Parasitic Flatworms

Molecular Biology, Biochemistry, Immunology and Physiology

Parasitic Flatworms

Molecular Biology, Biochemistry, Immunology and Physiology

Edited by

Aaron G. Maule

Parasitology Research Group
School of Biology and Biochemistry
Queen's University of Belfast
Belfast
UK

and

Nikki J. Marks

Parasitology Research Group
School of Biology and Biochemistry
Queen's University of Belfast
Belfast
UK

www.cabi.org

CABI is a trading name of CAB International

CABI Head Office
Nosworthy Way
Wallingford
Oxfordshire OX10 8DE
UK
Tel: +44 (0)1491 832111
Fax: +44 (0)1491 833508
E-mail: cabi@cabi.org
Website: www.cabi.org

CABI North American Office
875 Massachusetts Avenue
7th Floor
Cambridge, MA 02139
USA
Tel: +1 617 395 4056
Fax: +1 617 354 6875
E-mail: cabi-nao@cabi.org

A catalogue record for this book is available from the British Library, London, UK.

Library of Congress Cataloging-in-Publication Data

Parasitic flatworms : molecular biology, biochemistry, immunology and physiology / edited by Aaron G. Maule and Nikki J. Marks.
p. ; cm.
Includes bibliographical references and index.
ISBN-13: 978-0-85199-027-9 (alk. paper)
ISBN-10: 0-85199-027-4 (alk. paper)
1. Platyhelminthes.
[DNLM: 1. Platyhelminths. 2. Cestode Infections. QX 350 P224 2005] I. Maule, Aaron G. II. Marks, Nikki J. III. Tittle.

QL391.P7P368 2005
616.9'62--dc22

2005016094

ISBN-10: 0-85199-027-4
ISBN-13: 978-0-85199-027-9

Typeset by SPi, Pondicherry, India.
Printed and bound in the UK by Biddles Ltd, King's Lynn.

Contents

Contributors

John Barrett, Parasitology Research Group, Institute of Biological Sciences, Cledwyn Building, The University of Wales, Aberystwyth, Wales SY23 3DA, UK. Email, jzb@aber.ac.uk

David Blair, School of Tropical Biology, James Cook University, Townsville, Queensland 4811, Australia. Email, david.blair@jcu.edu.au

Sanaa Botros, Theodor Bilharz Research Institute, Cairo, Warrak El-Hadar, Imbaba, PO Box 30, Giza, 12411 Egypt. Email, sanaabotros@link.net

Paul J. Brindley, Department of Tropical Medicine, School of Public Health and Tropical Medicine, Tulane University Health Sciences Center, New Orleans, LA 70112, USA. Email, paul.brindley@tulane.edu

Peter M. Brophy, Parasitology Research Group, Institute of Biological Sciences, Edward Llwyd Building, University of Wales, Aberystwyth, Wales SY23 3DA, UK. Email, peter@brophy3218.fsnet.co.uk

Conor R. Caffrey, Tropical Disease Research Unit and Sandler Center for Basic Research in Parasitic Diseases, QB3 Building, PO Box 8 2550, 1700 4th Street, University of California, San Francisco, CA 94143, USA. Email, caffrey@cgl.ucsf.edu

Gerald C. Coles, Department of Clinical Veterinary Science, University of Bristol, Langford House, BS40 5DU, UK. Email, Gerald.C.Coles@bristol.ac.uk

Patricia S. Coulson, Department of Biology, University of York, York YO10 5YW, UK. Email, psc3@york.ac.uk

Carey O. Cunningham, FRS Marine Laboratory, PO Box 101, 375 Victoria Road, Aberdeen AB11 9DB, UK. Email, c.cunningham@marlab.ac.uk

John P. Dalton, Institute for the Biotechnology of Infectious Diseases, University of Technology Sydney, Westbourne Street, Gore Hill, Sydney, NSW 2065, Australia. Email, john.dalton@uts.edu.au

Tim A. Day, Department of Biomedical Sciences, Veterinary Medicine Building, Iowa State University, Ames, IA 50011, USA. Email, day@iastate.edu

Tegan Don, Helminth Biology Laboratory, Division of Infectious Diseases and Immunology, Queensland Institute of Medical Research, 300 Herston Road, Herston, Brisbane, Queensland 4006, Australia. Email, tegan.don@qimr.edu.au

Sheila Donnelly, Institute for the Biotechnology of Infectious Diseases, University of Technology Sydney, Westbourne Street, Gore Hill, Sydney, NSW 2065, Australia. Email, sheila.donnelly@uts.edu.au

Ricardo DeMarco, Universidade de São Paulo, Instituto de Quimica, Av. Prof. Lineu Prestes 748 sala 1200 05508-000 São Paulo, SP Brazil. Email, rdemarco@iq.usp.br

Sean G. Forrester, Faculty of Science, University of Ontario Institute of Technology, 2000 Simcoe St. North, Oshawa, ON, Canada, L1H 7K4. Email, Sean.Forrester@uoit.ca

Rudolf Geyer, Institute of Biochemistry, University of Giessen, Friedrichstrasse 24, 35392 Giessen, Germany. Email, Rudolf.Geyer@biochemie.med.uni-giessen.de

Robert M. Greenberg, Marine Biological Laboratory, 7 MBL Street, Woods Hole, MA 02543, USA. Email, rgreenberg@mbl.edu

Christoph G. Grevelding, Institute for Parasitology, Justus-Liebig-University, Rudolf-Buchheim-Strasse. 2, 35392 Giessen, Germany. Email, Christoph.Grevelding@vetmed.uni-giessen.de

David W. Halton, Parasitology Research Group, School of Biology and Biochemistry, Medical Biology Centre, Queen's University Belfast, 97 Lisburn Road, Belfast BT9 7BL, Northern Ireland, UK. Email, d.halton@qub.ac.uk

Judith E. Humphries, Department of Pathobiological Sciences, University of Wisconsin-Madison, School of Veterinary Medicine, 2015 Linden Drive, Madison, WI 53706, USA. Email, humphrij@svm.vetmed.wisc.edu

David A. Johnston, Wolfson-Wellcome Biomedical Laboratory, Department of Zoology, The Natural History Museum, Cromwell Road, London SW7 5BD, UK. Email, daj@nhm.ac.uk

Matty Knight, Biomedical Research Institute, 12111 Parklawn Drive, Rockville, MD 20852, USA. Email, mknight@afbr-bri.com

Marshall W. Lightowlers, Veterinary Clinical Centre, The University of Melbourne, 250 Princes Highway, Werribee, Victoria 3030, Australia. Email, marshall@unimelb.edu.au

D.T.J. Littlewood, Department of Zoology, The Natural History Museum, Cromwell Road, London SW7 5BD, UK. Email, T.Littlewood@nhm.ac.uk

Alex Loukas, Helminth Biology Laboratory, Division of Infectious Diseases and Immunology, Queensland Institute of Medical Research, 300 Herston Road, Herston, Brisbane, Queensland 4006, Australia. Email, alex.loukas@qimr.edu.au

Nikki J. Marks, Parasitology Research Group, Queen's University Belfast, 97 Lisburn Road, Belfast BT9 7BL, Northern Ireland, UK. Email, n.marks@qub.ac.uk

Iveta Matějusová, FRS Marine Laboratory, PO Box 101, 375 Victoria Road, Aberdeen AB11 9DB, UK. Email, I.Matejusova@marlab.ac.uk

Aaron G. Maule, Parasitology Research Group, Queen's University Belfast, 97 Lisburn Road, Belfast BT9 7BL, Northern Ireland, UK. Email, a.maule@qub.ac.uk

Derek M. Mckay, Intestinal Disease Research Programme, McMaster University, Hamilton, Ontario, Canada L8N 3Z5. Email, mckayd@mcmaster.ca

James McKerrow, Tropical Disease Research Unit and Sandler Center for Basic Research in Parasitic Diseases, QB3 Building, Box. 2550, 1700 4th Street, University of California, San Francisco, CA 94143, USA. Email, jmck@cgl.ucsf.edu

Don P. McManus, Molecular Parasitology Laboratory, Division of Infectious Diseases and Immunology, Australian Centre for International and Tropical Health and Nutrition, The Queensland Institute of Medical Research and The University of Queensland, Post Office Royal Brisbane Hospital, Queensland 4029, Australia. Email, Don.McManus@qimr.edu.au

Russell M. Morphew, Parasitology Research Group, Institute of Biological Sciences, Edward Llwyd Building, University of Wales, Aberystwyth, Wales SY23 3DA, UK. Email, rmm97@aber.ac.uk

Edward J. Pearce, Department of Pathobiology, School of Veterinary Medicine, University of Pennsylvania, Philadelphia, PA 19104-6008, USA. Email, ejpearce@mail.med.upenn.edu

Nithyakalyani Raghavan, Biomedical Research Institute, 12111 Parklawn Drive, Rockville, MD 20852, USA. Email, nkr@helix.nih.gov

Mohammed Sajid, Tropical Disease Research Unit and Sandler Center for Basic Research in Parasitic Diseases, QB3 Building, PO Box 2550, 1700 4th Street, University of California, San Francisco, CA 94143, USA. Email, sajid@cgl.ucsf.edu

Patrick J. Skelly, Department of Biomedical Sciences, Division of Infectious Diseases, Tufts University School of Veterinary Medicine, 200 Westboro Road, Grafton, MA 01536, USA. Email, Patrick.Skelly@Tufts.edu

Colin Stack, Institute for the Biotechnology of Infectious Diseases, University of Technology Sydney, Westbourne Street, Gore Hill, Sydney, NSW 2065, Australia. Email, colin.stack@uts.edu.au

Aloysius G.M. Tielens, Faculty of Veterinary Medicine, Department of Biochemistry and Cell Biology, Utrecht University, PO Box 80176, The Netherlands. Email, A.G.M. Tielens@vet.uu.nl

Jaap J. Van Hellemond, Faculty of Veterinary Medicine, Department of Biochemistry and Cell Biology, Utrecht University, PO Box 80176, The Netherlands. Email, j.vanhellemond@vet.uu.nl

Sergio Verjovski-Almeida, Instituto de Química, Universidade de São Paulo, Av. Prof. Lineu Prestes 748 sala 1200 05508-000 São Paulo, SP Brazil. Email, verjo@iq.usp.br

Jon J. Vermeire, Department of Pathobiological Sciences, University of Wisconsin-Madison, School of Veterinary Medicine, 2015 Linden Drive, Madison, WI 53706, USA. Email, vermeij@svm.vetmed.wisc.edu

Rodney A. Webb, Department of Biology, York University, Toronto, Ontario, Canada. Email, raw@yorku.ca

R. Alan Wilson, Department of Biology, University of York, York YO10 5YW, UK. Email, raw3@york.ac.uk

Manfred Wuhrer, Biomolecular Mass Spectrometry Unit, Department of Parasitology, Leiden University Medical Center, PO Box 9600, 2300 RC Leiden, The Netherlands. Email, m.wuhrer@lumc.nl

Timothy P. Yoshino, Department of Pathobiological Sciences, University of Wisconsin-Madison, School of Veterinary Medicine, 2015 Linden Drive, Madison, WI 53706, USA. Email, yoshinot@svm.vetmed.wisc.edu

Preface

Buoyed by the utility of research data on the model nematode *Caenorhabditis elegans* and the rapidly growing expressed sequence tag (EST) resources for roundworms, research on nematode parasites has progressed rapidly compared to that seen for flatworm parasites. This molecular research base has facilitated considerable advances in nematode phylogenetics, genomics, proteomics and biology that have often been the envy of many platyhelminthologists. Only in the last few years have we seen the benefits of concerted efforts to propel flatworm research into the 21st century with genome projects on schistosomes and EST projects on these and other parasitic flatworms. Even though EST data for schistosomes have only become available relatively recently, it is clear from many of the chapters of this book that they have served to invigorate parasitic flatworm research, providing a bountiful resource for almost all researchers in the field.

The undisputed lead role played by schistosomes in parasitic flatworm research is echoed by the content of this book – a reflection of the field as it stands today. However, it would be naïve to assume that all the significant progress is confined to schistosomes as much quality research is emanating from research laboratories that focusses on other trematodes, cestodes or monogeneans. All serve to inform us on flatworm biology and parasitology. We believe that flatworm parasitology is now entering a new era in which some of the considerable gaps between the frontiers of nematode and flatworm research are beginning to be bridged. In this respect, we believe that this book is timely.

The phylogenetic relationships amongst flatworms are only now being unravelled and are so complex that predicting how findings in one flatworm parasite species relate to another is almost impossible to decipher. In nematodes, *C. elegans* has proved a useful model to study the facets of biology of parasitic species. Although there is no such flatworm model, it is noteworthy that much significant progress has been made recently in planaria, especially in relation to genome-wide gene-silencing methods that have been developed in *Schmidtea mediterranea*. However, the value of such species as model organisms for diverse flatworm parasites is simply not known and there is limited evidence to suggest that they will be hugely beneficial in this respect. Indeed, comparative studies on flatworms are few and far between such that any judgement on the value of potential model flatworms is based on a very small dataset. Nevertheless, progress in the molecular manipulation of planarians has the potential to aid parasitologists, and more interaction and cross-fertilization between these research communities should be encouraged.

Although research on parasitic flatworms lags behind that on nematodes, progress is evident in almost all areas of focus, from phylogenetics and genomics to immunobiology and vaccine development, from cell signalling and physiology to gene silencing and transgenics. While this book addresses many of these areas, it does not attempt to be comprehensive. Unfortunately size constraints within a single volume preclude consideration of many other interesting aspects of flatworm research. Nevertheless, we hope the contents endow the reader with valuable insight into what we perceive as some of the most pivotal and exciting areas of parasitic flatworm biology.

The first section of this book provides a strong foundation for understanding the evolutionary interrelationships of flatworms, their phylogeny, genetics and transcriptomics; it informs on where we are today and highlights the molecular advances that are being made and where they are likely to lead. It finishes by charting recent progress in the development of transgenic flatworms; developments, which could have huge impact on many of our research goals. Section two focuses on the host–parasite relationship and parasite control. Within the host–parasite relationship there are contributions on immunobiology, the host–parasite interface and parasite-induced host transcriptome changes that have much merit well beyond the boundaries of parasitology. In relation to control, the chapters examine developments in flatworm chemotherapy and drug resistance. This section ends with chapters on vaccine development, believed by many to be the future of helminth parasite control. The third and final section turns to proteomics and the biology of flatworm parasites by exploring proteases, physiology, metabolism and glycoconjugates. Many of these chapters reveal how the new molecular resources for flatworms, discussed in earlier sections, have aided research on more focused aspects of flatworm biology. The final chapter examines gene silencing and the progress made in the induction of RNA interference in flatworm parasites as a tool to examine protein function. The authors are all leading researchers in their respective fields and have provided expert coverage of key areas in parasitic flatworm research as well as insight into the future directions for this work. As with the sister text on parasitic nematodes, the chapters included here are diverse in style and approach, but provide a broad insight into parasitic flatworms.

Without a highly characterized model species to underpin flatworm research, scientific progress will most likely rely on our ability to capitalize on evolving molecular advances in a range of flatworm species. We hope this book will provide clear evidence that significant advances are already being made across a wide spectrum of parasitic flatworm research, and will prove a useful resource for continued study.

Access to Colour Illustrations

In order to reduce the cost of this book, and thereby improve its accessibility, there are no colour reproductions. Some of the illustrations, however, can only properly be appreciated in colour (some of those in Chapters 7 and 17).

The colour illustrations can be viewed and downloaded from the following internet site: http://www.qub.ac.uk/bb/books/flatworms/index.html

If you have any suggestions or problems relating to the illustrations appearing in this book, then please feel free to contact Aaron Maule (a.maule@qub.ac.uk).

1 The Evolution of Parasitism in Flatworms

D.T.J. Littlewood
Department of Zoology, The Natural History Museum, Cromwell Road, London SW7 5BD, UK

Introduction

The phylum Platyhelminthes is comprised of an enormous diversity of species that occur in all seas, rivers and lakes, and on all continental land masses. With a soft body lacking any cuticle or protective covering, the majority of species are found in moist or aquatic environments. Generally bilaterally symmetrical, acoelomate, lacking an anus, possessing a low level of cephalization, usually hermaphroditic, and dorsoventrally flattened, flatworms are commonly small but may reach enormous lengths. These defining features are not unique to the phylum and are, e.g. multiciliated gastrodermal cells, best viewed with an electron microscope. However, there are many defining features for the various subgroups (Caira and Littlewood, 2001; Littlewood *et al.*, 2004).

As parasites, flatworms have extended their global presence by taking advantage of the adaptations of many diverse invertebrate and vertebrate hosts. Many parasitic forms are host-specific and many of these are site-specific within or on their host. Contemplate on the biodiversity of vertebrates, consider that many platyhelminths use one or more intermediate hosts and one may just begin to grasp the diversity of parasitic flatworms. From the microscopic interstitial free-living species that live between particles of mud to the enormously long tapeworms of blue whales, an estimate of 100,000 extant species, of which only about 20,000 have been formally described, still seems conservative when we consider the hosts and habitats yet to be surveyed. Even among those hosts and habitats already surveyed, many of the smaller parasitic species seem to have been overlooked and new species from commonly collected hosts are not uncommon. Small-bodied parasitic taxa are harder to find and are generally more specios than related large-bodied taxa (Poulin and Morand, 2000), suggesting a long future ahead for flatworm systematists.

As soft-bodied organisms with little or no protection from predation, flatworms tend to be cryptic with some notable exceptions; the very colourful marine polyclads found on coral reefs offer some of the most spectacular colour schemes exhibited by any organism (Newman and Cannon, 2003), and some terrestrial planarians (Tricladida) are also large and conspicuous. These brightly coloured free-living representatives tend to exhibit aposematic coloration, indicating their distastefulness, or in the case of some polyclads they mimic distasteful nudibranch molluscs. In contrast, parasitic forms are less noticeable and often have complex life cycles with only fleeting periods when they are not in or on a host. Usually they are microscopic during these free-living developmental stages; it takes an effort to find them in spite of their ubiquity.

Although lacking a fossil record, the deep-branching position of Platyhelminthes among the Lophotrochozoa in the tree of life, suggests a relatively ancient origin but perhaps not quite as ancient as once thought. For over 100 years, scientists have suggested that the relatively simple body plan of these flatworms renders them as ideal model organisms from which other phyla can be derived and it is common to see hypothetical ancestors of the Metazoa described as 'flatworm-like'. However, platyhelminth phylogenetics has progressed markedly over the last 10 years thanks to a sustained effort by morphologists and the advent of molecular systematics providing additional sources of phylogenetic data. We are now armed with more clearly resolved evolutionary trees with which to investigate the origins and subsequent radiation of flatworm groups. Recent advances, for example, indicate that acoelomorph flatworms (Acoela and Nemertodermatida) are now generally considered to be basal bilaterians (a role once occupied by the Platyhelminthes), occupying a pivotal position in the tree of metazoan life and likely representing one or two separate phyla (Ruiz-Trillo *et al.*, 1999, 2002; Telford *et al.*, 2000, 2003; Littlewood *et al.*, 2001; Jondelius *et al.*, 2002).

Whilst it is the inclusion of Acoelomorpha in the Platyhelminthes that contributed to a perception that the whole phylum is 'primitive', the remaining non-acoelomorph flatworms, the Catenulida + Rhabditophora (Platyhelminthes *sensu stricto*), may be more closely allied to phyla such as Gastrotricha (Giribet *et al.*, 2000) or at least as derived members of the Lophotrochozoa appearing sometime after the last protostome/deuterostome ancestor; the resolution of their true position among the Lophotrochozoa remains controversial and unresolved (Jenner, 2004; Valentine, 2004). Most molecular estimates of metazoan interrelationships recognize three main clades, the deuterostomes, the ecdysozoans and the lophotrochozoans, usually with the latter two protostome clades forming a monophyletic group (Adoutte, 1999). However, a recent analysis (Eernisse and Peterson, 2004), employing multiple genes and including morphology, places the deuterostomes and lophotrochozoans as sister groups, and the ectoprocts as a sister group to the Platyhelminthes. Clearly, many nodes require greater resolution from additional data.

Figure 1.1 illustrates a recent estimate of relationships among the Metazoa with an indication showing which phyla are parasitized by the main parasitic flatworm groups. Parasitic flatworms have had a marked impact on many other animal groups, most notably the Chordata. Most vertebrates are parasitized by at least one species of flatworm, and larger vertebrates tend to have richer parasite communities with many helminths (see, for example, Poulin, 1995). Considering the importance of parasitism as an 'engine of diversity', maintaining the diversity of the major histocompatibility complex, imposing selection favouring sexual reproduction in hosts and increasing speciation rates, among other factors (Summers *et al.*, 2003), the influence of the Platyhelminthes in evolution and ecology seems greater than the sum of its parts.

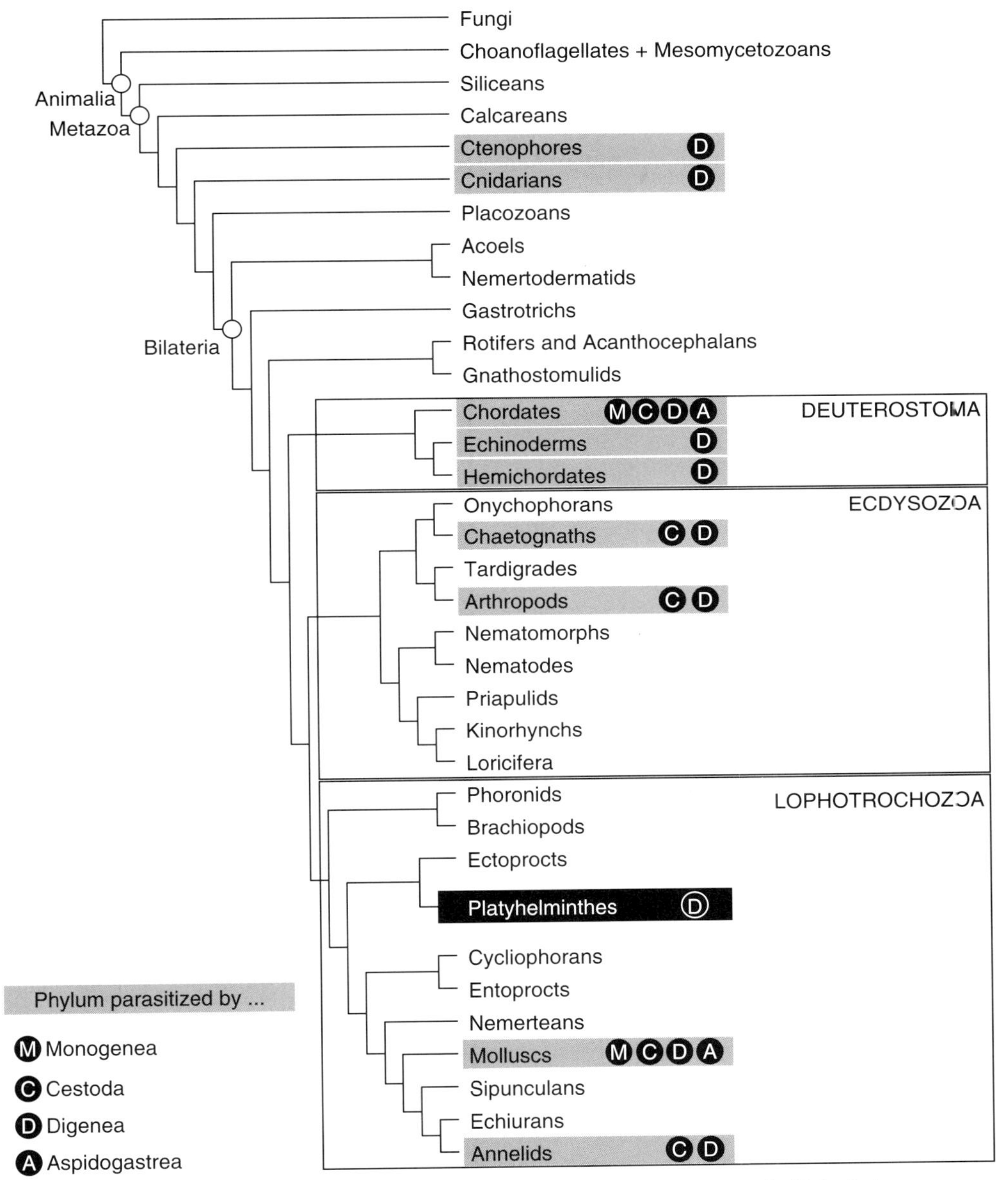

Fig. 1.1. Platyhelminthes and their position in the tree of life with an indication of which phyla are parasitized by neodermatan flatworms (Monogenea, Cestoda, Aspidogastrea, Digenea); basic tree adapted from Eernisse and Peterson (2004) who estimated this tree topology using a combined analysis of molecular (SSU rDNA and myosin II) and morphological data; monophyletic protostomes are shown as this remains the general consensus (Baldauf, 2003). Acoelomorph flatworms (Acoela and Nemertodermatida) are no longer members of the Platyhelminthes, but are instead recognized as basal bilaterians. True flatworms are members of the Lophotrochozoa but their relative position within this clade and the identity of their sister group is still debated. Digenea utilize the greatest diversity of metazoan phyla as hosts, including some free-living flatworms.

Figure 1.2 provides a consensus view of the major flatworm groups and a hypothesis of their interrelationships based on molecular and morphological estimates (Littlewood *et al.*, 1999b); individual estimates may be found in Brooks (1989), Carranza *et al.* (1997), Ehlers (1984, 1985), Littlewood *et al.* (1999b) and Zamparo *et al.* (2001). The 'Turbellaria' is a paraphyletic assemblage and the relationships between its constituent taxa and the Neodermata are yet to be satisfactorily resolved (see section on 'The Origins of Obligate Parasitism'). Catenulid flatworms employ a different mitochondrial genetic code than the remaining flatworms, the Rhabditophora (Telford *et al.*, 2000), and appear to have few, if any, convincing morphological synapomorphies with the Rhabditophora. Nevertheless, molecular data routinely resolve the Catenulida as the most basal members of the Platyhelminthes and they have long been considered flatworms (Ehlers, 1984). Most of the turbellarian groups include marine and freshwater examples and only some of the triclad turbellarians are terrestrial. The majority of the 'Turbellaria' are very small, and of these most are meiofaunal, except the Polycladida, which are often large and strictly marine. A number of turbellarian lineages include symbiotic and some parasitic species (see section on 'Parasitism in the Platyhelminthes'), but the obligate parasites (Neodermata) form a convincing monophyletic group.

Flatworms have no fossil record, beyond one or two trace fossils of uncertain origin or limited utility (Conway Morris, 1981; Upeniece, 2001; Valentine, 2004), including Quaternary turbellarian eggs (Binford, 1982), a typhloplanoid rhabdocoel in Eocene amber (Poinar, 2003) and shell pitting in marine bivalves caused by digeneans (Ruiz and Lindberg, 1989). This paucity of dateable evidence seriously hampers an understanding of the timing of key evolutionary events in the history of flatworm radiation (Littlewood and Donovan, 2003). To unpick the evolutionary history of parasite evolution and radiation objectively is no easy task and we must do it based on our knowledge of extant species, our

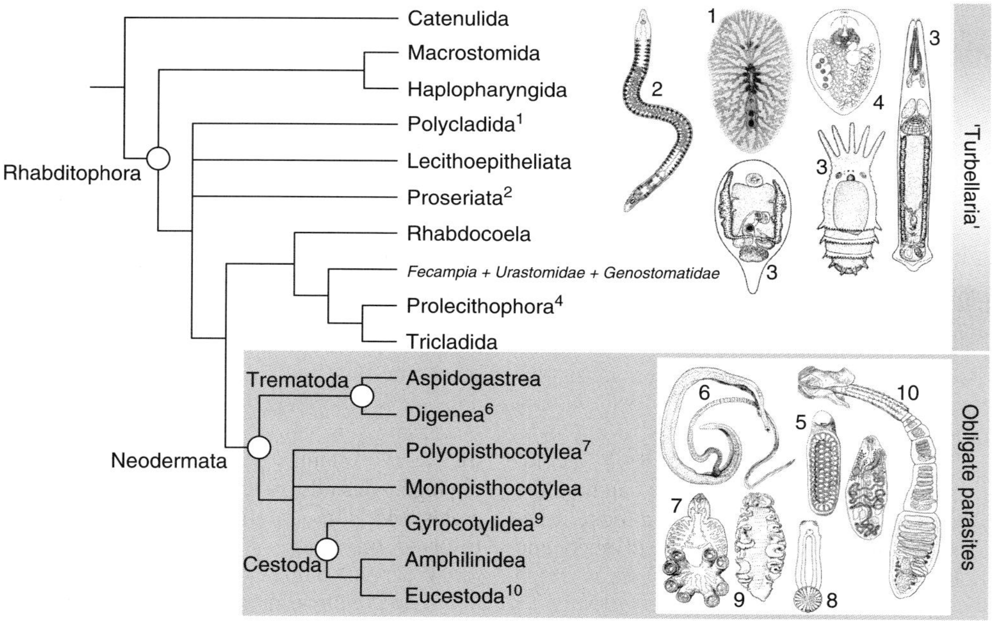

Fig. 1.2. Interrelationships of the major groups of Platyhelminthes based on a consensus of morphological and molecular estimates. Parasitic flatworms, the Neodermata, form a monophyletic group although their interrelationships are estimated differently by different molecular analyses (see Fig. 1.3).

ability to reconstruct phylogenies and the extent to which we can recognize shared ancestral characters. Considering present day host associations, which provides an insight into the diversity of adaptations exhibited by parasitic flatworms, and to some extent reference to the phylogeny of the various intermediate and definitive hosts may provide an understanding of the evolutionary history shared by the parasites and their hosts. This process depends on parasites having cospeciated with their hosts. However, strict cophyly, where parasite speciation events are unequivocally mapped on to a fully congruent host tree, requires coevolution to have occurred in the absence of host switching, and we know this to be relatively rare. Evolutionarily ancient lineages of hosts are not necessarily parasitized by evolutionarily ancient lineages of parasites; Szidat (1956) postulated that the more 'primitive' a host the more 'primitive' the parasites it harbours and there are many cases where this appears to hold among the parasitic flatworms. However, it is not a hypothesis that can be used unreservedly or uncritically to infer past host associations. In the absence of reliable phylogenies, hypotheses such as this have fuelled a storytelling approach to evolutionary parasitology that has led to erroneous interpretations of comparative data (Brooks and McLennan, 1993). Constructing scenarios that lead from ectoparasitism to endoparasitism or iteratively from one, to two, to three, to four hosts (and never, for example, from one to three, or three to one) may be compelling and persuasive, but in the absence of means by which we can test alternative hypotheses these stories remain of limited value. Although not phylogenetically based, Combes (2001) reviewed the bewildering range of interactions between hosts and parasites in an ecological and evolutionary context; flatworms provided many examples.

Phylogenetic analysis provides an evolutionary framework with which we can map the evolution of characters, including those pertinent to parasitism, that define the diversity and biology of extant flatworms, and from which ancestral conditions may be inferred. Alternative scenarios can be tested in terms of how parsimoniously they may map on to the evolutionary hypothesis to hand and through these means we may begin to reconstruct evolutionary history. Host phylogenies and the complicated field of cophylogeny mapping offer even greater opportunities to infer the nature of historical cospeciation and host switching events (Page, 2003), but there are few cases where these data have been suitably sampled among the parasitic flatworms (see also Klassen, 1992). Of course, phylogenies themselves are not without error or bias, depending on the source of data, the method by which it is coded, the models of evolution employed in their analysis, and to varying degrees the individual user's predelictions and/or competence. Furthermore, there are many evolutionary scenarios that cannot be distinguished unequivocally from phylogenies alone; it is the combination of phylogenies and comparative data that offer the best opportunities to differentiate between competing possibilities (Brooks and McLennan, 2002).

Presented here are recent estimates of flatworm phylogenies at various taxonomic levels from which one can infer the radiation of parasitic flatworm taxa and the evolution of parasitism through the mapping of comparative data. A phylogenetic approach to inferring the evolution of parasitism within the Platyhelminthes is relatively recent. Most effort has been expended on resolving the phylogenies and mapping the characters that are used to build them, rather than collating comparative data and viewing these data in the light of evolutionary ecology (Morand and Poulin, 2003).

Phylogenies are hypotheses and remain so until the weight of independent evidence provides overwhelming support for a particular evolutionary scenario to become accepted. Few platyhelminth phylogenies receive overwhelming support from helminth systematists, and whilst it is surely tiresome for comparative biologists to have to be aware of changing phylogenies, substantial progress in platyhelminth phylogenetics allows us to interpret recent estimates with some confidence. In other words, resolved phylogenies are a worthwhile goal, and biologists would do well to employ them to interpret comparative data in the light of evolution. However, first it is important to survey the scope of parasitism as it is found throughout the Platyhelminthes.

Parasitism in the Platyhelminthes

Parasitic flatworms, as detailed in this volume, form a monophyletic group known as the Neodermata, i.e. the tapeworms (Cestoda) and flukes (Trematoda) and Monogenea share a common ancestor. These taxa are obligate parasites, albeit with free-living stages that make up their life cycles, and they derive all their nutrition from their hosts. However, the ancestor of these worms and its many descendants has not been the only flatworm to take the path to parasitism. Many species within the paraphyletic 'Turbellaria' have become intimately associated with other organisms as commensals, and others may be classified as obligate parasites; Jennings (1971) provides a comprehensive summary of the diversity of parasitic and commensal turbellarians, and Rohde (1994) notes that only among the Catenulida, Macrostomida, Haplopharyngida and Lecithoepitheliata are there no known species that live in symbiosis with other larger Metazoa. The lesser-known parasitic and commensal turbellarian parasites are worthy of attention, not just from an evolutionary perspective, but because they may hold the key to understanding parasitism throughout the phylum, and how it began in the Neodermata. Differentiating between a commensal and a parasite is not only important ecologically, but considering the full gamut of host–flatworm interactions may provide a greater understanding of the parasitic forms' ability to have become so successful.

Feeding directly or indirectly from their hosts, parasites set themselves apart from commensals as their growth and/or development is arrested in the absence of a suitable host through lack of nutrition; more usually, parasitic species in the phase of their life cycle requiring a host will die if one is not available and without a host they are unable to complete their life cycle. Commensals, whilst often found in association with their hosts, are not so dependent on them; their nutrition may be facilitated by their hosts but they do not necessarily die in the absence of a host. There is obviously a continuum between the terms 'free-living' and 'parasitic', with 'commensal' somewhere between the two; where one draws the line between commensal and parasite is certainly subjective. This continuum has played a role in the early history of understanding the evolution of parasitism within the phylum with a number of cases of parasitic or commensal species of turbellarian being proposed as the sister group to the Neodermata. Such 'missing links' have the potential to provide useful insight into the evolution of parasitism as well as to confound an understanding; it would be wrong to propose evolutionary relationships based on ecological habit. Comparative studies have highlighted the various (frequently convergent) adaptations to parasitism found among the parasitic turbellarians. Functional genomics (Newmark and Sánchez Alvarado, 2002), including gene knockout through RNAi (Orii *et al.*, 2003), are established for planarian species and it will be interesting to see if similar studies with parasitic flatworms can add anything to our understanding of the evolution of parasitism within the phylum.

Notable parasites among the 'Turbellaria' include the Fecampiidae, Notenteridae, Genostomatidae and Urastomidae. Many of these species have been shown to produce very large numbers of eggs, presumably to offset the huge loss among those individuals unable to find a suitable host (Rohde, 1994). Entirely endoparasitic, fecampiids infect crustaceans (including crabs, amphipods and barnacles) and myzostomids; the one species of notenterid (*Notentera ivanovi*) infects a polychaete annelid; genostomatids infect crustaceans and fish, and urastomids are found only in bivalve molluscs. As a whole, these families infect the major groups that neodermatans are found in, and when it was discovered that certain features concerning sperm ultrastructure characterized these parasitic turbellarians and the Neodermata, it was suggested, not for the first time, that these parasitic turbellarians were the closest living relatives of the Neodermata (Joffe and Kornakova, 1998; Kornakova and Joffe, 1999); additional evidence linking the Fecampiidae with the Neodermata is also discussed in (Williams, 1988, 1994; Watson *et al.*, 1992; Watson and Rohde, 1993a,b; Rohde, 1994). However, molecular systematic studies have not confirmed this Revertospermata hypothesis and Rohde (1994) suggested that the morphological similarities were most likely due to convergent evolution. Molecular phylogenetic estimates have routinely placed the non-neodermatan revertospermatans in a clade with the prolecithophoran and triclad turbellarians (Littlewood *et al.*, 1999b; Norèn and Jondelius, 1999; Joffe and Kornakova, 2001; Littlewood and Olson, 2001; Lockyer *et al.*, 2003a); see section on 'The Origins of Obligate Parasitism'.

Few turbellarian groups do not have one or more commensal or parasitic species; for example, among the dalyellioids (Acholadidae, Graffillidae, Provorticidae, Pterastericolidae and Umagillidae) the lecithophorans (Typhloplanidae), the triclads (Bdellouridae, Procerodidae and Micropharyngidae), the polyclads (Apidioplanidae, Emprosthopharyngidae, Hoploplanidae, Latocestidae, Leptoplanidae, Stylochidae and Prosthiostomidae), the prolecithophorans (Cylindrostomatidae, Hypotrichinidae and Plagiostomidae) and all the temnocephalans. Together these taxa have managed to live on or in a bewildering array of other animal groups. Jennings (1971) and Cannon (1986) documented over 200 species belonging to 35 families of turbellarians living in permanent association with another organism. These species occupy various positions in and on their hosts depending on whether they are feeding with assistance from the host or feeding on the host themselves (Jennings, 1997). As Jennings noted, there is 'a continuous spectrum of often overlapping nutritional strategies from ecto- and entozoic predation (generally supplemented by opportunistic commensalism) through full commensalism (in which ingestion of the host's food may incidentally include ingestion of digestive enzymes) to obligate entoparasitism'. If we consider the platyhelminths as a whole, this latter stage is taken to its acme in the Cestoda, which have no gut and rely entirely on the passage of nutrients through their modified epidermis. Amongst the Neodermata this modified syncitial epidermis (i.e. the neodermis) performs this, including other, critical roles. For the most part, a ciliated, cellular epidermis characterizes turbellarians, but many of the commensal and parasitic forms have developed independently a syncitial epidermis; *Genostoma*, *Kronborgia*, and the temnocephalans offer well-documented examples and a review of epidermal morphology and development among parasitic and commensal turbellarians is provided by Tyler and Tyler (1997).

As Tyler and Hooge (2004) note, the developmental mechanism by which the ciliated epidermis is replaced by neodermis is not unique to the neodermatan flatworms. However, the Neodermata have specifically adapted this mechanism as they attack hosts in the life cycle. They speculate that the general phenomenon of epidermal-shedding and replacement with unciliated, syncitial, insunk (i.e. nuclei lie below the body wall musculature) teguments may have pre-adapted the group to parasitism. Clearly, the neodermis plays a critical role in defending the parasite against the host, whether it is from its immune system, or more frequently among enteric groups, its digestive system.

The Origins of Obligate Parasitism – The Appearance of the Neodermata

The Neodermata share a number of unique features (apomorphies), of which some have undoubtedly contributed to the successful radiation of the group and the main lineages.

Apomorphies that unite these taxa include:

- multiciliated ectoderm limited to 'larval' stages and shed later and replaced by a syncitial neodermis with sub-epidermal perikarya each separately connected to surface layer;
- protonephridia with a two-cell weir;
- epidermal locomotory cilia with single, cranial rootlet;
- epithelial sensory receptors with electron-dense collars;
- complete incorporation of both axonemes in sperm body;
- two long and one short insertions in nuclear small subunit (SSU) ribosomal DNA (rDNA).

Acknowledging the monophyly of the Neodermata as first proposed by Ehlers (1984) allows us to argue that obligate parasitism, as found in the constituent Trematoda, Monogenea and Cestoda, has its origins in a major single evolutionary event. This does not mean it happened quickly, or without intermediate forms, but the lineage that gave rise to the Neodermata surely began with a single speciation event and continued with a remarkable adaptive radiation. Not withstanding the many forays into parasitism made by turbellarian flatworms, it is without doubt that the emergence of the Neodermata that allowed the platyhelminths to engage successfully in a lifestyle that ties them so closely to the lives and evolution of so many other organisms. To infer the plesiomorphic ('primitive') characters of the common neodermatan ancestors we need to determine their interrelationships and to find the sister group to the Neodermata; which lineage of platyhelminths is most closely related to the obligate parasites? There have been a number of phylogenetic estimates from morphological and molecular data yielding a variety of solutions (see Fig. 1.3).

One group of turbellarians, which have provided a compelling link to the obligate parasites, is the Temnocephalida. With various degrees of adaptation towards their ectocommensal existence with freshwater crustaceans, including loss of locomotory cilia and possession of a distinct posterior sucker (Cannon and Joffe, 2001), it is easy to visualize a generalized transition towards obligate parasitism by placing the temnocephalans as sister group to the Neodermata. Indeed, an early cladistic analysis yielded just this answer (Brooks and McLennan, 1993; Fig. 1.3b). Other candidate lineages have been the Urastomidae and Genostomatidae, forming with the Neodermata the Mediofusata, as part of the Revertospermata hypothesis; Revertospermata = Fecampiida + Mediofusata (Kornakova and Joffe, 1999; Fig. 1.3c). Ehlers was the first to suggest another revertospermatan group, the Fecampiidae, as a possible sister group to the Neodermata, although he settled on a 'dalyellioid' clade, which included Temnocephalida, Fecampiidae and Udonellidae (Ehlers, 1985; Fig. 1.3a). Udonellidae has since been shown to be unequivocally a group of highly modified monopisthocotylean monogeneans (Littlewood *et al.*, 1998), but the morphologists have

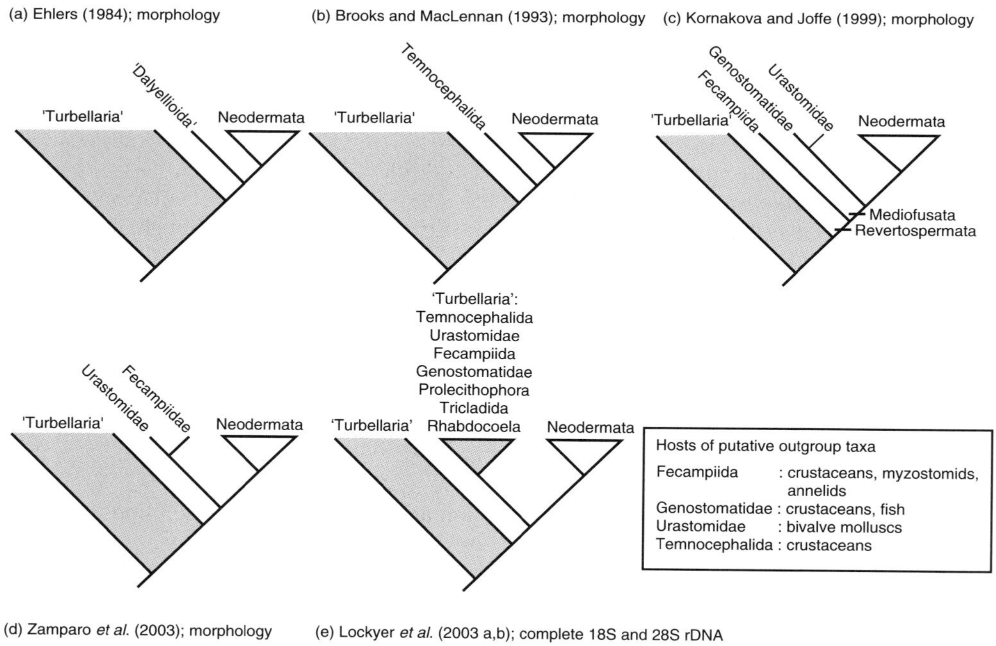

Fig. 1.3. Four morphological and one molecular estimate of the interrelationships of the Platyhelminthes with emphasis on identifying the sister group to the obligate parasites (Neodermata). Currently, the two main competing hypotheses concern (c) a revertospermatan clade and (e) a large clade of neophoran turbellarians.

frequently placed parasitic turbellarian groups as close relatives of the Neodermata (Fig. 1.3a–d). Certainly, if one includes the spermatological features in a morphological cladistic analysis, the Revertospermata are resolved (Littlewood *et al.*, 1999b) and a web-based listing of turbellarian taxonomy recognizes the integrity of this grouping (http://devbio.umesci.maine.edu/styler/turbellaria/). In contrast, molecular systematic estimates of flatworm phylogeny from nuclear ribosomal data are quite different, and consistently recognize a large clade of neoophoran turbellarians as the likely sister group to the Neodermata; the parasitic turbellarians along with the triclads and prolecithophorans form a well-supported sister group (Fig. 1.3e).

As with many other taxonomic groups, molecular phylogenetic estimates for the flatworms have suffered from poor sampling, single gene sampling, long-branch attraction and inadequate analyses, and although some of these issues have now been addressed, additional data are needed to support or refute the various morphological and molecular estimates; summaries of earlier molecular studies can be found in Littlewood *et al.* (1999a), Littlewood and Olson (2001), Lockyer *et al.* (2003a) and Baguñà and Riutort (2004). A review of the recent literature might suggest that there are too many polarized opinions and entrenched viewpoints on molecules versus morphology or characters being homologous or homoplasious for the evidence to speak for itself. However, if one can argue that molecular data do not come quite as loaded with *a priori* statements on homology as has been shown or claimed for morphology, as I would, then the two ribosomal genes sampled so far both suggest that a large clade of turbellarians is the most likely sister group to the Neodermata (Fig. 1.3e). Nevertheless, consensus of opinion based on congruence between independent data sets remains an elusive but necessary goal.

The solutions provided by neither molecular nor morphological data allow a simple story to be told concerning the stem group neodermatans, whether considered individually or collectively (Fig. 1.4). The sister group according to molecular data is so large that plesiomorphic characters for the clade are all but non-existent and hold little clue as to the origins of parasitism, except that the ancestral neodermatan was endoparasitic and adopted a vertebrate as its host first (Littlewood *et al.*, 1999a; Fig. 1.4d). Without being able to reconcile a parasitic or commensal sister group to the Neodermata, whether intuitive, attractive or indeed correct, the most parsimonious solution presently is that the vertebrates have always been host to neodermatans and so vertebrates must have been the first host. This scenario has generated disquiet amongst those workers who would argue that parasites with complex life cycles most likely parasitized the first intermediate hosts first, on an evolutionary as well as an ontogenetic scale, and thereafter recruited other hosts, i.e. trematodes first used molluscs and cestodes first used arthropods before they each ultimately involved vertebrates as their definitive hosts. A common argument is that as vertebrates were eating these molluscs and crustaceans, the parasites simply adapted to survive within the guts of the vertebrates. Was this a one-host life cycle with vertebrates, as it is among monogeneans? The argument will fail if it can be shown that the associations between the major parasitic lineages and vertebrates are not homologous. If the Revertospermata hypothesis is correct, and similar sperm morphology can be demonstrated to have arisen through common ancestry, rather than convergent evolution, then a two-host life cycle with an arthropod and a vertebrate may be the most parsimonious scenario; although the relative position of the Urastomidae could easily make a two-host life cycle with a mollusc and a vertebrate plausible. See Cribb *et al.* (2001a, 2003) for further studies on the origins and radiation of trematode life cycles.

The interrelationships among the Neodermata are just as important in reconciling this conundrum, and yet unfortunately they seem just as confused. In this case the confusion arises from alternative solutions provided by molecular data. Morphological analysis consistently resolves the Trematoda (Digenea and Aspidogastrea) as sister taxon to the Cercomeromorpha (Cestoda and Monogenea). Historically, the Monogenea have been resolved as a monophyletic group by most morphologists offering at least four synapomorphies: (i) larva with three ciliated zones; (ii) larva and adult with two pairs of pigmented eyes; (iii) one pair of ventral anchors and (iv) one egg filament (Boeger and Kritsky, 2001). In contrast, molecular data have provided two alternative options of a paraphyletic Monogenea (Justine, 1998; Littlewood *et al.*, 1999a) and a monophyletic Monogenea (Lockyer *et al.*, 2003a), but in the latter study where complete SSU and complete large subunit (LSU) ribosomal RNA (rRNA) genes were utilized, the Cestoda and Trematoda were also resolved as a monophyletic clade thus challenging a long-held theory uniting the Monogenea and Cestoda (Lockyer *et al.*, 2003a). Subsequent to the earlier molecular estimates Euzet and Combes (2003) reviewed the morphology of the two main constituent taxa, the Monopisthocotylea and the Polyopisthocotylea. The monophyly of each of these monogenean groups is not disputed, but Euzet and Combes (2003) emphasized the differences between them, most notably the morphology of the adult haptor, the structure of the genital system, mode of nutrition, the anatomy and morphology of the oncomiracidia (larvae) and the process of host invasion. They concluded that the Monogenea is a paraphyletic assemblage, although they did not argue for a specific topology for the

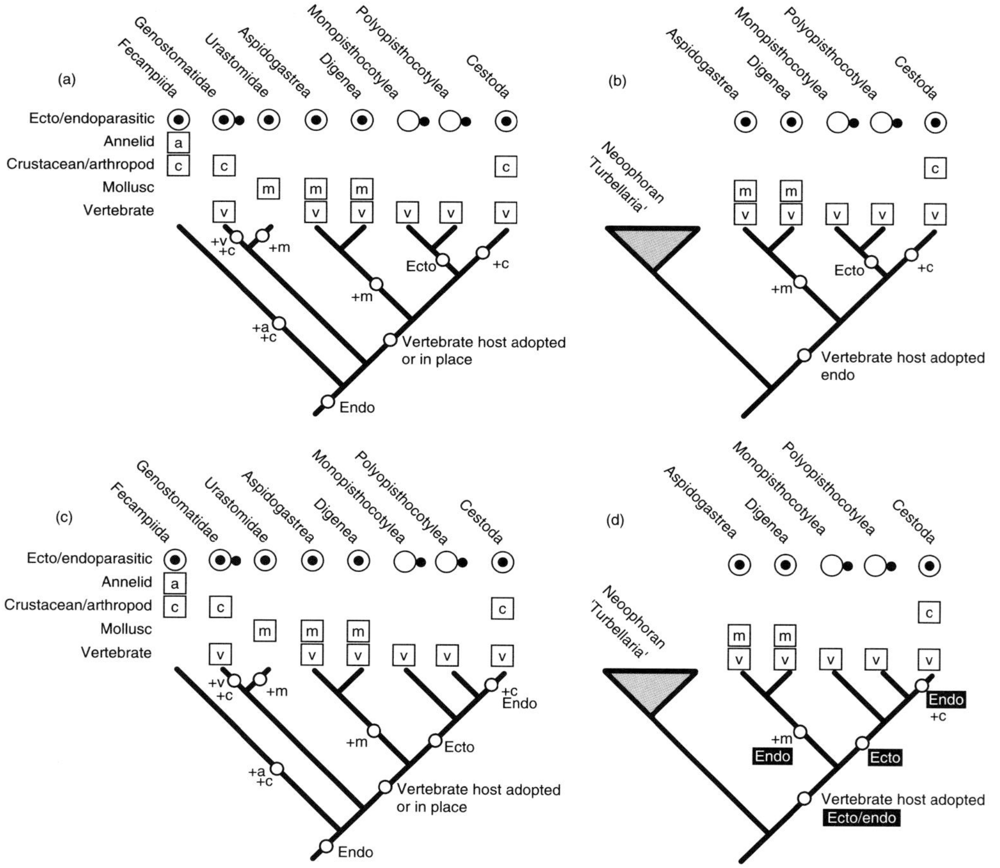

Fig. 1.4. Competing hypotheses on the interrelationships of the Platyhelminthes and the consequences for interpreting the evolution of parasitism. Four competing topologies for the interrelationships of the Neodermata have been proposed based on (a,b) morphology, (c,d) partial LSU rDNA, (e,f) complete SSU rDNA and (g,h) combined complete SSU and complete LSU rDNA.

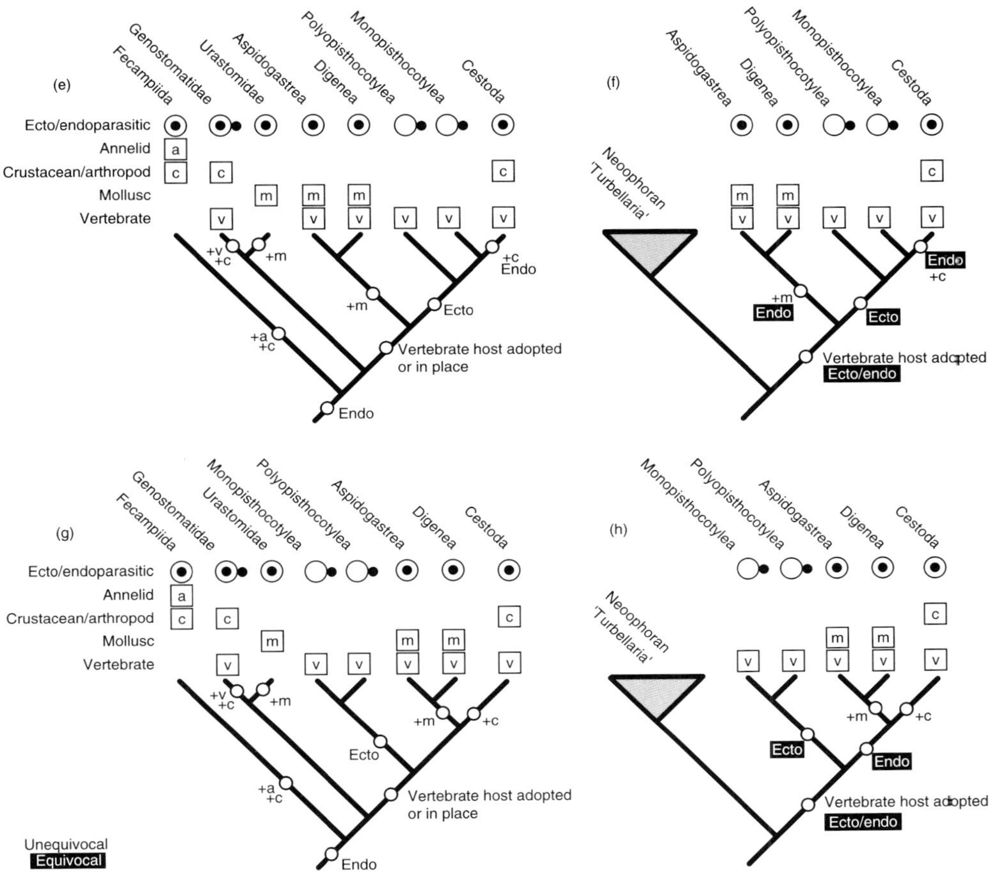

Fig. 1.4. (*cont'd*) Two main competing outgroups for the Neodermata are (a,c,e,g) a clade of parasitic turbellaria, which, with the obligate parasites, form the Mediofusata (see Fig. 1.3c) and (b,d,f,h) a clade of neoophoran turbellarians (see Fig. 1.3e). By mapping site of infection (ectoparasitic/endoparasitic) and identity of hosts used, few topologies offer unequivocal reconstructions of plesiomorphic states An interpretation of the origins of parasitism, the point at which vertebrates were adopted and the adoption of various invertebrate hosts depends on tree topology. Only a (Ehlers, 1985), d (Mollaret *et al.*, 1997), f (Littlewood *et al.*, 1999b) and h (Lockyer *et al.*, 2003a) represent topologies found in the literature, the remaining topologies are hybrid solutions between those offered by molecular and morphological results.

interrelationships of the Neodermata. Recently, Sopott-Ehlers *et al.* (2003) argued that the stem Neodermata must have had mitochondrial lenses, a feature found in larval monogeneans, apparently lost in endoparasitic Monogenea and not present in Cestoda or Trematoda, suggesting that the stem taxa must also have been ectoparasitic. However, whilst the presence/absence of such lenses may indicate ecto/endoparasitism, it says little about neodermatan relationships beyond possibly supporting a Trematoda + Cestoda clade. As with its sister group, until additional data are collected for the Neodermata, the precise interrelationships among the group remain unresolved.

Resolving the interrelationships of the Neodermata, and identifying its sister group, will allow us to narrow the likely evolutionary events that underscored the origins and radiation of the obligate parasitic flatworms, but just as phylogenies are important at this level they are also important to resolve within the major parasitic groups.

Figure 1.3 considers two of the most favoured possibilities concerning the likely sister group to the Neodermata from morphology (Joffe and Kornakova, 2001; Fig. 1.4d) and from a recent molecular estimate (Lockyer *et al.*, 2003a; Fig. 1.4e). In turn, each of these topologies are considered in the light of four recent estimates of the interrelationships of the Neodermata, one from morphology (Ehlers, 1985) and the others representing molecular estimates where the Monogenea are either paraphyletic or monophyletic and not a sister group to the Cestoda. By plotting whether parasites are essentially ecto- or endoparasitic and which hosts they use, simple parsimony character mapping can be used to predict at which points on each cladogram various events in the evolution of parasitism took place. Only five of the topologies reflect results found in the literature, the others combine alternative topologies for different ingroups and outgroups: Fig. 1.4a is the topology suggested by Joffe and Kornakova (1998), Fig. 1.4b by Littlewood *et al.* (1999b) using a combined molecular and morphological approach, Fig. 1.4d by Mollaret *et al.* (1997) using partial LSU rDNA, Fig. 1.4f by Littlewood *et al.* (1999b) using SSU rDNA and Fig. 1.4h using complete LSU and SSU rDNA by Lockyer *et al.* (2003a). The multifarious solutions should alert the reader that without a clear idea of the interrelationships of the Neodermata and without identifying its sister group, the plesiomorphic host(s) or site of parasitism of the ancestral obligate (neodermatan) parasite cannot be inferred. However, even if the phylogeny were resolved, there are few solutions that suggest an unequivocal interpretation. Furthermore, these inferences need to be reconciled with present day diversity and our knowledge of biology.

When did the Neodermata appear?

With no fossil record we can only use indirect evidence to address this question. As neodermatans are all obligate parasites of vertebrates it is to the fossil record of the hosts that we must turn. Among the extant Craniata, all lineages are parasitized by platyhelminths except the Myxini (hagfish) and Petromyzontida (lampreys). It is assumed then that neodermatans became parasites of the Vertebrata at least at the time when Chondrichthyes and Osteichthyes diverged or before (Littlewood *et al.*, 1999a).

The radiation and distribution of parasitic platyhelminths among the chordates is shown in Fig. 1.5. Brooks (1989) suggested that the divergence of the Aspidogastrea and Digenea and that of Gyrocotylidea from other Cestoda was associated with the divergence of the Chondrichthyes and Osteichthyes. Boeger and Kritsky (1997) suggested that among the Monogenea, the Polyopisthocotylea (Oligonchoinea) and the Monopisthocotylea (Heteronchoinea) each radiated with the divergence of the Chondrichthyes and Osteichthyes, suggesting they arose somewhere among the stem group Gnathostomata. Subsequently, based on present day host-associations and our knowledge of the fossil record, the origin of the Neodermata may be estimated to have occurred anytime during the history of the stem group Gnathostomata (540–480 Mya), but not the stem group Vertebrata (560–540 Mya) if extant host-associations reflect early history. To assume that parasitic platyhelminths have been around at least since the appropriate vertebrate and invertebrate hosts were all present provides only a crude estimate of the timing of the appearance of neodermatans. Whether or not the parasitic flatworms appeared earlier cannot be known without additional evidence. Furthermore, phylogenetic uncertainty is not just restricted to the Platyhelminthes; there are ongoing debates on the phylogenetic relationships among the various host groups too (Cracraft and Donoghue, 2004) and there are a wide range of estimates of divergence times for metazoan lineages (Benton and Ayala, 2003). A suitably calibrated molecular clock may provide better estimates of divergence times for the Platyhelminthes. Recent estimates place the divergence between the Mollusca and the Platyhelminthes as 538 Mya (Peterson *et al.*, 2004). However, a molecular-clock approach for estimating

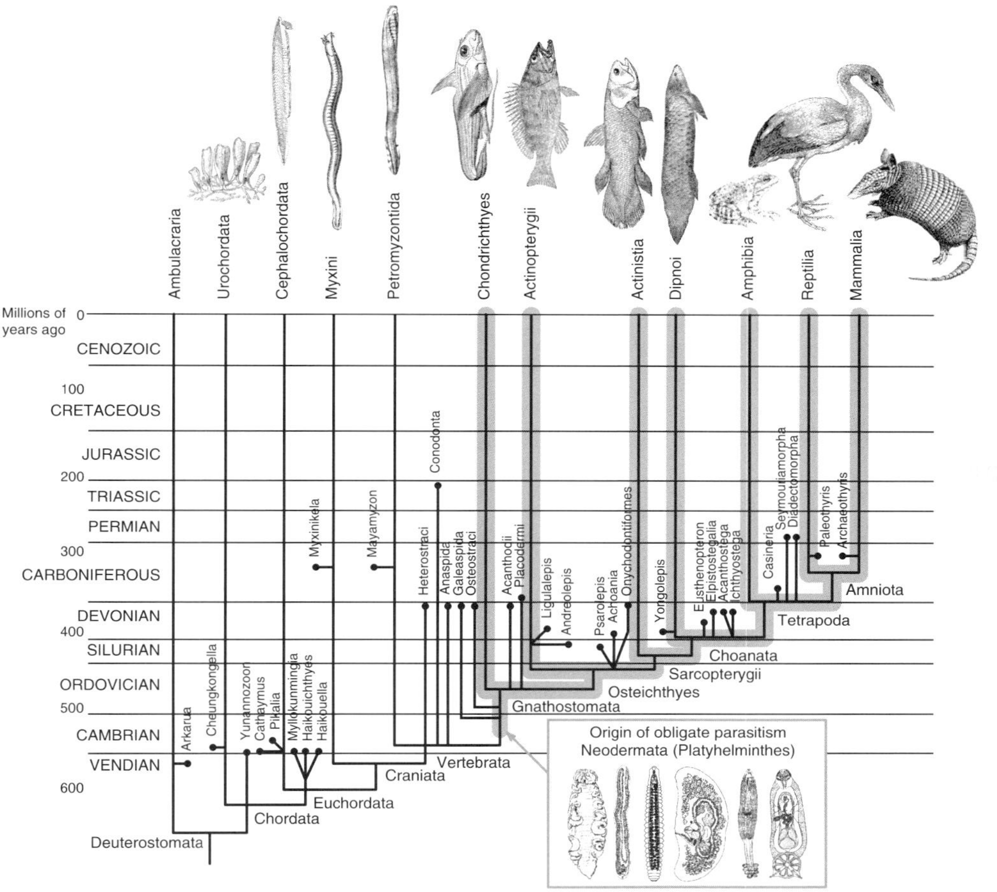

Fig. 1.5. Phylogeny of the deuterostomes indicating position of oldest fossil taxa against the geological record. All neodermatans parasitize vertebrates and, if present, host associations represent historical associations; the Neodermata must have parasitized the stem group Gnathostomata between the Cambrian and Ordovician. Tree redrawn from Rowe (2004).

divergence times within the Platyhelminthes, and in particular for that of the Neodermata, has yet to be attempted, presumably because neither a suitable single gene, a suite of genes, nor a reasonable calibration point for the platyhelminth phylogeny has been found yet, notwithstanding the inherent problems associated with molecular clocks and error estimation (Benton and Ayala, 2003; Douzery *et al.*, 2004). Expressed sequence tag and genome projects on platyhelminth taxa will undoubtedly provide sufficient protein coding genes to attempt these estimates in the near future.

The Neodermata – advantages of a new skin

The tegument of the Neodermata is a living structure that can presumably react dynamically to changes in environmental conditions, engage in defence and provide a means by which nutrients can be absorbed (Halton, 1997; Dalton *et al.*, 2004). Tyler and Tyler (1997) reviewed the four criteria characterized by the neodermatan tegument assumed to be adaptations to parasitism, namely that it: (i) shows features, which will promote nutrient absorption, such as an increase in surface area, best demonstrated by the microtriches found in the Cestoda; (ii) has lost ciliation; (iii) is syncitial, providing a continuous unbroken surface and (iv) that its nuclei lie below the basement membrane and the body wall musculature in so-called cytons or perikarya (described as 'insunk'), but connected to the surface layer by multiple branching bridges. That these features are common to the neodermatan tegument do not necessarily make them adapted to parasitism. No advantage has been suggested for a lack of cilia in adult Neodermata. Although common to the neodermatans and many commensal turbellarians, Tyler and Tyler (1997) noted that a syncitial tegument has also been reported for free-living turbellarians. However, they argue that an insunk tegument might have adaptive value in protecting the nucleated portion of the epidermis from host defences. Also, Dalton *et al.* (2004) noted that disturbance to the genetic control of differentiation and functioning of the tegument will be prohibited, or at least limited by having insunk nuclei. Nevertheless, as the site for nutrient absorption and defence against host attack mechanisms and the fact that the neodermis forms just as the parasite enters the host suggest that it has been critical to the success of the parasitic flatworms.

Tyler and Hooge (2004) reviewed the epidermis of each of the major parasitic platyhelminth groups (see Fig. 1.6) and noted some features unique to particular taxa: adult Aspidogastrea have microvilli shaped like minute, rounded tubercles, under a well-developed glycocalyx; adult Digenea have short, irregularly shaped microvilli under a usually well-developed glycocalyx or no microvilli but a relatively smooth surface with minute folds and canals, and spines formed from actin anchored to the basal membrane and piercing the apical membrane; adult Monogenea have an epidermis with irregular microvilli or folds, sometimes restricted in distribution, and some have rounder tubercles also restricted in distribution; adult Cestoda are perceived to have the most developed epidermis with cylindrical hollow microvilli with strong supports that maintain a high pH preventing the action of host gut enzymes, and that also form the microtrichs in certain body regions. Different life-cycle stages demonstrate different degrees of neodermis development, and these differences are quite distinct between the major neodermatan lineages. For instance, the epidermis is always syncitial in larval Cestoda, although realistically this 'character' is more a failure of the ability to recognize that 'larval' cestodes are in fact simply juvenile adults and so there is little surprise that the neodermis has already developed among these stages. Additionally, some species have unique features of the neodermis,

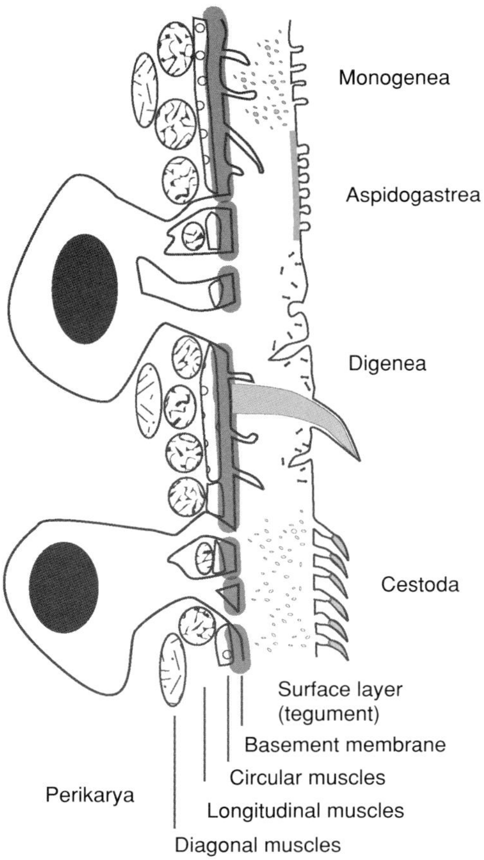

Fig. 1.6. Schematic representation of the adult body wall of neodermatan flatworms as seen in longitudinal section, and subdivided to show typical surface structure typical of major groups. Unlike turbellarian flatworms, diagonal muscle fibres lie innermost. Redrawn from Tyler and Hooge (2004); image kindly supplied by the authors.

presumably as adaptations to their particular parasitic lifestyle (Dalton *et al.*, 2004).

Whatever the functional advantages conferred by the neodermis might be, clearly these need to be elucidated further. The obligate parasitic flatworms share another important characteristic feature – with the exception of a few derived forms, they almost all parasitize vertebrates and use them as a host where they reach maturity, reproduce and complete their life cycle. Monogeneans are found predominantly 'on' their respective hosts and the Trematoda and Cestoda are found predominantly 'in' their hosts.

The Radiation of the Parasitic Flatworms

The functional morphology of the parasitic flatworms is well covered elsewhere (Coil, 1991; Fried and Haseeb, 1991), as are descriptions of the general biology of many of the major groups and taxa of biomedical, veterinary and economic importance (Burt, 1970; Arme and Pappas, 1983a,b; Williams and Jones, 1994; Kearn, 1998; Caira and Littlewood, 2001; Galaktionov and Dobrovolskij, 2003). Excellent taxonomic keys for the Cestoda (Khalil *et al.*, 1994) and the Trematoda (Gibson *et al.*, 2002; Jones *et al.*, 2005; Bray *et al.*, *Keys to the Trematoda*, 2006, in preparation) are available, but nothing of similar breadth is available for the Monogenea beyond what was produced in a classic series of monographs by Yamaguti (1963). Each of the major lineages is characterized by shared morphologies and developmental strategies that enable them to capture and utilize suitable hosts. Here, I consider the major features of each group and demonstrate the appearance of key innovations on recently estimated phylogenies for the major groups. As mentioned earlier, there is still the need to utilize comparative ecological (and biochemical) data in understanding the influences of phylogeny, host and life-history strategy on the evolution of parasitism in the flatworms.

The nature of parasite life cycles is known to shape parasite evolution as they predispose the parasite to different genetic structures (Criscione and Blouin, 2004). Thus, our knowledge of life cycles not only enables an understanding of origins but also evolutionary processes such as speciation, coevolution and host switching. However, although the field is replete with stories as to how complex life cycles may have arisen, hypotheses often remain untested and untestable. Even hypothetical model-driven studies remain divorced from reality in the absence of an evolutionary framework (e.g. Parker *et al.*, 2003).

Beveridge (2001) scored just 205 complete life cycles of Cestoda and 138 of these were for the more easily studied Cyclophyllidea. Generally, revealing life cycles in terms of host association is not a problem for the Monogenea, which entails simply identifying the host on or in which they are found; they use just one host although one species may utilize many different species of host. However, there is a gaping hole in our knowledge of the life cycles of Trematoda and Cestoda. The interpretation of the evolution of life cycles, larval development and transmission strategies are just some of the problems affected by incomplete data sets. Frequently, it is the well-known life cycles, which dominate the interpretation of development and the evolution of parasitism throughout the Trematoda and Cestoda, and more often than not, these life cycles, including those affecting humans, are peculiar to the species and not to the higher taxa they are purported to represent.

Understanding the radiation of the obligate parasites requires knowledge of parasite phylogeny, parasite life cycle, host phylogeny and parasite biology, particularly in relation to its role in the various host–parasite combinations. None of these elements are fully understood for any single group of parasites, although those parasites of biomedical, veterinary or economic importance are generally much better studied than others. Modern phylogenetic methods and the use of evolutionary trees depicting interrelationships to infer evolutionary patterns and processes are relatively the recent tools employed by evolutionary parasitologists (e.g. Littlewood, 2003 and chapters therein). The majority of comparative data available comes from a long history of morphological and ultrastructural studies, ecological assessments taking the form of faunal lists and biochemical studies of selected model or important parasites. From a wider biodiversity perspective, life-cycle studies are a much-neglected element of parasitology that until recently relied on laborious and difficult experimental manipulations. More recently, tracking species and identifying larvae or juveniles by molecular means are allowing accurate, albeit indirect, life-cycle inference. As a result, larval and juvenile hosts can be readily identified through gene sequencing in the absence of key morphological features and linked to their fully characterized adult counterparts (e.g. Dezfuli *et al.*, 2002). The growth of molecular systematics is providing a foundation for species identification and phylogenetic placement where morphology alone has previously failed. Presently, completed life cycles and an understanding of host–parasite associations and diversity remain very patchy and reflect more the sampling preferences and/or capabilities of the parasitologists who studied them. Patchiness of data sets limits on the conclusions we can draw from comparative evolutionary studies (e.g. Poulin, 2002). However, in the light of over 100 years of parasitology concerning platyhelminths we can illuminate a number of key elements underlying their successful radiation. Simple host–parasite associations are useful (e.g. http://www.nhm.ac.uk/research-curation/projects/host-parasites/), as is the understanding of the host–parasite interaction. Finally, but perhaps of greatest significance, parasites offer unique opportunities to study evolutionary processes that generate diversity (Brooks and McLennan, 1993; Poulin and Morand, 2000).

Radiation of the Monogenea

Monogenea are generally characterized by the possession of an anterior attachment structure, with one or two suckers and a posterior attachment structure (the haptor) armed variously with hardened structures including hooks, hooklets, hamuli (anchors) and suckers. Haptors may also possess glands, can be symmetrical or asymmetrical and may be divided into many loculi. The mouth of Monogenea is anterior and it generally leads to a muscular pharynx and then into a two-part blind-ending intestine. Male and female genital pores open together and most species are oviparous with short-lived larvae (oncomiracidia) hatching from eggs to find a suitable host. Adults are generally ectoparasites of cold-blooded vertebrates but a few occur in internal body cavities and rarely in the digestive tract of their hosts. With a direct life cycle, generally with high host-specificity and well-adapted attachment structures allowing them to secure themselves to all manner of structures (e.g. gills, fins, scales, eyes), this group is distributed widely globally and on or in most aquatic and some terrestrial vertebrates. No monogeneans infect humans although some are important parasites of commercial fish stocks.

Apomorphies that unite the Monogenea include:

- larva with three ciliated zones;
- larva and adult with two pairs of pigmented eyes;
- one pair of ventral anchors;
- one egg filament.

Although small and difficult to handle, experimental parasitologists and systematists have employed the Monogenea to address evolutionary questions concerning the radiation of parasites that cannot be achieved easily with the other parasitic flatworms with their more complex life cycles. Arguably, monogeneans remain a model only for one-host life cycles and cannot provide insight into more complex systems. As Poulin (2002) noted, Monogenea are an ideal group for investigating the processes behind their past diversification and present diversity because their phylogeny is reasonably well resolved at least to the family level, they are numerous, morphologically diverse and generally host-specific; for a review on host specificity in Monogenea see Whittington *et al.* (2000b). Coevolution studies, tracking host switching, measuring host specificity, determining modes of speciation and biogeographic studies have all been attempted with Monogenea (Bakke *et al.*, 2002; Desdevises *et al.*, 2002; Hoberg and Klassen, 2002; Poulin, 2002; Zietara and Lumme, 2002; Huyse *et al.*, 2003; Simkova *et al.*, 2004; Whittington, 2004).

Molecular estimates of monogenean phylogeny (Olson and Littlewood, 2002) are largely congruent with recent estimates based on morphology (Boeger and Kritsky, 2001), at least at the family level (Fig. 1.7), but greater sampling of taxa for molecular analysis is needed to fully test the hypotheses suggested by the morphological data.

All Monogenea (or Monogenoidea as some authors prefer; see Wheeler and Chisholm, 1995; Boeger and Kritsky, 2001) are dependent on an aquatic environment for the development of their eggs and the distribution of their larvae (Kearn, 1994). Although predominantly ectoparasitic (often on skin or gills), there are many successful endoparasitic Monogenea that tend to be restricted to the least hostile environments offered within the host. The mouth, pharynx, reproductive system, urinary system, body cavity, olfactory capsules and ocular orbits are the most common sites inhabited by endoparasitic Monogenea, although

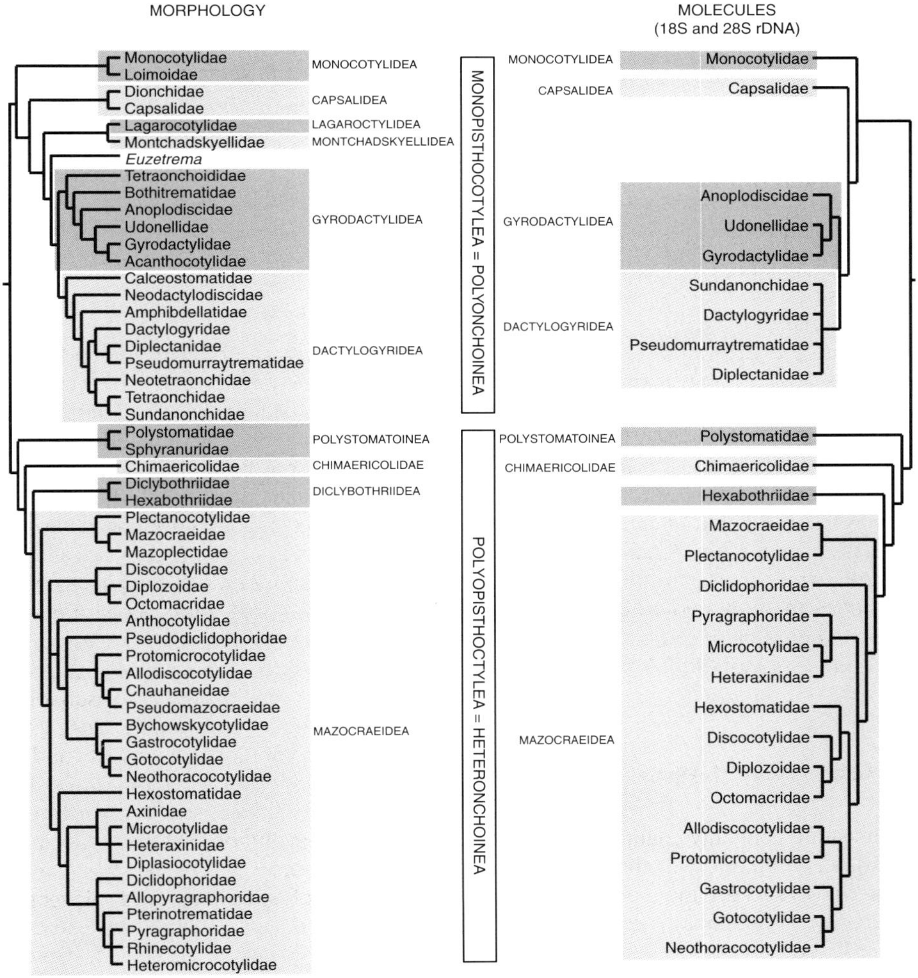

Fig. 1.7. Interrelationships of the Monogenea. An extensive morphological analysis by Boeger and Kritsky (2001) is largely supported by a preliminary molecular analysis by Olson and Littlewood (2002). In each case, the Monogenea were treated as monophyletic although it is possible that the Monopisthocotylea (Polyonchoinea) and Polyopisthocotylea (Heteronchoinea) form a paraphyletic Monogenea (see text).

some species inhabit the gut and stomach (Kearn, 1994). Collectively they are found in or on a diversity of freshwater, euryhaline and marine vertebrates including teleosts, chondrosteans (sturgeon and paddlefish), elasmobranchs, holocephalans, amphibians, reptiles, squid, marine crustaceans and even the eyes of hippopotamus. Whittington *et al.* (2000b) suggest that monogeneans are the most host-specific of parasites in general and may be the most host-specific of all fish parasites. Subsequently, their evolution is considered to be intimately connected to that of their hosts, with coevolution and strict cophyly playing important roles in their radiation (Kearn, 1994; Boeger and Kritsky, 1997; see Fig. 1.8).

Generally with both anterior and posterior attachment structures, monogeneans are capable of movement on and in their hosts, and they tend to migrate towards favoured sites of infection once settled. The posterior attachment structure, the haptor, may bear glands, suckers and may be divided into numerous loculi; the terminology of the various structures, hooks and sclerites is complicated and some researchers use these terms differently. Usually with an anterior mouth, most monogeneans possess a muscular pharynx and a blind-ending intestine with two, or sometimes just one, caeca. They feed usually on the skin, the mucus it produces or blood of their hosts. All monogeneans are hermaphroditic and most possess a ciliated larval stage known as an oncomiracidium. With a single host monogeneans have relatively simple life cycles. The oncomiracidium bears a posterior haptor that develops into the adult haptor. Asexual multiplication is uncommon within the group, although it occurs in the form of sequential polyembryony (hyperviviparity) in some Gyrodactylidea. Adult viviparous gyrodactylids can produce sequential generations of offspring without

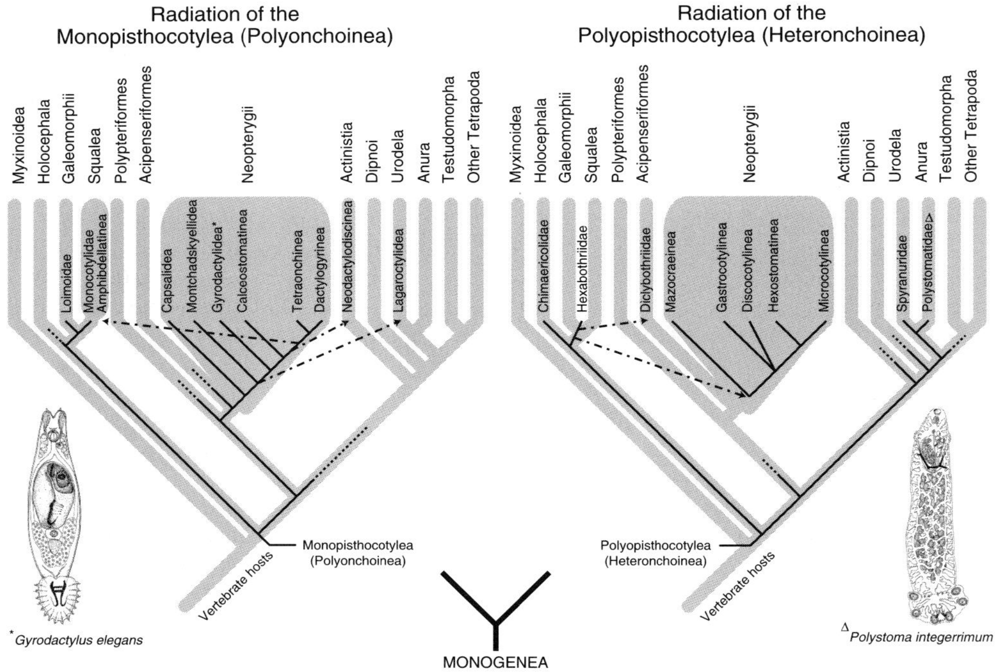

Fig. 1.8. The two main groups of Monogenea each diverged with the Chondrichthyes and Osteichthyes, suggesting a relatively ancient origin. Boeger and Kritsky (1997) proposed historical relationships of the Polyonchoinea and Heteronchoinea and their hosts; redrawn here. The parasite cladograms (solid lines) are superimposed on that of their plesiomorphic hosts (broad grey lines). Dotted lines indicate postulated extinctions; dashed–dotted lines indicate postulate primary host switching (dispersal) of parasites.

releasing them, resulting in a Russian doll effect until a single individual might harbour up to 20 successive generations *in utero* (Harris, 1983, 1985, 1993; Kearn, 1994). Other defining features of the Monogenea and the main constituent groups are detailed in Fried and Haseeb (1991), Kearn (1998) and Caira and Littlewood (2001). Whittington *et al.* (2000a) provide an excellent review of the larvae of Monogenea.

An account of the evolutionary expansion of the Monogenea was provided in some detail by Kearn (1994). The study is rich in comparative biology, but without a phylogenetic framework many of the conclusions remain untested. Boeger and Kritsky (1993, 1997, 2001) are responsible for comprehensive phylogenetic estimates of the Monogenea based on morphology and these estimates have been largely corroborated through a series of molecular studies using rRNA gene sequence data (Mollaret *et al.*, 1997; Olson and Littlewood, 2002). Boeger and Kritsky are also responsible for in-depth interpretations of the phylogenies in terms of evolutionary expansion, host use and dispersal, through comparison of host and parasite phylogenies (Boeger and Kritsky, 1997). Figure 1.8 summarizes the proposed historical relationships between the Monogenea and their hosts. The interpretation remains largely unaffected if the Monogenea are indeed paraphyletic. The authors conclude that the two main monogenean clades underwent sympatric speciation or dispersal while parasitic on stem-group (ancestral) Gnathostomata, resulting in the Monopisthocotylea and Polyopisthocotylea. These lineages in turn each cospeciated with the divergence of the Chondrichthyes and Osteichthyes. Thereafter the evolutionary histories of each group differ somewhat and whilst the authors rely heavily on cospeciation as the main evolutionary force explaining the radiation of Monogenea, they necessarily discuss host switching and extinction events as various extant vertebrates are not infected by monogeneans. Whereas most monogenean lineages track their vertebrate hosts throughout their subsequent radiations, with many basal groups such as Chimaericolidae, Loimoidae and Monocotylidae parasitizing Holocephala, Galeomorphii and Squalea, respectively, there are some notable host switching events (indicated with dashed–dotted lines, Fig. 1.8) that indicate major host changes during the radiation of the Monogenea. A number of hypotheses remain plausible, although not all are most parsimonious. Until the phylogenies of each host group are resolved fully, and the means by which cophylogeny mapping can be easily implemented (Page, 2003) it seems reasonable to accept the interpretations provided by Boeger and Kritsky (1997).

Monogeneans are not only frequently host-specific but are also typically site-specific. Phylogenetic studies of site-specific polystomes from freshwater turtles sharing ancient histories have revealed that morphological evolution in these groups is very slow. Many of these parasites occupy particular sites of infection such as oral cavities and urinary bladders. Additionally, studies of congeneric taxa have shown that species infecting a particular site in one host have given rise to species infecting the same site in different hosts; rather than one host speciating to occupy different sites in the same host (Littlewood *et al.*, 1997). Further studies are required to see to what extent site-specific speciation occurs across the Monogenea. Polystomes diverged at least 200 Mya and their hosts (mostly anurans and turtles, but also lungfish and the African hippopotamus) hold the key to understanding the phylobiogeography of host–parasite relationships within and between post-Gondwanan land masses (Bentz *et al.*, 2001, 2003; Sinnappah *et al.*, 2001; Verneau *et al.*, 2002). Among the polystomes it would be interesting to see where the only polystome parasite (*Oculotrema hippopotami*) of the African hippopotamus, which infects its eyes, fits into their radiation (du Preez and Moeng, 2004).

The one-host life cycle of Monogenea has enabled studies on the colonization and radiation of freshwater systems by South American sciaenid fishes (Boeger and Kritsky, 2003), and a test of whether the species rich Gyrodactylidae underwent an adaptive radiation through multiple vicariant events or adaptive modes of speciation (Boeger *et al.*, 2003). In the latter, rapid diversification of Gyrodactylidae was facilitated by, among other features, the development of hypervivparity and high dispersal to new host groups and geographical areas.

Phylogenetically based studies offer the opportunity to reveal additional micro- and macroevolutionary patterns in the Monogenea, but details on biogeographic ranges, larval development and dispersal, site specificity and phylogenies of the hosts are needed to complete a broader understanding of their widespread radiation.

Radiation of the Cestoda

All but the oldest tapeworm lineages completely lack organs of a digestive system; the anterior attachment organ of Amphilinidea and Gyrocotylidea are considered by some to be vestiges of such a system. Without a mouth, pharynx, oesophagus or intestine at any stage during their development the true tapeworms (Eucestoda) are quite different from the other platyhelminth groups. Instead, tapeworms gain their nutrition by absorption through the neodermis, often modified with microtriches that assist in this process. All cestodes have at least two hosts (except in cases with one host through secondary reduction of a two-host life cycle), adding considerable complexity to their life cycles compared with the Monogenea (see Fig. 1.9). Movement from host to host is always passive, except in amphilinideans. The first host is commonly an arthropod in which larvae develop into the metacestode stage. Having arisen among aquatic vertebrates they have successfully made the transition, along with their vertebrate hosts, to freshwater, land and back to the sea again (with marine mammals). Often long-lived and with the ability to increase their reproductive potential through the serial repetition of gonads by strobilation (usually accompanied by segmentation of the body) and, in some, asexually with the production of cysts, tapeworms are ubiquitous, globally distributed and species rich.

Apomorphies that unite the Cestoda include:

- all stages without an intestine;
- neodermis with distinct microtriches (also known as microvilli or microthrix);

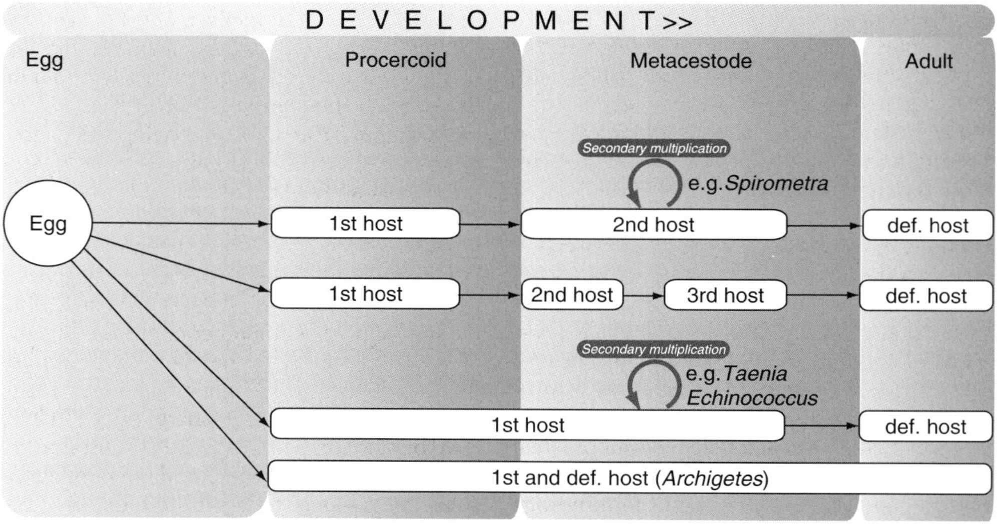

Fig. 1.9. Life-cycle variations among the Cestoda, indicating the position of the main developmental stages in relation to hosts. Usually with two or more hosts, cestodes develop through a number of stages between egg and adult, often distinct and with their own terminology for different tapeworm groups (Chervy, 2002). Asexual multiplication by metacestode proliferation occurs in about 20 species of polyzoic tapeworms through larval stages known as coeneri or hydatid cysts. The caryophyllaeid *Archigetes* undergoes all its development in a single host, an oligochaete annelid.

- first canal of protonephridium lacks cell gap and desmosome;
- reticulate protonephridial system in post-larvae;
- cell bodies of protonephridial canal cells under basal lamina;
- larval epidermis is syncitial, neodermal tissue does not reach body surface;
- ten larval hooks;
- large body dimensions;
- apical pit forms when in first host;
- male copulatory organ a cirrus;
- vertebrate host in life cycle.

Amphilinideans and gyrocotylideans possess a ciliated larval stage (lycophore) with ten hooks at the posterior end; see a review on these groups in Xylander (2001). Eaten by small crustaceans, or penetrating them directly, amphilinidean larvae develop further by shedding their ciliated epidermis and growing. Development into the adult stage takes place when the crustacean is eaten by a suitable definitive host. Gyrocotylideans may have a direct life cycle with their holocephalan hosts, although Dollfus (1923) reported a record of a larval form being found in the mollusc *Mulinia edulis*; certainly they are very strictly confined to parasitizing Holocephali and it is thought they simply failed to radiate with their hosts or host-switch (Xylander, 2001). Amphilinids are restricted to the coelomic cavities of their hosts (chondrosteans, some freshwater and marine teleosts and some freshwater turtles) but are more broadly distributed.

Eucestodes produce a larval stage (hexacanth) with three pairs of hooks that penetrate, after ingestion, the first intermediate host. The development of the juvenile (metacestode) stage takes place in the first intermediate host and may continue until the cestode reaches the definitive host. The nomenclature for 'larval' cestode (metacestode) forms remains confused as terminology has conflated developmental stage, life-cycle stage and morphology (Chervy, 2002); an improvement in terminology requires a better understanding of development within and between cestode lineages.

The majority of adult tapeworms have an attachment region, the scolex, from which a series of proglottids develop, giving the animal a segmented appearance; proglottids distal to the scolex are the oldest and most developed and frequently contain eggs. This is not true segmentation but serial repetition of proglottids through strobilation. The proglottids of the majority of tapeworms are hermaphroditic with one or more sets of male and female reproductive organs. Caryophyllideans are the most basal eucestodes and have a single set of reproductive organs. The spathebothriideans suggest an intermediary step in the body plan organization of cestodes, as they lack proglottids but possess multiple sets of reproductive organs arranged along the length of the body. All the more derived taxa exhibit serial repetition of reproductive organs (proglottization) and external subdivision of proglottides. The scolex is often armed with hooks, suckers or hooked tentacles.

The radiation of the tapeworms suggests an evolutionary trend towards higher fecundity and more complex scolex armature, the former (and possibly the latter) presumably intimately linked to the evolution of tapeworm life cycles involving different intermediate and vertebrate host taxa (Hoberg *et al.*, 1997; Olson *et al.*, 2001). Morphological adaptations to different sites of infection or host identity have not been studied in great detail (except see Williams, 1966, 1968) but certainly many lineages are classified according to these features, and lineages tend to be relatively narrowly confined to particular host groups (Khalil *et al.*, 1994; Hoberg *et al.*, 2001b). Finer phylogenetic resolution is required to test patterns of host association within the tapeworms.

Both morphologists and molecular systematists have contributed to estimating phylogenies for the Cestoda and although many general patterns of tapeworm evolution are accepted, the monophyly of a number of groups (e.g. the 'Trypanorhyncha') and the interrelationships of the tetrafossate clades remains problematic (Brooks *et al.*, 1991; Hoberg *et al.*, 1997, 2001b; Mariaux and Olson, 2001; Olson *et al.*, 2001). A generalized view of the interrelationships of cestode orders based on multiple sources is shown in Fig. 1.10, indicating the evolutionary trend towards serial repetition of reproductive organs (polyzooy, with monozooy as ancestral), proglottization and the differentiation of the scolex into a tetrapartite condition.

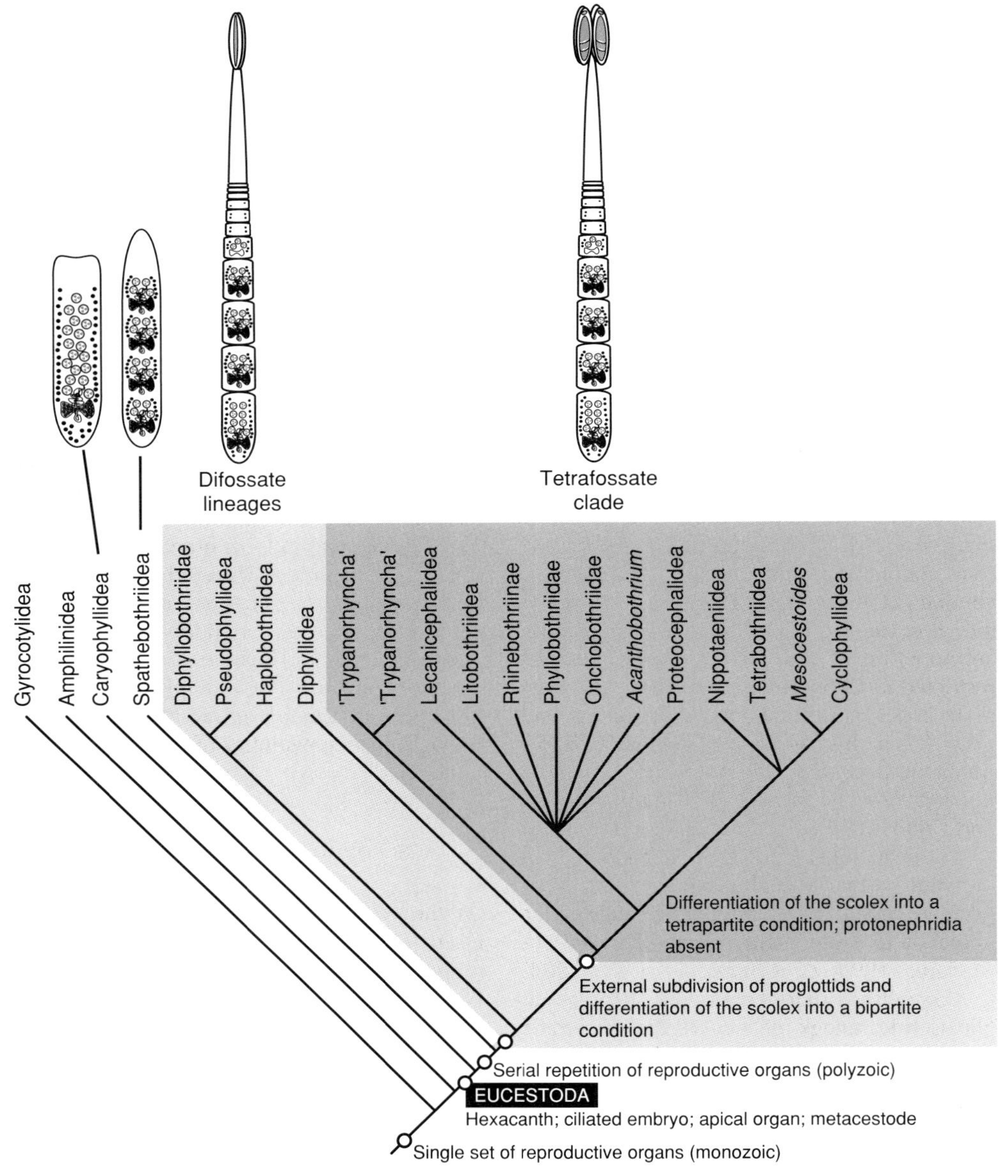

Fig. 1.10. Interrelationships of the Cestoda; adapted from Olson *et al.* (2001) after Hoberg *et al.* (2001b). The serial repetition of gonads appeared in cestode evolution with the appearance of the Spathebothriidea; proglottization immediately after that with the emergence of difossate lineages; and then the majority of lineages with a four-part scolex appearing thereafter.

Most eucestodes require at least two hosts to complete their life cycles but the use of paratenic hosts is common. Little or no development takes place in a paratenic host but cestodes may use them to bridge food-chain gaps so as to reach appropriate vertebrate hosts. By far the best-studied tapeworms are the Cyclophyllidea, as their hosts (commonly freshwater or terrestrial) are easier to sample than many other orders and it is by far the biggest group. There are some orders with little or no information about life cycles (e.g. Tetraphyllidea, Trypanorhyncha, Lecanicephalidea, Diphyllidea, Tetrabothriidea, Spathebothriidea), thus preventing an in-depth analysis of their radiation. Of the life cycles known for the Cestoda (Beveridge, 2001) and with reference to estimates of phylogeny, it seems that arthropods are the primitive intermediate hosts. Associations with definitive hosts show that many basal cestodes are present in basal vertebrate groups (e.g. gyrocotylideans in Holocephala), but as with most parasitic platyhelminths, it appears that host switching has been relatively common with many lineages parasitizing a diversity of vertebrates, e.g. Caryophyllideans and basal cestodes occur in siluriform and cypriniform freshwater fishes; Pseudophyllidea occur as adults in marine and freshwater fishes, some piscivorous mammals, lizards, snakes, anurans and caudates. Other groups are more restricted in their host use, e.g. Diphyllidea, Trypanorhyncha, Tetraphyllidea and Lecanicephalidea are restricted to Chondricthyes as adults, although many use marine teleosts as intermediate hosts. It is in these groups, which tend to be the dominant parasites of sharks and rays, where a much poorer diversity of Digenea exists (Bray and Cribb, 2003).

Higher level phylogenies tracking the radiation of individual cestode orders have been attempted for a number of groups, in particular the Proteocephalidea (Zehnder and Mariaux, 1999; de Chambrier *et al.*, 2004), Cyclophyllidea (Hoberg *et al.*, 1999; von Nickisch-Rosenegk *et al.*, 1999; Wickström *et al.*, 2003), Pseudophyllidea (Bray *et al.*, 1999), Tetraphyllidea (Caira *et al.*, 1999, 2001; Olson *et al.*, 1999; Caira and Jensen, 2001), Diphyllidea (Caira *et al.*, 1999) and Lecanicephalidea (Caira *et al.*, 1999, 2001). Many of these studies are systematic in nature although some consider the evolutionary history of the parasites and their hosts (e.g. Hoberg *et al.*, 1999; Olson *et al.*, 1999; Caira *et al.*, 2001). Lacking the simplicity of the monogenean life cycle, considering the relative ease in capturing and identifying all the intermediate host stages of digeneans (in particular with respect to first intermediate hosts), and the lack of synoptic studies on tapeworm life cycles, there is a relatively poor understanding of the radiation of cestodes with their hosts.

Much research effort has been expended on tapeworms infecting humans, but not in terms of their evolutionary radiation. Ashford and Crewe (2003) listed 54 species of cestodes (from one family of Diphyllobothriidea and six families of Cyclophyllidea) found in humans but only few of these are of major significance (e.g. *Hymenolepis nana*, *Taenia saginata*, *T. solium*, *T. asiatica*, *Echinococcus granulosus* and *E. multilocularis*) and these are all cyclophyllideans. All occurrences of cestode infection in humans, and indeed by all the definitive hosts infected by eucestodes, arise through the ingestion of larval stages in an intermediate host. Only among the Taeniidae has the origin and evolution of human tapeworms been closely studied; phylogenetic analyses of samples from humans and wildlife suggested that *Taenia* entered humans on at least two separate occasions in Africa during their evolutionary history (Hoberg *et al.*, 2000, 2001a). Similar phylogenies show that within the Taeniidae asexual multiplication appears to have been lost and recovered several times throughout their evolution (Trouvé *et al.*, 2003).

Radiation of the Aspidogastrea

Of the two trematode lineages, the Aspidogastrea have frequently been considered to be more 'primitive'. As the sister group to the Digenea this is in some respects erroneous; sister taxa diverge at the same time. However, there are features and behaviours among extant Aspidogastrea that are thought to reflect the likely habits of the proto-trematode. Interpretation of these features in the literature is somewhat biased towards accepting proto-trematodes use of mollusc hosts before adopting vertebrates (Gibson, 1987), but is also bolstered by their use of early branching vertebrates (holocephalans and elasmobranchs) as hosts, low host specificity among their mollusc hosts and a relatively simpler life cycle than Digenea (Rohde, 2001); Aspidogastrea lack multiplicative larval stages in the mollusc host and indeed lack any asexual reproductive strategy in their life cycle. Also, unlike Digenea, Aspidogastrea enter their molluscan hosts via external openings, as opposed to penetrating their tissue. Some species (e.g. *Aspidogaster conchicola*) reach sexual maturity in molluscs and the use of vertebrates in some taxa is facultative (e.g. Rohde, 2001). Rohde (1994) also suggested that the complex life cycle of the Digenea was readily derived from the simple life cycle found in Aspidogastrea. Complete life cycles are only known for the Aspidogastridae; life cycles for the other three families, the Multicalycidae, Stichocotylidae and Rugogastridae, remain elusive. The Rugogastridae (with two species infecting holocephalan fishes) and the Multicalycidae (from the gall bladder of holocephalans and elasmobranchs) each include only a single genus. Indeed, most of the families appear to be relictual with very few species. The Aspidogastridae include many more species, in ten genera, infecting teleosts and turtles, but it remains a small family in comparison to those of the Digenea. Only few attempts at reconstructing phylogenies for the group have been made. Detailed studies on aspidogastreans may be found in Rohde (1972, 2001) and Gibson (1987) and keys to genera in Rohde (2002). An excellent overview of the group, prepared by Klaus Rohde, is available online from the Tree of Life website at: http://tolweb.org/tree?group = Aspidogastrea& contgroup = Platyhelminthes#TOC3

Apomorphies that unite the Aspidogastrea include:

- larva with ventrocaudal sucker becoming alveolated adhesive organ in adults;
- few ciliated cells in larvae;
- neodermis with characteristic microvilli (= microtubercles).

Most sexual adult Aspidogastrea generally live in the intestines of vertebrates and release eggs in the faeces, but there are species restricted to gall bladders, rectal glands and other sites. These hatch in the external environment to release a swimming larva, the cotlyocidium that seeks out and attaches itself to a mollusc. The vertebrate is always infected by ingestion of the mollusc. Variations on this theme exist (Rohde, 1972). During development the posterior sucker of larval Aspidogastrea divides into smaller suckers (suckerlets, alveoli or rugae), each divided by a transverse septum. These rugae give the adult worm the appearance of being segmented. Such pseudosegmentation is not entirely superficial though as the nervous system and marginal glands also appear to show repeated organization in relation to the rugae (Rohde, 2001). Indeed, the nervous system of the Aspidogastrea is well developed (Rohde, 1971).

Gibson (1987) was the first to suggest a phylogeny for the Aspidogastrea and mapped on to its key host associations and morphological changes; a reanalysis of these morphological characters by Brooks *et al.* (1989) yielded a different solution, shown in Fig. 1.11. Rohde (2001) prepared an explicit morphological matrix for the group but these characters alone were unable to provide a resolved tree. However, the Aspidogastridae were resolved as the sister group to a polytomy uniting the Stichocotylidae, Multicalycidae and Rugogastridae. In contrast, a recent morphological analysis by Zamparo and Brooks (2003), coding 20 aspidogastrean taxa using 33 characters confirms the original general topology suggested by Brooks *et al.* (1989), and provides resolution within the Aspidogastridae among which only two constituent subfamilies were resolved as monophyletic. Preliminary molecular analyses (D.T.J. Littlewood, P.D. Olson and K. Rohde, unpublished data) support the basal position of the Rugogastridae but the Aspidogastridae are polyphyletic with the Multicalycidae nested within them (see also Olson *et al.*, 2003). Gene sequencing of all families is currently hampered by an inability to secure tissues of *Stichocotyle nephropis*, the single member of the Stichocotylidae.

There are no aspidogastreans of medical or economic importance, but their sister group status to the Digenea is critical in understanding their radiation.

Radiation of the Digenea

The Digenea is arguably the largest group of internal metazoan parasites with over 2500 nominal genera in over 150 families. As with the other obligate parasite flatworm groups, the Digenea have a distinctive morphology,

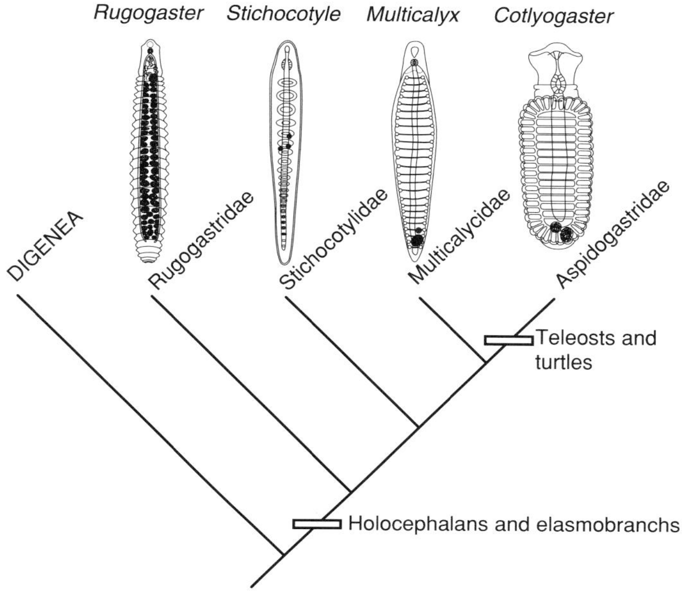

Fig. 1.11. Phylogeny of the Aspidogastrea at the family level, according to a reanalysis of characters in Gibson (1987) by Brooks *et al.* (1989) and recently from a broader analysis by Zamparo and Brooks (2003); figures of animals redrawn from Rohde (2001).

reproductive habit and attachment organs that are specialized for parasitism. Usually with three hosts – first a mollusc, second an invertebrate or vertebrate and third a vertebrate – in its complex life cycle an enormous diversity of host species are affected by Digenea making them important, although often overlooked, animals in many ecosystems (Cribb *et al.*, 2001b). Poulin (1998) noted that the complexity of a parasite life cycle may influence the number of opportunities for speciation and diversification. Indeed, this may have influenced the expansion of the Digenea and rendered it the most species-rich group of parasitic flatworms, although the idea remains to be tested formally.

Characterized by unique larval stages, including miracidia, sporocyst, redia, cercaria and metacercaria, digeneans are highly diverse in morphology and life habit as adults. Most adults possess anterior and mid-ventral suckers, an anterior mouth opening into a bulbous pharynx and a blind-ending intestine with one or two main branches and sometimes with diverticulae.

Apomorphies that unite the Digenea include:

- series of asexual generations in first intermediate (mollusc) host;
- ciliated epidermal cells of miracidium arranged in regular transverse rows;
- jawed vertebrates in complex life cycle;
- cercaria;
- miracidium and mother sporocyst without digestive system.

Yamaguti (1975) reviewed a wealth of literature on digenean life cycles concerning 69 families dealt with in over 1700 studies spanning the previous 90 years. Dr Tom Cribb (University of Queensland) has collated a database on digenean life cycles for approximately 1350 species and, combined with recent studies on digenean phylogenetics, this has provided the means by which patterns of host association, host use, modes of parasite transmission, life-cycle complexity and morphological evolution can be inferred. Cribb *et al.* (2003) described the evolution of the digenean life cycle in some detail. Digenea have developed alternation of sexual and asexual generations in their life cycle and the life cycle involves both free-living and parasitic stages. As adults they are primarily parasites of vertebrate intestines, but may also be found in the body cavity, the urinary bladder, the gall bladder, in the flesh, connective tissue, ovary, circulatory system, in fish under their scales, on their gills and in their swim bladder and in tetrapods in the oesophagus, liver and eye (Cribb *et al.*, 2003). In these various sites adult Digenea produce eggs that pass out to the external environment whereupon they hatch to release a non-feeding, ciliated, swimming larva, the miracidium, which penetrates a molluscan host. In this intermediate host, the digenean develops into a mother sporocyst, which reproduces asexually to produce either multiple daughter sporocysts or rediae. In each of these morphologically different forms, there is another round of asexual reproduction whereupon cercariae are produced. Usually with a tail, the cercariae actively emerge from the mollusc and either penetrate, encyst in, or are eaten by a second intermediate host. Frequently, the digenean develops a little further, as a metacercaria in this second host, which is itself ingested by the third and definitive host in which the sexual adult develops. In two-host life cycles the ingested/encysted cercariae mature through metacercariae to adults in the second (vertebrate) host. Multiple variations on the general theme exist with the addition or loss of intermediate hosts (see Cribb *et al.*, 2003, and references therein; Fig. 1.12).

Cribb *et al.* (2001b) reviewed the early views on the interrelationships of flukes and provided new estimates of their phylogeny based on an explicit character matrix and a molecular analysis based on SSU rDNA. Thirty adult and 26 life-cycle characteristics provided a tree largely congruent with the molecular estimate, and a combined morphological and molecular analysis provided a reasonably well-resolved phylogeny, in spite of a few polytomies. More recently, partial LSU and complete SSU rDNA has provided the foundation for a further resolved phylogenetic tree that has been the basis for the interpretation of morphology, a re-evaluation of relationships and nomenclature (Olson *et al.*, 2003) and a study

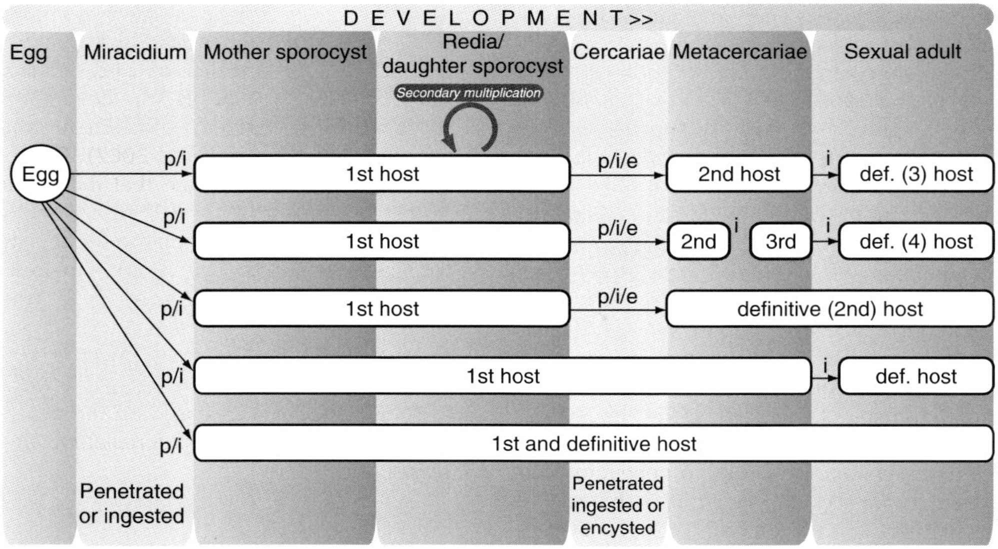

Fig. 1.12. Life-cycle variations among the Digenea, indicating the position of the main developmental stages in relation to hosts. Generally more complex than tapeworm life cycles, digeneans utilize more hosts and include more discrete developmental stages. Secondary multiplication is ubiquitous and takes place within sporocysts and rediae.

on the evolution of life cycles (Cribb *et al.*, 2003; see Fig. 1.13). The Digenea are split into two main groups, the Diplostomida and the Plagiorchiida (Olson *et al.*, 2003) comprised of 1 and 13 orders, respectively. A total of 22 superfamilies were recognized; but see Gibson *et al.* (2002).

The main host group identities used as first, second (where applicable, 17 of the 22 digenean superfamilies) and definitive hosts are mapped against a summary phylogeny for the Digenea in Fig. 1.14. A glance at the phylogeny indicates that the more derived Digenea have adopted more complex life cycles involving a greater diversity of host taxa. Choisy *et al.* (2003) argued, from simulation models, that selection favours a complex life cycle only if intermediate hosts are more abundant than definitive hosts, and that the selective advantage of a complex life cycle increased with predation rates by definitive hosts on intermediate hosts. Gastropods are by far the commonest group of molluscs used as first intermediate hosts with bivalve molluscs used by only eight superfamilies; two using bivalves exclusively, the Bucephaloidea and Gymnophalloidea. Some members of the Hemiuroidea use scaphopod molluscs and some of the Sanguinicolidae have been shown to use annelids. Parasitism of non-gastropods is thought to have been the result of host switching from gastropods (Cribb *et al.*, 2001a, 2003). Three-host life cycles have been adopted repeatedly and it is most parsimonious to infer that three-host life cycles are derived from two host life cycles. Indeed, with two-host life cycles characterizing the Aspidogastrea this seems to be the plesiomorphic (ancestral) state of the Digenea. The use of different hosts throughout the various life cycles characterize various adaptations of both the life cycles themselves and the evolutionary development of the intermediate life stages, e.g. cercariae exist with or without tails and with or without the ability to penetrate hosts, there are numerous behavioural strategies of cercariae that increase the likelihood of transmission, sporocysts are more common among bivalves than gastropods and the morphology (and even presence/absence) of oral and ventral suckers appears to be host related. Clearly, digeneans are most commonly parasites of teleosts. Digenea parasitize all classes of vertebrates although they are relatively rare in chondrichthyans (Bray and Cribb, 2003). Indeed, Bray and Cribb (2003) concluded that digeneans are primitively associated with teleosts and parasitism of elasmobranchs is a result of several host switches over a long geological period. Parasitism of tetrapods is less common in terms of numbers, but few superfamilies do not have representatives that employ them as definitive hosts.

Ashford and Crewe (2003) listed 130 species of digeneans that have been recorded from humans. Most records are of incidental occurrences when metacercariae were ingested with poorly cooked or raw second intermediate hosts. However, there are also a number of species that require humans as part of their life cycle (e.g. *Schistosoma mansoni*, *S. haematobium*, *S. intercalatum*, *S. guineensis*, *S. japonicum*, *S. mekongi*, *S. malayensis*, *Clonorchis sinensis*) and others where occurrences are frequent enough to make them medically important (e.g. *Paragonimus westermani*, *Fasciola hepatica*, *F. gigantica*). The major families involved are highlighted in Fig. 1.13.

Perhaps the most studied digeneans, in terms of general biology, are the schistosomes. Causative agents of schistosomiasis (among species of *Schistosoma*) and cercarial dermatitis (among species of bird schistosomes) the Schistosomatoidea include important parasites of humans (Rollinson and Simpson, 1987; Horak *et al.*, 2002). Recent phylogenetic studies indicate that the genus *Schistosoma* arose in Asia and subsequently spread into the Indian subcontinent and Africa and from Africa back out into India (Attwood *et al.*, 2002; Lockyer *et al.*, 2003b). The movement into humans as definitive hosts appears to have occurred on at least five different occasions (Lockyer *et al.*, 2003b, and unpublished results based on the phylogenetic placement of *S. intercalatum* and *S. guineensis*).

Concluding Remarks

There are many important life-history traits of parasitic flatworms that determine their success, as measured either as numbers, biomass,

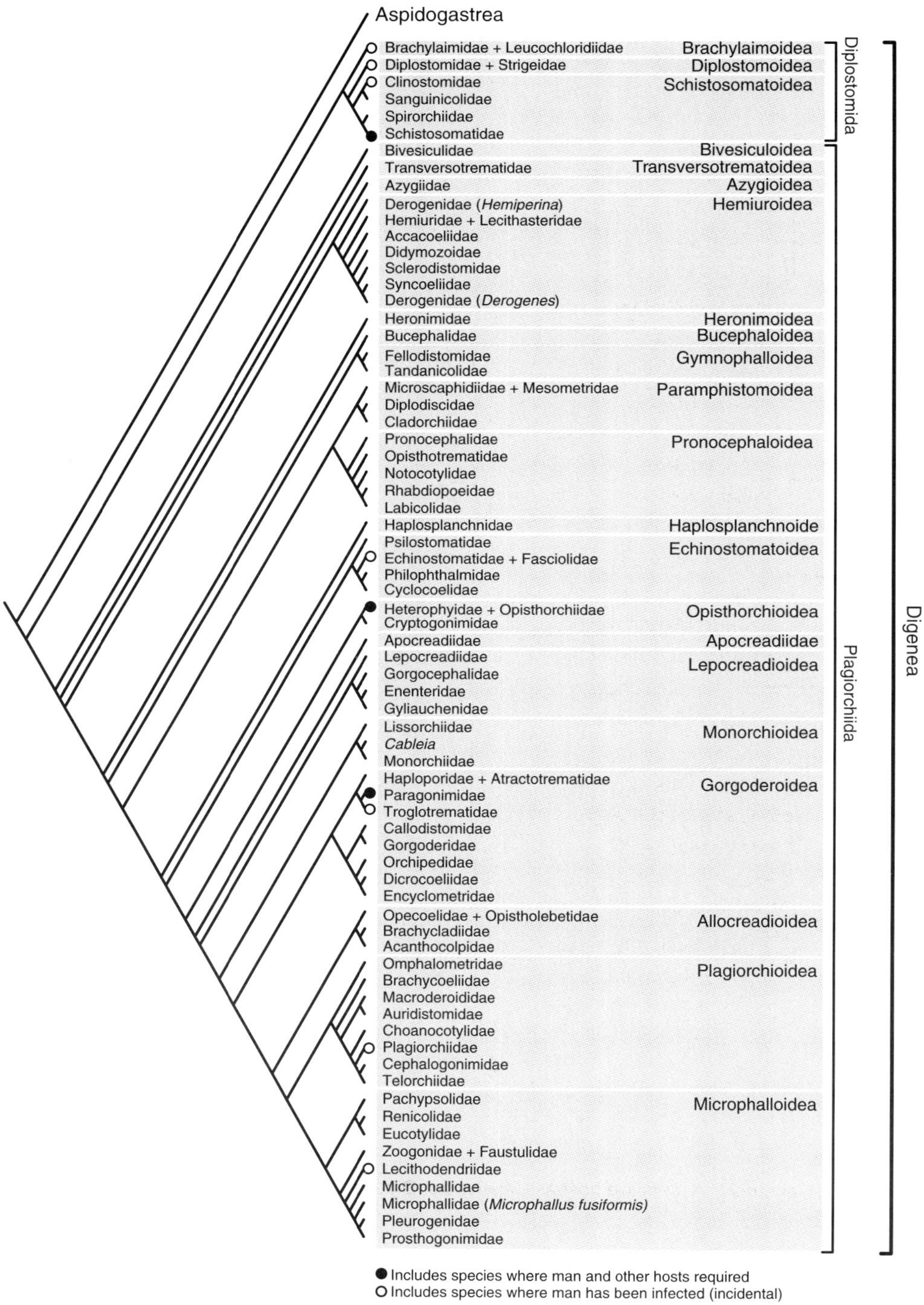

Fig. 1.13. Interrelationships of the Trematoda estimated from combined complete SSU and partial LSU rDNA with a revision of higher taxon nomenclature; redrawn from Olson *et al.* (2003) and with an indication of major families affecting humans (from Ashford and Crewe, 2003).

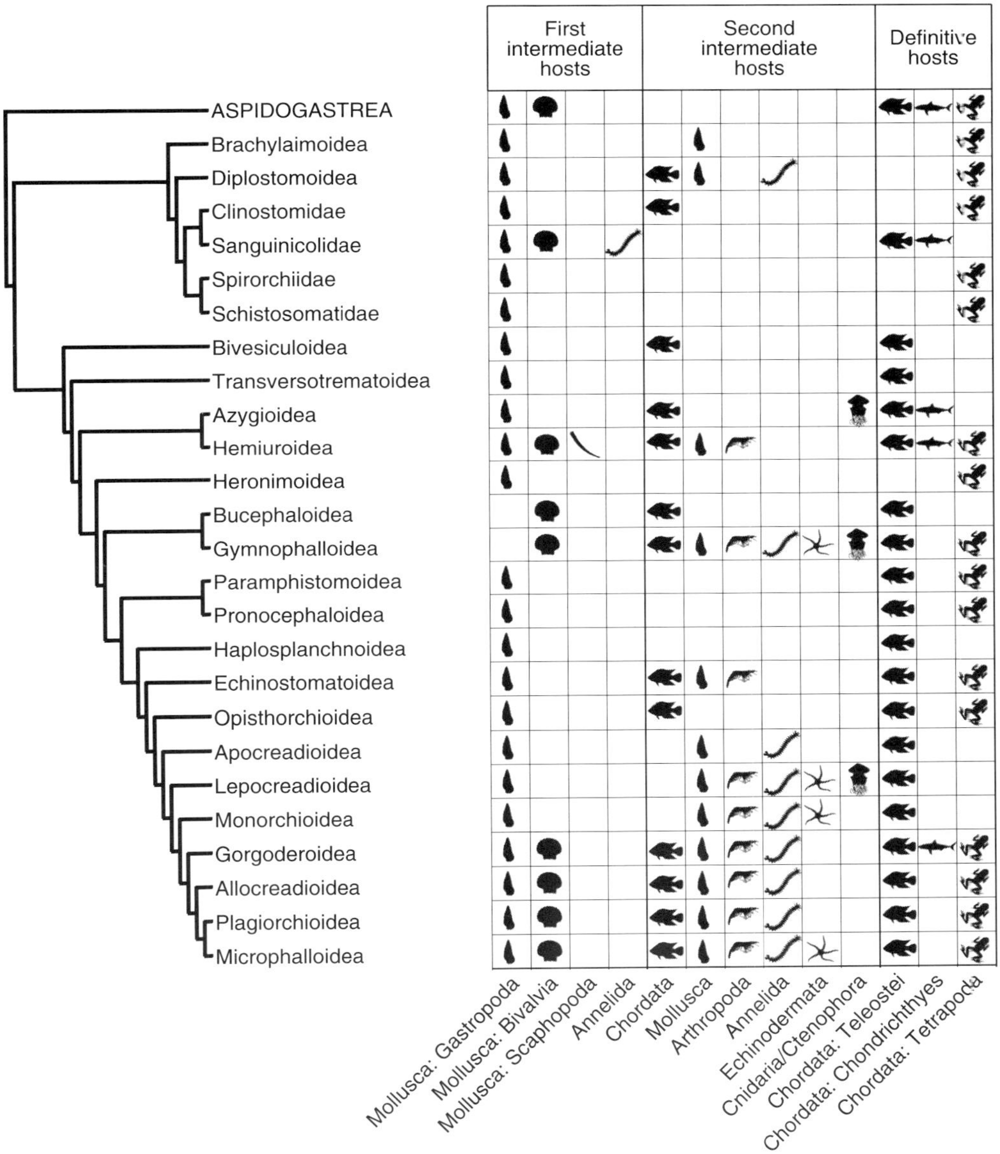

Fig. 1.14. First, second and definitive host associations of Digenea as viewed in a phylogenetic context; after Cribb *et al.* (2003).

distribution or any combination of these metrics. Host specificity determines the likelihood that a parasite will find a suitable host, move to new habitats or be accommodated by new hosts in new environments (Poulin and Mouillot, 2003). Body size affects fecundity and larger adults tend to produce larger offspring and have a relatively increased reproductive output (Trouvé *et al.*, 2003). Longevity affects total reproductive output, as does the extent to which parasites engage of asexual reproduction. Life-cycle complexity may enhance transmission particularly if, by encysting in the predator's prey, this facilitates a worm finding a host in which it can become adult (Choisy *et al.*, 2003). It seems to be a general rule that once a taxonomic lineage engages in parasitism it takes an irreversible evolutionary path. As a group the Neodermata have a secure future, and yet individually, parasitic flatworms are in a precarious balance with each of their hosts and the many environments they inhabit. As long as there are suitable hosts there will be parasitic platyhelminths and the evolutionary struggle to break (host), moderate (host), maintain (parasite) or facilitate (parasite) this relationship will continue. The highly specific host–parasite associations of some taxa surely make them more vulnerable to extinction than others and this seems to be the case amongst some of the basal groups of monogeneans, cestodes and trematodes. In contrast, the greatest diversity of species appears to exist at the crown (most derived) groups of these parasites; is this because only the derived (younger) lineages are capable of taking advantage of the new host opportunities? The premise and the question remain to be examined further.

Host and site specificity affect the ability to switch hosts and enter new habitats and intuitively it seems that maintaining the ability to switch hosts might be an integral component of being a successful parasite. However, strict site and host specificity are the hallmark of one of the most successful groups, the Monogenea. The influence of site and host specificity on success, in terms of species numbers at least, also requires further study.

Each of the major lineages of parasitic flatworms shows remarkable adaptations to parasitism. Whilst larval forms, mode of development, mode of reproduction, life-cycle complexity, method of attachment, site of infection and host use differ enormously between the monogeneans, cestodes, aspidogastreans and digeneans, each lineage has exploited this heterogeneity to a great effect.

The evolutionary fate of parasitic flatworms lies intimately intertwined with the taxa they currently utilize, the taxa they can but have not yet utilized and the selective pressures driving speciation within and between parasites and their hosts. As with any other biological system, if time and conditions permit, natural selection will maintain diversity. Host switching, host speciation, vicariant events affecting parasites and/or hosts, an increase in host population size and dispersal and drug resistance are just some of the factors that will maintain or even promote the success of parasites and subsequently drive diversity in the hosts.

Acknowledgements

Like a parasite, I have fed from a number of hosts (colleagues, collaborators, researchers and institutions), their ingestions (data) and digestions (publications, ideas and insights) in providing this overview. Many thanks to all concerned and particular thanks to Tom Cribb for much help, guidance and tolerance, to Seth Tyler for Fig. 1.5 and to Aaron Maule for the opportunity to contribute. My research is sponsored through a Wellcome Trust Senior Research Fellowship (043965) and an NERC project grant (NER/A/S/2003/00313) hosted and facilitated by the Natural History Museum.

References

Adoutte, A. (1999) Animal evolution – the end of the intermediate taxa? *Trends in Genetics*, 15, 104–108.

Arme, C. and Pappas, P.W. (1983a) *Biology of the Eucestoda*, vol. 1. Academic Press, London.

Arme, C. and Pappas, P.W. (1983b) *Biology of the Eucestoda*, vol. 2. Academic Press, London.

Ashford, R.W. and Crewe, W. (2003) *The Parasites of Homo sapiens – An Annotated Checklist of the Protozoa, Helminths and Arthropods for Which We Are Home*. Taylor & Francis, London.

Attwood, S.W., Upatham, E.S., Meng, X.H., Qiu, D.C. and Southgate, V.R. (2002) The phylogeography of Asian *Schistosoma* (Trematoda: Schistosomatidae). *Parasitology* 125, 99–112.

Baguñà, J. and Riutort, M. (2004) Molecular phylogeny of the Platyhelminthes. *Canadian Journal of Zoology* 82, 168–193.

Bakke, T.A., Harris, P.D. and Cable, J. (2002) Host specificity dynamics: observations on gyrodactylid monogeneans. *International Journal for Parasitology* 32, 281–308.

Baldauf, S.L. (2003) The deep roots of eukaryotes. *Science* 300, 1703–1706.

Benton, M.J. and Ayala, F.J. (2003) Dating the tree of life. *Science* 300, 1698–1700.

Bentz, S., Leroy, S., du Preez, L., Mariaux, J., Vaucher, C. and Verneau, O. (2001) Origin and evolution of African *Polystoma* (Monogenea: Polystomatidae) assessed by molecular methods. *International Journal for Parasitology* 31, 697–705.

Bentz, S., Combes, C., Euzet, L., Riutord, J.J. and Verneau, O. (2003) Evolution of monogenean parasites across vertebrate hosts illuminated by the phylogenetic position of *Euzetrema* Combes, 1965 within the Monopisthocotylea. *Biological Journal of the Linnean Society* 80, 727–734.

Beveridge, I. (2001) The use of life-cycle characters in studies of the evolution of cestodes. In: Littlewood, D.T.J. and Bray, R.A. (eds) *Interrelationships of the Platyhelminthes*. Taylor & Francis, London, pp. 250–256

Binford, M.W. (1982) Ecological history of Lake Valencia, Venezuela: interpretation of animal microfossils and some chemical, physical, and geological features. *Ecological Monographs* 52, 307–337.

Boeger, W.A. and Kritsky, D.C. (1993) Phylogeny and a revised classification of the Monogenoidea Bychowsky, 1937 (Platyhelminthes). *Systematic Parasitology* 26, 1–32.

Boeger, W.A. and Kritsky, D.C. (1997) Coevolution of the Monogenoidea (Platyhelminthes) based on a revised hypothesis of parasite phylogeny. *International Journal for Parasitology* 27, 1495–1511.

Boeger, W.A. and Kritsky, D.C. (2001) Phylogenetic relationships of the Monogenoidea. In: Littlewood, D.T.J. and Bray, R.A. (eds) *Interrelationships of the Platyhelminthes*. Taylor & Francis, London, pp. 92–102.

Boeger, W.A. and Kritsky, D.C. (2003) Parasites, fossils and geologic history: historical biogeography of the South American freshwater croakers, *Plagioscion* spp. (Teleostei, Sciaenidae). *Zoologica Scripta* 32, 3–11.

Boeger, W.A., Kritsky, D.C. and Pie, M.R. (2003) Context of diversification of the viviparous Gyrodactylidae (Platyhelminthes, Monogenoidea). *Zoologica Scripta* 32, 437–448.

Bray, R.A. and Cribb, T.H. (2003) The digeneans of elasmobranchs – distribution and evolutionary significance. In: Combes, C. and Jourdane, J. (eds) *Taxonomy, Ecology and Evolution of Metazoan Parasites. Tome I*. Perpignan University Press, Perpignan, France, pp. 67–96.

Bray, R.A., Jones, A. and Hoberg, E.P. (1999) Observations on the phylogeny of the cestode order Pseudophyllidea Carus, 1863. *Systematic Parasitology* 42, 13–20.

Brooks, D.R. (1989) The phylogeny of the Cercomeria (Platyhelminthes: Rhabdocoela) and general evolutionary principles. *Journal of Parasitology* 75, 606–616.

Brooks, D.R. and McLennan, D.A. (1993) *Parascript: Parasites and the Language of Evolution*. Smithsonian Institution Press, Washington, DC.

Brooks, D.R. and McLennan, D.A. (2002) *The Nature of Diversity – An Evolutionary Voyage of Discovery*. The University of Chicago Press, Chicago, Illinois.

Brooks, D.R., Bandoni, S.M. and O'Grady, R.T. (1989) Aspects of the phylogeny of the Trematoda Rudolphi, 1808 (Platyhelminthes: Cercomeria). *Canadian Journal of Zoology* 67, 2609–2624.

Brooks, D.R., Hoberg, E.P. and Weekes, P. (1991) Preliminary phylogenetic systematic analysis of the major lineages of the Eucestoda (Platyhelminthes: Cercomeria). *Proceedings of the Biological Society of Washington* 104, 651–668.

Burt, D.R.R. (1970) *Platyhelminthes and Parasitism*. The English Universities Press, London.

Caira, J.N. and Jensen, K. (2001) An investigation of the co-evolutionary relationships between onchobothriid tapeworms and their elasmobranch hosts. *International Journal for Parasitology* 31, 960–975.

Caira, J.N. and Littlewood, D.T.J. (2001) Diversity of Platyhelminthes. In: Levin, S.A. (ed.) *Encyclopedia of Biodiversity*. Academic Press, New York, pp. 863–899.

Caira, J.N., Jensen, K. and Healy, C.J. (1999) On the phylogenetic relationships among tetraphyllidean, lecanicephalidean and diphyllidean tapeworm genera. *Systematic Parasitology* 42, 77–151.
Caira, J.N., Jensen, K. and Healy, C.J. (2001) Interrelationships among tetraphyllidean and lecanicephalidean cestodes. In: Littlewood, D.T.J. and Bray, R.A. (eds) *Interrelationships of the Platyhelminthes*. Taylor & Francis, London, pp. 135–158.
Cannon, L.R.G. (1986) *Turbellaria of the World. A Guide to Families and Genera*. Queensland Museum, Brisbane, Australia.
Cannon, L.R.G. and Joffe, B.I. (2001) The Temnocephalida. In: Littlewood, D.T.J. and Bray, R.A. (eds) *Interrelationships of the Platyhelminthes*. Taylor & Francis, London, pp. 83–91.
Carranza, S., Baguñà, J. and Riutort, M. (1997) Are the Platyhelminthes a monophyletic primitive group? An assessment using 18S rDNA sequences. *Molecular Biology and Evolution* 14, 485–497.
Chervy, L. (2002) The terminology of larval cestodes or metacestodes. *Systematic Parasitology* 52, 1–33.
Choisy, M., Brown, S.P., Lafferty, K.D. and Thomas, F. (2003) Evolution of trophic transmission in parasites: why add intermediate hosts? *The American Naturalist* 162, 172–181.
Coil, W.C. (1991) Platyhelminthes: Cestoidea. In: Harrison, F.W. and Bogitsh, B.J. (eds) *Microscopic Anatomy of Invertebrates*. Wiley-Liss, New York, pp. 211–283.
Combes, C. (2001) *Parasitism – The Ecology and Evolution of Intimate Interactions*. The University of Chicago Press, Chicago, Illinois.
Conway Morris, S. (1981) Parasites and the fossil record. *Parasitology* 82, 489–509.
Cracraft, J. and Donoghue, M.J. (2004) *Assembling the Tree of Life*. Oxford University Press, Oxford, UK.
Cribb, T.H., Bray, R.A. and Littlewood, D.T.J. (2001a) The nature and evolution of the association among digeneans, molluscs and fishes. *International Journal for Parasitology* 31, 997–1011.
Cribb, T.H., Bray, R.A., Littlewood, D.T.J., Pichelin, S. and Herniou, E.A. (2001b) Relationships of the Digenea – evidence from molecules and morphology. In: Littlewood, D.T.J. and Bray, R.A. (eds) *Interrelationships of the Platyhelminthes*. Taylor & Francis, London, pp. 186–193.
Cribb, T.H., Bray, R.A., Olson, P.D. and Littlewood, D.T.J. (2003) Life cycle evolution in the Digenea: a new perspective from phylogeny. *Advances in Parasitology*, 54, 197–254.
Criscione, C.D. and Blouin, M.S. (2004) Life cycles shape parasite evolution: comparative population genetics of salmon trematodes. *Evolution* 58, 198–202.
Dalton, J.P., Skelly, P. and Halton, D.W. (2004) Role of the tegument and gut in nutrient uptake by parasitic platyhelminths. *Canadian Journal of Zoology* 82, 211–232.
de Chambrier, A., Zehnder, M., Vaucher, C. and Mariaux, J. (2004) The evolution of the Proteocephalidea (Platyhelminthes, Eucestoda) based on an enlarged molecular phylogeny, with comments on their uterine development. *Systematic Parasitology* 57, 159–171.
Desdevises, Y., Morand, S. and Legendre, P. (2002) Evolution and determinants of host specificity in the genus *Lamellodiscus* (Monogenea). *Biological Journal of the Linnean Society* 77, 431–443.
Dezfuli, B.S., Capuano, S. and Congiu, L. (2002) Identification of life cycle stages of *Cyathocephalus truncatus* (Cestoda: Spathebothriidea) using molecular techniques. *Journal of Parasitology* 88, 632–634.
Dollfus, R.-P. (1923) Énumération des cestodes du plancton et des invertébrés marins. I. Plancton marin. *Annales de Parasitologie Humaine et Comparée* 1, 276–300.
Douzery, E.J.P., Snell, E.A., Bapteste, E., Delsuc, F. and Philippe, H. (2004) The timing of eukaryotic evolution: does a relaxed molecular clock reconcile proteins and fossils? *Proceedings of the National Academy of Sciences USA* 101, 15386–15391.
du Preez, L.H. and Moeng, I.A. (2004) Additional morphological information on *Oculotrema hippopotami* Stunkard, 1924 (Monogenea: Polystomatidae) parasitic on the African hippopotamus. *African Zoology* 39, 225–233.
Eernisse, D.J. and Peterson, K.J. (2004) The history of animals. In: Cracraft, J. and Donoghue, M.J. (eds) *Assembling the Tree of Life*. Oxford University Press, Oxford, UK, pp. 197–208.
Ehlers, U. (1984) Phylogenetisches System der Plathelminthes. *Verhandlungen des Naturwissenschaftlichen Vereins In Hamburg (NF)* 27, 291–294.
Ehlers, U. (1985) Phylogenetic relationships within the Platyhelminthes. In: Conway Morris, S., George, J.D., Gibson, R. and Platt, H.M. (eds) *The Origins and Relationships of Lower Invertebrates*. Oxford University, Oxford, UK, pp. 143–158.
Euzet, L. and Combes, C. (2003) Some controversial questions regarding Monogenea. In: Combes, C. and Jourdane, J. (eds) *Taxonomy, Ecology and Evolution of Metazoan Parasites*. Presses Universitaires de Perpignan, Perpignan, France, pp. 303–320.
Fried, B. and Haseeb, M.A. (1991) Platyhelminthes: Aspidogastrea, Monogenea, and Digenea. In: Harrison, F.W. and Bogitsh, B.J. (eds) *Microscopic Anatomy of Invertebrates*. Wiley-Liss, New York, pp. 141–209.

Galaktionov, K.V. and Dobrovolskij, A.A. (2003) *The Biology and Evolution of Trematodes*. Kluwer Academic Publishers, London.

Gibson, D.I. (1987) Questions in digenean systematics and evolution. *Parasitology* 95, 429–460.

Gibson, D.I., Jones, A. and Bray, R.A. (2002) *Keys to the Trematoda*, vol. 1. CAB International, Wallingford, UK.

Giribet, G., Distel, D.L., Polz, M., Sterrer, W. and Wheeler, W.C. (2000) Triploblastic relationships with emphasis on the acoelomates and the position of Gnathostomulida, Cycliophora, Plathelminthes, and Chaetognatha: a combined approach of 18S rDNA sequences and morphology. *Systematic Biology* 49, 539.

Halton, D.W. (1997) Nutritional adaptations to parasitism within the Platyhelminthes. *International Journal for Parasitology* 27, 693–704.

Harris, P.D. (1983) The morphology and life cycle of the oviparous *Oögyrodactylus farlowellae* gen. et sp. nov. (Monogenea, Gyrodactylidea). *Parasitology* 87, 405–420.

Harris, P.D. (1985) Observations on the development of the male reproductive system in *Gyrodactylus gasterostei* Gläser, 1974 (Monogenea, Gyrodactylidea). *Parasitology* 91, 519–529.

Harris, P.D. (1993) Interactions between reproduction and population biology in gyrodactylid monogeneans – a review. *Bulletin français de la Pêche et de la Pisciculture* 328, 47–65.

Hoberg, E.P. and Klassen, G.J. (2002) Revealing the faunal tapestry: co-evolution and historical biogeography of hosts and parasites in marine systems. *Parasitology* 124, S3–S22.

Hoberg, E.P., Mariaux, J., Justine, J.-L., Brooks, D.R. and Weekes, P. (1997) Phylogeny of the orders of the Eucestoda (Cercomeromorphae) based on comparative morphology: historical perspectives and a new working hypothesis. *Journal of Parasitology* 83, 1128–1147.

Hoberg, E.P., Jones, A. and Bray, R.A. (1999) Phylogenetic analysis among the families of the Cyclophyllidea (Eucestoda) based on comparative morphology, with new hypotheses for co-evolution in vertebrates. *Systematic Parasitology* 42, 51–73.

Hoberg, E.P., Jones, A., Rausch, R.L., Eom, K.S. and Gardner, S.L. (2000) A phylogenetic hypothesis for species of the genus *Taenia* (Eucestoda: Taeniidae). *Journal of Parasitology* 86, 89–98.

Hoberg, E.P., Alkire, N.L., de Queiroz, A. and Jones, A. (2001a) Out of Africa: origins of the *Taenia* tapeworms in humans. *Proceedings of the Royal Society of London Series B – Biological Sciences* 268, 781–787.

Hoberg, E.P., Mariaux, J. and Brooks, D.R. (2001b) Phylogeny among orders of the Eucestoda (Cercomeromorphae): integrating morphology, molecules and total evidence. In: Littlewood, D.T.J. and Bray, R.A. (eds) *Interrelationships of the Platyhelminthes*. Taylor & Francis, London, pp. 112–126.

Horak, P., Kolarova, L. and Adema, C.M. (2002) Biology of the schistosome genus *Trichobilharzia*. *Advances in Parasitology* 52, 155–233.

Huyse, T., Audenaert, V. and Voickaert, F.A.M. (2003) Speciation and host–parasite relationships in the parasite genus *Gyrodactylus* (Monogenea, Platyhelminthes) infecting gobies of the genus *Pomatoschistus* (Gobiidae, Teleostei). *International Journal for Parasitology* 33, 1679–1689.

Jenner, R.A. (2004) Towards a phylogeny of the Metazoa: evaluating alternative phylogenetic positions of Platyhelminthes, Nemertea, and Gnathostomulida, with a critical reappraisal of cladistic characters. *Contributions to Zoology* 73, 3.

Jennings, J.B. (1971) Parasitism and commensalism in the Turbellaria. *Advances in Parasitology* 9, 1–32.

Jennings, J.B. (1997) Nutritional and respiratory pathways to parasitism exemplified in the Turbellaria. *International Journal for Parasitology* 27, 679–691.

Joffe, B.I. and Kornakova, E.E. (1998) *Notentera ivanovi* Joffe *et al.*, 1997: a contribution to the question of phylogenetic relationships between 'turbellarians' and the parasitic Plathelminthes (Neodermata). *Hydrobiologia* 383, 245–250.

Joffe, B.I. and Kornakova, E.E. (2001) Flatworm phylogeneticist: between molecular hammer and morphological anvil. In: Littlewood, D.T.J. and Bray, R.A. (eds) *Interrelationships of the Platyhelminthes*. Taylor & Francis, London, pp. 279–291.

Jondelius, U., Ruiz-Trillo, I., Baguñà, J. and Riutort, M. (2002) The Nemertodermatida are basal bilaterians and not members of the Platyhelminthes. *Zoologica Scripta* 31, 201–215.

Jones, A., Bray, R.A. and Gibson, D.I. (2005) *Keys to the Trematoda*, vol. 2. CAB International, Wallingford, UK.

Justine, J.-L. (1998) Non-monophyly of the monogeneans? *International Journal for Parasitology* 28, 1653–1657.

Kearn, G.C. (1994) Evolutionary expansion of the Monogenea. *International Journal for Parasitology* 24, 1227–1271.

Kearn, G.C. (1998) *Parasitism and the Platyhelminthes*. Chapman & Hall, London.

Khalil, L.F., Jones, A. and Bray, R.A. (1994) *Keys to the Cestode Parasites of Vertebrates*. CAB International, Wallingford, UK.

Klassen, G.J. (1992) Coevolution – a history of the macroevolutionary approach to studying host–parasite associations. *Journal of Parasitology* 78, 573–587.

Kornakova, E.E. and Joffe, B.I. (1999) A new variant of the neodermatan-type spermiogenesis in a parasitic 'turbellarian', *Notentera ivanovi* (Platyhelminthes) and the origin of the Neodermata. *Acta Zoologica* 80, 135–151.

Littlewood, D.T.J. (2003) *The Evolution of Parasitism – A Phylogenetic Approach,* Vol. 54, *Advances in Parasitology*. Elsevier Academic Press, London.

Littlewood, D.T.J. and Donovan, S.K. (2003) Fossil parasites: a case of identity. *Geology Today* 19, 136–142.

Littlewood, D.T.J. and Olson, P.D. (2001) SSU rDNA and the Platyhelminthes: signal, noise, conflict and compromise. In: Littlewood, D.T.J. and Bray, R.A. (eds) *Interrelationships of the Platyhelminthes*. Taylor & Francis, London, pp. 292–301.

Littlewood, D.T.J., Rohde, K. and Clough, K.A. (1997) Parasite speciation within or between host species? Phylogenetic evidence from site-specific polystome monogeneans. *International Journal for Parasitology* 27, 1289–1297.

Littlewood, D.T.J., Rohde, K. and Clough, K.A. (1998) The phylogenetic position of *Udonella* (Platyhelminthes). *International Journal for Parasitology* 28, 1241–1250.

Littlewood, D.T.J., Rohde, K., Bray, R.A. and Herniou, E.A. (1999a) Phylogeny of the Platyhelminthes and the evolution of parasitism. *Biological Journal of the Linnean Society* 68, 257–287.

Littlewood, D.T.J., Rohde, K. and Clough, K.A. (1999b) The interrelationships of all major groups of Platyhelminthes: phylogenetic evidence from morphology and molecules. *Biological Journal of the Linnean Society* 66, 75–114.

Littlewood, D.T.J., Olson, P.D., Telford, M.J., Herniou, E.A. and Riutort, M. (2001) Elongation factor 1-alpha sequences alone do not assist in resolving the position of the Acoela within the Metazoa. *Molecular Biology and Evolution* 18, 437–442.

Littlewood, D.T.J., Telford, M.J. and Bray, R.A. (2004) Protostomes and Platyhelminthes – the worm's turn. In: Cracraft, J. and Donoghue, M.J. (eds) *Assembling the Tree of Life*. Oxford University Press, New York, pp. 209–234.

Lockyer, A.E., Olson, P.D. and Littlewood, D.T.J. (2003a) Utility of complete large and small subunit rRNA genes in resolving the phylogeny of the Neodermata (Platyhelminthes): implications and a review of the cercomer theory. *Biological Journal of the Linnean Society* 78, 155–171.

Lockyer, A.E., Olson, P.D., Østergaard, P., Rollinson, D., Johnston, D.A., Attwood, S.A., Southgate, V.R., Horak, P., Snyder, S.D., Le, T.H., Agatsuma, T., McManus, D.P., Carmichael, A.C., Naem, S. and Littlewood, D.T.J. (2003b) The phylogeny of the Schistosomatidae based on three genes with emphasis on the interrelationships of *Schistosoma* Weinland, 1858. *Parasitology* 126, 203–224.

Mariaux, J. and Olson, P.D. (2001) Cestode systematics in the molecular era. In: Littlewood, D.T.J. and Bray, R.A. (eds) *Interrelationships of the Platyhelminthes*. Taylor & Francis, London, pp. 127–134.

Mollaret, I., Jamieson, B.G.M., Adlard, R.D., Hugall, A., Lecointre, G., Chombard, C. and Justine, J.-L. (1997) Phylogenetic analysis of the Monogenea and their relationships with Digenea and Eucestoda inferred from 28S rDNA sequences. *Molecular and Biochemical Parasitology* 90, 433–438.

Morand, S. and Poulin, R. (2003) Phylogenies, the comparative method and parasite evolutionary ecology. *Advances in Parasitology* 54, 281–302.

Newman, L.J. and Cannon, L.R.G. (2003) *Marine Flatworms: The World of Polyclads*. CSIRO, Collingwood, Victoria, Australia.

Newmark, P.A. and Sánchez Alvarado, A. (2002) Not your father's planarian: a classic model enters the era of functional genomics. *Nature Reviews Genetics* 3, 210–219.

Norèn, M. and Jondelius, U. (1999) Phylogeny of the Prolecithophora (Platyhelminthes) inferred from 18S rDNA sequences. *Cladistics* 15, 103–112.

Olson, P.D. and Littlewood, D.T.J. (2002) Phylogenetics of the Monogenea – evidence from a medley of molecules. *International Journal for Parasitology* 32, 233–244.

Olson, P.D., Ruhnke, T.R., Sanney, J. and Hudson, T. (1999) Evidence for host-specific clades of tetraphyllidean tapeworms (Platyhelminthes: Eucestoda) revealed by analysis of 18S ssrDNA. *International Journal for Parasitology* 29, 1465–1476.

Olson, P.D., Littlewood, D.T.J., Bray, R.A. and Mariaux, J. (2001) Interrelationships and evolution of the tapeworms (Platyhelminthes: Cestoda). *Molecular Phylogenetics and Evolution* 19, 443–467.

Olson, P.D., Cribb, T.H., Tkach, V.V., Bray, R.A. and Littlewood, D.T.J. (2003) Phylogeny and classification of the Digenea (Platyhelminthes: Trematoda). *International Journal for Parasitology* 33, 733–755.

Orii, H., Mochii, M. and Watanabe, K. (2003) A simple "soaking method" for RNA interference in the planarian *Dugesia japonica*. *Development Genes and Evolution* 213, 138–141.

Page, R.D.M. (2003) *Tangled Trees: Phylogeny, Cospeciation, and Coevolution*. The University of Chicago Press, Chicago, Illinois.
Parker, G.A., Chubb, J.C., Ball, M.A. and Roberts, G.N. (2003) Evolution of complex life cycles in helminth parasites. *Nature* 425, 480–484.
Peterson, K.J., Lyons, J.B., Nowak, K.S., Takacs, C.M., Wargo, M.J. and McPeek, M.A. (2004) Estimating metazoan divergence times with a molecular clock. *Proceedings of the National Academy of Sciences USA* 101, 6536–6541.
Poinar, G. (2003) A rhabdocoel turbellarian (Platyhelminthes, Typhlopanoida) in Baltic amber with a review of fossil and sub-fossil platyhelminths. *Invertebrate Biology* 122, 308–312.
Poulin, R. (1995) Phylogeny, ecology, and the richness of parasite communities in vertebrates. *Ecological Monographs* 65, 283–302.
Poulin, R. (1998) *Evolutionary Ecology of Parasites*. Chapman and Hall, London.
Poulin, R. (2002) The evolution of monogenean diversity. *International Journal for Parasitology* 32, 245–254.
Poulin, R. and Morand, S. (2000) The diversity of parasites. *Quarterly Review of Biology* 75, 277–293.
Poulin, R. and Mouillot, D. (2003) Parasite specialization from a phylogenetic perspective: a new index of host specificity. *Parasitology* 126, 473–480.
Rohde, K. (1971) Untersuchungen an *Multicotyle purvisi* Dawes, 1941 (Trematoda: Aspidogastrea). I. Entwicklung un Morphologie. *Zoologischer Jahrbuch der Anatmoie* 88, 320–363.
Rohde, K. (1972) The Aspidogastrea, especially *Multicotyle purvisi* Dawes, 1941. *Advances in Parasitology* 10, 77–151.
Rohde, K. (1994) The origins of parasitism in the Platyhelminthes. *International Journal for Parasitology* 24, 1099–1115.
Rohde, K. (2001) The Aspidogastrea: an archaic group of Platyhelminthes. In: Littlewood, D.T.J. and Bray, R.A. (eds) *Interrelationships of the Platyhelminthes*. Taylor & Francis, London, pp. 159–167.
Rohde, K. (2002) Subclass Aspidogastrea Faust & Tang, 1936. In: Gibson, D.I., Jones, A. and Bray, R.A. (eds) *Keys to the Trematoda*, vol. 1. CAB International, Wallingford, UK, pp. 5–14.
Rollinson, D. and Simpson, A.J.G. (1987) *From Genes to Latrines*. Academic Press, London.
Rowe, T. (2004) Chordate phylogeny and development. In: Cracraft, J. and Donoghue, M.J. (eds) *Assembling the Tree of Life*. Oxford University Press, Oxford, UK, pp. 384–409.
Ruiz, G. and Lindberg, D.R. (1989) A fossil record for trematodes: extent and potential uses. *Lethaia* 22, 431–438.
Ruiz-Trillo, I., Riutort, M., Littlewood, D.T.J., Herniou, E.A. and Baguñà, J. (1999) Acoel flatworms: earliest extant bilaterian metazoans, not members of Platyhelminthes. *Science* 283, 1919–1923.
Ruiz-Trillo, I., Paps, J., Loukota, M., Ribera, C., Jondelius, U., Baguñà, J. and Riutort, M. (2002) A phylogenetic analysis of myosin heavy chain type II sequences corroborates that Acoela and Nemertodermatida are basal bilaterians. *Proceedings of the National Academy of Sciences USA* 99, 11246–11251.
Simkova, A., Morand, S., Jobet, E., Gelnar, M. and Verneau, O. (2004) Molecular phylogeny of congeneric monogenean parasites (*Dactylogyrus*): a case of intrahost speciation. *Evolution* 58, 1001–1018.
Sinnappah, N.D., Lim, L.H.S., Rohde, K., Tinsley, R., Combes, C. and Verneau, O. (2001) A paedomorphic parasite associated with a neotenic amphibian host: phylogenetic evidence suggests a revised systematic position for Sphyranuridae within anuran and turtle Polystomatoineans. *Molecular Phylogenetics and Evolution* 18, 189–201.
Sopott-Ehlers, B., Haas, W. and Ehlers, U. (2003) Ultrastructure of pigmented and unpigmented photoreceptors in cercariae of *Trichobilharzia ocellata* (Plathelminthes, Trematoda, Schistosomatidae): evidence for the evolution of parasitism within the Neodermata. *Parasitology Research* 91, 109–116.
Summers, K., McKeon, S., Sellars, J., Keusentothen, M., Morris, J., Gloeckner, D., Pressley, C., Price, B. and Snow, H. (2003) Parasitic exploitation as an engine of diversity. *Biological Reviews* 78, 639–675.
Szidat, L. (1956) Der marine charakter der Parasitenfauna der Susswässerfische des Stromsystems des Rio de la Plata und ihre Deutung als Reliktfauna des Tertaren Tethys-Meeres. *Proceedings of the XIV International Congress of Zoology* 1953, 128–138.
Telford, M.J., Herniou, E.A., Russell, R.B. and Littlewood, D.T.J. (2000) Changes in mitochondrial genetic codes as phylogenetic characters: two examples from the flatworms. *Proceedings of the National Academy of Sciences USA* 97, 11359–11364.
Telford, M.J., Lockyer, A.E., Cartwright-Finch, C. and Littlewood, D.T.J. (2003) Combined large and small subunit ribosomal RNA phylogenies support a basal position of the acoelomorph flatworms. *Proceedings of the Royal Society of London Series B – Biological Sciences* 270, 1077–1083.
Trouvé, S., Morand, S. and Gabrion, C. (2003) Asexual multiplication of larval parasitic worms: a predictor of adult life-history traits in Taeniidae? *Parasitology Research* 89, 81–88.

Tyler, S. and Hooge, M. (2004) Comparative morphology of the body wall in flatworms (Platyhelminthes). *Canadian Journal of Zoology* 82, 194–210.

Tyler, S. and Tyler, M.S. (1997) Origin of the epidermis in parasitic platyhelminths. *International Journal for Parasitology* 27, 715–738.

Upeniece, I. (2001) The unique fossil assemblage from the Lode Quarry (Upper Devonian, Latvia). *Mitteilungen aus dem Museum für Naturkunde in Berlin – Geowissenschaftliche Reihe* 4, 101–119.

Valentine, J.W. (2004) *On the Origin of Phyla*. The University of Chicago Press, Chicago, Illinois.

Verneau, O., Bentz, S., Sinnappah, N.D., du Preez, L., Whittington, I. and Combes, C. (2002) A view of early vertebrate evolution inferred from the phylogeny of polystome parasites (Monogenea: Polystomatidae). *Proceedings of the Royal Society of London Series B – Biological Sciences* 269, 535–543.

von Nickisch-Rosenegk, M., Lucius, R. and Loos-Frank, B. (1999) Contributions to the phylogeny of the Cyclophyllidea (Cestoda) inferred from mitochondrial 12S rDNA. *Journal of Molecular Evolution* 48, 586–596.

Watson, N.A. and Rohde, K. (1993a) Ultrastructure of sperm and spermiogenesis of *Kronborgia isopodicola* (Platyhelminthes, Fecampiidae). *International Journal for Parasitology* 23, 737–744.

Watson, N.A. and Rohde, K. (1993b) Ultrastructural evidence for an adelphotaxon (sister group) to the Neodermata (Platyhelminthes). *International Journal for Parasitology* 23, 285–289.

Watson, N.A., Williams, J.B. and Rohde, K. (1992) Ultrastructure and development of the eyes of larval *Kronborgia isopodicola* (Platyhelminthes, Fecampiidae). *Acta Zoologica* 73, 95–102.

Wheeler, T.A. and Chisholm, L.A. (1995) Monogenea versus Monogenoidea: the case for stability in nomenclature. *Systematic Parasitology* 30, 159–164.

Whittington, I.D. (2004) The Capsalidae (Monogenea: Monopisthocotylea): a review of diversity, classification and phylogeny with a note about species complexes. *Folia Parasitologica* 51, 109–122.

Whittington, I.D., Chisholm, L.A. and Rohde, K. (2000a) The larvae of Monogenea (Platyhelminthes). *Advances in Parasitology* 44, 139–232.

Whittington, I.D., Cribb, B.W., Hamwood, T.E. and Halliday, J.A. (2000b) Host specificity of monogenean (platyhelminth) parasites: a role for anterior adhesive areas? *International Journal for Parasitology* 30, 305–320.

Wickström, L.M., Haukisalmi, V., Varis, S., Hantula, J., Fedorov, V.B. and Henttonen, H. (2003) Phylogeography of the circumpolar *Paranoplocephala arctica* species complex (Cestoda: Anoplocephalidae) parasitizing collared lemmings (*Dicrostonyx* spp.). *Molecular Ecology* 12, 3359–3371.

Williams, H.H. (1966) The ecology, functional morphology and taxonomy of *Echeneibothrium* Beneden, 1849 (Cestoda: Tetraphyllidea), a revision of the genus and comments on *Discobothrium* Beneden, 1870 *Pseudanthobothrium* Baer, 1956 and *Phormobothrium* Alexander, 1963. *Parasitology* 56, 227–285.

Williams, H.H. (1968) *Phyllobothrium piriei* p. nov. (Cestoda: Tetraphyllida) from *Raja naevus* with a comment on its habitat and mode of attachment. *Parasitology* 58, 929–937.

Williams, J.B. (1988) Ultrastructural studies on *Kronborgia* (Platyhelminthes, Neoophora): the spermatozoon. *International Journal for Parasitology* 18, 477–483.

Williams, J.B. (1994) Comparative study of sperm morphology and reproductive biology of *Kronborgia* and *Temnocephala* (Platyhelminthes, Neoophora) – implications for platyhelminth phylogeny. *New Zealand Journal of Zoology* 21, 179–194.

Williams, H.H. and Jones, A. (1994) *Parasitic Worms of Fishes*. Taylor & Francis, London.

Xylander, W.E.R. (2001) The Gyrocotylidea, Amphilinidea and the early evolution of the Cestoda. In: Littlewood, D.T.J. and Bray, R.A. (eds) *Interrelationships of the Platyhelminthes*. Taylor & Francis, London, pp. 103–111.

Yamaguti, S. (1963) Monogenea and Aspidocotylea, Vol. 4. In: *Systema Helminthum*. John Wiley & Sons, New York.

Yamaguti, S. (1975) *A Synoptical Review of Digenetic Trematodes*. Keigaku Publishers, Tokyo.

Zamparo, D. and Brooks, D.R. (2003) Phylogenetic systematic assessment of the Aspidobothrea (Platyhelminthes, Neodermata, Trematoda). *Zoologica Scripta* 32, 83–93.

Zamparo, D., Brooks, D.R., Hoberg, E.P. and McLennan, D.A. (2001) Phylogenetic analysis of the Rhabdocoela (Platyhelminthes) with emphasis on the Neodermata and relatives. *Zoologica Scripta* 30, 59–77.

Zehnder, M.P. and Mariaux, J. (1999) Molecular systematic analysis of the order Proteocephalidea (Eucestoda) based on mitochondrial and nuclear rDNA sequences. *International Journal for Parasitology* 29, 1841–1852.

Zietara, M.S. and Lumme, J. (2002) Speciation by host switch and adaptive radiation in a fish parasite genus *Gyrodactylus* (Monogenea, Gyrodactylidae). *Evolution* 56, 2445–2458.

2 Genomes and Genomics of Parasitic Flatworms

D.A. Johnston

Wolfson-Wellcome Biomedical Laboratory, Department of Zoology, The Natural History Museum, Cromwell Road, London SW7 5BD, UK

Introduction

For the purposes of this review, 'flatworm' will be considered to encompass the phylum Platyhelminthes as presently understood from the molecular phylogenetic viewpoint, i.e. excluding the Acoelomorpha (Acoela and Nemertodermatida), which have recently been removed from the phylum on the combined evidence of small and large subunit (SSU; LSU) ribosomal RNA (rRNA) and myosin II gene sequence data (Ruiz-Trillo *et al.*, 1999, 2002; Jondelius *et al.*, 2002; Telford *et al.*, 2003) as well as mitochondrial genetic code (Telford *et al.*, 2000) and gene order data (Ruiz-Trillo *et al.*, 2004; see Baguñà and Riutort, 2004, and Chapter 1, this volume, for reviews). For convenience, 'Turbellaria' will be used to encompass the paraphyletic grouping of (largely) free-living flatworms that fall outside of the parasitic Neodermata (Carranza *et al.*, 1997). As flatworm genomics, in the broad sense in which it is considered here, is an emerging discipline, reference will be made to data from both free-living and parasitic taxa in the hope that knowledge of, and method development for each will help stimulate research in the other.

As of March 2005, the public sequence databases contain some 330,000 platyhelminth DNA sequences and 14,000 platyhelminth protein sequences (Table 2.1). With the exception of data generated for molecular phylogenetic purposes, these show a very heavy bias towards only four genera; the important medical and veterinary parasites *Schistosoma* (by far the largest single contributor) and *Echinococcus*, and the 'model' planarians *Dugesia* and *Schmidtea* (Table 2.1). This bias simply reflects the scientific, social and economic priorities of

Table 2.1. Distribution of platyhelminth molecular data on GenBank, March 2005.

		Number of entries				
		DNA	Protein	Structure	Genome	Popset
Platyhelminthes		329,880	13,837	22	10	118
Cestoda		12,690	1,560	2	5	42
	Cestodaria	18	–	–	–	2
	Eucestoda	12,671	1,560	2	5	42
	inc. *Echinococcus*	10,793	659	1	1	3
Monogenea		753	124	–	–	22
	Monopisthocotylea	559	100	–	–	18
	Polyopisthocotylea	194	24	–	–	6
Trematoda		306,264	11,747	20	5	45
	Aspidogastrea	26	1	–	–	3
	Digenea	306,238	11,746	20	5	45
	inc. *Schistosoma*	302,786	10,697	16	3	14
'Turbellaria'		10,173	406	–	–	14
	Catenulida	22	4	–	–	3
	Lecithoepitheliata	14	–	–	–	1
	Macrostomorpha	24	14	–	–	1
	Mediofusata	8	–	–	–	1
	Polycladida	54	31	–	–	9
	Prolecithophora	47	–	–	–	1
	Rhabdocoela	60	4	–	–	1
	Seriata	9,943	353	–	–	8
	inc. *Schmidtea*	3,089	99	–	–	3
	inc. *Dugesia*	6,586	104	–	–	1
	Unclassified	1	–	–	–	–

Note: Taxonomic groupings taken from GenBank's Taxonomy Division (http://www.ncbi.nlm.nih.gov/Taxonomy).

a developing discipline and will change dramatically in the future as attention shifts to other, currently under-represented taxa. By necessity, much of this review will consider lessons learned from *Schistosoma* and its close relatives. However, at the fundamental level, all flatworm genomes are comparable, and methods and analyses that have been applied to *Schistosoma* can equally be applied elsewhere in the future. The reader is encouraged to exploit the current knowledge base, and especially the eagerly anticipated results of the first platyhelminth genome sequencing projects, as a foundation from which to develop projects for their own systems.

On a wider taxonomic canvas, recent taxonomic revisions of the animal kingdom identify three major clades within the bilateral Metazoa; the Deuterostoma, the Ecdysozoa (the moulting animals) and the Lophotrochozoa (animals bearing a lophophore or displaying trochophore larvae) (Adoutte *et al.*, 2000). The major 'model' invertebrates (*Drosophila*, *Caenorhabditis*, etc.) all fall within the Ecdysozoa and the mammals within the Deuterostoma, whilst platyhelminths are lophotrochozoans. The sequence databases therefore currently also display a massive bias towards the Deuterostoma and Ecdysozoa; the huge biological diversity encompassed by the Lophotrochozoa contributes relatively little (Table 2.2). Amongst the Lophotrochozoa, the Platyhelminthes make the largest contribution and so, de facto, serve as models for the clade.

Table 2.2. Platyhelminth contribution to DNA sequence data for the bilateral Metazoa on GenBank, March 2005.

Taxonomic group[a]	DNA sequences
Deuterostomes	30,308,602
Vertebrates	28,908,567
Cephalochordates	16,772
Urochordates	862,324
Hemichordates	112
Echinoderms	520,827
Ecdysozoans	3,695,260
Gastrotrichs	32
Nematodes	922,845
Priapulids	34
Kinorhynchs	4
Onychophorans	236
Tardigrades	6,460
Arthropods	2,765,649
Lophotrochozoans	412,479
Bryozoans	551
Entoprocts	6
Platyhelminths	330,856
Pogonophorans	313
Brachiopods	165
Phoronids	22
Nemerteans	321
Annelids	17,698
Echiurans	85
Molluscs	61,775
Sipunculans	131
Gnathostomulids	14
Rotifers	540

[a]Taxonomic groups taken from Adoutte *et al.* (2000).

Genome Features

Genome size

Genome size has been calculated for a variety of flatworm species (see Table 2.3 for summary data and references).

For the 'Turbellaria', direct measurement of the DNA content of Feulgen-stained nuclei by densitometry or image analysis reveals a huge variation in estimated haploid genome size from 50 megabases (Mb) in *Stenostomum arevaloi* to 18,390 Mb in *Otomesostoma auditivum* (>350-fold variation) (Gregory *et al.*, 2000). For comparison, the human genome is approximately 3200 Mb. Within the 'Turbellaria', genome size appears to correlate with both taxonomic order and body size, although considerable variation is seen at both large and small taxonomic scales (Gregory *et al.*, 2000). Discontinuous, step-like variation in genome size is also observed within certain 'turbellarian' genera (e.g. *Dugesia*, *Phaenocora* and *Mesostoma*) and is suggested to reflect historic genome duplication events, which have lead to rapid speciation (Gregory *et al.*, 2000). However, as polyploidy has also been extensively reported in the 'Turbellaria' (see section entitled 'Ploidy'), such observations must be interpreted with caution unless the karyotypes have also been determined.

For the Trematoda and Cestoda, both reassociation kinetics/Cot analysis and direct measurement of the DNA content of Feulgen-stained nuclei have been used to estimate genome size. As relatively few taxa have been examined, it is difficult to draw any firm conclusions although the available data suggest much less extreme variation than in the 'Turbellaria' and the absence of step-like intrageneric variation (Table 2.3).

Table 2.3. Platyhelminth genome size estimates.

	Haploid genome size[c]		
Taxon[a,b]	Picograms	Megabases	Reference
Cestoda–Eucestoda			
Echinococcus granulosus	0.17	150	7
Hymenolepis diminuta	1.7	1,500	12
Spirometra mansonoides	1.79	1,600	8
Trematoda–Digenea			
Diplostomum pseudospathaceum	0.89	800	5
Schistosoma mansoni	0.30	270	10
Trichobilharzia spp.	1.06–1.31	950–1,120	5
'Turbellaria'–Seriata			
Archilina spp.	0.75	670	2
Archiloa petiti	1.15	1,030	2
Archilopsis marifuga	0.95	850	2
Archimonocelis spp.	2.4	2,160	2
Archotoplana holotriche	1.2	1,080	2
Bothrioplana semperi	4.7	4,210	1
Carenscoilia biforamen	1.75	1,570	2
Crenobia alpina	1.1	990	6
Dendrocoelum lacteum	1.2	1,080	6
Dugesia spp.	1.66–2.40	1,490–2,150	6, 11
Duploperaclistus circocirrus	0.6	540	2
Foviella affinis	1.8	1,610	1
Girardia tigrina	1.40–1.88	1,260–1,690	1, 6

Table 2.3. Platyhelminth genome size estimates. (*cont'd*)

Taxon[a,b]	Haploid genome size[c] Picograms	Megabases	Reference
Hymanella retenuova	4.46	4,000	1
Monocelis lineata	0.5	450	2
Otomesostoma auditivum	20.52	18,390	1
Otoplana truncaspina	2.05	1,840	2
Phagocata woodworthi	3.76	3,370	1
Polycelis spp.	1.05–2.06	940–1,850	1, 6
Procerodes litoralis	1.06	950	1
Procotyla fluviatilis	2.62	2,350	1
Schmidtea spp.	0.53–1.32	480–1,180	1, 4, 6, 9
Uteriporus vulgaris	1.14	1,020	1
'Turbellaria'–Prolecithophora			
Hydrolimax grisea	2.55	2,290	1
'Turbellaria'–Lecithoepitheliata			
Prorhynchus stagnalis	4.28	3,840	1
'Turbellaria'–Catenulida			
Myostenostomum tauricum	0.19	170	1
Rhynchoscolex simplex	1.68	1,510	1
Stenostomum spp.	0.06–0.75	50–670	1
'Turbellaria'–Rhabdocoela			
Bothromesostoma spp.	3.77	3,380	1
Castrella pinguis	1.34	1,200	1
Dalyellia viridis	1.03	920	1
Gieysztoria sp.	0.63	570	1
Gyratrix hermaphroditus	2.74	2,460	1
Krumbachia hiemalis	1.67	1,500	1
Mesostoma spp.	5.10–16.36	4,570–14,660	1, 3
Olisthanella truncula	0.12	110	1
Phaenocora sp.	0.72–1.61	650–1,440	1
Rhynchomesostoma spp.	2.16	1,940	1
Strongylostoma spp.	0.65–0.83	580–740	1
Typhloplana viridata	0.98	880	1
'Turbellaria'–Macrostomorpha			
Macrostomum spp.	0.17–0.58	150–520	1

[a]Higher level taxonomy taken from GenBank's Taxonomy Division (http://www.ncbi.nlm.nih.gov/Taxonomy).
[b]Intrageneric ranges are given for genera where data for more than one species are available.
[c]Where the primary literature cites genome size in picograms of DNA, size in megabases (Mb) has been estimated on the basis that 1 Mb of double stranded DNA = 0.001116 pg (assuming the mean molecular weight of a base pair to be 672 Daltons and an unbiased genome base composition).
Note: 1, Gregory *et al.* (2000); 2, Martens *et al.* (1989); 3, Hebert and Beaton (1990); 4, Pellicciari *et al.* (1983); 5, T.R. Gregory, unpublished data, cited on http://www.genomesize.com; 6, J. Baguñà, unpublished data, cited on http://www.genomesize.com; 7, Rishi and McManus (1988); 8, Cox *et al.* (1990); 9, www.genome.gov/Pages/Research/Sequencing/SeqProposals/PlanarianSEQ.pdf; 10, Simpson *et al.* (1982); 11, Garcia-Fernandez *et al.* (1995); and 12, Searcy and MacInnis (1970).

Genome composition

To date, very few detailed analyses of overall genome composition have been reported for platyhelminths. The tapeworm *Spirometra mansonoides* contains 10% highly repetitious, 35% moderately repetitious, and at least 25% single copy sequence (Cox *et al.*, 1990). It has an overall (G+C) content of 44% (Cox *et al.*, 1990) compared to 36% in *Hymenolepis diminuta* (Carter *et al.*, 1972). *Schistosoma mansoni* possess 6% highly repetitive sequence, 34% moderately repetitive sequence and 60% single copy sequence (Simpson *et al.*, 1982). The (G+C) content of the genomes of *S. mansoni, Schistosoma haematobium* and *Schistosoma japonicum* was initially estimated at 34% (Hillyer, 1974), but more recent studies give values of 29.4%, 28.5% and 26.1%, respectively (Marx *et al.*, 2000). The lower (G+C) content in *S. japonicum* suggests a greater bias in nucleotide DNA replication and repair in this species (Marx *et al.*, 2000). Such a bias has also been detected in coding regions in *S. mansoni*, where over-represented codons tend to have A or T at the third codon position, while under-represented codons have C or G at both the second and third positions (Ellis and Morrison, 1995; Musto *et al.*, 1998). Moreover, there is much variation in (G+C) content between coding regions, and a strong, but skewed correlation between the (G+C) content of a coding region and that of its flanking regions and introns, such that in (G+C) poor regions, G and C are even more under-represented in third codon positions, whilst in (G+C) rich regions they are proportionally over-represented (Musto *et al.*, 1995, 1998). These observations suggest that the schistosome genome is arranged into 'isochore-like structures' of varying (G+C) content (Musto *et al.*, 1995, 1998). Codon biases have also been detected in coding sequences of *Echinococcus* species where a subset of codons (almost all of which end in G or C) show increased usage in highly expressed genes (Fernandez *et al.*, 2001). As this runs against the mutational bias of an (A+T)-rich genome, it suggests that selection pressures are acting at the level of translation of highly expressed genes to maintain the bias. To date, this is the only observation of translational selection for codon usage amongst the Platyhelminthes (Fernandez *et al.*, 2001).

DNA methylation

Several platyhelminth genomes have been examined for the presence of methylated DNA by chromatography, differential restriction with methylation sensitive/insensitive isoschizomeric endonucleases or restriction with methyl-dependent endonucleases. Taxa examined include the cestode *S. mansonoides* (Cox et *al.*, 1990) and the digeneans *S. mansoni* (Simpson *et al.*, 1982; Fantappie *et al.*, 2001), *Opisthorchis viverrini, Fasciola gigantica* and *Gigantocotyle siamensis* (Sermswan *et al.*, 1991b) and *Fasciola hepatica* (Musto *et al.*, 1994b). Neither cytosine nor adenine methylation has been detected in any of these taxa and it would appear that DNA methylation does not occur in parasitic flatworms. Two main hypotheses have been proposed to explain the evolution, distribution and extent of DNA methylation in animal genomes. First, that methylation initially served to silence newly introduced 'genomic parasites', subsequently becoming a general repressor of endogenous repeated sequences and, eventually, of tissue-specific genes in the genomes of higher eukaryotes (Bestor, 1990). In multicellular invertebrates, DNA methylation would therefore be expected to correlate with genomic parasite load and genome size. Second, that DNA methylation correlates with high levels of mitotic turnover between meioses and serves as a long-term 'cell memory system', which facilitates the transmission of non-obligatory sequences to daughter cells (Jablonka *et al.*, 1992; Jablonka and Lamb, 1995). The absence of DNA methylation in parasitic flatworms, which possess both extensive asexual reproductive stages and genomes with significant amounts of repetitive DNA, appears to contradict both hypotheses (Regev *et al.*, 1998). Analysis of dinucleotide frequencies in coding and flanking regions, introns and repetitive sequences of *S. mansoni* suggest an under-representation of CpG dinucleotides and over-representation of TpG and CpA in the genome, with coding regions being more similar in bias to their flanking regions than to their introns (Musto *et al.*, 1994a,b). As such biases are characteristic of methylated genomes, the absence of DNA methylation in parasitic flatworms could be a recent, derived characteristic (Musto *et al.*, 1994b). It could also simply reflect an isochore genome organization (Musto *et al.*, 1994b, 1995, 1998). Interestingly, initial automated analysis of both expressed sequence tag (EST) and genomic sequence data from the *S. mansoni* genome project suggests that sequences with putative homology to DNA methyltransferase and DNA methyltransferase-associated protein genes exist. If confirmed, the functionality and biological role of these sequences remain to be determined. Methylated, host microsatellite-related sequences have also been reported to occur in *S. mansoni* and *S. japonicum* (Iwamura *et al.*, 1991) although their origins are disputed (Simpson and Pena, 1991); see section entitled 'Integration of Host DNA Sequences' for more detailed comment. It will be interesting to determine whether free-living flatworms possess methylated genomes (Regev *et al.*, 1998).

Repeat sequences

Studies of overall genome composition based on reassociation kinetics (Simpson *et al.*, 1982; Cox *et al.*, 1990; Marx *et al.*, 2000) and analysis of fully sequenced bacterial artificial chromosome (BAC) clones from the *S. mansoni* genome project show that platyhelminth genomes contain abundant highly and moderately repetitive sequence (Fig. 2.1). Much of the repetitive DNA comprises two classes of integrated mobile elements; class I elements, which include long terminal repeat (LTR) retrotransposons and retroviruses, non-LTR retrotransposons and short interspersed nuclear elements (SINES) and transpose via an RNA intermediate, and class II elements (transposons), which transpose as DNA (Brindley *et al.*, 2003). Additionally, small dispersed or tandemly repeated sequences are common. A wide variety of these sequences have been isolated and characterized from a variety of taxa (Table 2.4).

Individual copies of integrated mobile elements may either be degenerate relics or

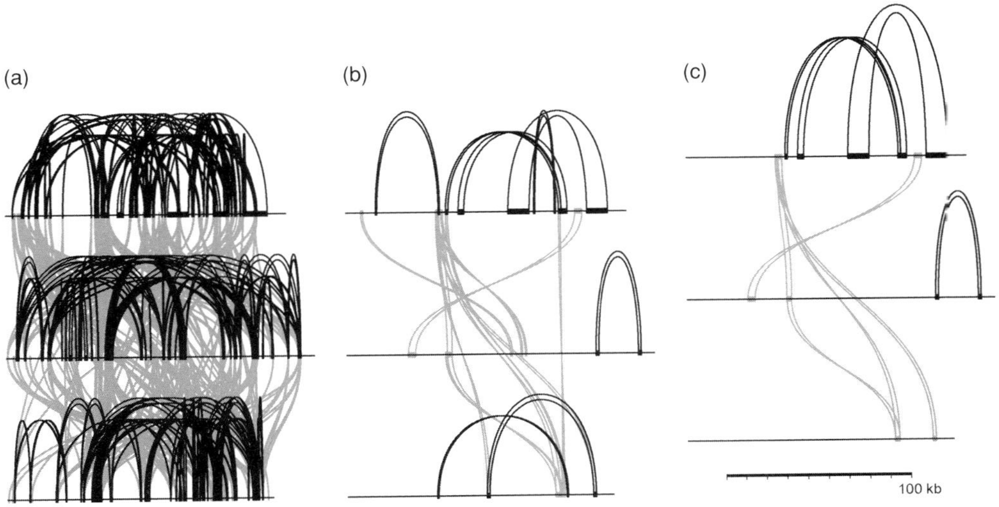

Fig. 2.1. Flatworm genomes possess large amounts and large numbers of repetitive DNA elements. Diagrams show the distribution of repeat elements of different scales that occur within (black), or are shared between (grey), three fully sequenced *Schistosoma mansoni* bacterial artificial chromosome clones. (a) 100 bp repeat size threshold; (b) 500 bp repeat size threshold; and (c) 1000 bp repeat size threshold. Each diagram shows: top = SmBAC-53J5 (126 kb insert); middle = SmBAC-54H2 (133 kb insert); and bottom = SmBAC-44D11 (127 kb insert). Figures compiled from postscript output of Miropeats software (Parsons, 1995), freely available from http://www.littlest.co.uk/software/bioinf/, using sequence data available by anonymous FTP from ftp.sanger.ac.uk/pub/databases/Trematode/S.mansoni/BACs/

Table 2.4. Mobile genetic elements and other repetitive sequences in platyhelminth genomes.

Species	Repeat	Repeat type	Estimated frequency in genome[a]	Transcribed	Reference
Digenea					
Schistosoma mansoni	Boudicca	LTR retrotransposon (Ty3/Gypsy)	>1,000/1,000–10,000 copies	Yes	Copeland *et al.* (2003)
	Saci-1	LTR retrotransposon (BEL)	70–700 copies	Yes	DeMarco *et al.* (2004)
	Saci-2	LTR retrotransposon (Ty3/Gypsy)	85–850 copies	Yes	DeMarco *et al.* (2004)
	Saci-3	LTR retrotransposon (Ty3/Gypsy)	150–1,500 copies	Yes	DeMarco *et al.* (2004)
	Sinbad	LTR retrotransposon (Pao/BEL)	50 copies	Yes	Copeland *et al.* (2005)
	Fugitive	LTR retrotransposon (Ty3/Gypsy)	>2000 copies	Yes	Laha *et al.* (2004)
	Perere	Non-LTR retrotransposon (CR1)	250–2,500 copies	Yes	DeMarco *et al.* (2004)
	SR1	Non-LTR retrotransposon (CR1)	200/2,000–20,000 copies	Yes	Drew and Brindley (1997)
	SR2	Non-LTR retrotransposon (RTE-1)	2,600–26,000 copies	Yes	Drew *et al.* (1999)
	Sm-alpha family	Tandemly repeated and dispersed SINE	7,000–10,000/20,000–200,000 copies	Yes	Spotila *et al.* (1989)
	Sm-alpha fem-1	Female-specific, tandemly repeated SINE	ND	ND	Drew and Brindley (1997)
	Sm750	Polymorphic repeat	ND	Yes	Spotila *et al.* (1991)
	W1	476 base, female-specific repeat	>500 copies	ND	Webster *et al.* (1989)
	W2	715 base, female-specific tandem repeat	ND	ND	Drew and Brindley (1997)
	SM23	Female-specific repeat	75 copies	No	Walker *et al.* (1989)
	pSm1-7	121 base, species-specific tandem repeat	250,000 copies	ND	Hamburger *et al.* (1991)
Schistosoma japonicum	Gulliver	LTR retrotransposon (Ty3/Gypsy)	500 copies	Yes	Laha *et al.* (2001)
	SjR2	Non-LTR retrotransposon (RTE-1)	10,000 copies	Yes	Laha *et al.* (2002a)
	Pido	Non-LTR retrotransposon (CR1)	>1000 copies	Yes	Laha *et al.* (2002b)
	Sj-alpha family	SINE	10,000 copies	ND	Laha *et al.* (2000)
	Sjrh1.0/1.3	>1000 and >1300 base, species-specific tandem repeats	100–200 copies	ND	Drew *et al.* (1998)
Schistosoma haematobium	*Dra*1	121 base tandem repeat	>300,000 copies	ND	Hamburger *et al.* (1991)
Clonorchis sinensis	CsRn1	LTR-retrotransposon (Ty3/Gypsy)	>100 copies	Yes	Bae *et al.* (2001), Bae and Kong (2003b)

Paragonimus westermani	Multiple	LTR-retrotransposon (Ty3/Gypsy)	Highly repetitive	Yes	Bae and Kong (2003a)
	Multiple	LTR-retrotransposon (Pao/BEL)	Highly repetitive	Yes	Bae and Kong (2003a)
	Multiple	LTR-retrotransposon (Xena)	ND	No	Bae and Kong (2003a)
Fasciola hepatica	–	124 base, genus-specific repeat	15% of genome/300,000 copies	ND	Kaplan *et al.* (1995), Kramer and Schnieder (1998)
	Fhr-I	Species-specific repeat	–	–	Heussler *et al.* (1993)
	Fhr-II	Species-specific repeat	–	–	Heussler *et al.* (1993)
	Fhr-III	Species-specific repeat	–	–	Heussler *et al.* (1993)
Dicrocoelium dendriticum	Ddr-IV	Species-specific repeat	ND	ND	Heussler *et al.* (1998)
Opisthorchis viverrini	pOV-A6	334 base, species-specific tandem repeat	–	–	Sermswan *et al.* (1991a)
Cestoda					
Echinococcus granulosus	–	477–489 base repeat	Middle repetitive	ND, unlikely	Marin *et al.* (1993)
	TREg	186 base, genus-specific, polymorphic, tandem repeat	120–23,000 copies (strain-dependent)	ND, unlikely	Rosenzvit *et al.* (1997, 2001)
	EgG1 Hae III	269 base, species-specific, dispersed tandem repeat	6,900 copies	ND	Abbasi *et al.* (2003)
Taenia solium	–	158 base, species-specific tandem and dispersed repeat	ND	ND	Chapman *et al.* (1995)
Taenia saginata	HDP1	53 base, species-specific tandem repeat	ND	ND, unlikely	Gonzalez *et al.* (2002)
'Turbellaria'					
Dugesia etrusca	DEL	≥1.4 kb dispersed LINE	1% of genome	ND, unlikely	Batistoni *et al.* (1998)
Giradia tigrina	PEVE	Helicase-encoding, extra chromosomal, virus-like element	–	–	Rebrikov *et al.* (2002)
	–	Mariner-like transposon	8,000 copies	Not proven	Garcia-Fernandez *et al.* (1995)

[a]Data from original references: Brindley *et al.* (2003) and DeMarco *et al.* (2004).
ND = not determined.

retain some potential functionality. Virtually all platyhelminth integrated mobile elements described to date appear to be transcribed to some degree. Indeed, the overall transcription level can be very high; in *S. mansoni* some 14% of cercarial transcripts encode sequences with putative homology to reverse transcriptase and there appear to be low copy number elements with high transcriptional activity and vice versa (DeMarco *et al.*, 2004). Direct evidence for translation of these transcripts to create functional mobile elements has yet to be demonstrated. However, reverse transcriptase activity has been detected in cultured *S. mansoni* (Ivanchenko *et al.*, 1999) and the reverse transcriptase domain of the *S. japonicum* SjR2 non-LTR retrotransposon has been expressed in insect cells and shown to prime reverse transcription of SjR2 mRNA *in vitro* (Laha *et al.*, 2002a). Therefore, at least indirect evidence exists for functional mechanisms. Such activity would certainly help explain reports of genome instability (see section entitled 'Genome Instability').

LTR retrotransposons usually possess one or two open reading frames with putative homology to retroviral *gag* and *pol* proteins, but some, including the Gypsy family possess a third open reading frame with homology to the retroviral envelope (*env*) gene. The *env* gene product is required for the production of infective viral particles; by binding to specific receptors on the host cell membrane it both allows penetration and confers specificity (Hoffman and Doms, 1999). The identification of schistosome LTR retrotransposons containing *env* genes therefore paves the way for the potential development of specific retroviral transfection systems. A potential third open reading frame has been identified in the *S. mansoni* LTR retrotransposons Boudicca and Saci-3 (Copeland *et al.*, 2003; DeMarco *et al.*, 2004) and the sequence and structure of putative, transcriptionally active versions of the retrotransposons have been reconstructed by assembling sequences derived from *gag* and *pol* transcripts with raw data from the *S. mansoni* genome sequencing project (Copeland *et al.*, 2004; DeMarco *et al.*, 2004). Similarly, seqences with homology to the promiscuous transposon Mariner have been detected in *Girardia tigrina* (Garcia-Fernandez *et al.*, 1995). Both Mariner and other promiscuous transposons (*Hermes* and *piggyBac*) have successfully been used for transgenesis in this species (Gonzalez-Estevez *et al.*, 2003). Whilst *Hermes* and *piggyBac* produce stable transfectants, the Mariner constructs were unstable; possibly as a result of activity of the endogenous elements. It may prove possible to reconstruct a functional planarian Mariner from these sequences, as has been achieved for the salmonid transposon *Sleeping Beauty* (Ivics *et al.*, 1997). An extrachromosomal, virus-like element has also been reported in *G. tigrina* and found to be distributed nonuniformly through the planarian body (Rebrikov *et al.*, 2002). Its significance remains to be determined.

In addition to the mobile genetic elements, a wide variety of short, non-transcribed, tandem and dispersed repeat sequences have been described in digeneans and cestodes. These are often species- or genus-specific and present in high copy numbers, allowing the development of sensitive, specific diagnostic molecular assays for *S. mansoni* (Hamburger *et al.*, 1998a,b), *S. haematobium* (Hamburger *et al.*, 2001), *F. hepatica* (Heussler *et al.*, 1993; Kaplan *et al.*, 1995), *Dicrocoelium dendriticum* (Heussler *et al.*, 1998), *Echinococcus granulosus* (Abbasi *et al.*, 2003), *Taenia solium* (Chapman *et al.*, 1995) and *Taenia saginata* (Gonzalez *et al.*, 2002). Additionally, for *Schistosoma*, which has chromosomally determined sex in which the female is heterogametic (ZW), female-specific tandem repeats have also been characterized (Walker *et al.*, 1989; Webster *et al.*, 1989; Drew and Brindley, 1995; see section entitled 'Sex Chromosomes').

Integration of host DNA sequences

Despite the hostility of the environment in which they make their home, adult *Schistosoma* are remarkably long lived; frequently surviving for years, and regularly for decades (Warren *et al.*, 1974; Garcia, 1976; Basch, 1991). It has long been recognized that schistosomes acquire a cloak of host antigens including immunoglobulins, erythrocyte antigens and major histocompatibility complex (MHC) products as a means of avoiding immune attack (Salzet *et al.*, 2000; see also Chapter 8, this volume). Early experiments failed to detect sequences homologous to host class 1 MHC genes within schistosome DNA and it was concluded that the MHC antigens expressed on the parasite surface are acquired from the host as intact glycoproteins, rather than synthesized from sequences within the parasite genome (Simpson *et al.*, 1983). Despite this, a series of papers published over the past 15 years have reported the existence, developmental plasticity, rapid integration and vertical transmission of host DNA sequences in the schistosome genome (Tanaka *et al.*, 1989; Iwamura *et al.*, 1991, 1995; Irie and Iwamura, 1993; Imase *et al.*, 1999, 2000, 2001, 2003, 2004). These authors have used a wide variety of methods of increasing sophistication and sensitivity, including immunolocalization, immunoblotting, Southern blotting, polymerase chain reaction (PCR), nested PCR, direct sequencing of PCR products, *in situ* hybridization with radiolabelled probes and *in situ* PCR with localization with radiolabelled or digoxygenin-labelled probes to test for a variety of sequences in adult worms, eggs, miracidia and cercariae of *S. mansoni* and *S. japonicum*. The host sequences studied comprise different Type A and C murine retroviruses and different genes thereof, type 1 Alu and type 2 Alu (B2) repeat sequences, various mouse microsatellites and different regions of MHC-H2. In total, some 200 different combinations of host sequence, schistosome target and experimental method have been examined. The overall picture revealed is incredibly complex, and it is impossible to make any broad generalizations to summarize the observations. It is not appropriate to collate all the experimental data here; the interested reader is referred to the original publications and to a recent review (Imase *et al.*, 2003) for the full information. However, what is apparent is that there is huge variation in recorded behaviour shown by the different host sequences: (i) some show vertical transmission, whilst others do not; (ii) patterns of presence/absence throughout the life cycle differ greatly (some are detected in adult worms of both sexes and eggs, but not in miracidia or cercariae, some only in adult worms, some in adult females but not males and vice versa, some in eggs but not adults, some in miracidia but not cercariae, etc.; (iii) the timing of acquisition by adult worms varies; (iv) some show tissue-specific localization including the germ cells (but then do not display vertical transmission); (v) some localize to the cytoplasm, others in the nucleus; (vi) different regions of the same sequence behave differently; (vii) some behave similarly in both *S. mansoni* and *S. japonicum*, others show significant differences in presence/absence in different developmental stages, in the timing of acquisition in adults and in tissue localization; (viii) some sequences are identical to those seen in the host, some show homology but not identity, and some even vary between developmental stages and sexes; and (ix) some appear to become methylated (see comments in section entitled 'DNA Methylation'). In some of these experiments, *F. hepatica* (from cattle) and *Paragonimus ohirai* (from rat) were used as 'controls' and host-related sequences were also occasionally detected in the latter (Iwamura *et al.*, 1991, 1995). It is difficult to conceive of any molecular mechanism, which could explain such a diverse range of observed behaviours and, despite the presence of seemingly valid controls, these reports have proved controversial, with other authors claiming that the observations must be artefacts resulting from host DNA contamination in schistosome DNA preparations (Simpson and Pena, 1991; Clough *et al.*, 1996). Clearly, independent confirmation of results would help to settle the argument.

Genome instability

Irrespective of the debate surrounding the reported acquisition of host sequences by *Schistosoma* spp., a significant body of evidence supports the idea that genome plasticity/instability is a common feature in the genus. The first evidence came from studies of inducible drug resistance in *S. mansoni*; when immature worms of particular strains are exposed to the anthelmintic hycanthone *in vivo*, hereditable drug resistance appears within a single generation (Jansma *et al.*, 1977). A restriction fragment length polymorphism

(RFLP) within the rRNA gene complex was consistently found to be associated with induced resistance (Brindley *et al.*, 1989) and was subsequently shown to involve a 732 bp duplicative insertion within the SSU rRNA gene, which affected a small number of copies in the tandemly arrayed complex (Brindley *et al.*, 1991). This genomic change does not appear to be the actual basis of hycanthone resistance, but rather to arise from genomic alterations related to those that are responsible. Although hycanthone intercalates into DNA and causes frameshift mutations, such random mutagenic effects could not cause a consistent genomic change across independent experiments and individuals and the mechanism responsible remains to be determined.

When total miracidial DNA and total adult worm DNA from *S. mansoni* are used as probes against blots of restricted total DNA from the same sources, one would predict the two probes to give identical results; showing equal and strong hybridization to both life-cycle stages. However, very different results have been reported where both the probes give a massively stronger hybridization signal with DNA from the same life-cycle stage, than with DNA from the other. Moreover, the total miracidial DNA probe shows significant hybridization with both adult and miracidial DNA from *S. japonicum*, whereas the total adult DNA probe shows minimal hybridization with either (Nara *et al.*, 1990). Similarly, when random, cloned genomic fragments are used as probes, hybridization patterns appear to suggest both stage specificity and the occurrence of developmentally regulated genome rearrangements (Nara *et al.*, 1990). Experimental controls appear to rule out the possibility that host DNA contamination of adult worm DNA preparations alters its composition relative to miracidial preparations. Therefore, results have been interpreted as suggesting that major DNA amplification and deletion events occur during the *S. mansoni* life cycle, resulting in the DNA from the two stages having completely different compositions. This then affects their behaviour when used as probes or targets (Nara *et al.*, 1990). Such unexpected results clearly call for independent verification, but until proven otherwise should be accepted as valid.

The *S. mansoni* W1 tandem repeat was originally isolated from Puerto Rican parasites and shown to hybridize exclusively to female DNA of this isolate (Webster *et al.*, 1989). It localizes to a euchromatic gap within the heterochromatin of the long arm of the female-specific W chromosome, where suppression of recombination in the ZW bivalent by surrounding heterochromatin has been suggested to explain its sex-specific distribution (Hirai *et al.*, 1993). However, studies in other isolates revealed that copies of the sequence can also occur in male worms (Grevelding, 1995). Analysis of W1 distribution in the supposedly clonal progeny that arise from monomiracidial snail infections reveals significant inter- and even intra-clonal variation. Moreover, breeding experiments revealed the presence of variable patterns of W1 elements in the male progeny of crosses involving males that lacked W1. These data are indicative of genome instability and suggest that unusual DNA recombination events occur during both mitosis and meiosis (Grevelding, 1999). Subsequently, analysis of clonal daughter sporocyst populations arising from single mother sporocysts *in vitro* confirmed the existence of intra-clonal variation in W1 distribution, supporting the hypothesis that mitotic recombination events occur during the asexual development of schistosomes (Bayne and Grevelding, 2003).

Rapid rearrangement of a polymorphic minisatellite within the D-loop/control region of the *S. mansoni* mitochondrial genome has also been reported by various authors (Minchella *et al.*, 1994; Bieberich and Minchella, 2001; Jannotti-Passos *et al.*, 2001). These observations are discussed in detail in the section entitled 'Maternal Inheritance'.

Taken together, these diverse data suggest that a high rate of non-meiotic genome rearrangement occurs in *S. mansoni*. Whether this is a result of, or the driving mechanism behind, the relatively recent radiation of African *Schistosoma* species (Le *et al.*, 2001b; Lockyer *et al.*, 2003) remains to be determined, as do the mechanisms responsible.

Gene processing

Spliced leader RNA *trans*-splicing is a form of RNA editing in which a short RNA exon (the spliced leader) is spliced from the 5′ end of a non-polyadenylated RNA, on to the 5′ end of a polyadenylated pre-mRNA in order to form a mature mRNA. Spliced leader sequences have been reported in a variety of platyhelminths (Table 2.5) including all digeneans examined to date, some cestodes and primitive (polyclad) 'turbellarians' but not in the single monogenean that has been studied (Rajkovic *et al.*, 1990; Davis *et al.*, 1994; Davis, 1997; Brehm *et al.*, 2000, 2002). Flatworm spliced leaders show considerable diversity in size (34–51 nucleotides) and in sequence. In particular, the *Schistosoma* spliced leader, whilst totally conserved between *S. mansoni*, *S. haematobium* and *S. japonicum*, is markedly different to that of the other trematodes (Davis, 1997). However, there is sufficient homology across the phylum to suggest a common ancestry (Davis, 1997; Brehm *et al.*, 2000). Such diversity is in marked contrast to the situation in nematodes where a 22-nucleotide spliced leader is highly conserved (Stratford and Shields, 1994). Only a small proportion of platyhelminth transcripts appear to acquire a spliced leader (Davis, 1997) (again in contrast to nematodes where the majority do so, McCarter *et al.*, 2003) but amongst those that do, the enolase gene seems particularly favoured. Consequently, 5′-rapid amplification of cDNA ends (RACE) methods targeting the 5′ end of the enolase gene have been used with great success to identify spliced leaders in flatworm species (Davis, 1997). There appears to be neither conservation between cestodes and trematodes in the spectrum of mRNAs that are

Table 2.5. Platyhelminth spliced leader RNAs/genes.

	Spliced leader (length)	Enolase *trans*-splicing	Reference
Digenea			
Schistosoma mansoni	Yes (36)	Yes	1, 3
Fasciola hepatica	Yes (37)	Yes	2, 3
Echinostoma caproni	Yes (37)	Yes	3
Haematolechus spp.	Yes (34)	Yes	3
Stephanostomum spp.	Yes (36)	Yes	3
Cestoda			
Echinococcus multilocularis	Yes (36)	No	4
Echinococcus granulosus	Putative identification	No	4
Taenia solium	Yes	No	5
Hymenolepis diminuta	No	No	3
Calliobothrium spp.	No	No	3
Lacistorynchus tenuis	No	No	3
Monogenea			
Megacotyle tributa	No	No	3
'Turbellaria'			
Stylochus zebra	Yes (51)	Yes	3
Notoplana spp.	Yes (51)	Yes	3
Dugesia dorotocephala	No	No	3
Bdelloura candida	No	No	3
Phagocata morgani	No	No	3
Syndesmis franciscana	No	No	3

Note: 1, Rajkovic *et al.* (1990); 2, Davis *et al.* (1994); 3, Davis (1997); 4, Brehm *et al.* (2000); and 5, Brehm *et al.* (2002).

trans-spliced (Brehm *et al.*, 2000) nor any discernible pattern or apparent sex, tissue or cell specificity in the genes that get *trans*-spliced within a species (Davis *et al.*, 1995), and the functional significance of flatworm spliced leaders remains to be determined. There is a conserved AUG (methionine) codon immediately before the splice site but this does not necessarily serve as the translation initiation site and can be out of frame with the following open reading frame (ORF) (Davis, 1997). However, 43% of *trans*-spliced *Echinococcus* sequences potentially utilize the donated AUG (Fernandez *et al.*, 2002). In other taxonomic groups (e.g. nematodes) spliced leaders are involved in the processing of poly-cistronic transcription units (Spieth *et al.*, 1993). In *S. mansoni* the gene encoding the ubiquinol-binding protein, UbCRBP, is situated only 54 nucleotides upstream of the enolase gene and RNAs, which significantly overlap both the genes are observed, suggesting that they form a poly-cistronic unit. A close linkage of the two genes is also seen in *F. hepatica*, which also *trans*-splices its enolase gene (Davis and Hodgson, 1997). Such evidence strongly suggests that spliced leaders are involved in processing poly-cistronic units in flatworms.

Karyotype Features

Chromosome number

Platyhelminths have typical metazoan chromosomes that condense during mitosis and their genome size is such that chromosomes are visible under the light microscope. Karyotype analysis has therefore been undertaken in a very wide variety of parasitic and free-living species. Data for 230 species of trematode have been collated (Barsiene, 1993), but no comparable reviews are available for the other groups. In general, platyhelminths appear to possess a relatively small chromosome compliment, ranging from $n = 2$ in the planarian *Polycelis auriculata* (Nishitan *et al.*, 1998) to $n = 28$ in *Clonorchis sinensis* (Park *et al.*, 2000), with $n < 10$ as a general 'rule' (Barsiene, 1993) and $n = 9$ most common in cestodes (Petkeviciute, 1996).

Ploidy

Except for the gametes of sexually reproducing species, platyhelminths are generally diploid ($2n$) throughout their life cycles. However, a vast array of 'non-standard' chromosome complements have been reported, including polyploid taxa, supernumerary chromosomes and aneuploids.

Naturally haploid taxa have not generally been reported. However, in the dioecous trematode, *Schistosomatium douthitti*, unpaired females regularly mature and produce eggs that hatch and release miracidia capable of undergoing asexual development in the snail host to generate single sex cercariae. Karyotype analysis reveals these to be parthenogenetic haploids (Short and Menzel, 1959). Likewise, *Schistosoma*, which is normally diploid, can be forced into a haploid, parthenogenetic state by experimental pairing with *S. douthitti*, which provides a stimulus for reproduction but no genetic contribution. If mice are co-infected with male *S. douthitti* and female *S. mansoni*, the two species pair up and the female worms produce typical lateral-spined eggs containing haploid embryos. These hatch to release miracidia that can infect the snail intermediate hosts of *S. mansoni* in which they give rise to haploid sporocysts. These subsequently produce haploid, male or female cercariae, which can infect mice. Diploidy is restored when parthenogenetic female *S. mansoni* mate with either parthenogenetic (haploid) or normal (diploid) males of their own species. However, if paired with male *S. douthitti*, they again reproduce parthenogenetically, giving rise to further haploid sporocysts in the snail host (Basch and Basch, 1984). Parthenogenetic reproduction has also been shown to occur in both diploid and triploid *Fasciola* (diploid *Fasciola* are normally sexual) (Agatsuma *et al.*, 1994a; Fletcher *et al.*, 2004) and in triploid *Paragonimus westermani* (van Herwerden *et al.*, 1999; Blair, 2000). In both genera triploidy appears to have arisen on multiple occasions. Triploid *P. westermani* are larger and much more pathogenic to humans than their diploid counterparts and so have been a focus of study (Blair, 2000). Their origins are disputed; some authors report that the C-banding patterns of all copies of a particular chromosome are identical, suggesting autotriploidy (Tan and Li, 1990; Terasaki *et al.*, 1996), whilst others report differences, suggesting allotriploidy as a result of interspecific hybridization (Hirai *et al.*, 1985). Rare tetraploids may be autotetraploids resulting from insemination of parthenogenic triploids with haploid gametes from sexual diploids and appear to be capable of sexual reproduction (Terasaki *et al.*, 1995). Triploidy has also been reported in one population of the digenean *Ichthyocotylurus platycephalus* (Staneviciute and Kiseliene, 2001).

Polyploidy, especially triploidy, is not unusual in the 'Turbellaria' (Benazzi and Benazzi-Lentati, 1976) and is usually associated with asexual lineages; it has been reported in *Dugesia japonica* (where triploids lack genital organs and are presumed to be asexual (Bessho *et al.*, 1997)), *Dugesia subtentaculata* (Roca *et al.*, 1992), *Polycelis nigra* (where polyploid (typically triploid) individuals are pseudogamous parthenogens (Beukeboom *et al.*, 1998)) and in *Crenobia alpina* (where chromosome number suggests hexaploidy but individuals behave as functional diploids, possessing only two copies of particular chromosome morphologies rather than six (Van der Linden, 1969; Roca *et al.*, 1992)). In the Proseriata it is suggested that gene duplication as a result of autotetraploidy leads to genetic diversification and speciation (Martens *et al.*, 1989). The most extreme polyploid state reported to date is in the planarian *Phagocata ullala* where a polyploid series of $4n = 68$, $8n = 136$, $10n = 170$ and $14n = 238$ is seen (Roca *et al.*, 1992; Sluys *et al.*, 1995). The occurrence of polyploidy also appears to be related to life in harsh, cold habitats (Roca *et al.*, 1992).

Although asexual/parthenogenetic species may be considered as evolutionary dead ends, some are widespread and genetically diverse, suggesting that cryptic sexuality could be contributing to their success. The planarian *Schmidtea polychroa* is a sperm-dependent, parthenogenetic hemaphrodite in which copulation and sperm exchange is needed to initiate embryogenesis but where sperm do not contribute genetic material. It has recently been shown that limited genetic exchange involving chromosome addition and loss occurs in *S. polychroa*, resulting in up to 12% of offspring receiving novel genetic makeups (D'Souza *et al.*, 2004). Such processes, no doubt, help to explain the success of some parthenogenetic species.

Mixoploidy, in which individuals are a mosaic of different ploidy levels, has a functional significance in several planarians. A diploid/triploid mosaicism of somatic and germ cells is found in *Dugesia aethiopica* (Stocchino *et al.*, 2004) whilst in polyploid *Dugesia lugubris* embryonic and somatic cells are triploid ($3n = 12$), male germ cells are diploid ($2n = 8$) and female germ cells are hexaploid ($6n = 24$) (Gremigni *et al.*, 1980). Regeneration studies show that diploid male germ cells can recreate both triploid somatic cells and hexaploid oocytes (Gremigni *et al.*, 1982). Certain populations of *D. japonica* are also ($3n + 2n$) mixoploids, lack genital organs and are assumed to be assexual (Bessho *et al.*, 1997). Mixoploidy has also been recorded in the intramolluscan stages of the digeneans *S. mansoni*, *Calicophoron microbothrium* and *Carmyerius gregarius* (Hirai and LoVerde, 1989; Ashour *et al.*, 1995) and in the cestode *Cyathocephalus truncatus* (Petkeviciute, 1996). Its significance, if any, for these species is unknown.

In addition to gross changes in ploidy level, more subtle variations in chromosome complement are also known to occur in flatworms. Aneuploidy, the gain of extra copies of individual chromosomes of the normal complement, or the loss of individual chromosomes from the normal complement, has been described in larvae of the ORF strain of the cestode *Taenia crassiceps*. Compared to the standard $2n = 16$ karyotype, the ORF strain displays a $2n = 14$ karyotype in which both copies of chromosome 2 are missing (Smith *et al.*, 1972). Despite this major loss of genetic material, ORF strain metacestodes are still viable and can be maintained continuously by serial inoculations in mice, although they appear unable to infect their definitive hosts. In the 'turbellarian' *P. nigra*, 66% of individuals showed positive aneuploidy (Beukeboom *et al.*, 1998). In addition to true aneuploidy, B chromosomes have also been recorded in various platyhelminth taxa. These are supernumerary chromosomes that are morphologically distinct from the normal chromosome complement. They appear to be derived from autosomes or sex chromosomes following intra- and interspecific crosses and subsequently evolve in a manner similar to univalent sex chromosomes, with heterochromatin formation, gene silencing and the accumulation of repetitive sequences and transposons (Camacho *et al.*, 2000; Camacho, 2004). B chromosomes are reported to occur in some 15% of eukaryote species (Camacho, 2004) but have only been recorded in a limited number of flatworms, including the digeneans *Trichobilharzia regenti* (2*n* + 1–2 B in >60% of cells in the sporocyst stage, with female sporocysts being affected more frequently than males) (Spakulova *et al.*, 2001), *Ichthyocotylurus platycephalus (2n + 1–2B), Apatemon, Notocotylus, Diplodiscus* and *Echinostoma* (Staneviciute and Kiseliene, 2001) and the turbellarians *D. aethiopica* (diploid = 2*n* + 0–1 B, triploid = 3*n* + 1–2 B) (Stocchino *et al.*, 2004) and *Dugesia sicula* (3*n* + 2–3 B) (Pala *et al.*, 1995). Given the large number of karyotype studies that have been performed on flatworms, this seems a disproportionately small number of recordings (approximately 5% frequency in trematodes (Staneviciute and Kiseliene, 2001)) and could indicate that strong selection forces exist to eliminate B chromosomes from the phylum.

Sex chromosomes

The vast majority of platyhelminths are hermaphroditic. However, separate sexes/dioecy occurs in a few, unrelated taxa amongst the Eucestoda (*Dioecotaenia, Dioecocestus, Shipleya, Gyrocoelia*) and 'Turbellaria' (Fecampiidae) (Zamparo *et al.*, 2001). Mechanisms of sex determination in these taxa are not yet well understood and may vary from taxon to taxon. Karyotype analysis in *Shipleya inermis* reveals that females possess one pair of chromosomes with heteromorphic heterochromatin regions that result in non-homologous/non-pairing segments. Consequently, female *S. inermis* produce two types of gametes (with respect to heterochromatic DNA), while males are homogametic and it is suggested that this could provide a molecular basis for sex determination (Rausch and Rausch, 1990). However, an alternative hypothesis, based on population analysis, suggests that *S. inermis* is actually a protogynous hermaphrodite, which can become regionally dioecious, possibly in response to interaction with other individuals of the same species (Didyk and Burt, 1998). In *Dioecocestus*, initial evidence suggests that sex is determined genetically as two morphologically distinct metacestode forms are seen, one of which matures into male worms, the other into female worms (Ryzhikov and Tolkacheva, 1981). In *Dioecotaenia*, adult worms are morphologically distinct (Schmidt, 1969), but mechanisms have yet to be determined. More detailed investigations of these taxa are clearly warranted.

True, chromosome-based sex determination and marked sexual dimorphism does occur in the Schistosomatidae (Trematoda, Digenea) and has been extensively studied. The female is the heterogametic sex (ZW) and the male homogametic (ZZ) and sex is determined at fertilization in the mammalian host, with the asexual stages in the molluscan host giving rise to either male or female cercariae (Atkinson, 1980; Grossman *et al.*, 1980, 1981a,b; Short, 1983; Barshene *et al.*, 1989; Spakulova *et al.*, 1996, 1997, 2001). The schistosome Z and W chromosomes are large and distinguished by the presence of extensive, centromeric heterochromatin blocks on the W chromosome (Grossman *et al.*, 1980). Molecular studies using fluorescent *in situ* hybridization (FISH) of DNA probes to schistosome chromosome spreads reveal both regions of homology between Z and W, and regions of unique, sex chromosome-specific sequence (Hirai and LoVerde, 1995; Tanaka *et al.*, 1995). The only reported exception to the schistosome ZW system is in isolates of *Heterobilharzia americana* from Louisiana, which, in contrast to other populations of the parasite, have a ZZ/ZWA mechanism in which the female only has one normal chromosome 5; the other number 5 and most of the W having fused to form a WA chromosome (Short and Grossman, 1986; Short *et al.*, 1987).

Detailed comparative studies of the chromosomes of schistosomes and of their hermaphrodite relatives, the sanguinicolids and spirorchids, have produced hypotheses for the evolution of schistosome sex chromosomes from the autosomes of hermaphrodite ancestors. The hermaphroditic spirorchid *Spirorchis* also possesses heteromorphic chromosomes, suggesting that 'pre-sex' chromosomes existed in flatworms before the splitting away of the lineage that lead to the Schistosomatidae with their distinct, differentiated sex chromosomes (Grossman *et al.*, 1981a). It is interesting to note that chromosomal polymorphism, suggestive of sex chromosomes/sex determination has also been reported in the digeneans *Megalodiscus temperatus* (Grossman and Cain, 1981) and *Rubenstrema exasperatum* (where one copy of chromosome 1 is approximately 50% larger than the other, Mutafova and Kanev, 1996). As there is no evidence of dioecy in these species, it has been suggested that heteromorphic chromosomes of hermaphroditic species may carry genes involved in gonad differentiation (Grossman and Cain, 1981; Mutafova and Kanev, 1996). Current efforts to generate and analyse the full genome sequence of *S. mansoni* and *S. japonicum* (see section entitled 'Genome Sequencing') will reveal the composition of schistosome sex chromosomes and autosomes and greatly facilitate our understanding of the evolution, and molecular basis, of sex in these fascinating organisms. What is already apparent from *Schistosoma* is that genes relating to female sexual differentiation are not restricted to the W chromosome as 'hermaphroditic' male *S. mansoni* (ZZ) showing variable degrees of development of the female reproductive system have been observed (Short, 1948). Hermaphroditic female *S. douthitti* have also been recorded (Short, 1951).

The existence of true sex chromosomes in schistosomes permits the development of methods for determining the sex of morphologically indistinguishable life-cycle stages. Being able to determine the sex of the clonal cercarial populations that result from monomiracidial snail infections allows the efficient conduct of breeding experiments and facilitates genetic analysis. At the gross level, simple DNA staining and microscopy has been used to sex cercariae in *S. mansoni*. Staining with carbol fuchsin or thionin reveals two prominent, densely staining bodies in the nuclei of male cercariae and only one in those from female cercariae, which allows sexing with >80% accuracy (Raghunathan and Bruckner, 1975). This phenomenon was initially suggested to indicate sex chromosome inactivation/'Barr bodies' in both sexes in a process akin to dosage compensation-related X inactivation in humans (Raghunathan and Bruckner, 1975). However, it was subsequently demonstrated that both schistosome sex chromosomes contain euchromatic regions (Grossman *et al.*, 1980). Therefore, total inactivation cannot be occurring and the biological basis of this staining pattern remains to be determined. The presence of a persistent heterochromatin block on the W chromosome of interphase cells of female *S. mansoni* does, however, allow this feature to be stained with conventional C-banding techniques, providing a simple, rapid and reliable marker for female parasites (Liberatos and Short, 1983). Despite the fact that the W chromosomes of other *Schistosoma* species also contain large amounts of heterochromatin, the technique appears to not work for them (Liberatos and Short, 1983). At the molecular level, several female *S. mansoni*-specific DNA probes have been reported including Sm23 (Walker *et al.*, 1989), D9 (Spotila *et al.*, 1987), W1 (Webster *et al.*, 1989), SM alpha fem-1 and W2 tandem repeats (Drew and Brindley, 1995). W1 is a degenerative repeat present in more than 500 copies/female genome (Webster *et al.*, 1989), which maps to an euchromatic gap within the heterochromatin block on the long arm of the W chromosome (Hirai *et al.*, 1993). It has been used to develop sexing assays based on dot blot (Penschow *et al.*, 1993), specific PCR (Gasser *et al.*, 1991; Grevelding *et al.*, 1997) and low-stringency PCR (Dias Neto *et al.*, 1993). However, more recent reports indicate that W1's sex specificity may be restricted to particular isolates of *S. mansoni* (Grevelding, 1995).

Recombination frequency

The chiasma graph method (Wada and Imai, 1995) has been used to examine the frequency and chromosomal distribution of chiasma on the chromosomes of three digenean flukes, *S. mansoni*, *S. japonicum* and *P. ohirai*. These studies indicate that recombination is suppressed within a minimum distance of other chiasmata, of the centromere and of the telomere (equivalent to approximately 1.8% of the length of the bivalent concerned). In female schistosomes, recombination is also known to be suppressed in the region of the ZW bivalent corresponding to the heterochromatic region of the W chromosome (Hirai *et al.*, 1989). Outside of these distances, the distribution of chiasmata on the chromosomes of *S. mansoni* and *P. ohirai* is essentially uniform and random, and chiasma frequencies are similar to those observed in other taxa. Up to five interstitial chiasmata or four interstitial chiasmata and one terminal chiasma are seen per chromosome in *S. mansoni*, and up to three interstitial chiasmata, with zero or one terminal chiasma per chromosome in *P. ohirai* (Hirai *et al.*, 1996). In marked contrast, chiasma frequency in *S. japonicum* is over fivefold lower and a maximum of one interstitial chiasma plus one terminal chiasma is seen per chromosome (Hirai *et al.*, 1996). These observations have been confirmed in other strains of *S. japonicum*. Interstrain variation is observed, with some strains showing increased chiasma frequency, but this is never greater than about 50% of that seen in *S. mansoni* (Hirai *et al.*, 2000). Two other Asian schistosomes, *Schistosoma malayensis* and *Schistosoma mekongi* also show reduced chiasma frequency relative to *S. mansoni*, but at least in *S. mekongi*, multiple chiasma are seen in interstitial regions (Hirai *et al.*, 2000). A similar suppression of chiasma formation has been observed during male meiosis in the planarian *Mesostoma ehrenbergii* where three pairs of chromosomes form bivalents with a single, distant chiasma whilst the remaining two pairs of chromosomes remain as unpaired univalents (Croft and Jones, 1989). The mechanisms behind, and true genetic effects of, such phenomena remain to be determined.

Historically, *S. japonicum* has been considered to be a genetically diverse parasite with different geographic isolates showing variation in snail host specificity (Chiu, 1967), pathogenicity and development (Hsu and Hsu, 1962), morphology (Sobhon *et al.*, 1986) and protein/isoenzyme markers (Ruff *et al.*, 1973; Woodruff *et al.*, 1987). In support of these observations, the organization of high copy number repeat sequences and of the rRNA gene complex appears to be variable within and between isolates (Drew *et al.*, 1998; van Herwerden *et al.*, 1998). However, contradictory evidence has come from comparative molecular analyses of Chinese and Philippine isolates in which very little variation was seen in marker gene sequences by RFLP and sequence analysis or at the gross, whole genome level by randomly amplified polymorphic DNA (RAPD) profiling (Bowles *et al.*, 1993; Qiu *et al.*, 2002).

Telomeres

Using primed *in situ* hybridization techniques (PRINS), the terminal regions of chromosomes of the 'turbellarian' *Polycelis tenuis* and of the schistosomes *S. mansoni*, *S. haematobium*, *S. japonicum* and *Schistosoma sinensium* have all been shown to possess sequences that hybridize with the core telomere repeat sequence $TTAGGG_n$, suggesting that flatworm chromosomes have typical telomeres (Hirai and LoVerde, 1996; Joffe *et al.*, 1996; Hirai *et al.*, 2000). Curiously, the telomeric repeat was also detected in centromeric and some interstitial regions in the African *Schistosoma* species, but not in the Asian ones (Hirai *et al.*, 2000). The significance of this observation will be revealed once full genome sequence becomes available for *S. mansoni* and *S. japonicum* (see section entitled 'Genome Sequencing').

Mitochondrial Genomes

General characteristics

In addition to the nuclear (chromosomal) genome, flatworms also possess a closed circular mitochondrial genome containing a limited repertoire of genes. Mitochondrial genes are present in high copy numbers per cell, facilitating molecular analysis and tend to show higher mutation rates than nuclear genes. Consequently, they have been extensively studied for phylogenetic purposes and to investigate intra- and interspecific genetic variation (e.g. Qiu *et al.*, 2002; Lockyer *et al.*, 2003). The higher mutation rate was long thought to result from the combined effects of a highly reactive chemical environment, which promotes DNA mutation, and a lack of proofreading activity during mitochondrial DNA replication by DNA polymerase (pol) gamma, the nuclear genome-encoded polymerase responsible for mitochondrial DNA replication. The latter argument no longer holds true as pol gamma has now been shown to display a high fidelity in nucleotide incorporation combined with proofreading based on a 3′–5′ exonuclease activity (Kaguni, 2004).

In recent years numerous groups have undertaken sequencing of platyhelminth mitochondrial genomes. One of the initial aims was to identify mitochondrial genes in schistosomes in order to provide 'quality control' for EST data-sets. However, those studies revealed unexpected results, which have prompted a much more extensive taxonomic coverage. At the time of writing, complete, near complete or significant coding region sequences are available for 7 genera and 13 species of flatworm (summarized in Table 2.6), with taeniid tapeworms and schistosomes especially well represented.

Mitochondrial genome sequences have been generated by a combination of direct sequencing of short intra- and longer intergene PCR products (for which conserved primer sequences are often available) and primer walking on cloned, long PCR products. Transposon-mediated sequencing speeds the process of mitochondrial genome-sequencing although the high (A+T) content of most platyhelminth mitochondrial genomes (McManus *et al.*, 2004b) appears to bias incorporation of transposons towards vector sequence (D.A. Johnston and A. Waeschenbach, London, 2005, unpublished data). Gene regions are initially identified by comparison with other flatworm sequences on the database. For protein-coding sequences, this is coupled to ORF analysis using the appropriate genetic code, taking into account the potential use of non-standard initiation and termination sequences (see section entitled 'Genetic Code'). As all mitochondria-encoded proteins are integral membrane proteins, hydrophobicity plots can be used to assist in localizing the less conserved genes (NAD3, NAD4l, NAD6) (Le *et al.*, 2001a). The rRNA gene sequences are identified by homology, coupled to their ability to form appropriate secondary structures; their ends have not been determined experimentally. Transfer RNA (tRNA) genes are identified by a combination of automated prediction using the tRNAscan-SE server at http://www.genetics.wustl.edu/eddy/tRNAscan-SE/ (Lowe and Eddy, 1997), manual inspection, homology and the ability to form appropriate secondary structures. In some cases it is not possible to definitively identify the limits of tRNA genes as alternative, overlapping sequences fulfilling these criteria can be identified, e.g. tRNA-cystine in *S. mansoni* and *S. japonicum* (Le *et al.*, 2001b).

The original publications describing the individual mitochondrial genome sequences provide comprehensive annotation of specific features such as the length of gene and inferred amino acid sequences, base composition, codon bias, start and stop codon usage, tRNA structure, etc. to which the interested reader is referred (see Le *et al.*, 2004 and Table 2.6 for references). For those sequences which have not been formally described in the literature, initial annotation is provided by the original database submissions, by curated GenBank Genome and RefSeq entries and through the Organellar Genome Retrieval System (OGRe) at McMaster University (http://drake.physics.mcmaster.ca/ogre/) (Jameson *et al.*, 2003). General characteristics of platyhelminth mitochondrial genomes can be summarized as follows:

1. They all possess a common set of genes, all coded on the same strand. These comprise: 12 protein coding genes (three subunits of the cytochrome *c* oxidase complex (COI/II/III), seven subunits of the nicotinamide dehydrogenase complex (NAD1/2/3/4/4L/5/6), cytochrome B (CytB) and ATP-synthase subunit 6 (ATP6), two rRNA genes (12S (small) and 16S (large)), and 22 tRNA genes (two copies of tRNA-serine and tRNA-leucine). The only apparent exception to this is *P. westermani* in which a second copy of tRNA-glycine has been predicted in

Table 2.6. Progress in sequencing platyhelminth mitochondrial genomes.

Species	Length	Complete genome/coding region	Accession and reference
Cestoda; Eucestoda; Cyclophyllidea; Taeniidae			
Taenia asiatica	13,703	Yes/yes	NC_004826 – K.S. Eom and H.K. Jeon (unpublished)
Taenia solium	13,709	Yes/yes	NC_004022 – Nakao *et al.* (2003)
Taenia crassiceps	13,503	Yes/yes	NC_002547 – Le *et al.* (2002)
Echinococcus multilocularis	13,738	Yes/yes	NC_000928 – Nakao *et al.* (2002)
Echinococcus granulosus G1 genotype (sheep–dog strain)	13,588	Yes/yes	AF297617 – Le *et al.* (2002)
Echinococcus granulosus G4 genotype (horse–dog strain)	13,598	Yes/yes	AF346403 – Le *et al.* (2002)
Cestoda; Eucestoda; Cyclophyllidea; Hymenolepididae			
Hymenolepis diminuta	13,900	Yes/yes	NC_002767 – von Nickisch-Rosenegk *et al.* (2001)
Trematoda; Digenea; Strigeidida; Schistosomatoidea; Schistosomatidae			
Schistosoma mansoni	14,415	No/yes	NC_002545 – Le *et al.* (2001b)
Schistosoma japonicum	14,085	No/yes	NC_002544 – Le *et al.* (2001b)
Schistosoma mekongi	14,072	No/yes	NC_002529 – Le *et al.* (2001b)
Schistosoma malayensis	9,581	No/no	AF295106 – Le *et al.* (2001b)
Trematoda; Digenea; Echinostomida; Echinostomata; Fascioloidea; Fasciolidae			
Fasciola hepatica	14,462	Yes/yes	NC_002546 – Le *et al.* (2001a)
Trematoda; Digenea; Plagiorchiida; Troglotremata; Paragonimidae			
Paragonimus westermani (triploid)	14,965	Yes/yes	NC_002354 – T. Agatsuma and M. Iwagami (unpublished)
Paragonimus westermani (diploid)	14,244	No/no	AF540958 – M. Iwagami, Y. Sato, A. Iwashita, T. Abe, S.J. Hong and T. Agatsuma (unpublished)
'Turbellaria'; Macrostomorpha; Macrostomida; Microstomidae			
Microstomum lineare	6,882	No/no	AY228756 – Ruiz-Trillo *et al.* (2004)

both diploid and triploid forms (annotation of two tRNA-cysteine genes in *S. mansoni* and *S. japonicum* in some sources reflects alternate models covering the same genomic region (Le *et al.*, 2001b)). In addition to the genes, there is a D-loop/control region.

2. All platyhelminth genomes examined to date lack the gene for ATP-synthase subunit 8, suggesting that this is a characteristic of the phylum, or at least of the Rhabditophora. This gene is commonly found in metazoan mitochondrial genomes, but is also known to be missing in some bivalve molluscs (*Crassostrea gigas, Mytilus edulis, Venerupis* (*Ruditapes*) *philippinarum*), some tunicates (*Halocynthia roretzi* and *Ciona savignyi*, but interestingly present in *Ciona intestinalis*), both chaetognaths sequenced to date (*Paraspadella gotoi* and *Spadella cephaloptera*), one old world tree frog (*Polypedates megacephalus*) and some parasitic and free-living nematodes (*Ascaris suum, Caenorhabditis elegans, Onchocerca volvulus, Strongyloides stercoralis, Dirofilaria immitis, Brugia malayi, Steinernema carpocapsae, Ancylostoma duodenale, Necator americanus, Xiphinema americanum, Cooperia oncophora*). Whether the gene is present in flatworm nuclear genomes has yet to be determined. Initial examination of data from the *S. mansoni* genome project indicates that it has not yet been detected in either cDNA or genomic DNA sequences. However, because the gene is absent from the whole phylum, there is no good, closely matching homologue to search the data-sets with, and it may simply have been overlooked. Adult schistosomes living in the venous blood system of their mammalian hosts have access to effectively unlimited supplies of oxygen and glucose and it might be expected that they would derive most of their energy from ATP synthesized through oxidative phosphorylation. Surprisingly, available data suggest that in both the definitive and intermediate host they are primarily homolactic fermenters; exploiting the anaerobic glycolytic pathway with lactate as the main end product, although some oxidative phosphorylation is also known to occur (van Oordt *et al.*, 1985; Tielens *et al.*, 1992). In contrast, both free-living stages of the life cycle utilize aerobic metabolism (Van Oordt *et al.*, 1989; Tielens *et al.*, 1992). Schistosomes must, therefore, possess a functional ATP-synthase. Whether it functions with or without subunit 8 awaits more detailed analysis. In the yeast *Saccharomyces cerevisiae*, mutations affecting ATP8 are not lethal, but produce a slow growing, 'petite' phenotype that is unable to survive without a fermentable carbon source (Marzuki *et al.*, 1989).

3. None of the genes possess introns and, with the exception of the D-loop/control region, intergenic distances are only a few nucleotides long. Some genes, especially tRNA genes, may abutt and some overlap, especially ND4L and ND4 (by up to 40 bases in all taxa examined apart from *T. solium* and *Echinococcus multilocularis*) and COI and tRNA-threonine.

4. Excepting *P. westermani*, which has an (A+T) content of 51.6%, all the genomes show a significant (A+T) bias, ranging from 63.6% in *F. hepatica* to 74% in *T. crassiceps* (McManus *et al.*, 2004b). Additionally, codon bias, skew (the unequal representation of complementary bases on a single strand) and convergence and divergence in nucleotide and amino acid composition are observed across taxa (Le *et al.*, 2004).

5. The D-loop/control regions are non-coding regions of the mitochondrial genomes and may contain repeated sequences and stable stem-loop structures involved in replication. Usually there is a short and a long non-coding region, separated by one or more genes. The control regions of digeneans are generally much longer than those of cestodes and their structure makes them difficult to sequence through. Hence, the majority of available digenean sequences represent complete coding regions, rather than complete genomes (Table 2.5). The full size of the mitochondrial genome can be estimated by RFLP studies. For *P. westermani* these suggest an overall size of 21 kb versus 15 kb for the sequenced coding region (Agatsuma *et al.*, 1994b) and for *F. hepatica* of up to 19 kb (Zurita *et al.*, 1988) (although a genome of 14.5 kb has been completely sequenced (Le *et al.*, 2001a)). For *S. mansoni* considerable inter-isolate variation is apparent, with sizes ranging from approximately 16.5 to 24.9 kb (Despres *et al.*, 1991, 1993) compared to 14.4 kb for the sequenced coding region (Le *et al.*, 2000). The detected polymorphism is as great between different African isolates as it is between African and South American isolates, providing evidence that the introduction of *S. mansoni* into the New World was a recent historical event (likely as a result of the slave trade), rather than an ancient one due to tectonic separation (Despres *et al.*, 1993). The polymorphism localizes to a single region, which does not hybridize with genic probes, suggesting that it affects the control region. A complex, tandemly arrayed minisatellite (F21) comprising a 558 base sequence and a variable-length, subarray of a 62 bp repeat motif (effectively forming a nested variable number of tandem repeat (VNTR) polymorphism) has since been identified in the *S. mansoni* mitochondrial genome (Pena *et al.*, 1995) and is a likely cause of the polymorphism. Both the 62-base sequence and parts of the 558-base sequence are also found in the SM750 repeat found in the nuclear genome (Spotila *et al.*, 1991) and the relative organization of the two versions suggests either that the element has moved from nuclear to mitochondrial genome (a vanishingly rare occurrence in nature) or that both genomes acquired it simultaneously from a mobile element (Pena *et al.*, 1995).

6. Mitochondrial genome sequence is available for two different populations of both *P. westermani* and *E. granulosus*. For *P. westermani*, both diploid (sexual) and triploid (parthenogenetic) forms have been sequenced (T. Agutsuma, M. Iwagami, Y. Sato, A. Iwashita, T. Abe and S.-J. Hong, unpublished; sequence accessions AF540958 (2n), NC_002354/AF219379 (3n)). Gene order is identical in the two sequences and, apart from a 163-base deletion in the short non-coding region of the triploid, the sequences are very highly conserved (99% at the nucleotide level across the whole genome, 99.5% at the amino acid level in the protein coding regions), suggesting a recent divergence. In contrast, the mitochondrial genomes of the G1 (sheep–dog) and G4 (horse–dog) strains of *E. granulosus* are almost as different from each other (12% divergence at both DNA and amino acid levels in protein coding regions) as each is from *E. multilocularis* (13–15% divergence at DNA level, 12–14% at amino acid level). This supports suggestions that the G1 and G4 genotypes should be designated as separate species (Bowles *et al.*, 1995; Thompson *et al.*, 1995; Le *et al.*, 2002).

Gene order

Perhaps the most unexpected finding from these studies concerns the relative positions of genes within the mitochondrial genome. Mitochondrial gene order is generally regarded as a highly stable characteristic, conserved across large phylogenetic distances and appropriate to use in investigations of deep phylogenies (Boore, 1999; Saccone *et al.*, 2002). When the mitochondrial genomes of flatworms are compared, gene order is seen to be very highly conserved across all the cestode species together with *F. hepatica*, *P. westermani* and the Asian *Schistosoma* species (Fig. 2.2). The order of the 12-protein coding genes plus the two rRNA genes is completely conserved across all these taxa, with only minor rearrangements of a few tRNA genes and of the two non-coding regions seen; these largely occurring at the cestode–trematode 'divide'. However, complex, large-scale gene rearrangements (which also involve protein coding genes) are observed between the Asian *Schistosoma* species and *S. mansoni*, an African species. The fact that Asian *Schistosoma* species possess a 'conserved neodermatan' gene order, whilst an African *Schistosoma* species has a highly derived gene order provides further evidence that the genus originated in Asia and subsequently spread to Africa (Hirai *et al.*, 2000; Snyder and Loker, 2000; Agatsuma, 2003; Lockyer *et al.*, 2003). This recent hypothesis reverses the long accepted idea that *Schistosoma* originated in Africa and spread to Asia (Davis, 1980, 1992). Work is currently underway to sequence mitochondrial genomes from a variety of other flatworm taxa (D.T.J. Littlewood, London, 2005, personal communication). These include: (i) *S. haematobium*, *Schistosoma incognitum*, *Schistosoma spindale*, *Schistosoma guiniensis* and *Orientobilharzia turkestanicum* (which molecular phylogenetic analysis places within the genus *Schistosoma*, Snyder and Loker, 2000) in order to map gene order rearrangements on to an extensive and robust phylogeny (Lockyer *et al.*, 2003) and thus identify how and when they occurred during the radiation from Asia and (ii) representatives of other major flatworm groups including the 'Turbellaria' (*Bdelloura candida*, *Pseudostylochus inter-*

medius), Cestoda (*Gigantolina magna, Caryophyllaeus laticeps*) and Monogenea (*Diclidophora denticulata, Dictyocotyle coeliaca*) in order to examine gene order variation across the phylum. A partial sequence is already available for *Microstomum lineare* and indicates that gene order is not at all conserved between the parasitic Neodermata and the free-living 'Turbellaria' (Ruiz-Trillo *et al.*, 2004).

The small scale, localized swapping of gene order, seen (i) with tRNA-Ser and tRNA-Leu between *H. diminuta* and the other cestodes; (ii) with tRNA-Glu and the small non-coding region and with tRNA-Gly and the large non-coding region between *F. hepatica* and *P. westermani*; and (iii) with tRNA-Tyr and tRNA-Leu and with tRNA-Leu and the large non-coding region between *S. japonicum* and *S. mekongi* (Fig. 2.2) cannot be explained by simple inversion of a segment of the mitochondrial genome. In all of the platyhelminth mitochondrial genomes examined to date, all of the genes always occur on the same strand and inversions would produce strand switching. Consequently other mechanisms must be responsible. All platyhelminth mitochondrial genomes examined to date posses two copies of the serine and leucine tRNA genes and, in the majority of cases, both tRNA-Leu genes and one of the tRNA-Ser genes lie between, or immediately adjacent to, the non-coding regions. With very few exceptions, gene order rearrangements also affect sequences in these areas. It is tempting to speculate that mutation or duplication events involving the 'redundant', multi-copy tRNA-Ser and tRNA-Leu genes could allow new tRNA specificities to arise, which are then moved around the mitochondrial genome by recombination events associated with repeat sequences in the non-coding regions. It is therefore interesting to note that the reported extra copy of tRNA-Gly in *P. westermani* also lies next to the larger of the non-coding regions.

Genetic code

Reassignment of the platyhelminth mitochondrial genetic code from the standard invertebrate mitochondrial code (GenBank code 5: ftp://ftp.ncbi.nih.gov/entrez/misc/data/gc.prt) is based on analysis of patterns of codon usage at conserved positions in mitochondrial DNA-encoded protein coding genes (primarily COX1); i.e. the observation of novel codons at positions that are known to represent highly conserved amino acids across sequences from other organisms, or as a result of intraspecific sequence variation, coupled with the lack of use of that codon at conserved positions, which code for the amino acid stipulated by the standard code. For stop codons, reassignment results from the presence of standard-code stop codons in the middle of known coding sequences and the occurrence of novel codons at conserved ends of coding regions. However, this approach is only as reliable as the sampling density that identifies the conserved positions and the range of amino acids occurring at them. Moreover, it fails to consider the possibility that conservative amino acid substitutions, which might have minimal impact on protein function, could have occurred. A more sophisticated analysis method, which weights the probability of code change with respect to known amino acid change/substitution preferences derived from empirical data (protein alignment matrices) has therefore been developed (Telford *et al.*, 2000). Multiple versions of the platyhelminth mitochondrial genetic code have been proposed and are summarized in Table 2.7. Currently GenBank lists three different genetic codes for the Platyhelminthes, codes 9, 14 and 21 (ftp://ftp.ncbi.nih.gov/entrez/misc/data/gc.prt). Of the approximately 2100 platyhelminth mitochondrial sequences on the database, 1800 have a genetic code assigned to them (the remainder comprise 12S or 16S rRNA gene sequences, tRNA genes, non-coding region sequences or putative nuclear pseudogenes). Code 9 (Telford *et al.*, 2000) is specified by the vast majority of entries (92%) and has been used to define a major clade within the flatworms, the Rhabditophora, which comprises the majority of the free-living turbellarian taxa together with the parasitic Neodermata, but excludes three classes of turbellarian flatworm, the Acoela, Catenulida and Nemertodermatida. Code 14 is still retained by 8% of entries as GenBank curation staff report that

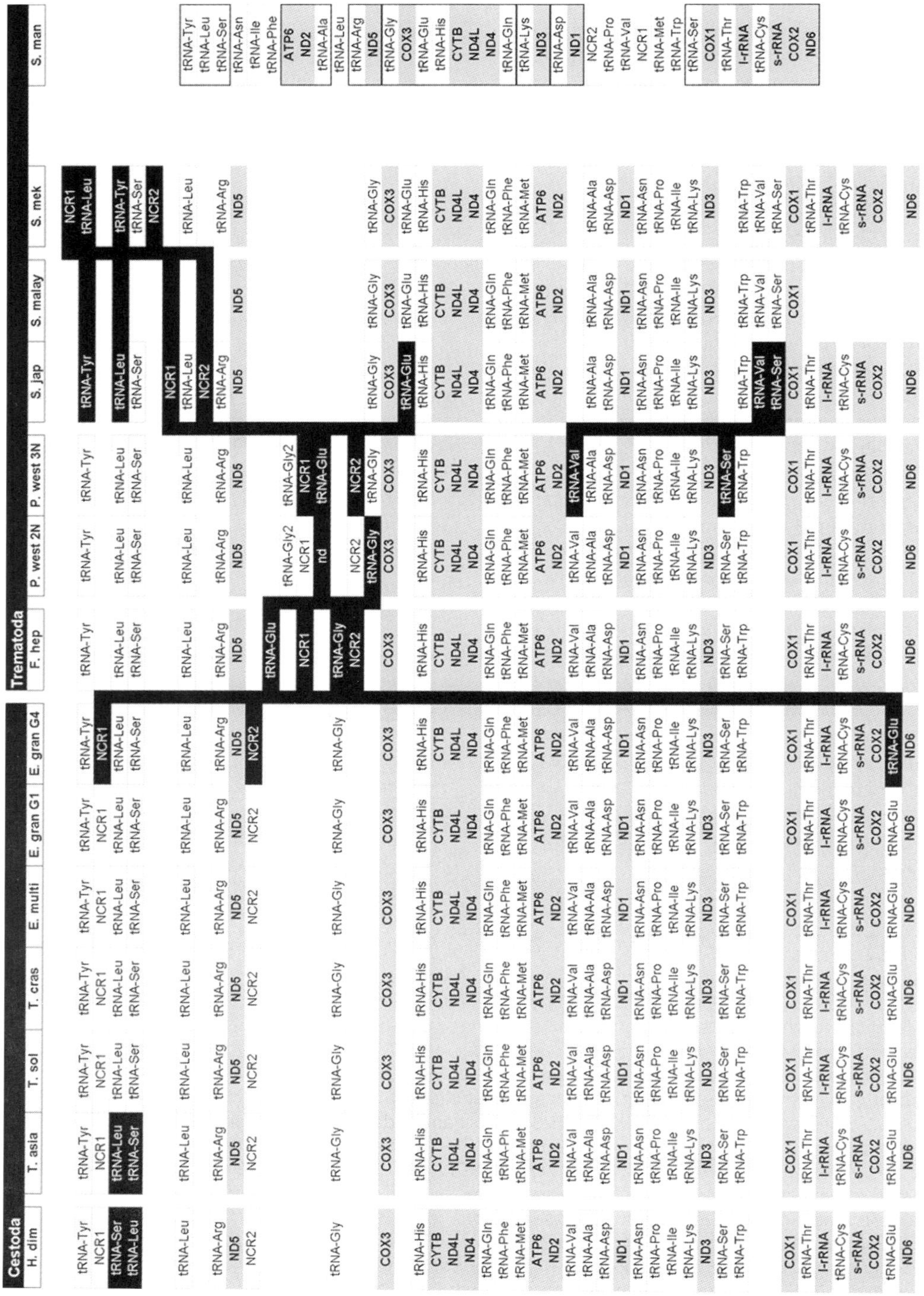
Cestoda
Trematoda
H. dim: tRNA-Tyr NCR1 tRNA-Ser tRNA-Leu tRNA-Leu tRNA-Arg ND5 NCR2 tRNA-Gly COX3 tRNA-His CYTB ND4L ND4 tRNA-Gln tRNA-Phe tRNA-Met ATP6 ND2 tRNA-Val tRNA-Ala tRNA-Asp ND1 tRNA-Asn tRNA-Pro tRNA-Ile tRNA-Lys ND3 tRNA-Ser tRNA-Trp COX1 tRNA-Thr l-rRNA tRNA-Cys s-rRNA COX2 tRNA-Glu ND6
T. asia: tRNA-Tyr NCR1 tRNA-Leu tRNA-Ser tRNA-Leu tRNA-Arg ND5 NCR2 tRNA-Gly COX3 tRNA-His CYTB ND4L ND4 tRNA-Gln tRNA-Ph tRNA-Met ATP6 ND2 tRNA-Val tRNA-Ala tRNA-Asp ND1 tRNA-Asn tRNA-Pro tRNA-Ile tRNA-Lys ND3 tRNA-Ser tRNA-Trp COX1 tRNA-Thr l-rRNA tRNA-Cys s-rRNA COX2 tRNA-Glu ND6
T. sol: tRNA-Tyr NCR1 tRNA-Leu tRNA-Ser tRNA-Leu tRNA-Arg ND5 NCR2 tRNA-Gly COX3 tRNA-His CYTB ND4L ND4 tRNA-Gln tRNA-Phe tRNA-Met ATP6 ND2 tRNA-Val tRNA-Ala tRNA-Asp ND1 tRNA-Asn tRNA-Pro tRNA-Ile tRNA-Lys ND3 tRNA-Ser tRNA-Trp COX1 tRNA-Thr l-rRNA tRNA-Cys s-rRNA COX2 tRNA-Glu ND6
T. cras: tRNA-Tyr NCR1 tRNA-Leu tRNA-Ser tRNA-Leu tRNA-Arg ND5 NCR2 tRNA-Gly COX3 tRNA-His CYTB ND4L ND4 tRNA-Gln tRNA-Phe tRNA-Met ATP6 ND2 tRNA-Val tRNA-Ala tRNA-Asp ND1 tRNA-Asn tRNA-Pro tRNA-Ile tRNA-Lys ND3 tRNA-Ser tRNA-Trp COX1 tRNA-Thr l-rRNA tRNA-Cys s-rRNA COX2 tRNA-Glu ND6
E. multi: tRNA-Tyr NCR1 tRNA-Leu tRNA-Ser tRNA-Leu tRNA-Arg ND5 NCR2 tRNA-Gly COX3 tRNA-His CYTB ND4L ND4 tRNA-Gln tRNA-Phe tRNA-Met ATP6 ND2 tRNA-Val tRNA-Ala tRNA-Asp ND1 tRNA-Asn tRNA-Pro tRNA-Ile tRNA-Lys ND3 tRNA-Ser tRNA-Trp COX1 tRNA-Thr l-rRNA tRNA-Cys s-rRNA COX2 tRNA-Glu ND6
E. gran G1: tRNA-Tyr NCR1 tRNA-Leu tRNA-Ser tRNA-Leu tRNA-Arg ND5 NCR2 tRNA-Gly COX3 tRNA-His CYTB ND4L ND4 tRNA-Gln tRNA-Phe tRNA-Met ATP6 ND2 tRNA-Val tRNA-Ala tRNA-Asp ND1 tRNA-Asn tRNA-Pro tRNA-Ile tRNA-Lys ND3 tRNA-Ser tRNA-Trp COX1 tRNA-Thr l-rRNA tRNA-Cys s-rRNA COX2 tRNA-Glu ND6
E. gran G4: tRNA-Tyr NCR1 tRNA-Leu tRNA-Ser tRNA-Leu tRNA-Arg ND5 NCR2 tRNA-Gly COX3 tRNA-His CYTB ND4L ND4 tRNA-Gln tRNA-Phe tRNA-Met ATP6 ND2 tRNA-Val tRNA-Ala tRNA-Asp ND1 tRNA-Asn tRNA-Pro tRNA-Ile tRNA-Lys ND3 tRNA-Ser tRNA-Trp COX1 tRNA-Thr l-rRNA tRNA-Cys s-rRNA COX2 tRNA-Glu ND6
F. hep: tRNA-Tyr tRNA-Leu tRNA-Ser tRNA-Leu tRNA-Arg ND5 tRNA-Glu NCR1 tRNA-Gly NCR2 COX3 tRNA-His CYTB ND4L ND4 tRNA-Gln tRNA-Phe tRNA-Met ATP6 ND2 tRNA-Val tRNA-Ala tRNA-Asp ND1 tRNA-Asn tRNA-Pro tRNA-Ile tRNA-Lys ND3 tRNA-Ser tRNA-Trp COX1 tRNA-Thr l-rRNA tRNA-Cys s-rRNA COX2 ND6
P. west 2N: tRNA-Tyr tRNA-Leu tRNA-Ser tRNA-Leu tRNA-Arg ND5 tRNA-Gly2 NCR1 nd NCR2 tRNA-Gly COX3 tRNA-His CYTB ND4L ND4 tRNA-Gln tRNA-Phe tRNA-Met ATP6 ND2 tRNA-Val tRNA-Ala tRNA-Asp ND1 tRNA-Asn tRNA-Pro tRNA-Ile tRNA-Lys ND3 tRNA-Ser tRNA-Trp COX1 tRNA-Thr l-rRNA tRNA-Cys s-rRNA COX2 ND6
P. west 3N: tRNA-Tyr tRNA-Leu tRNA-Ser tRNA-Leu tRNA-Arg ND5 tRNA-Gly2 NCR1 tRNA-Glu NCR2 tRNA-Gly COX3 tRNA-His CYTB ND4L ND4 tRNA-Gln tRNA-Phe tRNA-Met ATP6 ND2 tRNA-Val tRNA-Ala tRNA-Asp ND1 tRNA-Asn tRNA-Pro tRNA-Ile tRNA-Lys ND3 tRNA-Ser tRNA-Trp COX1 tRNA-Thr l-rRNA tRNA-Cys s-rRNA COX2 ND6
S. jap: tRNA-Tyr tRNA-Leu tRNA-Ser NCR1 tRNA-Leu NCR2 tRNA-Arg ND5 tRNA-Gly COX3 tRNA-Glu tRNA-His CYTB ND4L ND4 tRNA-Gln tRNA-Phe tRNA-Met ATP6 ND2 tRNA-Ala tRNA-Asp ND1 tRNA-Asn tRNA-Pro tRNA-Ile tRNA-Lys ND3 tRNA-Trp tRNA-Val tRNA-Ser COX1 tRNA-Thr l-rRNA tRNA-Cys s-rRNA COX2 ND6
S. malay: ND5 tRNA-Gly COX3 tRNA-Glu tRNA-His CYTB ND4L ND4 tRNA-Gln tRNA-Phe tRNA-Met ATP6 ND2 tRNA-Ala tRNA-Asp ND1 tRNA-Asn tRNA-Pro tRNA-Ile tRNA-Lys ND3 tRNA-Trp tRNA-Val tRNA-Ser COX1
S. mek: NCR1 tRNA-Leu tRNA-Tyr tRNA-Ser NCR2 tRNA-Leu tRNA-Arg ND5 tRNA-Gly COX3 tRNA-Glu tRNA-His CYTB ND4L ND4 tRNA-Gln tRNA-Phe tRNA-Met ATP6 ND2 tRNA-Ala tRNA-Asp ND1 tRNA-Asn tRNA-Pro tRNA-Ile tRNA-Lys ND3 tRNA-Trp tRNA-Val tRNA-Ser COX1 tRNA-Thr l-rRNA tRNA-Cys s-rRNA COX2 ND6
S. man: tRNA-Tyr tRNA-Leu tRNA-Ser tRNA-Asn tRNA-Ile tRNA-Phe ATP6 ND2 tRNA-Ala tRNA-Leu tRNA-Arg ND5 tRNA-Gly COX3 tRNA-Glu tRNA-His CYTB ND4L ND4 tRNA-Gln tRNA-Lys ND3 tRNA-Asp ND1 NCR2 tRNA-Pro tRNA-Val NCR1 tRNA-Met tRNA-Trp tRNA-Ser COX1 tRNA-Thr l-rRNA tRNA-Cys s-rRNA COX2 ND6

test translations of these sequences using code 9 can cause premature termination. However, of the 59 source species from 37 genera that retain code 14, 16 species from six genera also have database entries that use code 9. Code 21 is no longer used by any platyhelminth sequences, but is retained on the GenBank listing. Code 5 is reported for one *Clonorchis* (digenean) COX1 sequence, which has an identical ORF under codes 9 and 5, and for 13 acoel and catenulid sequences. The Acoela, together with the Nemertodermatida, have recently been removed from the Platyhelminthes on the basis of evidence from phylogenetic sequence analysis (Ruiz-Trillo *et al.*, 1999, 2002; Telford *et al.*, 2003), mitochondrial gene order (Ruiz-Trillo *et al.*, 2004) and mitochondrial genetic code (Telford *et al.*, 2000), and the catenulids are basal to the code 9-using Rhabditophora in the platyhelminth tree (Telford *et al.*, 2003). Direct experimental verification of platyhelminth genetic codes awaits the application of proteomic analysis to mitochondrial membrane fractions.

The abbreviated stop codon T has been proposed for two cestode mitochondrial genes: *H. diminuta* COX1 (von Nickisch-Rosenegk *et al.*, 2001) and *T. solium* ND1 gene (Nakao *et al.*, 2003); and for COX1 in the free-living Rhabditophoran *M. lineare* (Ruiz-Trillo *et al.*, 2004). It is suggested that post-transcriptional polyadenylation converts this to a full TAA stop codon and for *T. solium* ND1, 3′-RACE analysis has confirmed this suggestion (Nakao *et al.*, 2003).

Maternal inheritance

The *S. mansoni* mitochondrial microsatellite (Pena *et al.*, 1995) is highly polymorphic, both within and between strains, and has been used as a marker for breeding experiments (Minchella *et al.*, 1994). These appear to show a predominantly maternal pattern of inheritance, with the appearance of occasional, non-maternal bands, consistent with the generally accepted notion of maternal inheritance of mitochondria (Minchella *et al.*, 1994). However, subsequent experiments by other groups were interpreted as suggesting biparental mitochondrial inheritance (Jannotti-Passos *et al.*, 2001) and so the issue has been re-examined. Monomiracidial snail infections were generated and the resulting (theoretically clonal) cercariae crossed, grown to maturity and then typed to determine their mitochondrial SM750 haplotype. Eggs from these crosses were then also used to establish a multi-miracidial F1 population. Haplotypes were visualized as multi-band profiles that revealed the number of copies of the 62-base repeat in the different units of the tandem array (Pena *et al.*, 1995). Within a theoretically clonal, single-sex population, all individuals shared a core pattern of bands, but 34–53% also displayed novel bands, indicating that a high mutation rate was generating multiple, non-maternal haplotypes. Importantly, a subset of the new female haplotypes (18–30%) matched paternal haplotypes by chance and thus the female worm population contained individuals who would pass on a haplotype, which could be

Fig. 2.2. Diagrammatic Representation of Mitochondrial Gene Orders in Parasitic Flatworms. The orders of genes around the circular mitochondrial genomes of parasitic platyhelminths are shown in vertical columns. Protein-coding and rRNA genes are shown with grey shading and bold text. White text with black shading indicates gene order rearrangements between adjacent sequences. The *Schistosoma mansoni* data are shown unlinked to the other species because it shows very significant rearrangements; black boxes around subsets of the *S. mansoni* genes indicate localized regions of gene order conservation with Asian *Schistosoma* species. Abbreviations: COX1/2/3 = Cytochrome *c* oxidase complex subunits one, two and three; ND1/2/3/4/4L/5/6 = nicotinamide dehydrogenase complex subunits one, two, three, four, four-L five, six; CYTB = cytochrome B; ATP6 = ATP-synthase subunit six; s-rRNA = 12S rRNA (small subunit); l-rRNA = 16S rRNA (large subunit); NCR1 = small non-coding region; NCR2 = large non-coding region; nd = not determined (note also that the *Schistosoma malayensis* sequence represents a partial coding region); H. dim = *Hymenolepis diminuta*; T. asia = *Taenia asiatica*; T. sol = *Taenia solium*; T. cras = *Taenia crassiceps*; E. multi = *Echinococcus multilocularis*; E. gran G1/G4 = *Echinococcus granulosus* genotypes G1/G4; F. hep = *Fasciola hepatica*; P. west 2N/3N = *Paragonimus westermani* diploid/triploid; S. jap = *Schistosoma japonicum*; S. malay = *Schistosoma malayensis*; S. mek = *Schistosoma mekongi*; S. man = *Schistosoma mansoni.*

Table 2.7. Platyhelminth mitochondrial genetic code assignment.

Genetic code	Codon identity	Derivation	Reference	Comments
GenBank Code 1: standard code	Base1 = TTTTTTTTTTTTTTTTCCCCCCCCCCCCCCCCAAAAAAAAAAAAAAAAGGGGGGGGGGGGGGGG Base2 = TTTTCCCCAAAAGGGGTTTTCCCCAAAAGGGGTTTTCCCCAAAAGGGGTTTTCCCCAAAAGGGG Base3 = TCAGTCAGTCAGTCAGTCAGTCAGTCAGTCAGTCAGTCAGTCAGTCAGTCAGTCAGTCAGTCAG AAs = FFLLSSSSYY**CC*WLLLLPPPPHHQQRRRRIIIMTTTTNNKKSSRRVVVVAAAADDEEGGGG Starts = ---i---------------i---------------i----------------------------	–	–	
GenBank Code 5: invertebrate mitochondrial code	AAs = FFLLSSSSYY**CCWWLLLLPPPPHHQQRRRRIIMMTTTTNNKKSSSSVVVVAAAADDEEGGGG Starts = ---i----------------------------iiii---------------i------------	COX1 from catenulid flatworms and Acoelomorpha	Telford *et al.* (2000)	
	Starts = --------------------------------i-ii------------i---------------	Partial mitochondrial genomes of two Acoelomorpha	Ruiz-Trillo *et al.* (2004)	Truncated T and TA stop codons proposed
GenBank Code 9: Rhabditophora mitochondrial code	AAs = FFLLSSSSYY**CCWWLLLLPPPPHHQQRRRRIIIMTTTTNNNKSSSSVVVVAAAADDEEGGGG Starts = -----------------------------------i---------------i------------	Unification of: (1) COX1 from 24 species in 10 Rhabditophoran classes (2) Code 14	Telford *et al.* (2000)	No evidence found to confirm TAA as Tyrosine (Y)
GenBank Code 14: alternative platyhelminth mitochondrial code	AAs = FFLLSSSSYYY*CCWWLLLLPPPPHHQQRRRRIIIMTTTTNNNKSSSSVVVVAAAADDEEGGGG Starts = -----------------------------------i----------------------------	Unification of: (1) *Dugesia japonica* COX1, from multiple isolates showing sequence variation (2) Code 21	Bessho *et al.* (1992)	(1) AAA incorrectly abbreviated as Asp (= Aspartic Acid, D) rather than Asn (= asparagine, N) in original paper (2) Only canonical ATG (Methionine) assigned as start

GenBank Code 21: Trematode mitochondrial code	AAs = --------------W-------------------M-------N---SS---------------- Starts = -----------------------------------i---------------i------------	*Fasciola hepatica* ND1, ND3, COX1	Unification of data from the following two sources, below:	–
	AAs = --------------W-------------------M-----------SS---------------- Starts = -----------------------------------i---------------i------------	*Fasciola hepatica* ND1, ND3, COX1	Garey and Wolstenholme (1989)	ATA = methionone is proposed but not confirmed experimentally
	AAs = --N---------------------	*Fasciola hepatica* ND1, COX1	Ohama *et al.* (1990)	Re-examination of sequences from (Garey and Wolstenholme, 1989)
Cestode mitochondrial code	AAs = FFLLSSSSYY**CCWWLLLLPPPPHHQQRRRRIIIMTTTTNNNKSSSSVVVVAAAADDEEGGGG Starts = -----------------------------------i---------------i------------	COX1, ND2, ND4, ATP6 from eight species of cestode	Nakao *et al.* (2000; 2003)	No evidence found to confirm TAA as Tyrosine (Y)

Adapted from: http://www.ncbi.nlm.nih.gov/Taxonomy and primary literature.

misinterpreted as being of paternal origin (Bieberich and Minchella, 2001).

Flatworm Genomics

Rationale

Largely as a result of the competition between the public and private consortia to sequence the human genome, the past decade and a half has witnessed huge advances in DNA sequencing, laboratory automation and scientific computing technologies as well as dramatic increases in the capacity of the publicly funded sequencing centres. Over the same period, the cost of acquiring and exploiting these technologies has also fallen significantly and automated DNA sequencers are now common place. Consequently, it is now possible to rapidly acquire an amount of genomic information for any organism deemed worthy of the effort through laboratory, institutional or collaborative efforts.

As a general rule, parasitic platyhelminths are amongst the least tractable of experimental systems to work with. Compared to traditional 'model organisms' like *Drosophila* or *Caenorhabditis*: (i) they can be difficult or impossible to maintain in the laboratory, either *in vitro* or *in vivo*, with complex and protracted developmental cycles that yield limited amounts of biological material; (ii) they rarely exist as clonal populations/defined strains with well-characterized variation in morphology, biochemistry, behaviour, development, etc.; (iii) complexities of the life cycle make genetic analysis very hard, if it is even possible; and (iv) they have limited potential for experimental manipulation. Consequently, as organisms of choice for genome analysis, parasitic platyhelminths fall well short of the conventional benchmarks. However, as reviewed elsewhere (see Chapters 12, 15 and 16, this volume), our choices for current and short-term future, pharmacological or immunological intervention strategies for the major biomedical species are very limited. What if resistance to praziquantel develops? What if current vaccine candidates fail to deliver? As parasitic platyhelminths cannot be obtained in the kinds of numbers required for mass screening studies, rational drug and vaccine design strategies based on a sound understanding, of how they develop, survive and reproduce in their different hosts, of parasite–parasite and host–parasite interactions and of the factors, which determine pathogenicity, drug resistance, antigenic variation and behaviour, will provide the best prospect for identifying new therapeutic approaches. The genome is the 'blueprint' for all of these characteristics, and gene discovery and genome sequencing efforts will reveal them. Projects have already been reported, or are currently underway, for both parasitic and free-living flatworms, which should all be regarded as 'works in progress'. Currently, the most advanced projects involve *Schistosoma* spp. for which progress has regularly been reported and reviewed in the scientific literature (Franco *et al.*, 1995, 1997, 2000; Hirai and LoVerde, 1995; Tanaka *et al.*, 1995, 1997; Dias Neto *et al.*, 1996; Rabelo *et al.*, 1997; Fan *et al.*, 1998; Zouain *et al.*, 1998; Johnston *et al.*, 1999; Rollinson and Johnston, 1999; Williams and Johnston, 1999; Le Paslier *et al.*, 2000; Oliveira and Johnston, 2001; Foster and Johnston, 2002; Prosdocimi *et al.*, 2002; Hoffmann and Dunne, 2003; Hu *et al.*, 2003; Merrick *et al.*, 2003; Verjovski-Almeida *et al.*, 2003; El-Sayed *et al.*, 2004; LoVerde *et al.*, 2004; McManus *et al.*, 2004a,b) and in reports from annual meetings of the Schistosome Genome Network (http://www.nhm.ac.uk/hosted_sites/schisto/network). It is not proposed to revisit already well-worked territory, rather to give some pointers to some general lessons learned and methods exploited to help design future projects on other flatworm species. Of course, each new genome will have its own unique characteristics that will throw up equally unique challenges to be overcome.

Gene discovery and transcriptome analysis

ESTs are partial cDNA sequences generated by single pass, end-sequencing of randomly selected cDNA clones that are assigned a putative identity by comparison with the sequence databases (Adams *et al.*, 1991). They provide a quick and cheap means of cataloging the genes expressed in the biological source of the mRNA from which the cDNA library was constructed. For flatworms, where the solid body plan means that individual organs cannot be isolated, this tends to mean the whole body at different developmental stages, or in response to specific experimental manipulations, although with larger specimens at least partial enrichment for particular body areas is possible (Mineta *et al.*, 2003). Near-complete gene catalogues for both *S. mansoni* and *S. japonicum* have recently been reported (Hu *et al.*, 2003; Verjovski-Almeida *et al.*, 2003) and reveal that schistosomes possess most, if not all, of the physiological and cellular systems seen in higher animals. However, 35–55% of schistosome EST sequences match nothing on the public databases and their functions remain unknown. As the databases are populated largely by deuterostome and ecdysozoan sequences (Table 2.2), it is tempting to think that a large, clade-specific gene complement awaits characterization. These data are reviewed elsewhere (Chapter 6, this volume).

For *Schistosoma*, transcriptome analysis was initially restricted to developmental stages for which meaningful quantities of biological material were available for cDNA library construction. Novel methods have been developed for *Schistosoma* gene discovery, which use arbitrarily primed reverse transcriptase PCR to generate cDNA (Dias Neto *et al.*, 1996, 1997). These have been refined as the ORESTES (open reading frame EST) method of gene discovery (de Souza *et al.*, 2000), which has since been widely exploited for gene discovery in many systems including *S. mansoni* (Verjovski-Almeida *et al.*, 2003). It has the twin advantages of allowing cDNA analysis from minute quantities of biological material, allowing the transcriptome of the *S. mansoni* germ ball to be investigated for the first time (Verjovski-Almeida *et al.*, 2003), and of preferentially targeting the middle of open reading frames (de Souza *et al.*, 2000), to complement conventional 5′ and 3′ EST sequences and to increase the coverage of individual cDNAs by clustering. This is particularly valuable for proteomics where the identification of proteins from their peptide mass fingerprints requires long cDNA sequences to match against (Ashton *et al.*, 2001).

EST data-sets have also been generated for other flatworms including approximately 9700 sequences for *E. granulosus* (C. Fernandez and R.M. Maizels, unpublished data, available from http://www.ncbi.nlm.nih.gov/dbEST/), 6400 for *D. japonica* (Mineta *et al.*, 2003), 4500 for the planarian *Schmidtea mediterranea* (Sanchez Alvarado *et al.*, 2002) and several thousand for the macrostomid turbellarian *Macrostomum* (V. Hartenstein, UCLA, unpublished data). Projects are currently underway for *S. haematobium*, *F. hepatica* and *F. gigantica* (D.A. Johnston, London, 2005, unpublished data, sequences available by anonymous FTP from ftp.sanger.ac.uk/pub/databases/Trematode/) and for further sequences from *E. granulosus* and *E. multilocularis* (http://www.sanger.ac.uk/Projects/Echinococcus/). Taken together, these EST data-sets will allow comparative genomic analyses to be undertaken additively across several biologically significant ranges. *Schistosoma* spp. and *Fasciola* spp. represent averagely derived members of the digenean orders Strigeidida and Plagiorchiida, respectively, so comparative analysis between them will reasonably reflect gene diversity across the Digenea. Comparisons of these taxa with *Echinococcus* sp., a highly derived Cyclophyllidean cestode, will then encompass diversity across the parasitic Neodermata and possibly identify targets for pan-helminthic therapeutics. *S. mediterranea*, and *D. japonica* as triclad flatworms, fall within the sister clade to the Neodermata, so comparisons between them and the parasitic taxa may provide fundamental insights into the molecular basis of parasitism within the phylum. Finally, *Macrostomum* represents a 'primitive', early branching flatworm lineage allowing comparisons across the entire phylum (Chapter 1, this volume).

What is clear from all these EST initiatives is that the scientific impact of generating a few thousand sequences for a new species significantly outweighs the relatively small amount of time, effort and finance required to do it. It provides a dramatic stimulus to the research community – by revealing novel genes to work on, by identifying genes that had long been sought by other means and by providing probes or sequences to use to search for gene homologues in other taxa. Even with approximately 50% of flatworm EST sequences unidentified by database homology searches such data can also immediately be exploited for functional studies through the fabrication of small scale cDNA microarrays in order to analyse expression profiles and reveal yet more research opportunities (Hoffmann *et al.*, 2002; Nakazawa *et al.*, 2003; Fitzpatrick *et al.*, 2004).

Reliable gene prediction from genomic sequence requires an extensive training set of full length cDNA sequences to map on to the sequence and develop robust gene models from. Therefore, whilst EST projects provide valuable data-sets for new taxa, if genome sequencing is an ultimate goal, efforts also need to be directed towards full sequencing of full-length cDNA clones or to generating full-length cDNA sequences through 5′- and 3′-RACE methods.

Physical mapping

The physical map of a genome is an organized 'tile path' of clones, which cover the entire genome in known order and with minimum overlap. It thus provides a resource to facilitate full sequencing and assembly of a genome. Numerous methods are available to identify clone overlaps in large fragment genomic DNA libraries and these have been successfully exploited in different parasite genome projects, including cosmid fingerprint mapping in *Leishmania* (Ivens *et al.*, 1998) and PCR-pool mapping in *Plasmodium* (Dame *et al.*, 1996). To date, the only flatworm for which physical mapping has been attempted is *S. mansoni* where a 'map as you go' approach was tested in the initial phases of genome sequencing (El-Sayed *et al.*, 2004). This involved first designing PCR primers to amplify the end sequence of a selected seed clone from a BAC library (Le Paslier *et al.*, 2000) and using the resulting PCR product to screen BAC library filters and identify overlapping BAC clones, then designing PCR primers for the end sequences of each of the overlapping BAC clones and testing these against all the other identified clones to identify the one which represents minimal overlap and maximum contig extension. The cycle is then repeated using the new contig end to generate a new PCR probe and so walk out from the seed point. This approach needs the ends of the cloned fragments to be unique and evenly distributed across the genome. Unfortunately, the restriction enzyme, which had been used for *S. mansoni* BAC library construction, cut within a previously unidentified, large, highly abundant, dispersed repeat element. Consequently, this sequence was significantly over represented in the BAC ends and mapping yielded many false positives, which had to be eliminated by extensive testing. It rapidly became apparent that the approach was not in the least cost-effective (at least with the available library resources) and was soon abandoned in favour of diverting the funding into a whole genome shotgun (WGS) sequencing effort. Any future physical mapping initiatives for other taxa should certainly involve significant efforts to identify and characterize repetitive elements and to fully sequence some large insert genomic clones to analyse repeat distribution and genomic structure in order to optimize mapping strategies.

Chromosome mapping

A chromosome map is a collection of reference markers, which have been assigned to specific chromosomal locations across the karyotype and thus serve to anchor and orientate both physical maps and assembled genome sequences. Procedures for flatworm chromosome preparation are long established (Lamatsch *et al.*, 1998) and have been refined to allow probe localization by light microscopy using FISH and PRINS (Hirai and LoVerde, 1995; Tanaka *et al.*, 1995; Batistoni *et al.*, 1998; Imai *et al.*, 1999). Such methods are technically demanding and laborious to conduct, but as flatworm chromosomes are too large to be separated by pulse field electrophoresis, and, in general, too small to be separated by flow cytometry, this remains the only viable method for mapping sequences on to the karyotype. Such analyses would be greatly facilitated by the availability of flatworm cell lines, which could provide a routine, reliable source of mitotic cells. However, with the noticeable exception of established, putative germ cell cell lines from *E. multilocularis* (Furuya, 1991) and *E. granulosus* (Liu *et al.*, 1998), this remains something of a 'Holy Grail' for the flatworm community (Bayne *et al.*, 1994; Bayne, 1998). Such resources would also hugely facilitate the development of RNA interference and transfection systems (Chapters 7 and 22, this volume).

Genome sequencing

As the initial round of genome projects for selected 'model organisms' reach completion, the genome centres have been able to release sequencing capacity for use on other projects and the funding agencies have begun to allocate funding for genomic analyses on a wider diversity of (neglected) taxa including flatworms (http://www.genome.gov/ 10002154). As of March 2005, three flatworm genome sequencing projects are underway: (i) a 8–10x genome coverage WGS project for *S. mansoni* being performed jointly by The Institute for Genomic Research (http://www.tigr.org/tdb/e2k1/sma1/) and at The Wellcome Trust Sanger Institute (http://www.sanger.ac.uk/Projects/S_mansoni/); (ii) a 6x WGS project for *S. japonicum* at The Chinese National Human Genome Center in Shanghai (http://schistosoma.chgc.sh.cn); and (iii) a 6–8x WGS project for the planarian *S. mediterranea* at the Genome Sequencing Center of Washington University (St. Louis) (ftp://genome.wustl.edu/pub/seqmgr/planarian/). The *Schistosoma* projects are end-sequencing from libraries that have been generated from size-selected, randomly sheared genomic DNA to try to minimize the problems of bias associated with repetitive DNA in restriction-based libraries, and to provide scaffolding information for assembly purposes; the assembly algorithms take into account the fact that sequences from either end of an individual cloned insert are paired, are oriented in opposite directions and are located a (roughly) known distance apart in the genome. The main sequencing phase of the *Schistosoma* projects is nearing completion and initial assembly attempts are underway. For *S. mansoni*, these already suggest that more, large-scale scaffolding information is needed to link together thousands of small contigs that possess repeat sequences at either end. End-sequencing of 38–42 kb inserts in fosmid libraries and of 130 kb BAC clones from a new BAC library is currently underway to provide this additional information (M. Quail, Wellcome Trust Sanger Institute, Hinxton, 2005, personal communication). Even a early draft genome assembly can provide valuable scientific data and an auto-annotated database of the *S. mansoni* data has already been released (http://www. genedb.org/genedb/smansoni/). Ultimately it is hoped that resources will become available to allow proteomic, microarray, comparative EST, literature and community-based annotation data to be integrated with the genome assembly data to provide a 'one stop' curated reference system.

None of the three flatworm WGS projects has yet secured funding for genome closure, the highly labour-intensive (and thus slow and costly) process of using PCR to amplify across the large number of individual, small gaps that will be present in the completed assembly and sequencing the PCR products to fill in those gaps. Only when robust assemblies are in place will decisions be made as to the scientific need for, and cost-effectiveness of, closure.

Concluding Remarks

A review of platyhelminth genomes written just 10 years ago would have painted a very different picture to the one which we can write today and we would never then have predicted that we would be in the position that we currently find ourselves in, with full genome sequencing projects underway for several flatworm species. Even if one's scientific interest falls outside the currently targeted species, these efforts should not just be viewed as 'stamp collecting'; they are revealing fundamental information of potential interest and value to the entire flatworm research community. Moreover, the methods of analysis are transferable across taxa. Given the ever-falling costs of high-throughput sequencing and the lessons that have been learned from the current projects, there is no reason why, with sufficient community support and collaboration to justify the efforts, a similar review undertaken 10 years from now should not chronicle even more dramatic advances across a broader range of flatworm taxa. We will all be scientifically richer for it.

Acknowledgements

This chapter is dedicated to all members of the Schistosome Genome Network, past and present, and to WHO-TDR, without whose combined vision and willingness to collaborate for the common good, we would not be where we are today, and to the funding agencies and sequencing centres for having the faith to turn that vision into reality.

References

Abbasi, I., Branzburg, A., Campos-Ponce, M., Abdel Hafez, S.K., Raoul, F., Craig, P.S. and Hamburger, J. (2003) Copro-diagnosis of *Echinococcus granulosus* infection in dogs by amplification of a newly identified repeated DNA sequence. *American Journal of Tropical Medicine and Hygiene* 69, 324–330.

Adams, M.D., Kelley, J.M., Gocayne, J.D., Dubnick, M., Polymeropoulos, M.H., Xiao, H., Merril, C.R., Wu, A., Olde, B. and Moreno, R.F. (1991) Complementary DNA sequencing: expressed sequence tags and human genome project. *Science* 252, 1651–1656.

Adoutte, A., Balavoine, G., Lartillot, N., Lespinet, O., Prudhomme, B. and de Rosa, R. (2000) The new animal phylogeny: reliability and implications. *Proceedings of the National Academy of Sciences USA* 97, 4453–4456.

Agatsuma, T. (2003) Origin and evolution of *Schistosoma japonicum*. *Parasitology International* 52, 335–340.

Agatsuma, T., Terasaki, K., Yang, L. and Blair, D. (1994a) Genetic variation in the triploids of Japanese *Fasciola* species, and relationships with other species in the genus. *Journal of Helminthology* 68, 181–186.

Agatsuma, T., Yang, L., Kim, D. and Yonekawa, H. (1994b) Mitochondrial DNA differentiation of Japanese diploid and triploid *Paragonimus westermani*. *Journal of Helminthology* 68, 7–11.

Ashour, A.A., Hassan, N.H., Ibrahim, A.M. and el-Kabbany, A.I. (1995) The karyotypes of *Calicophoron microbothrium* and *Carmyerius gregarius* from Egyptian ruminants. *Journal of Helminthology* 69, 107–111.

Ashton, P.D., Curwen, R.S. and Wilson, R.A. (2001) Linking proteome and genome: how to identify parasite proteins. *Trends in Parasitology* 17, 198–202.

Atkinson, K.H. (1980) Chromosome analysis of *Schistosoma rodhaini* (Trematoda: Schistosomatidae). *Canadian Journal of Genetics and Cytology* 22, 143–147.

Bae, Y.A. and Kong, Y. (2003a) Divergent long-terminal-repeat retrotransposon families in the genome of *Paragonimus westermani*. *Korean Journal of Parasitology* 41, 221–231.

Bae, Y.A. and Kong, Y. (2003b) Evolutionary course of Csrn1 long-terminal-repeat retrotransposon and its heterogeneous integrations into the genome of the liver fluke, *Clonorchis sinensis*. *Korean Journal of Parasitology* 41, 209–219.

Bae, Y.A., Moon, S.Y., Kong, Y., Cho, S.Y. and Rhyu, M.G. (2001) Csrn1, a novel active retrotransposon in a parasitic trematode, *Clonorchis sinensis*, discloses a new phylogenetic clade of Ty3/gypsy-like LTR retrotransposons. *Molecular Biology and Evolution* 18, 1474–1483.

Baguñà, J. and Riutort, M. (2004) The dawn of bilaterian animals: the case of acoelomorph flatworms. *Bioessays* 26, 1046–1057.

Barshene, V., Staniavichiute, G.I. and Orlovskaia, O.M. (1989) Karyologic research on trematodes of the family Schistosomatidae from northwestern Chukota. *Parazitologiia* 23, 496–503.

Barsiene, J.V. (1993) The chromosome sets of trematodes. *Parazitologiia* 27, 336–353.

Basch, P.F. (1991) *Schistosomes: Development, Reproduction, and Host Relations*, 1st edn. Oxford University Press, Oxford, UK.

Basch, P.F. and Basch, N. (1984) Intergeneric reproductive stimulation and parthenogenesis in *Schistosoma mansoni*. *Parasitology* 89, 369–376.

Batistoni, R., Filippi, C., Salvetti, A., Cardelli, M. and Deri, P. (1998) Repeated DNA elements in planarians of the *Dugesia gonocephala* group (Platyhelminthes, Tricladida). *Hydrobiologia* 383, 139–146.

Bayne, C.J. (1998) Invertebrate cell culture considerations: insects, ticks, shellfish, and worms. *Methods in Cell Biology* 57, 187–201.

Bayne, C.J. and Grevelding, C.G. (2003) Cloning of *Schistosoma mansoni* sporocysts *in vitro* and detection of genetic heterogeneity among individuals within clones. *Journal of Parasitology* 89, 1056–1060.

Bayne, C.J., Menino, J.S., Hobbs, D.J. and Barnes, D.W. (1994) *In vitro* cultivation of cells from larval *Schistosoma mansoni*. *Journal of Parasitology* 80, 29–35.

Benazzi, M. and Benazzi-Lentati, G. (1976) *Animal Cytogenetics*, Vol. 1: Platyhelminthes, 1st edn. Gebr. Borntraeger Verlagsbuchhandlung, Stuttgart, Germany.

Bessho, Y., Ohama, T. and Osawa, S. (1992) Planarian mitochondria. ii. The unique genetic code as deduced from cytochrome c oxidase subunit 1 gene sequences. *Journal of Molecular Evolution* 34, 331–335.

Bessho, Y., Tamura, S., Hori, H., Tanaka, H., Ohama, T. and Osawa, S. (1997) Planarian mitochondria sequence heterogeneity: relationships between the type of cytochrome c oxidase subunit 1 gene sequence, karyotype and genital organ. *Molecular Ecology* 6, 129–136.

Bestor, T.H. (1990) DNA methylation: evolution of a bacterial immune function into a regulator of gene expression and genome structure in higher eukaryotes. *Philosophical Transactions of the Royal Society of London. Series B, Biological Sciences* 326, 179–187.

Beukeboom, L.W., Sharbel, T.F. and Michiels, N.K. (1998) Reproductive modes, ploidy distribution, and supernumerary chromosome frequencies of the flatworm *Polycelis nigra* (Platyhelminthes: Tricladida). *Hydrobiologia* 383, 277–285.

Bieberich, A.A. and Minchella, D.J. (2001) Mitochondrial inheritance in *Schistosoma mansoni*: mitochondrial variable number tandem repeat mutation produces noise on top of the signal. *Journal of Parasitology* 87, 1011–1015.

Blair, D. (2000) Genomes of *Paragonimus westermani* and related species: current state of knowledge. *International Journal for Parasitology* 30, 421–426.

Boore, J.L. (1999) Animal mitochondrial genomes. *Nucleic Acids Research* 27, 1767–1780.

Bowles, J., Hope, M., Tiu, W.U., Liu, X. and McManus, D.P. (1993) Nuclear and mitochondrial genetic markers highly conserved between Chinese and Philippine *Schistosoma japonicum*. *Acta Tropica* 55, 217–229.

Bowles, J., Blair, D. and McManus, D.P. (1995) A molecular phylogeny of the genus *Echinococcus*. *Parasitology* 110, 317–328.

Brehm, K., Jensen, K. and Frosch, M. (2000) mRNA *trans*-splicing in the human parasitic cestode *Echinococcus multilocularis*. *Journal of Biological Chemistry* 275, 38311–38318.

Brehm, K., Hubert, K., Sciutto, E., Garate, T. and Frosch, M. (2002) Characterization of a spliced leader gene and of *trans*-spliced mRNAs from *Taenia solium*. *Molecular and Biochemical Parasitology* 122, 105–110.

Brindley, P.J., Lewis, F.A., McCutchan, T.F., Bueding, E. and Sher, A. (1989) A genomic change associated with the development of resistance to hycanthone in *Schistosoma mansoni*. *Molecular and Biochemical Parasitology* 36, 243–252.

Brindley, P.J., Heath, S., Waters, A.P., McCutchan, T.F. and Sher, A. (1991) Characterization of a programmed alteration in an 18S ribosomal gene that accompanies the experimental induction of drug resistance in *Schistosoma mansoni*. *Proceedings of the National Academy of Sciences USA* 88, 7754–7758.

Brindley, P.J., Laha, T., McManus, D.P. and Loukas, A. (2003) Mobile genetic elements colonizing the genomes of metazoan parasites. *Trends in Parasitology* 19, 79–87.

Camacho, J.P.M. (2004) *B Chromosomes in the Eukaryote Genome*, 1st edn. Karger, Basel, Switzerland.

Camacho, J.P., Sharbel, T.F. and Beukeboom, L.W. (2000) B-chromosome evolution. *Philosophical Transactions of the Royal Society of London. Series B, Biological Sciences* 355, 163–178.

Carranza, S., Baguna, J. and Riutort, M. (1997) Are the platyhelminthes a monophyletic primitive group? An assessment using 18S rDNA sequences. *Molecular Biology and Evolution* 14, 485–497.

Carter, C.E., Wells, J.R. and MacInnis, A.J. (1972) DNA from anaerobic *Ascaris lumbricoides* and *Hymenolepis diminuta* mitochondria isolated by zonal centrifugation. *Biochimica et Biophysica Acta* 262, 135–144.

Chapman, A., Vallejo, V., Mossie, K.G., Ortiz, D., Agabian, N. and Flisser, A. (1995) Isolation and characterization of species-specific DNA probes from *Taenia solium* and *Taenia saginata* and their use in an egg detection assay. *Journal of Clinical Microbiology* 33, 1283–1288.

Chiu, J.K. (1967) Susceptibility of *Oncomelania hupensis chiui* to infection with *Schistosoma japonicum*. *Malacologia* 6, 145–153.

Clough, K.A., Drew, A.C. and Brindley, P.J. (1996) Host-like sequences in the schistosome genome. *Parasitology Today* 12, 283–286.

Copeland, C.S., Brindley, P.J., Heyers, O., Michael, S.F., Johnston, D.A., Williams, D.L., Ivens, A.C. and Kalinna, B.H. (2003) Boudicca, a retrovirus-like long terminal repeat retrotransposon from the genome of the human blood fluke *Schistosoma mansoni*. *Journal of Virology* 77, 6153–6166.

Copeland, C.S., Heyers, O., Kalinna, B.H., Bachmair, A., Stadler, P.F., Hofacker, I.L. and Brindley, P.J. (2004) Structural and evolutionary analysis of the transcribed sequence of Boudicca, a *Schistosoma mansoni* retrotransposon. *Gene* 329, 103–114.

Copeland, C.S., Mann, V.H., Morales, M.E., Kalinna, B.H. and Brindley, P.J. (2005) The Sinbad retrotransposon from the genome of the human blood fluke, *Schistosoma mansoni*, and the distribution of related Pao-like elements. *BioMed Central (BMC) Evolutionary Biology* 5, 20.

Cox, G.S., Phares, C.K. and Schmidt, R.A. (1990) Molecular characterization of the *Spirometra mansonoides* genome: renaturation kinetics, methylation, and hybridization to human cDNA probes. *Biochimica et Biophysica Acta* 1049, 134–144.

Croft, J.A. and Jones, G.H. (1989) Meiosis in *Mesostoma ehrenbergii ehrenbergii*. iv. Recombination nodules in spermatocytes and a test of the correspondence of late recombination nodules and chiasmata. *Genetics* 121, 255–262.

Dame, J.B., Arnot, D.E., Bourke, P.F., Chakrabarti, D., Christodoulou, Z., Coppel, R.L., Cowman, A.F., Craig, A.G., Fischer, K., Foster, J., Goodman, N., Hinterberg, K., Holder, A.A., Holt, D.C., Kemp, D.J., Lanzer, M., Lim, A., Newbold, C.I., Ravetch, J.V., Reddy, G.R., Rubio, J., Schuster, S.M., Su, X.-Z., Thompson, J.K., Vital, F., Wellems, T.E. and Werner, E.B (1996) Current status of the *Plasmodium falciparum* genome project. *Molecular and Biochemical Parasitology* 79, 1–12.

Davis, G.M. (1980) Snail hosts of Asian *Schistosoma* infesting man: evolution and coevolution. *Malacological Reviews* 2, 195–238.

Davis, G.M. (1992) Evolution of prosobranch snails transmitting Asian *Schistosoma*; coevolution with *Schistosoma*: a review. *Progress in Clinical Parasitology* 3, 145–204.

Davis, R.E. (1997) Surprising diversity and distribution of spliced leader RNAs in flatworms. *Molecular and Biochemical Parasitology* 87, 29–48.

Davis, R.E. and Hodgson, S. (1997) Gene linkage and steady state RNAs suggest *trans*-splicing may be associated with a polycistronic transcript in *Schistosoma mansoni*. *Molecular and Biochemical Parasitology* 89, 25–39.

Davis, R.E., Singh, H., Botka, C., Hardwick, C., Ashraf el Meanawy, M. and Villanueva, J. (1994) RNA *trans*-splicing in *Fasciola hepatica*. Identification of a spliced leader (sl) RNA and sl sequences on mRNAs. *Journal of Biological Chemistry* 269, 20026–20030.

Davis, R.E., Hardwick, C., Tavernier, P., Hodgson, S. and Singh, H. (1995) RNA *trans*-splicing in flatworms. Analysis of *trans*-spliced mRNAs and genes in the human parasite, *Schistosoma mansoni*. *Journal of Biological Chemistry* 270, 21813–21819.

DeMarco, R., Kowaltowski, A.T., Machado, A.A., Soares, M.B., Gargioni, C., Kawano, T., Rodrigues, V., Madeira, A.M., Wilson, R.A., Menck, C.F., Setubal, J.C., Dias Neto, E., Leite, L.C. and Verjovski-Almeida, S. (2004) Saci-1, -2, and -3 and Perere, four novel retrotransposons with high transcriptional activities from the human parasite *Schistosoma mansoni*. *Journal of Virology* 78, 2967–2978.

de Souza, S.J., Camargo, A.A., Briones, M.R., Costa, F.F., Nagai, M.A., Verjovski-Almeida, S., Zago, M.A., Andrade, L.E., Carrer, H., El-Dorry, H.F., Espreafico, E.M., Habr-Gama, A., Giannella-Neto, D., Goldman, G.H., Gruber, A., Hackel, C., Kimura, E.T., Maciel, R.M., Marie, S.K., Martins, E.A., Nobrega, M.P., Paco-Larson, M.L., Pardini, M.I., Pereira, G.G., Pesquero, J.B., Rodrigues, V., Rogatto, S.R., da Silva, I.D., Sogayar, M.C., de Fatima Sonati, M., Tajara, E.H., Valentini, S.R., Acencio, M., Alberto, F.L., Amaral, M.E., Aneas, I., Bengtson, M.H., Carraro, D.M., Carvalho, A.F., Carvalho, L.H., Cerutti, J.M., Correa, M.L., Costa, M.C., Curcio, C., Gushiken, T., Ho, P.L., Kimura, E., Leite, L.C., Maia, G., Majumder, P., Marins, M., Matsukuma, A., Melo, A.S., Mestriner, C.A., Miracca, E.C., Miranda, D.C., Nascimento, A.N., Nobrega, F.G., Ojopi, E.P., Pandolfi, J.R., Pessoa, L.G., Rahal, P., Rainho, C.A., da Ros, N., de Sa, R.G., Sales, M.M., da Silva, N.P., Silva, T.C., da Silva, W., Simao, D.F. Jr, Sousa, J.F., Stecconi, D., Tsukumo, F., Valente, V., Zalcbeg, H., Brentani, R.R., Reis, F.L., Dias Neto, E. and Simpson, A.J. (2000) Identification of human chromosome 22 transcribed sequences with ORF expressed sequence tags. *Proceedings of the National Academy of Sciences USA* 97, 12690–12693.

Despres, L., Imbert-Establet, D., Combes, C., Bonhomme, F. and Monnerot, M. (1991) Isolation and polymorphism in mitochondrial DNA from *Schistosoma mansoni*. *Molecular and Biochemical Parasitology* 47, 139–142.

Despres, L., Imbert-Establet, D. and Monnerot, M. (1993) Molecular characterization of mitochondrial DNA provides evidence for the recent introduction of *Schistosoma mansoni* into America. *Molecular and Biochemical Parasitology* 60, 221–229.

Dias Neto, E., Rodrigues Santos, F., Pena, S.D.J. and Simpson, A.J.G. (1993) Sex determination by low stringency PCR (ls-PCR). *Nucleic Acids Research* 21, 763–764.

Dias Neto, E., Harrop, R., Correa-Oliveira, R., Pena, S.D., Wilson, R.A. and Simpson, A.J. (1996) The schistosome genome project: RNA arbitrarily primed PCR allows the accelerated generation of expressed sequence tags. *Memorias do Instituto Oswaldo Cruz* 91, 655–657.

Dias Neto, E., Harrop, R., Correa-Oliveira, R., Wilson, R.A., Pena, S.D. and Simpson, A.J. (1997) Minilibraries constructed from cDNA generated by arbitrarily primed rt-PCR: an alternative to normalized libraries for the generation of ESTs from nanogram quantities of mRNA. *Gene* 186, 135–142.

Didyk, A.S. and Burt, M.D. (1998) Geographical, seasonal, and sex dynamics of *Shipleya inermis* (Cestoidea: Dioecocestidae) in *Limnodromus griseus gmelin* (Aves: Charadriiformes). *Journal of Parasitology* 84, 931–934.

Drew, A.C. and Brindley, P.J. (1995) Female-specific sequences isolated from *Schistosoma mansoni* by representational difference analysis. *Molecular and Biochemical Parasitology* 71, 173–181.

Drew, A.C. and Brindley, P.J. (1997) A retrotransposon of the non-long terminal repeat class from the human blood fluke *Schistosoma mansoni*. Similarities to the chicken-repeat-1-like elements of vertebrates. *Molecular Biology and Evolution* 14, 602–610.

Drew, A.C., Brindley, P.J., Lewis, F.A., Liang, Y.S. and Minchella, D.J. (1998) Tandemly repeated genomic sequence demonstrates inter- and intrastrain genetic variation in *Schistosoma japonicum*. *Tropical Medicine and International Health* 3, 373–380.

Drew, A.C., Minchella, D.J., King, L.T., Rollinson, D. and Brindley, P.J. (1999) Sr2 elements, non-long terminal repeat retrotransposons of the RTE-1 lineage from the human blood fluke *Schistosoma mansoni*. *Molecular Biology and Evolution* 16, 1256–1269.

D'Souza, T.G., Storhas, M., Schulenburg, H., Beukeboom, L.W. and Michiels, N.K. (2004) Occasional sex in an 'asexual' polyploid hermaphrodite. *Proceedings of the Royal Society of London. Series B, Biological Sciences* 271, 1001–1007.

Ellis, J.T. and Morrison, D.A. (1995) *Schistosoma mansoni*: patterns of codon use and bias. *Parasitology* 110, 53–60.

El-Sayed, N.M., Bartholomeu, D., Ivens, A., Johnston, D.A. and LoVerde, P.T. (2004) Advances in schistosome genomics. *Trends in Parasitology* 20, 154–157.

Fan, J., Minchella, D.J., Day, S.R., McManus, D.P., Tiu, W.U. and Brindley, P.J. (1998) Generation, identification, and evaluation of expressed sequence tags from different developmental stages of the Asian blood fluke *Schistosoma japonicum*. *Biochemical and Biophysical Research Communications* 252, 348–356.

Fantappie, M.R., Gimba, E.R. and Rumjanek, F.D. (2001) Lack of DNA methylation in *Schistosoma mansoni*. *Experimental Parasitology* 98, 162–166.

Fernandez, V., Zavala, A. and Musto, H. (2001) Evidence for translational selection in codon usage in *Echinococcus* spp. *Parasitology* 123, 203–209.

Fernandez, C., Gregory, W.F., Loke, P. and Maizels, R.M. (2002) Full-length-enriched cDNA libraries from *Echinococcus granulosus* contain separate populations of oligo-capped and *trans*-spliced transcripts and a high level of predicted signal peptide sequences. *Molecular and Biochemical Parasitology* 122, 171–180.

Fitzpatrick, J.M., Johansen, M.V., Johnston, D.A., Dunne, D.W. and Hoffmann, K.F. (2004) Gender-associated gene expression in two related strains of *Schistosoma japonicum*. *Molecular and Biochemical Parasitology* 136, 191–209.

Fletcher, H.L., Hoey, E.M., Orr, N., Trudgett, A., Fairweather, I. and Robinson, M.W. (2004) The occurrence and significance of triploidy in the liver fluke, *Fasciola hepatica*. *Parasitology* 128, 69–72.

Foster, J.M. and Johnston, D.A. (2002) Helminth genomics: from gene discovery to genome sequencing. *Trends in Parasitology* 18, 241–242.

Franco, G.R., Adams, M.D., Soares, M.B., Simpson, A.J., Venter, J.C. and Pena, S.D. (1995) Identification of new *Schistosoma mansoni* genes by the EST strategy using a directional cDNA library. *Gene* 152, 141–147.

Franco, G.R., Rabelo, E.M., Azevedo, V., Pena, H.B., Ortega, J.M., Santos, T.M., Meira, W.S., Rodrigues, N.A., Dias, C.M., Harrop, R., Wilson, A., Saber, M., Abdel-Hamid, H., Faria, M.S., Margutti, M.E., Parra, J.C.

and Pena, S.D. (1997) Evaluation of cDNA libraries from different developmental stages of *Schistosoma mansoni* for production of expressed sequence tags (ESTs). *DNA Research* 30, 231–240.

Franco, G.R., Valadao, A.F., Azevedo, V. and Rabelo, E.M. (2000) The *Schistosoma* gene discovery programme: state of the art. *International Journal for Parasitology* 30, 453–463.

Furuya, K. (1991) An established cell line of larval *Echinococcus multilocularis*. *International Journal for Parasitology* 21, 233–240.

Garcia, E.G. (1976) The biology of *Schistosoma japonicum*, Philippine strain: a review. *Southeast Asian Journal of Tropical Medicine and Public Health* 7, 190–196.

Garcia-Fernandez, J., Bayascas-Ramirez, J.R., Marfany, G., Munoz-Marmol, A.M., Casali, A., Baguna, J. and Salo, E. (1995) High copy number of highly similar Mariner-like transposons in planarian (platyhelminthe): evidence for a *trans*-phyla horizontal transfer. *Molecular Biology and Evolution* 12, 421–431.

Garey, J.R. and Wolstenholme, D.R. (1989) Platyhelminth mitochondrial DNA: evidence for early evolutionary origin of a tRNA(ser AGN) that contains a dihydrouridine arm replacement loop, and of serine-specifying AGA and AGG codons. *Journal of Molecular Evolution* 28, 374–387.

Gasser, R.B., Morahan, G. and Mitchell, G.F. (1991) Sexing single larval stages of *Schistosoma mansoni* by polymerase chain reaction. *Molecular and Biochemical Parasitology* 47, 255–258.

Gonzalez, L.M., Montero, E., Sciutto, E., Harrison, L.J., Parkhouse, R.M. and Garate, T. (2002) Differential diagnosis of *Taenia saginata* and *Taenia solium* infections: from DNA probes to polymerase chain reaction. *Transactions of the Royal Society of Tropical Medicine and Hygiene* 96, S243–S250.

Gonzalez-Estevez, C., Momose, T., Gehring, W.J. and Salo, E. (2003) Transgenic planarian lines obtained by electroporation using transposon-derived vectors and an eye-specific gfp marker. *Proceedings of the National Academy of Sciences USA* 100, 14046–14051.

Gregory, T.R., Hebert, P.D. and Kolasa, J. (2000) Evolutionary implications of the relationship between genome size and body size in flatworms and copepods. *Heredity* 84, 201–208.

Gremigni, V., Miceli, C. and Puccinelli, I. (1980) On the role of germ cells in planarian regeneration. I. A karyological investigation. *Journal of Embryology and Experimental Morphology* 55, 53–63.

Gremigni, V., Nigro, M. and Puccinelli, I. (1982) Evidence of male germ cell redifferentiation into female germ cells in planarian regeneration. *Journal of Embryology and Experimental Morphology* 70, 29–36.

Grevelding, C.G. (1995) The female-specific W1 sequence of the Puerto Rican strain of *Schistosoma mansoni* occurs in both genders of a Liberian strain. *Molecular and Biochemical Parasitology* 71, 269–272.

Grevelding, C.G. (1999) Genomic instability in *Schistosoma mansoni*. *Molecular and Biochemical Parasitology* 101, 207–216.

Grevelding, C.G., Kampkotter, A. and Kunz, W. (1997) *Schistosoma mansoni*: sexing cercariae by PCR without DNA extraction. *Experimental Parasitology* 85, 99–100.

Grossman, A.I. and Cain, G.D. (1981) Karyotypes and chromosome morphologies of *Megalodiscus temperatus* and *Philophthalamus gralli*. *Journal of Helminthology* 55, 71–78.

Grossman, A.I., McKenzie, R. and Cain, G.D. (1980) Sex heterochromatin in *Schistosoma mansoni*. *Journal of Parasitology* 66, 368–370.

Grossman, A.I., Short, R.B. and Cain, G.D. (1981a) Karyotype evolution and sex chromosome differentiation in schistosomes (Trematoda, Schistosomatidae). *Chromosoma* 84, 413–430.

Grossman, A.I., Short, R.B. and Kuntz, R.E. (1981b) Somatic chromosomes of *Schistosoma rodhaini*, *S. mattheei*, and *S. intercalatum*. *Journal of Parasitology* 67, 41–44.

Hamburger, J., Turetski, T., Kapeller, I. and Deresiewicz, R. (1991) Highly repeated short DNA sequences in the genome of *Schistosoma mansoni* recognized by a species-specific probe. *Molecular and Biochemical Parasitology* 44, 73–80.

Hamburger, J., He, N., Xin, X.Y., Ramzy, R.M., Jourdane, J. and Ruppel, A. (1998a) A polymerase chain reaction assay for detecting snails infected with bilharzia parasites (*Schistosoma mansoni*) from very early prepatency. *American Journal of Tropical Medicine and Hygiene* 59, 872–876.

Hamburger, J., Xu, Y.X., Ramzy, R.M., Jourdane, J. and Ruppel, A. (1998b) Development and laboratory evaluation of a polymerase chain reaction for monitoring *Schistosoma mansoni* infestation of water. *American Journal of Tropical Medicine and Hygiene* 59, 468–473.

Hamburger, J., He, N., Abbasi, I., Ramzy, R.M., Jourdane, J. and Ruppel, A. (2001) Polymerase chain reaction assay based on a highly repeated sequence of *Schistosoma haematobium*: a potential tool for monitoring schistosome-infested water. *American Journal of Tropical Medicine and Hygiene* 65, 907–911.

Hebert, P.D.N. and Beaton, M.J. (1990) Breeding system and genome size of the rhabdocoel turbellarian *Mesostoma ehrenbergii*. *Genome* 33, 719–724.

Heussler, V., Kaufmann, H., Strahm, D., Liz, J. and Dobbelaere, D. (1993) DNA probes for the detection of *Fasciola hepatica* in snails. *Molecular and Cellular Probes* 7, 261–267.

Heussler, V., Kaufmann, H., Glaser, I., Ducommun, D., Muller, C. and Dobbelaere, D. (1998) A DNA probe for the detection of *Dicrocoelium dendriticum* in ants of *Formica* spp. and *Lasius* spp. *Parasitology Research* 84, 505–508.

Hillyer, G.V. (1974) Buoyant density and thermal denaturation profiles of schistosome DNA. *Journal of Parasitology* 60, 725–727.

Hirai, H. and LoVerde, P.T. (1989) Triploid cells found in intramolluscan stages of *Schistosoma mansoni*. *Journal of Parasitology* 75, 800–802.

Hirai, H. and LoVerde, P.T. (1995) FISH techniques for constructing physical maps on schistosome chromosomes. *Parasitology Today* 11, 310–314.

Hirai, H. and LoVerde, P.T. (1996) Identification of the telomeres on *Schistosoma mansoni* chromosomes by FISH. *Journal of Parasitology* 82, 511–512.

Hirai, H., Sakaguchi, Y., Habe, S. and Imai, H.T. (1985) C-banding analysis of six species of lung flukes, *Paragonimus* spp. (Trematoda: Platyhelminthes), from Japan and Korea. *Zeitschrift fur Parasitenkunde* 71, 617–629.

Hirai, H., Spotila, L.D. and LoVerde, P.T. (1989) *Schistosoma mansoni*: chromosomal localization of DNA repeat elements by *in situ* hybridization using biotinylated DNA probes. *Experimental Parasitology* 69, 175–188.

Hirai, H., Tanaka, M. and LoVerde, P.T. (1993) *Schistosoma mansoni*: chromosomal localization of female-specific genes and a female-specific DNA element. *Experimental Parasitology* 76, 175–181.

Hirai, H., Hirata, M., Aoki, Y., Tanaka, M. and Imai, H.T. (1996) Chiasma analyses of the parasite flukes, *Schistosoma* and *Paragonimus* (Trematoda), by using the chiasma distribution graph. *Genes and Genetic Systems* 71, 181–188.

Hirai, H., Taguchi, T., Saitoh, Y., Kawanaka, M., Sugiyama, H., Habe, S., Okamoto, M., Hirata, M., Shimada, M., Tiu, W.U., Lai, K., Upatham, E.S. and Agatsuma, T. (2000) Chromosomal differentiation of the *Schistosoma japonicum* complex. *International Journal for Parasitology* 30, 441–452.

Hoffman, T.L. and Doms, R.W. (1999) HIV-1 envelope determinants for cell tropism and chemokine receptor use. *Molecular Membrane Biology* 16, 57–65.

Hoffmann, K.F. and Dunne, D.W. (2003) Characterization of the *Schistosoma* transcriptome opens up the world of helminth genomics. *Genome Biology* 5, 203.

Hoffmann, K.F., Johnston, D.A. and Dunne, D.W. (2002) Identification of *Schistosoma mansoni* gender-associated gene transcripts by cDNA microarray profiling. *Genome Biology* 3, RESEARCH0041.

Hsu, H.F. and Hsu, S.Y.L. (1962) *Schistosoma japonicum* in Formosa: a critical review. *Experimental Parasitology* 12, 459–462.

Hu, W., Yan, Q., Shen, D.K., Liu, F., Zhu, Z.D., Song, H.D., Xu, X.R., Wang, Z.J., Rong, Y.P., Zeng, L.C., Wu, J., Zhang, X., Wang, J.J., Xu, X.N., Wang, S.Y., Fu, G., Zhang, X.L., Wang, Z.Q., Brindley, P.J., McManus, D.P., Xue, C.L., Feng, Z., Chen, Z. and Han, Z.G. (2003) Evolutionary and biomedical implications of a *Schistosoma japonicum* complementary DNA resource. *Nature Genetics* 35, 139–147.

Imai, H.T., Wada, M.Y., Hirai, H., Matsuda, Y. and Tsuchiya, K. (1999) Cytological, genetic and evolutionary functions of chiasmata based on chiasma graph analysis. *Journal of Theoretical Biology* 198, 239–257.

Imase, A., Kumagai, T., Ohmae, H., Irie, Y. and Iwamura, Y. (1999) Localization of mouse type 2 alu sequence in schistosomes. *Parasitology* 119, 315–321.

Imase, A., Kobayashi, K., Ohmae, H., Irie, Y. and Iwamura, Y. (2000) Horizontal and vertical transfer of mouse endogenous retroviral DNA sequences in schistosomes. *Parasitology* 121, 155–162.

Imase, A., Kobayashi, K., Ohmae, H., Matsuda, H. and Iwamura, Y. (2001) Horizontal and vertical transmission of mouse class 1 MHC sequence in *Schistosoma mansoni*. *Parasitology* 123, 163–168.

Imase, A., Matsuda, H., Irie, Y. and Iwamura, Y. (2003) Existence of host DNA sequences in schistosomes – horizontal and vertical transmission. *Parasitology International* 52, 369–373.

Imase, A., Ohmae, H., Iwamura, Y., Kirinoki, M. and Matsuda, H. (2004) A comparative study on mouse MHC class 1 sequences detected in *Schistosoma japonicum* recovered from balb/c (H-2d) and c57bl/6 (H-2b) mice. *Southeast Asian Journal of Tropical Medicine and Public Health* 35, 10–18.

Irie, Y. and Iwamura, Y. (1993) Host-related DNA sequences are localized in the body of schistosome adults. *Parasitology* 107, 519–528.

Ivanchenko, M.G., Lerner, J.P., McCormick, R.S., Toumadje, A., Allen, B., Fischer, K., Hedstrom, O., Helmrich, A., Barnes, D.W. and Bayne, C.J. (1999) Continuous *in vitro* propagation and differentiation of cultures of the intramolluscan stages of the human parasite *Schistosoma mansoni*. *Proceedings of the National Academy of Sciences USA* 96, 4965–4970.

Ivens, A.C., Lewis, S.M., Bagherzadeh, A., Zhang, L., Chan, H.M. and Smith, D.F. (1998) A physical map of the *Leishmania major* Friedlin genome. *Genome Research* 8, 135–145.

Ivics, Z., Hackett, P.B., Plasterk, R.H. and Izsvak, Z. (1997) Molecular reconstruction of sleeping beauty, a TC1-like transposon from fish, and its transposition in human cells. *Cell* 91, 501–510.

Iwamura, Y., Irie, Y., Kominami, R., Nara, T. and Yasuraoka, K. (1991) Existence of host-related DNA sequences in the schistosome genome. *Parasitology* 102, 397–403.

Iwamura, Y., Yonekawa, H. and Irie, Y. (1995) Detection of host DNA sequences including the H-2 locus of the major histocompatibility complex in schistosomes. *Parasitology* 110, 163–170.

Jablonka, E. and Lamb, M.J. (1995) *Epigenetic Inheritance and Evolution,* 1st edn. Oxford University Press, Oxford, UK.

Jablonka, E., Lachmann, M. and Lamb, M.J. (1992) Evidence, mechanisms and models for the inheritance of acquired characters. *Journal of Theoretical Biology* 158, 245–268.

Jameson, D., Gibson, A.P., Hudelot, C. and Higgs, P.G. (2003) OGRe: a relational database for comparative analysis of mitochondrial genomes. *Nucleic Acids Research* 31, 202–206.

Jannotti-Passos, L.K., Souza, C.P., Parra, J.C. and Simpson, A.J. (2001) Biparental mitochondrial DNA inheritance in the parasitic trematode *Schistosoma mansoni*. *Journal of Parasitology* 87, 79–82.

Jansma, W.B., Rogers, S.H., Liu, C.L. and Bueding, E. (1977) Experimentally produced resistance of *Schistosoma mansoni* to hycanthone. *American Journal of Tropical Medicine and Hygiene* 26, 926–936.

Joffe, B.I., Solovei, I.V. and Macgregor, H.C. (1996) Ends of chromosomes in *Polycelis tenuis* (Platyhelminthes) have telomere repeat TTAGGG. *Chromosome Research* 4, 323–324.

Johnston, D.A., Blaxter, M.L., Degrave, W.M., Foster, J.M., Ivens, A.C. and Melville, S.E. (1999) Genomics and the biology of parasites. *Bioessays* 21, 131–147.

Jondelius, U., Ruiz-Trillo, I., Baguna, J. and Riutort, M. (2002) The nemertodermatida are basal bilaterians and not members of the Platyhelminthes. *Zoologica Scripta* 31, 201–215.

Kaguni, L.S. (2004) DNA polymerase gamma, the mitochondrial replicase. *Annual Review of Biochemistry* 73, 293–320.

Kaplan, R.M., Dame, J.B., Reddy, G.R. and Courtney, C.H. (1995) A repetitive DNA probe for the sensitive detection of *Fasciola* hepatica infected snails. *International Journal for Parasitology* 25, 601–610.

Kramer, F. and Schnieder, T. (1998) Sequence heterogeneity in a repetitive DNA element of *Fasciola*. *International Journal for Parasitology* 28, 1923–1929.

Laha, T., McManus, D.P., Loukas, A. and Brindley, P.J. (2000) Sj-alpha elements, short interspersed element-like retroposons bearing a hammerhead ribozyme motif from the genome of the oriental blood fluke *Schistosoma japonicum*. *Biochimica et Biophysica Acta* 1492, 477–482.

Laha, T., Loukas, A., Verity, C.K., McManus, D.P. and Brindley, P.J. (2001) Gulliver, a long terminal repeat retrotransposon from the genome of the oriental blood fluke *Schistosoma japonicum*. *Gene* 264, 59–68.

Laha, T., Brindley, P.J., Smout, M.J., Verity, C.K., McManus, D.P. and Loukas, A. (2002a) Reverse transcriptase activity and untranslated region sharing of a new RTE-like, non-long terminal repeat retrotransposon from the human blood fluke, *Schistosoma japonicum*. *International Journal for Parasitology* 32, 1163–1174.

Laha, T., Brindley, P.J., Verity, C.K., McManus, D.P. and Loukas, A. (2002b) Pido, a non-long terminal repeat retrotransposon of the chicken repeat 1 family from the genome of the oriental blood fluke, *Schistosoma japonicum*. *Gene* 284, 149–159.

Laha, T., Loukas, A., Smyth, D.J., Copeland, C.S. and Brindley, P.J. (2004) The fugitive LTR retrotransposon from the genome of the human blood fluke, *Schistosoma mansoni*. *International Journal for Parasitology* 34, 1365–1375.

Lamatsch, D.K., Sharbel, T.F., Martin, R. and Bock, C. (1998) A drop technique for flatworm chromosome preparation for light microscopy and high-resolution scanning electron microscopy. *Chromosome Research* 6, 654–656.

Le, T.H., Blair, D., Agatsuma, T., Humair, P.F., Campbell, N.J., Iwagami, M., Littlewood, D.T., Peacock, B., Johnston, D.A., Bartley, J., Rollinson, D., Herniou, E.A., Zarlenga, D.S. and McManus, D.P. (2000)

Phylogenies inferred from mitochondrial gene orders – a cautionary tale from the parasitic flatworms. *Molecular Biology and Evolution* 17, 1123–1125.

Le, T.H., Blair, D. and McManus, D.P. (2001a) Complete DNA sequence and gene organization of the mitochondrial genome of the liver fluke, *Fasciola hepatica* (Platyhelminthes; Trematoda). *Parasitology* 123, 609–621.

Le, T.H., Humair, P.F., Blair, D., Agatsuma, T., Littlewood, D.T. and McManus, D.P. (2001b) Mitochondrial gene content, arrangement and composition compared in African and Asian schistosomes. *Molecular and Biochemical Parasitology* 117, 61–71.

Le, T.H., Pearson, M.S., Blair, D., Dai, N., Zhang, L.H. and McManus, D.P. (2002) Complete mitochondrial genomes confirm the distinctiveness of the horse–dog and sheep–dog strains of *Echinococcus granulosus*. *Parasitology* 124, 97–112.

Le, T.H., McManus, D.P. and Blair, D. (2004) Codon usage and bias in mitochondrial genomes of parasitic platyhelminthes. *Korean Journal of Parasitology* 42, 159–167.

Le Paslier, M.C., Pierce, R.J., Merlin, F., Hirai, H., Wu, W., Williams, D.L., Johnston, D., LoVerde, P.T. and Le Paslier, D. (2000) Construction and characterization of a *Schistosoma mansoni* bacterial artificial chromosome library. *Genomics* 65, 87–94.

Liberatos, J.D. and Short, R.B. (1983) Identification of sex of schistosome larval stages. *Journal of Parasitology* 69, 1084–1089.

Liu, J., Gu, H., Wang, A. and Chen, Y. (1998) Establishment of germinal cell line of *Echinococcus granulosus*. *Zhongguo Ji Sheng Chong Xue Yu Ji Sheng Chong Bing Za Zhi* 16, 353–356.

Lockyer, A.E., Olson, P.D., Ostergaard, P., Rollinson, D., Johnston, D.A., Attwood, S.W., Southgate, V.R., Horak, P., Snyder, S.D., Le, T.H., Agatsuma, T., McManus, D.P., Carmichael, A.C., Naem, S. and Littlewood, D.T. (2003) The phylogeny of the Schistosomatidae based on three genes with emphasis on the interrelationships of *Schistosoma* Weinland, 1858. *Parasitology* 126, 203–224.

LoVerde, P.T., Hirai, H., Merrick, J.M., Lee, N.H. and El-Sayed, N. (2004) *Schistosoma mansoni* genome project: an update. *Parasitology International* 53, 183–192.

Lowe, T.M. and Eddy, S.R. (1997) tRNAscan-Se: a program for improved detection of transfer RNA genes in genomic sequence. *Nucleic Acids Research* 25, 955–964.

Marin, M., Garat, B., Pettersson, U. and Ehrlich, R. (1993) Isolation and characterization of a middle repetitive DNA element from *Echinococcus granulosus*. *Molecular and Biochemical Parasitology* 59, 335–338.

Martens, P.M., Curini-Galletti, M.C. and Van Oostveldt, P. (1989) Polyploidy in Proseriata (Platyhelminthes) and its phylogenetical implications. *Evolution* 43, 900–907.

Marx, K.A., Bizzaro, J.W., Blake, R.D., Hsien Tsai, M. and Feng Tao, L. (2000) Experimental DNA melting behavior of the three major *Schistosoma* species. *Molecular and Biochemical Parasitology* 107, 303–307.

Marzuki, S., Watkins, L.C. and Choo, W.M. (1989) Mitochondrial H+-ATPase in mutants of *Saccharomyces cerevisiae* with defective subunit 8 of the enzyme complex. *Biochimica et Biophysica Acta* 975, 222–230.

McCarter, J.P., Mitreva, M.D., Martin, J., Dante, M., Wylie, T., Rao, U., Pape, D., Bowers, Y., Theising, B., Murphy, C.V., Kloek, A.P., Chiapelli, B.J., Clifton, S.W., Bird, D.M. and Waterston, R.H. (2003) Analysis and functional classification of transcripts from the nematode *Meloidogyne incognita*. *Genome Biology* 4, R26.

McManus, D.P., Hu, W., Brindley, P.J., Feng, Z. and Han, Z.G. (2004a) Schistosome transcriptome analysis at the cutting edge. *Trends in Parasitology* 20, 301–304.

McManus, D.P., Le, T.H. and Blair, D. (2004b) Genomics of parasitic flatworms. *International Journal for Parasitology* 34, 153–158.

Merrick, J.M., Osman, A., Tsai, J., Quackenbush, J., LoVerde, P.T. and Lee, N.H. (2003) The *Schistosoma mansoni* gene index: gene discovery and biology by reconstruction and analysis of expressed gene sequences. *Journal of Parasitology* 89, 261–269.

Minchella, D.J., Lewis, F.A., Sollenberger, K.M. and Williams, J.A. (1994) Genetic diversity of *Schistosoma mansoni*: quantifying strain heterogeneity using a polymorphic DNA element. *Molecular and Biochemical Parasitology* 68, 307–313.

Mineta, K., Nakazawa, M., Cebria, F., Ikeo, K., Agata, K. and Gojobori, T. (2003) Origin and evolutionary process of the CNS elucidated by comparative genomics analysis of planarian ESTs. *Proceedings of the National Academy of Sciences USA* 100, 7666–7671.

Musto, H., Alvarez, F., Tort, J. and Maseda, H.R. (1994a) Dinucleotide biases in the platyhelminth *Schistosoma mansoni*. *International Journal for Parasitology* 24, 277–283.
Musto, H., Rodriguez-Maseda, H., Alvarez, F. and Tort, J. (1994b) Possible implications of CpG avoidance in the flatworm *Schistosoma mansoni*. *Journal of Molecular Evolution* 38, 36–40.
Musto, H., Rodriguez-Maseda, H. and Alvarez, F. (1995) Compositional correlations in the nuclear genes of the flatworm *Schistosoma mansoni*. *Journal of Molecular Evolution* 40, 343–346.
Musto, H., Romero, H. and Rodriguez-Maseda, H. (1998) Heterogeneity in codon usage in the flatworm *Schistosoma mansoni*. *Journal of Molecular Evolution* 46, 159–167.
Mutafova, T. and Kanev, I. (1996) Cytological study of *Rubenstrema exasperatum* (Trematoda: Omphalometridae). *International Journal for Parasitology* 26, 1405–1406.
Nakao, M., Sako, Y., Yokoyama, N., Fukunaga, M. and Ito, A. (2000) Mitochondrial genetic code in cestodes. *Molecular and Biochemical Parasitology* 111, 415–424.
Nakao, M., Yokoyama, N., Sako, Y., Fukunaga, M. and Ito, A. (2002) The complete mitochondrial DNA sequence of the cestode *Echinococcus multilocularis* (Cyclophyllidea: Taeniidae). *Mitochondrion* 1, 497–509.
Nakao, M., Sako, Y. and Ito, A. (2003) The mitochondrial genome of the tapeworm *Taenia solium*: a finding of the abbreviated stop codon U. *Journal of Parasitology* 89, 633–635.
Nakazawa, M., Cebria, F., Mineta, K., Ikeo, K., Agata, K. and Gojobori, T. (2003) Search for the evolutionary origin of a brain: planarian brain characterized by microarray. *Molecular Biology and Evolution* 20, 784–791.
Nara, T., Iwamura, Y., Tanaka, M., Irie, Y. and Yasuraoka, K. (1990) Dynamic changes of DNA sequences in *Schistosoma mansoni* in the course of development. *Parasitology* 100, 241–245.
Nishitan, S., Niimura, S., Ishida, S. and Teshirogi, W. (1998) Chromosomal polymorphism of a freshwater planarian, *Polycelis auriculata* in Mt. Apoi and its adjacent area, Hokkaido. *Zoological Science* 15, 40.
Ohama, T., Osawa, S., Watanabe, K. and Jukes, T.H. (1990) Evolution of the mitochondrial genetic code. IV. AAA as an asparagine codon in some animal mitochondria. *Journal of Molecular Evolution* 30, 329–332.
Oliveira, G. and Johnston, D.A. (2001) Mining the schistosome DNA sequence database. *Trends in Parasitology* 17, 501–503.
Pala, M., Vacca, R.A., Casu, S. and Stocchino, G. (1995) The freshwater planarian *Dugesia sicula* Lepori from Sardinia (Platyhelminthes, Tricladida). *Hydrobiologia* 310, 151–156.
Park, G.M., Im, K., Huh, S. and Yong, T.S. (2000) Chromosomes of the liver fluke, *Clonorchis sinensis*. *Korean Journal of Parasitology* 38, 201–206.
Parsons, J.D. (1995) Miropeats: graphical DNA sequence comparisons. *Computer Applications in the Biosciences* 11, 615–619.
Pellicciari, C., Redi, C.A. and Garagna, S. (1983) Genome size in the genus *Dugesia*. *Archives D'Anatomie Microscopique et de Morphologie Experimentale* 72, 243.
Pena, H.B., de Souza, C.P., Simpson, A.J. and Pena, S.D. (1995) Intracellular promiscuity in *Schistosoma mansoni*: nuclear transcribed DNA sequences are part of a mitochondrial minisatellite region. *Proceedings of the National Academy of Sciences USA* 92, 915–919.
Penschow, J.D., Gasser, R.B., Darling, P.E. and Mitchell, G.F. (1993) Female-specific genomic PCR sequences from *Schistosoma mansoni* as probes for hybridization histochemistry. *International Journal for Parasitology* 23, 407–409.
Petkeviciute, R. (1996) A chromosome study in the progenetic cestode *Cyathocephalus truncatus* (Cestoda: Spathebothriidea). *International Journal for Parasitology* 26, 1211–1216.
Prosdocimi, F., Faria-Campos, A.C., Peixoto, F.C., Pena, S.D., Ortega, J.M. and Franco, G.R. (2002) Clustering of *Schistosoma mansoni* mRNA sequences and analysis of the most transcribed genes: implications in metabolism and biology of different developmental stages. *Memorias do Instituto Oswaldo Cruz* 97, 61–69.
Qiu, C.P., Chris, S., Xia, M.Y., Chen, Q., Tong, X.M., Gong, X., Davis, G. and Feng, Z. (2002) Study on genomic diversity among different isolates of *Schistosoma japonicum* in China. *Zhongguo Ji Sheng Chong Xue Yu Ji Sheng Chong Bing Za Zhi* 20, 79–82.
Rabelo, E.M., Franco, G.R., Azevedo, V.A., Pena, H.B., Santos, T.M., Meira, W.S., Rodrigues, N.A., Ortega, J.M. and Pena, S.D. (1997) Update of the gene discovery program in *Schistosoma mansoni* with the expressed sequence tag approach. *Memorias do Instituto Oswaldo Cruz* 92, 625–629.
Raghunathan, L. and Bruckner, D. (1975) Identification of sex in *Schistosoma mansoni* cercariae. *Journal of Parasitology* 61, 66–68.

Rajkovic, A., Davis, R.E., Simonsen, J.N. and Rottman, F.M. (1990) A spliced leader is present on a subset of mRNAs from the human parasite *Schistosoma mansoni*. *Proceedings of the National Academy of Sciences USA* 87, 8879–8883.

Rausch, R.L. and Rausch, V.R. (1990) Reproductive anatomy and gametogenesis in *Shipleya inermis* (Cestoda: Dioecocestidae). *Annals de Parasitologie Humaine et Comparée* 65, 229–237.

Rebrikov, D.V., Bogdanova, E.A., Bulina, M.E. and Luk'ianov, S.A. (2002) A new planarian extrachromosomal virus-like element revealed by subtraction hybridization. *Molekuliarnaia Genetika Mikrobiologia I Virusologa (Moskva)* 36, 1002–1011.

Regev, A., Lamb, M.J. and Jablonka, E. (1998) The role of DNA methylation in invertebrates: developmental regulation or genome defense? *Molecular Biology and Evolution* 15, 880–891.

Rishi, A.K. and McManus, D.P. (1988) Molecular cloning of *Taenia solium* genomic DNA and characterization of taeniid cestodes by DNA analysis. *Parasitology* 97, 161–176.

Roca, J.R., Ribas, M. and Baguna, J. (1992) Distribution, ecology, mode of reproduction and karyology of freshwater planarians (Platyhelminthes, Turbellaria, Tricladida) in the springs of the Central Pyrenees. *Ecography* 15, 373–384.

Rollinson, D. and Johnston, D.A. (1999) Genome analysis: helping to solve the molecular mysteries of schistosomes. *Helminthologia* 36, 147–158.

Rosenzvit, M.C., Canova, S.G., Kamenetzky, L., Ledesma, B.A. and Guarnera, E.A. (1997) *Echinococcus granulosus*: cloning and characterization of a tandemly repeated DNA element. *Experimental Parasitology* 87, 65–68.

Rosenzvit, M.C., Canova, S.G., Kamenetzky, L. and Guarnera, E.A. (2001) *Echinococcus granulosus*: intraspecific genetic variation assessed by a DNA repetitive element. *Parasitology* 123, 381–388.

Ruff, M.D., Davis, G.M. and Werner, J.K. (1973) *Schistosoma japonicum*: disc electrophoretic protein patterns of the Japanese, Philippine, and Formosan strains. *Experimental Parasitology* 33, 437–446.

Ruiz-Trillo, I., Riutort, M., Littlewood, D.T., Herniou, E.A. and Baguna, J. (1999) Acoel flatworms: earliest extant bilaterian metazoans, not members of platyhelminthes. *Science* 283, 1919–1923.

Ruiz-Trillo, I., Paps, J., Loukota, M., Ribera, C., Jondelius, U., Baguna, J. and Riutort, M. (2002) A phylogenetic analysis of myosin heavy chain type II sequences corroborates that Acoela and Nemertodermatida are basal bilaterians. *Proceedings of the National Academy of Sciences USA* 99, 11246–11251.

Ruiz-Trillo, I., Riutort, M., Fourcade, H.M., Baguna, J. and Boore, J.L. (2004) Mitochondrial genome data support the basal position of Acoelomorpha and the polyphyly of the platyhelminthes. *Molecular Phylogenetics and Evolution* 33, 321–332.

Ryzhikov, K.M. and Tolkacheva, L.M. (1981) *Acoleata – Tapeworms of Birds*, 1st edn. Nauka, Moscow.

Saccone, C., Gissi, C., Reyes, A., Larizza, A., Sbisa, E. and Pesole, G. (2002) Mitochondrial DNA in metazoa: degree of freedom in a frozen event. *Gene* 286, 3–12.

Salzet, M., Capron, A. and Stefano, G.B. (2000) Molecular crosstalk in host–parasite relationship: schistosome– and leech–host interactions. *Parasitology Today* 16, 536–540.

Sanchez Alvarado, A., Newmark, P.A., Robb, S.M. and Juste, R. (2002) The *Schmidtea mediterranea* database as a molecular resource for studying platyhelminthes, stem cells and regeneration. *Development* 129, 5659–5665.

Schmidt, G.D. (1969) *Dioecotaenia cancellata* (Linton, 1890) gen. et. comb. n. a dioeceous cestode (Tetraphyllidea) from the cow-nosed ray *Rhinopterabonasus* (Mitchell) in Chesapeake Bay, with the proposal of a new family, Dioecotaeniidae. *Journal of Parasitology* 55, 271–275.

Searcy, D.G. and MacInnis, A.J. (1970) Measurements by DNA renaturation of the genetic basis of parasite reduction. *Evolution* 24, 796–806.

Sermswan, R., Mongkolsuk, S., Panyim, S. and Sirisinha, S. (1991a) Isolation and characterization of *Opisthorchis viverrini* specific DNA probe. *Molecular and Cellular Probes* 5, 399–407.

Sermswan, R., Mongkolsuk, S. and Sirisinha, S. (1991b) Characterization of the *Opisthorchis viverrini* genome. *Journal of Helminthology* 65, 51–54.

Short, R.B. (1948) Hermaphrodites in a Puerto Rican strain of *Schistosoma mansoni*. *Journal of Parasitology* 34, 240–242.

Short, R.B. (1951) Hermaphroditic female *Schistosomatium douthitti* (Trematoda: Schistosomatidae). *Journal of Parasitology* 37, 547–555.

Short, R.B. (1983) Presidential address: sex and the single schistosome. *Journal of Parasitology* 69, 3–22.

Short, R.B. and Grossman, A.I. (1986) Chromosomes of *Heterobilharzia americana* (Digenea: Schistosomatidae) from Texas. *Journal of Parasitology* 72, 807–809.

Short, R.B. and Menzel, M.Y. (1959) Chromosomes in parthenogenetic miracidia and embryonic cercariae of *Schistosomatium douthitti*. *Experimental Parasitology* 8, 249–264.

Short, R.B., Teehan, W.H. and Liberatos, J.D. (1987) Chromosomes of *Heterobilharzia americana* (Digenea: Schistosomatidae), with ZWA sex determination, from Louisiana. *Journal of Parasitology* 73, 941–946.

Simpson, A.J.G. and Pena, S.D.J. (1991) Host-related DNA sequences in the schistosome genome? *Parasitology Today* 7, 266.

Simpson, A.J., Sher, A. and McCutchan, T.F. (1982) The genome of *Schistosoma mansoni*: isolation of DNA, its size, bases and repetitive sequences. *Molecular and Biochemical Parasitology* 6, 125–137.

Simpson, A.J., Singer, D., McCutchan, T.F., Sacks, D.L. and Sher, A. (1983) Evidence that schistosome MHC antigens are not synthesized by the parasite but are acquired from the host as intact glycoproteins. *Journal of Immunology* 131, 962–965.

Sluys, R., Ribas, M. and Baguna, J. (1995) Taxonomy, ecology, and karyology of a new species of *Phagocata* from Spain, with a discussion on the phylogenetic systematics of the genus *Phagocata* Sl (Platyhelminthes, Tricladida, Paludicola). *Canadian Journal of Zoology* 73, 557–568.

Smith, J.K., Esch, G.W. and Kuhn, R.E. (1972) Growth and development of larval *Taenia crassiceps* (Cestoda). I. Aneuploidy in the anomalous ORF strain. *International Journal for Parasitology* 2, 261–263.

Snyder, S.D. and Loker, E.S. (2000) Evolutionary relationships among the Schistosomatidae (Platyhelminthes: Digenea) and an Asian origin for *Schistosoma*. *Journal of Parasitology* 86, 283–288.

Sobhon, P., Koonchornboon, T., Yuan, H.C., Upatham, E.S., Saitongdee, P., Krautrachue, M., Bubphaniroj, P. and Vongpayabal, P. (1986) Comparison of the surface morphology of adult *Schistosoma japonicum* (Chinese, Philippine and Indonesian strains) by scanning electron microscopy. *International Journal for Parasitology* 16, 205–216.

Spakulova, M., Horak, P. and Kral'ova, I. (1996) Karyotype of an avian schistosome *Trichobilharzia szidati* (Digenea: Schistosomatidae). *International Journal for Parasitology* 26, 783–785.

Spakulova, M., Horak, P. and Muller, V. (1997) The karyotype of *Trichobilharzia franki* Muller et Kimmig, 1994 (Digenea: Schistosomatidae), a new European schistosome agent of swimmers' itch. *Parasite* 4, 63–66.

Spakulova, M., Horak, P. and Dvorak, J. (2001) The karyotype of *Trichobilharzia regenti* Horak, Kolarova et Dvorak, 1998 (Digenea: Schistosomatidae), a nasal avian schistome in Central Europe. *Parasitology Research* 87, 479–483.

Spieth, J., Brooke, G., Kuersten, S., Lea, K. and Blumenthal, T. (1993) Operons in *C. elegans*: polycistronic mRNA precursors are processed by *trans*-splicing of SL2 to downstream coding regions. *Cell* 73, 521–532.

Spotila, L.D., Rekosh, D.M., Boucher, J.M. and LoVerde, P.T. (1987) A cloned DNA probe identifies the sex of *Schistosoma mansoni* cercariae. *Molecular and Biochemical Parasitology* 26, 17–20.

Spotila, L.D., Hirai, H., Rekosh, D.M. and Lo Verde, P.T. (1989) A retroposon-like short repetitive DNA element in the genome of the human blood fluke, *Schistosoma mansoni*. *Chromosoma* 97, 421–428.

Spotila, L.D., Rekosh, D.M. and LoVerde, P.T. (1991) Polymorphic repeated DNA element in the genome of *Schistosoma mansoni*. *Molecular and Biochemical Parasitology* 48, 117–120.

Staneviciute, G. and Kiseliene, V. (2001) Chromosome studies of *Ichthyocotylurus platycephalus* (Creplin, 1825) Odening 1969 with description of triploid variant and comparative karyology of the genus *Ichthyocotylurus*. *Parasite* 8, 137–145.

Stocchino, G.A., Manconi, R., Corso, G. and Pala, M. (2004) Karyology and karyometric analysis of an afrotropical freshwater planarian (Platyhelminthes, Tricladida). *Italian Journal of Zoology* 71, 89–93.

Stratford, R. and Shields, R. (1994) A *trans*-spliced leader RNA sequence in plant parasitic nematodes. *Molecular and Biochemical Parasitology* 67, 147–155.

Tan, Q. and Li, D. (1990) Study on chromosome C-banding in *Paragonimus westermani*: comparison of the diploid and triploid types. *Zhongguo Ji Sheng Chong Xue Yu Ji Sheng Chong Bing Za Zhi* 8, 35–37.

Tanaka, M., Iwamura, Y., Amanuma, H., Irie, Y., Watanabe, M., Watanabe, T., Uchiyama, Y. and Yasuraoka, K. (1989) Integration and expression of murine retrovirus-related sequences in schistosomes. *Parasitology* 99, 31–38.

Tanaka, M., Hirai, H., LoVerde, P.T., Nagafuchi, S., Franco, G.R., Simpson, A.J. and Pena, S.D. (1995) Yeast artificial chromosome (YAC)-based genome mapping of *Schistosoma mansoni*. *Molecular and Biochemical Parasitology* 69, 41–51.

Tanaka, M., Tanaka, T., Inazawa, J., Nagafuchi, S., Mutsui, Y., Kaukas, A., Johnston, D.A. and Rollinson, D. (1997) Proceedings of the schistosome genome project. *Memorias do Instituto Oswaldo Cruz* 92, 829–834.

Telford, M.J., Herniou, E.A., Russell, R.B. and Littlewood, D.T. (2000) Changes in mitochondrial genetic codes as phylogenetic characters: two examples from the flatworms. *Proceedings of the National Academy of Sciences USA* 97, 11359–11364.

Telford, M.J., Lockyer, A.E., Cartwright-Finch, C. and Littlewood, D.T. (2003) Combined large and small subunit ribosomal RNA phylogenies support a basal position of the acoelomorph flatworms. *Proceedings of the Royal Society of London. Series B, Biological Sciences* 270, 1077–1083.

Terasaki, K., Habe, S., Ho, L., Jian, H., Agatsuma, T., Shibahara, T., Sugiyama, H. and Kawashima, K. (1995) Tetraploids of the lung fluke *Paragonimus westermani* found in China. *Parasitology Research* 81, 627–630.

Terasaki, K., Shibahara, T., Noda, Y. and Kayano, H. (1996) The oocyte of triploid fluke receiving intrusion of sperm from a diploid fluke – evidence for the origin of tetraploids in *Paragonimus westermani*. *Journal of Parasitology* 82, 947–950.

Thompson, R.C., Lymbery, A.J. and Constantine, C.C. (1995) Variation in *Echinococcus*: towards a taxonomic revision of the genus. *Advances in Parasitology* 35, 145–176.

Tielens, A.G., Horemans, A.M., Dunnewijk, R., van der Meer, P. and van den Bergh, S.G. (1992) The facultative anaerobic energy metabolism of *Schistosoma mansoni* sporocysts. *Molecular and Biochemical Parasitology* 56, 49–57.

Van der Linden, A.G. (1969) Chromosome number and ploidy level in a dutch population of *Crenobia alpina* Dana (Planaria). *Genetica* 40, 61–64.

van Herwerden, L., Blair, D. and Agatsuma, T. (1998) Intra- and inter-specific variation in nuclear ribosomal internal transcribed spacer 1 of the *Schistosoma japonicum* species complex. *Parasitology* 116, 311–317.

van Herwerden, L., Blair, D. and Agatsuma, T. (1999) Genetic diversity in parthenogenetic triploid *Paragonimus westermani*. *International Journal for Parasitology* 29, 1477–1482.

van Oordt, B.E., van den Heuvel, J.M., Tielens, A.G. and van den Bergh, S.G. (1985) The energy production of the adult *Schistosoma mansoni* is for a large part aerobic. *Molecular and Biochemical Parasitology* 16, 117–126.

Van Oordt, B.E., Tielens, A.G. and van den Bergh, S.G. (1989) Aerobic to anaerobic transition in the carbohydrate metabolism of *Schistosoma mansoni* cercariae during transformation *in vitro*. *Parasitology* 98, 409–415.

Verjovski-Almeida, S., DeMarco, R., Martins, E.A., Guimaraes, P.E., Ojopi, E.P., Paquola, A.C., Piazza, J.P., Nishiyama, M.Y. Jr, Kitajima, J.P., Adamson, R.E., Ashton, P.D., Bonaldo, M.F., Coulson, P.S., Dillon, G.P., Farias, L.P., Gregorio, S.P., Ho, P.L., Leite, R.A., Malaquias, L.C., Marques, R.C., Miyasato, P.A., Nascimento, A.L., Ohlweiler, F.P., Reis, E.M., Ribeiro, M.A., Sa, R.G., Stukart, G.C., Soares, M.B., Gargioni, C., Kawano, T., Rodrigues, V., Madeira, A.M., Wilson, R.A., Menck, C.F., Setubal, J.C., Leite, L.C. and Dias Neto, E. (2003) Transcriptome analysis of the acoelomate human parasite *Schistosoma mansoni*. *Nature Genetics* 35, 148–157.

von Nickisch-Rosenegk, M., Brown, W.M. and Boore, J.L. (2001) Complete sequence of the mitochondrial genome of the tapeworm *Hymenolepis diminuta*: gene arrangements indicate that platyhelminths are eutrochozoans. *Molecular Biology and Evolution* 18, 721–730.

Wada, M.Y. and Imai, H.T. (1995) Theoretical analysis of chiasmata using a novel chiasma graph method applied to Chinese hamsters, mice, and dog. *Japanese Journal of Genetics* 70, 233–265.

Walker, T.K., Rollinson, D. and Simpson, A.J. (1989) A DNA probe from *Schistosoma mansoni* allows rapid determination of the sex of larval parasites. *Molecular and Biochemical Parasitology* 33, 93–100.

Warren, K.S., Mahmoud, A.A., Cummings, P., Murphy, D.J. and Houser, H.B. (1974) *Schistosomiasis mansoni* in Yemeni in California: duration of infection, presence of disease, therapeutic management. *American Journal of Tropical Medicine and Hygiene* 23, 902–909.

Webster, P., Mansour, T.E. and Bieber, D. (1989) Isolation of a female-specific, highly repeated *Schistosoma mansoni* DNA probe and its use in an assay of cercarial sex. *Molecular and Biochemical Parasitology* 36, 217–222.

Williams, S.A. and Johnston, D.A. (1999) Helminth genome analysis: the current status of the filarial and schistosome genome projects. *Parasitology* 118, S19–S38.

Woodruff, D.S., Merenlender, A.M., Upatham, E.S. and Viyanant, V. (1987) Genetic variation and differentiation of three *Schistosoma* species from the Philippines, Laos, and peninsular Malaysia. *American Journal of Tropical Medicine and Hygeine* 36, 345–354.

Zamparo, D., Brooks, D.R., Hoberg, E.P. and McLennan, D.M. (2001) Phylogenetic analysis of the Rhabdocoela (Platyhelminthes) with emphasis on the Neodermata and relatives. *Zoologica Scripta* 30, 59–77.

Zouain, C.S., Azevedo, V.A., Franco, G.R., Pena, S.D. and Goes, A.M. (1998) Identification of genes encoding *Schistosoma mansoni* antigens using an antigenic sequence tag strategy. *Journal of Parasitology* 84, 1307–1310.

Zurita, M., Bieber, D., Ringold, G. and Mansour, T.E. (1988) cDNA cloning and gene characterization of the mitochondrial large subunit (LSU) rRNA from the liver fluke *Fasciola hepatica*. Evidence of heterogeneity in the fluke mitochondrial genome. *Nucleic Acids Research* 16, 7001–7012.

3 Genetic Discrimination of *Echinococcus* Species and Strains

D.P. McManus
Molecular Parasitology Laboratory, Division of Infectious Diseases and Immunology, Australian Centre for International and Tropical Health and Nutrition, The Queensland Institute of Medical Research and The University of Queensland, Post Office Royal Brisbane Hospital, Queensland 4029, Australia

Introduction

Echinococcosis (hydatid disease) is a near-cosmopolitan zoonosis caused by adult or larval stages of dog cestodes belonging to the genus *Echinococcus* (family Taeniidae). The two major species of medical and public health importance are *Echinococcus granulosus* and *E. multilocularis* that, respectively, cause cystic echinococcosis (CE) and alveolar echinococcosis (AE) (McManus *et al.*, 2003). Both CE and AE are serious and severe diseases, the latter especially so, with high fatality rates and poor prognosis if managed incorrectly (McManus *et al.*, 2003). Hydatid cysts of *E. granulosus* develop in internal organs (mainly liver and lungs) of humans and other intermediate hosts as unilocular fluid-filled bladders. The cyst (metacestode) of *E. multilocularis* is a tumour-like infiltrating structure consisting of numerous

 Parasitic Flatworms: Molecular Biology, Biochemistry, Immunology and Physiology (eds A.G. Maule and N.J. Marks)

small vesicles embedded in stroma of connective tissue. The metacestode mass usually contains a semisolid matrix rather than fluid. The initial phase of primary *Echinococcus* infection is always asymptomatic. The incubation period of CE is unclear but probably lasts for many months to years. The infection may become symptomatic if the cysts either rupture or exert a mass-effect. Recurrence may occur following surgery on primary cysts. CE has been reported in subjects ranging in age from below 1 year to more than 75 years old. The mortality rate is estimated to be 0.2/100,000 with a fatality rate of 2.2%. Over 90% of cysts occur in the liver, lungs or both. Symptomatic cysts have been reported occasionally in other organs.

AE typically presents later than CE. Cases of AE are characterized by an initial asymptomatic incubation period of 5–15 years duration and a subsequent chronic course. Untreated or inadequately managed cases have high fatality rates. The peak age group for infection is from 50 to 70 years. The larval stage develops almost exclusively in the liver (99% of cases). Parasite lesions in the liver can vary from small foci of a few millimetre in size to large (15–20 cm in diameter) areas of infiltration. Early diagnosis of CE and AE can provide significant improvements in the quality of the management and treatment of both diseases. The definitive diagnosis for most human cases of CE and AE is by physical imaging methods such as ultrasonography. Immunodiagnosis complements the clinical picture while molecular techniques have been recently developed and adapted to advance laboratory diagnosis of AE and CE. Molecular approaches are proving of value in the study of both CE and AE, in particular for investigating genetic variation in the causative organisms, for the detection of parasite nucleic acids in clinical samples, and in the identification of *Echinococcus* eggs. Detecting minute amounts of parasite DNA and mRNA, not only to identify but also to characterize the biological status of parasite material, is becoming recognized as a powerful identification method.

Genetic Variation in *Echinococcus*

An important feature of the biology of *E. granulosus* is the fact that it comprises a number of intraspecific variants or strains that exhibit considerable variation at the genetic level (Thompson and McManus, 2001, 2002). By contrast, there appears to be very limited genetic variation within *E. multilocularis* (McManus and Bryant, 1995; Haag *et al.*, 1997; Rinder *et al.*, 1997; Kedra *et al.*, 2000). There are no available data to indicate that either *Echinococcus vogeli* or *E. oligarthrus* is variable. The term strain is used here to describe variants that differ from other groups of the same species in gene frequencies or DNA sequences, and in one or more characters of actual or potential significance to the epidemiology and control of echinococcosis (Thompson and Lymbery, 1990; Bowles *et al.*, 1995).

The extensive intraspecific variation in nominal *E. granulosus* may be associated with life-cycle patterns, host specificity, development rate, antigenicity, transmission dynamics, sensitivity to chemotherapeutic agents and pathology (Thompson and Lymbery, 1990; Thompson, 1995; Thompson and McManus, 2001, 2002). This may have important implications for the design and development of vaccines, diagnostic reagents and drugs impacting on the epidemiology and control of echinococcosis. For example, the adult parasite of the cattle strain of *E. granulosus* exhibits a precocious development in the definitive host with a short pre-patent period of only 33–35 days, nearly a week less than that of the common sheep strain (Thompson, 1995). This complicates control efforts where drug treatment of definitive hosts is utilized as a means of breaking the cycle of transmission, as it necessitates an increase in frequency of adult cestocidal treatment.

A number of well-characterized strains are now recognized (Table 3.1) that all appear to be adapted to particular life-cycle patterns and host assemblages (Thompson and McManus, 2001, 2002; McManus, 2002). Up to the time of the publication by McManus (2002), molecular studies, using mainly mitochondrial DNA (mtDNA) sequences, had identified nine distinct genetic types (genotypes (G) 1–9) within *E. granulosus*. This categorization followed very closely the pattern of strain variation emerging based on biological characteristics. Another genotypic group (G10), named the Fennoscandian cervid strain, was recently described infecting reindeer and moose in northeastern Finland (Lavikainen

Table 3.1. Genotypes/strains of *Echinococcus granulosus* categorized by DNA analysis with their host and geographical range. (Modified from McManus and Thompson, 2003[a].)

Genotype (strain)	Host origin	Geographic origin
G1 (common sheep strain)	Sheep	UK, Spain, China, Australian mainland, Tasmania, Kenya, Uruguay, Turkey, Jordan, Lebanon, Italy, Argentina, Brazil, Iran, Nepal, North Algeria, South Algeria, Bulgaria, Portugal
	Cattle	UK, Spain, Kenya, Tasmania, Jordan, China, North Algeria, Bulgaria, Australia
	Human	Australian mainland, Tasmania, Jordan, Lebanon, Holland, Kenya, China, Argentina, Spain, Algeria, Austria
	Goat	Kenya, China, Nepal, Spain
	Buffalo	India, Nepal
	Camel	Kenya, China, Tunisia
	Pig	China, Kenya, Bulgaria
	Wild boar	Spain
	Kangaroo	Australian mainland
	Dog (adult)	Kenya
	Dingo (adult)	Australian mainland
	Wolf (adult)	Bulgaria
	Jackal (adult)	Bulgaria
G2 (Tasmanian sheep strain)	Sheep	Tasmania, Argentina
	Human	Argentina
G3 (buffalo strain?)	Buffalo	India
G4 (horse strain; [b]*Echinococcus equinus*)	Horse	UK, Ireland, Switzerland, Spain
	Donkey	Ireland
G5 (cattle strain; [b]*Echinococcus ortleppi*)	Sheep	Nepal
	Goat	Nepal
	Cattle	Switzerland, Holland, Brazil, Sudan
	Buffalo	India, Nepal
	Pig	Kenya, Sudan
	Human	Holland, Argentina
	Zebra	Namibia
G6 (camel strain)	Camel	Kenya, Somalia, Kenya, Sudan, China, Iran, Mauritania, Egypt, South Algeria, Tunisia
	Cattle	China, Iran, Mauritania, Kenya, Sudan
	Human	Argentina, Nepal, Iran, Mauritania, Kenya, Egypt
	Sheep	Iran, Sudan
	Goat	Kenya
G7 (pig strain)	Pig	Poland, Slovakia, Ukraine, Argentina, Spain, Egypt (?), Kenya, Austria (G6/G7)
	Wild boar	Ukraine, Spain
	Goat	Spain
	Beaver	Poland
	Cattle	Slovakia
	Human	Poland, Slovakia, Austria (G6/G7)
G8 (cervid strain)	Moose	USA
	Human	USA
G9	Human	Poland
G10 (Fennoscandian cervid strain)	Moose, Reindeer	Northeast Finland

[a]Additional references: Bardonnet *et al.* (2003), Lavikainen *et al.* (2003), Azab *et al.* (2004), Bart *et al.* (2004), Breyer *et al.* (2004), Daniel-Mwambete *et al.* (2004), Dinkel *et al.* (2004), Haag *et al.* (2004), Lahmar *et al.* (2004) and Obwaller *et al.* (2004)
[b]Proposed in Thompson and McManus (2002).

et al., 2003). Recently, species status was suggested for two of these strains as *Echinococcus equinus* (horse–dog strain, G4) and *Echinococcus ortleppi* (cattle–dog strain, G5), respectively (Thompson and McManus, 2002) although formal reinstatement of the names was not made. The case for this revised taxonomy is further explored later.

Identification of *Echinococcus* Isolates using Molecular Genetic Techniques

Genetic variation in *Echinococcus* has been investigated using sequences from both the nuclear and mitochondrial genomes. The advent of the polymerase chain reaction (PCR) has provided a highly sensitive approach that is now widely used for *Echinococcus* identification purposes, including discrimination of eggs. The various techniques applied for studies of this genetic variation have been described in detail (McManus, 2002; McManus and Thompson, 2003). A brief synopsis follows.

RFLP/RAPD analysis

Earlier studies of molecular genetic variation in *Echinococcus* involved restriction fragment length polymorphism (RFLP) analysis using conventional Southern blotting. It was able to distinguish between several distinct strains of *E. granulosus* and extensive study showed that RFLP patterns were stable within a particular strain. The conventional RFLP procedure was simplified, without loss of resolution or accuracy, by linking RFLP analysis with PCR targeting the nuclear ribosomal DNA (rDNA) internal transcribed spacer 1 (ITS-1) region. The random amplified polymorphic (RAPD)-PCR (RAPD-PCR) has also been used under carefully controlled conditions for distinguishing the four recognized *Echinococcus* species and genetically distinct forms of *E. granulosus*. This technique was used recently to show that human cases in Egypt were of the camel–dog strain, and that camels are important hosts for the transmission of human hydatidosis there (Azab *et al.*, 2004).

PCR-amplified DNA sequences

Direct comparison of the nucleotide sequences of defined DNA segments between organisms provides a highly diagnostic and sensitive means of detecting genetic variation. Mitochondrial sequences, particularly fragments of the mitochondrial protein-coding genes, *cox1* and *nad1*, have proved invaluable for *E. granulosus* strain identification. Recently, Dinkel *et al.* (2004) have devised a specific and sensitive PCR/semi-nested PCR system for the rapid diagnosis of *E. granulosus* genotype G1, *E. granulosus* genotype G6/7 and *E. ortleppi* (G5). Diagnosis of G1 and the group G5/6/7 is performed by a simple PCR, while discrimination between *E. ortleppi* (G5) and G6/7 involves a subsequent semi-nested PCR step. The target sequence for amplification is part of the mitochondrial 12S ribosomal RNA (rRNA) gene. Specificity of the PCRs was 100% when evaluated with isolates of 16 species of cestodes, including *E. multilocularis, E. equinus* (G4), *E. ortleppi* (G5) and three strains of *E. granulosus* (G1, G6 and G7). This approach can be readily applied for the screening of large numbers of samples in epidemiological surveys. Preliminary epidemiological results obtained with this PCR approach included the detection of a camel strain (G6) infection for the first time in a human patient from eastern Africa, and the first reports of *E. ortleppi* (G5) in livestock from Kenya and the Sudan. It is noteworthy that human CE in Central Sudan appears to be relatively rare and this may result from the absence or limited distribution of the sheep–dog strain there (Elmahdi *et al.*, 2004), and/or the fact that in Sudan, sheep and goats seem to harbour calcified or infertile cysts of *E. granulosus* (Saad and Magzoub, 1989; Elmahdi *et al.*, 2004).

Daniel-Mwambete *et al.* (2004) analysed *cox1* and *nad1* sequences in combination with RAPD-PCR analysis to genotype isolates of *E. granulosus* from a range of hosts from Spain. Previously, three strains (sheep–dog, horse–dog and pig–dog) of *E. granulosus* had been identified in Spain. The new study confirmed that the Spanish ‘sheep’ strain corresponded to genotype 1 (G1), infecting sheep, cattle, goats, pigs, wild boar and human; the ‘horse’ strain corresponded to genotype 4 (G4), only infecting horses; and the ‘pig’ strain corresponded with genotype 7 (G7), infecting goats, pigs and wild boar. Goats, pigs and wild boar can be infected by two genotypes, G1 and G7. The authors pointed out that this circumstance, and especially the possibility of sylvatic intermediate hosts serving as reservoirs of the G1 genotype of the parasite, should be taken into consideration by authorities to develop and evaluate effective control programmes for echinococcosis.

Mutation scanning methods

Mutation scanning methods provide alternatives to DNA sequencing for the high resolution analysis of PCR-amplified fragments (Gasser and Zhu, 1999). One such method is single strand conformation polymorphism (SSCP) that has been used to rapidly screen large numbers of *Echinococcus* isolates. Another useful mutation scanning method is dideoxy fingerprinting (ddF), which is a hybrid between SSCP and conventional dideoxysequencing. The technique has been used reproducibly for the direct display of sequence variation in the *cox1* gene to type and differentiate all of the *Echinococcus* genotypes examined.

Microsatellite markers

A virtually untapped area for studying diversity in *Echinococcus* is the use of microsatellite DNA. Some microsatellite markers are available for *E. multilocularis*, following the earlier studies of Bretagne *et al.* (1996) who were able to use microsatellite DNA to divide isolates of *E. multilocularis* into three groups: European, North American (Montana) and Japanese. More recently Nakao *et al.* (2003) isolated two microsatellite loci that were used to demonstrate population-level polymorphisms in *E. multilocularis* adult worms derived from wild red foxes collected from a limited geographical area of Hokkaido, Japan.

Bartholomei-Santos *et al.* (2003) described for the first time the isolation and characterization of microsatellites from *E. granulosus*. This group searched the *E. granulosus* genome for microsatellites using eight different oligonucleotides containing particular repeats as probes. Southern blot experiments revealed that DNA regions containing GT, CAA, CATA and CT repeats are the most frequent in the *E. granulosus* genome. Two loci containing CA/GT (Egmsca1 and Egmsca2) and 1 locus containing GA/CT (Egmsga1) repeats were cloned and sequenced. The locus Egmsca1 was analysed in 73 isolates from Brazil and Argentina whose genotypes had been characterized previously. Brazilian isolates of the cattle strain and Argentinean isolates of the camel strain were monomorphic and fixed for the allele (CA)7. Argentinean isolates of sheep and Tasmanian sheep strains shared two alleles ((CA)8 and (CA)10) with the Brazilian isolates of sheep strain. The allele (CA)11 was found only in Brazilian isolates of the sheep strain at a low frequency. The Brazilian, but not the Argentinean, sheep strain populations were in agreement with the Hardy–Weinberg equilibrium. No polymorphism was found among individual protoscoleces from a single hydatid cyst, validating the procedure of pooling protoscoleces from one cyst, grouped as an isolate, in population studies. The availability of the current microsatellite markers for *E. granulosus* and *E. multilocularis*, and the identification of additional microsatellites in the future, will provide exquisitely sensitive tools for studying the population genetics and transmission biology of *Echinococcus* species.

Molecular Identification of *E. granulosus* Strains

The various genotypes of *E. granulosus* that have been identified using DNA analysis together with their host range and geographical distribution are presented in Table 3.1. A brief description of the major findings only is presented here as the area has been reviewed extensively in Thompson and McManus (2001), McManus (2002) and McManus and Thompson (2003).

Sheep–dog (Genotype 1) and horse–dog (Genotype 4 – *E. equinus*) strains

Discrete horse–dog and sheep–dog forms of *E. granulosus* occur that differ in a wide spectrum of biological criteria. Of public health significance is the fact that the sheep strain is infective to humans but, probably, non-infective to horses. The horse strain appears to be poorly infective to sheep and may be non-infective to humans. This is borne out by extensive DNA data as, to date, the horse strain (G4) has not been reported in sheep or humans, and the sheep strain (G1) has not been identified in horses. Pairwise differences among genes can give a measure of relative levels of divergence among taxa. Complete mitochondrial genomes for the sheep–dog and horse–dog strains were recently obtained and compared with those of *E. multilocularis* and *Taenia crassiceps* (Le *et al.*, 2002). Comparisons of the concatenated nucleotide sequences of the mitochondrial protein-encoding genes, and of the two subunits (small: *rrnS* and large: *rrnL*) of rRNA are shown in Table 3.2. The data indicate that the sheep and horse strains differ by 12.4% (nucleotides (nt)), and 11.6% (amino acids (aa)), a level similar to differences between these two genotypes and *E. multilocularis* (13–15% nt and 11.5–13.5% aa). As expected, divergence was considerably higher (26–30% nt and aa differences) when any member of the genus *Echinococcus* was compared with *T. crassiceps*. Sequences for the variable genes *atp6* and *nad3* were obtained from additional genotypes of *E. granulosus*, from *E. vogeli* and *E. oligarthrus*. Again, pairwise comparisons showed the distinctiveness of the G1 and G4 genotypes.

Another approach to investigating levels of divergence is by means of phylogenetic trees. Careful phylogenetic analysis of mitochondrial sequence data, in combination with additional nuclear sequence data (Bowles *et al.*, 1995), formally demonstrated the evolutionary distinctiveness of the sheep and horse strains of *E. granulosus*. Additional phylogenetic analyses (Fig. 3.1) of concatenated *atp6*, *nad1* (partial) and *cox1* (partial) genes from *E. multilocularis*, *E. vogeli*, *E. oligarthrus*, five genotypes of *E. granulosus*, and using *T. crassiceps* as an outgroup, yielded the same results (Le *et al.*, 2002). Overall, the comparisons suggested that the horse and sheep strains are as distinct from each other as either is from *E. multilocularis*. In addition, data have accumulated from throughout the world demonstrating the conserved nature of the horse and sheep strains/genotypes of *E. granulosus*. Furthermore, the fact that their genetic characteristics are maintained in sympatry in endemic areas such as UK or Spain where the life cycles overlap, reinforces the argument

Table 3.2. Divergence (%) in 12 mitochondrial protein-coding genes (nucleotide sequence (above diagonal) and amino acid sequence (below diagonal)) and in nucleotide sequences of the mitochondrial large (*rrnL*) (above diagonal) and small (*rrnS*) (below diagonal) ribosomal RNA (rRNA) subunits between the sheep–dog (*Eg*G1) and horse–dog (*Eg*G4) strains of *Echinococcus granulosus*, *E. multilocularis* (Em) and *Taenia crassiceps* (Tc). (After Le *et al.*, 2002.)

	Protein-coding sequences[a]				rrnL and rrnS sequences[b]			
	EgG1	EgG4	Em	Tc	EgG1	EgG4	Em	Tc
EgG1		12.37	14.97	27.01		8.76	11.05	23.73
EgG4	11.57		13.01	26.37	8.18		11.24	24.47
Em	13.67	11.53		25.73	11.20	10.24		25.41
Tc	30.60	30.78	29.58		22.45	22.56	22.25	

[a] *E. granulosus* G1 differs from G4 by 12.4% (nucleotides (nt)) and 11.6% (amino acids (aa)), a level similar to differences between these two genotypes and *E. multilocularis* (13–15% nt and 11.5–13.5% aa). As expected, divergence is considerably higher when any member of the genus *Echinococcus* is compared with *T. crassiceps* (26–30% nt and aa differences), suggesting that saturation has not been reached within *Echinococcus*.
[b] *E. granulosus* G1 and G4 differ by 11% from *E. multilocularis* and differ from each other by 8%. The differences between the two *E. granulosus* genotypes are noteworthy as rRNAs are known to be conserved among related taxa. Overall, the comparisons suggest that *E. granulosus* G1 and G4 are as distinct from each other as either is from *E. multilocularis*.

that the two forms are separate species. Accordingly, considering the new biological, epidemiological and molecular evidence that has accumulated since the original description and subspecies taxonomic status given by Williams and Sweatman (1963) for the form of *Echinococcus* that uses horses as its intermediate host, Thompson and McManus (2002) have suggested that the horse–dog strain of *E. granulosus* be recognized as a distinct species, *E. equinus*.

It has been recognized for some time that *E. granulosus* is maintained in two cycles of transmission on mainland Australia. One principally involves domestic sheep, while the other involves numerous species of macropod marsupials (kangaroos and wallabies). There is interaction between these cycles through a range of carnivores (domestic dogs, feral dogs, dingoes and red foxes), which are definitive hosts. Early evidence led to their proposed designation as distinct strains but subsequent molecular analysis indicates that only the common sheep strain is present. In biological, epidemiological and molecular features the common sheep strain can be regarded as homogeneous except that in Tasmania, morphological distinctiveness, a significantly shortened pre-patency period and molecular evidence has indicated that a variant of the common *E. granulosus* G1 genotype (designated genotype G2) occurs. The G2 genotype is also present in Argentina, possibly having been introduced with Merino sheep exported from Australia to Argentina (Rosenzvit *et al.*, 1999; Haag *et al.*, 2004).

Cattle–dog strain (Genotype 5 – *E. ortleppi*)

Until the early 1990s, all surgically obtained human isolates of *E. granulosus* examined by isoenzyme and DNA analysis conformed to the common domestic sheep strain. However, a partly calcified hydatid cyst, removed from a young Dutch patient and typed by PCR/RFLP analysis and *cox1* and *nad1* sequences, was shown to belong to the genetically distinct bovine or cattle strain (G5 genotype) of *E. granulosus*. This strain has also been shown to infect humans in Argentina (Kamenetzky *et al.*, 2002; Haag *et al.*, 2004). Thus, in regions where the bovine strain occurs, cattle may act as reservoirs of human infection.

Cattle are commonly found to harbour hydatid cysts throughout the world but the

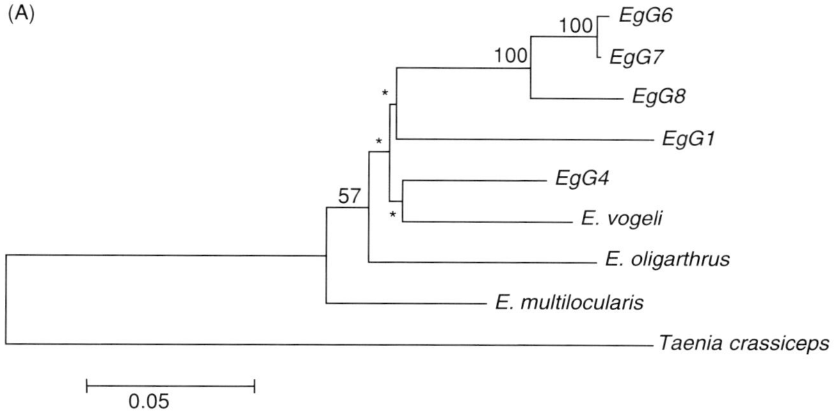

Fig. 3.1. Inferred relationships among species and genotypes of *Echinococcus*, using concatenated sequences of the mitochondrial protein-encoding genes *atp6*, *nad1* (partial) and *cox1* (partial) and *Taenia crassiceps* as an outgroup (modified from Le *et al.*, 2002). *Eg*G1, *Eg*G4, *Eg*G6, *Eg*G7, *Eg*G8 are different genotypes of *Echinococcus granulosus* (Table 3.1). Units on scale bar: changes per site. (A) A distance matrix was constructed from the inferred amino acid sequences using a Poisson correction for multiple hits and the tree constructed using the minimum evolution approach. Five hundred bootstrap resamplings were carried out. Branches with bootstrap support values less than 50% are indicated with an asterisk.

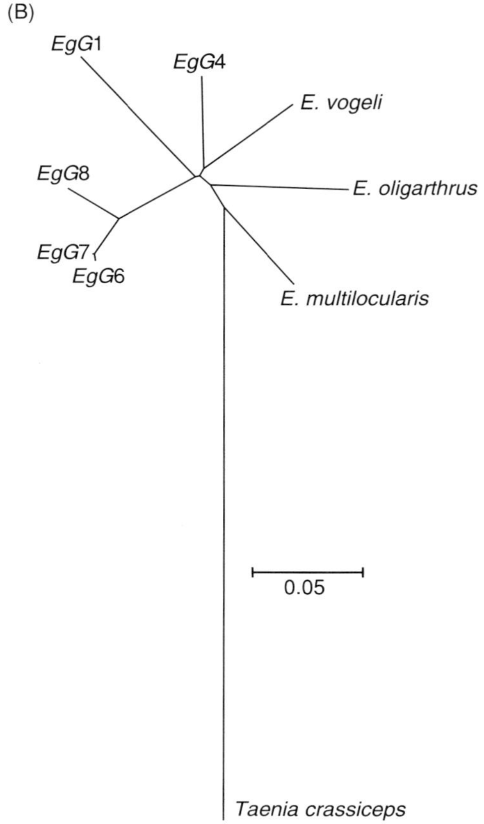

Fig. 3.1. (*cont'd*) (B) A distance matrix was constructed from the nucleotide sequences using the Kimura 2-parameter correction for multiple hits and the unrooted tree constructed using the minimum evolution approach. It is clear that *Eg*G4, *Eg*G1, *Echinococcus vogeli* and *E. oligarthrus* are almost equidistant from each other in terms of mitochondrial sequences. Furthermore, the *E. granulosus* G1 and G4 genotypes are also almost equidistant from the G6–G8 genotype cluster, although there is some structure in this latter group. *Echinococcus multilocularis* appears as basal within the genus.

aetiological agent is usually the sheep strain of *E. granulosus*, and infected cattle are an accidental host with resultant cysts rarely fertile. The form of *Echinococcus*, which is adapted to cattle as its intermediate host is characterized by the nature of its pulmonary metacestode development with the production of predominantly fertile cysts, the unusual strobilar morphology and rapid rate of development of the adult worm. In addition, although the molecular data are not as rich as those available for comparing the horse and sheep strains, there is no question of its genetic distinctiveness as clearly shown by pairwise distance matrix and phylogenetic analysis using nuclear and mitochondrial genes (Bowles *et al.*, 1995). The cattle-adapted form has a widespread geographical distribution that includes parts of Central Europe, South Africa, India, Sri Lanka, Nepal and possibly South America (Thompson and McManus, 2001, 2002). As with *Echinococcus* of horse origin, the cattle form of *Echinococcus* was originally given specific status as *E. ortleppi* (Lopez-Neyra and Soler Planas, 1943) based on a re-evaluation of an earlier description (Ortlepp, 1934) of *Echinococcus* in South Africa where the cattle-adapted form is now known to occur. The validity of *E. ortleppi* as a separate species was not accepted by Rausch and Nelson (1963) but subsequent studies (Verster, 1965) showed that previous taxonomic considerations were based only on a limited appraisal of the morphological features, which characterize this form of *Echinococcus*. Accordingly, Thompson and McManus (2002) suggested that *E. ortleppi* should be reinstated and recognized as the cattle-adapted form of *Echinococcus*. Recently, genotype 5 was shown by *cox1* and *nad1* sequence analysis to infect a Namibian zebra (Obwaller *et al.*, 2004). There have been no molecular genetic studies undertaken on *Echinococcus* isolates from South African hosts and clearly this would be a fruitful area for future research.

Camel–dog strain (Genotype 6)

The camel–dog strain of *E. granulosus* has been identified by molecular methods in East Africa where it infects camels, goats, cattle and pigs. DNA analysis has also shown that the G6 genotype of *E. granulosus* occurs in several other countries including Iran, Argentina and China (Table 3.1). DNA studies have indicated that the camel strain is infectious to humans and circulates between intermediate hosts including camels and cattle in Mauritania; if the sheep strain is present there, it is rarely found. As well as Mauritania, molecular studies of *E. granulosus* from Argentina, Iran, Egypt and Nepal have reported the presence of the G6 genotype in several human subjects; recently the camel strain was reported for the first time in a human patient from Kenya (Dinkel *et al.*, 2004). This has potentially important implications for public health and implementation of echinococcosis control programmes where the camel strain is involved in *E. granulosus* transmission. The camel strain has a shorter maturation time in dogs compared with the common sheep strain, which is the form generally associated with human infection. However, as is evident from scrutiny of Table 3.1, DNA analysis indicates that the majority of the characterized *E. granulosus* genotypes (G1, G2, G5, G6, G7, G8 and G9) have been shown to occur in humans.

Pig–dog strains (Genotypes 7 and 9)

Analysis of *E. granulosus* from Polish patients indicated that they were not infected with the common sheep strain (G1 genotype) of *E. granulosus*. Instead, the isolates shared similar DNA sequence with the previously characterized pig–dog strain (G7) genotype but exhibited some clear differences. Accordingly, it was proposed that these human isolates represented a distinct *E. granulosus* genotype (designated G9) (Scott *et al.*, 1997). Subsequent studies of human and pig isolates from Poland, Slovakia and Ukraine failed to confirm the existence of this genotype but have provided evidence for the almost exclusive presence of the G7 genotype (Kedra *et al.*, 1999; Pawlowski and Stefaniak, 2003a). Confirmation of the existence of the G9 genotype and identification of the major reservoir(s) of human echinococcosis in Poland (almost certainly pigs; Pawlowski and Stefaniak, 2003b) and other countries in Central and Eastern Europe remain to be resolved.

Cervid strains (Genotypes 8 and 10)

The northern biotype of *E. granulosus* occurs in North America and northern Eurasia in life cycles involving cervids (moose and reindeer) with the wolf as the principal definitive host; cycles involving sled dogs and domesticated reindeer also occur. Based on its unique *nad1* sequence and ITS-1 PCR-RFLP pattern, a cervid strain obtained from Alaskan moose appeared to represent a distinct genotype of *E. granulosus*, which was designated G8 (Bowles *et al.*, 1994).

Case-based data have suggested that the course of sylvatic disease is less severe than that of domestic disease, which led to the recommendation to treat CE patients in the Arctic, presumed to be infected with the cervid strain, by careful medical management rather than by aggressive surgery. The first two documented *E. granulosus* human cases in Alaska with accompanying severe sequelae in the liver were recently reported (Castrodale *et al.*, 2002). The results of molecular analysis of the cyst material of one of the subjects supported identification of the parasite as the sylvatic (cervid) strain (G8) and not as the domestic (common sheep) strain, which was initially thought to be implicated in these unusually severe cases (McManus *et al.*, 2002). The adverse outcomes could have been rare complications that are part of the clinical spectrum of sylvatic CE, an indication that the sylvatic form of *E. granulosus*, especially when affecting the liver, has potential for severe clinical consequences.

In a recent study, Lavikainen *et al.* (2003) provided molecular evidence for the existence of a new variant of the cervid strain. Five isolates of *E. granulosus* were collected from four reindeer and one moose in northeastern Finland. DNA sequences within regions of the *cox1* and *nad1* genes and the rDNA ITS-1 were analysed. The mitochondrial nucleotide sequences were identical in all isolates, but high sequence variation was found in cloned fragments of the ITS-1 region. Mitochondrial and nuclear sequences of the Finnish cervid *E. granulosus* and the camel strain (G6) of *E. granulosus* closely resembled each other. Phylogenetic analysis indicated the Finnish isolates had close relationship also with the pig (G7) and cattle (G5) strains. Although some similarities were found with G8 cervid strain, particularly in the *nad1* and some of the ITS-1 clones, the Finnish *E. granulosus* form appeared to represent a distinct, previously undescribed genotype of *E. granulosus*, which the authors named as the Fennoscandian cervid strain (G10). It is important to evaluate the geographical distribution of this new genotype and to determine whether it is infective to humans as has been shown for the G8 form.

Detection of *Echinococcus* Nucleic Acids in Clinical Samples

Clinical findings require sophisticated investigations to confirm echinococcosis. Usually primary identification and characterization of echinococcal lesions occur by imaging techniques. However, the diagnostic potential of such techniques is sometimes limited by the atypical appearance of the visualized lesions that may also be insufficient in providing information about the involved species or about the viability of the parasite. Immunodiagnosis is a useful complementary diagnostic tool for the identification of infection and disease (Zhang *et al.*, 2003). Recent efforts have been undertaken to apply molecular identification methods for direct detection of parasite RNA or DNA in clinical samples by Southern/Northern blotting or PCR (Siles-Lucas and Gottstein, 2001). These molecular methods have mainly been used in the clinical context for the primary diagnostic identification of parasite materials in biological specimens resected or biopsied from patients, and also for the assessment of the viability of parasite samples after chemotherapy or other treatment. Detection of parasite nucleic acids in clinical samples from AE or CE patients has been substantially improved by the use of the PCR. Generation of specific primers and their use in PCR and reverse transcriptase (RT)-PCR (RT-PCR) allow the detection of minute amounts of parasite RNA/DNA collected during surgical removal of cyst material or by fine needle aspiration biopsy (FNAB) (Pawlowski and Stefaniak, 2003a).

These nucleic acid-based assays are highly sensitive and can provide exquisite levels of specificity for absolute and unambiguous identification of alcohol-fixed and appropriately stored *Echinococcus* samples. Such assays incorporating appropriate PCR primers could readily provide confirmation of the first report of a human case of polycystic echinococcosis due to *E. vogeli* from Peru (Somocurcio *et al.*, 2004), and whether two cases of echinococcosis reported from India were caused by *E. oligarthrus* (Kini *et al.*, 1997; Sahni *et al.*, 2000) or *E. granulosus* as argued by D'Alessandro and Rausch (2004).

DNA Detection of Infection in Definitive and Intermediate Hosts

Two approaches for diagnosis of intestinal *Echinococcus* infection, the detection of *E. multilocularis-/E. granulosus*-specific coproantigens in enzyme-linked immunosorbent assay (ELISA) and of copro-DNA by PCR, have been successfully implemented. These methods are valuable for the post-mortem and the intravitam diagnosis of *Echinococcus* infection in definitive hosts (Dinkel *et al.*, 1998; Cabrera *et al.*, 2002; Mathis and Deplazes, 2002; Abbasi *et al.*, 2003;

Deplazes *et al.*, 2003; Stefanic *et al.*, 2004). They provide also the tools to study the transmission biology of *E. multilocularis* and *E. granulosus* as they allow detection of an infection in faecal samples collected in the environment. Coproantigen detection is the diagnostic method of choice as it is sensitive, fast and cheap. The use of PCR for routine diagnostic or large-scale purposes is hampered by the fact that DNA extraction from faecal material is a very laborious task (Deplazes *et al.*, 2003). Nevertheless, the copro-PCR is a valuable method for confirmation of positive coproantigen results by ELISA and for diagnosis in individual animals. As taeniid eggs cannot further be differentiated morphologically, PCR is the method of choice to identify *Echinococcus* infections in faecal or environmental samples containing taeniid eggs. The sensitivity of an *E. granulosus* G1-specific coprodiagnostic PCR was recently shown to give a positive result with a single egg (A. Dinkel, University of Hohenheim, Stuttgart, Germany, personal communication).

In intermediate rodent hosts, the PCR is routinely used in epidemiological studies to identify *E. multilocularis* from liver lesions, which are often very small, atypical or calcified. The PCR using proteinase K-digested lesion material has been successfully employed for monitoring the prevalence of *E. multilocularis* in rodents (Dinkel *et al.*, 1998; Gottstein *et al.*, 2001; Stieger *et al.*, 2002).

PCR amplification/sequencing of mtDNA sequences (*cox1*; *cytb* (cytochrome *b*)) has been used to show that yaks (*Bos grunniens*) seem inadequate and dead-end hosts for the sheep–dog strain of *E. granulosus*, and also for *E. multilocularis* in western Sichuan, PR China. Both unilocular (*E. granulosus* in yaks, sheep and goats) and multilocular (alveolar) (*E. multilocularis* in hares and rodents) hydatids are common in this region. Humans and dogs are equally infected with both species. Previous studies had shown that many yaks harbour unilocular (probably *E. granulosus*) and multilocular cysts that visually were deemed to be *E. multilocularis*. However, two studies involving histological and PCR-based mitochondrial genotyping of echinococcal lesions from the infected yaks showed that they were actually the G1 genotype of *E. granulosus* (Xiao *et al.*, 2003; Heath *et al.*, 2005).

In the study by Heath *et al.* (2005) no viable cysts were found in 125 necropsied yaks taken from four townships of the Datangba Flatlands, Ganzi County, Sichuan. The convoluted laminated membranes in the multilocular cysts were lined on the inside by germinal membrane, but showed no budding to the exterior, and there were no protoscoleces present. These multilocular cysts were probably a manifestation of an immune response to *E. granulosus* that walls off the developing cyst so that the laminated and germinal membranes continue to proliferate within a confined space. The unilocular cysts also had no evidence of protoscoleces or developing brood capsules.

The results from these two studies have important public health consequences and implications for echinococcosis control. A hydatid control and community health project underway in Sichuan is intended to provide guideline information for the development of future hydatid control programmes in China, and is focused on interrupting the life cycle of the hydatid parasite by dosing dogs with praziquantel and vaccinating intermediate hosts (Heath *et al.*, 2003). An understanding of the tapeworm life cycle and how people can avoid becoming infected with hydatid disease is included as part of the community and health education activities.

At the beginning of the project it was thought that Datangba Flatlands yaks, sheep and goats could all produce cysts that would be able to reinfect dogs with the parasite. Vaccination of all these animals would prevent infections becoming established and reduce the chances of dogs becoming reinfected by eating animal organs containing hydatid cysts. Now that it has been shown that this local Datangba sheep strain of *E. granulosus* usually only produces protoscoleces in sheep and goats, and not in yaks, the control of hydatid disease caused by *E. granulosus* can concentrate there on sheep and goats, while putting less emphasis on yaks.

Concluding Remarks

Nucleic acid-based approaches and the use of microsatellites are proving of considerable value for investigating genetic variation, the molecular epidemiology and population genetics of *Echinococcus* taxa, and for identification and discrimination of *Echinococcus* eggs from those of other taeniid species in definitive hosts. As well, the range of DNA techniques now available for genetic discrimination of *Echinococcus* species and strains is impressive and much valuable information on the molecular categorization of the different genotypes is now at hand. In many cases, molecular techniques have validated the genetic basis of important morphological and other biological differences that can now be used with confidence as a reliable and simple means of identifying and differentiating between *Echinococcus* taxa. The publication of the complete sequences of the mitochondrial genomes of the horse and sheep strains of *E. granulosus* and *E. multilocularis* and the availability of mtDNA sequences for a number of other *Echinococcus* genotypes (Le *et al.*, 2002; Nakao *et al.*, 2002) provide additional genetic information that can be used for even more in-depth strain characterization and phylogeny construction of *Echinococcus* spp. As discussed earlier, the availability of this rich genetic information has provided a solid molecular basis for revising the taxonomy of the genus *Echinococcus* (Thompson and McManus, 2002).

Acknowledgements

The author acknowledges the National Health and Medical Research Council of Australia, the Australian Research Council and The Wellcome Trust for financial support of his research on the molecular epidemiology of *Echinococcus* and echinococcosis. He would also like to thank numerous collaborators, especially Li Hua Zhang, Josephine Bowles, Robin Gasser, Thanh Hoa Le, David Blair, Mark Pearson and Mara Rosenzvit, for their significant contributions to the DNA studies.

References

Abbasi, I., Branzburg, A., Campos-Ponce, M., Abdel Hafez, S.K., Raoul, F., Craig, P.S. and Hamburger J. (2003) Copro-diagnosis of *Echinococcus granulosus* infection in dogs by amplification of a newly identified repeated DNA sequence. *American Journal of Tropical Medicine and Hygiene* 69, 324–330.

Azab, M.E., Bishara, S.A., Helmy, H., Oteifa, N.M., El-Hoseiny, L.M., Ramzy, R.M. and Ahmed, MA. (2004) Molecular characterization of Egyptian human and animal *Echinococcus granulosus* isolates by RAPD-PCR technique. *Journal of the Egyptian Society of Parasitology* 34, 83–96.

Bardonnet, K., Benchikh-Elfegoun, M.C., Bart, J.M., Harraga, S., Hannache, N., Haddad, S., Dumon, H., Vuitton, D.A. and Piarroux R. (2003) Cystic echinococcosis in Algeria: cattle act as reservoirs of a sheep strain and may contribute to human contamination. *Veterinary Parasitology* 116, 35–44.

Bart, J.M., Bardonnet, K., Elfegoun, M.C., Dumon, H., Dia, L., Vuitton, D.A. and Piarroux, R. (2004) *Echinococcus granulosus* strain typing in North Africa: comparison of eight nuclear and mitochondrial DNA fragments. *Parasitology* 128, 229–234.

Bartholomei-Santos, M.L., Heinzelmann, L.S., Oliveira, R.P., Chemale, G., Gutierrez, A.M., Kamenetzky, L., Haag, K.L. and Zaha A. (2003) Isolation and characterization of microsatellites from the tapeworm *Echinococcus granulosus. Parasitology* 126, 599–605.

Bowles, J., Blair, D. and McManus, D.P. (1994) Molecular genetic characterisation of the cervid strain ('northern form') of *Echinococcus granulosus. Parasitology* 109, 215–221.

Bowles, J., Blair, D. and McManus D.P. (1995) A molecular phylogeny of the genus *Echinococcus. Parasitology* 110, 317–328.

Bretagne, S., Assouline, B., Vidaud, D., Houin, R. and Vidaud, M. (1996) *Echinococcus multilocularis*: microsatellite polymorphism in U1 snRNA genes. *Experimental Parasitology* 82, 324–328.

Breyer, I., Georgieva, D., Kurdova, R. and Gottstein, B. (2004) *Echinococcus granulosus* strain typing in Bulgaria: the G1 genotype is predominant in intermediate and definitive wild hosts. *Parasitology Research* 93, 127–130.

Cabrera, M., Canova, S., Rosenzvit, M. and Guarnera, E. (2002) Identification of *Echinococcus granulosus* eggs. *Diagnostic Microbiology and Infectious Disease* 44, 29–34.

Castrodale, L.J., Beller, M., Wilson, J.F., Schantz, P.M., McManus, D.P., Zhang, L.H., Fallico, F.G. and Sacco, F.D. (2002) Two atypical cases of cystic echinococcosis (*Echinococcus granulosus*) in Alaska, 1999. *American Journal of Tropical Medicine and Hygiene* 66, 325–327.

D'Alessandro, A. and Rausch, R.L. (2004) Erroneous reports of the neotropical *Echinococcus oligarthrus* as a cause of echinococcosis in India. *Journal of Parasitology* 90, 202–203.

Daniel-Mwambete, K., Ponce-Gordo, F. and Cuesta-Bandera, C. (2004) Genetic identification and host range of the Spanish strains of *Echinococcus granulosus*. *Acta Tropica* 91, 87–93.

Deplazes, P., Dinkel, A. and Mathis, A. (2003) Molecular tools for studies on the transmission biology of *Echinococcus multilocularis*. *Parasitology* 127, S53–61.

Dinkel, A., von Nickisch-Rosenegk, M., Bilger, B., Merli, M., Lucius, R. and Romig, T. (1998) Detection of *Echinococcus multilocularis* in the definitive host: coprodiagnosis by PCR as an alternative to necropsy. *Journal of Clinical Microbiology* 36, 1871–1876.

Dinkel, A., Njoroge, E.M., Zimmermann, A., Walz, M., Zeyhle, E., Elmahdi, I.E., Mackenstedt, U. and Romig, T. (2004) A PCR system for detection of species and genotypes of the *Echinococcus granulosus*-complex, with reference to the epidemiological situation in eastern Africa. *International Journal for Parasitology* 34, 645–653.

Elmahdi, I.E., Ali, Q.M., Magzoub, M.M., Ibrahim, A.M., Saad, M.B. and Romig, T. (2004) Cystic echinococcosis of livestock and humans in central Sudan. *Annals of Tropical Medicine and Parasitology* 98, 473–479.

Gasser, R.B. and Zhu, X.Q. (1999) Sequence-based analysis of enzymatically amplified DNA fragments by mutation detection techniques. *Parasitology Today* 15, 462–465.

Gottstein, B., Saucy, F., Deplazes, P., Reichen, J., Demierre, G., Busato, A., Zuercher, C. and Pugin, P. (2001) Is high prevalence of *Echinococcus multilocularis* in wild and domestic animals associated with disease incidence in humans? *Emerging Infectious Diseases* 7, 408–412.

Haag, K.L., Zaha, A., Araujo, A.M. and Gottstein, B. (1997) Reduced genetic variability within coding and noncoding regions of the *Echinococcus multilocularis* genome. *Parasitology* 115, 521–529.

Haag, K.L., Ayala, F.J., Kamenetzky, L., Gutierrez, A.M. and Rosenzvit, M. (2004) Livestock trade history, geography, and parasite strains: the mitochondrial genetic structure of *Echinococcus granulosus* in Argentina. *Journal of Parasitology* 90, 234–239.

Heath, D.D., Jensen, O. and Lightowlers, M.W. (2003) Progress in control of hydatidosis using vaccination – a review of formulation and delivery of the vaccine and recommendations for practical use in control programmes. *Acta Tropica* 85, 133–143.

Heath, D.D., Zhang, L.H. and McManus, D.P. (2005) Short report: inadequacy of yaks as hosts for the sheep–dog strain of *Echinococcus granulosus* or *E. multilocularis*. *American Journal of Tropical Medicine and Hygiene* 72, 289–290.

Kamenetzky, L., Gutierrez, A.M., Canova, S.G., Haag, K.L., Guarnera, E.A., Parra, A., Garcia, G.E. and Rosenzvit, M.C. (2002) Several strains of *Echinococcus granulosus* infect livestock and humans in Argentina. *Infection, Genetics and Evolution* 2, 129–136.

Kedra, A.H., Swiderski, Z., Tkach, V.V., Dubinsky, P., Pawlowski, Z., Stefaniak, J. and Pawlowski, J. (1999) Genetic analysis of *Echinococcus granulosus* from humans and pigs in Poland, Slovakia and Ukraine. A multicenter study. *Acta Parasitologica* 44, 248–254.

Kedra, A.H., Swiderski, Z., Tkach, V.V., Rocki, B., Pawlowski, J. and Pawlowski, Z. (2000) Variability within NADH dehydrogenase sequences of *Echinococcus multilocularis*. *Acta Parasitologica* 45, 353–355.

Kini, U., Shariff, S. and Nirmala, V. (1997) Aspiration cytology of *Echinococcus oligarthrus*. A case report. *Acta Cytologica* 41, 544–548.

Lahmar, S., Debbek, H., Zhang, L.H., McManus, D.P., Souissi, A., Chelly, S. and Torgerson, P.R. (2004) Transmission dynamics of the *Echinococcus granulosus* sheep–dog strain (G1 genotype) in camels in Tunisia. *Veterinary Parasitology* 121, 151–156.

Lavikainen, A., Lehtinen, M.J., Meri, T., Hirvela-Koski, V. and Meri, S. (2003) Molecular genetic characterization of the Fennoscandian cervid strain, a new genotypic group (G10) of *Echinococcus granulosus*. *Parasitology* 127, 207–215.

Le, T.H., Pearson, M.S., Blair, D., Dai, N., Zhang, L.H. and McManus, D.P. (2002) Complete mitochondrial genomes confirm the distinctiveness of the horse–dog and sheep–dog strains of *Echinococcus granulosus*. *Parasitology* 124, 97–112.

Lopez-Neyra, C.R. and Soler Planas, M.A. (1943) Revision del genero *Echinococcus* rud y description de una especie nueva parasita intestinal del porro en Almeria. *Revista Iberica Parasitologia* 3, 169–194.

Mathis, A. and Deplazes, P. (2002) Role of PCR-DNA detection of *Echinococcus multilocularis.* In: Craig, P. and Pawlowski, Z. (eds) *Cestode Zoonoses: Echinococcosis and Cysticercosis.* IOS Press, Amsterdam, pp. 195–204.

McManus, D.P. (2002) The molecular epidemiology of *Echinococcus granulosus* and cystic hydatid disease. *Transactions of the Royal Society of Tropical Medicine and Hygiene* 96, S151–157.

McManus, D.P. and Bryant, C. (1995) Biochemistry, physiology and molecular biology of *Echinococcus.* In: Thompson, R.C.A. and Lymbery, A.J. (eds) *The Biology of Echinococcus and Hydatid Disease.* CAB International, Wallingford, UK, pp. 355–410.

McManus, D.P. and Thompson, R.C.A. (2003) Molecular epidemiology of cystic echinococcosis. *Parasitology* S37–51.

McManus, D.P., Zhang, L.H., Castrodale, L.J., Le, T.H., Pearson, M. and Blair, D. (2002) Short report: molecular genetic characterization of an unusually severe case of hydatid disease in Alaska caused by the cervid strain of *Echinococcus granulosus. American Journal of Tropical Medicine and Hygiene* 67, 296–298.

McManus, D.P., Zhang, W., Li, J. and Bartley, P.B. (2003) *Echinococcosis.* Lancet 362, 1295–1304.

Nakao, M., Yokoyama, N., Sako, Y., Fukunaga, M. and Ito, A. (2002) The complete mitochondrial DNA sequence of the cestode *Echinococcus multilocularis* (Cyclophyllidea: Taeniidae). *Mitochondrion* 1, 497–509.

Nakao, M., Sako, Y. and Ito, A. (2003) Isolation of polymorphic microsatellite loci from the tapeworm *Echinococcus multilocularis. Infection Genetics and Evolution* 3, 159–163.

Obwaller, A., Schneider, R., Walochnik, J., Gollackner, B., Deutz, A., Janitschke, K., Aspock, H. and Auer, H. (2004) *Echinococcus granulosus* strain differentiation based on sequence heterogeneity in mitochondrial genes of cytochrome *c* oxidase-1 and NADH dehydrogenase-1. *Parasitology* 128, 569–575.

Ortlepp, R.J. (1934) *Echinococcus* in dogs from Pretoria and vicinity. *Onderstepoort Journal of Veterinary Science and Animal Industry* 3, 97–108.

Pawlowski, Z.S. and Stefaniak, J. (2003a) Cystic echinococcosis (CE) in the Poznan region in the years 1990–2000. *Przeglad Epidemiologiczny (Warsawa)* 57, 579–586. (In Polish.)

Pawlowski, Z.S. and Stefaniak, J. (2003b) The pig strain of *Echinococcus granulosus* in humans: a neglected issue? *Trends in Parasitology* 19, 439.

Rausch, R.L. and Nelson, G.S. (1963) A review of the genus *Echinococcus* Rudolphi, 1801. *Annals of Tropical Medicine and Parasitology* 57, 127–135.

Rinder, H., Rausch, R.L., Takahashi, K., Kopp, H., Thomschke, A. and Löscher T. (1997) Limited range of genetic variation in *Echinococcus multilocularis. Journal of Parasitology* 183, 1045–1050.

Rosenzvit, M.C., Zhang, L.H., Kamenetzky, L., Canova, S.G., Guarnera, E.A. and McManus, D.P. (1999) Genetic variation and epidemiology of *Echinococcus granulosus* in Argentina. *Parasitology* 118, 523–530.

Saad, M.B. and Magzoub, M. (1989) Hydatidosis in sheep and goats in the Sudan. *Sudan Journal of Veterinary Science and Animal Husbandry* 28, 33–37.

Sahni, J.K., Jain, M., Bajaj, Y., Kumar, V. and Jain, A. (2000) Submandibular hydatid cyst caused by *Echinococcus oligarthrus. Journal of Laryngology and Otology* 114, 473–476.

Scott, J.C., Stefaniak, J., Pawlowski, Z.S. and McManus, D.P. (1997) Molecular genetic analysis of human cystic hydatid cases from Poland: identification of a new genotypic group (G9) of *Echinococcus granulosus. Parasitology* 114, 37–43.

Siles-Lucas, M.M. and Gottstein, B.B. (2001) Molecular tools for the diagnosis of cystic and alveolar echinococcosis. *Tropical Medicine and International Health* 6, 463–475.

Somocurcio, J.R., Sanchez, E.L., Naquira, C., Schilder, J., Rojas, F., Chacon, P. and Yabar, A. (2004) First report of a human case of polycystic echinococcosis due to *Echinococcus vogeli* from a neotropical area of Peru, South America. *Revista do Instituto de Medicina Tropical de São Paulo* 46, 41–42.

Stefanic, S., Shaikenov, B.S., Deplazes, P., Dinkel, A., Torgerson, P.R. and Mathis, A. (2004) Polymerase chain reaction for detection of patent infections of *Echinococcus granulosus* ('sheep strain') in naturally infected dogs. *Parasitology Research* 92, 347–351.

Stieger, C., Hegglin, D., Schwarzenbach, G., Mathis, A. and Deplazes, P. (2002) Spatial and temporal aspects of urban transmission of *Echinococcus multilocularis. Parasitology* 124, 631–40.

Thompson, R.C.A. (1995) Biology and systematics of *Echinococcus.* In: Thompson, R.C.A. and Lymbery, A.J. (eds) *Echinococcus and Hydatid Disease.* CAB International, Wallingford, UK, pp. 1–50.

Thompson, R.C.A. and Lymbery, A. (1990) *Echinococcus*: biology and strain variation. *International Journal for Parasitology* 20, 457–470.

Thompson, R.C.A. and McManus, D.P. (2001) Aetiology: parasites and life cycles. In: Eckert, J., Gemmell, M.A., Meslin, F.-X. and Pawlowski, Z.S. (eds) *WHO/OIE Manual on Echinococcosis in Humans and Animals a Public Health Problem of Global Concern.* World Organisation for Animal Health, Paris, pp. 1–19.

Thompson, R.C.A. and McManus, D.P. (2002) Towards a taxonomic revision of the genus *Echinococcus. Trends in Parasitology* 18, 452–457.

Verster, A.J.M. (1965) Review of *Echinococcus* species in South Africa. *Onderstepoort Journal of Veterinary Research and Animal Industry* 32, 7–118.

Williams, R.J. and Sweatman, G.K. (1963) On the transmission, biology and morphology of *Echinococcus granulosus equinus,* a new subspecies of hydatid tapeworm in horses in Great Britain. *Parasitology* 53, 391–407.

Xiao, N., Qiu, J., Nakao, M., Nakaya, K., Yamasaki, H., Sako, Y., Mamuti, W., Schantz, P.M., Craig, P.S. and Ito, A. (2003) Short report: identification of *Echinococcus* species from a yak in the Qinghai-Tibet plateau region of China. *American Journal of Tropical Medicine and Hygiene* 69, 445–446.

Zhang, W., Li, J. and McManus, D.P. (2003) Concepts in immunology and diagnosis of hydatid disease. *Clinical Microbiology Reviews* 16, 18–36.

4 Ribosomal DNA Variation in Parasitic Flatworms

D. Blair
School of Tropical Biology, James Cook University, Townsville, Queensland 4811, Australia

Introduction

Ribosomes, the intracellular particles on which proteins are assembled, are highly complex and dynamic entities. The structural framework of ribosomes is provided by ribosomal RNA (rRNA) molecules with which many proteins are associated (summarized in Capowski and Tracy, 2003). Homologous rRNA genes occur in all prokaryotes and eukaryotes. The mitochondrial and chloroplast rRNA genes in eukaryotes clearly have prokaryote affinities (Pace *et al.*, 1986). The genomic DNA from which ribosomal genes are transcribed, along with any associated spacers, is collectively termed ribosomal DNA (rDNA). Sequences and other data from rDNA and its products, both nuclear and mitochondrial, have a long history of use in phylogenetic, systematic and related studies across all kingdoms of life (summarized for parasitic flatworms in Table 4.1). Most studies on rDNA of parasitic flatworms (Neodermata – the trematodes, monogeneans and cestodes) have had a focus on phylogeny and/or systematics, a focus that will flavour this review. They have also revealed that this region of the genome is a dynamic and fascinating one.

Terminology applied to the rRNAs can be confusing. Individual rRNAs are often referred to according to their sedimentation coefficient (*S* value). However, *S* values vary among orthologous genes according to the cellular location (nucleus or mitochondrion in metazoans) and taxon in which each is found. Thus, the bacterial

Table 4.1. Summary of research areas for which ribosomal sequences and related data have been used. This is not comprehensive. Several studies tabulated here could be listed in more than one category.

Research area	Gene region	Taxon and notes	Reference
Broad-scale phylogeny of flatworms	nSSU	Platyhelminthes	Littlewood and Olson (2001)
	nSSU/LSU	Platyhelminthes and Neodermata	Lockyer *et al.* (2003a)
Molecular taxonomy and discovery of cryptic species	ITS	Trematodes of marine fish	Jousson *et al.* (2000)
	ITS2 and mtLSU	Cryptic species of trematodes in marine gastropods	Donald *et al.* (2004)
	ITS1	Cryptic species of trematode in freshwater salmonids	Criscione and Blouin (2004)
Identification of life-cycle stages	ITS	Matched cercariae and adults of mesometrids (Trematoda)	Jousson *et al.* (1998)
	ITS1	Opecoelids (Trematoda) in marine fish and snails	Jousson and Bartoli (2000)
	ITS2	Bivesiculids (Trematoda) have three-host life cycle	Cribb *et al.* (1998)
Evaluation of morphological features	nSSU	Affinities of Gyliauchenidae (Trematoda)	Blair and Barker (1993)
Host specificity and speciation	ITS	*Gyrodactylus* species	Zietara and Lumme (2002)
	nSSU and ITS	*Gyrodactylus* species	Huyse *et al.* (2003)
	nSSU and ITS1	Sympatric speciation in *Dactylogyrus*	Simkova *et al.* (2004)
	nLSU	Polystomes in freshwater turtles	Littlewood *et al.* (1997)
Coevolution	nSSU	Host–parasite cospeciation not typical in a monogenean/marine fish system	Desdevises *et al.* (2002)
Biogeography	ITS1	Cestodes of lemmings	Haukisalmi *et al.* (2001)
	nLSU and ITS	Lung flukes of frogs	Snyder and Tkach (2001)
	nSSU/LSU and mtSSU	Asian *Schistosoma* species	Attwood *et al.* (2002)
	nSSU/LSU and ITS and mtLSU	*Schistosoma* species	Morgan *et al.* (2003a)
	nSSU/LSU	*Schistosoma* species	Lockyer *et al.* (2003b)
Host evolutionary history	nLSU	Polystomes (Monogenea) and early vertebrates	Verneau *et al.* (2002)
Molecular diagnosis	nSSU	*Hymenolepis nana* formed cystic mass in AIDS patient	Olson *et al.* (2003)
	mtSSU	*Echinococcus granulosus* strains	Dinkel *et al.* (2004)

16S rRNA is orthologous with the mitochondrial (mt) 12S and the nuclear (n) small subunit (SSU) (nSSU) or 18S rRNAs in eukaryotes. Similarly, the bacterial 23S rRNA is orthologous with the mt16S and the nuclear (n) large subunit (LSU) (nLSU) or 28S rRNAs of eukaryotes. Two other small rRNAs transcribed from the eukaryote nuclear genome are universally referred to as the 5S and 5.8S molecules. Both form a complex with the nLSU, which is the scaffolding for the LSU of the ribosome. The nSSU forms the scaffolding for the SSU of the ribosome.

In this review, the organization and genomic location of rDNA in both the nuclear and mitochondrial genomes of neodermatans will be discussed. There will be an emphasis on the kinds and causes of variation in rDNA and its products. Variation occurs at many levels. Sequence differences may be due to point mutations, insertions, deletions and translocations, and to repeated and inverted regions. Variants can be specific to certain taxa within the Neodermata. Variation may also occur within a species, at the level of populations and even within an individual. Variation in the non-transcribed portion of the nuclear intergenic spacers (IGSs) can affect promoters and other signals. Variation at the nucleotide level can affect secondary and higher level structures of the rRNAs and their transcribed spacers, and hence influence their function. It has often been noted that secondary structure features are far more conserved than the nucleotide strings that interact to form such features (e.g. Després *et al.*, 1992). Consequently, secondary structure features have often been used as an aid to sequence alignment, and as a guide to the 'anatomy' of rRNA molecules.

Nuclear Ribosomal Operons

The gene encoding the smallest structural RNA component of the nuclear ribosome is the 5S rRNA gene. This gene is not co-located with the remaining nuclear rRNA genes. Neither sequence nor chromosomal location of the 5S gene is known for any neodermatan and it will not be discussed further here.

The remaining three nuclear rRNA genes, those for the nSSU, 5.8S and nLSU rRNAs, are arranged in a linear manner within an operon. This in turn forms an array of tandem repeats, on one or more chromosomes, visible in chromosome spreads as a 'nucleolar organizer region'. In the two neodermatans, both trematodes (the lung fluke, *Paragonimus ohirai*, and the blood fluke, *Schistosoma mansoni*), for which data are available, the array of operons occurs only on the short arm of a single chromosome (Hirai, 1988; Hirai *et al.*, 1989). Numbers of copies in the tandem array can be considerable: about 100 per haploid genome in *S. mansoni* (Simpson *et al.*, 1984). In the oriental liver fluke (*Opisthorchis viverrini*), ribosomal operons constitute about 6.1% of the genome (Korbsrisate *et al.*, 1992).

The structure of the operon seems very similar in all metazoans and has been reviewed for *S. mansoni* by Capowski and Tracy (2003). Transcription from the chromosomal DNA by RNA polymerase I initially produces a long precursor RNA molecule in which the future nSSU and 5.8S rRNAs are separated by the first internal transcribed spacer (ITS)1 and the future 5.8S and nLSU rRNAs are separated by ITS2 (Fig. 4.1). External transcribed spacers (ETSs) occur immediately 5′ of the nSSU gene and 3′ of the nLSU gene (it is unclear whether the latter is invariably present in neodermatans). Individual operons are separated by the IGS, which comprises the ETSs and the non-transcribed spacer (NTS) (terminology as used by Collins and Cunningham, 2000). Successive cleavages of the full-length precursor eventually yield the mature rRNA molecules. The process in schistosomes seems to be similar to that in well-studied model organisms such as yeast (Capowski and Tracy, 2003).

Early studies on neodermatans used restriction mapping to estimate the size and general structure of the nuclear ribosomal operon (e.g. Walker *et al.*, 1986 and Kaukas *et al.*, 1994 – *Schistosoma* species ≈10,000 bp; Blair and McManus, 1989 – fasciolids ~12,000–16,000 bp; Korbsrisate *et al.*, 1992 – *O. viverrini* ~13,000 bp; and Silva *et al.*, 1991 – *Echinococcus granulosus*, considerable length variation noted). Subsequently, sequences spanning the entire operon have been assembled for the trematode *S. mansoni* and the monogenean *Gyrodactylus salaris* (see below and also Fig. 4.1).

In addition to the tandem repetition of the operon, diverse repeats of varying length often occur within each operon, mostly within spacers.

	nSSU = 18S	ITS1	5.8S	ITS2	nLSU = 28S	3′ ETS + IGS
Monogenea						
Gyrodactylus salaris[1]	1991	623	157	428	3818	2168
Digenea						
Schistosoma mansoni[2]	1994	456	155	309	4127	2558
Fasciola hepatica[3]	~1900	433		364	~4 kb	~5–9 kb
Opisthorchis viverrini[4]	~2000				~4.2 kb	~5.8 kb
Cestoda						
Gigantolina magna[5]	2873				5503	
Echinococcus granulosus[6]	2394	251–957		~250	>3.6 kb	? - >13 kb

[1] Matejusova and Cunningham (2004); [2] Kane and Rollinson (1998), Capowski and Tracy (2003); [3] Blair and McManus (1989), Mas-Coma *et al.* (2001); [4] Korbsrisate *et al.* (1992); [5] Lockyer *et al.* (2003a); [6] This study and Picón *et al.* (1996).

Fig. 4.1. Lengths of various components (diagram not to scale) of the nuclear rRNA operon in selected neodermatans. Mature rRNA molecules are shown as shaded boxes. Literature sources are listed. Lengths of some nSSU and nLSU are slightly underestimated because conserved primer sites at each end are not always included in the GenBank accessions.

Long tracts of tandemly repeated sequences, such as ribosomal operons are expected to undergo concerted evolution (reviewed by Elder and Turner, 1995) whereby sequences are homogenized among repeats. That is, every sequence in a tandem array is more similar to all others than would be expected by chance. The same process should homogenize sequences among members of a population. Mechanisms responsible for concerted evolution are still somewhat mysterious. Biased gene conversion and unequal crossing over have been implicated (Hillis *et al.*, 1991; Elder and Turner, 1995), both phenomena that work best where repeat sequences are abundant, as they are in ribosomal operons.

Concerted evolution apparently operates among neodermatan ribosomal operons (Collins and Cunningham, 2000). However, intraspecific and intra-individual variation is common in neodermatans and other eukaryotes, especially in spacers, and indicates that the force of concerted evolution cannot completely overcome contrary influences. Such influences include the effects of replication slippage and unequal crossing over, both of which can occur in regions of 'simple' sequence, which is abundant in eukaryotic genomes (Tautz and Renz, 1984). Simple sequence is often biased in nucleotide composition and includes runs of short, tandem repeats (typically no longer than tetranucleotides), or non-tandem repetitions of such motifs in close proximity to one another ('cryptic simplicity' – Tautz *et al.*, 1986), to a far greater extent than would be expected at random. Cryptic simplicity is common in sequences of ribosomal spacers and the most variable portions of rRNA molecules. Identical short motifs throughout a spacer can be generated again and again convergently in an individual (Dover and Tautz, 1986; Hancock and Vogler, 2000) and presumably also in different individuals and species. Such regions appear to have escaped some of the homogenizing effects of concerted evolution. Regions of cryptic simplicity, where replication slippage and unequal crossing over generate variety, are usually those varying within and among individuals (Gerbi, 1996). By contrast, in conserved portions of the operon, point mutations may be the main source of variation (Hancock and Vogler, 2000).

Most of the studies reported here used the polymerase chain reaction (PCR) to amplify portions of the ribosomal operon for subsequent manipulation, usually sequencing. The PCR has the potential to generate artefacts, especially if the template DNA contains repeats (e.g. Vartapetian, 1994). Although acknowledged as a problem in some of the reviewed papers (e.g. Kane and Rollinson, 1998), there have been few attempts to rule out the possibility of PCR artefacts. This should be remembered when reading what follows.

Intergenic spacers (IGSs)

The IGS region is the least conserved in the operon (although it may be rivalled by the ITS1 in some taxa). For example, Cunningham *et al.* (2003) found intra- and inter-individual sequence variation in the IGS among taxa that had identical sequences at other variable regions of the operon, such as the V4 region of the nSSU rRNA gene and the ITS. Intra- and inter-individual variation in the IGS was noted in the early restriction-mapping studies (Walker *et al.*, 1986 for *Schistosoma margrebowiei*; Blair and McManus, 1989 for fasciolids; and Korbsrisate *et al.*, 1992 for *O. viverrini*). Subsequent sequencing studies have provided further information. Complete IGS sequences are available for the monogenean *G. salaris* (see Collins and Cunningham, 2000) and three species of the trematode genus *Schistosoma* (see Kane and Rollinson, 1998). Partial sequence has been obtained for additional species of *Gyrodactylus* (see Cunningham *et al.*, 2003) and for the cestode *E. granulosus* (see Silva *et al.*, 1991; Picón *et al.*, 1996).

PCR amplification of the IGS region of *Schistosoma intercalatum* and *S. haematobium*, followed by cloning of PCR products, yielded two size variants in each case (Kane and Rollinson, 1998). Similarly, three bands were obtained for the monogenean *G. salaris* (Collins and Cunningham, 2000). Analyses reported by these authors were mostly based on the largest sequenced products. The IGSs of these *Schistosoma* species and of *E. granulosus* resembled one another in containing many repeats and AT-rich tracts. In *Schistosoma* species, diverse repeats (19–379 bp long) were scattered throughout the IGS and copies could be direct or inverted. Sequences of the entire IGS region from three species of *Schistosoma* could be aligned with each other with reasonable ease, provided gaps were inserted to accommodate species-specific repeats and insertions (Kane and Rollinson, 1998) (Fig. 4.2). Differences among species were due to differences in numbers and types of identifiable repeats, as well as to insertions/deletions and sequence variation within non-repetitive regions (Fig. 4.2). The smaller sequenced PCR product from *S. intercalatum* lacked a large repeat present in the larger product, and differed somewhat in sequence. In the case of

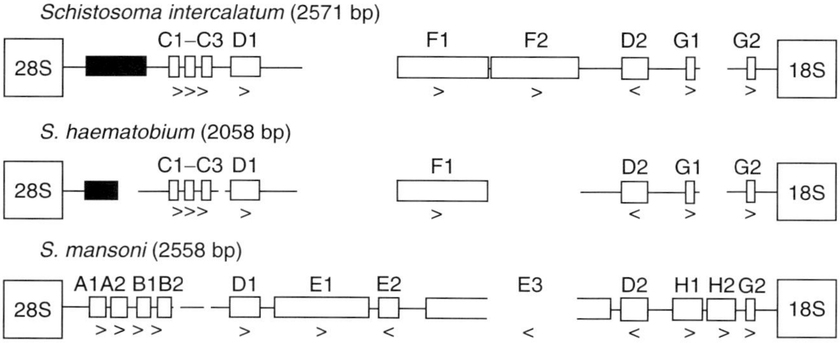

Fig. 4.2. Schematic diagram of the components of the IGSs for three species of *Schistosoma*, after Kane and Rollinson (1998). Major repeats are shown as open boxes, with an arrowhead indicating whether each is direct or inverted. Repeats coded by the same letter are similar in sequence. The AT rich region (present in *Schistosoma intercalatum* and *S. haematobium*) is shown as a filled box. Breaks in the horizontal line indicate large alignment gaps.

S. haematobium, the smaller product was very short and surmised to be a non-functional relic or a PCR artefact. Intra-individual variation was apparently present in the AT-rich regions occurring near the 5′ end of the IGS (Kane and Rollinson, 1998).

In the case of *G. salaris*, repeats were concentrated in two tandem arrays somewhat 3′ of the middle of the spacer (Fig. 4.3). The arrays were separated by 81 bp of unique sequence and an AT-rich region occurred just before and after each. Each repeat was 23 bp in length, and those within a single cluster were similar to one another but not necessarily identical. Cloned PCR products spanning the portion of the IGS containing these clusters were sequenced from numerous *Gyrodactylus* specimens from several fish species (Cunningham *et al.*, 2003). Although the pattern was similar to that seen in the original IGS characterized from *G. salaris*, sequence variation in individual repeats was found within and among individuals as well as among nominal species.

AT-rich regions, such as those flanking the arrays of repeats in the IGS of *G. salaris*, and those lying near the 5′ end of the IGS in

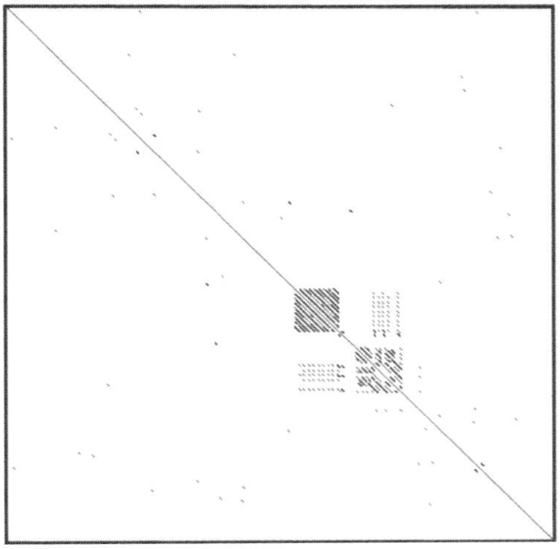

Fig. 4.3. Dotplot of the IGS of *Gyrodactylus salaris* (AJ276032 – 2619 bp) to show locations of repeats. This was made using the resource at http://arbl.cvmbs.colostate.edu/molkit/dnadot/. A moving window of 11 and a mismatch limit of 2 were specified. Two blocks of repeats are indicated – see text for details.

S. intercalatum and *S. haematobium*, but not found in *S. mansoni*, might represent DNA unwinding elements (DUEs). Such regions, typically ranging in length from 30 to 100 bp, are associated with origins of DNA replication in both prokaryotes and eukaryotes (discussed in Kane and Rollinson, 1998).

Other motifs and repeats, suspected to be functional in a number of model organisms, have been found in IGSs of eukaryotes (e.g. Fujiwara and Ishikawa, 1987; Safrany and Hidvegi, 1989). Similar motifs have been noted in all neodermatan IGS regions that have been sequenced to date. However, no experimental work has been done to demonstrate whether these are truly functional or simply represent random assortments of bases. Putative functional motifs identified include chi-like sites, transcription start and termination signals and transcription factor binding sites (tabulated and summarized in Collins and Cunningham, 2000). One problem in seeking functional motifs in ribosomal operons is that promoters for RNA polymerase I are unusually variable, even differing among closely related taxa (summarized in Picón *et al.*, 1996). Chi-like sites resemble motifs experimentally determined to be recombination hot spots in prokaryotes, and a similar function has been assumed in some eukaryotes (Kane *et al.*, 1996). Recombination, especially unequal crossing over, mediated by these and other motifs might be a source of repeats and may partly drive IGS variation within and among individuals and species.

The length to which transcription extends 3′ of the nLSU and 5′ of the nSSU is uncertain for *Schistosoma* species and for *G. salaris*. More definite information is available for the 5′ ETS of the cestode *E. granulosus*. The region of the IGS immediately upstream of the nSSU gene in this species was investigated by Picón *et al.* (1996). These workers used experimental methods to place the transcription start point at 1100 bp upstream of the 5′ end of the nSSU gene. Regions close to this contained a variety of interesting motifs, notably a poly-T tract and short inverted repeats. Copies of a 12 bp repeat were scattered in the remainder of the ETS closer to the nSSU gene. The availability of this sequence data should make it possible to explore variation in sequence and patterns of repeats in the 3′ end of the IGS of other *Echinococcus* species.

Internal transcribed spacers (ITSs)

Conserved sequences at the 3′ end of the nSSU gene and the 5′ end of the nLSU gene have made it easy to design primers to span the ITS region (Fig. 4.1). The conserved 5.8S gene lies between the two spacers and can also be used as a target for PCR primers. The spacers themselves are poorly conserved (in both sequence and length) and can rarely be aligned at or above the level of family. ITS1 is generally the more variable in sequence and length (e.g. Luton *et al.*, 1992; Morgan and Blair, 1995; Bell *et al.*, 2001; and Galazzo *et al.*, 2002 for the trematode genera *Dolichosaccus*, *Echinostoma*, *Diplostomum* and *Ichthyocotylurus*, respectively and Cunningham *et al.*, 2000 for the monogenean genus *Gyrodactylus*). However, León-Règagnon *et al.* (1999) found ITS2 to be the more variable spacer in *Haematoloechus* species and Luo *et al.* (2002) and Olson *et al.* (2002) found the same for the cestode genera *Bothriocephalus* and *Ligula*, respectively. Figure 4.4 shows an edited screen shot of the result from a BLAST search in GenBank. The ITS sequence (including parts of the flanking nSSU and nLSU genes) of the heterophyid trematode *Haplorchis taichui* was submitted to BLAST using default parameters and is shown as the first 'hit' in the figure. Solid lines, indicating highly conserved regions, are shown for all 'hits' in the conserved nSSU, 5.8S and nLSU rRNA sequences. The second, third and fourth hits represent other genera of heterophyids and show that the portions of ITS1 and ITS2 flanking the 5.8S are more conserved (relative to the first hit) than the remainder of these spacers. Finally, sequences 5–7 are from more distantly related trematodes and demonstrate the low sequence conservation in the spacers beyond the level of family.

The first cleavage of the primary rRNA transcript occurs within ITS1 (Capowski and Tracy, 2003). Possibly associated with this, ITS1 contains features reminiscent of the IGS; repeats, simple-sequence tracts and possible promoter sites and recombination hot spots. Secondary structure models for the ITS1 are tentative. The ITS2, on the other hand, functionally resembles the expansion segments found within the rRNA genes (Gerbi, 1996). It folds so as to bring the 3′ end of the 5.8S rRNA and the 5′ end of the nLSU rRNA together to facilitate their base pairing in a conserved secondary structure. The ITS2 differs from almost all expansion segments in neodermatan rRNA genes, however, in that it is excised from the ribosomal transcript (i.e. it could be regarded as a processed expansion segment (Gerbi, 1996)).

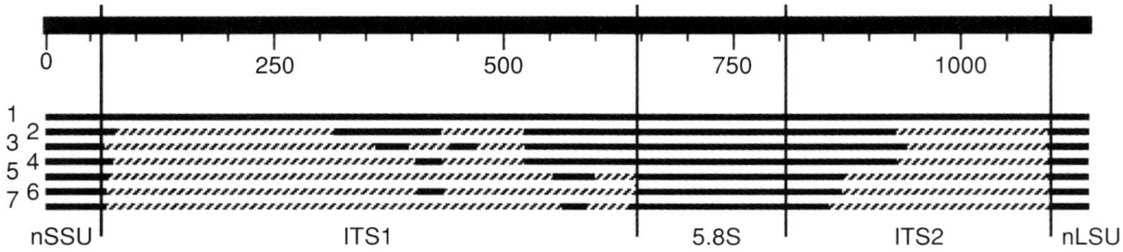

Fig. 4.4. Edited screen shot of results of a BLAST search (default parameters) in GenBank using as the query a portion of the rDNA of *Haplorchis taichui* (AY245705)(fam. Heterophyidae) spanning the ITS region and including the 3′ 60 bp of the nSSU and 5′ 50 bases of the nLSU genes. All 'hits' not spanning the entire ITS region have been edited out. Regions of relatively high similarity are shown as a solid line, those of lower similarity as a broken line. 1, *Haplorchis taichui*; 2, *Centrocestus* sp. (AY245699)(fam. Heterophyidae); 3, *Pygidiopsis genata* (AY245710)(fam. Heterophyidae); 4, *Haplorchis pumilio* (AY245706)(fam. Heterophyidae); 5, *Zoogonus rubellus* (AJ241804)(fam. Zoogonidae); 6, *Cainocreadium labracis* (AJ241795)(fam. Opecoelidae); 7, *Nicolla elongata* (AJ241792)(fam. Opecoelidae).

Intra-individual variation and repeat sequences

Repeats are infrequently reported from ITS2. A tract of 50 bp was repeated once in the ITS2 of a number of *Echinostoma* species (Morgan and Blair, 1995). Morgan and Blair (1998) noted a tract of 35–45 bp occurring three times in the ITS2 of *G. salaris*. Intra-individual variation in this region seems not to have been reported, even among species in which such variation occurs in ITS1 (e.g. *Paragonimus* species and *Schistosoma japonicum*; see below).

Intra-individual variation in ITS1 is common and is often due to the presence of differing numbers of repeat regions. Other differences can be due to sequence variation, including insertions and deletions, outside the repeats. Differences between individuals can also have such a basis, and mention of this will be made in this section.

The first published digenean ITS1 sequences were from *Dolichosaccus symmetrus* (GenBank accession L01631) and *Dolichosaccus helocirrus* (L01630) (Luton *et al.*, 1992). Cloned PCR products were sequenced. Respective lengths of the ITS1 were 525 and 868 bp, with almost all of the difference in length being due to the presence of a number of repeats close to the 5′ end of the ITS1 in *D. helocirrus* of which only a single example appeared in *D. symmetrus* (Fig. 4.5). Each repeat consisted of two sequence tracts, each about 50 bp long, here arbitrarily designated as *a* and *b*. In *D. helocirrus*, the repeats occurred tandemly as $(a,b)_4a$ and in *D. symmetrus* as *b,a* only. Sequences of *a* and *b* were not obviously similar and different copies of *a* or *b* were not 100% identical. Some geographical variation in sequence was noted.

A similar pattern of repeats has been found near the 5′ end of ITS1 in other trematodes unrelated to *Dolichosaccus*: a *b* tract is flanked on either side by an *a* tract (van Herwerden *et al.*, 1998). Sequences of *a* and *b* within a species are not similar. Apart from close congeners, neither *a* nor *b* regions are similar among species. In the lung fluke, *Paragonimus westermani*, the longest sequence derived from a cloned PCR product had the pattern of repeats $(a,b)_2a$. In this species, *a* was 82 bp long and *b* 38. Shorter sequences containing fewer repeats were also obtained from PCR products from the same individual worm (van Herwerden *et al.*, 1999). Similarly, in blood flukes of the *S. japonicum* group, variable numbers of near-identical repeats are present within and among individuals. The repeats are again bipartite – *a* (about 90 bp) and *b* (15–53 bp depending on the species) (van Herwerden *et al.*, 1998). Cloned and sequenced PCR products from a single *S. japonicum* specimen yielded variants with a single *a*, with $(a,b)a$ and with $(a,b)_2a$. In neither of these examples did the

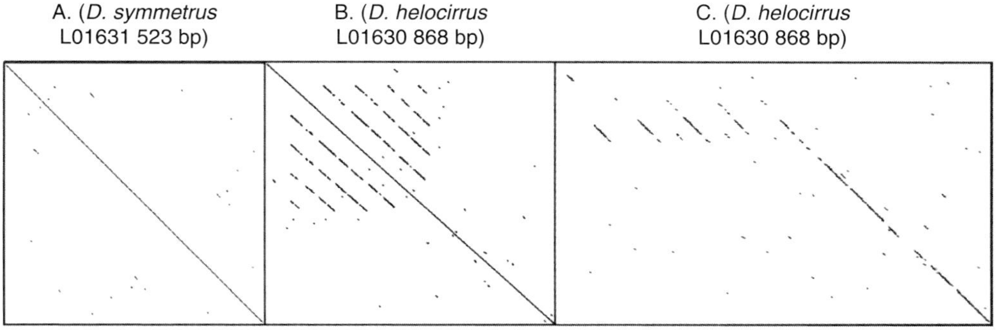

Fig. 4.5. Dotplots (constructed as for Fig. 4.3) to show repeats in the ITS1 of *Dolichosaccus helocirrus* and *Dolichosaccus symmetrus*. In panels A and C, the sequence on the *y*-axis is that of *D. symmetrus*. In panel B, the sequence on the *y*-axis is that of *D. helocirrus*. Identity of the sequence on the *x*-axis is indicated on the figure in each case.

sequenced PCR products represent all the variants present. By probing Southern blots of genomic digests using a radiolabelled ITS fragment, van Herwerden *et al.* (1999) demonstrated that most variants of ITS1 in *P. westermani* individuals were much longer than the longest cloned PCR product. Patterns on autoradiographs were very different for each individual of *P. westermani* from different localities. Other species in the genus were relatively monomorphic. A similar approach applied to *S. japonicum* suggested that some copies of the ITS1 might contain as many as $(a,b)_7a$ and that, at least in Chinese populations of this parasite, most copies of the ITS1 had $(a,b)_4a$ or more. Copies of *a* at the same position share more similarity across clones than they do within a given clone (van Herwerden *et al.*, 1998).

The pattern of repeats in ITS1 found in members of the *S. japonicum* group and among some unrelated trematodes is not, however, seen in remaining members of the genus *Schistosoma* investigated to date. In the phylogenetic tree of Morgan *et al.* (2003a), one major lineage in *Schistosoma* includes the *S. japonicum* group and *Schistosoma sinensium* found only in Asia. The remaining members of the genus are here referred to as the African lineage – even though some members occur in Asia – because the basal species in this lineage occur in Africa. These basal African species (*Schistosoma hippopotami*, a form thought to represent *Schistosoma edwardiense* and an unidentified species) lack repeat elements at the 5′ end of the ITS1 whereas at least some of the more derived species in this lineage possess them (Kane and Rollinson, 1994; Kane *et al.*, 1996). *S. mansoni*, *S. intercalatum* and *S. haematobium* all have two copies of a 70–80 bp tract. These repeats are not bipartite in the manner described above, but each does contain three copies of a 9 bp subrepeat. Nor are the repeats arranged in tandem in the species listed above: one copy occurs near each end of the ITS1. *S. margrebowiei* has four copies and *Schistosoma mattheei* has four or five copies (depending on geographical origin). In the last two species, one copy occurs near the 3′ end of ITS1 and the remaining copies are in tandem near the 5′ end. In *Schistosoma spindale* from Sri Lanka, there are seven copies of the repeat near the 5′ end of ITS1, but sequence is not available for the 3′ end (Kane and Rollinson, 1994; Kane *et al.*, 1996). Sequences of each copy are similar, occasionally identical, among species, but they cannot be aligned with repeats in the *S. japonicum* group (van Herwerden *et al.*, 1998). Interestingly, dotplots reveal (D. Blair, unpublished data) that the copy of the repeat at the 3′ end of ITS1 in the African lineage can also be recognized in a similar location in members of the *S. japonicum* group and even in species of a distantly related schistosome genus *Trichobilharzia* (Fig. 4.6). Repeats have apparently evolved independently in the African lineage and the *S. japonicum* group,

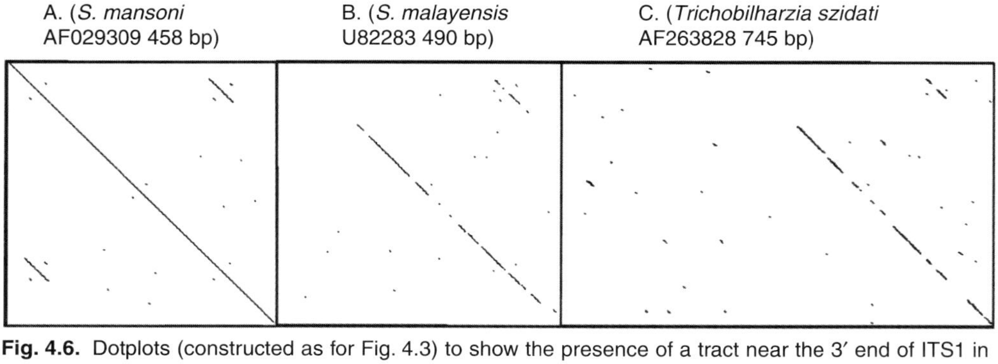

Fig. 4.6. Dotplots (constructed as for Fig. 4.3) to show the presence of a tract near the 3′ end of ITS1 in *Schistosoma* and *Trichobilharzia* species that has been repeated near the 5′ end of ITS1 in many members of the African lineage of *Schistosoma*. In each case the sequence on the *y*-axis is that of *Schistosoma mansoni* (AF029309). Identity of the sequence on the *x*-axis is indicated on the figure in each case.

with the African lineage using a tract common to ITS1 of schistosomes as the template for repeats. Intra-individual and intraspecific variation in ITS1, including numbers of repeats, is very limited in members of the African lineage relative to the *S. japonicum* group.

Not all trematodes possess repeats in the ITS1. They are absent from members of the Echinostomatidae (see Morgan and Blair, 1995) and the genus *Diplostomum* (e.g. Galazzo *et al.*, 2002). Similarly, large repeats have not been reported from the ITS1 of monogeneans and cestodes, but it must be noted that a search for repeats has not always been mentioned in published work.

Non-repetitive regions of ITS1 of trematodes, especially towards the 3′ end of the spacer, have been investigated for phylogenetic utility. von der Schulenburg *et al.* (1999) reported no intra-individual variation among a range of species and contended that the 3′ region of ITS1 could be aligned at considerable phylogenetic depths and be used to determine relationships among trematode families. They aided alignment by use of a secondary structure model (see section on 'Secondary Structure'). By contrast, van Herwerden *et al.* (1999) found intra-individual variation in the 3′ portion (post-repeat) of the ITS1 in members of the *P. westermani* species complex that was sufficient to obscure phylogenetic relationships in this complex. However, intra-individual variation was not noted in members of the *P. ohirai* species group (van Herwerden *et al.*, 1999).

Intra-individual variation in ITS1 is absent or slight in monogeneans (e.g. Zietara *et al.*, 2002) and most cestodes (e.g. Hancock *et al.*, 2001 who noted variation in *Taenia solium* due to differing numbers of a tetranucleotide repeat). A striking exception to this rule is among members of the cestode genus *Echinococcus*. Many sequence variants can be found within a single isolate, especially in the case of *E. granulosus*, a species in which multiple strains are known, each with a degree of mammalian-host specificity (McManus *et al.*, 2003). In *Echinococcus* species, numerous small (typically di- and trinucleotide) repeats occur, in short tandem tracts as well as scattered throughout the ITS1 (van Herwerden *et al.*, 2000). These sequences therefore exhibit simplicity and cryptic simplicity. Variant sequences are generally not markedly different from one another. Phylogenies constructed from these sequences frequently show a number of clusters, each containing one or more ITS1 variants from each isolate investigated (Bowles *et al.*, 1995; Kedra *et al.*, 1999; van Herwerden *et al.*, 2000; Lavikainen *et al.*, 2003). At least two of the species (*E. granulosus* and *Echinococcus multilocularis*) were paraphyletic for ITS1 (van Herwerden *et al.*, 2000). This variability is illustrated graphically in Fig. 4.7. Most available ITS1 sequences from *Echinococcus* species (four short variant sequences were omitted) were aligned and a tree constructed in MEGA v3 (Kumar *et al.*, 2004) using the minimum evolution method and complete deletion (any column containing a gap in any sequence was omitted). The locations on the tree of sequences from certain isolates are indicated. Note that unambiguous alignment of cryptically simple sequences with many insertions/deletions can be difficult. Inferred trees may vary according to choice of alignment parameters, decisions concerning inclusion of sites with gaps and the method of tree construction. However, any tree will demonstrate the variability of sequences from individual isolates. Intraspecific and intra-individual variation occurs in other *Echinococcus* species (Bowles *et al.*, 1995), but is not always extensive (e.g. see Rinder *et al.*, 1997 for *E. multilocularis*).

Explanations for the sequence heterogeneity observed in *E. granulosus* fall into several non-exclusive categories. It might be that arrays of rRNA operons occur on more than one chromosome (a possibility mentioned by van Herwerden *et al.*, 2000). If so, concerted evolution may not homogenize sequences among arrays, or this process may only occur slowly. A second class of explanation involves hybridization (e.g. Bowles *et al.*, 1995; van Herwerden *et al.*, 2000). The presence of sequences from different phylogenetic clusters in a single isolate/sample of *E. granulosus* suggests this. If so, then these events might have taken place relatively recently, and concerted evolution has not had time to homogenize the array(s). Another possibility is suggested by visual inspection of an alignment of available sequences (D. Blair, unpublished data). The ITS1 region of *Echinococcus* species is obviously one of simple and cryptically simple sequence with

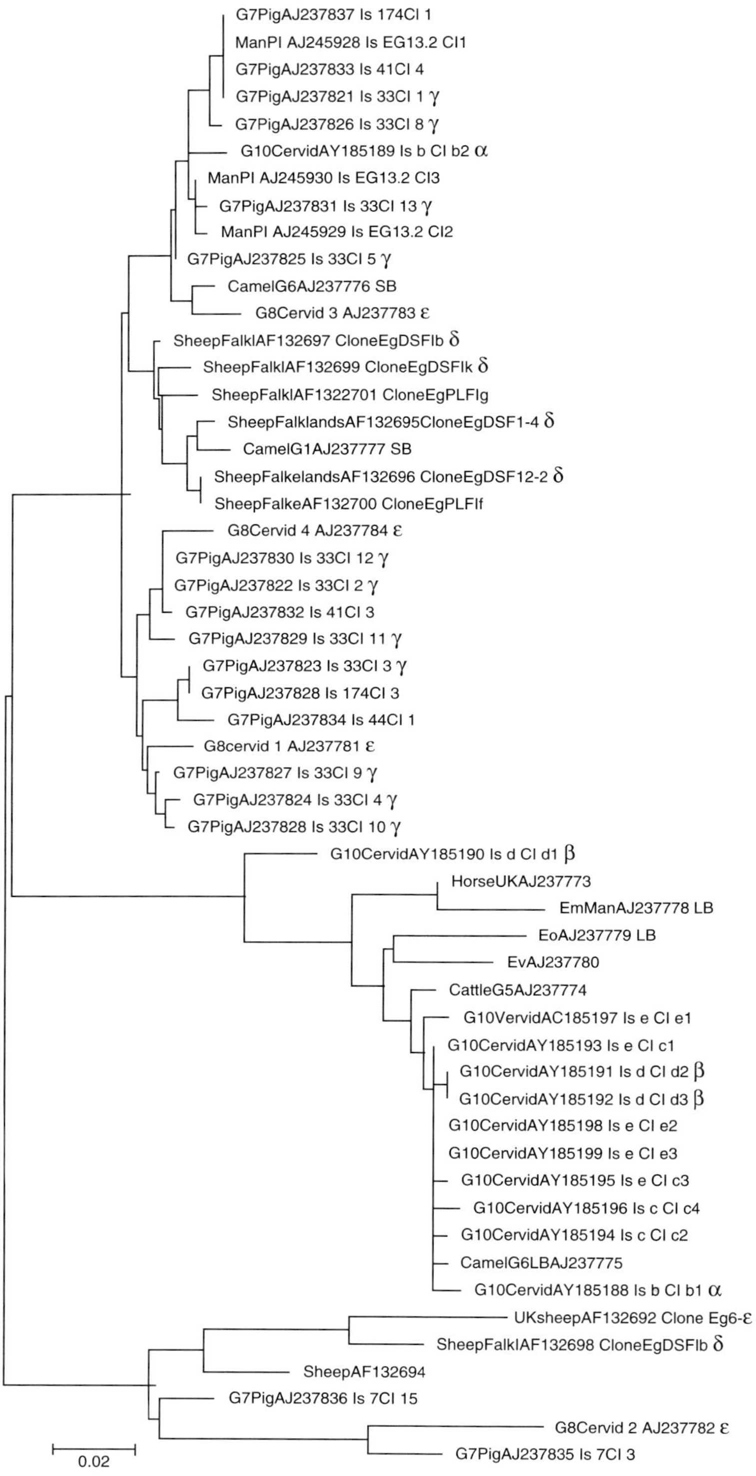

Fig. 4.7. Phylogenetic tree of most available ITS1 sequences from *Echinococcus* species. Minimum evolution tree found in MEGAv3 as described in the text. Identical Greek letters indicate sequences of cloned PCR products from the same isolate. α and β – Lavikainen *et al.* (2003); γ – Kedra *et al.* (1999); δ – van Herwerden *et al.* (2000); ε – Bowles *et al.* (1995).

trinucleotide motifs being particularly common. Simplicity is confirmed using the SIMPLE v 3.0 algorithm (Hancock and Armstrong, 1994) (available at: http://www.biochem.ucl.ac.uk/bsm/SIMPLE/index.html). Score parameters were set at 1 for dinucleotide repeats, 3 for tri- and 3 for tetranucleotides. A relative simplicity factor (RSF) score of 1.42 was obtained for a typical *E. granulosus* ITS1 sequence (AF132697) – a significant departure from random expectation based on the scores for ten permuted sequences with the same base composition as the test sequence. For comparison, the nSSU sequence for *E. granulosus* (U27015) had an RSF score of 1.18 using the same parameters. This score was also significant but implies a lower degree of simplicity. Hancock and Vogler (2000) investigated sequence evolution within an expansion segment (V4) of the nSSU rRNA in tiger beetles. Like ITS sequences, expansion segments often consist of simple and cryptically simple sequences (Gerbi, 1996). Hancock and Vogler (2000) concluded that replication slippage often produced similar sequence motifs in different lineages – a parallelism rendering questionable the use of such regions for phylogeny. ITS1 sequences in *Echinococcus* species could well be experiencing slippage at a rate that outruns concerted evolution. The result is extensive microheterogeneity within individuals and sequence convergence/parallelism among copies, limiting the value of ITS for molecular taxonomy in this genus (Kedra *et al.*, 1999).

Intra- and interspecific variation

ITS sequences of neodermatans are generally regarded as difficult or impossible to align beyond the level of genus or family (Morgan and Blair, 1998). However, phylogenetic depths at which alignment is possible vary. Desdevises *et al.* (2000) noted that ITS1 sequences were difficult to align among members of the monogenean genus *Lamellodiscus*. Similarly, Matejusova *et al.* (2003) found it difficult to align some portions of the ITS region, especially the 5′ end of ITS1, among species of *Gyrodactylus* and related genera. However, ITS1 sequences of *Dactylogyrus* species were apparently easy to align but could not be aligned outside the genus with other dactylogyrids (Simkova *et al.*, 2004). Morgan and Blair (1998) found that only about the first 130 bases of ITS2 could be aligned with any confidence across several families of trematodes.

The relatively high rates of evolutionary change in the ITS region have made this the target of choice for many taxonomic studies. In some neodermatans, intra-individual and intraspecific variation, especially in ITS1, is too great to permit such uses (see above). However, in many others, intraspecific or intra-individual variation is limited: sequences from either or both of the spacers have been used for investigating relationships among closely related species and to aid recognition of synonyms and of cryptic species. Problems of interpretation must be acknowledged. When ITS sequences of different nominal species are identical, can this be taken as evidence that the species are synonymous? Lack of variation in ITS was used by Morgan and Blair (1995) to propose that three species of the trematode genus *Echinostoma* from widely separated parts of Africa were synonymous. In this case, there was other evidence to support this view (Morgan and Blair, 1995). A contrasting example was reported by Niewiadomska and Laskowski (2002) who found identical partial ITS1 sequences in the trematodes *Diplostomum spathaceum* and *D. parviventosum*, species that are morphologically distinguishable at all major stages of the life cycle.

Where there is little doubt concerning the identity of a species, intraspecific differences in ITS sequences are generally few, even across considerable geographic distances. Cable *et al.* (1999 – note added in proof) found identical ITS2 sequences in specimens of *Gyrodactylus kobayashii* from Europe and Australia. Huyse and Volckaert (2002) found no differences in ITS1 and ITS2 sequences of specimens of *Gyrodactylus rugiensis* and related species from France, Norway, Belgium and the Netherlands. Lo *et al.* (2001) found no differences in ITS2 sequences from digenean species inhabiting fish from sites 6000 km apart in the Pacific. Differences were few in ITS2 of *Dolichosaccus* species from localities 340 km apart in northern Australia (Luton *et al.*, 1992). ITS2 sequences of *Fasciola hepatica* from Australia, Hungary and New Zealand were identical (Adlard *et al.*, 1993) and differed from a sequence from a Mexican specimen by a single transition. Some variation in ITS2 was noted from several samples of *Fasciola gigantica* from Zambia and between these and southern Asian specimens (Itagaki and Tsutsumi, 1998). Sorensen *et al.* (1998) found differences at six sites in ITS sequences between two samples of *Echinostoma trivolvis* from the same pond in Indiana.

Variation in ITS sequences has sometimes been used as evidence for the existence of cryptic species, especially if variants differ in host specificity or other biological attributes. For example, geographic strains of *P. westermani* differ in ITS2 sequence, consistent with evidence from other sources that some strains are distinct at the level of species (Iwagami *et al.*, 2000). Although Morgan and Blair (1995) found no differences in ITS sequences of *Echinostoma revolutum* specimens from Germany and Indonesia, Sorensen *et al.* (1998) did detect an apparent variant sequence in North America. Subsequent study led to the conclusion that *E. revolutum* of Morgan and Blair (1995) in fact represented a different species (Kostadinova *et al.*, 2003). Among cestodes, *Bothriocephalus acheilognathi* from freshwater fish on several continents exhibited limited variation (<2%) in the ITS region, but this variation was correlated both with geography and host species (Luo *et al.*, 2002). Olson *et al.* (2002) found host-correlated ITS sequence variation in *Ligula intestinalis* plerocercoids within the same lake, consistent with proposals that two strains or species might be present. Jousson and Bartoli (2001) used a combination of ITS sequences and morphometric approaches to characterize cryptic species in the trematode genus *Cainocreadium* from marine fish. Huyse and Volckaert (2002) used a similar approach to characterize cryptic species of *Gyrodactylus* inhabiting gobies in northern Europe.

Functional motifs

Putative functional motifs are commonly reported in ITS1 sequences of neodermatans. In processing motifs such as chi-like sites (possible hot spots for recombination), the ITS1 resembles the IGS. Chi-like sites were first noted by Kane *et al.* (1996) in ITS1 of *Schistosoma* species (African lineage) in which they are associated with repeats. Dvorák *et al.* (2002) found similar sites within repeats in avian blood flukes of the genus *Trichobilharzia* and they have also been noted in ITS1 of other trematodes, but not associated with repeats (Kane *et al.*, 1996). Perhaps surprisingly, chi-like sites are absent from ITS1 of the *S. japonicum* group (van Herwerden *et al.*, 1998). These authors speculated that the greater degree of intra-individual variation in ITS1 in the *S. japonicum* group relative to the African lineage of *Schistosoma* might be due to differences in rates of recombination. The chiasma frequency in *S. japonicum* is only a fifth of that which is seen in *S. mansoni* (see Hirai *et al.*, 1996). Lower rates of recombination/gene conversion are likely to lead to slower homogenization within the ribosomal array.

van Herwerden *et al.* (2003) examined ITS1 sequence data from 18 species of trematodes and cestodes for motifs associated with transcription, such as TATAAT, CCAAT and GC boxes. One or (usually) more of these was found in each species. Simple statistical tests demonstrated that these motifs occurred significantly more frequently in many species than might be expected by random assortment of bases. Although this suggests that these motifs might be maintained by selection pressure, it remains important to emphasize that there is as yet no experimental evidence that they are actually functional in ITS1.

van Herwerden *et al.* (1998) noted AT-rich regions, reminiscent of those seen in the IGS of other *Schistosoma* species, in some variant ITS1 sequences from *S. japonicum*. The possibility of functional motifs occurring in the ITS2 of neodermatans has not been investigated.

Secondary structure

Phylogenetic and functional studies require alignments where there is high confidence in positional homology of bases. Secondary structural features of rRNA can provide reference points to assist alignment (e.g. Morgan and Blair, 1998; Zietara *et al.*, 2002). Secondary structural features can also be used as phylogenetic characters in their own right (Caetano-Anollés, 2002). The only attempts at developing a secondary structural model for the ITS1 of a neodermatan are for the monogenean *G. salaris* by Cunningham *et al.* (2000) and by von der Schulenburg *et al.* (1999) for a portion of the ITS1 in a range of digeneans. Among three species of *Gyrodactylus* many differences were found in ITS sequence, primarily in the 5′ half (Cunningham *et al.*, 2000), and many different secondary structures of similar free-energy values could be inferred. von der Schulenburg *et al.* (1999) aligned the 3′ portions of most ITS1 sequences from digeneans then available, refining their alignment using inferred secondary structures. They postulated the existence of seven helices in this region, and could produce phylogenetic trees from their alignment that recovered monophyly of each family for which multiple sequences were available.

It increasingly appears that transcribed ITS2 forms a similar secondary structure in many – perhaps all – eukaryotes (Joseph *et al.*, 1999; Coleman, 2003), the function of which is to juxtapose the 5.8S and nLSU rRNAs, after which the spacer is excised. Researchers are increasingly converging on the view that the ITS2 contains four domains/helices (Coleman, 2003). Secondary structure models for ITS2 of neodermatans have been proposed by Michot *et al.* (1993), Morgan and Blair (1998) and Cunningham *et al.* (2000). The first two studies used sequences from trematodes and the last, monogeneans. All models are similar, resembling the 'ring' model proposed for yeast by Joseph *et al.* (1999). The repeats in ITS2 of *G. salaris* are each capable of forming a short helix, adding to the total number of helices in that structure (Cunningham *et al.*, 2000). Semi-conserved sequence tracts were found in all neodermatan taxa investigated by Morgan and Blair (1998) (Fig. 4.8). The first of these, in domain B, appears to represent a processing site identified in yeast and rat (Morgan and Blair, 1998). Sequences could be aligned across trematode families for domains A and B, but not beyond this (Morgan and Blair, 1998).

Small and Large Subunit Genes

The nLSU rRNA and the nSSU rRNA are by far the most studied portions of the ribosomal operon across all domains of life. There are about 347 complete or near-complete neodermatan nSSU and about 50 nLSU in the public databases (February 2005). When partial sequences are included, the numbers are much greater (Table 4.2). The nSSU sequences (typically ~ 2000 bp) were among the first to be used in phylogenetic studies, probing the earliest divergences of life and the relationships among major kingdoms (e.g. Pace *et al.*, 1986). They have also been used to infer phylogenies of metazoans, asking, for example, where the Platyhelminthes nest within the metazoans (Ruiz-Trillo *et al.*, 1999), whether the phylum is monophyletic (e.g. Telford *et al.*, 2003) and to explore the interrelationships of the main groups of Platyhelminthes (Littlewood and Olson, 2001). The nLSU has been less used in phylogenetic studies, partly because its greater length (3200–5500 bp in flatworms – Fig. 4.1) has been an impediment to obtaining full-length sequences. However, partial nLSU sequences are now accumulating in the databases at an increasing rate (Table 4.2). The two subunits have been combined for analysis in some recent phylogenetic studies (Mallatt and Winchell, 2002 for metazoans; Lockyer *et al.*, 2003a for neodermatans).

Like the ribosomal transcription unit itself, nSSU and nLSU rRNAs are a patchwork of variable and conserved domains. The former can vary in length as well as sequence and are often termed expansion segments. By contrast, the conserved regions of the molecules form universal helices (Wuyts *et al.*, 2001b) recognizable across all domains of life (some absent in mt rRNAs). In mature ribosomes, these portions of the rRNA genes are found near the centre of the ribosomal particle, and variable regions are peripheral (Wuyts *et al.*, 2001b) where they form distinct helices at least some of which have the capacity to interact to form tertiary structures (Alkemar and Nygård, 2004), implying that they are functional. Schematic diagrams of the secondary structures of the nSSU and nLSU of neodermatans, indicating conserved and variable regions, are shown in Figs 4.9 and 4.10 (modified from Wuyts *et al.*, 2001b and based on a figure kindly provided by Dr Jan Wuyts). Note that both nSSU and nLSU rRNAs in cestodes are much longer than in trematodes and monogeneans

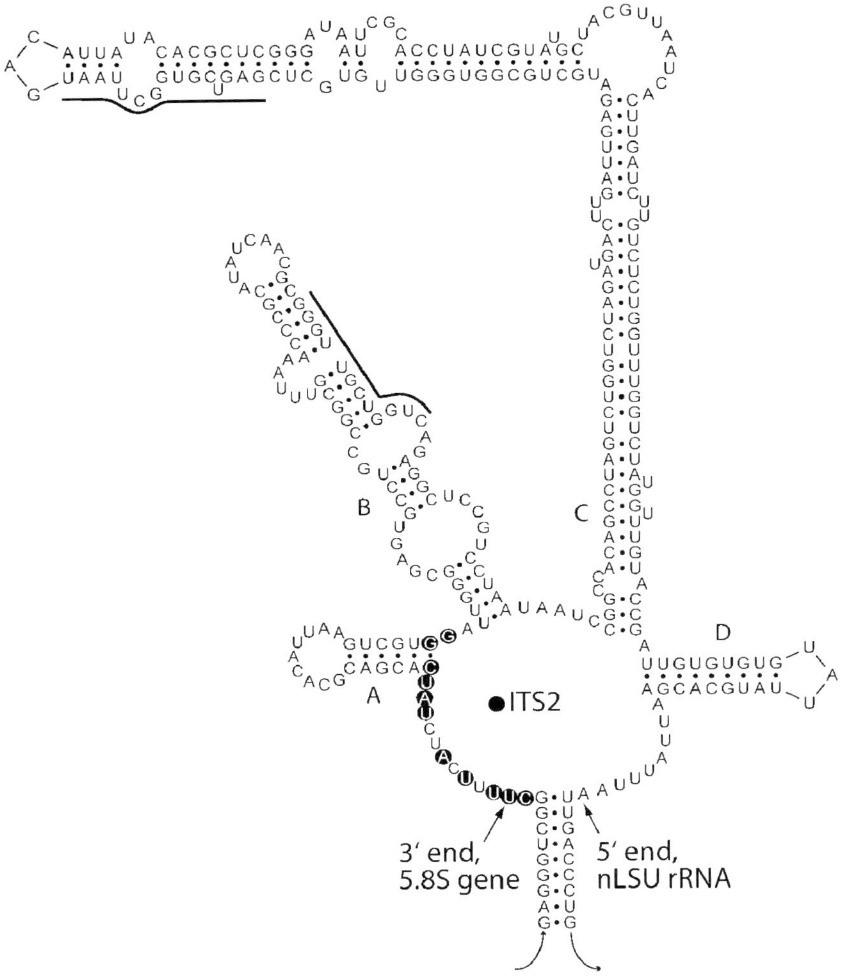

Fig. 4.8. ITS2 folding of *Schistosoma mansoni* (U22168) based largely on Morgan and Blair (1998) and showing the four domains, A–D. Bases conserved in domains A and B across all 42 sequences from trematodes of eight families studied by Morgan and Blair (1998) are indicated (white letters on black circles). Lines along parts of domains B and C indicate the position of semi-conserved sequence tracts (see text).

Table 4.2. Numbers of nSSU and nLSU sequences for neodermatans found on 28 February 2005 by submitting a series of queries to GenBank modelled on this example: '18S rRNA gene AND Trematoda[Orgn]'. This recovers all relevant accessions including partial sequences and multiple accessions for the same species.

	Trematoda	Cestoda	Monogenea
nSSU	402	476	260
nLSU	625	380	176
mtSSU	70	27	0
mtLSU	107	66	11

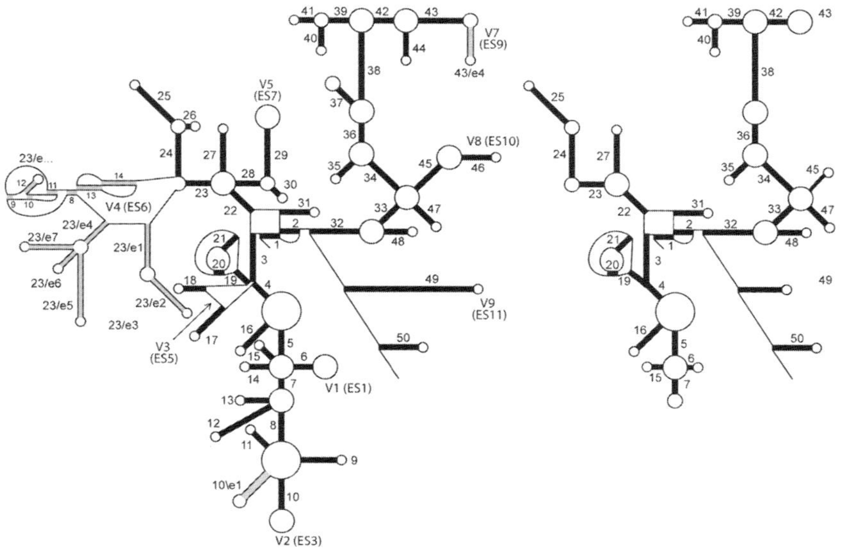

Fig. 4.9. Diagrammatic representations of the secondary structures of the nSSU (left) and mtSSU (right) rRNAs. 'Universal' helices are shown in solid black. Helices present in eukaryotic expansion segments of neodermatans are not shaded. Expansion segments are numbered as V1 to V9, following De Rijk *et al.* (1992). The equivalent numbering scheme used by Gerbi (1996) is indicated in parentheses. Note that the V7 region adjacent to helix 43 might have a different, as yet unknown, structure in cestodes. The mtSSU rRNA structure is based on that for *Fasciola hepatica* (Le *et al.*, 2001). Diagram modified from one supplied by Dr Jan Wuyts.

(Fig. 4.1). Secondary structure models for the neodermatan nSSU have been published by Omer Ali *et al.* (1991) – *Schistosoma*; Johnston *et al.* (1993) – *Schistosoma*; Picón *et al.* (1996) – *Echinococcus*; Liu *et al.* (1997) – *Spirometra*; and for the nLSU by Matejusova and Cunningham (2004) – *Gyrodactylus*. A model for the nSSU of the trematode *Calicophoron calicophorum* is available at http://www.psb.ugent.be/rRNA/secmodel/Ccal_SSU.html

Intraspecific variation is slight or non-existent in the nSSU and nLSU genes. Sequences from the conserved helices can be aligned among all flatworms and indeed many portions can be aligned at far greater phylogenetic depths than that. Lockyer *et al.* (2003a) have explored variation in these genes among neodermatans and flatworms in general. They found that, across the Platyhelminthes, the genes had probably not experienced substitution saturation and could therefore be used with confidence for phylogenetic studies. However, a large proportion of sites in both genes could not be aligned with confidence across the Platyhelminthes and were omitted from analyses. These variable regions (presumably largely in expansion segments, although this was not stated explicitly) had accumulated substitutions at such a rate that their alignment remained a problem even when only sequences from the Trematoda were considered. Figure 4.2 in Lockyer *et al.* (2003a) shows this clearly. Thus, sequences from expansion segments are difficult to use in phylogenetic studies across, e.g. the trematodes (but see below).

Matejusova and Cunningham (2004) reported primary and secondary structures of the nLSU rRNA in the monogenean *G. salaris*. Like Lockyer *et al.* (2003a), they considered that the LSU gene may be useful in phylogenetic analysis of the Monogenea or Platyhelminthes and comparisons with other eukaryotes. They suggested that the variable domains C and H (terminology of Wuyts *et al.*, 2001a) may be the most suitable for this purpose.

Variable regions have been termed 'expansion segments' (insertions relative to the sequence in *Escherichia coli*), 'variable domains' and 'divergent domains' by different workers

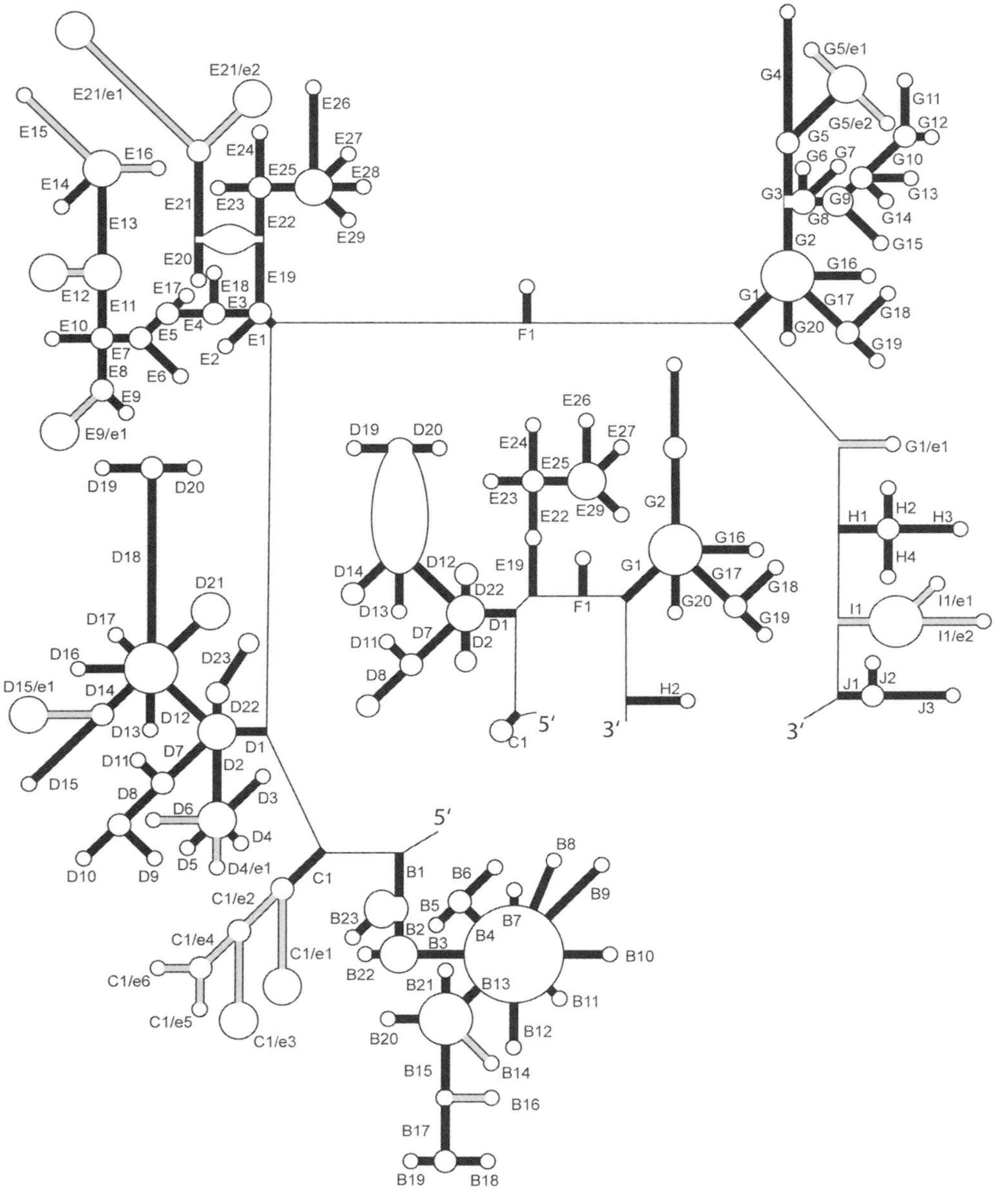

Fig. 4.10. Diagrammatic representations of the secondary structures of the nLSU and mtLSU rRNAs. The latter is shown in the centre of the figure. 'Universal' helices are shown in solid black. Helices present in eukaryotic expansion segments of neodermatans are not shaded. The mtLSU rRNA structure is based on that for *Fasciola hepatica* in Le *et al.* (2001). Diagram modified from one supplied by Dr Jan Wuyts.

(reviewed in Gerbi, 1996). Similarly, different numbering schemes for helices in both conserved and variable regions have been proposed. Gerbi (1996) has tried to reconcile these for expansion segments. In the nSSU Gerbi (1996) recognized 12 expansion segments, not all of which are present in neodermatans. The two largest in the nSSU of most neodermatans are ES6 of Gerbi (1996) (= V4 of De Rijk *et al.*, 1992) and ES9 of Gerbi (1996) (= V7). Cestodes have much longer expansion segments than other neodermatans (Fig. 4.1; Table 4.3) and additional, as yet undetermined, helices may be present in these.

Gerbi (1996) recognized 41 expansion segments in nLSU sequences. There has been little analysis of secondary structures and expansion segments in nLSU from neodermatans. The only published secondary structure model is that for *G. salaris* published by Matejusova and Cunningham (2004). Attempts to fit nLSU sequence of *S. mansoni* to this model met difficulties (D. Blair, unpublished data) and a revised model for domains B and D (terminology of Wuyts *et al.*, 2001a (broadly equivalent to D1-D3 of Michot *et al.*, 1984)), based on the general model for eukaryotes (Wuyts *et al.*, 2001a), is presented in Fig. 4.11. Although secondary structures can be deduced for domain C using programs such as MFOLD (Zuker, 2000) this region is so variable that comparative methods (Gutell *et al.*, 2002) are also required before we can be confident as to which helices occur in neodermatans. Most of the numerous available partial nLSU sequences for neodermatans (Table 4.2) span domains B-D, including the highly variable domain C.

Expansion segments accumulate mutations at 5–20 times the rate for the conserved core (reviewed in Gerbi, 1996 and see comments in Wuyts *et al.*, 2001b) and have therefore been a target for systematic studies on neodermatans at shallow phylogenetic levels. For example, Barker and Blair (1996), Kaukas and Rollinson (1997) and Blair *et al.* (1998) have used the nSSU V4 region in such studies on trematodes. Huyse *et al.* (2003) used this region for a species-level phylogeny within *Gyrodactylus*. The comments by Hancock and Vogler (2000), mentioned earlier, cautioning against unreserved use of such regions for phylogenetic work, need to be considered. However, at shallow taxonomic levels, such as among members of a family or order, expansion segments can be aligned with relative ease and will contain most of the variable and phylogenetically informative sites within the molecule. Blair and Barker (1993) pointed this

Table 4.3. Range of lengths of the nSSU of neodermatans and V4 and V7 expansion segments found in the molecule. Lengths of some nSSU rRNAs are slightly underestimated because conserved primer sites at each end are not always included in the GenBank accessions.

	Total length of nSSU	Length of V4	Length of V7	GenBank accession
Trematoda: Aspidobothrea				
Aspidogaster conchicola	1957	326	122	AJ287478
Lobatostoma manteri	1976	316	122	L16911
Trematoda: Digenea				
Opisthorchis viverrini	1992	324	116	X55357
Tetracerasta blepta	1973	316	117	L06670
Monogenea				
Gyrodactylus salaries	1965	335	100	Z26942
Cestoda				
Schizochoerus linguloideus[a]	2382	585	352	AF124454
Gigantolina magna	2873	474	845	AJ243681
Echinococcus granulosus	2394	560	222	U27015
Echinobothrium fautleyi[a]	1844	372	181	AF124464
Gyrocotyle rugosa[a]	2209	455	174	AF124455

[a]Data from Olson and Caira (1999).

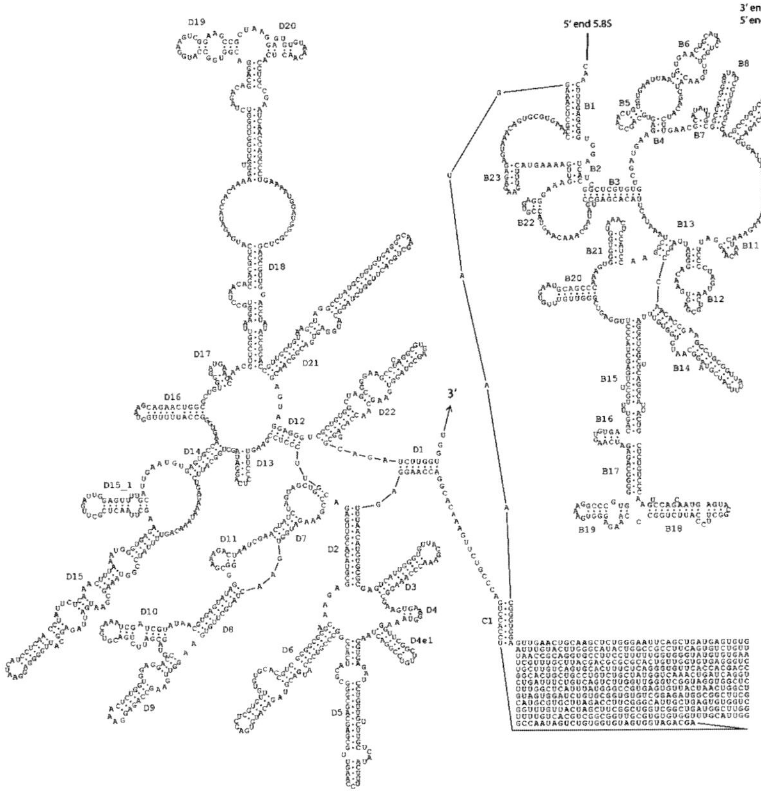

Fig. 4.11. Secondary structure model for the 5′ half of the nLSU of *Schistosoma mansoni*. This is based on the conservation diagram for yeast, *Saccharomyces cerevisiae* in Ben Ali *et al.* (1999) and the helix numbering of Wuyts *et al.* (2001a).

out for an alignment of SSU sequences from eight trematodes belonging to both subclasses (Digenea and Aspidobothrea). In an alignment of 1970 bp, 1107 bp were located in conserved core sites and of these only 104 were variable. The eight expansion segments combined contributed 863 sites of which 331 were variable.

Alignment of expansion segment sequences can be aided by knowledge of their secondary structures. Blair *et al.* (1998) investigated relationships among hemiuroid trematodes using sequences from the V4 region (ES6 of Gerbi, 1996). The secondary structure of this region is similar among all neodermatans and is shown for a monogenean and a cestode in Fig. 4.12. Although the V4 of the cestode *E. granulosus* has the same numbers of helices as that of the monogenean *G. salaris*, the lengths of these, and especially E23-1/2, differ considerably (Fig. 4.12). Among digeneans, almost all of the sequence variation occurs in the stems at the 5′ end of the region; the 3′ end, including two pseudoknots (Wuyts *et al.*, 2000), is almost invariant at this level. Blair *et al.* (1998) included the bases of stems E23-1/2, E23-5 and E23-6 in their alignment, but not the distal portions of these stems that were very variable in both sequence and length.

One of the expansion segments in the nLSU of at least some neodermatans is removed from the transcript before the mature rRNA molecules have been formed (Mertz *et al.*, 1991; Korbsrisate *et al.*, 1992). This short 'gap' region, located midway along the gene, varies in length (54–67 bp) among members of the

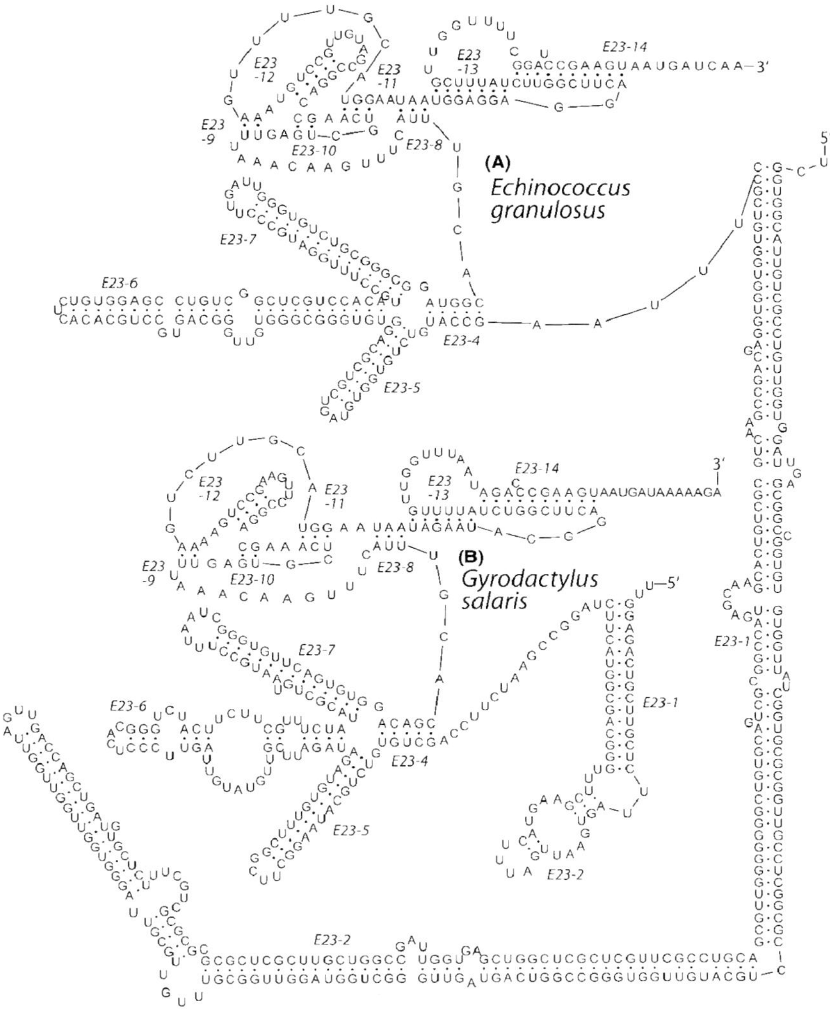

Fig. 4.12. Secondary structures for the nSSU V4 region of a cestode, *Echinococcus granulosus* (A), and a monogenean, *Gyrodactylus salaris* (B), showing the considerable differences in lengths of the helices in the 5′ half of the region. The figure for *G. salaris* is based partly on Cunningham *et al.* (2000).

genus *Schistosoma* (Mertz *et al.*, 1991). This is the only 'processed expansion segment' (Gerbi, 1996) seen in neodermatan rRNA although, as mentioned earlier, ITS2 could also be interpreted as one (Gerbi, 1996).

Mitochondrial Ribosomal Genes

Sequences of the entire coding portions of mitochondrial genomes are known for 12 species of trematodes and cestodes (McManus *et al.*, 2004). In all of these, the mtSSU and mtLSU genes are transcribed in the same direction and are separated only by a single tRNA gene.

Mitochondrial ribosomal genes are much smaller than even their bacterial homologues. Expansion segments found in nuclear orthologues are, of course, absent. Gerbi (1996) referred to 'contraction segments' as features found in the bacterium *E. coli* that are absent from other rRNAs, particularly mt rRNAs. The

SSU rRNA of *E. coli* is 1522 bp long and the LSU is 2904 bp. The mitochondrial equivalents in neodermatans are ~750 bp and ~1000 bp, respectively (Le *et al.*, 2001). The mt rRNAs of cestodes are similar in length to those of trematodes despite the considerable lengths of cestode nuclear rRNAs. Many of the 'universal' helices found in both prokaryotes and eukaryotes are absent or shorter than in bacteria or eukaryotes. Particularly in the case of the mtLSU rRNA, missing helices are among those identified as the most variable in bacterial LSU genes (Van de Peer *et al.*, 1996) (Figs 4.9 and 4.10). Despite the absence of the expansion segments that are so variable in their nuclear orthologs, the mt rRNAs exhibit plenty of variability. Inspection of an alignment of available complete mtSSU genes shows that a variability map of this gene in neodermatans would closely resemble that published for bacteria by Van de Peer *et al.* (1996) (latest version available at: http://www.psb.ugent.be/rRNA/varmaps/index.html). Le *et al.* (2001) presented a secondary structure for the mtLSU of *F. hepatica*. On this, they highlighted nucleotides also conserved in at least 90% of available molluscan mtLSU rRNAs (study by Lydeard *et al.*, 2000). Nucleotides conserved among 90% or more of available neodermatan mtLSU rRNAs map to almost exactly the same positions (D. Blair, unpublished data): there are only a few more sites conserved among neodermatans than are conserved between neodermatans and molluscs. Neodermatan-wide conserved nucleotides cluster mostly in helices D11, D19, E24, E25 and parts of G2 and G16 (Fig. 4.10).

Relatively few phylogenetic studies on neodermatans have used mt rRNA genes. The earliest study was by Després *et al.* (1992) who determined the sequence from the 3′ end of the mtLSU rRNA (3′ side of E22 through to G18) for several *Schistosoma* species. They estimated the rate of divergence of the partial mtLSU gene to be three times that of the nuclear ITS2 region. They also noted some intraspecific variation in *S. mansoni* and *Schistosoma bovis*. Morgan *et al.* (2003a) sequenced the same region for a larger number of *Schistosoma* species, but did not report results in detail. This region was also used by Donald *et al.* (2004) who noted up to 17.5% sequence difference between samples of a single cercarial morphotype from marine snails in New Zealand and Australia, suggesting the presence of cryptic species.

The mtSSU rRNA gene has also been used in a few phylogenetic studies. von Nickisch-Rosenegk *et al.* (1999) sequenced a portion of this gene from 21 species of cestodes spanning the 3′ side of helix 27 through to helix 48. As was the case with the mtLSU gene, some intraspecific variation was noted. Between samples of *Taenia taeniaeformis* the degree of difference was consistent with the presence of cryptic species. Variation in the same portion of the mtSSU gene among species and strains of *Echinococcus* was sufficient to enable Dinkel *et al.* (2004) to design PCR-based diagnostic tests for several species and strains. Attwood *et al.* (2002) also used the same region of the mtSSU gene in a phylogenetic study on *Schistosoma* species. They found the mtSSU sequences to be much more variable than the orthologous nSSU sequences. However, they did not compare the same regions of the orthologs, so interpretation should be cautious.

Hybrids

Hybridization plays a role in evolution and speciation (Arnold, 1997) and is known to occur in some neodermatans (e.g. *Schistosoma* spp. – reviewed in Rollinson *et al.*, 1990). In the F1 generation, ribosomal operons from each parent species should be represented. Rollinson *et al.* (1990) demonstrated this experimentally for various crosses of pairs of *Schistosoma* species: bands on Southern blots of genomic DNA from hybrids included those expected from both parent species. With the passing of generations, the processes of concerted evolution might be expected to homogenize the ribosomal operons, obscuring the patterns seen in the F1 generation (e.g. Hillis *et al.*, 1991 – parthenogenetic lizards). Nevertheless, there are several reports of neodermatans apparently bearing ribosomal sequences derived from two species. Parthenogenetic diploid and triploid forms of *Fasciola* in Japan and adjacent regions have long presented taxonomic difficulties (reviewed in Blair and McManus, 1989; Terasaki *et al.*,

2000). Early genetic studies suggested identity with *F. gigantica* (Blair and McManus, 1989; Adlard *et al.*, 1993; Hashimoto *et al.*, 1997). However, Itagaki and Tsutsumi (1998) found ITS2 sequences typical of both *F. gigantica* and *F. hepatica* in Japanese triploid worms, but not co-occurring in the same individual. They suggested a hybrid origin for Japanese worms. Agatsuma *et al.* (2000) sequenced ITS2 and a portion of the nLSU, as well as portions of mitochondrial protein-coding genes, from five individual worms from Korea. All had mitochondrial sequences typical of *F. gigantica*. However, for both ribosomal sequences, two worms were close to *F. gigantica*, one was close to *F. hepatica* and two had sequences of both types. Similarly, Huang *et al.* (2004) found both types of sequence present in individual fasciolids from NE China. The most likely explanation of these findings is hybridization between the two *Fasciola* species.

Patterns of repeats in the IGS region of a *G. salaris* strain specific to rainbow trout suggested to Cunningham *et al.* (2003) that this form might represent a hybrid between *G. salaris* and *Gyrodactylus thymalli*, although other explanations were also considered by Cunningham *et al.* (2003).

Morgan *et al.* (2003b) found evidence of natural hybridization between *S. mansoni* and *Schistosoma rodhaini*. The mtSSU and mtLSU sequences matched those of the former species whereas ITS1 and ITS2 sequences matched those of the latter. As ITS sequences of both parents were not present in the hybrid, this is probably not an F1 but rather a result of backcrossing of a hybrid form with *S. rodhaini*.

Concluding Remarks

Ribosomal operons contain genes with primary responsibility for one of the most fundamental processes within the cell: protein synthesis. They are, therefore, inevitably a source of fascination for biologists. Our knowledge of rDNA variation in neodermatans has come a long way since the first sequences were obtained about 20 years ago by Qu *et al.* (1986). This increased knowledge has been in step with information on the structure of the nuclear operon from other realms of life. Various, often antagonistic, forces such as natural selection, concerted evolution, gene conversion, unequal crossing over and replication slippage have forged a complex, dynamic and fascinating region that we are still very far from fully understanding. In particular, detection of the presence and appreciation of the nature of functional motifs in spacers present a challenge for the future. The roles, if any, of repeated sequences also await clarification. Portions of the operon, especially the nSSU and nLSU will continue to be used for taxonomic and phylogenetic work, as will the mitochondrial homologues of these genes. Increased understanding of their evolution and secondary (and higher-level) structure will substantially assist such work in the future. But we have come a long way in 20 years.

References

Adlard, R.D., Barker, S.C., Blair, D. and Cribb, T.H. (1993) Comparison of the second internal transcribed spacer (Ribosomal DNA) from populations and species of Fasciolidae (Digenea). *International Journal for Parasitology* 23, 423–425.

Agatsuma, T., Arakawa, Y., Iwagami, M., Honzako, Y., Cahyaningsih, U., Kang, S.Y. and Hong, S.J. (2000) Molecular evidence of natural hybridization between *Fasciola hepatica* and *F. gigantica*. *Parasitology International* 49, 231–238.

Alkemar, G. and Nygård, O. (2004) Secondary structure of two regions in expansion segments ES3 and ES6 with the potential of forming a tertiary interaction in eukaryotic 40S ribosomal subunits. *RNA* 10, 403–411.

Arnold, M.L. (1997) *Natural Hybridization and Evolution*. Oxford University Press, Oxford, UK.

Attwood, S.W., Upatham, E.S., Meng, X.H., Qiu, D.C. and Southgate, V.R. (2002) The phylogeography of Asian *Schistosoma* (Trematoda: Schistosomatidae). *Parasitology* 125, 99–112.

Barker, S.C. and Blair, D. (1996) Molecular phylogeny of *Schistosoma* species supports traditional groupings within the genus. *Parasitology* 82, 292–298.

Bell, A.S., Sommerville, C. and Valtonen, E.T. (2001) A molecular phylogeny of the genus *Ichthyocotylurus* (Digenea, Strigeidae). *International Journal for Parasitology* 31, 833–842.
Ben Ali, A., Wuyts, J., De Wachter, R., Meyer, A. and Van de Peer, A. (1999) Construction of a variability map for eukaryotic large subunit ribosomal RNA. *Nucleic Acids Research* 27, 2825–2831.
Blair, D. and Barker, S.C. (1993) Affinities of the Gyliauchenidae: utility of the 18S rRNA gene for phylogenetic inference in the Digenea (Platyhelminthes). *International Journal for Parasitology* 23, 527–532.
Blair, D. and McManus, D.P. (1989) Restriction enzyme mapping of ribosomal DNA can distinguish between fasciolid (liver fluke) species. *Molecular and Biochemical Parasitology* 36, 201–208.
Blair, D., Bray, R.A. and Barker, S.C. (1998) Molecules and morphology in phylogenetic studies of the Hemiuroidea (Digenea: Trematoda: Platyhelminthes). *Molecular Phylogenetics and Evolution* 9, 15–25.
Bowles, J., Blair, D. and McManus, D.P. (1995) A molecular phylogeny of the genus *Echinococcus*. *Parasitology* 110, 317–328.
Cable, J., Harris, P.D., Tinsley, R.C. and Lazarus, C.M. (1999) Phylogenetic analysis of *Gyrodactylus* spp. (Platyhelminthes: Monogenea) using ribosomal DNA sequences. *Canadian Journal of Zoology* 77, 1439–1449.
Caetano-Anollés, G. (2002) Tracing the evolution of RNA structure in ribosomes. *Nucleic Acids Research* 30, 2575–2587.
Capowski, E.E. and Tracy, J.W. (2003) Ribosomal RNA processing and the role of SmMAK16 in ribosome biogenesis in *Schistosoma mansoni*. *Molecular and Biochemical Parasitology* 132, 67–74.
Coleman, A.W. (2003) ITS2 is a double-edged tool for eukaryote evolutionary comparisons. *Trends in Genetics* 19, 370–375.
Collins, C.M. and Cunningham, C.O. (2000) Characterization of the *Gyrodactylus salaris* Malmberg, 1957 (Platyhelminthes: Monogenea) ribosomal intergenic spacer (IGS) DNA. *Parasitology* 121, 555–563.
Cribb, T.H., Anderson, G.R., Adlard, R.D. and Bray, R.A. (1998) A DNA-based demonstration of a three-host life cycle for the Bivesiculidae (Platyhelminthes: Digenea). *International Journal for Parasitology* 28, 1791–1795.
Criscione, C.D. and Blouin, M.S. (2004) Life cycles shape parasite evolution: comparative population genetics of salmon trematodes. *Evolution* 58, 198–202.
Cunningham, C.O., Aliesky, H. and Collins, C.M. (2000) Sequence and secondary structure variation in the *Gyrodactylus* (Platyhelminthes: Monogenea) ribosomal RNA gene array. *Journal of Parasitology* 86, 567–576.
Cunningham, C.O., Collins, C.M., Malmberg, G. and Mo, T.A. (2003) Analysis of ribosomal RNA intergenic spacer (IGS) sequences in species and populations of *Gyrodactylus* (Platyhelminthes: Monogenea) from salmonid fish in northern Europe. *Diseases of Aquatic Organisms* 57, 237–246.
De Rijk, P., Neefs, J.M., Van de Peer, Y. and De Wachter, R. (1992) Compilation of small ribosomal subunit RNA sequences. *Nucleic Acids Research* 20, 2075–2089.
Desdevises, Y., Jovelin, R., Jousson, O. and Morand, S. (2000) Comparison of ribosomal DNA sequences of *Lamellodiscus* spp. (Monogenea, Diplectanidae) parasitising *Pagellus* (Sparidae, Teleostei) in the North Mediterranean Sea: species divergence and coevolutionary interactions. *International Journal for Parasitology* 30, 741–746.
Desdevises, Y., Morand, S., Jousson, O. and Legendre, P. (2002) Coevolution between *Lamellodiscus* (Monogenea: Diplectanidae) and Sparidae (Teleostei): the study of a complex host–parasite system. *Evolution* 56, 2459–2471.
Després, L., Imbert-Establet, D., Combes, C. and Bonhomme, F. (1992) Molecular evidence linking hominoid evolution to recent radiation of schistosomes (Platyhelminthes: Trematoda). *Molecular Phylogenetics and Evolution* 1, 295–304.
Dinkel, A., Njoroge, E.M., Zimmermann, A., Walz, M., Zeyhle, E., Almahdi, I.E., Mackenstedt, U. and Romig, T. (2004) A PCR system for detection of species and genotypes of the *Echinococcus granulosus*-complex, with reference to the epidemiological situation in eastern Africa. *International Journal for Parasitology* 34, 645–653.
Donald, K.M., Kennedy, M., Poulin, R. and Spencer, H.G. (2004) Host specificity and molecular phylogeny of larval Digenea isolated from New Zealand and Australian topshells (Gastropoda: Trochidae). *International Journal for Parasitology* 34, 557–568.
Dover, G.A. and Tautz, D. (1986) Conservation and divergence in multigene families: alternatives to selection and drift. *Philosophical Transactions of the Royal Society of London Series B* 312, 275–289.
Dvorák, J., Horák, P., Hampl, V., Flegr, J. and Horák, P. (2002) Comparison of European *Trichobilharzia* species based on ITS1 and ITS2 sequences. *Parasitology* 124, 307–313.

Elder, J.F. and Turner, B.J. (1995) Concerted evolution of repetitive DNA sequences in eukaryotes. *Quarterly Review of Biology* 70, 297–320.

Fujiwara, H. and Ishikawa, H. (1987) Structure of the *Bombyx mori* rDNA: initiation site for its transcription. *Nucleic Acids Research* 15, 1245–1258.

Galazzo, D.E., Dayanandan, S., Marcogliese, D.J. and McLaughlin, J.D. (2002) Molecular systematics of some North American species of *Diplostomum* (Digenea) based on rDNA-sequence data and comparisons with European congeners. *Canadian Journal of Zoology* 80, 2207–2217.

Gerbi, S.A. (1996) Expansion segments: regions of variable size that interrupt the universal core secondary structure of ribosomal RNA. In: Dahlberg, A.E. and Zimmermann, R.A. (eds) *Ribosomal RNA: Structure, Evolution, Processing and Function in Protein Synthesis.* CRC Press, Boca Raton, Florida, pp. 71–87.

Gutell, R.R., Lee, J.C. and Cannone, J.J. (2002) The accuracy of ribosomal RNA comparative structure models. *Current Opinion in Structural Biology* 12, 301–310.

Hancock, J.M. and Armstrong, J.S. (1994) SIMPLE34: an improved and enhanced implementation for VAX and SUN computers of the SIMPLE algorithm for analysis of clustered repetitive motifs in nucleotide sequences. *Cabios* 10, 67–70.

Hancock, J.M. and Vogler, A.P. (2000) How slippage-derived sequences are incorporated into rRNA variable-region secondary structure: implications for phylogeny reconstruction. *Molecular Phylogenetics and Evolution* 14, 366–374.

Hancock, K., Broughel, D.E., Moura, I.N., Khan, A., Pieniazek, N.J., Gonzalez, A.E., Garcia, H.H., Gilman, R.H. and Tsang, V.C. (2001) Sequence variation in the cytochrome oxidase I, internal transcribed spacer 1 and Ts14 diagnostic antigen sequences of *Taenia solium* isolates from South and Central America, India, and and Asia. *International Journal for Parasitology* 31, 1601–1607.

Hashimoto, K., Watanobe, T., Liu, C.X., Init, I., Blair, D., Ohnishi, S. and Agatsuma, T. (1997) Mitochondrial DNA and nuclear DNA indicate that the Japanese *Fasciola* species is *F. gigantica*. *Parasitology Research* 83, 220–225.

Haukisalmi, V., Wickstrom, L.M., Hantula, J. and Henttonen, H. (2001) Taxonomy, genetic differentiation and Holarctic biogeography of *Paranoplocephala* spp. (Cestoda: Anoplocephalidae) in collared lemmings (*Dicrostonyx*; Arvicolinae). *Biological Journal of the Linnean Society* 74, 171–196.

Hillis, D.M., Moritz, C., Porter, C.A. and Baker, R.J. (1991) Evidence for biased gene conversion in concerted evolution of ribosomal DNA. *Science* 251, 308–310.

Hirai, H. (1988) *Paragonimus ohirai*: identification of nucleolar organizer regions (NORs) and silver nitrate staining pattern in spermatogenesis. *Experimental Parasitology* 67, 281–286.

Hirai, H., Spotila, L.D. and LoVerde, P.T. (1989) *Schistosoma mansoni*: chromosomal localization of DNA repeat elements by *in situ* hybridization using biotinylated DNA probes. *Experimental Parasitology* 69, 175–188.

Hirai, H., Hirata, M., Aoki, Y., Tanaka, M. and Imai, H.T. (1996) Chiasma analyses of the parasite flukes, *Schistosoma* and *Paragonimus* (Trematoda), by using the chiasma distribution graph. *Genes and Genetic Systems* 71, 181–188.

Huang, W.Y., He, B., Wang, C.R. and Zhu, X.Q. (2004) Characterisation of *Fasciola* species from Mainland China by ITS-2 ribosomal DNA sequence. *Veterinary Parasitology* 120, 75–83.

Huyse, T. and Volckaert, F.A.M. (2002) Identification of a host-associated species complex using molecular and morphometric analyses, with the description of *Gyrodactylus rugiensoides* n. sp (Gyrodactylidae, Monogenea). *International Journal for Parasitology* 32, 907–919.

Huyse, T., Audenaert, V. and Volckaert, F.A.M. (2003) Speciation and host–parasite relationships in the parasite genus *Gyrodactylus* (Monogenea, Platyhelminthes) infecting gobies of the genus *Pomatoschistus* (Gobiidae, Teleostei). *International Journal for Parasitology* 33, 1679–1689.

Itagaki, T. and Tsutsumi, K.I. (1998) Triploid form of *Fasciola* in Japan: genetic relationships between *Fasciola hepatica* and *Fasciola gigantica* determined by ITS-2 sequence of nuclear rDNA. *International Journal for Parasitology* 28, 777–781.

Iwagami, M., Ho, L.Y., Su, K., Lai, P.F., Fukushima, M., Nakano, M., Blair, D., Kawashima, K. and Agatsuma, T. (2000) Molecular phylogeographic studies on *Paragonimus westermani* in Asia. *Journal of Helminthology* 74, 315–322.

Johnston, D.A., Kane, R.A. and Rollinson, D. (1993) Small subunit (18S) ribosomal RNA gene divergence in the genus *Schistosoma*. *Parasitology* 107, 147–156.

Joseph, N., Krauskopf, E., Vera, M.I. and Michot, B. (1999) Ribosomal internal transcribed spacer 2 (ITS2) exhibits a common core of secondary structure in vertebrates and yeast. *Nucleic Acids Research* 27, 4533–4540.

Jousson, O. and Bartoli, P. (2000) The life cycle of *Opecoeloides columbellae* (Pagenstecher, 1863) n. comb, (Digenea, Opecoelidae): evidence from molecules and morphology. *International Journal for Parasitology* 30, 747–760.

Jousson, O. and Bartoli, P. (2001) Molecules, morphology and morphometrics of *Cainocreadium labracis* and *Cainocreadium dentecis* n. sp. (Digenea: Opecoelidae) parasitic in marine fishes. *International Journal for Parasitology* 31, 706–714.

Jousson, O., Bartoli, P. and Pawlowski, J. (1998) Molecular phylogeny of Mesometridae (Trematoda, Digenea) with its relation to morphological changes in parasites. *Parasite* 5, 365.

Jousson, O., Bartoli, P. and Pawlowski, J. (2000) Cryptic speciation among intestinal parasites (Trematoda: Digenea) infecting sympatric host species (Sparidae). *Journal of Evolutionary Biology* 13, 778–785.

Kane, R.A. and Rollinson, D. (1994) Repetitive sequences in the ribosomal DNA internal transcribed spacer of *Schistosoma haematobium*, *Schistosoma intercalatum* and *Schistosoma mattheei*. *Molecular and Biochemical Parasitology* 63, 153–156.

Kane, R.A. and Rollinson, D. (1998) Comparison of the intergenic spacers and 3′ end regions of the large subunit (28S) ribosomal RNA gene from three species of *Schistosoma*. *Parasitology* 117, 235–242.

Kane, R.A., Ridgers, I.L., Johnston, D.A. and Rollinson, D. (1996) Repetitive sequences within the first internal transcribed spacer of ribosomal DNA in schistosomes contain a Chi-like site. *Molecular and Biochemical Parasitology* 75, 265–269.

Kaukas, A. and Rollinson, D. (1997) Interspecific variation within the hypervariable region of the 18S ribosomal RNA gene among species of *Schistosoma* Weinland, 1858 (Digenea). *Systematic Parasitology* 36, 157–160.

Kaukas, A., Johnston, D.A., Kane, R.A. and Rollinson, D. (1994) Restriction enzyme mapping of ribosomal DNA of *Schistosoma spindale* and *S. leiperi* (Digenea) and its application to interspecific differentiation. *Systematic Parasitology* 27, 13–17.

Kedra, A.H., Swiderski, Z., Tkach, V.V., Dubinsky, P., Pawlowski, Z., Stefaniak, J. and Pawlowski, J. (1999) Genetic analysis of *Echinococcus granulosus* from humans and pigs in Poland, Slovakia and Ukraine. A mulicenter study. *Acta Parasitologica* 44, 248–254.

Korbsrisate, S., Mongkolsuk, S., Haynes, J.R., Wong, Q. and Sirisinha, S. (1992) Cloning and characterization of ribosomal RNA genes from *Opisthorchis viverrini*. *Parasitology* 104, 323–329.

Kostadinova, A., Herniou, E.A., Barrett, J. and Littlewood, D.T.J. (2003) Phylogenetic relationships of *Echinostoma* Rudolphi, 1809 (Digenea: Echinostomatidae) and related genera re-assessed via DNA and morphological analyses. *Systematic Parasitology* 54, 159–176.

Kumar, S., Tamura, K. and Nei, M. (2004) MEGA3: integrated software for molecular evolutionary genetics analysis and sequence alignment. *Briefings in Bioinformatics* 5, 150–163.

Lavikainen, A., Lehtinen, M.J., Meri, T., Hirvelä-Koski, V. and Meri, S. (2003) Molecular genetic characterization of the Fennoscandian cervid strain, a new genotypic group (G10) of *Echinococcus granulosus*. *Parasitology* 127, 207–215.

Le, T.H., Blair, D. and McManus, D.P. (2001) Complete DNA sequence and gene organization of the mitochondrial genome of the liver fluke, *Fasciola hepatica* L. (Platyhelminthes; Trematoda). *Parasitology* 123, 609–621.

León-Règagnon, V., Brooks, D.R. and Pérez-Ponce de León, G. (1999) Differentiation of Mexican species of Haematoloechus Looss, 1899 (Digenea: Plagiorchiformes): molecular and morphological evidence. *Journal of Parasitology* 85, 935–946.

Littlewood, D.T.J. and Olson, P.D. (2001) SSU rDNA and the Platyhelminthes: signal, noise, conflict and compromise. In: Littlewood, D.T.J. and Bray, R.A. (eds) *Interrelationships of the Platyhelminthes*. Taylor & Francis, London, pp. 262–278.

Littlewood, D.T., Rohde, K. and Clough, K.A. (1997) Parasite speciation within or between host species? Phylogenetic evidence from site-specific polystome monogeneans. *International Journal for Parasitology* 27, 1289–1297.

Liu, D., Kato, H. and Sugane, K. (1997) The nucleotide sequence and predicted secondary structure of small subunit (18S) ribosomal RNA from *Spirometra erinaceieuropaei*. *Gene* 184, 221–227.

Lo, C.M., Morgan, J.A.T., Galzin, R. and Cribb, T.H. (2001) Identical digeneans in coral reef fishes from French Polynesia and the Great Barrier Reef (Australia) demonstrated by morphology and molecules. *International Journal for Parasitology* 31, 1573–1578.

Lockyer, A.E., Olson, P.D. and Littlewood, D.T.J. (2003a) Utility of complete large and small subunit rRNA genes in resolving the phylogeny of the Neodermata (Platyhelminthes): implications and a review of the cercomer theory. *Biological Journal of the Linnean Society* 78, 155–171.

Lockyer, A.E., Olson, P.D., Østergaard, P., Rollinson, D., Johnson, D.A., Attwood, S.W., Southgate, V.R., Horak, P., Snyder, S.D., Le, T.H., Agatsuma, T., McManus, D.P., Carmichael, A.C., Naem, S. and Littlewood, D.T.J. (2003b) The phylogeny of the Schistosomatidae based on three genes with emphasis on the interrelationships of *Schistosoma* Weinland, 1858. *Parasitology* 126, 203–224.

Luo, H.Y., Nie, P., Zhang, Y.A., Wang, G.T. and Yao, W.J. (2002) Molecular variation of *Bothriocephalus acheilognathi* Yamaguti, 1934 (Cestoda: Pseudophyllidea) in different fish host species based on ITS rDNA sequences. *Systematic Parasitology* 52, 159–166.

Luton, K., Walker, D. and Blair, D. (1992) Comparisons of ribosomal internal transcribed spacers from 2 congeneric species of flukes (Platyhelminthes, Trematoda, Digenea). *Molecular and Biochemical Parasitology* 56, 323–327.

Lydeard, C., Holznagel, W.E., Schnare, M.N. and Gutell, R.R. (2000) Phylogenetic analysis of molluscan mitochondrial LSU rDNA sequences and secondary structures. *Molecular Phylogenetics and Evolution* 15, 83–102.

Mallatt, J. and Winchell, C.J. (2002) Testing the new animal phylogeny: first use of combined large-subunit and small-subunit rRNA gene sequences to classify the protostomes. *Molecular Biology and Evolution* 19, 289–301.

Mas-Coma, S., Funatsu, I.R. and Bargues, M.D. (2001) *Fasciola hepatica* and lymnaeid snails occurring at very high altitude in South America. *Parasitology* 123, 115–127.

Matejusova, I. and Cunningham, C.O. (2004) The first complete Monogenean ribosomal RNA gene operon: sequence and secondary structure of the *Gyrodactylus salaris* Malmberg, 1957, large subunit ribosomal RNA gene. *Journal of Parasitology* 90, 146–151.

Matejusova, I., Gelnar, M., Verneau, O., Cunningham, C.O. and Littlewood, D.T.J. (2003) Molecular phylogenetic analysis of the genus *Gyrodactylus* (Platyhelminthes: Monogenea) inferred from rDNA ITS region: subgenera versus species groups. *Parasitology* 127, 603–611.

McManus, D.P., Zhang, W.B., Li, J. and Bartley, P.B. (2003) Echinococcosis. *The Lancet* 362, 1295–1304.

McManus, D.P., Le, T.H. and Blair, D. (2004) Genomics of parasitic flatworms. *International Journal for Parasitology* 34, 153–158.

Mertz, P.M., Bobek, L.A., Rekosh, D.M. and LoVerde, P.T. (1991) *Schistosoma haematobium* and *S. japonicum* – analysis of the ribosomal RNA genes and determination of the gap boundaries and sequences. *Experimental Parasitology* 73, 137–149.

Michot, B., Hassouna, N. and Bachellerie, J.P. (1984) Secondary structure of mouse 28S rRNA and general model for the folding of the large rRNA in eukaryotes. *Nucleic Acids Research* 12, 4259–4279.

Michot, B., Després, L., Bonhomme, F. and Bachellerie, J.P. (1993) Conserved secondary structures in the ITS2 of trematode pre-rRNA. *FEBS Letters* 316, 247–252.

Morgan, J.A.T. and Blair, D. (1995) Nuclear rDNA ITS sequence variation in the trematode genus *Echinostoma*: an aid to establishing relationships with the 37-collar-spine group. *Parasitology* 111, 609–615.

Morgan, J.A.T. and Blair, D. (1998) Trematode and monogenean rRNA ITS2 secondary structures support a four-domain model. *Journal of Molecular Evolution* 47, 406–419.

Morgan, J.A.T., DeJong, R.J., Kazibwe, F., Mkoji, G.M. and Loker, E.S. (2003a) A newly identified lineage of *Schistosoma*. *International Journal for Parasitology* 33, 977–985.

Morgan, J.A.T., DeJong, R.J., Lwambo, N.J.S., Mungai, B.N., Mkoji, G.M. and Loker, E.S. (2003b) First report of a natural hybrid between *Schistosoma mansoni* and *S. rodhaini*. *Journal of Parasitology* 89, 416–418.

Niewiadomska, K. and Laskowski, K. (2002) Systematic relationships among six species of *Diplostomum* Nordmann, 1832 (Digenea) based on morphological and molecular data. *Acta Parasitologica* 47, 20–28.

Olson, P.D. and Caira, J.N. (1999) Evolution of the major lineages of tapeworms (Platyhelminthes: Cestoidea) inferred from 18S ribosomal DNA and elongation factor-1alpha. *Journal of Parasitology* 85, 1134–1159.

Olson, P.D., Littlewood, D.T.J., Griffiths, D., Kennedy, C.R. and Arme, C. (2002) Evidence for the co-existence of separate strains or species of *Ligula* in Lough Neagh, Northern Ireland. *Journal of Helminthology* 76, 171–174.

Olson, P.D., Yoder, K., Fajardo, L.F., Marty, A.M. and Van De Pas, S. (2003) Lethal invasive cestodiasis in immunosuppressed patients. *Journal of Infectious Diseases* 187, 1962–1966.

Omer Ali, P.O., Simpson, A.J., Allen, R., Waters, A.P., Humphries, C.J., Johnston, D.A. and Rollinson, D. (1991) Sequence of a small subunit rRNA gene of *Schistosoma mansoni* and its use in phylogenetic analysis. *Molecular and Biochemical Parasitology* 46, 201–208.

Pace, N.R., Olsen, G.J. and Woese, C.R. (1986) Ribosomal RNA phylogeny and the primary lines of evolutionary descent. *Cell* 45, 325–326.

Picón, M., Gutell, R.R., Ehrlich, R. and Zaha, A. (1996) Characterisation of a flatworm ribosomal RNA-encoding gene: promoter sequence and small subunit rRNA secondary structure. *Genetics* 171, 215–220.

Qu, L.H., Hardman, J.P., Gill, N., Chappell, L., Nicoloso, L. and Bachellerie, M. (1986) Phylogeny of helminths determined by rRNA sequence comparison. *Molecular and Biochemical Parasitology* 20, 93–99.

Rinder, H., Rausch, R.L., Takahashi, K., Kopp, H., Thomschke, A. and Löscher, T. (1997) Limited range of genetic variation in *Echinococcus multilocularis*. *Journal of Parasitology* 83, 1045–1050.

Rollinson, D., Walker, T.K., Knowles, R.J. and Simpson, A.J. (1990) Identification of schistosome hybrids and larval parasites using rRNA probes. *Systematic Parasitology* 15, 65–74.

Ruiz-Trillo, I., Riutort, M., Littlewood, D.T., Herniou, E.A. and Baguñà, J. (1999) Acoel flatworms: earliest extant bilaterian metazoans, not members of Platyhelminthes. *Science* 283, 1919–1922.

Safrany, G. and Hidvegi, E.J. (1989) New tandem repeat unit in the non-transcribed spacer of human ribosomal RNA gene. *Nucleic Acids Research* 17, 3013–3036.

Silva, S.C., Rodrigues, J.J.S. and Zaha, A. (1991) Molecular cloning and characterization of the ribosomal RNA genes of the cestode *Echinococcus granulosus*. *Brazilian Journal of Medical and Biological Research* 24, 345–357.

Simkova, A., Morand, S., Jobet, E., Gelnar, M. and Verneau, O. (2004) Molecular phylogeny of congeneric monogenean parasites (*Dactylogyrus*): a case of intra-host speciations. *Evolution* 58, 1001–1008.

Simpson, A.J., Dame, J.B., Lewis, F.A. and McCutchan, T.F. (1984) The arrangement of ribosomal RNA genes in *Schistosoma mansoni*: identification of polymorphic structural variants. *European Journal of Biochemistry* 139, 41–45.

Snyder, S.D. and Tkach, V.V. (2001) Phylogenetic and biogeographical relationships among some holarctic frog lung flukes (Digenea: Haematoloechidae). *Journal of Parasitology* 87, 1433–1440.

Sorensen, R.E., Curtis, J. and Minchella, D.J. (1998) Intraspecific variation in the rDNA ITS loci of 37-collar-spined echinostomes from North America: implications for sequence-based diagnoses and phylogenetics. *Journal of Parasitology* 84, 992–997.

Tautz, D. and Renz, M. (1984) Simple sequences are ubiquitous repetitive components of eukaryotic genomes. *Nucleic Acids Research* 12, 4127–4138.

Tautz, D., Trick, M. and Dover, G.A. (1986) Cryptic simplicity in DNA is a major source of genetic variation. *Nature* 322, 652–656.

Telford, M.J., Lockyer, A.E., Cartwright-Finch, C. and Littlewood, D.T.J. (2003) Combined large and small subunit ribosomal RNA phylogenies support a basal position of the acoelomorph flatworms. *Proceedings of the Royal Society of London Series B* 270, 1077–1083.

Terasaki, K., Noda, Y., Shibahara, T. and Itagaki, T. (2000) Morphological comparisons and hypotheses on the origin of polyploids in parthenogenetic *Fasciola* sp. *Journal of Parasitology* 86, 724–729.

Van de Peer, Y., Chapelle, S. and De Wachter, R. (1996) A quantitative map of nucleotide substitution rates in bacterial rRNA. *Nucleic Acids Research* 24, 3381–3391.

van Herwerden, L., Blair, D. and Agatsuma, T. (1998) Intra- and interspecific variation in nuclear ribosomal internal transcribed spacer 1 of the *Schistosoma japonicum* species complex. *Parasitology* 116, 311–317.

van Herwerden, L., Blair, D. and Agatsuma, T. (1999) Intra- and interindividual variation in ITS1 of *Paragonimus westermani* (Trematoda: Digenea) and related species: implications for phylogenetic studies. *Molecular Phylogenetics and Evolution* 12, 67–73.

van Herwerden, L., Gasser, R.B. and Blair, D. (2000) ITS-1 ribosomal DNA sequence variants are maintained in different species and strains of *Echinococcus*. *International Journal for Parasitology* 30, 157–169.

van Herwerden, L., Caley, M.J. and Blair, D. (2003) Regulatory motifs are present in the ITS1 of some flatworm species. *Journal of Experimental Zoology Part B – Molecular and Developmental Evolution* 296B, 80–86.

Vartapetian, A.B. (1994) DNA recombination in the course of PCR. In: Griffin, H.G. and Griffin, A.M. (eds) *PCR Technologies: Current Innovations*. CRC Press, Boca Raton, Florida, pp. 215–218.

Verneau, O., Bentz, S., Sinnappah, N.D., du Preez, L., Whittington, I. and Combes, C. (2002) A view of early vertebrate evolution inferred from the phylogeny of polystome parasites (Monogenea: Polystomatidae). *Proceedings of the Royal Society of London Series B* 269, 535–543.

von der Schulenburg, J.H.G., Englisch, U. and Wägele, J.W. (1999) Evolution of ITS1 rDNA in the Digenea (Platyhelminthes: Trematoda): 3′ end sequence conservation and its phylogenetic utility. *Journal of Molecular Evolution* 48, 2–12.

von Nickisch-Rosenegk, M., Lucius, R. and Loos-Frank, B. (1999) Contributions to the phylogeny of the Cyclophyllidea (Cestoda) inferred from mitochondrial 12S rDNA. *Journal of Molecular Evolution* 48, 586–596.

Walker, T.K., Rollinson, D. and Simpson, A.J. (1986) Differentiation of *Schistosoma haemotobium* from related species using cloned ribosomal RNA gene probes. *Molecular and Biochemical Parasitology* 20, 123–131.

Wuyts, J., De Rijk, P., Van de Peer, Y., Pison, G., Rousseeuw, P. and De Wachter, R. (2000) Comparative analysis of more than 3000 sequences reveals the existence of two pseudoknots in area V4 of eukaryotic small subunit ribosomal RNA. *Nucleic Acids Research* 28, 4698–4708.

Wuyts, J., De Rijk, P., Van de Peer, Y., Winkelmans, T. and Wachter, R.D. (2001a) The European large subunit ribosomal RNA database. *Nucleic Acids Research* 29, 175–177.

Wuyts, J., Van de Peer, Y. and Wachter, R.D. (2001b) Distribution of substitution rates and location of insertion sites in the tertiary structure of ribosomal RNA. *Nucleic Acids Research* 29, 5017–5028.

Zietara, M.S. and Lumme, J. (2002) Speciation by host switch and adaptive radiation in a fish parasite genus *Gyrodactylus* (Monogenea, Gyrodactylidae). *Evolution* 56, 2445–2458.

Zietara, M.S., Huyse, T., Lumme, J. and Volckaert, F.A. (2002) Deep divergence among subgenera of *Gyrodactylus* inferred from rDNA ITS region. *Parasitology* 124, 39–52.

Zuker, M. (2000) Calculating nucleic acid secondary structure. *Current Opinion in Structural Biology* 10, 303–310.

5 Genetic Studies on Monogeneans with Emphasis on *Gyrodactylus*

C.O. Cunningham and I. Matějusová
FRS Marine Laboratory, PO Box 101, 375 Victoria Road, Aberdeen AB11 9DB, UK

Introduction

Our knowledge of the molecular biology of monogeneans is currently lagging some way behind that of digeneans or cestodes, described in other chapters in this volume. Nevertheless, this group presents a number of characteristics that make them attractive candidates for molecular study. No doubt we shall see significant developments in our knowledge of the immunology, physiology and biochemistry of monogeneans in the near future. Many of the studies on the biology of other flatworms will guide the experiments required to analyse monogeneans, as the molecular studies carried out to date have been enhanced by previous work on other platyhelminths.

The first studies on the molecular biology of monogeneans could be said to be the determination of partial small subunit (SSU) ribosomal RNA (rRNA) genes, or 18S genes, from *Dictyocotyle coeliaca*, *Diclidophora merlangi* and *Anoplodiscus cirrusspiralis* by Baverstock *et al.* (1991). These sequences were obtained and used in phylogenetic analysis of a range of parasitic platyhelminths. Sequences for use in phylogenetic analysis form the majority of the over 700 monogenean nucleotide sequences available in GenBank at the time of writing.

The genetic characterization of monogeneans began in earnest in the 1990s with the search for molecular markers of species. This work continues with a variety of monogenean families and now encompasses a wide range of techniques and target regions of the genome. This chapter provides a brief overview of the development of molecular characterization of monogeneans, notable successes and some current work. Many of the techniques applied have been used with other classes or phyla. Molecular research on Monogenea has been dominated by a focus on the Gyrodactylidae, and therefore this chapter necessarily concentrates on the genus *Gyrodactylus*. Many of the methods used in molecular studies of Monogenea have been described by Gasser (2001) in another book in this series, so here, only their current and future applications are included.

The first intensive molecular genetic study of monogenean species began in 1991 with the aim of developing novel and objective methods of identifying *Gyrodactylus salaris* Malmberg, 1957. *G. salaris* had caused serious losses of wild Atlantic salmon in Norway since its introduction to that country in the 1970s (Mo, 1994). It was listed in the Office International des Epizooties (OIE) diseases of fish and European Community (EC) Fish Health legislation. To operate systems such as the EC regulations effectively and protect fish health while not unnecessarily impeding trade, unambiguous methods of diagnosing pathogens are required. The identification of some monogenea is notoriously difficult. Malmberg (1970) proposed the use of the excretory system and the hard parts of the attachment apparatus located on the haptor for *Gyrodactylus* species discrimination (see Fig. 5.1A). However, difficulties have been experienced with some closely related species, particularly those found on salmonid fish. For example, *G. salaris* exhibits an extremely wide range of morphology and measurements within the haptoral hard parts, related mainly to seasonal changes in water temperature (Malmberg, 1987; Mo, 1991a,b,c; Malmberg and Malmberg, 1993). Moreover, this species is very difficult to discriminate from *Gyrodactylus thymalli* Zitnan, 1960 on the basis of the shape and size of haptoral hard parts, and for these reasons work was instigated to develop alternative methods of identifying *G. salaris*.

Difficulties in Molecular Studies of Monogenea

The generally small size of monogenea is not in itself a major problem for molecular techniques. After all, polymerase chain reaction (PCR) can amplify DNA from tiny amounts of starting material. It has proven possible to amplify sufficient DNA from even partial gyrodactylids. However, the small size of specimens does present some complications. Many studies of genetic characteristics of monogenea are currently producing molecular markers for species previously identified by morphological characteristics. In order to connect sequence and morphology definitively in a single description, the ideal situation is to have microscopic and molecular examination of the same specimen. This is possible via dissection or digestion of the monogenean (Cunningham *et al.*, 1995b; Harris *et al.*, 1999), but the additional manipulations increase the risk of contaminating or losing what may be precious specimens.

Contamination may also arise from host tissue within or attached to the specimen. Host DNA may be amplified alongside that from the monogenean in reactions such as random amplified polymorphic DNA (RAPD) and can complicate interpretation of results, particularly if only very small amounts of monogenean material are available and the proportion of host nucleic acid in the sample is therefore rather large. This problem is of course greatly reduced when methods with greater stringency are applied, using primers or probes with specificity for at least the phylum Platyhelminthes.

Combined molecular and morphological analysis involves compromise, usually in the preservative solution used. Molecular analyses perform best on ethanol-preserved material, but the alcohol renders the specimens rather hard and suboptimal for microscopic examination. The use of solutions that assist subsequent microscopic examinations, such as formalin, can result in specimens that are difficult or impossible to use for molecular work, so the

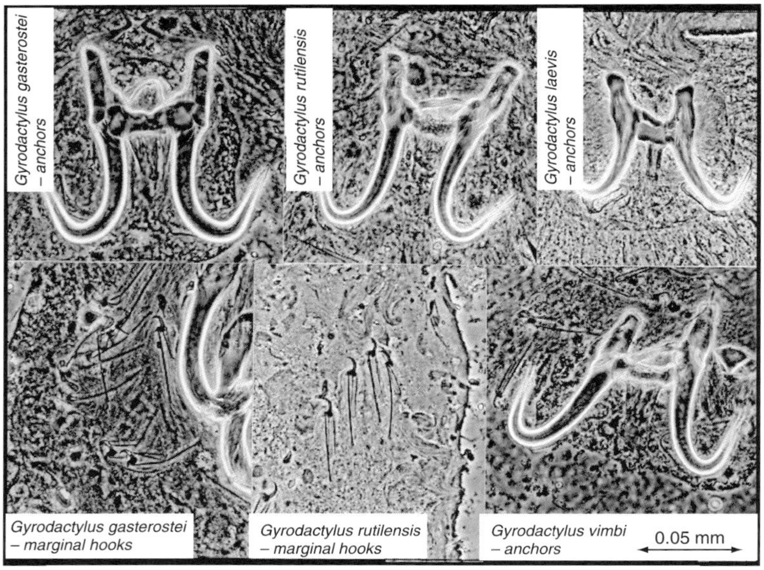

(A)

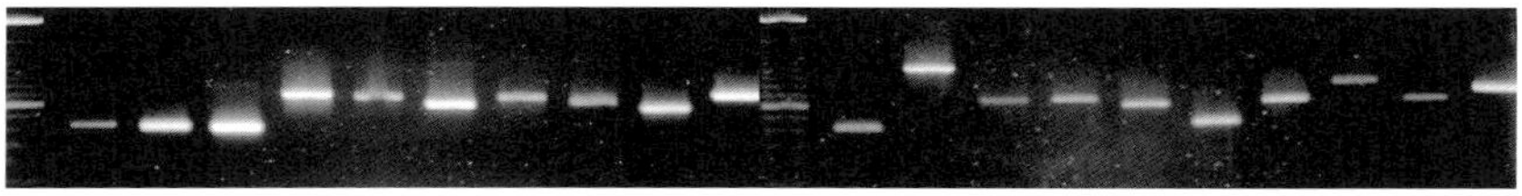

(B)

Fig. 5.1. Hard parts of the opisthaptor of *Gyrodactylus* species viewed under phase contrast microscopy (A) and agarose gel electrophoresis of ribosomal RNA (rRNA) gene internal transcribed spacer 1 (ITS1) (B) from: 1, *Gyrodactylus carassii*; 2, *G.* sp.1; 3, *G. elegans*; 4, *G. gobiensis*; 5, *G. gobii*; 6, *G. gracilihamatus*; 7, *G. gurleyi*; 8, *G. hronosus*; 9, *G. katharineri*; 10, *G. kobayashii*; 11, *G. laevis*; 12, *G. lomi*; 13, *G. longoacuminatus*; 14, *G.* sp.2; 15, *G. macronychus*; 16, *G. markakulensis*; 17, *G. rhodei*; 18, *G.* sp.3; 19, *G. rutilensis*; 20, *G. vimbi* (M-100 bp DNA ladder), illustrating the diversity in morphological and genetic features used in species discrimination.

compromise in favour of nucleic acid preservation is necessary.

Difficulties in the morphological identification of parasites from other monogenean groups are expected because recent preservation methods favour molecular analyses. Alcohol preservation of larger monogeneans such as diplozoids increases the difficulty in species identification because the attachment clamps used for species discrimination of such preserved individuals are deeply embedded in parasite tissue and the accuracy of species identification is reduced.

Molecular Markers for Monogenean Species

The rRNA genes and spacers

The region of the genome encoding rRNA genes and spacers has a common arrangement in eukaryotes (Long and Dawid, 1980) and the genes and RNA occur at high abundance in cells (Waters and McCutchan, 1990). Within the rRNA there are conserved regions that contain very similar nucleotide sequences in all the eukaryotes studied, interspersed with regions that show considerable sequence variation, even between closely related organisms (Gerbi *et al.*, 1982; Hillis and Dixon, 1991). The rRNA genes are suitable for the purposes of species probe or primer development from organisms that have not previously been characterized at the molecular level.

The SSU (18S) rRNA gene

As mentioned above, partial 18S rRNA gene sequences were obtained from some monogeneans by Baverstock *et al.* (1991) in phylogenetic analysis of platyhelminths. Molecular investigations of monogeneans specifically began with the PCR amplification of the SSU, or 18S, rRNA gene from DNA extracted from a pool of many *G. salaris* specimens. At that time, this gene had been used to develop methods of identifying other parasite groups (Cameron *et al.*, 1988; Blair and McManus, 1989; Rollinson *et al.*, 1990). The 18S gene sequence from *G. salaris* was determined and was used to examine the phylogenetic relationship of this species to other platyhelminths (Cunningham *et al.*, 1995a). As this 18S sequence was being completed, work on the most closely related organism to *Gyrodactylus* that had been studied in any detail using molecular methods (i.e. *Schistosoma* spp.) showed that one region of the 18S gene, V4, was more variable than others (Johnston *et al.*, 1993). The conservation of sequence in regions of the 18S gene enabled the use of alignments of the nucleotide sequences to identify the homologous V4 region of the 18S gene to be identified in *G. salaris*. Methods were developed to amplify and analyse this region from even single specimens of *Gyrodactylus* and sequences were obtained from *Gyrodactylus truttae* and *G. derjavini*, two other species commonly recorded on salmonids in Europe. These sequences did indeed show species variation and the V4 region was used in restriction fragment length polymorphism (RFLP), oligonucleotide probe and DNA sequence methods of distinguishing these three species (Cunningham *et al.*, 1995b,c). Later, the probes were shown to produce potentially confusing results, with cross reaction of the probe designed to detect *G. salaris* V4 DNA with the V4 of *Gyrodactylus teuchis* (Cunningham *et al.*, 2001).

More recently, the complete 18S genes have been determined from *Gyrodactylus gobiensis*, *G. rhodei*, *G. rutilensis*, *G. carassi*, *G. sedelnikovi*, *Gyrodactyloides bychowskii* and *Macrogyrodactylus polypteri*. They were used to identify the sister group to *Gyrodactylus*, and found that *G. bychowskii* was resolved as the sister group to *Gyrodactylus* and *Macrogyrodactylus* (Matějusová *et al.*, 2003) (Fig. 5.2). The 18S sequences have also been used in phylogenetic analysis of other Monogenea (Littlewood *et al.*, 1998; Desdevises, 2001; Littlewood and Olson, 2001; Verneau *et al.*, 2002; Bentz *et al.*, 2003; Simková *et al.*, 2003, 2004).

The rRNA internal transcribed spacer (ITS)

Although the variable region V4 of the 18S rRNA gene proved useful for distinguishing *G. salaris*, *G. derjavini* and *G. truttae*, it was not suitable for separating *G. salaris* and *G. thymalli*, which have the same nucleotide sequence in this region of the gene. Therefore work began to analyse a more variable part of the rRNA gene operon, the ITS, including the 5.8S gene. The ITS also proved useful for identifying *G. salaris*, *G. derjavini* and *G. truttae* by way of RFLP or sequence, but again sequences from *G. salaris* and *G. thymalli* were identical (Cunningham, 1997; Cunningham *et al.*, 2000).

The development of molecular markers that could be used to distinguish *Gyrodactylus* species more objectively than by microscopic examination has been the aim of a large number of studies (Harris and Cable, 2000; Ziętara *et al.*, 2000, 2002; Cunningham *et al.*, 2001; Matějusová *et al.*, 2001a; Huyse and Volckaert, 2002; Ziętara and Lumme, 2002, 2003; Huyse *et al.*, 2003, 2004; Huyse and Malmberg, 2004). This work has diverged from members of the genus *Gyrodactylus* only, and ITS sequences from Diplozoinae (Matějusová *et al.*, 2001b, 2004), *Gyrdicotylus* (Cable *et al.*, 1999b) and *Gyrodactyloides* (Bruno *et al.*, 2001) have also become available through their application as species markers.

Experience in analysis of many specimens of *Gyrodactylus* from various host species,

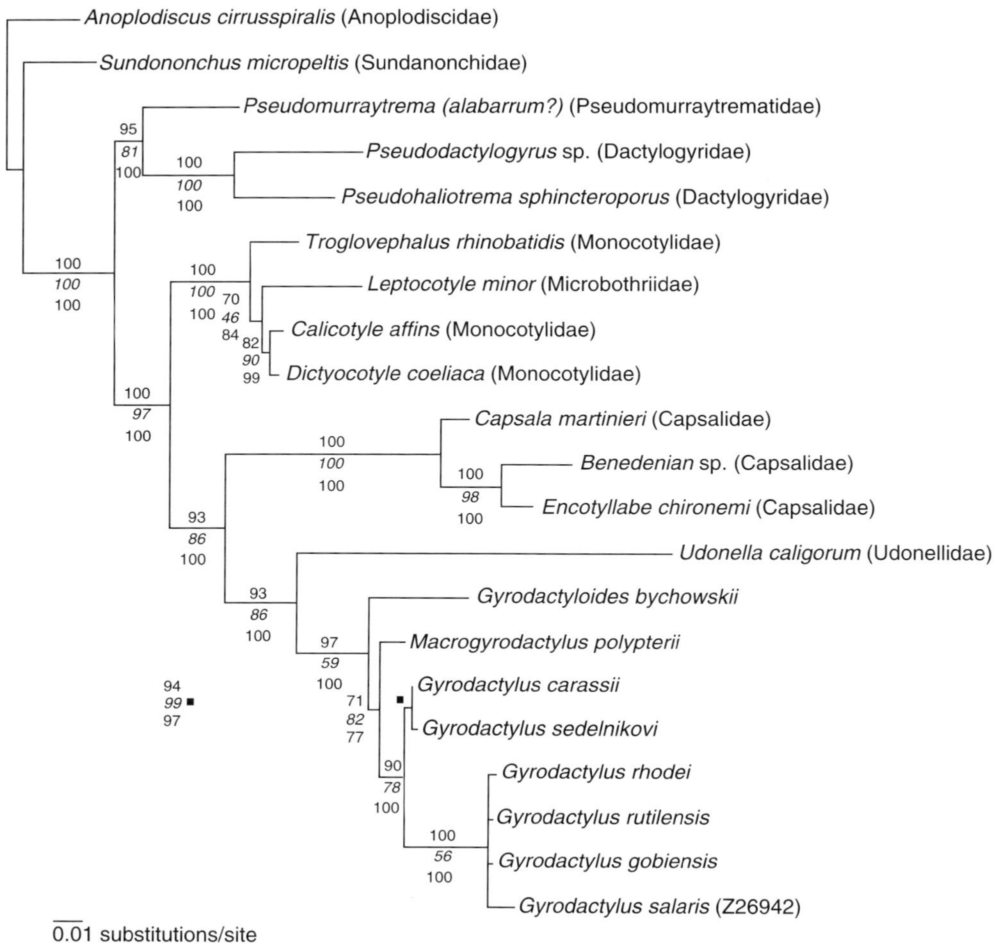

Fig. 5.2. Phylogeny of monopisthocotylean Monogenea based on SSU rDNA. The tree topology is from a Bayesian analysis with nodal support indicated, from top to bottom, for maximum likelihood (bootstrap%, n = 100), maximum parsimony (bootstrap%, n = 1000) and Bayesian inference (posterior probabilities). Figure from Matějusová *et al.* (2003).

varied locations and sampling times, has shown that the ITS is a stable region of the genome, with intra-specific variation of less than five nucleotides. This variation is of the same order as expected between different sequencing methodologies or laboratories. Thus, ITS sequence appears to be a robust species marker. Sequencing this region of the genome is specified, in preference to analysis by RFLP with single restriction enzyme or probe hybridization. This is in the light of experience with application of short DNA probes to the 18S rRNA of *Gyrodactylus*. A probe designed to hybridize to DNA of *G. salaris* also binds to the same region of *G. teuchis* DNA and this cross reaction could lead to erroneous identification (Cunningham *et al.*, 2001). The ITS, with greater length and variability of sequence than the 18S V4 region, presents less opportunity for misidentification in sequence analysis. The inclusion of ITS sequence along with morphological analysis is a useful addition to new *Gyrodactylus* species descriptions (Cunningham *et al.*, 2001; Ziętara and Lumme, 2003; Huyse and Malmberg, 2004).

The ITS region is useful to reveal species differences (Fig. 5.1B) but in the case of sequencing being the diagnostic tool, it becomes costly as additional sequencing reactions with internal primers are usually necessary to ensure clear overlap between forward and reverse reactions. Thus, only partial ITS sequences, the second ITS (ITS2) rDNA, were used to differentiate species of another group of monogeneans, members of the Diplozoinae. Molecular markers based on the ITS2 region only were developed for eight species of diplozoids from three genera: *Diplozoon*, *Eudiplozoon* and *Paradiplozoon* (Matějusová *et al.*, 2001b, 2004). However in this case, RFLP with multiple enzymes has to be applied for species identification, as the ITS2 region is relatively conserved and it becomes more difficult to obtain species-specific restriction patterns for closely related species.

RFLP can also be used to differentiate the variant of *G. salaris* described by Lindenstrøm *et al.* (2003b). Interestingly, this variant displayed heterogeneity in the ITS of the rRNA genes, varying at three positions compared to *G. salaris sensu* Malmberg, 1957.

There is still the possibility that confusion could arise if oligonucleotide probes or RFLP are the only techniques applied, as these methods only examine a very small portion of the ITS. Therefore, nucleotide sequencing of the ITS is at present the most reliable molecular marker of monogenean species.

To date, the ITS is the most commonly sequenced part of the monogenean genome. In addition to the application of ITS sequences to develop species markers, there are an increasing number of studies where the ITS is used to infer phylogeny. This region has been proven valuable within *Gyrodactylus* (Cable *et al.*, 1999b; Ziętara and Lumme, 2002, 2003; Ziętara *et al.*, 2002; Kritsky and Boeger, 2003; Matějusová *et al.*, 2003) and has demonstrated the potential to resolve the phylogeny of other monogenean groups such as diplozoids (Sicard *et al.*, 2001, 2002; Matějusová *et al.*, 2004) or polystomes (Bentz *et al.*, 2001).

The rRNA intergenic spacer (IGS)

The intergenic spacer includes both non-transcribed and external transcribed spacers of rRNA. Compared to the ITS region, the IGS is poorly characterized within the Platyhelminthes and the only known IGS of monogeneans is that of *G. salaris* (Collins and Cunningham, 2000). Characterization of the *G. salaris* intergenic spacer revealed the presence of repetitive DNA resembling minisatellites. Within these repetitive regions, variations in sequence were found that distinguish *G. salaris* and *G. thymalli* (Sterud *et al.*, 2002). Length heterogeneity in the IGS has been commonly observed for other parasite groups (Kane and Rollinson, 1998) and various free-living organisms (Morales-Hojas *et al.*, 2002), and further IGS data from other species may reveal if this also occurs in monogeneans.

Large subunit (LSU) (28S) rRNA gene

The LSU rRNA gene of *G. salaris* has now been characterized, and together with the regions of rRNA discussed above, provides the first complete sequence of the rRNA operon for the Monogenea (Matějusová and Cunningham, 2004). As there are no other 28S gene sequences available from species of *Gyrodactylus* at present, the potential use of this region for species molecular markers within this genus is only suggested based on the experience from other platyhelminth groups. The 28S gene has mainly been studied to infer phylogeny of various parasitic groups, and for this purpose partial sequences of this gene are sufficient. The D1-D3 regions are most frequently used to infer phylogeny, and these regions correspond to the B, C and D domains of the 28S defined for *Saccharomyces cerevisiae* (Wuyts *et al.*, 2001). This region of the 28S gene has been sequenced from members of the families Ancyrocephalidae, Dactylogyridae, Pseudodactylogyridae and Tetraonchidae (I. Matějusová, FRS Marine Laboratory, 2004, unpublished data). When the secondary structure of *G. salaris* and *S. cerevisiae* 28S genes were compared, domains C and H were revealed to be variable regions and therefore could potentially be used in phylogenetic analyses and also for the development of molecular markers for closely related species (Matějusová and Cunningham, 2004).

The 28S gene, often a partial sequence, has been widely used in phylogenetic analysis within the Monogenea (Mollaret *et al.*, 1997, 2000a,b; Litvaitis and Rohde, 1999; Chisholm *et al.*, 2001; Jovelin and Justine, 2001; Justine *et al.*, 2002; Olson and Littlewood, 2002; Lockyer *et al.*, 2003; Whittington *et al.*, 2004).

Other Regions of the Monogenean Genome

Other regions of the monogenean genome are poorly studied compared to the genome of some human parasite groups. The main driving force to study other genes has so far been the search for molecular species markers or regions suitable for phylogenetic analysis. The elongation factor 1-α gene of *Neomicrocotyle pacifica* was sequenced for use in phylogenetic analysis (Littlewood *et al.*, 2001) but has not yet been used as a species marker.

The β-tubulin gene of *G. salaris* has been examined (Collins *et al.*, 2004). This gene has previously been described as a molecular species marker in *Echinococcus, Babesia* and *Theileria* (Brehm *et al.*, 2000; Cacciò *et al.*, 2000; Robinson *et al.*, 2001). As this gene has only been sequenced from *G. salaris* to date, its potential as a species marker for monogeneans is not yet known, but the sequence of this gene is thought to be less variable than regions of the ribosomal genes or spacers. Interesting questions remain regarding the role of the β-tubulin gene as a tool for monitoring anthelminthic resistance and its potential role in this resistance has still to be studied (see 'Concluding Remarks').

The cytochrome oxidase subunit I (COI) gene of the mitochondrial genome has been isolated and sequenced from *G. salaris* and *G. thymalli* (Meinilä *et al.*, 2002; Hansen *et al.*, 2003). The COI gene was proven to be polymorphic and different haplotypes were found in both species. Monophyletic origins of the COI variants for *G. salaris* and *G. thymalli* were not confirmed. This gene has also been used to study the population genetics of *G. salaris* (see section on 'Molecular Analysis within Monogenean Species').

Mitochondrial COI and 16S rRNA sequences have been used in phylogenetic analysis (Littlewood *et al.*, 1997; Telford *et al.*, 2000; Plaisance *et al.*, 2005).

A project aimed at sequencing the remainder of the mitochondrial genome of the genus *Gyrodactylus* and determining variable regions for further intraspecific analyses is underway (D.T.J. Littlewood, Natural History Museum, London, personal communication).

Monogenean Species Concept and Species Complexes

Monogenean species descriptions have hitherto relied upon information on the host, geographical sampling location and microscopic examination and comparison with drawings from other authors. Because of this, and the diversity of morphological characters found in *G. salaris*, it was expected that molecular investigation might result in the synonymization of forms previously described as separate species. To our surprise, the converse has occurred. Based on sequence of the V4 region of the 18S rRNA gene and the ITS, *G. teuchis* was separated from *G. salaris* (Cunningham *et al.*, 2001) although initial morphological examination considered them the same species.

The traditional species concept is based on differences in morphological or life-history characteristics. However, the existence of so-called cryptic species is common and can be resolved by several genetic or molecular biology methods. Allozyme electrophoresis has been used but largely superseded by faster and more accurate approaches based on differences in various regions of the ribosomal or mitochondrial DNA to detect cryptic species in parasitic groups (Chilton *et al.*, 1995; Overstreet *et al.*, 2002). The use of nucleic acid techniques rather than allozymes is an obvious advantage in analysis of monogeneans, when their small size often prevents allozyme analysis of individual specimens. The monogenean parasites are no exception to the occurrence of cryptic species, as illustrated in the existence of *G. teuchis* mentioned above. With recent extensive sampling and sequencing of *Gyrodactylus* parasites from a wide range of fish, this could become a serious problem. Even species with apparently clear morphological characteristics have caused difficulties when sequences have been obtained following morphological identification of species. The discovery of more than one distinctly different sequence from specimens that are extremely similar in morphology reveals the existence of more cryptic species. This has been found in *Gyrodactylus macronychus*-like, *Gyrodactylus longoacuminatus*-like and *Gyrodactylus jiroveci*-like species groups (Ziętara and Lumme, 2002, 2003; I. Matějusová, FRS Marine Laboratory, Aberdeen, 2004, unpublished data).

It is apparent that the application of molecular biology to species redescriptions or to complement descriptions based on morphology is not always straightforward. Ziętara and Lumme (2002) proposed 'DNA tag priority' and a guide of 1% variation in ITS sequence to separate *Gyrodactylus* species. They also suggested that sequences obtained from the type locality should take priority. These suggestions are unique and have not been widely adopted in any field of parasitology, let alone the study of Monogenea. *Gyrodactylus* species, for example, can be notoriously difficult to separate or identify on the basis of morphology, and several species can occur in the same locality or even on the same host as shown for *G. teuchis* (Cunningham *et al.*, 2001). Therefore, redescriptions, new descriptions and assigning DNA sequences to species should as far as possible take a holistic approach, involving as much information as possible on the location, host identity, range and preference and thorough morphological and molecular analysis. In the study of monogeneans, when morphological analysis can throw up many examples of sibling or cryptic species, and the species concept is itself a matter for debate, the addition of DNA sequence data can often create more difficulties than it solves. Each study should use as much data as possible and it is perhaps too early in the molecular analysis of Monogenea to be able to lay down rules or guidelines on the degree of sequence variation that occurs between species.

One of the possible solutions for the time being is to include new 'species', determined solely on the basis of DNA sequence variation, into species groups or complexes as suggested by Malmberg (1970), who grouped *Gyrodactylus* spp. on the basis of morphological features. These species groups, e.g. the *Gyrodactylus wageneri* or *Gyrodactylus elegans* groups, were constructed consisting of species with haptoral hard parts of similar morphology. This species-group approach could prevent the premature description of new species before detailed morphological study of all related *Gyrodactylus* species from all related hosts is performed, considering even those parasites recorded on the hosts not typical for the locality or geographical region studied. However, such extensive comparative studies cannot be accomplished without collaboration and exchange of material between research groups.

Molecular Analysis within Monogenean Species

Increasing attention is being paid to the analysis of relationships of monogeneans within a species. These sub-species groups may be cryptic species, strains or different specimens, e.g. from different locations. Again, this work stems from the *G. salaris* problem. Initially, the question of separating *G. salaris*, which is pathogenic to Atlantic salmon, from *G. thymalli*, which is not pathogenic, arose. The two species are very similar morphologically and their validity as separate species has been questioned. Host preference, pathogenicity and rRNA intergenic spacer DNA analysis has provided good evidence that *G. salaris* and *G. thymalli* are valid species (Sterud *et al.*, 2002).

Within the species, molecular analysis of *G. salaris* strains or populations is of interest in addressing the questions of whether different strains have differing pathogenicity and whether population genetics can provide epidemiological information or inform knowledge of reproductive systems. Lindenstrøm *et al.* (2003b) described a variant of *G. salaris* that exhibited unusual preference for rainbow trout (*Oncorhynchus mykiss*) rather than Atlantic salmon (*Salmo salar*). This variant was less virulent to Atlantic salmon than other forms, such as those that have caused salmon mortality in Norway. The variant form differs from *G. salaris sensu* Malmberg, 1957 at three positions in the

ITS. As the ITS sequence has been found to be very stable in specimens from very different geographical locations or sampling times, the variation found raises some questions regarding the species status of this form. Both the morphology and DNA sequence results place this variant clearly in the same species group as *G. salaris* and yet its pathogenicity may be very different. It will be interesting to see if this variant is found elsewhere or if its distribution is very restricted.

Mitochondrial COI sequences seem to be useful in providing epidemiological evidence for introduction of *G. salaris* to Norway (Hansen *et al.*, 2003), but are probably too conserved to provide detailed information on the spread of the parasite within a country. Nevertheless, Meinilä *et al.* (2004) used COI sequences in phylogenetic analysis of *G. salaris* and *G. thymalli* from Norway, Sweden, Finland, Russia and Latvia, concentrating on hypotheses of taxonomy and speciation.

The rRNA intergenic spacer region was also found to be difficult to use for detailed epidemiological analysis (Cunningham *et al.*, 2003). The RAPD and the related amplified fragment length polymorphism (AFLP) have also been applied to *G. salaris* (Cunningham and Mo, 1997; Cable *et al.*, 1999a). Large numbers of DNA fragments were revealed by these studies but the fragments have not been characterized. These methods have the potential to yield intraspecific differences but the application of the technique to gyrodactylid samples creates difficulty in analysis and interpretation and the methods have not been widely used.

Microsatellite DNA appears at present to be the best choice for revealing population differences within monogenean species. Work is underway to isolate suitable microsatellites by different groups and using different methods (J. Cable, University of Cardiff, 2004, personal communication; C.M. Collins, Fisheries Research Services, 2004, personal communication). It is hoped that suitable markers may be discovered in the near future, opening up a large field of population genetics study. Microsatellite DNA markers were also considered for screening populations of diplozoid species (Diplozoinae). Members of this subgenus usually show high host specificity and most of the species parasitize only one host species. The notable exception is *Paradiplozoon homoion*, which has been recorded from more than 15 species of cyprinid fish in Europe. Molecular study of diplozoids carried out to date has included sequencing the ITS2 rDNA and morphological analysis of *P. homoion* specimens from the gills of ten cyprinid species. This did not provide any proof of differences between specimens of *P. homoion* collected from different hosts (Matějusová *et al.*, 2002). Polymorphic microsatellite DNA markers might be useful tools to screen for possible variation between specimens collected from different hosts.

Other Genetic Studies in Monogenea

Although not studying monogenean genetics directly, work has begun to examine the genetic response of the host to monogenean parasite infection. Differential display reverse transcription PCR and suppressive subtractive hybridization are being used to examine differences in gene expression between infected and uninfected fish and between fish of differing susceptibility to *G. salaris* infection (Gilbey *et al.*, 2003). The expression of some genes related to the host immune response are also being studied, by screening complimentary DNA (cDNA) libraries of infected and uninfected fish using PCR with primers for these genes (Lindenstrøm *et al.*, 2003a). These studies may reveal which monogenean genes are important in stimulating host response and the same molecular techniques of differential display of gene expression that are being applied to host material could indicate variations in gene expression between monogeneans that may be related to differences in pathogenicity. Although the same problems of sample size and purity that can complicate other molecular studies also apply to this work, molecular and gene identification techniques are improving so fast that it cannot be long before the genes expressed by these tiny animals are uncovered and studied.

Concluding Remarks

The reproduction and developmental biology of some monogeneans, such as gyrodactylids or diplozoons, is one area that will derive enormous benefit from developments in the study of the genetics of these organisms. Cable and Harris (2002) pointed out the value of analysing hox gene expression in pattern formation during development. Fingerprinting-type examinations could shed some light on the relative contribution of sexual and asexual reproduction in different reproductive modes of gyrodactylids (Harris, 1993, 1998; Harris *et al.*, 1994), or the relative contribution of each individual in diplozoon reproduction.

The host–parasite interaction is another area where molecular biology can be applied to great advantage. Analysis of proteins and expression profiles of both proteins and nucleic acids can be used to determine the nature of molecules used in adhesion of the parasite to the host, their expression and induction. Discovery and analysis of the host molecules towards which monogeneans are attracted to or used as cues, as well as the parasite molecules that trigger immune responses in the host, will also be important areas for further application of molecular biology.

The farming of species new to such culture offers ideal opportunities for the emergence of pathogens that are not serious problems in wild populations. This is particularly true in aquaculture, where monogeneans can pose serious difficulties (Mo and Lile, 1998; Ogawa, 2002). The use of anthelminthic treatment can be complicated by the development of resistance. Resistance to benzimidazole has already been identified in *Pseudodactylogyrus* in eel farms (Waller and Buchmann, 2001). Benzimidazole and derivatives target β-tubulin and mutations in the β-tubulin gene are associated with resistance to these drugs. Analysis of the β-tubulin gene, or any other gene found to be associated with resistance to these drugs in nematodes, offers a means of monitoring potential resistance problems in monogeneans (Collins *et al.*, 2004). Differential display of gene expression in resistant and susceptible types offers a means of discovering more about the pathways and genes involved in the mechanisms of anthelminthic resistance, and will inform future selection or development of control treatments.

References

Baverstock, P.R., Fielke, R., Johnson, A.M., Bray, R.A. and Beveridge, I. (1991) Conflicting phylogenetic hypotheses for the parasitic platyhelminths tested by partial sequencing of 18S ribosomal RNA. *International Journal for Parasitology* 21, 329–339.

Bentz, S., Leroy, S., du Preez, L., Mariaux, J., Vaucher, C. and Verneau, O. (2001) Origin and evolution of African Polystoma (Monogenea: Polystomatidae) assessed by molecular methods. *International Journal for Parasitology* 31, 697–705.

Bentz, S., Combes, C., Euzet, L., Riutord, J.J. and Verneau, O. (2003) Evolution of monogenean parasites across vertebrate hosts illuminated by the phylogenetic position of *Euzetrema* Combes, 1965 within the Monopisthocotylea. *Biological Journal of the Linnean Society* 80, 727–734.

Blair, D. and McManus, D.P. (1989) Restriction enzyme mapping of ribosomal DNA can distinguish between fasciolid (liver fluke) species. *Molecular and Biochemical Parasitology* 36, 201–208.

Brehm, K., Kronthaler, K., Jura, H. and Frosch, M. (2000) Cloning and characterization of β-tubulin genes from *Echinococcus multilocularis*. *Molecular and Biochemical Parasitology* 107, 297–302.

Bruno, D.W., Collins, C.M., Cunningham, C.O. and MacKenzie, K. (2001) Combined morphological and molecular analysis of *Gyrodactyloides bychowskii*. *Diseases of Aquatic Organisms* 45, 191–196.

Cable, J. and Harris, P.D. (2002) Gyrodactylid developmental biology: historical review, current status and future trends. *International Journal for Parasitology* 32, 255–280.

Cable, J., Harris, P.D., Tinsley, R.C. and Lazarus, C.M. (1999a) Detection of genetic variation in strains of *Gyrodactylus turnbulli* using AFLP PCR. British Society for Parasitology Spring Meeting, 11–14 April 1999, University of Warwick, UK.

Cable, J., Harris, P.D., Tinsley, R.C. and Lazarus, C.M. (1999b) Phylogenetic analysis of *Gyrodactylus* spp. (Platyhelminthes: Monogenea) using ribosomal DNA sequences. *Canadian Journal of Zoology* 77, 1439–1449.

Cacciò, S., Cammà, C., Onuma, M. and Severini, C. (2000) The β-tubulin gene of *Babesia* and *Theileria* parasites is an informative marker for species discrimination. *International Journal for Parasitology* 30, 1181–1185.

Cameron, M.L., Levy, P., Nutman, T., Vanamala, C.R., Narayanan, P.R. and Rajan, T.V. (1988) Use of restriction fragment length polymorphisms (RFLPs) to distinguish between nematodes of pathogenic significance. *Parasitology* 96, 381–390.

Chilton, N.B., Gasser, R.B. and Beveridge, I. (1995) Differences in a ribosomal DNA sequence of morphologically indistinguishable species within the *Hypodontus macropi* complex (Nematoda: Strongyloidea). *International Journal for Parasitology* 25, 647–651.

Chisholm, L.A., Morgan, J.A.T., Adlard, R.D. and Whittington, I.D. (2001) Phylogenetic analysis of the Monocotylidae (Monogenea) inferred from 28S rDNA sequences. *International Journal for Parasitology* 31, 1537–1547.

Collins, C.M. and Cunningham, C.O. (2000) Characterisation of the *Gyrodactylus salaris* Malmberg, 1957 (Platyhelminthes: Monogenea) ribosomal intergenic spacer (IGS) DNA. *Parasitology* 121, 555–563.

Collins, C.M., Miller, K. and Cunningham, C.O. (2004) Characterisation of a β-tubulin gene from the monogenean parasite, *Gyrodactylus salaris* Malmberg, 1957. *Parasitology Research* 92, 390–399.

Cunningham, C.O. (1997) Species variation within the internal transcribed spacer (ITS) region of *Gyrodactylus* (Monogenea; Gyrodactylidae) ribosomal RNA genes. *Journal of Parasitology* 83, 215–219.

Cunningham, C.O. and Mo, T.A. (1997) Random amplified polymorphic DNA (RAPD) analysis of three Norwegian populations of *Gyrodactylus salaris* (Monogenea; Gyrodactylidae). *Journal of Parasitology* 83, 311–314.

Cunningham, C.O., McGillivray, D.M. and MacKenzie, K. (1995a) Phylogenetic analysis of *Gyrodactylus salaris* Malmberg, 1957 based on the small subunit (18S) ribosomal RNA gene. *Molecular and Biochemical Parasitology* 71, 139–142.

Cunningham, C.O., McGillivray, D.M., MacKenzie, K. and Melvin, W.T. (1995b) Discrimination between *Gyrodactylus salaris, G. derjavini* and *G. truttae* (Platyhelminthes: Monogenea) using restriction fragment length polymorphisms and an oligonucleotide probe within the small subunit ribosomal RNA gene. *Parasitology* 111, 87–94.

Cunningham, C.O., McGillivray, D.M., MacKenzie, K. and Melvin, W.T. (1995c) Identification of *Gyrodactylus* (Monogenea) species parasitizing salmonid fish using DNA probes. *Journal of Fish Diseases* 18, 539–544.

Cunningham, C.O., Aliesky, H. and Collins, C.M. (2000) Sequence and secondary structure variation in the *Gyrodactylus* (Platyhelminthes: Monogenea) ribosomal RNA gene array. *Journal of Parasitology* 86, 567–576.

Cunningham, C.O., Mo, T.A., Collins, C.M., Buchmann, K., Thiery, R., Blanc, G. and Lautraite, A. (2001) Redescription of *Gyrodactylus teuchis* Lautraite, Blanc, Thiery, Daniel and Vigneulle, 1999 (Monogenea: Gyrodactylidae), a species identified by ribosomal RNA sequence. *Systematic Parasitology* 48, 141–150.

Cunningham, C.O., Collins, C.M., Malmberg, G. and Mo, T.A. (2003) Analysis of ribosomal RNA intergenic spacer (IGS) sequences in species and populations of *Gyrodactylus* (Platyhelminthes: Monogenea) from salmonid fish in Northern Europe. *Diseases of Aquatic Organisms* 57, 237–246.

Desdevises, Y. (2001) The phylogenetic position of *Furnestinia echeneis* (Monogenea, Diplectanidae) based on molecular data: a case of morphological adaptation? *International Journal for Parasitology* 31, 205–208.

Gasser, R.B. (2001) Identification of parasitic nematodes and study of genetic variability using PCR approaches. In: Kennedy, M. and Harnett, W. (eds) *Parasitic Nematodes: Molecular Biology, Biochemistry and Immunology*. CAB International, Wallingford, UK, pp. 53–82.

Gerbi, S.A., Gourse, R.L. and Clark, C.G. (1982) Conserved regions within ribosomal DNA: locations and some possible functions. In: Busch, H. and Rothblum, L. (eds) *The Cell Nucleus*. Academic Press, New York, pp. 351–386.

Gilbey, J., Matějusová, I., Sorsa-Leslie, T., Collins, C.M., Cunningham, C.O., Verspoor, E., Noble, L.R. and Jones, C.S. (2003) The genetic basis of *Gyrodactylus salaris* resistance in Atlantic salmon (*Salmo salar*). International Society for Developmental and Comparative Immunology (ISDCI) 9th International Congress, 29 June–4 July 2003, University of St. Andrews, Scotland, UK.

Hansen, H., Bachmann, L. and Bakke, T.A. (2003) Mitochondrial DNA variation of *Gyrodactylus* spp. (Monogenea, Gyrodactylidae) populations infecting Atlantic salmon, grayling, and rainbow trout in Norway and Sweden. *International Journal for Parasitology* 33, 1471–1478.

Harris, P.D. (1993) Interactions between reproduction and population biology in Gyrodactylid monogeneans – a review. *Bulletin Français de la Pêche et de la Pisciculture* 328, 47–65.

Harris, P.D. (1998) Ecological and genetic evidence for clonal reproduction in *Gyrodactylus gasterostei* Gläser, 1974. *International Journal for Parasitology* 28, 1595–1607.

Harris, P.D. and Cable, J. (2000) *Gyrodactylus poeciliae* n. sp. and *G. milleri* n. sp. (Monogenea: Gyrodactylidae) from *Poecilia caucana* (Steindachner) in Venezuela. *Systematic Parasitology* 47, 79–85.

Harris, P.D., Jansen, P.A. and Bakke, T.A. (1994) The population age structure and reproductive biology of *Gyrodactylus salaris* Malmberg (Monogenea). *Parasitology* 108, 167–173.

Harris, P.D., Cable, J., Tinsley, R.C. and Lazarus, C.M. (1999) Combined ribosomal DNA and morphological analysis of individual gyrodactylid monogeneans. *Journal of Parasitology* 85, 188–191.

Hillis, D.M. and Dixon, M.T. (1991) Ribosomal DNA: molecular evolution and phylogenetic inference. *Quarterly Review of Biology* 66, 411–453.

Huyse, T. and Malmberg, G. (2004) Molecular and morphological comparisons between *Gyrodactylus ostendicus* n. sp. (Monogenea: Gyrodactylidae) on *Pomatoschistus microps* (Krøyer) and *G. harengi* Malmberg, 1957 on *Clupea harengus membras* L. *Systematic Parasitology* 58, 105–113.

Huyse, T. and Volckaert, F.A.M. (2002) Identification of a host-associated species complex using molecular and morphometric analyses, with the description of *Gyrodactylus rugiensoides* n. sp. (Gyrodactylidae, Monogenea). *International Journal for Parasitology* 32, 907–919.

Huyse, T., Audenaert, V. and Volckaert, F.A.M. (2003) Speciation and host–parasite relationships in the parasite genus *Gyrodactylus* (Monogenea, Platyhelminthes) infecting gobies of the genus *Pomatoschistus* (Gobiidae, Teleostei). *International Journal for Parasitology* 33, 1679–1689.

Huyse, T., Malmberg, G. and Volckaert, F.A.M. (2004) Four new species of *Gyrodactylus* von Nordmann, 1832 (Monogenea, Gyrodactylidae) on gobiid fishes: combined DNA and morphological analyses. *Systematic Parasitology* 59, 103–120.

Johnston, D.J., Kane, R.A. and Rollinson, D. (1993) Small subunit (18S) ribosomal RNA gene divergence in the genus *Schistosoma*. *Parasitology* 107, 147–156.

Jovelin, R. and Justine, J.L. (2001) Phylogenetic relationships within the polyopisthocotylean monogeneans (Platyhelminthes) inferred from partial 28S rDNA sequences. *International Journal for Parasitology* 31, 393–401.

Justine, J.L., Jovelin, R., Neifar, L., Mollaret, I., Lim, L.H.S., Hendrix, S.S. and Euzet, L. (2002) Phylogenetic positions of the Bothitrematidae and Neocalceostomatidae (Monopisthocotylean monogeneans) inferred from 28S rDNA sequences. *Comparative Parasitology* 69, 20–25.

Kane, R.A. and Rollinson, D. (1998) Comparison of the intergenic spacers and 3′ end regions of the large subunit (28S) ribosomal RNA gene from three species of *Schistosoma*. *Parasitology* 117, 235–242.

Kritsky, D.C. and Boeger, W.A. (2003) Phylogeny of the Gyrodactylidae and the phylogenetic status of *Gyrodactylus* Nordmann, 1832 (Platyhelminthes: Monogenoidea). In: Combes, C. and Jourdane, J. (eds) *Taxonomy, Ecology and Evolution of Metazoan Parasites*. Presses Universitaires de Perpignan, Perpignan, France, pp. 37–58.

Lindenstrøm, T., Buchmann, K. and Secombes, C.J. (2003a) *Gyrodactylus derjavini* infection elicits IL-1beta expression in rainbow trout skin. *Fish and Shellfish Immunology* 15, 107–115.

Lindenstrøm, T., Collins, C.M., Bresciani, J., Cunningham, C.O. and Buchmann, K. (2003b) Characterisation of a *Gyrodactylus salaris* variant: infection biology, morphology and molecular genetics. *Parasitology* 127, 165–178.

Littlewood, D.T.J. and Olson, P.D. (2001) SSU rDNA and the Platyhelminthes: signal, noise, conflict and compromise. In: Littlewood, D.T.J. and Bray, R.A. (eds) *Interrelationships of the Platyhelminthes*. Taylor & Francis, London, pp. 262–278.

Littlewood, D.T.J., Rohde, K. and Clough, K.A. (1997) Site-specific parasites: speciation in one host? *International Journal for Parasitology* 27, 1289–1297.

Littlewood, D.T.J., Rohde, K. and Clough, K.A. (1998) The phylogenetic position of *Udonella* (Platyhelminthes). *International Journal for Parasitology* 28, 1241–1250.

Littlewood, D.T.J., Olson, P.D., Telford, M.J., Herniou, E.A. and Riutort, M. (2001) Elongation factor 1-alpha sequences alone do not assist in resolving the position of the acoela within the metazoa. *Molecular and Biological Evolution* 18, 437–442.

Litvaitis, M. and Rhode, K. (1999) A molecular test of platyhelminth phylogeny: inferences from partial 28S rDNA sequences. *Invertebrate Biology* 118, 42–56.

Lockyer, A.E., Olson, P.D., Ostergaard, P., Rollinson, D., Johnston, D.A., Attwood, S.W., Southgate, V.R., Horak, P., Snyder, S.D., Le, T.H., Agatsuma, T., McManus, D.P., Carmichael, A.C., Naem, S. and

Littlewood, D.T.J. (2003) Utility of complete large and small subunit rRNA genes in resolving the phylogeny of the Neodermata (Platyhelminthes): implications and a review of the cerconmer theory. *Biological Journal of the Linnean Society of London* 78, 155–171.

Long, E.O. and Dawid, I.B. (1980) Repeated genes in eukaryotes. *Annual Review of Biochemistry* 49, 727–764.

Malmberg, G. (1970) The excretory systems and the marginal hooks as a basis for the systematics of *Gyrodactylus* (Trematoda, Monogenea). *Arkiv for Zoologi Serie 2* 23, 235.

Malmberg, G. (1987) *Gyrodactylus salaris* Malmberg 1957 and *G. truttae* Gläser 1974 – two problematic species. *Abo Akademi Information* 19, 34.

Malmberg, G. and Malmberg, M. (1993) Species of *Gyrodactylus* (Platyhelminthes, Monogenea) on salmonids in Sweden. *Fisheries Research* 17, 59–68.

Matějusová, I. and Cunningham, C.O. (2004) The first complete monogenean ribosomal RNA gene operon: sequence and secondary structure of the *Gyrodactylus salaris* Malmberg, 1957 large subunit ribosomal RNA gene. *Journal of Parasitology* 90, 146–151.

Matějusová, I., Gelnar, M., McBeath, A.J.A., Collins, C.M. and Cunningham, C.O. (2001a) Molecular markers for gyrodactylids (Gyrodactylidae: Monogenea) from five fish families (Teleostei). *International Journal for Parasitology* 31, 738–745.

Matějusová, I., Koubková, B., D'Amelio, S. and Cunningham, C.O. (2001b) Genetic characterisation of six species of diplozoids (Oligonchoinea: Diplozoidae). *Parasitology* 123, 465–474.

Matějusová, I., Koubková, B., Gelnar, M. and Cunningham, C.O. (2002) *Paradiplozoon homoion* Bychowsky and Nagibina 1959 versus *Paradiplozoon gracile* Reichenbach-Klinke 1961: two species or phenotypic plasticity? *Systematic Parasitology* 53, 39–47.

Matějusová, I., Gelnar, M., Verneau, O., Cunningham, C.O. and Littlewood, D.T.J. (2003) Molecular phylogenetic analysis of the genus Gyrodactylus (Platyhelminthes: Monogenea) inferred from rDNA ITS region: subgenera versus species groups. *Parasitology* 127, 603–611.

Matějusová, I., Koubková, B. and Cunningham, C.O. (2004) Identification of European diplozoids (Diplozoidae, Monogenea) by restriction digestion of the ribosomal RNA internal transcribed spacer (ITS2). *Journal of Parasitology* 90, 817–822.

Meinilä, M., Kuusela, J., Ziętara, M. and Lumme, J. (2002) Primers for amplifying approximately 820 bp of highly polymorphic mitochondrial COI gene of *Gyrodactylus salaris*. *Hereditas* 137, 72–74.

Meinilä, M., Kuusela, J., Ziętara, M. and Lumme, J. (2004) Initial steps of speciation by geographic isolation and host switch in salmonid pathogen *Gyrodactylus salaris* (Monogenea: Gyrodactylidae). *International Journal for Parasitology* 34, 515–526.

Mo, T.A. (1991a) Seasonal variations of opisthaptoral hard parts of *Gyrodactylus salaris* Malmberg, 1957 (Monogenea: Gyrodactylidae) on parr of Atlantic salmon *Salmo salar* L. in the river Batnfjordselva, Norway. *Systematic Parasitology* 19, 231–240.

Mo, T.A. (1991b) Variations of opisthaptoral hard parts of *Gyrodactylus salaris* Malmberg, 1957 (Monogenea: Gyrodactylidae) on parr of Atlantic salmon *Salmo salar* L. in laboratory experiments. *Systematic Parasitology* 20, 11–19.

Mo, T.A. (1991c) Variations of opisthaptoral hard parts of *Gyrodactylus salaris* Malmberg, 1957 (Monogenea: Gyrodactylidae) on rainbow trout *Oncorhynchus mykiss* (Walbaum, 1792) in a fish farm, with comments on the spreading of the parasite in south-eastern Norway. *Systematic Parasitology* 20, 1–9.

Mo, T.A. (1994) Status of *Gyrodactylus salaris* problems and research in Norway. In: Pike, A.W. and Lewis, J.W. (eds) *Parasitic Diseases of Fish*. Samara Publishing, Dyfed, pp. 43–56.

Mo, T.A. and Lile, N.K. (1998) *Gyrodactylus anarhichatis* n. sp. and *G. microanchoratus* n. sp. (Monogenea: Gyrodactylidae) from the Atlantic wolffish *Anarhichas lupus* (Anarhichadidae) off the coast of Norway. *Systematic Parasitology* 40, 213–220.

Mollaret, I., Jamieson, B.G.M., Adlard, R.D., Hugall, A., Lecointre, G., Chombard, C. and Justine, J.L. (1997) Phylogenetic analysis of the Monogenea and their relationships with Digenea and Eucestoda inferred from 28S rDNA sequences. *Molecular and Biochemical Parasitology* 90, 433–438.

Mollaret, I., Jamieson, B.G.M. and Justine, J.-L. (2000a) Phylogeny of the Monopisthocotylea and Polyopisthocotylea (Platyhelminthes) inferred from 28S rDNA sequences. *International Journal for Parasitology* 30, 171–185.

Mollaret, I., Lim, L.H.S. and Justine, J.L. (2000b) Phylogenetic position of the monogeneans *Sundanonchus, Thaparocleidus,* and *Cichlidogyrus* inferred from 28S rDNA sequences. *International Journal for Parasitology* 30, 659–662.

Morales-Hojas, R., Post, R.J., Cheke, R.A. and Wilson, M.D. (2002) Assessment of rDNA IGS as a molecular marker in the *Simulium damnosum* complex. *Medical and Veterinary Entomology* 16, 395–403.

Ogawa, K. (2002) Impacts of diclidophorid monogenean infections on fisheries in Japan. *International Journal for Parasitology* 32, 373–380.

Olson, P.D. and Littlewood, D.T.J. (2002) Phylogenetics of the Monogenea – evidence from a medley of molecules. *International Journal for Parasitology* 32, 233–244.

Overstreet, R.M., Curran, S.S., Pote, L.M., King, D.T., Blend, C.K. and Grater, W.D. (2002) *Bolbophorus damnificus* n. sp. (Digenea: Bolbophoridae) from the channel catfish *Ictalurus punctatus* and American white pelican *Pelecanus erythrorhynchos* in the USA based on life-cycle and molecular data. *Systematic Parasitology* 52, 81–96.

Plaisance, L., Littlewood, D.T.J., Olson, P.D. and Morand, S. (2005) Molecular phylogeny of gill monogeneans (platyhelminthes, Monogenea, Dactylogyridae) and colonization of Indo-west Pacific butterfly fish hosts (Perciformes, Chaetodontidae). *Zoologica scripta* 34, 425–436.

Robinson, M.W., Hoey, E.M., Fairweather, I., Dalton, J.P., McGonigle, S. and Trudgett, A. (2001) Characterisation of a β-tubulin gene from the liver fluke, *Fasciola hepatica*. *International Journal for Parasitology* 31, 1264–1268.

Rollinson, D., Walker, T.K., Knowles, R.J. and Simpson, A.J.G. (1990) Identification of *Schistosoma* hybrids and larval parasites using rRNA probes. *Systematic Parasitology* 15, 65–73.

Sicard, M., Desmarais, E. and Lambert, A. (2001) Molecular characterisation of Diplozoidae populations on five Cyprinidae species: consequences for host specificity. *Comptes Rendus De L Academie Des Sciences Serie III-Sciences De La Vie-Life Sciences* 324, 709–717.

Sicard, M., Desmarais, E. and Lambert, A. (2002) Is the Octomacridae the sister family of the Diplozoidae? *Parasite-Journal De La Societe Francaise De Parasitologie* 9, 85–87.

Simková, A., Plaisance, L., Matějusová, I., Morand, S. and Verneau, O. (2003) Phylogenetic relationships of the Dactylogyridae Bychowsky, 1933 (Monogenea: Dactylogyridea): the need for the systematic revision of the Ancyrocephalinae Bychowsky, 1937. *Systematic Parasitology* 54, 1–11.

Simková, A., Morand, S., Jobet, E., Gelnar, M. and Verneau, O. (2004) Molecular phylogeny of congeneric monogenean parasites (Dactylogyrus): a case of intrahost speciation. *Evolution* 58, 1001–1018.

Sterud, E., Mo, T.A., Collins, C.M. and Cunningham, C.O. (2002) The use of host specificity, pathogenicity, and molecular markers to differentiate between *Gyrodactylus salaris* Malmberg, 1957 and *G. thymalli* Zitnan, 1960 (Monogenea: Gyrodactylidae). *Parasitology* 124, 203–214.

Telford, M.J., Herniou, E.A., Russell, R.B. and Littlewood, D.T.J. (2000) Changes in mitochondrial genetic codes as phylogenetic characters: two examples from the flatworms. *Proceedings of the National Academy of the Sciences USA* 97, 11359–11364.

Verneau, O., Bentz, S., Sinnappah, N.D., du Preez, L., Whittington, I. and Combes, C. (2002) A view of early vertebrate evolution inferred from the phylogeny of polystome parasites (Monogenea: Polystomatidae). *Proceedings of the Royal Society of London Series B – Biological Sciences* 269, 535–543.

Waller, P.J. and Buchmann, K. (2001) Anthelmintic resistance and parasite control in commercial eel farms: consequences for producers. *Veterinary Record* 148, 783–784.

Waters, A.P. and McCutchan, T.F. (1990) Ribosomal RNA: nature's own polymerase-amplified target for diagnosis. *Parasitology Today* 6, 56–59.

Whittington, I.D., Deveney, M.R., Morgan, J.A.T., Chisholm, L.A. and Adlard, R.D. (2004) A preliminary phylogenetic analysis of the Capsalidae (Platyhelminthes: Monogenea: Monopisthocotylea) inferred from large subunit rDNA sequences. *Parasitology* 128, 511–519.

Wuyts, J., Van de Peer, Y. and De Wachter, R. (2001) Distribution of substitution rates and location of insertion sites in the tertiary structure of ribosomal RNA. *Nucleic Acids Research* 29, 5017–5028.

Ziętara, M.S. and Lumme, J. (2002) Speciation by host switch and adaptive radiation in a fish parasite genus *Gyrodactylus* (Monogenea, Gyrodactylidae). *Evolution* 56, 2445–2458.

Ziętara, M.S. and Lumme, J. (2003) The crossroads of molecular, typological and biological species concepts: two new species of *Gyrodactylus* Nordmann, 1832 (Monogenea: Gyrodactylidae). *Systematic Parasitology* 55, 39–52.

Ziętara, M.S., Arndt, A., Geets, A., Hellemans, B. and Volckaert, F.A.M. (2000) The nuclear rDNA region of *Gyrodactylus arcuatus* and *G. branchicus* (Monogenea: Gyrodactylidae). *Journal of Parasitology* 86, 1368–1373.

Ziętara, M.S., Huyse, T., Lumme, J. and Volckaert, F.A. (2002) Deep divergence among subgenera of *Gyrodactylus* inferred from rDNA ITS region. *Parasitology* 124, 39–52.

6 The Schistosome Transcriptome

S. Verjovski-Almeida and R. DeMarco
Instituto de Quimica, Universidade de São Paulo, Av. Prof. Lineu Prestes 748 sala 1200 05508-000 São Paulo, SP, Brazil

Introduction

Schistosomes are one of the most complex parasites infecting humans, being dioecious multicellular platyhelminths. They present a life cycle with a phase of asexual reproduction in the intermediate host (snail) and a phase of sexual reproduction in the definitive host (human). There are also two intermediate aquatic free-living phases (cercaria and miracidium) with a short lifespan linking the two phases of reproduction, thus completing the life cycle. The parasite is subjected to a significant amount of stress, which is triggered by the defence systems of the hosts or the external environment of the free-living stages. The significant differences in morphology and physiology that are displayed by the parasite during its life cycle probably reflect a constant adaptation to the different environments in which the parasite resides (Hoffmann and Dunne, 2003).

Different species of *Schistosoma* such as *S. mansoni*, *S. japonicum* and *S. haematobium* infect the human host (Hoffmann and Dunne, 2003). The adult stage of *Schistosoma* is adapted to life in human blood vessels, being capable of surviving for years in the host despite a significant immune response (Pearce and MacDonald, 2002). The ability of *Schistosoma* to foil host defences depends on a series of mechanisms such as modulation of the immune system of the host by either interference in cytokine synthesis (Angeli *et al.*, 2001) or by the induction of defence cell apoptosis (Carneiro-Santos *et al.*, 2000; Chen *et al.*, 2002). The intimate interaction between *Schistosoma* and the human immune system is also stressed by modulation of *Schistosoma* development and reproduction by signals from the host immune system (Amiri *et al.*, 1992; Davies and McKerrow, 2003). Additionally, endocrine signals from the host have been shown to influence *Schistosoma* development and viability (de Mendonca *et al.*, 2000; Davies and McKerrow, 2003), suggesting molecular cross-talk between host and parasite. These facts demonstrate a highly sophisticated level of interaction and adaptation such that the schistosome would be expected to possess a large transcriptome with complex and intricate protein expression patterns through the various life stages.

The Transcriptome Projects

For years, efforts to generate expressed sequence tags (ESTs) of *Schistosoma* have resulted in the production of a modest repository of approximately 16,000 sequences, mostly from the adult stage (Franco *et al.*, 2000; Fietto *et al.*, 2002; Fung *et al.*, 2002), which along with the few hundred full-length sequences deposited in the public databases by several groups of investigators allowed a partial description of the *Schistosoma* transcriptome, with an estimated coverage of 15–20% of the gene complement for *S. mansoni* (Franco *et al.*, 2000). In the beginning of this decade, two major independent efforts were initiated, aiming to obtain a more extensive sequencing of the transcriptome from two species, *S. mansoni* and *S. japonicum*. These efforts were launched concomitantly by a Brazilian and a Chinese sequencing consortium, respectively. Both projects were able to produce a significant amount of data, which have been publicly released, and two articles were published in the same issue of *Nature Genetics* by late 2003. The Chinese project generated 43,707 ESTs, focused on the transcriptome of the adult (either males and females separately, or mixed samples) and egg stages of *S. japonicum* (Fig. 6.1) using poly-dT primed cDNA libraries generated from these two stages (Hu *et al.*, 2003). The Brazilian effort used material from six different stages of the parasite life cycle, namely adults, cercariae, miracidia, germ balls, schistosomula and eggs (Fig. 6.1), for the generation of 163,586 ESTs of *S. mansoni* (Verjovski-Almeida *et al.*, 2003); 151,684 of the ESTs were produced with the ORESTES technique (Dias-Neto *et al.*, 2000) from the six stages, and 11,902 ESTs were generated from a normalized poly-dT primed library (Soares *et al.*, 1994) derived from adult worms. Additional contributions to the EST repository are expected to include ESTs of *S. mansoni* generated by the Genome Network of State of Minas Gerais and 15,000 ESTs of *S. hematobium* that the Welcome Trust Sanger Institute is planning to produce (McManus *et al.*, 2004).

Transcriptome Comparisons

The analysis resulting from both studies indicated that the sequenced ESTs effectively sampled the majority of messages from the transcriptome of these organisms. The Brazilian study estimated a 92% sampling of messages and obtained 395 novel ortholog genes with their entire coding region sequenced. A gene complement of 14,000 genes was estimated for the six stages of the *S. mansoni* life cycle using two independent methods (Verjovski-Almeida *et al.*, 2003). The first was a clustering method with an additional analysis to eliminate redundancy due to alternate splicing, bad clustering of sequences due to low sequencing quality and sequencing of different non-overlapped portions of the same message; the second method called for determining the total number of non-redundant bases acquired in the project followed by computing the extrapolated maximal number of non-redundant bases and dividing it by the mean gene size of *S. mansoni*, estimated as 2063 bp (Verjovski-Almeida *et al.*, 2003). The estimated number of 14,000 genes in *S. mansoni* (Verjovski-Almeida *et al.*, 2003) is comparable to the 14,000–19,000 predicted genes from other fully sequenced invertebrates, such as *Caenorhabditis elegans* (The *C. elegans* Sequencing Consortium, 1998) and *Ciona intestinalis* (Dehal *et al.*, 2002). In contrast, the

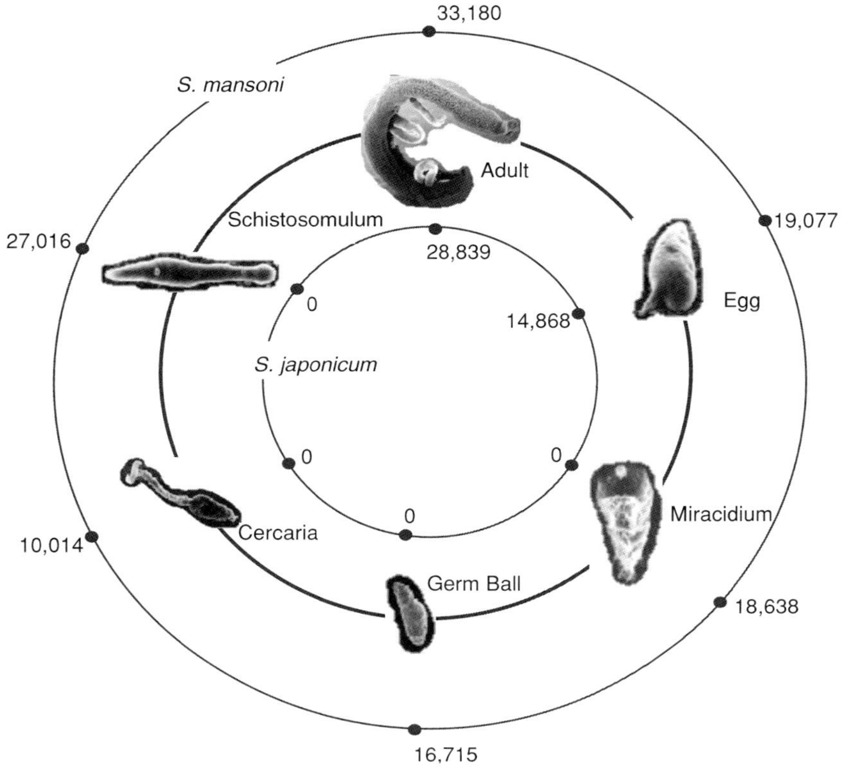

Fig. 6.1. Summary data for the large-scale expressed sequence tag (EST) sequencing projects for *Schistosoma mansoni* and *S. japonicum*. Numbers refer to ESTs obtained for each of the six stages of the parasite's life cycle.

Chinese work estimated that approximately 13,000 gene clusters were expressed in adults and eggs, which would correspond to the majority of the estimated gene complement and indicate that most of the transcripts of *S. japonicum* are present in just these two stages (Hu *et al.*, 2003; McManus *et al.*, 2004). We believe that this is probably an overestimation, as the analysis consisted of a simple counting of the total number of gene clusters generated by the assembly of ESTs and did not account for redundancy that originated from alternate splicing, bad clustering of sequences due to low sequencing quality and sequencing of different non-overlapped portions of the same message. In fact, an independent experimental determination of the number of genes expressed in *S. mansoni* was obtained (Verjovski-Almeida *et al.*, 2003), using serial analysis of gene expression (SAGE) and counting the tags from a SAGE library constructed with *S. mansoni* adult stage mRNA. Counting reached a plateau with approximately 7200 tags, indicating that only half of the *S. mansoni* genes are expressed in the adult stage (Verjovski-Almeida *et al.*, 2003), giving a very different picture to that proposed for *S. japonicum*. Moreover, several of the new genes identified either in *S. japonicum* or in *S. mansoni* were seen to have a stage or gender-specific variation in their expression levels, indicating a possible role of these genes in adaptation and differentiation of the parasite (Hu *et al.*, 2003; Verjovski-Almeida *et al.*, 2003). These facts indicate that several of the schistosome genes are transiently expressed during the parasite life cycle such that it is unlikely that sequencing of adult and egg messages only would permit sampling of the entire parasite transcriptome. It seems unlikely that such closely related organisms would present such an overwhelming difference in the number of expressed genes in a given stage of their life cycle – further analysis is warranted to explain this discrepancy.

Evolutionary Implications

As a platyhelminth, schistosomes diverged from other eubilaterian metazoa at the early stages in bilaterian evolution, more than a billion years ago (Hausdorf, 2000). Recent analysis utilizing novel schistosome genes such as actin, cathepsin, XPA protein and polymerase delta confirmed this placement (Hu *et al.*, 2003; Verjovski-Almeida *et al.*, 2003). Currently, *S. mansoni* and *S. japonicum* are the only two platyhelminths whose transcriptomes have been extensively sequenced, and the nucleotide sequences for these two organisms represent 92% of the platyhelminth nucleotides deposited in GenBank. Therefore they are unique examples of well-described transcriptomes from an evolutionary branch long diverged from other organisms for which the full genome or transcriptome have been described. Moreover, *Schistosoma* has a long evolutionary history of parasitism, its ancestor probably parasitized fishes and reptilians before migrating to vertebrate hosts such as mammals and birds (Lenzi *et al.*, 1997; Rollinson *et al.*, 1997). Schistosomes parasitize a wide range of snails within the Gastropoda (Blair *et al.*, 2001) and it is hypothesized that the two-host life cycle of the schistosome is a truncation of the three-host cycle of the typical digenean trematodes (Poulin and Cribb, 2002). It would be expected that such an early diverging organism, with a long history of co-evolution with several different hosts and diverse environments that exert distinct evolutionary pressures, must present several unique mechanisms of adaptation that evolved along with the organism.

Novel Transcripts

In fact, comparison of EST-translated sequences with all other gene sequences from other species available in the public databases showed that 70% and 55% of the clusters obtained from *S. japonicum* and *S. mansoni*, respectively, did not display significant similarity and were unique (using an expected value significance cut-off of $E \leq 10^{-10}$ in the Blastx comparisons) (Hu *et al.*, 2003; Verjovski-Almeida *et al.*, 2003). This difference between the numbers of clusters with no similarity to any other database entries in the two organisms probably reflects the different methodologies used to sample sequences in the two studies, namely the use of either poly-dT primed cDNA libraries or ORESTES libraries. In the former method, the non-coding 3′-end of messages is preferentially cloned and sequenced. In the latter method, two low-stringency, arbitrarily primed events, namely reverse transcription and polymerase chain reaction (PCR), are sequentially combined, resulting in the statistically determined preferential sampling of the central, protein-coding portion of genes (Dias-Neto *et al.*, 2000). Nonetheless, it is possible to state that a considerable portion of the schistosome transcriptome codes for proteins with primary structures that are unrelated to those seen in other well-described organisms such as *C. elegans* and *Drosophila melanogaster*. It is a major challenge to determine the exact function of these novel proteins and unravelling their mechanisms of action in schistosomes will require a tremendous amount of effort, especially because homology-based screens provide no clues to their function.

Several of the transcript fragments described in these new sets of *Schistosoma* genes could have their function inferred by similarity with known proteins. The overall view of the schistosome obtained by analysis of these data is of a complex organism, sharing several sophisticated mechanisms with other distantly related organisms. Approximately 1443 *S. mansoni* genes are conserved among all fully sequenced eukaryotes, including the multicellular eukaryotes *Homo sapiens*, *D. melanogaster*, *C. elegans* and *C. intestinalis,* and the unicellular eukaryotes *Saccharomyces cerevisiae* and *Plasmodium falciparum*, probably representing proteins with basic cellular or 'housekeeping' functions (Verjovski-Almeida *et al.*, 2003). Approximately 645 genes appear to be conserved only among metazoans. The metazoa-specific protein set reveals a noticeable enrichment of protein functions essential for the establishment of a complex multicellular organism compared to the eukarya-conserved protein set, indicating that several cellular and physiological systems of higher animals were established before the divergence of platyhelminths (Verjovski-Almeida *et al.*, 2003).

Receptors and Host–Parasite Interaction

The description of schistosome receptors is of particular interest as they play an essential role in the process of parasite development and differentiation. Additionally, they could participate in the complex interaction of the parasite and host (de Mendonca *et al.*, 2000) and in the molecular cross-talk between adult males and females (Kunz, 2001) that are known to influence parasite development; parasites lacking either of these stimuli remain sexually immature (Davies *et al.*, 2001; Kunz, 2001). Previous studies have detected the presence of nuclear receptors of the COUP-TF, RXR and FTZ-F1 families in *Schistosoma*, with the characterization of cDNA clones for RXR and FTZ-F1 receptors (Escriva *et al.*, 1997), and a possible implication of the RXR receptors for female-specific gene expression in vitelline cells (Fantappie *et al.*, 2001). Another receptor that may play a significant role in the host–parasite relationship is SMTβR-I, which was shown to be activated by human TGFβ (Beall and Pearce, 2001) and to be present on the surface of the parasite (Davies *et al.*, 1998). The identification of fragments of additional schistosome proteins that have homology to host endocrine signalling system receptors (e.g. a progesterone receptor component, thyroid-associated proteins and insulin receptors) suggests that the parasite could use these components as sensors for host signals that could ultimately regulate cell proliferation and development (Hu *et al.*, 2003; Verjovski-Almeida *et al.*, 2003). The discovery of components with homology to host receptors opens up the field for novel studies to further characterize the host–parasite interaction, and for possible treatments to revert the parasite adaptation capability. Some of these components might be part of an endogenous endocrine system of the parasite, in which diffusion may be facilitated by the syncytial nature of most schistosome tissues (Kunz, 2001). However, the physiological architecture of the schistosome, as an acoelomate, limits the distribution of signalling molecules such that such factors may only act over short distances or through the neuroendocrine system (Verjovski-Almeida *et al.*, 2003). Taken together, these data reveal that the schistosome transcriptome encodes components of multiple signalling pathways that are likely to form a highly evolved system for the intricate regulation of *Schistosoma* cell functions.

Immune Evasion

Several mechanisms of schistosome evasion from the human immune system such as antigenic variation, antigen mimicry and modification of the host immune response have been proposed (Salzet *et al.*, 2000). Analysis of the schistosome transcriptome data supports this notion and indicates that mechanisms of interference with the host immune system are likely to be present. Indeed, homologues of proteins that are known to be involved in the control of immune processes have been described both in *S. japonicum* and *S. mansoni*. The activities of these parasite-secreted proteins appear to include: (i) the modulation of host immune responses; (ii) the simple inhibition of host immune responses; and (iii) mimicking of the host antigen presentation system. The presence of proteins such as Phospholipase A2-activating protein, tumour necrosis factor-associated protein and B-cell receptor-associated protein suggests the production of, and sensitivity for, host immune system signals (Hu *et al.*, 2003; Verjovski-Almeida *et al.*, 2003). Additionally, expression of tropomyosin presenting immunological cross reactivity and the high similarity in amino acid primary sequence to the *Biomphalaria glabrata* tropomyosins suggests molecular mimicry (Dissous and Capron, 1995). Highly variable gene families, which offer one possible mechanism that could be used to generate antigenic variation, were not detected in the *S. mansoni* transcriptome. However, the presence of several paralogues of a number of genes suggests that parasite evolution may have favoured the presence of multiple enzyme paralogues with redundant functions in similar pathways, a mechanism that may have enabled the parasite to evade some host defences (Verjovski-Almeida *et al.*, 2003).

Sex Differences

The Schistosomatidae is the only family within the trematodes to be dioecious; as such it represents a unique example of this adaptation within this class of organisms. It is known that permanent and intimate contact between male and female is necessary for reproduction to occur. Females that mature in the absence of males are considerably smaller and do not produce eggs (Kunz, 2001). This is likely due to molecular cross-talk between male and female and it is known that differential expression of some genes is dependent on contact between male and female worms (Grevelding *et al.*, 1997). As expected, males and females have several genes that are differentially expressed between the sexes, some of them involved in reproductive functions, such as the production of eggs (Franco *et al.*, 2000; Boag *et al.*, 2001). In fact, recent experiments using a 576-element cDNA microarray identified 12 new female-associated and 4 new male-associated gene transcripts in adult worms (Hoffmann *et al.*, 2002) and semi-quantitative RT-PCR also revealed several genes with different expression levels between the sexes (Hu *et al.*, 2003). Analysis of schistosome ESTs allowed the description of fragments from sex determination pathway proteins (fox-1, mog-1, mog-4, tra-2 and fem-1) and 19 proteins characteristic of sexual chromosomes (Hu *et al.*, 2003; Verjovski-Almeida *et al.*, 2003), indicating mechanisms that could impact sexual differentiation. The presence of such proteins points to a shared ancient origin of mechanisms of sexual determination.

Genome Data

Schistosoma has a large genome of approximately 2.7×10^8 base pairs and it is composed of 4–8% highly repetitive sequences and 40% of middle repetitive sequences (Simpson *et al.*, 1982). This genome is organized in eight pairs of chromosomes and heteromorphic sexual chromosomes. It is significantly larger than most human pathogens and only one-tenth the size of the human genome, reflecting the complexity of this organism. Recently, the completion of a 7.5-fold coverage of the *S. mansoni* genome by whole genome shotgun sequencing has been announced (El-Sayed *et al.*, 2004). Based on Lander–Waterman model this sequencing would achieve 99.5% coverage of the *S. mansoni* genome and it is expected that a draft of the genome will be released soon. A four- to fivefold coverage sequencing of the *S. japonicum* genome will also be available soon (El-Sayed *et al.*, 2004). With completion of the *S. mansoni* and *S. japonicum* genomes it will be possible to analyse the gene content of these genomes, a task known to be difficult and prone to misinterpretation in complex organisms. Although the use of computer algorithms in gene prediction in a genome is widespread, it is known that they are still prone to make inaccurate gene predictions with overpredictions and false negatives. The mapping of sequences generated from mRNAs is still the best way to identify genes in complex eukaryote genomes, as it is based upon physical evidence (Dunham *et al.*, 1999). In this context, the existing schistosome ESTs can be useful to help map the genomic locus of thousands of genes, although they unveil only part of the structure of each gene in the genome. Computer programs may be trained using the existing data set of expressed sequences to make more accurate predictions of gene structure in platyhelminth genomes.

Novel Drug Targets

The use of new transcriptome information for discovery of novel drug candidates is desirable, as the control of schistosomiasis relies mostly on the use of a single drug, praziquantel, a heterocyclic pyrazino-isoquinolone. This drug exerts multiple effects, such as damage to the tegumental membrane, changes in calcium flux and muscular contractions in the parasite by a mechanism or mechanisms that are not fully understood (Hagan *et al.*, 2004). There are reports of treatment failure in Senegal and Egypt, where verification of isolates with reduced susceptibility to praziquantel in the schistosome population raised concern about the development of resistance (Hagan *et al.*, 2004). However, evidence of drug resistance is considered to be limited and the expansion of the use of praziquantel has been proposed due to the relative success of control programmes in Egypt, China, Brazil and Morocco (Fenwick *et al.*, 2003; Hagan *et al.*, 2004). Nonetheless, the lack of knowledge of the mechanism of action of praziquantel limits the analysis of the mechanisms of resistance. It is not known if the drug acts on a single or multiple targets, and if multiple targets are involved analysis of the heritability of resistance would be complex (Hagan *et al.*, 2004). These facts create an uncertainty regarding the probability of the emergence and spread of novel resistant strains. In this context, novel drug target candidates, mostly novel receptors of neurotransmitters and ion channels, have been identified using the recent database of protein fragments of *S. mansoni* (Verjovski-Almeida *et al.*, 2003). It is expected that further trawls for 'druggable' targets would improve selection of the most appropriate candidates and subsequent screens for specific agonists/antagonists would uncover novel anti-schistosome drugs. In fact, one novel nicotinic acetylcholine receptor has been identified and located on the surface of the parasite – a convenient location for drug intervention (Bentley *et al.*, 2004). Several novel fragments of proteins of the multi-drug resistance (MDR) family of ABC transporters have been described, revealing potential effectors of drug resistance in the parasite (Verjovski-Almeida *et al.*, 2003). Additional studies of the substrate specificity and the identification of possible inhibitors of such transporters may contribute to a more rational design of drugs. Although the discovery, validation and development of novel classes of schistomicidal agents will be a long and slow process, it is imperative that efforts are made to improve the current drug arsenal and so ensure the long-term control of schistosome parasites of man. The transcriptome provides a rich and untapped resource for novel drug targets.

Vaccines

One of the most prized goals in schistosomiasis research is the development of an effective vaccine. An effective vaccine would enable the protection of a large proportion of exposed populations with only a single or a few treatments and, importantly, would prevent their re-infection; this would negate the need for continuous re-treatment that is seen with chemotherapeutics that require constant redistribution (Hagan *et al.*, 2000). The concept of protection against schistosomiasis is different from the sterile protection commonly associated with vaccines. A putative vaccine that protects against the accumulation of eggs in the body or a vaccine conferring a reduction on the worm burden could have a useful application (Bergquist *et al.*, 2002). Although there is still significant controversy over the feasibility of developing an effective vaccine against schistosomiasis using the currently available technology (Gryseels, 2000; Hagan *et al.*, 2000), the efficacy of irradiated cercariae in immunization tests provides 'proof-of-principle' that an efficient vaccine is achievable (Hagan *et al.*, 2000). Recently, six vaccine candidate proteins, namely paramyosin, IrV-5, triose phosphate isomerase, glutathione-*S*-transferase and Sm14, were chosen for further development based on protective capability, experience and availability (Bergquist *et al.*, 2002). The clinical trials using these candidates have been disappointing, particularly because none of the antigens achieved a protection of at least 40% against challenge infection such that currently the vaccine effort is stalled (Hagan and Sharaf, 2003). The description of a large set of novel genes allowed the proposition of novel vaccine candidates, namely characteristic surface exposed and exported proteins expressed in the inter-mammalian stages (Verjovski-Almeida *et al.*, 2003). Despite the speculative nature of such candidates, it offers the opportunity for large-scale trials to discover a few novel effective formulations for further experimentation, providing new hopes for the development of a vaccine. Additionally, proteomic analyses utilizing the transcriptome information will permit the identification of novel secreted proteins that would be strong vaccine candidates (Verjovski-Almeida *et al.*, 2004).

Trans-splicing

A subset of mRNA sequences of *S. mansoni* has been shown to possess a splice leader sequence. Apparently, the RNA processing mechanisms involved in *trans*-splicing are similar to those described in trypanosomatid protozoa and nematodes. In trypanosomatids *trans*-splicing is associated with the maturation of constituent genes in polycistronic products and provides a 5′ cap for mature RNA species; in nematodes and schistosomes, the biological significance of splice-leader *trans*-splicing remains unknown (Davis and Hodgson, 1997). Splice-leader *trans*-splicing has a sporadic evolutionary occurrence and it is not known if this process arose on multiple occasions or originated once and was subsequently lost in multiple lineages (Nilsen, 2001). In this respect, the presence of conserved splice-leader sequence elements in flatworms suggests that they are an ancestral feature within the phylum (Davis and Hodgson, 1997). The messages presenting the splice-leader appear to be restricted to less than 10% of the cDNA transcripts (Williams and Johnston, 1999), although there are reports that suggest a higher proportion of splice-leader-led transcripts in the schistosome transcriptome (Hamdan *et al.*, 2002). Nonetheless, addition of the splice-leader sequence to immature mRNAs is not a general mechanism in schistosome transcription, and the extent of its influence to adaptation and development of *Schistosoma* remains to be determined.

Retrotransposable Elements

The expression of several retrotransposable elements has been described in *S. mansoni* and *S. japonicum*, indicating that these elements are part of the transcriptome ecology of these parasites. Most of them belong to non-LTR retrotransposon families CR1 and RTE and to the LTR retrotransposon family Ty3/Gypsy, although there is also a member of the BEL family described (Brindley *et al.*, 2003; Copeland *et al.*, 2003; DeMarco *et al.*, 2004). The detection of expressed retrotransposons with high transcriptional activity and evidence for their insertion into active genes suggests that retrotransposition may represent a remarkable driving force in the *Schistosoma* genome (DeMarco *et al.*, 2004). Apparently, the dynamics of these elements in the schistosome genome is complex with different elements presenting very distinct strategies in the colonization of the genome (DeMarco *et al.*, 2004). It is not obvious which retrotransposons are still active, but reports of reverse transcriptase activity in *S. mansoni* extracts (Ivanchenko *et al.*, 1999) suggest that there still are transposition events in the schistosome genome. New insights of retrotransposon biology will improve understanding of its dynamics and possible new synergic mechanisms between the retroelements and the parasite that may improve the capacity for adaptations to a parasitic lifestyle. Description of retrotransposons is also a valuable resource that should help genome assembly and annotation, especially as it is estimated that 15–55% of the schistosome genome has mobile genetic elements (Brindley *et al.*, 2003). It is expected that a considerable fraction of these mobile elements must be of retrotransposons. A new superfamily of DNA transposons, called Merlin-IS1016, which is common in many eubacterial and animal genomes, has been identified (Feschotte, 2004). There is evidence that these transposons have been recently active in several animal species, particularly in *S. mansoni*, where Merlin is also the first described DNA transposon family (Feschotte, 2004).

Functional Genomics

The prospects for exploitation of these new sequences for functional genomics studies in schistosomes are exciting. It is expected that results from large-scale experiments using the new repositories of schistosome genetic information should emerge soon, as several groups are initiating efforts in this area (Verjovski-Almeida *et al.*, 2004). Microarray studies of gene expression should provide additional information about the different transcriptome profiles throughout the parasite life cycle and the differential activation of genes by diverse environmental stimuli. These studies should highlight genes involved in several processes of adaptation and development of the parasite during its life cycle. The microarray technique may also be used to obtain a more detailed view of parasite responses to future vaccine and novel drug treatments. The use of RNA interference (RNAi) technique for large-scale studies is feasible because recently this technique has been successfully used to manipulate gene expression in sporocysts (Boyle *et al.*, 2003). Accumulation of EST information is also useful for proteomic analyses, as mRNA sequence information can be linked to the information obtained from peptide mass fingerprints (Ashton *et al.*, 2001). In this way, the transcriptome data will help to provide a more complete description of the protein content of the parasite during its various stages and, in particular, will aid characterization of the secreted protein complement; correlation between protein and RNA message levels may provide additional insight into the rate of protein turnover and the rate of translation among the various schistosome proteins.

Concluding Remarks

The enormous potential for further experimental analyses that has been uncovered using the schistosome transcriptome data is overwhelming. It seems highly probable that much more rapid progress will be made in molecular-based studies of schistosomes. Furthermore, the forthcoming data from transcriptome, genome, post-genomic and proteomic analyses will permit a more integrated approach to the molecular biology of schistosomes. Comparative studies of the *S. mansoni* and *S. japonicum* transcriptomes and genomes will also provide new information on their evolution. All these analyses should permit the evolution of novel approaches in efforts to design more effective treatments for schistosomiasis.

References

Amiri, P., Locksley, R.M., Parslow, T.G., Sadick, M., Rector, E., Ritter, D. and McKerrow, J.H. (1992) Tumour necrosis factor alpha restores granulomas and induces parasite egg-laying in schistosome-infected SCID mice. *Nature* 356, 604–607.

Angeli, V., Faveeuw, C., Delerive, P., Fontaine, J., Barriera, Y., Franchimont, N., Staels, B., Capron, M. and Trottein, F. (2001) *Schistosoma mansoni* induces the synthesis of IL-6 in pulmonary microvascular endothelial cells: role of IL-6 in the control of lung eosinophilia during infection. *European Journal of Immunology* 31, 2751–2761.

Ashton, P.D., Curwen, R.S. and Wilson, R.A. (2001) Linking proteome and genome: how to identify parasite proteins. *Trends in Parasitology* 17, 198–202.

Beall, M.J. and Pearce, E.J. (2001) Human transforming growth factor-beta activates a receptor serine/threonine kinase from the intravascular parasite *Schistosoma mansoni*. *Journal of Biological Chemistry* 276, 31613–31619.

Bentley, G.N., Jones, A.K., Oliveros Parra, W.G. and Agnew, A. (2004) ShAR1alpha and ShAR1beta: novel putative nicotinic acetylcholine receptor subunits from the platyhelminth blood fluke Schistosoma. *Gene* 329, 27–38.

Bergquist, R., Al-Sherbiny, M., Barakat, R. and Olds, R. (2002) Blueprint for schistosomiasis vaccine development. *Acta Tropica* 82, 183–192.

Blair, D., Davis, G.M. and Wu, B. (2001) Evolutionary relationships between trematodes and snails emphasizing schistosomes and paragonimids. *Parasitology* 123, S229–S243.

Boag, P.R., Newton, S.E. and Gasser, R.B. (2001) Molecular aspects of sexual development and reproduction in nematodes and schistosomes. *Advances in Parasitology* 50, 153–198.

Boyle, J.P., Wu, X.J., Shoemaker, C.B. and Yoshino, T.P. (2003) Using RNA interference to manipulate endogenous gene expression in *Schistosoma mansoni* sporocysts. *Molecular and Biochemical Parasitology* 128, 205–215.

Brindley, P.J., Laha, T., McManus, D.P. and Loukas, A. (2003) Mobile genetic elements colonizing the genomes of metazoan parasites. *Trends in Parasitology* 19, 79–87.

Carneiro-Santos, P., Martins-Filho, O., Alves-Oliveira, L.F., Silveira, A.M., Coura-Filho, P., Viana, I.R., Wilson, R.A. and Correa-Oliveira, R. (2000) Apoptosis: a mechanism of immunoregulation during human *Schistosomiasis mansoni*. *Parasite Immunology* 22, 267–277.

Chen, L., Rao, K.V., He, Y.X. and Ramaswamy, K. (2002) Skin-stage schistosomula of *Schistosoma mansoni* produce an apoptosis-inducing factor that can cause apoptosis of T cells. *Journal of Biological Chemistry* 277, 34329–34335.

Copeland, C.S., Brindley, P.J., Heyers, O., Michael, S.F., Johnston, D.A., Williams, D.L., Ivens, A.C. and Kalinna, B.H. (2003) Boudicca, a retrovirus-like long terminal repeat retrotransposon from the genome of the human blood fluke *Schistosoma mansoni*. *Journal of Virology* 77, 6153–6166.

Davies, S.J. and McKerrow, J.H. (2003) Developmental plasticity in schistosomes and other helminths. *International Journal for Parasitology* 33, 1277–1284.

Davies, S.J., Shoemaker, C.B. and Pearce, E.J. (1998) A divergent member of the transforming growth factor beta receptor family from *Schistosoma mansoni* is expressed on the parasite surface membrane. *Journal of Biological Chemistry* 273, 11234–11240.

Davies, S.J., Grogan, J.L., Blank, R.B., Lim, K.C., Locksley, R.M. and McKerrow, J.H. (2001) Modulation of blood fluke development in the liver by hepatic CD4+ lymphocytes. *Science* 294, 1358–1361.

Davis, R.E. and Hodgson, S. (1997) Gene linkage and steady state RNAs suggest *trans*-splicing may be associated with a polycistronic transcript in *Schistosoma mansoni*. *Molecular and Biochemical Parasitology* 89, 25–39.

de Mendonca, R.L., Escriva, H., Bouton, D., Laudet, V. and Pierce, R.J. (2000) Hormones and nuclear receptors in schistosome development. *Parasitology Today* 16, 233–240.

Dehal, P., Satou, Y., Campbell, R.K., Chapman, J., Degnan, B., De Tomaso, A., Davidson, B., Di Gregorio, A., Gelpke, M., Goodstein, D.M., Harafuji, N., Hastings, K.E., Ho, I., Hotta, K., Huang, W., Kawashima, T., Lemaire, P., Martinez, D., Meinertzhagen, I.A., Necula, S., Nonaka, M., Putnam, N., Rash, S., Saiga, H., Satake, M., Terry, A., Yamada, L., Wang, H.G., Awazu, S., Azumi, K., Boore, J., Branno, M., Chin-Bow, S., DeSantis, R., Doyle, S., Francino P., Keys, D.N., Haga, S., Hayashi, H., Hino, K., Imai, K.S., Inaba, K., Kano, S., Kobayashi, K., Kobayashi, M., Lee, B.I., Makabe, K.W., Manohar, C., Matassi, G., Medina, M., Mochizuki, Y., Mount, S., Morishita, T., Miura, S., Nakayama, A., Nishizaka, S., Nomoto, H., Ohta, F., Oishi, K., Rigoutsos, I., Sano, M., Sasaki, A., Sasakura, Y., Shoguchi, E., Shin-i, T., Spagnuolo, A., Stainier, D., Suzuki, M.M., Tassy, O., Takatori, N., Tokuoka, M., Yagi, K., Yoshizaki, F., Wada, S., Zhang, C., Hyatt, P.D., Larimer, F., Detter, C., Doggett, N., Glavina, T., Hawkins, T., Richardson, P., Lucas, S., Kohara, Y., Levine, M., Satoh, N. and Rokhsar D.S. (2002) The draft genome of *Ciona intestinalis*: insights into chordate and vertebrate origins. *Science* 298, 2157–2167.

DeMarco, R., Kowaltowski, A.T., Machado, A.A., Soares, M.B., Gargioni, C., Kawano, T., Rodrigues, V., Madeira, A.M.B.N., Wilson, R.A., Menck, C.F.M., Setubal, J.C., Dias-Neto, E., Leite, L.C.C. and Verjovski-Almeida, S. (2004) Saci-1, -2 and -3 and Perere, four novel retrotransposons with high transcriptional activities from the human parasite *Schistosoma mansoni*. *Journal of Virology* 78, 2967–2978.

Dias-Neto, E., Correa, R.G., Verjovski-Almeida, S., Briones, M.R., Nagai, M.A., da Silva, W. Jr, Zago, M.A., Bordin, S., Costa, F.F., Goldman, G.H., Carvalho, A.F., Matsukuma, A., Baia, G.S., Simpson, D.H., Brunstein, A., de Oliveira, P.S., Bucher, P., Jongeneel, C.V., O'Hare, M.J., Soares, F., Brentani, R.R., Reis, L.F., de Souza, S.J. and Simpson, A.J. (2000) Shotgun sequencing of the human transcriptome with ORF expressed sequence tags. *Proceedings of the National Academy of Sciences USA* 97, 3491–3496.

Dissous, C. and Capron, A. (1995) Convergent evolution of tropomyosin epitopes. *Parasitology Today* 11, 45–46.

Dunham, I., Shimizu, N., Roe, B.A., Chissoe, S., Hunt, A.R., Collins, J.E., Bruskiewich, R., Beare, D.M., Clamp, M., Smink, L.J., Ainscough, R., Almeida, J.P., Babbage, A., Bagguley, C., Bailey, J., Barlow, K., Bates, K.N., Beasley, O., Bird, C.P., Blakey, S., Bridgeman, A.M., Buck, D., Burgess, J., Burrill, W.D. and O'Brien, K.P. (1999) The DNA sequence of human chromosome 22. *Nature* 402, 489–495.

El-Sayed, N.M., Bartholomeu, D., Ivens, A., Johnston, D.A. and LoVerde, P.T. (2004) Advances in schistosome genomics. *Trends in Parasitology* 20, 154–157.

Escriva, H., Safi, R., Hanni, C., Langlois, M.C., Saumitou-Laprade, P., Stehelin, D., Capron, A., Pierce, R. and Laudet, V. (1997) Ligand binding was acquired during evolution of nuclear receptors. *Proceedings of the National Academy of Sciences USA* 94, 6803–6808.

Fantappie, M.R., Freebern, W.J., Osman, A., LaDuca, J., Niles, E.G. and LoVerde, P.T. (2001) Evaluation of *Schistosoma mansoni* retinoid X receptor (SmRXR1 and SmRXR2) activity and tissue distribution. *Molecular and Biochemical Parasitology* 115, 87–99.

Fenwick, A., Savioli, L., Engels, D., Robert Bergquist, N. and Todd, M.H. (2003) Drugs for the control of parasitic diseases: current status and development in schistosomiasis. *Trends in Parasitology* 19, 509–515.

Feschotte, C. (2004) Merlin, a new superfamily of DNA transposons identified in diverse animal genomes and related to bacterial IS1016 insertion sequences. *Molecular Biology and Evolution* 21, 1769–1780.

Fietto, J.L., DeMarco, R. and Verjovski-Almeida, S. (2002) Use of degenerate primers and touchdown PCR for construction of cDNA libraries. *Biotechniques* 32, 1404–1411.

Franco, G.R., Valadao, A.F., Azevedo, V. and Rabelo, E.M. (2000) The *Schistosoma* gene discovery program: state of the art. *International Journal for Parasitology* 30, 453–463.

Fung, M.C., Lau, M.T. and Chen, X.G. (2002) Expressed sequence tag (EST) analysis of a *Schistosoma japonicum* cercariae cDNA library. *Acta Tropica* 82, 215–224.

Grevelding, C.G., Sommer, G. and Kunz, W. (1997) Female-specific gene expression in *Schistosoma mansoni* is regulated by pairing. *Parasitology* 115, 635–640.

Gryseels, B. (2000) Schistosomiasis vaccines: a devils' advocate view. *Parasitology Today* 16, 46–48.

Hagan, P. and Sharaf, O. (2003) Schistosomiasis vaccines. *Expert Opinion on Biological Therapy* 3, 1271–1278.

Hagan, P., Doenhoff, M.J., Wilson, R.A., Al-Sherbiny, M. and Bergquist, R. (2000) Schistosomiasis vaccines: a response to a devils' advocate's view. *Parasitology Today* 16, 322–323.

Hagan, P., Appleton, C.C., Coles, G.C., Kusel, J.R. and Tchuem-Tchuente, L.A. (2004) Schistosomiasis control: keep taking the tablets. *Trends in Parasitology* 20, 92–97.

Hamdan, F.F., Abramovitz, M., Mousa, A., Xie, J., Durocher, Y. and Ribeiro, P. (2002) A novel *Schistosoma mansoni* G protein-coupled receptor is responsive to histamine. *Molecular and Biochemical Parasitology* 119, 75–86.

Hausdorf, B. (2000) Early evolution of the bilateria. *Systematic Biology* 49, 130–142.

Hoffmann, K.F. and Dunne, D.W. (2003) Characterization of the Schistosoma transcriptome opens up the world of helminth genomics. *Genome Biology* 5, 203.

Hoffmann, K.F., Johnston, D.A. and Dunne, D.W. (2002) Identification of *Schistosoma mansoni* gender-associated gene transcripts by cDNA microarray profiling. *Genome Biology* 3, research 0041.1–0041.12.

Hu, W., Yan, Q., Shen, D.K., Liu, F., Zhu, Z.D., Song, H.D., Xu, X.R., Wang, Z.J., Rong, Y.P., Zeng, L.C., Wu, J., Zhang, X., Wang, J.J., Xu, X.N., Wang, S.Y., Fu, G., Zhang, X.L., Wang, Z.Q., Brindley, P.J., McManus, D.P., Xue, C.L., Feng, Z., Chen, Z. and Han, Z.G. (2003) Evolutionary and biomedical implications of a *Schistosoma japonicum* complementary DNA resource. *Nature Genetics* 35, 139–147.

Ivanchenko, M.G., Lerner, J.P., McCormick, R.S., Toumadje, A., Allen, B., Fischer, K., Hedstrom, O., Helmrich, A., Barnes, D.W. and Bayne, C.J. (1999) Continuous *in vitro* propagation and differentiation of cultures of the intramolluscan stages of the human parasite *Schistosoma mansoni*. *Proceedings of the National Academy of Sciences USA* 96, 4965–4970.

Kunz, W. (2001) Schistosome male–female interaction: induction of germ-cell differentiation. *Trends in Parasitology* 17, 227–231.

Lenzi, H.L., Pacheco, R.G., Pelajo-Machado, M., Panasco, M.S., Romanha, W.S. and Lenzi, J.A. (1997). Immunological system and *Schistosoma mansoni*: co-evolutionary immunobiology. What is the eosinophil role in parasite–host relationship? *Memorias do Instituto Oswaldo Cruz* 92 (Suppl. 2), 19–32.

McManus, D.P., Le, T.H. and Blair, D. (2004) Genomics of parasitic flatworms. *International Journal for Parasitology* 34, 153–158.

Nilsen, T.W. (2001) Evolutionary origin of SL-addition *trans*-splicing: still an enigma. *Trends in Genetics* 17, 678–680.

Pearce, E.J. and MacDonald, A.S. (2002) The immunobiology of schistosomiasis. *Nature Reviews in Immunology* 2, 499–511.

Poulin, R. and Cribb, T.H. (2002) Trematode life cycles: short is sweet? *Trends in Parasitology* 18, 176–183.

Rollinson, D., Kaukas, A., Johnston, D.A., Simpson, A.J. and Tanaka, M. (1997) Some molecular insights into schistosome evolution. *International Journal for Parasitology* 27, 11–28.

Salzet, M., Capron, A. and Stefano, G.B. (2000) Molecular crosstalk in host–parasite relationships: schistosome– and leech–host interactions. *Parasitology Today* 16, 536–540.

Simpson, A.J., Sher, A. and McCutchan, T.F. (1982) The genome of *Schistosoma mansoni*: isolation of DNA, its size, bases and repetitive sequences. *Molecular and Biochemical Parasitology* 6, 125–137.

Soares, M.B., Bonaldo, M.F., Jelene, P., Su, L., Lawton, L. and Efstratiadis, A. (1994) Construction and characterization of a normalized cDNA library. *Proceedings of the National Academy of Sciences USA* 91, 9228–9232.

The *C. elegans* Sequencing Consortium (1998) Genome sequence of the nematode *C. elegans*: a platform for investigating biology. *Science* 282, 2012–2018.

Verjovski-Almeida, S., DeMarco, R., Martins, E.A., Guimaraes, P.E., Ojopi, E.P., Paquola, A.C., Piazza, J.P., Nishiyama, M.Y., Kitajima, J.P., Adamson, R.E., Ashton, P.D., Bonaldo, M.F., Coulson, P.S., Dillon, G.P., Farias, L.P., Gregorio, S.P., Ho, P.L., Leite, R.A., Malaquias, L.C., Marques, R.C., Miyasato, P.A., Nascimento, A.L., Ohlweiler, F.P., Reis, E.M., Ribeiro, M.A., Sa, R.G., Stukart, G.C., Soares, M.B., Gargioni, C., Kawano, T., Rodrigues, V., Madeira, A.M., Wilson, R.A., Menck, C.F., Setubal, J.C., Leite, L.C. and Dias-Neto, E. (2003) Transcriptome analysis of the acoelomate human parasite *Schistosoma mansoni*. *Nature Genetics* 35, 148–157.

Verjovski-Almeida, S., Leite, L.C., Dias-Neto, E., Menck, C.F. and Wilson, R.A. (2004) Schistosome transcriptome: insights and perspectives for functional genomics. *Trends in Parasitology* 20, 304–308.

Williams, S.A. and Johnston, D.A. (1999) Helminth genome analysis: the current status of the filarial and schistosome genome projects. Filarial Genome Project. Schistosome Genome Project. *Parasitology* 118, S19–S38.

7 Transgenic Flatworms

C.G. Grevelding
Institute for Parasitology, Justus-Liebig-University, Rudolf-Buchheim-Strasse 2, 35392 Giessen, Germany

Introduction

Flatworms (platyhelminthes) represent a phylum that includes the classes Turbellaria (planarians), Trematoda (flukes) and Cestoda (tapeworms). The Turbellaria consist of mostly free-living aquatic flatworms with some insignificant parasitic forms. The Trematoda, in contrast, comprise platyhelminths that are exclusively adapted to parasitism. Within the trematodes, the Monogenea consist of ectoparasites with a rapid life cycle that involves a single host, whereas the Digenea represent endoparasites with a complex life cycle including more than one host. The latter feature also characterizes members of the Cestoda, which are obligate parasites.

For various reasons, flatworms have come into the focus of scientific interest. They occupy a key position in the evolution of metazoans and are widely acknowledged as being among the simplest organisms possessing three tissue layers (triploblasts), bilateral symmetry, cephalization and complex organ systems (Brusca and Brusca, 2002). Furthermore, flatworms exhibit remarkable features, some of which are unique in nature as will be discussed later on.

Along with the major advances in DNA technology, methods for gene discovery and their functional characterization have been developed in model organisms during the last decades. These methods have also been applied to non-model organisms such as parasites (Prichard and Tait, 2001; Ellis *et al.*, 2003; Ersfeld, 2003). Among these, the protozoa register greatest progress with established *in vitro* culture systems permitting experimental manipulations, genomic and expressed sequence tag (EST) databases and transgenic systems for the purposes of functional gene analyses. Gene insertion and knockout strategies can be applied to parasitic protozoa as well as gene silencing by RNA interference (RNAi) (Kelly, 1997; Clayton, 1999; Beverly, 2003; Kim and Weiss, 2004). In contrast, research on parasitic and non-parasitic flatworms, as with other parasitic helminths, has lagged well behind the protozoa because *in vitro* culture systems have not been established for whole life cycles, cell lines are not available and no transformation systems exist that enable genetic modification of the worm of interest. Cell lines are still missing for most flatworms (Bayne, 1998), although for planarians and schistosomes the first approaches to isolate and maintain cells have been reported (Hobbs *et al.*, 1993; Asami, *et al.*, 2002; Dong *et al.*, 2003). Furthermore, *in vitro* culture systems have been developed for specific life stages of trematodes (Basch, 1981a,b, 1984; Nollen, 1990; Grevelding *et al.*, 1997; Kook *et al.*, 1997; Ivanchenko *et al.*, 1999; Laursen and Yoshino, 1999; Coustau and Yoshino, 2000; Chaithong *et al.*, 2001; Fried and Robert, 2002; Gorbushin and Shaposhnikova, 2002; Stewart *et al.*, 2003) and cestodes (Hemphill *et al.*, 2002; Siles-Lucas and Hemphill, 2002; Spiliotis *et al.*, 2004). Beyond this, remarkable progress has been made in the field of genomics. In view of the increasing amount of sequence information available for flatworms (Foster and Johnston, 2002; McManus *et al.*, 2004; Verjovski-Almeida *et al.*, 2004), it is crucial now to develop methods for post-genomic analyses in these organisms to allow functional genomic studies. This chapter concentrates on recent progress towards the establishment of transgenesis systems for flatworms. These new tools have the potential to help flatworm research catch up with the protozoan research and to facilitate progress in both basic and applied research, especially concerning the development of novel strategies to control parasitic flatworms.

The Planarian Model

Within the Turbellaria the freshwater planarians have long been known to possess remarkable regenerative abilities. As far back as 1898 Morgan (1898) demonstrated that a very small planarian fragment, corresponding to only 1/279th of the intact organism, was capable of regenerating a complete individual. Such a striking regenerative capacity has attracted generations of biologists to investigate this phenomenon (Salo and Baguna, 2002). The most peculiar cell type in planarians is the totipotent stem cells, called neoblasts. These cells possess an extraordinary morphological plasticity as well as the potential to divide and to differentiate into all cell types, and to regulate body size in the adult organism (Baguna, 1998).

Despite their attractiveness as a model system, studies of planarians in the past have mainly been descriptive or phenomenological. In the recent decade, examinations were also performed at the molecular level (Newmark and Sanchez Alvarado, 2002; Sanchez Alvarado, 2003). Different efforts have been made to unravel the genetic programme of regeneration. However, functional analyses were hampered by the fact that a transformation system was not available. Towards this end, remarkable progress has been made using planarians. Sanchez Alvarado and Newmark (1999) reported the successful application of RNAi, a technique that allows the abrogation of gene function at the post-transcriptional level. This method was originally developed for *Caenorhabditis elegans* (Fire *et al.*, 1998) and was subsequently shown to be applicable for a variety of invertebrates and vertebrates. The introduction of double-stranded RNA (dsRNA) specifically inhibits the expression of the gene that codes for the same sequence, leading to a loss-of-function phenotype. Patrick Skelly discusses details of this technique in Chapter 22 of this volume. The approach used for planarians comprised the *in vitro* synthesis of dsRNA and its microinjection into planarian fragments of 1–2 mm. At different time points after microinjection, gene-silencing effects were observed, which proved the applicability of the RNAi principle in this organism. Injection into either the gut or the parenchyma produced defects in other parts of the planarian body. This indicated the existence of dsRNA-transport mechanisms, as had been reported for *C. elegans* (Timmons *et al.*, 2003). Here, the site of dsRNA injection may also differ from the site of action. Injection into the gut of this nematode produced germ-line defects, and injection into the gonads led to somatic defects (Fire *et al.*, 1998). In addition to RNAi, it is also possible to stably transform planarians with vector–DNA constructs leading to gain-of-function phenotypes. Gonzalez-Estevez *et al.* (2003) recently introduced a method based on mobile genetic elements to shuttle reporter-gene constructs into the genome of this organism. The authors made use of the transposable elements *mariner*, *Hermes* or *piggyBac*, genome vagabonds that exhibit distinct insertion specificities in a variety of organisms. A classical two-component strategy for insertional mutagenesis was performed using truncated versions of these transposons and separate helper plasmids containing the transposase functions of the appropriate transposable elements. The truncated versions, still capable to be mobilized in *trans*, were combined with green fluorescent protein (GFP) reporter-gene constructs and cloned into plasmid–DNA backbones. To transform planarians, both the transposon reporter-gene constructs and the helper plasmids were simultaneously introduced by microinjection into adults or, alternatively, into posterior or anterior regenerating individuals. In both cases, microinjection was followed by electroporation. Supported by the transposase function on the helper plasmids, the transposon reporter-gene constructs were mobilized and integrated into the genome. Subsequent analyses demonstrated that *Hermes* and *piggyBac* vectors transposed at a high frequency and integrated stably into chromosomes (Fig. 7.1), whereas *mariner* constructs were less stable. This was explained by the presence of endogenous *mariner*-type transposons in the planarian genome that may interfere with the exogenously applied copies.

With the transposon-based gain-of-function system and the RNAi-based loss-of-function system, powerful tools now exist for the molecular analyses of developmental mechanisms in planarians such as regeneration, one of the oldest unresolved phenomena of this organism.

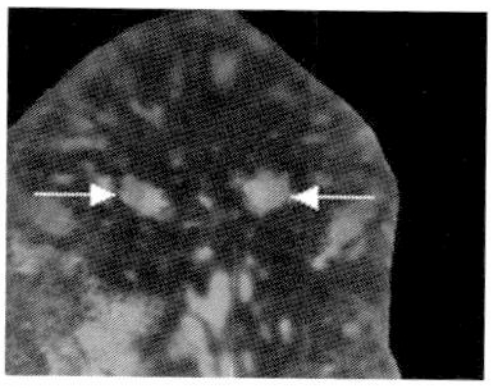

Fig. 7.1. A planarian individual of a line transformed with a *Hermes* transposon-derived vector 12 months after microinjection and subsequent electroporation. Besides the transposon part, the vector contained the E(enhanced)GFP reporter gene under the control of an artificial promoter that responds to the transcription factor Pax6 (Gonzalez-Estevez *et al.*, 2003). Binding sites for this transcription factor are found in rhodopsin and other photoreceptor-specific genes. Reporter-gene activity was found in the eyes (arrows) of transformed individuals or in regenerated heads as a mosaic of EGFP-positive photoreceptor cells. Figure courtesy of Dr E. Salo.

Trematodes and Cestodes

The increasing interest in flukes and tapeworms is based upon their impact on public health (Graczyk and Fried, 1998; Blair *et al.*, 1999; Deplazes and Eckert, 2001; Carpio, 2002; Hoberg, 2002; Ross *et al.*, 2002; Craig, 2003; Gonzalez *et al.*, 2003; McManus *et al.*, 2003; Savioli *et al.*, 2004). Beyond their medical importance for humans, parasitic flatworms are also of economical and ecological significance affecting livestock and wild animals (DeBont and Vercruysse, 1998; Hurtrez-Bousses *et al.*, 2001; Horak *et al.*, 2002; Otranto and Traversa, 2002). Tapeworms are endoparasitic organisms that successively infect intermediate and definitive hosts through oral ingestion of eggs or larvae. Serious diseases of humans or other mammalian hosts are mainly caused by larval stages. Echinococcosis, for example, is a zoonosis caused by larval stages of a dog tapeworm of the genus *Echinococcosus*. It is a major public health problem in many parts of the world (Jenkins and Macpherson, 2003). *Echinococcosus multilocularis* is the causative agent of alveolar echinococcosis, which is one of the most dangerous helminth infections of man (Rausch, 1995). The larval stage of the pork tapeworm (*Taenia solium*) causes neurocysticercosis in infected human, a widespread disease affecting the nervous system (Garcia *et al.*, 2003). Due to their relevance, *Echinococcus* spp. and indeed *Taenia* spp. have been well investigated. Besides these, *Hymenolepis* spp.

and *Mesocestoides* spp. parasites have been studied, although infections are rare in humans. But these species can be easily maintained in the laboratory (Siles-Lucas and Hemphill, 2002).

Although the biology of these parasites has been well studied, they still keep many secrets that enable these organisms to thrive in the face of remarkable efforts to eradicate them. During evolution, parasites have been equipped with impressive qualities that allow these animals to conquer their hosts and to modify the host's environment in their interest. One example of such remarkable features is the ability of schistosomes to live for decades in the blood system of their vertebrate definitive host, surviving the attacks of the immune system. The consequence of this parasitism is schistosomiasis, one of the most prevalent parasitic infections affecting millions of people in the tropics and subtropics (Rollinson and Simpson, 1987; Ross *et al.*, 2002; Savioli *et al.*, 2004). Because of the worldwide significance of this disease, programmes have been established to control the spread of schistosomes. Based on the classical knowledge of the life cycle of this digenean trematode, ecological programmes have been undertaken to eradicate the intermediate snail host with molluscicides (Ohmae *et al.*, 2003; Blas *et al.*, 2004). These approaches are expensive and questionable considering geography, climate, hydrographic conditions and ecological and biological consequences (Sturrock, 2001; Kiesecker, 2002). The safe and effective drug praziquantel is commonly used to fight schistosomes pharmacologically, but drug treatment does not prevent reinfection. Beyond that, the introduction of mass treatment programmes has provided the first evidence of drug resistance (Doenhoff *et al.*, 2002; Cioli and Pica-Mattoccia, 2003), leading to an acknowledged need for novel drugs or prophylactics. There have been considerable advances in vaccine discovery, but the current model vaccines are not perceived to be sufficiently protective (Bergquist *et al.*, 2002; Bergquist, 2004; Lebens *et al.*, 2004), and there is a continuing search for additional candidate molecules (Fenwick *et al.*, 2003; Pearce, 2003). Up to now, drug and vaccine discovery have been dependent on empirical approaches, largely because of the lack of substantial genomic and biological information. To identify novel targets with control implications, sequencing projects have been initiated, supported by the World Health Organization, governmental institutions or private companies. Towards this end, much progress has been made during the last decade (Foster and Johnson, 2002; Hoffmann and Dunne, 2003; LoVerde *et al.*, 2004a), and first data-sets of the transcriptomes of *Schistosoma mansoni* and *S. japonicum* were recently published (Hu *et al.*, 2003; Verjovski-Almeida *et al.*, 2003). These transcript data, which probably represent >90% of the protein-coding part of the genome, are being supplemented by genomic data generated at the Sanger and TIGR institutes. Details of these studies are summarized in Chapters 2 and 6 of this volume. Besides the enormous sequencing efforts, diverse molecular approaches of the pre-genomic sequencing era did lead to the identification of genes that may play important roles during the development of this parasite or during the host–parasite interaction (Ramachandran *et al.*, 1996; Davies *et al.*, 1998; Mastroberardino *et al.*, 1998; Skelly and Shoemaker, 1998; De Mendonca *et al.*, 2000; Inal and Sim, 2000; Rao and Ramaswamy, 2000; Fantappie *et al.*, 2001; Osman *et al.*, 2001; Caffrey *et al.*, 2002; Hoffmann *et al.*, 2002; Knobloch *et al.*, 2002; Schramm *et al.*, 2003; Vicogne *et al.*, 2003; Kapp *et al.*, 2004).

With the increasing amount of sequence information, there is now a pressing need to develop methods that enable exploitation of the genome data, and investigation of the function of genes of interest. However, schistosomes as well as other multicellular parasites are, for several reasons, not ideal organisms for laboratory work. Among these are the relatively low numbers of parasites that can be generated. Furthermore, life cycles have to be maintained in intermediate and definitive hosts such that access to appropriate life stages is limited. *In vitro* cultivation methods that allow maintaining life cycles completely *in vitro* have not yet been established. This complicates classical genetic approaches to identify genes such as the induction of mutations and the subsequent screening for interesting phenotypes that could be characterized by genetic hybridization and molecular methods. Due to these obstacles, the development of reverse genetic approaches appears to be the only possible way to study genes in these organisms. However, a prerequisite for such approaches is the availability of transformation systems that allow the introduction of transgenes into the appropriate parasite. In this respect, some progress has been made for trematodes in recent years.

Transformation Techniques for Schistosomes

For transformation experiments in general, biochemical or physical methods such as microinjection, electroporation or lipofection are available. These methods have been successfully applied for a variety of organisms including protozoan parasites (Clayton, 1999; De Koning-Ward *et al.*, 2000) and were consequently tested for their ability to generate transiently transformed schistosomes by different laboratories.

With GFP reporter-gene constructs, lipofection experiments were performed with adults or sporocysts cultured *in vitro* using different commercially available reagents. However, reporter-gene activity could be detected in neither case (Kapp, 2000). This was not due to the vector construct as it later turned out to be functional. In a parallel approach fluorescently labelled reagents for lipofection revealed signals at the schistosome surface after treatment, but there was no indication of the uptake of these compounds into subtegumental areas (J. Kusel, Glasgow, 1999, personal communication).

Electroporation was tested with miracidia, the first free-living larval stage. But, depending on the conditions, miracidia either did not survive, or they were biologically inactive after treatment (Kapp, 2000). However, the latest studies support the use of this technique for the introduction of RNA into schistosomula (Correnti and Pearce, 2004). This life stage represents a pre-adult form that develops from cercariae shortly after they penetrate the final host. Schistosomula can be generated from cercariae *in vitro* by chemically or physically induced transformation processes (Salafsky *et al.*, 1988). Correnti and Pearce (2004) used schistosomula for electroporation experiments and luciferase mRNA. Enzyme activity assays showed that electroporation targeted the majority of the schistosomula. Immunolocalization studies indicated that the RNA was delivered to tegumental and subtegumental tissues. However, the RNA was unstable, and luciferase activity declined 24 h post-electroporation and was not detectable at 72 h. Nevertheless, the results may open the perspective to test electroporation for dsRNA delivery to schistosomula for RNA-silencing experiments. Furthermore, DNA delivery can be envisaged for gain-of-function studies. The biological effect of such approaches could be studied in two ways: (i) *in vitro*, because schistosomula can be cultured for several weeks (Basch, 1981a) and (ii) as schistosomula can be intravenously injected into a final host to complete the life cycle (Stirewalt, 1974; James and Taylor, 1976), it is possible to investigate long-term effects.

Microinjection was found to be difficult due to the soft structure of the surface tegument. Tests with adults and sporocysts demonstrated the potential of microinjection using selected dyes, which diffused through the body after treatment (L. Gohr, Düsseldorf, 1995, PhD thesis at the University of Düsseldorf; Wippersteg, unpublished). However, only a very low number of individuals could be successfully manipulated within 1 day, and most of these did not survive in culture afterwards.

Finally, we tested the particle bombardment method, a biolistic approach that was shown before to work when other approaches have failed. In particular, particle bombardment has allowed the generation of transgenic plants such as crop species, which are not susceptible to *Agrobacterium tumefaciens* or cannot be regenerated from protoplasts. Furthermore, particle bombardment has facilitated organelle transformation in intact cells and the genetic modification of cultured cells that were not accessible to other transfection techniques.

Particle bombardment

The basic principle of this technology is to use DNA-coated microparticles that are accelerated by gas pressure to velocities that enable these particles to penetrate cell layers in order to introduce the genetic material into cells (Yang and Christou, 1994). Gold or tungsten particles are commonly used as carriers for the DNA. Researches at the Cornell University were the first to show that accelerated tungsten particles (1–4 μm) are able to penetrate membranes and cell layers (Sanford *et al.*, 1987). It was recognized that biological substances such as proteins, synthetic macromolecules, stains and genetic material could be transported into cells by particle bombardment. Work in subsequent years achieved the transient expression of reporter genes in epidermal cells of onions and the stable transformation of maize, soybean and tobacco (Klein *et al.*, 1990). This opened the door for biolistic approaches with other plant species. Motivated by the success for plant systems, the repertoire of target organisms was enlarged, and particle bombardment protocols were developed for algae, animals, bacteria, cells, fungi, yeast and other organisms (Klein *et al.*, 1992).

Particle bombardment includes a sequence of physical, chemical and biolistic interactions involving accelerated microprojectiles, DNA and living target tissues. The frequency of cells transiently expressing reporter genes after bombardment is influenced by different factors, the composition and size of the microprojectiles, the DNA-coating of the microprojectiles before bombardment, the velocity of the coated particles and the degree of tissue damage following bombardment. Considering the small diameter of a cell, preferred sizes of microprojectiles are in the range of 0.6–1.6 μm. The microprojectiles should be chemically inert to avoid reactivity with DNA or toxic effects in cells after penetration. Tungsten and gold particles meet these requirements and have been widely used. However, there is evidence that tungsten may have toxic effects reducing the recovery of transformants in some species (Russel *et al.*, 1992). The optimal size and velocity of the microprojectiles depend on the properties of the target cell such as size, penetrability and resilience. A study investigating the relationships of particle diameter, velocity, vacuum, target distances and transformation frequencies was published by Klein *et al.* (1992). It was demonstrated that large particles, small target distances and high vacuum conditions generate high particle velocities. Slightly different DNA-coating protocols have been developed that are probably universally applicable to all cell types regardless of other bombardment conditions. After the first attempts to precipitate DNA on to gold particles have been performed by ethanol, modified protocols were established that included polyethylene glycol (PEG) precipitation and $CaCl_2$/spermidine coprecipitation followed by an ethanol wash (Yang and Christou, 1994). Tungsten particles, in contrast, are loaded with DNA by $CaCl_2$/spermidine coprecipitation only. Although these protocols are accepted as standard, the importance of a consistent protocol to precipitate DNA on to the appropriate microprojectiles, the preparation of the DNA-coated particles, and the loading of the acceleration apparatus have considerable influence on the efficiency of the gene transfer. The generation of reproducible data depends on parameters such as oxidation, aggregation or settling of particles before DNA precipitation, purity and concentration of the DNA, altered precipitation kinetics with varying temperatures during precipitation, aggregation of particles following the precipitation process, degradation of DNA or loss of DNA from the coated particles by post-precipitation treatments such as sonication to disperse particle aggregates. Besides these parameters, the pressure used to accelerate the particles has a decisive influence on the transformation success. Size and velocity determine the deepness of the penetration of particles into the target organism (Sanford *et al.*, 1993).

Bombardment with microprojectiles is a physical trauma for target cells. Although bombarded cells have the ability to repair damage, e.g. membranes, some bombardment conditions may lead to the disruption of tissues and cell death, e.g. during gold particle preparation aggregates can form that act as macroprojectiles damaging the target tissue. This may result in a central area of tissue damage, which is surrounded by an area of unaffected, transgene-expressing cells. Aggregate formation can be prevented by a short sonication step before loading the DNA on to a macrocarrier.

Although not proven, it is assumed that following bombardment, gold particles enter the cytoplasm of a cell and release the vector

DNA, which is actively transported into the nucleus to be transcribed. In some cases, the nucleus may be a target of the microprojectile, but due to the size proportions, the nucleus may be destroyed by the impact of the microprojectile.

The PDS-1000/He particle bombardment system

This vacuum-based system has been developed by the BioRad Laboratories and uses gas to accelerate microparticles (see Fig. 7.2). By helium gas pressure a shock wave is generated, first passing a plastic disc (rupture disc) that closes the end of an acceleration tube. The rupture disc bursts at a defined pressure (optionally between 450 and 2200 psi) and releases the gas into the lower part of the tube, which is sealed by a second plastic disc carrying DNA-coated gold or tungsten particles (macrocarrier). This carrier sheet is propelled a short distance and immediately stops at a metal grid serving as a stopping screen that retains the plastic disc, but allows the DNA-coated microcarriers to move on through the pores of the grid. Accelerated by the energy of the shock wave, the microparticles finally reach their targets, which reside in Petri dishes at variable positions below the stopping screen area. The target distance determines the spread of the particles over the targets. During the procedure, rupture disc, macrocarrier and stopping screen are subjected to violent forces and therefore are replaced by new ones for the next experiment.

Based on previously used protocols (Wippersteg *et al.*, 2002a), the following slightly modified parameters are actually used in our laboratory to bombard schistosomes: (i) gold sizes of 0.6–1.6 μm; (ii) pressures of 1350–1800 psi; (iii) vacuum of 15 in Hg; (iv) target distance stage 1 (3 cm); and (v) 5–7 μg DNA/600 ng gold. An excellent dispersion of gold particles for optimal loading on to the macrocarrier sheet is accomplished by a short sonication step for 10 s at 50 kHz. These parameters allow a good dispersion of the particles on the macrocarrier, which leads to a wide distribution of particles during bombardment. Small gold particles and/or lower pressures lead to an accumulation of particles within the tegumental area and the parenchyma. Big microprojectiles and/or high

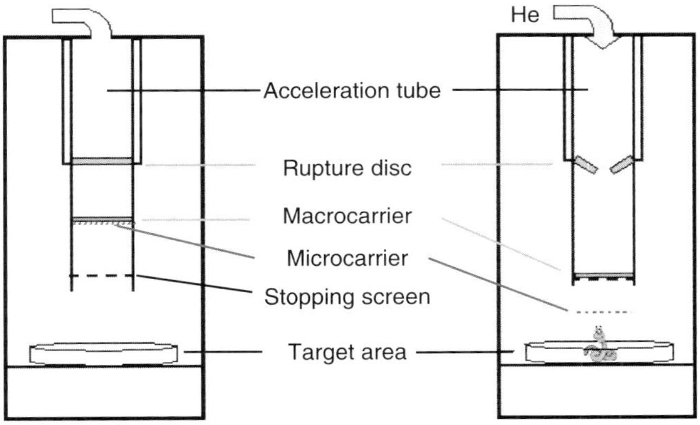

Fig. 7.2. Diagram of the PDS-1000/He, a stationary particle bombardment machine that is connected to a helium gas container. Controlled by adjustable valves, the gas stream (He) terminates in an acceleration tube, which is mounted on the top of a target chamber. This chamber is closed by a door and set under vacuum shortly before bombardment. When gas flows into the acceleration tube, the rupture disc bursts releasing the shock wave into the lower part of the tube. The gas pressure then accelerates the macrocarrier sheet containing the microprojectiles on its lower surface. The net-like stopping screen holds the macrocarrier sheet back and serves to block the shock wave, while the microprojectiles slip through the pores of the grid and continue on towards their final target.

pressures are needed to reach deeper tissue areas. Sections of bombarded worms revealed gold particles from the tegument to the gastrodermis, indicating that under these conditions nearly every tissue is hit. For schistosomes, bombardment under vacuum was found to be the only suitable way to introduce transgenes without damaging the worm too much. Using non-vacuum particle bombardment systems such as the gene gun, which works under normal atmospheric conditions, caused severe damages on schistosomes that probably resulted from the combination of (i) the air/gas blast used to accelerate the microprojectiles and (ii) the soft structure of the tegument surrounding the worm (Wippersteg *et al.*, 2002b). In contrast, the cuticle of filarial parasites, such as *Litosomoides sigmodontis*, resisted physical stress allowing bombardment experiments with the gene gun without damaging the worms significantly (Jackstadt *et al.*, 1999). In the meantime, the PDS-1000 system was successfully used for bombardment experiments with the parasites *Leishmania tarantolae* (Sbicego *et al.*, 1998), *Brugia malayi* (Higazi *et al.*, 2002), *Trypanosoma brucei* (Hara *et al.*, 2002) and with the free-living nematode *C. elegans* (Wilm *et al.*, 1999; Berezikov *et al.*, 2004). Compared to the microinjection technique commonly used for this model nematode, the biolistic approach allows the manipulation of a higher amount of individuals in a significantly shorter time period. Furthermore, it even allows the design of gene knockout approaches using substitution vectors that are designed for homologous recombination (Berezikov *et al.*, 2004).

Reporter-gene activity in bombarded schistosomes

When we started to elaborate the particle bombardment technique for *S. mansoni*, two questions were asked. First, is it generally possible to use the biolistic approach to genetically modify schistosomes? Second, if yes, does the method allow the generation of reliable and reproducible results? To approach these questions, we first designed test vectors for transformation experiments. These vectors should fulfil two different criteria. First, they should contain reporter genes whose expression can be satisfactorily demonstrated. Second, regulatory regions of genes that had already been identified and characterized in schistosomes should control the expression of the reporter genes. Thus, we wanted to confirm the reliability of the results obtained.

Parallel to our study Davies *et al.* (1999) started a similar approach with the PDS-1000 system to transform parasitic helminths. For initial experiments, a marked spliced leader (SL) RNA gene was biolistically introduced into 32- or 64-cell *Ascaris* embryos and was shown to be expressed accurately. Subsequently, vectors were used containing the luciferase reporter gene under the control of the regulatory elements from different genes such as a translation elongation factor, fert-I, rRNA or the SL-RNA promoter. In transformed *Ascaris*, the fert-I and SL-RNA promoters were able to significantly elevate luciferase activity up to 200-fold, whereas the rRNA promoter produced low levels similar to the promoter-less control. In addition, it was tested whether luciferase RNA instead of DNA can be introduced to *Ascaris* embryos and expressed following particle bombardment. Lyophilized on to gold particles, the biolistically introduced RNA led to significant levels of luciferase activity in a dose- and time-dependent manner. Besides *Ascaris* embryos, schistosomes were also chosen as targets for particle bombardment using DNA- or RNA-coated projectiles and pressures between 1500 and 2000 psi. With a DNA-construct containing the luciferase gene and the *S. mansoni* SL-RNA promoter, luciferase activity was found to be elevated 20-fold in adults after particle bombardment, indicating that the transgene is expressed in this organism. However, neither for *Ascaris* nor for *S. mansoni* were molecular or microscopical data provided, demonstrating the level of transgene expression or the quality of the worms after bombardment, or the distribution of the gold particles/constructs within the worms.

In our experiments a slightly modified form of the GFP gene from the jellyfish *Aequorea victoria* (Reichel *et al.*, 1996) and the β-glucoronidase (GUS) gene from *Escherichia coli* (Jefferson *et al.*, 1986) were chosen as reporter genes. GFP was selected because its expression generates strong visible bioluminescence that can be easily detected by fluorescent microscopy (Chalfie *et al.*, 1994; Kendall and Badminton, 1998). GUS is a hydrolase that catalyses the cleavage of a wide variety of β-glucoronides, many of which are available commercially as spectrophotometric, fluorometric or histochemical substrates providing alternative means for monitoring reporter-gene activity. Similar to the well-known β-galactosidase (β-Gal) system with X-Gal as a substrate, GUS activity can be microscopically visualized using the substrate X-Gluc (*p*-nitrophenyl β-D-glucoronide). In addition, 4-methyl umbelliferyl glucoronide (MUG) or *p*-nitrophenyl-β-D-galactopyranoside (PNPG) can be used as a substrate to determine values of GUS activity by spectrophotometric analyses. The classical β-Gal system was not considered as a reporter as there is an endogenous β-Gal activity in schistosomes that has been localized in the tegument by azo-dye substrates during immunolocalization experiments (W. Kunz, Düsseldorf, 1996, personal communication). Recently, schistosome β-Gal homologues were identified in *S. mansoni* (Franco *et al.*, 1997; Parra, Genbank Accession # AA525623) and in *S. japonicum* (Hu *et al.*, 2003). Preliminary experiments with the X-Gluc substrate on living worms or worm extracts indicated no endogenous GUS activity in schistosomes (C.G. Grevelding, unpublished data).

As regulatory elements to control reporter-gene activity, we used promoter and terminator regions of a variety of known genes. Among these was the heat-shock (HS) protein (Hsp) gene Hsp70, which was originally identified in a Puerto Rican strain of *S. mansoni* and carefully characterized. Expression studies demonstrated that the schistosome Hsp70 gene is developmentally regulated and stress inducible (Neumann *et al.*, 1993). Promoter and terminator regions of the Hsp70 gene were cloned by polymerase chain reaction (PCR) from a Liberian schistosome isolate and fused to the GFP reporter gene. As Hsp70 is highly conserved in nature, we first tested the Hsp70–GFP construct in an established cell culture system to confirm the functionality of the construct. The transfection results demonstrated HS-dependent expression of GFP, and thus confirmed the predicted promoter activity. As a next step, adult male schistosomes were used for particle bombardment experiments with this vector using pressure of 1550–1800 psi. Molecular analyses demonstrated the presence of the Hsp70–GFP vector, its transcription and translation (Wippersteg *et al.*, 2002b). Confocal microscopy finally confirmed the correct folding of the GFP protein because fluorescing signals were detected. In adults, signals were obtained following HS corresponding to published data. Neumann *et al.* (1993) had shown that there is a weak basal level of Hsp70 transcriptional activity that was elevated sevenfold following HS. In our experiments, this basal activity was not detectable with the GFP constructs at the microscopic level in non-stressed worms. However, in subsequent experiments some basal transcription of the GFP transgene was found by RT-PCR in bombarded adults (V. Wippersteg, Düsseldorf, 2002, unpublished data). But this basal level of activity was probably not sufficient to detect GFP fluorescence by confocal microscopy. In heat-shocked worms, GFP signals predominantly concentrated in the tegument including the tubercles (Fig. 7.3A). This was not a consequence of a technical problem such as the insufficient penetrance of gold particles into deeper tissue areas because previous histochemical analyses by methylene-blue staining of 5 μm sections indicated the presence of gold particles in nearly all tissues of bombarded worms.

There is supportive evidence for the assumption that the tegumental dominance of GFP may indicate a biological function of Hsp70 in this part of the worm following HS. First, Hsp70 represents a major immunogen in schistosomes (Hedstrom *et al.*, 1987; Moser *et al.*, 1990; F. Mutapi and R. Maizels, Edinburgh, 2004, personal communication) indicating that it may be present at the surface area of adults, or that it may be secreted. Second, immunohistochemical studies in *S. japonicum* showed the tegumental expression of the Hsp70 ortholog (Scallon *et al.*, 1987). Finally, to provide additive evidence for the tegumental expression pattern in *S. mansoni*, we performed immunohistochemical analyses with adult males. After HS for 4 h at 42°C, sections were prepared and incubated with an antibody

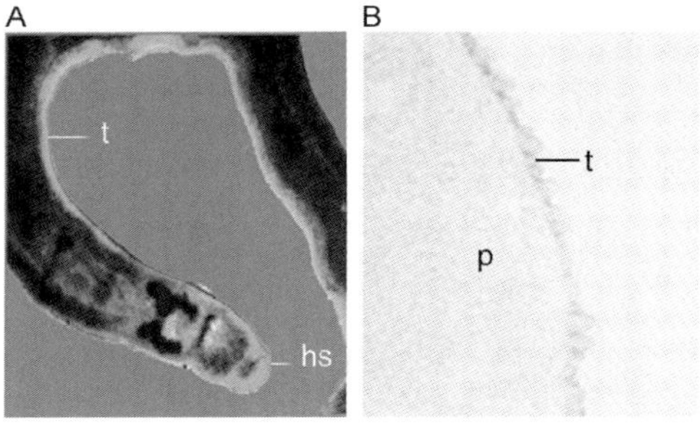

Fig. 7.3. A. Confocal laser scanning micrograph of a male schistosome bombarded with the Hsp70–GFP construct after heat-shock (HS) showing signals along the tegument (t) including the head sucker (hs) (partly reprinted from Wippersteg *et al.*, 2002b with permission from Elsevier). B. Bright-field microscope image of a 5 μm section of a wild-type male worm after HS. A section of the dorsal part is shown (p = parenchyma; t = tegument). Immunolocalization was performed with an antibody against *Schistosoma mansoni* Hsp70 (Moser *et al.*, 1990). Detection was performed using an alkaline-phosphatase conjugated secondary antibody, naphthol–AS–phosphate and Fast Red TR (Finken *et al.*, 1994).

against *S. mansoni* Hsp70 (Moser *et al.*, 1990), which was kindly provided by Mo Klinkert (Bernhard-Nocht-Institut, Hamburg). As shown in Fig. 7.3B, Hsp70 is predominantly expressed in the tegument of stressed worms after HS. These data correspond to the results obtained in *S. japonicum* and provide further supportive evidence for the reliability of the particle bombardment results with transgenic *S. mansoni*, although the role of Hsp70 in the schistosome tegument after HS still remains elusive.

As the biolistic approach worked well with adults, we also tried to perform bombardment experiments with larval schistosomes. As targets, we used sporocysts, which develop *in vivo* within the intermediate snail host following infection by the miracidium, the first free-living larval stages hatching from the egg. With the *Biomphalaria glabrata* embryonic cell (*Bge*-cell) co-cultivation method developed by Coustau *et al.* (1997) it is possible to transform miracidia into sporocysts *in vitro* and to maintain this developmental stage in culture for some time. Groups of 2000–4000 sporocysts were generated this way and used for bombardment experiments. Due to the more sensitive nature of these larvae, lower pressures (900 or 1350 psi) were applied that guaranteed the survival of the individuals after bombardment. The presence and the transcription of the Hsp70–GFP vector after bombardment were confirmed by PCR and RT-PCR analyses. Transcripts were detected in sporocysts without HS induction. This again corresponded to previous findings as Northern blot analyses had demonstrated that Hsp70 is constitutively active within the larval stages of schistosomes (Neumann *et al.*, 1993). In addition to the transcriptional data, we detected fluorescing signals by confocal microscopy in different tissues except the tegument. This indicated a more ubiquitous activity in sporocysts. Although the obviously different roles of Hsp70 during schistosome development are not clear yet, the results of these gene-transfer experiments demonstrated the possibility of transient transformation of different schistosome stages, and the expression of GFP reporter-gene constructs in this organism. Furthermore, data were obtained, which corresponded to previous findings.

A second test vector consisted of GFP fused to the regulatory elements of the ER60 gene, which codes for a cysteine protease in schistosomes (Finken-Eigen and Kunz, 1997). Proteases play important roles in all developmental stages of schistosomes (Trap and Boireau, 2000). In eggs, proteases are produced as excretory/secretory (ES) molecules that allow tissue migration. In miracidia, these enzymes are important for the penetration process into the intermediate snail hosts and the transformation to mother sporocysts. The ability of cercariae to penetrate the skin of a final host is also mediated by proteases. In adults, finally, proteases, among other tasks, manage digestion and nutrient uptake. Due to their essential functions for the biology of parasites, these enzymes are regarded as promising targets for the development of antiparasitic drugs or vaccines (Sajid and McKerrow, 2002). Database analyses classified ER60 as a member of the protein disulfide isomerases, and it was hypothesized that the enzyme has a function within the endoplasmic reticulum (ER). The ER60 gene is gender-independently transcribed, and localization experiments in adults demonstrated its expression in ES tissues such as the gastrodermis or the protonephridia (Finken-Eigen and Kunz, 1997). For particle bombardment experiments, part of the 5′ untranslated region (UTR) and the 3′ UTR of the ER60 gene were cloned and fused to GFP. Transfection experiments with COS-7 cells confirmed the functionality of the promoter, although only 276 bp of the 5′ UTR of ER60 were used. After transformation of adult schistosomes with an ER60–GFP vector, significant GFP signals were detected in the gastrodermis. In addition, fluorescence was also observed as stripe-like structures within the parenchyma (Wippersteg *et al.*, 2003). Due to the tissue-preferential expression in the gastrodermis of schistosomes and the activity in COS-7 cells, it was concluded that the used 5′ UTR contained most of the elements important for mediating tissue specificity. Within this short fragment, a TATA box occurs and in addition one S8 homeodomain-binding site followed by an inverted GATA element, a tandem element that was identified later (Rossi *et al.*, 2003). The homeobox protein S8 belongs to the *paired* homeodomain subfamily, which includes *S. mansoni* homologues (Webster and Mansour, 1992), while Cys_4 zinc finger factors bind to GATA elements (Patient and McGhee, 2002). The S8 and GATA sequence motifs occur in the promoter regions of tissue-specifically and developmentally expressed genes. To find out whether the stripe-like activity represented expression in the protonephridial system, a colocalization approach was undertaken (Wippersteg *et al.*, 2003). For this, we made use of Texas Red-BSA (TxR-BSA), a fluorescent dye that enters the ES system, especially the excretory tubules (Tan *et al.*, 2003). These tubules are part of the well-structured protonephridium. They are widely distributed along the worm and ramify in most tissues. For colocalization experiments, adult male schistosomes were first bombarded with the ER60–GFP construct and subsequently stained with TxR-BSA. Confocal laser scanning microscopy was done using the excitation wavelengths for GFP (488 nm) and TxR-BSA (568 nm). GFP-signals were detected in several tubular structures between the gut and the surface area. In the same structures, TxR-BSA-derived red fluorescence was identified (Fig. 7.4). The merged image of these areas

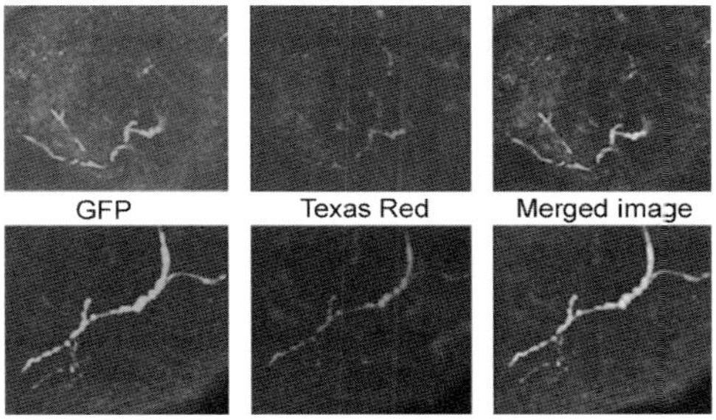

Fig. 7.4. Confocal laser scanning microscopy of parts of male schistosomes that were first bombarded with ER60–GFP and then stained with TxR-BSA. The upper and lower lanes show close-ups of different excretory tubules embedded within the parenchyma. Wavelengths of 488 nm for excitation of GFP and 568 nm for excitation of TxR-BSA were used (partly reprinted from Wippersteg *et al.*, 2003 with permission from Elsevier).

revealed yellow signals, which confirmed the occurrence of GFP and TxR-BSA in the same tissue (Wippersteg *et al.*, 2003). These results indicated that the cysteine protease ER60 is also expressed in the excretory tubules, which are part of the protonephridia of adult schistosomes. The obtained tissue-specific expression pattern of GFP under the control of the regulatory elements of the ER60 gene within the ES system corresponded to the expectation.

Because no data existed on ER60 expression in larval stages at this time point, the analysis was extended. Following RT-PCR analyses, which indicated that ER60 is transcribed in miracidia, mother sporocysts and cercariae, bombardment experiments were performed with mother sporocysts generated from miracidia *in vitro*. As the ES tissues of these larval stages are well known, it was expected to identify ER60 promoter-induced GFP activity here. After bombardment, bright field microscopy showed a wide distribution of gold particles inside the larvae. Molecular analyses confirmed that GFP is transcribed and translated. By confocal laser scanning microscopy, fluorescence was specifically observed in ES tissues, while the neural mass or the germ cells did not show fluorescence. Most significantly, two parallel structures at the anterior part exhibited clear signals (Fig. 7.5). According to ultrastructural data, these structures represent

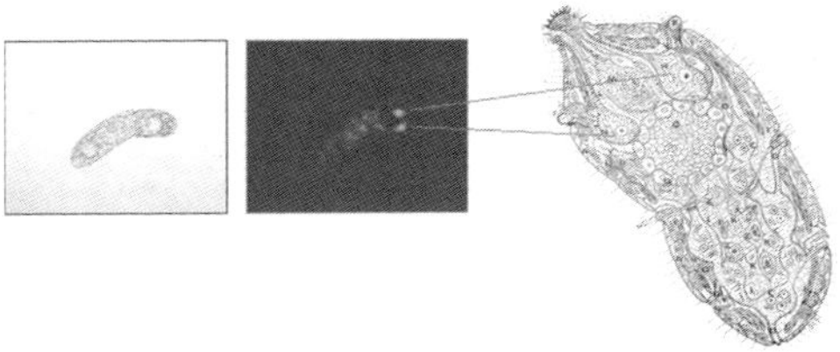

Fig. 7.5. Bright field (left side) and confocal laser scanning microscopy (middle part) of a sporocyst bombarded with ER60–GFP. The black lines indicate the lateral gland cells (N), which flank the medium gland cell (M) in the anterior part of the sporocyst as indicated by the morphological overview (right side; partly reprinted from Pan, 1980 and Wippersteg *et al.*, 2002c with permission from Elsevier).

the lateral gland cells flanking the apical gland (Pan, 1980). These cells contain ES products, which are supposed to be important for the penetration of miracidia into the snail. In addition to these cells, GFP activity was identified in the median and posterior part of the sporocysts where the ridge cytons, interstitial cytons, the cytons of the common excretory tubule and the protonephridia are located (Pan, 1980). These tissues belong to the ES system of larvae. From these data it was concluded that in miracidia/sporocysts ER60 also functions tissue-specifically in cells of the ES system.

The colocalization experiments using fluorescent molecules expanded the possibilities to characterize tissue-specific gene activity in living adults. The usefulness of TxR-BSA to specifically determine reporter-gene activity in the ES system was subsequently used to characterize the tissue-specific expression of the calcineurin (CN) gene from schistosomes. CN is a calcium/calmodulin-dependent serine–threonine protein phosphatase that is composed of a catalytic A subunit (CNA) and a regulatory B subunit (CNB). Homologues of both CN subunits were isolated from *S. mansoni*, and CNA protein was localized by immunofluorescence in the tegument, the excretory tubules and in the flame cells of all life cycle stages (Mecozzi *et al.*, 2000). Therefore, it was speculated that CN has a role in ES processes in schistosomes. As a first step to characterize gene regulation by the colocalization approach, the 5′- and 3′-regions of the SmCNA gene were cloned and ligated to the GFP gene. The resulting vector was used for particle bombardment of adult worms, which were analysed by confocal microscopy after the additional TxR-BSA staining. GFP expression was observed in the tegument and in the ES system. Similar to the results obtained for ER60–GFP, green fluorescing signals and TxR-BSA-derived red signals colocalized in the same Y-shaped tubules of the ES system (Rossi *et al.*, 2003). Besides the coinciding tubular expression observed for ER60- or CNA-induced GFP activity, computer-based sequence analyses of the cloned 5′ regions of the CNA or ER60 genes revealed another interesting congruence. Among the identified elements in the CNA promoter region is the tandem S8 homeodomain/GATA binding site that is also present in the ER60 promoter. Because these elements were thought to be involved in the regulation of tissue specificity, their corresponding occurrence in the promoter regions of both genes may indicate a common regulatory pathway leading to tissue specificity in the ES system. To elucidate the hypothesized function of these elements, deletion variants of the appropriate promoter sequences have to be generated and fused with appropriate reporter genes such as GFP, GUS, or luciferase to determine their function in more detail.

In addition to these transformation vectors, further constructs were designed for bombardment experiments to characterize gene activity by fusing the regulatory elements of a gene of interest with the GFP reporter gene. Among these is an Actin–GFP vector consisting of the 5′- and 3′-regions of a schistosome actin gene fused to GFP. Confocal microscopy demonstrated GFP activity specifically in the muscles, parenchyma, the tegument and the tubercles of bombarded male schistosomes (Wippersteg *et al.*, 2004). This pattern perfectly corresponded to immunolocalization results obtained with an anti-*S. mansoni* actin antibody (MacGregor and Shore, 1990). In a similar approach with the promoters of the genes for cathepsin L or cathepsin B2 from *S. mansoni* regulating GFP activity, fluorescence was detected exclusively in the gut or the tegument (Wippersteg *et al.*, 2005). Again, the specific activities corresponded to the expectations as previous immunolocalization studies demonstrated activity in exactly these tissues (Brady *et al.*, 1999; Caffrey *et al.*, 2002). The interpretation of the specificity of the results is assisted by light microscopical analyses, which indicated a wide distribution of gold particles in the schistosomes used for these experiments.

The data obtained so far confirmed that the method employed was able to generate reproducible data. Therefore, particle bombardment is at present the method of choice to genetically modify schistosomes, although there are a couple of obstacles. Among these are two forms of auto-fluorescence that occur at two different levels. First, worms cultured *in vitro* sometimes show a pale fluorescence within the gut lumen after bombardment (Fig. 7.6A). This auto-fluorescence occurs over a wide spectrum of wavelengths (488/520 nm and 568/590 nm). As 568 nm is not suitable to induce GFP fluorescence, this wavelength can help to distinguish between non-specific background and GFP-specific fluorescence. The second handicap is the strong auto-fluorescence within the vitellarium of mature females (Fig. 7.6B), which is detectable over a wide spectrum. In schistosomes, as in other plathyhelminths, the female gonad is divided into the ovary and the vitellarium (reviewed in Kunz, 2001). The ovary produces oocytes and the vitellarium produces vitelline cells that have a nurse-cell character providing vitelline and a variety of other substances for embryonic development and eggshell precursor proteins (Schüßler *et al.*, 1995). The ovary and vitellarium are needed to make up a composite egg consisting of one oocyte and 30–40 vitelline cells. To accomplish the daily egg production of about 300 eggs, more than 10,000 vitelline cells are generated. Due to this tremendous requirement the vitellarium is the largest organ in schistosome adults representing about two-thirds of the body of a mature female. Eggshell precursor proteins are synthesized within vitelline cells and stored as proteinaceous granules, which are exocytosed from vitelline cells to become crosslinked for eggshell formation (Smith and Clegg, 1959). During crosslinking, tyrosine-rich precursor proteins are oxidized, a reaction that is accompanied by auto-fluorescence (Cordingly, 1987). The emergence of auto-fluorescence in schistosome females is probably due to a premature oxidation of eggshell precursor proteins within vitelline cells under *in vitro* culture conditions. This assumption is supported by the observation that auto-fluorescence is not observed in mature females freshly isolated from a final host (W. Kunz, Düsseldorf, 1994, personal communication).

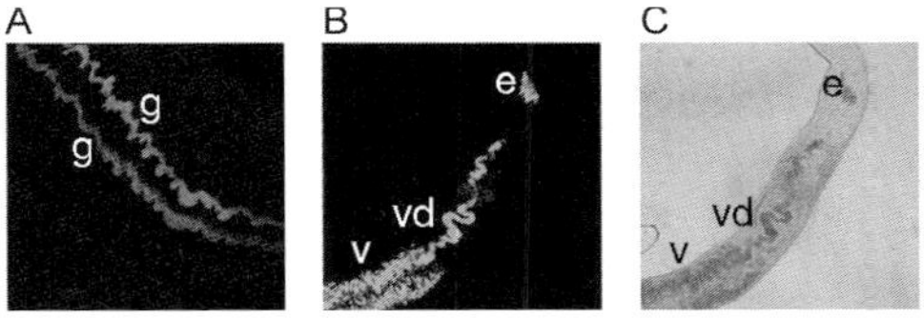

Fig. 7.6. A and B. Confocal laser scanning microscographs of schistosome adults showing auto-fluorescence; 488 nm was used for excitation and 520 nm for emission. A. Mid-body of a male schistosome with the focus on the lumen of the bipartite gut (g). In bombarded worms, a pale auto-fluorescence sometimes occurs under *in vitro* culture conditions. This auto-fluorescence occurs over a wide spectrum of wavelengths. B. Upper part of a female schistosome; auto-fluorescence can be observed over a wide spectrum within the vitellarium (v), the vitelloduct (vd) and the egg (e). C. Bright-field image of the female part shown in B.

With respect to sexual maturation, schistosomes are unique in nature because a continuous pairing-contact with a male is essential for the development of the reproductive organs of the female (Popiel and Basch, 1984; Basch, 1997; Kunz, 2001; Grevelding, 2004; LoVerde *et al.*, 2004a,b). Only after pairing, mitoses and differentiation processes are initiated in the female, leading to the differentiation of the reproductive organs, especially the vitellarium (Den Hollander and Erasmus, 1984). In females that have not been paired with a male, only undifferentiated precursor vitelline cells occur that have not yet started eggshell precursor protein synthesis. Therefore, the strong form of auto-fluorescence is not observed in virgin-like, immature females, which can be used for transformation experiments with GFP reporter-gene constructs to monitor gene expression.

Due to the auto-fluorescence of mature females, other reporter genes should be envisaged for transformation experiments in this gender such as luciferase whose applicability for schistosomes had already been demonstrated

(Davies *et al.*, 1999; Correnti and Pearce, 2004), or GUS whose applicability for parasites was shown for *Leishmania* (Sbicego *et al.*, 1998). In our laboratory, preliminary bombardment experiments have been performed using GUS reporter-gene constructs under the control of the regulatory elements of the ER60 promoter (Kapp, 2000). With X-Gluc as the substrate, in living worms only a faint blue staining was observed by bright field microscopy, which was not sufficient for a detailed characterization. In contrast to organisms like *C. elegans*, schistosomes are not translucent, and their pigmented structure impedes these kinds of analysis. However, the weak microscopical indication of GUS activity was confirmed by RT-PCR analyses demonstrating that the reporter gene was transcribed. Although X-Gluc staining is not the way of choice to monitor GUS activity, this reporter gene still can be used because of the wide variety of potential substrates that allow GUS activity to be monitored. With MUG or PNPG, for example, a spectroscopical determination of GUS activity can be performed, which allows studies such as promoter activity assays.

Stable transformation

Recent results demonstrate the feasibility of transgene introduction and expression in schistosomes. So far, adult worms have been the main targets for transformation experiments because they can be easily harvested from their final hosts and cultured *in vitro* for several weeks providing enough time for post-bombardment analyses. Reporter-gene constructs were based on plasmid backbones, not specifically designed for genome integration. Therefore, it has been assumed that the genetic modification is temporary and thus designated transient transformation. Although unexamined, the plasmid constructs probably reside as episomes in the nuclei of transfected cells. This kind of extrachromosomal coexistence was shown for plasmid transformation vectors in transgenic *C. elegans*. Following microinjection the plasmids occur as extrachromosomal concatemers in target cells (Mello and Fire, 1995).

In some publications the term transfection is used to describe the genetic modification of multicellular parasites such as schistosomes. Classically, transfection means the genetic modification of cultured eukaryotic cells by introducing DNA. The term transfection has also been used for the genetic modification of protozoa, which have a single cell character. For schistosomes as for other multicellular parasites, it would seem more appropriate to use the term transient transformation because following particle bombardment many cells or even tissue complexes exhibit the genetic modification contributing to a change of the phenotype of the multicellular organism.

To stably transform schistosomes, the used vectors must integrate into the chromosomes of germ cells. This is the prerequisite for the transmission of the transgenes to the progeny in a Mendelian fashion. Although considered to be unlikely, it cannot be excluded that the plasmid constructs are transferred to germ cells of adults during bombardment, and that they are integrated into chromosomes by recombination events. However, due to limitations of the *in vitro* culture system, it has not yet been possible to obtain miracidia from bombarded schistosomes in culture to investigate the first generation (F1) after transformation. Due to these obstacles, alternative approaches for stable transformation have to be envisaged.

Among these is the use of schistosome life stages that are rich in germ cells and suitable for *in vitro* cultivation. This condition is met by the first two larval stages, miracidia and sporocysts. Miracidia can be maintained in culture after they have been harvested in large amounts from eggs isolated from livers of infected hosts. In their posterior part, miracidia contain about 20 germinal cells, each with a diameter of 5 μm and a large nucleus of 3 μm. Following penetration in the intermediate host, each miracidium transforms into a mother sporocyst, which produces between 200 and 400 germ balls by asexual multiplication (Pan, 1980). The germ balls develop to daughter sporocysts generating cercariae, which are released from the daughter sporocysts to leave the snail. As it is possible to produce sporocysts from miracidia *in vitro* and to maintain this stage for some time in culture, another source for germ-cell-rich schistosome material is available, which may serve as a target for transformation approaches. To return to the life cycle, *in vitro* generated sporocysts can be transplanted into snails (Kapp *et al.*, 2003). Due to the asexual multiplication steps within the snail, it is theoretically possible to obtain hundreds of clonal cercariae from a single sporocyst. The technique for transplantation is based on previous work demonstrating that mother sporocysts (Chernin, 1966) or daughter sporocysts (DiConza and Hansen, 1972; Jourdane and Theron, 1980) can be recovered from donor snails and introduced into recipient snails. This technique allows the cloning of schistosomes to maintain strains or genotypes for unlimited time periods (reviewed in Jourdane, 1990). However, the technique is laborious, and it requires self-made glass equipment, manual skills and experience. Although high numbers of sporocysts can be generated *in vitro*, the number of transplantations, which can be conducted in 1 day, is limited (about 50 in our hands). Furthermore, the infection efficiencies vary significantly (16–43%) depending on the *in vitro* culture conditions applied to generate and maintain sporocysts (Kapp *et al.*, 2003).

An alternative to this approach is to use miracidia. Recently conducted experiments showed that miracidia are suitable for biolistics under low-pressure conditions (Wippersteg *et al.*, 2004). In spite of the significant physical influence, miracidia, which had been bombarded immediately after hatching from the egg, were still able to move and to penetrate the intermediate snail host. This finding reveals the potential of this first larval stage as a target for transformation experiments and subsequent reintroduction into the life cycle by its natural capacity for infection. Compared to the sporocyst approach, miracidia require less technical input, and a higher number of individuals can be generated.

With miracidia as targets we performed the first biolistic experiments towards a germ-line transformation using the reporter-gene construct ER60–GFP in circular or linearized forms, with pressure settings between 900 and 1350 psi and 1.6 μm gold particles. After bombardment, miracidia were used for intermediate-host infection, and cercariae were obtained from several snails. From a batch of the cercariae, DNA was extracted and used for PCR analyses that confirmed the presence of the vector. As germ cells are the only continuous cell line in the life cycle, the PCR result indirectly proved the presence of the transgenes within these cells. This demonstrated the possibility of penetrating the germ line of this organism (Wippersteg *et al.*, 2004). Furthermore, the result indicated that the germ cell tolerates gold particles whose size corresponds to approximately one-third of the size of this cell. This assumption was confirmed by an independent approach of Heyers *et al.* (2004) who biolistically transformed miracidia with an Hsp70-E(enhanced)–GFP reporter-gene construct and subsequently infected snails with the bombarded larvae. Two weeks post-infection, sections of snail tissue were histochemically analysed. Gold particles were identified within the sporocyst tissue in close proximity to the germ cells. In addition, EGFP gene transcription was confirmed by RT-PCR using total RNA from infected snails. However, neither the resulting cercariae nor adults were analysed.

To extend the analyses of the germ-line transformation approach in our laboratory, another batch of the putatively transformed cercariae were used for final-host infection to generate transgenic adults. Molecular analyses demonstrated the presence of the constructs in this life stage and, beyond that, even in adults of second generation after transformation. The microscopic analyses, however, revealed two unexpected findings. First, only a fraction of the worms showed reporter-gene activity. Second, in putatively transformed individuals, distinct fluorescing areas were detected, like tesseras of a mosaic, only in parts of the gastrodermis but not in all cells that belong to the ES system. Biological processes during larval development can explain the first finding. After penetration into the snail host, each of the 20 germ cells of a miracidium is able to transform into a mother sporocyst, which produces several ranks of daughter-sporocyst generations as a normal mode of asexual, larval multiplication (Jourdane *et al.*, 1980). Each rank of daughter sporocysts is able to continue sporocystogenesis or to enter cercariogenesis (Jourdane and Theron, 1987). This results in multiple generations of cercariae, which differ in their germ-cell origin and in their time of maturation. If only one or a few germ cells of a miracidium are hit during bombardment, a mosaic population of genetically modified and wild-type germ cells is present. Consequently, transformed and non-transformed mother and daughter sporocysts coexist in infected snails, which can be of monomiracidial or polymiracidial origin. Finally, a mixed population of cercariae is obtained and forms the basis for final-host infection and adult generation, some of these contain the transgenes and others do not. The second finding, the occurrence of distinct fluorescing areas, can be explained by an extrachromosomal maintenance of the plasmid constructs. If these would have been integrated following bombardment, every cell should contain and express the transgenes depending on the control regions – in the case of ER60–GFP, all cells of the ES system. If the plasmids occur extrachromosomally, some cells may maintain and express the plasmids, whereas other cells would lose them during development. This explains the distinct signal patterns observed by microscopy.

Taken together, these results demonstrate the feasibility of using particle bombardment to facilitate entry into the germ line and to enable the expression of reporter-gene constructs in developmental stages that are distinct from the stage of transformation. Although this is a significant step forward, it is still not the absolute breakthrough because a stable integration into the genome has not been achieved yet. To attain this final goal, new strategies for the design of transformation vectors should be developed. One possibility is to use linearized vector molecules that contain a higher amount of homologous sequences that may serve as better substrates for recombination events. If homologous recombination is envisaged, the portion of homologous DNA sequences within a transformation vector is a critical parameter. It was shown for *C. elegans* that 4.4 kb of homologous sequences were sufficient to achieve homologous recombination (Berezikov *et al.*, 2004), whereas 14 kb of homology are necessary for efficient gene targeting in murine embryonic stem cells (Deng and Capecchi, 1992). In our experiments, the lengths of homologous sequences varied between 1 and 1.3 kb, and these may not be sufficient to bring about integration events. A self-evident modification of transformation vectors for schistosomes is the addition of sequences from mobile genetic elements known to support genome integration. Recently, there have been many studies aimed at the identification and characterization of mobile genetic elements from schistosomes. A number of LTR and non-LTR retrotransposable elements were isolated, and their sequences analysed (Drew *et al.*, 1999; Laha *et al.*, 2001, 2002a,b; Copeland *et al.*, 2003; DeMarco *et al.*, 2004). One possibility would be to use this sequence information for the modification of vector constructs to enhance the probability of integration. In schistosomes, a DNA transposon has not yet been identified, which would be the favoured tool to shuttle transgenes into the genome. But it should be feasible to use well-characterized heterologous DNA-transposon systems that are known to work in a variety of organisms. Examples are *mariner*, *Hermes*, or *piggyBac* that have successfully been used to shuttle transgenes into planarians (Gonzalez-Estevez *et al.*, 2003). Besides heterologous transposons, pantropic retroviruses can also be envisaged as potential vectors for integration (Yee *et al.*, 1994). These viruses are able to randomly integrate into the genome of dividing cells of various vertebrates and invertebrates. Similar to other mobile genetic elements such as transposons, these viruses can be used as vehicles for the stable integration of DNA constructs. With these tools, strategies can be conceived to ectopically express genes to investigate gain-of-function effects, or to express dsRNA hairpin structures, which lead to loss-of-function effects. Both approaches facilitate the functional characterization of interesting genes. As recent studies confirmed the applicability of RNAi technology to schistosomes, albeit only for larvae or schistosomula (Boyle *et al.*, 2003; Skelly *et al.*, 2003), silencing approaches seem to be feasible for this organism.

Cestodes, on the Way to Transgenesis

Because of their significance for human or animal health, cestode worms have been intensively studied. The complexity of the host–parasite interaction, however, makes it difficult to investigate factors that regulate the differentiation of specific life stages. To overcome this problem, protocols have been developed to easily maintain tapeworms in the laboratory.

This is achieved by life cycling in suitable animal hosts or by *in vitro* culture techniques. *Hymenolepis* spp. and *Mesocestoides* spp., for example, have been used as models because their life-cycle stages can be maintained in animal hosts and some life stages can additionally be maintained *in vitro*, thereby allowing *ex vivo* studies of aspects of cestode biology (Siles-Lucas and Hemphill, 2002). For *Echinococcus* protocols have been developed to maintain metacestodes in the laboratory by serial transplantation passages into susceptible gerbils, enabling the parasite to proliferate asexually. In addition, *in vitro* culture techniques have been established that allow the long-term cultivation of metacestodes under low oxygen conditions, and even the differentiation to protoscoleces in the presence of host factors (Spiliotis *et al.*, 2004). Today, these culture models are used to study various aspects of *Echinococcus* biology such as (i) metacestode *in vitro* proliferation and differentiation; (ii) the ultrastructural analysis of the acellular laminated layer, a structure that represents the physical interface between parasite and host tissues; and (iii) immunological characteristics (reviewed in Hemphill *et al.*, 2002).

Although *in vitro* culture systems exist, not much has been done at the genomic or post-genomic levels to analyse gene function in cestodes. Some molecular approaches have been performed to isolate (i) immunologically relevant genes by screening of expression libraries with sera from patients (Ortona *et al.*, 2003) or by characterizing gene expression products from eggs (Lightowlers *et al.*, 2003), or (ii) genes with putative functions during development by homology-based identification methods (Brehm *et al.*, 2003; Konrad *et al.*, 2003; Spiliotis *et al.*, 2003). But the functional analysis of candidate genes is still in its infancy. First attempts have been made to genetically modify metacestodes cultured *in vitro* by particle bombardment using reporter-gene constructs. However, the fragile metacestodes did not survive biolistics (K. Brehm, Würzburg, 2004, personal communication). Further approaches are planned to test lipofection or virus-based systems as alternative transformation methods.

Concluding Remarks

Recent years have seen a revolutionary change in the study of organisms at the molecular level. Catalysed by the improvement of automated sequencing technologies, comprehensive genome information has been generated for various organisms including parasites. Today, we mark the beginning of the post-genomic era, which necessitates interpretation and exploitation of the sequence data. In this context it is crucial to develop technologies to perform functional genetics, i.e. to study a gene of interest within its natural genomic environment. To overcome existing limitations, heterologous expression systems such as *C. elegans*, or mammalian cell lines, or frog oocytes have been used to characterize parasite genes (Britton *et al.*, 1999; Hashmi *et al.*, 2001; Brooks and Isaac, 2002; Boyle and Yoshino, 2003). Although interesting aspects of gene regulation and function were obtained by these approaches, heterologous systems may not always allow conclusions on the function of a gene in the homologous system. Therefore, it is inescapable to establish transformation protocols for non-model organisms, too.

For planarians and trematodes, but not yet for cestodes, transformation protocols have been developed that demonstrate the possibility of generating genetically modified flatworms. The methods employed are largely based on existing techniques with some organism-specific modifications. In principle these methods can be applied to other flatworms too as well as to any other organism of interest, provided that *in vitro* culture systems are available. However, organism-specific features may complicate the applicability of some of these techniques. The results obtained for the development of transgenic planarians are outstanding and strengthen their position as model organisms for the class Turbellaria. The progress with transiently transformed schistosomes is not yet comparable to that of planarians, but encouraging results were obtained showing that post-genomic analyses towards gene regulation are possible. Although this is an important step forward, the final breakthrough of stable germ-line integration has not yet been achieved. Due to the many advantages of the schistosome system, however, such as the existence of *in vitro* culture protocols for larval and adult life stages, the transplantation techniques for the reintegration of larval stages or schistosomula into the natural life cycle, or an established transformation procedure that allows germ-line entry, it may be only a matter of time before this goal is realized. A few remaining obstacles remain to be overcome:

1. Protocols have to be established that allow the integration of transgenes into the germ line.
2. Strategies should be developed to effectively select transformants by drug treatment.
3. *In vitro* culture techniques should be improved to keep different life stages for unlimited time in culture and to have an immediate access to all developmental stages.

If this succeeds, schistosomes may be close to becoming a model for parasitic trematodes. This would allow researchers to further develop and optimize techniques for the genetic modification of other parasitic flatworms of medical or economic relevance. The combination of modern molecular approaches such as genomics, proteomics and transgenics, with classical disciplines such as systematics, cytology and physiology will lead to a more fundamental understanding of biological processes within these parasites and to the development of novel control strategies.

Acknowledgements

All post-docs, PhD and Diploma students are gratefully acknowledged for their enthusiasm, engagement and valuable inputs to the research in our laboratory. The author would like to point out the substantial contributions of Volker Wippersteg and Katja Kapp to the establishment of the transgenesis system. Particularly, the author expresses his gratitude to his former mentor Professor Werner Kunz. Furthermore, the author appreciates the fruitful collaborations with Horst Zahner and Peter Jackstadt, John Kusel and Fabio Ribeiro, Mo Klinkert and Alessandro Rossi as well as Connor Caffrey and Jim McKerrow concerning transgenic schistosomes. This part of the author's research has been mainly supported by grants of the Deutsche Forschungsgemeinschaft (GR1549) and in part by the UNDP/World Bank/WHO Special Program for Research and Training in Tropical Diseases.

References

Asami, M., Nakatsuka, T., Hayashi, T., Kou, K., Kagawa, H. and Agata, K. (2002) Cultivation and characterization of planarian neuronal cells isolated by fluorescence-activated cell sorting (FACS). *Zoological Science* 19, 1257–1265.

Baguna, J. (1998) From invertebrates to humans. In: Ferreti, P. and Geraudi, J. (eds) *Cellular and Molecular Basis for Regeneration*. John Wiley & Sons, New York, pp. 135–165.

Basch, P. (1981a) Cultivation of *Schistosoma mansoni in vitro*. I. Establishment of cultures from cercariae and development until pairing. *Journal of Parasitology* 67, 179–185.

Basch, P. (1981b) Cultivation of *Schistosoma mansoni in vitro*. II. Production of infertile eggs by worm pairs cultured from cercariae. *Journal of Parasitology* 67, 186–190.

Basch, P. (1984) Development and behavior of cultured *Schistosoma mansoni* fed on human erythrocyte ghosts. *The American Journal of Tropical Medicine and Hygiene* 33, 911–917.

Basch, P. (1997) *Schistosomes: Development, Reproduction, and Host Relations*. Oxford University Press, Oxford, UK.

Bayne, C.J. (1998) Invertebrate cell culture considerations: insects, ticks, shellfish, and worms. *Methods in Cell Biology* 57, 187–201.

Berezikov, E., Bargmann, C.I. and Plasterk, H.A. (2004) Homologous gene targeting in *Caenorhabditis elegans* by biolistic transformation. *Nucleic Acids Research* 32(4), 1–7.

Bergquist, R. (2004) Prospects for schistosomiasis vaccine development. *TDR News* 71, 6–7.

Bergquist, R., Al-Sherbiny, M., Barakat, R. and Olds, R. (2002) Blueprint for schistosomiasis vaccine development. *Acta Tropica* 82, 183–192.

Beverly, S. (2003) Protozomics: trypanosomatid parasite genetics comes of age. *Nature Reviews Genetics* 4, 11–19.

Blair, D., Xu, Z.B. and Agatsuma, T. (1999) Paragonimiasis and the genus *Paragonimus*. *Advances in Parasitology* 42, 113–222.
Blas, B.L., Rosales, M.I., Lipayon, I.L., Yasuraoka, K., Matsuda, H. and Hayashi, M. (2004) The schistosomasis problem in the Philippines: a review. *Parasitology International* 53, 127–134.
Boyle, J.P., and Yoshino, T.P. (2003) Gene manipulation in parasitic helminths. *International Journal for Parasitology* 33, 1259–1268.
Boyle, J.P., Wu, X.J., Shoemaker, C.B. and Yoshino, T.P. (2003) Using RNA interference to manipulate endogenous gene expression in *Schistosoma mansoni* sporocysts. *Molecular and Biochemical Parasitology* 128, 205–215.
Brady, C.P., Dowd, A.J., Brindley, P.J., Ryan, T., Day, S.R. and Dalton, J.P. (1999) Recombinant expression and localization of *Schistosoma mansoni* cathepsin L1 support its role in the degradation of host hemoglobin. *Infection and Immunity* 67, 368–374.
Brehm, K., Wolf, M., Beland, H., Kroner, A. and Frosch, M. (2003) Analysis of differential gene expression in *Echinococcus multilocularis* larval stages by means of spliced leader differential display. *International Journal for Parasitology* 33, 1145–1159.
Britton, C., Redmond, D.L., Knox, D.P., McKerrow, J.H. and Barry, J.D. (1999) Identification of promoter elements of parasite nematode genes in transgenic *Caenorhabditis elegans*. *Molecular and Biochemical Parasitology* 103, 171–181.
Brooks, D. and Isaac, R.E. (2002) Functional genomics of parasite worms: the dawn of a new era. *Parasitology International* 51, 319–325.
Brusca, R.C. and Brusca, G.J. (2002) *Invertebrates*, 2nd edn. Sinauer, Sunderland, Massachusetts.
Caffrey, C.R., Salter, J.P., Lucas, K.D., Khiem, D., Hsieh, I., Lim, K.C., Ruppel, A., McKerrow, J.H. and Sajid, M. (2002) SmCB2, a novel tegumental cathepsin B from adult *Schistosoma mansoni*. *Molecular and Biochemical Parasitology* 121, 49–61.
Carpio, A (2002) Neurocysticercosis: an update. *Lancet Infectious Diseases* 2, 751–762.
Chaithong, U., Sukontason, K., Boonsriwong, N., Sukontason, K.L. and Piangjai, S. (2001) *In vitro* development of *Haplorchis taichui* (Trematoda: Heterophyidae). *Southeast Asian Journal of Tropical Medicine and Public Health* 32, 31–35.
Chalfie, M., Tu, Y., Euskirchen, G., Ward, W.W. and Prasher, D.C. (1994) Green fluorescent protein as a marker for gene expression. *Science* 263, 802–805.
Chernin, E. (1966) Transplantation of larval *Schistosoma mansoni* from infected to uninfected snails. *Journal for Parasitology* 52, 473–482.
Cioli, D. and Pica-Mattoccia, L. (2003) Praziquantel. *Parasitology Research* 90, 3–9.
Clayton, C.E. (1999) Genetic manipulation of kinetoplastida. *Parasitology Today* 15, 372–377.
Copeland, C.S., Brindley, P.J., Heyers, O., Michael, S.F., Johnston, D.A., Williams, D.L., Ivens, A.C. and Kalinna, B.H. (2003) *Boudicca*, a retrovirus-like long terminal repeat retrotransposon from the genome of the human blood fluke *Schistosoma mansoni*. *The Journal of Virology* 77, 6153–6166.
Cordingly, J.S. (1987) Trematode eggshells: novel protein biopolymers. *Parasitology Today* 11, 341–344.
Correnti, J.M. and Pearce, E.J. (2004) Transgene expression in *Schistosoma mansoni*: introduction of RNA into schistosomula by electroporation. *Molecular and Biochemical Parasitology* 137, 75–79.
Coustau, C. and Yoshino, T.P. (2000) Flukes without snails: advances in the *in vitro* cultivation of intramollluscan stages of trematodes. *Experimental Parasitology* 94, 62–66.
Coustau, C., Ataev, G., Jourdane, J. and Yoshino, T.P. (1997) *Schistosoma japonicum*: *in vitro* cultivation of miracidium to daughter sporocyst using a *Biomphalaria glabrata* embryonic cell line. *Experimental Parasitology* 87, 77–87.
Craig, P. (2003) *Echinococcus multilocularis*. *Current Opinion in Infectious Diseases* 16, 437–444.
Davies, S.J., Shoemaker, C.B. and Pearce, E.J. (1998) A divergent member of the transforming growth factor beta receptor family from *Schistosoma mansoni* is expressed on the parasite surface membrane. *Journal of Biological Chemistry* 273, 11234–11240.
Davies, R.E., Parra, A., LoVerde, P.T., Ribeiro, E., Glorioso, G. and Hodgson, S. (1999) Transient expression of DNA and RNA in parasitic helminths by using particle bombardment. *Proceedings of the National Academy of Sciences USA* 96, 8687–8692.
DeBont, J. and Vercruysse, J. (1998) Schistosomiasis in cattle. *Advances in Parasitology* 41, 285–364.
DeMarco, R., Kowaltowski, A.T., Machado, A.A., Soares, M.B., Gargioni, C., Kawano, T., Rodrigues, V., Madeira, A.M., Wilson, R.A., Menck, C.F., Setubal, J.C., Dias-Neto, E., Leite, L.C. and Verjovski-Almeida, S. (2004) Saci-1, -2, and -3 and Perere, four novel retrotransposons with high transcriptional activities from the human parasite *Schistosoma mansoni*. *The Journal of Virology* 78, 2967–2978.

De Koning-Ward, T.F., Janse, C.J. and Waters, A.P. (2000) The development of genetic tools for dissecting the biology of malaria parasites. *Annual Review of Microbiology* 54, 157–185.

De Mendonca, R.L., Escriva, H., Bouton, D., Laudet, V. and Pierce, R.J. (2000) Hormones and nuclear receptors in schistosome development. *Parasitology Today* 16, 233–240.

Deng, C.X. and Capecchi, M.R. (1992) Re-examination of gene targeting frequency as a function of the extent of homology between the targeting vector and the target locus. *Molecular and Cellular Biology* 12, 3365–3371.

Den Hollander, J.E. and Erasmus, D.A. (1984) *Schistosoma mansoni*: DNA synthesis in males and females from mixed- and single-sex infections. *Parasitology* 88, 463–476.

Deplazes, P. and Eckert, J. (2001) Veterinary aspects of alveolar echinococcosis – a zoonosis of public health significance. *Veterinary Parasitology* 98, 65–87.

DiConza, J.J. and Hansen, E.L. (1972) Multiplication of transplanted *Schistosoma mansoni* daughter sporocysts. *Journal for Parasitology* 58, 181–182.

Doenhoff, M.J., Kusel, J.R., Coles, G.C. and Cioli, D. (2002) Resistance of *Schistosoma mansoni* to praziquantel: is there a problem? *Transactions of the Royal Society of Tropical Medicine and Hygiene* 96, 465–469.

Dong, H.F., Chen, X.B., Ming, Z.P., Zhong, Q.P. and Jiang, M.S. (2003) Ultrastructure of cultured cells from *Schistosoma japonicum*. *Acta Tropica* 82, 225–234.

Drew, A.C., Minchella, D.J., King, L.T., Rollinson, D. and Brindley, P.J. (1999) SR2 elements, non-long terminal repeat retrotransposons of the RTE-1 lineage from the human blood fluke *Schistosoma mansoni*. *Molecular Biology and Evolution* 16, 1256–1269.

Ellis, J.T., Morrison, D.A. and Reichel, M.P. (2003) Genomics and its impact on parasitology and the potential for development of new parasite control methods. *DNA and Cell Biology* 22, 395–403.

Ersfeld, K. (2003) Genomes and genome projects of protozoan parasites. *Current Issues in Molecular Biology* 5, 61–74.

Fantappie, M.R., Freebern, W.J., Osman, A., LaDuca, J., Niles, E.G. and LoVerde, P.T. (2001) Evaluation of *Schistosoma mansoni* retinoid X receptor (SmRXR1 and SmRXR2) activity and tissue distribution. *Molecular and Biochemical Parasitology* 115, 87–99.

Fenwick, A., Savioli, L., Engels, D., Bergquist, R. and Todd, M.H. (2003) Drugs for the control of parasitic diseases: current status and development in schistosomiasis. *Trends in Parasitology* 19, 509–515.

Finken-Eigen, M. and Kunz, W. (1997) *Schistosoma mansoni*: gene structure and localization of a homologue to cysteine protease ER60. *Experimental Parasitology* 86, 1–7.

Finken, M., Sobek, A., Symmons, P. and Kunz, W. (1994) Characterization of the complete protein disulfide isomerase gene of *Schistosoma mansoni* and identification of the tissues of its expression. *Molecular and Biochemical Parasitology* 64, 135–144.

Fire, A., Xu, S., Montgomery, M.K., Kostas, S.A., Driver, S.E. and Mello, C.C. (1998) Potent and specific genetic interference by double-stranded RNA in *Caenorhabditis elegans*. *Nature* 391, 806–811.

Foster, J.M. and Johnston, D.A. (2002) Helminth genomics: from gene discovery to genome sequencing. *Trends in Parasitology* 18, 241–242.

Franco, G.R., Rabelo, E.M., Azevedo, V., Pena, H.B., Ortega, J.M., Santos, T.M., Meira, W.S., Rodrigues, N.A., Dias, C.M., Harrop, R., Wilson, A., Saber, M., Abdel-Hamid, H., Faria, M.S., Margutti, M.E., Parra, J.C. and Pena, S.D. (1997) Evaluation of cDNA libraries from different developmental stages of *Schistosoma mansoni* for production of expressed sequence tags (ESTs). *DNA Research* 4, 231–240.

Fried, B. and Robert, L. (2002) *In vitro* and *in vivo* encystment of the cercariae of *Echinostoma caproni*. *Journal of Parasitology* 88, 1124–1129.

Garcia, H.H., Gonzalez, A.E., Evans, C.A. and Gilman, R.H. (2003) *Taenia solium* cysticercosis. *Lancet* 362, 547–556.

Gonzalez, A.E., Garcia, H.H., Gilman, R.H. and Tsang, V.C. (2003) Control of *Taenia solium*. *Acta Tropica* 87, 103–109.

Gonzalez-Estevez, C., Momose, T., Gehring, W.J. and Salo, E. (2003) Transgenic planarian lines obtained by electroporation using transposon-derived vectors and an eye-specifc GFP marker. *Proceedings of the National Academy of Sciences USA* 24, 14046–14051.

Gorbushin, A.M. and Shaposhnikova, T.G. (2002) *In vitro* culture of the avian echinostome *Himasthla elongata*: from redia to marita. *Experimental Parasitology* 101, 234–239.

Graczyk, T.K. and Fried, B. (1998) Echinostomiasis: a common but forgotten food-borne disease. *The American Journal of Tropical Medicine and Hygiene* 58, 501–504.

Grevelding, C.G. (2004) *Schistosoma*. *Current Biology* 14, R545.

Grevelding, C.G., Sommer, G. and Kunz, W. (1997) Female-specific gene expression in *Schistosoma mansoni* is regulated by pairing. *Parasitology* 115, 635–640.
Hara, T., Yasuda, K. and Fukuma, T. (2002) Effective gene transfer into *Trypanosoma brucei* bloodstream forms by particle bombardment. *Molecular and Biochemical Parasitology* 119, 117–119.
Hashmi, S., Tawe, W. and Lustigman, S. (2001) *Caenorhabditis elegans* and the study of gene function in parasites. *Trends in Parasitology* 17, 387–393.
Hedstrom, R., Culpepper, J., Harrison, R.A., Agabian, N. and Newport, G.J. (1987) A major immunogen in *Schistosoma mansoni* infections is homologous to the heat shock protein Hsp70. *The Journal of Experimental Medicine* 165, 1430–1435.
Hemphill, A., Stettler, M., Walker, M., Siles-Lucas, M., Fink, R. and Gottstein, B. (2002) Culture of *Echinococcus multilocularis* metacestodes: an alternative to animal use. *Trends in Parasitology* 18, 445–451.
Heyers, O., Walduck, A.K., Brindley, P.J., Bleiß, W., Lucius, R., Dorbic, T., Wittig, B. and Kalinna, B.H. (2004) *Schistosoma mansoni* miracidia transformed by particle bombardment infect *Biomphalaria glabrata* snails and develop into transgenic sporocysts. *Experimental Parasitology* 105, 174–178.
Higazi, T.B., Merriweather, A., Shu, L., Davis, R. and Unnasch, T.R. (2002) *Brugia malayi*: transient transfection by microinjection and particle bombardment. *Experimental Parasitology* 100, 95–102.
Hobbs, D.J., Fryer, S.E., Duimstra, J.R., Hedstrom, O.R., Brodie, A.E., Collodi, P.A., Menino, J.S., Bayne, C.J. and Barnes, D.W. (1993) Culture of cells from juvenile worms of *Schistosoma mansoni*. *Journal of Parasitology* 79, 913–921.
Hoberg, E.P. (2002) *Taenia* tapeworms: their biology, evolution and socioeconomic significance. *Microbes and Infection* 4, 859–866.
Hoffmann, K.F. and Dunne, D.W. (2003) Characterization of the *Schistosoma* transcriptome opens up the world of helminth genomics. *Genome Biology* 5, 203.
Hoffmann, K.F., Johnston, D.A. and Dunne, D.W. (2002) Identification of *Schistosoma mansoni* gender-associated gene transcripts by cDNA microarray profiling. *Genome Biology* 25, 41.
Horak, P., Kolarova, L. and Adema, C.M. (2002) Biology of the schistosome genus *Trichobilharzia*. *Advances in Parasitology* 52, 155–233.
Hu, W., Yan, Q., Shen, D.K., Liu, F., Zhu, Z.D., Song, H.D., Xu, X.R., Wang, Z.J., Rong, Y.P., Zeng, L.C., Wu, J., Zhang, X., Wang, J.J., Xu, X.N., Wang, S.Y., Fu, G., Zhang, X.L., Wang, Z.Q., Brindley, P.J., McManus, D.P., Xue, C.L., Feng, Z., Chen, Z. and Han, Z.G. (2003) Evolutionary and biomedical implications of a *Schistosoma japonicum* complementary DNA resource. *Nature Genetics* 35, 139–147.
Hurtrez-Bousses, S., Meunier, C., Durand, P. and Renaud, F. (2001) Dynamics of host–parasite interactions: the example of population biology of the liver fluke (*Fasciola hepatica*). *Microbes and Infection* 3, 841–849.
Inal, J.M. and Sim, R.B. (2000) A *Schistosoma* protein, Sh-TOR, is a novel inhibitor of complement which binds human C2. *FEBS Letters* 470, 131–134.
Ivanchenko, M.G., Lerner, J.P., McCormick, R.S., Toumadje, A., Allen, B., Fischer, K., Hedstrom, O., Helmrich, A., Barnes, D.W. and Bayne, C.J. (1999) Continuous *in vitro* propagation and differentiation of cultures of the intramolluscan stages of the human parasite *Schistosoma mansoni*. *Proceedings of the National Academy of Sciences USA* 96, 4965–4970.
Jackstadt, P., Wilm, T.P., Zahner, H. and Hobom, G. (1999) Transformation of nematodes by ballistic DNA transfer. *Molecular and Biochemical Parasitology* 103, 261–266.
James, E.R. and Taylor, M.G. (1976) Transformation of cercariae to schistosomula: a quantitative comparison of transformation techniques and of infectivity by different injection routes of the organisms produced. *Journal of Helminthology* 50, 223–233.
Jefferson, R.A., Burgess, S.M. and Hirsch, D. (1986) β-glucoronidase from *Escherichia coli* as a gene-fusion marker. *Proceedings of the National Academy of Sciences USA* 83, 8447–8451.
Jenkins, D.J. and Macpherson, C.N. (2003) Transmission ecology of *Echinococcus* in wild-life in Australia and Africa. *Parasitology* 127, 63–72.
Jourdane, J. (1990) Cloning of schistosomes. *Parasitology Today* 6, 363–365.
Jourdane, J. and Theron, A. (1980) *Schistosoma mansoni*: cloning by microsurgical transplantation of sporocysts. *Experimental Parasitology* 50, 349–357.
Jourdane, J. and Theron, A. (1987) Larval development: eggs to cercariae. In: Rollinson, D. and Simpson, A.J.G (eds) *The Biology of Schistosomes: From Genes to Latrines*. Academic Press, London, pp. 83–106.
Jourdane, J., Theron, A. and Combes, C. (1980) Demonstration of several sporocysts generations as a normal pattern of reproduction of *Schistosoma mansoni*. *Acta Tropica* 37, 177–182.
Kapp (2000) *Schistosoma mansoni:* Charakterisierung von zwei tyrosinkinasen und ansätze zur etablierung eines gentransfersystems. PhD thesis, The University of Düsseldorf, Düsseldorf, Germany.

Kapp, K., Coustau, C., Wippersteg, V., Jourdane, J., Kunz, W. and Grevelding, C.G. (2003) Transplantation of *in vitro*-generated *Schistosoma mansoni* mother sporocysts into *Biomphalaria glabrata*. *Experimental Parasitology* 91, 482–485.

Kapp, K., Knobloch, J., Schüßler, P., Sroka, S., Lammers, R., Kunz, W. and Grevelding, C.G. (2004) The *Schistosoma mansoni* Src kinase TK3 is expressed in the gonads and likely involved in cytoskeletal organization. *Molecular and Biochemical Parasitology* 138, 171–182.

Kelly, J.M. (1997) Genetic transformation of parasitic protozoa. *Advances in Parasitology* 39, 227–270.

Kendall, J.M. and Badminton, M.N. (1998) *Aequorea victoria* bioluminescence moves into an exciting new era. *Trends in Biotechnology* 16, 216–224.

Kiesecker, J.M. (2002) Synergism between trematode infection and pesticide exposure: a link to amphibian limb deformities in nature? *Proceedings of the National Academy of Sciences USA* 99, 9900–9904.

Kim, K. and Weiss, L.M. (2004) *Toxoplasma gondii*: the model apicomplexan. *International Journal for Parasitology* 34, 423–432.

Klein, T.M., Goff, S.A., Roth, B.A. and Fromm, M.E. (1990) Applications for the particle gun in plant biology. In: Nijkamp, H.J.J., Van der Plas, L.W.W. and Wan Aartrijk, J. (eds) *Progress in Plant Cellular and Molecular Biology*. Kluwer Academic Publishers, Boston, Massachusetts.

Klein, T.M., Arebtzen, R., Lewis, P.A. and Fitzpatrick-McElligot, S. (1992) Transformation of microbes, plants and animals by particle bombardment. *Biotechnology* 10, 286–291.

Knobloch, J., Winnen, R., Quack, M., Kunz, W. and Grevelding, C.G. (2002) A novel Syk-family tyrosine kinase from *Schistosoma mansoni*, which is preferentially transcribed in reproductive organs. *Gene* 294, 87–97.

Konrad, C., Kroner, A., Spiliotis, M., Zavala-Gongora, R. and Brehm, K. (2003) Identification and molecular characterisation of a gene encoding a member of the insulin receptor family in *Echinococcus multilocularis*. *International Journal for Parasitology* 33, 301–312.

Kook, J., Lee, S.H. and Chai, J.Y. (1997) *In vitro* cultivation of *Gymnophalloides seoi* metacercariae (Digenea: *Gymnophallidae*). *The Korean Journal of Parasitology* 35, 25–29.

Kunz, W. (2001) Schistosome male–female interaction: induction of germ-cell differentiation. *Trends in Parasitology* 17, 227–231.

Laha, T., Loukas, A., Verity, C.K., McManus, D.P. and Brindley P.J. (2001) Gulliver, a long terminal repeat retrotransposon from the genome of the oriental blood fluke *Schistosoma japonicum*. *Gene* 264, 59–68.

Laha, T., Brindley, P.J., Verity, C.K., McManus, D.P. and Loukas, A. (2002a) pido, a non-long terminal repeat retrotransposon of the chicken repeat 1 family from the genome of the oriental blood fluke *Schistosoma japonicum*. *Gene* 284, 149–159.

Laha, T., Brindley, P.J., Smout, M.J., Verity, C.K., McManus, D.P. and Loukas, A. (2002b) Reverse transcriptase activity and untranslated region sharing of a new RTE-like, non-long terminal repeat retrotransposon from the human blood fluke *Schistosoma japonicum*. *International Journal for Parasitology* 32, 1163–1174.

Laursen, J.R. and Yoshino, T.P. (1999) *Biomphalaria glabrata* embryonic (*Bge*) cell line supports *in vitro* miracidial transformation and early larval development of the deer liver fluke, *Fascioloides magna*. *Parasitology* 118, 187–194.

Lebens, M., Sun, J.B., Czerkinsky, C. and Holmgren, J. (2004) Current status and future prospects for a vaccine against schistosomiasis. *Expert Review of Vaccines* 3, 315–328.

Lightowlers, M.W., Colebrook, A.L., Gauci, C.G., Gauci, S.M., Kyngdon, C.T., Monkhouse, J.L., Vallejo Rodriquez, C., Read, A.J., Rolfe, R.A. and Sato, C. (2003) Vaccination against cestode parasites: anti-helminth vaccines that work and why. *Veterinary Parasitology* 115, 83–123.

LoVerde, P.T., Hirai, H., Merrick, J.M., Lee, N.H. and El-Sayed, N. (2004a) *Schistosoma mansoni* genome project: an update. *Parasitology International* 53, 183–192.

LoVerde, P.T., Niles, E.G., Osman, A. and Wu, W. (2004b) *Schistosoma mansoni* male–female interactions. *Canadian Journal of Zoology* 82, 357–374.

MacGregor, A.N. and Shore, S.J. (1990) Immunocytochemistry of cytoskeletal proteins in adult *Schistosoma mansoni*. *International Journal for Parasitology* 20, 279–284.

Mastroberardino, L., Spindler, B., Pfeiffer, R., Skelly, P.J., Loffing, J., Shoemaker, C.B. and Verrey, F. (1998) Amino-acid transport by heterodimers of 4F2hc/CD98 and members of a permease family. *Nature* 395, 288–291.

McManus, D.P., Zhang, W., Li, J. and Bartley, P.B. (2003) *Echinococcosis*. *Lancet* 362, 1295–1304.

McManus, D.P., Hu, W., Brindley, P.J., Feng, Z., Han, Z.G. (2004) Schistosome transcriptome analysis at the cutting edge. *Trends in Parasitology* 20, 301–304.

Mecozzi, B., Rossi, A., Lazzaretti, P., Kady, M., Kaiser, S., Valle, C., Cioli, D. and Klinkert, M.Q. (2000) Molecular cloning of *Schistosoma mansoni* calcineurin subunits and immunolocalization to the excretory system. *Molecular and Biochemical Parasitology* 110, 333–343.

Mello, C. and Fire, A. (1995) DNA transformation. In: Epstein, H.F. and Shakes, D. (eds) *Caenorhabditis elegans: Modern Biological Analysis of an Organism*. Academic Press, New York, pp. 451–482.

Morgan, T.H. (1898) Experimental studies of the regeneration in *Planaria maculata*. *Archiv für Entwicklungsmechanik der Organe (Wilhelm Roux)* 7, 364–397.

Moser, D., Doumbo, O. and Klinkert, M.Q. (1990) The humoral response to heat shock protein 70 in human and murine *Schistosomiasis mansoni*. *Parasite Immunology* 12, 341–352.

Neumann, S., Ziv, E., Lantner, F. and Schechter, I. (1993) Regulation of Hsp70 gene expression during the life cycle of the parasitic helminth *Schistosoma mansoni*. *European Journal of Biochemistry* 212, 589–596.

Newmark, P.A. and Sanchez Alvarado, A. (2002) Not your father's planarian: a classic model enters the era of functional genomics. *Nature Reviews Genetics* 3, 210–219.

Nollen, P.M. (1990) Escape of rediae from miracidia of *Philophthalmus megalurus* and *Philophthalmus gralli* during *in vitro* culture. *Journal of Parasitology* 76, 725–729.

Ohmae, H., Iwanaga, Y., Nara, T., Matsuda, H. and Yasuraoka, K. (2003) Biological characteristics and control of intermediate snail host of *Schistosoma japonicum*. *Parasitology International* 52, 409–417.

Ortona, E., Margutti, P., Delunardo, F., Vaccari, S., Rigano, R., Profumo, E., Buttari, B., Teggi, A. and Siracusano, A. (2003) Molecular and immunological characterization of the C-terminal region of a new *Echinococcus granulosus* Heat Shock Protein 70. *Parasite Immunology* 25, 119–126.

Osman, A., Niles, E.G. and LoVerde, P.T. (2001) Identification and characterization of a Smad2 homologue from *Schistosoma mansoni*, a transforming growth factor-beta signal transducer. *Journal of Biological Chemistry* 276, 10072–10082.

Otranto, D. and Traversa, D. (2002) A review of dicrocoeliosis of ruminants including recent advances in the diagnosis and treatment. *Veterinary Parasitology* 107, 317–335.

Pan, S.C-T. (1980) The fine structure of the miracidium of *Schistosoma mansoni*. *Journal of Invertebrate Pathology* 36, 307–373.

Patient, R.K. and McGhee, J.D. (2002) The GATA family (vertebrates and invertebrates). *Current Opinion in Genetics and Development* 12, 416–422.

Pearce, E.J. (2003) Progress towards a vaccine for schistosomiasis. *Acta Tropica* 86, 309–313.

Popiel, I. and Basch, P.F. (1984) Reproductive development of female *Schistosoma mansoni* (Digenea: *Schistosomatidae*) following bisexual pairing of worms and worm segments. *Journal of Experimental Zoology* 232, 141–150.

Prichard, R. and Tait, A. (2001) The role of molecular biology in veterinary parasitology. *Veterinary Parasitology* 98, 169–194.

Ramachandran, H., Skelly, P.J. and Shoemaker, C.B. (1996) The *Schistosoma mansoni* epidermal growth factor receptor homologue, SER, has tyrosine kinase activity and is localized in adult muscle. *Molecular and Biochemical Parasitology* 83, 1–10.

Rao, K.V. and Ramaswamy, K. (2000) Cloning and expression of a gene-encoding Sm16, an anti-inflammatory protein from *Schistosoma mansoni*. *Molecular and Biochemical Parasitology* 108, 101–108.

Rausch, R.L. (1995) Life cycle patterns and geographical distribution of *Echinoccocus* species. In: Thompson, R.C.A. and Lymbery, A.J. (eds) *Echinoccocus and Hydatid Disease*. CAB International, Wallingford, UK, pp. 89–134.

Reichel, C., Mathur, J., Eckes, P., Langenkemper, K., Koncz, C., Schell, J., Reiss, B. and Maas, C. (1996) Enhanced green fluorescence by the expression of an *Aequoria victoria* green fluorescent protein mutant in mono- and dicotyledonous plant cells. *Proceedings of the National Academy of Sciences USA* 93, 5888–5893.

Rollinson, D. and Simpson, A.J.G. (1987) The biology of schistosomes. In: *From Genes to Latrines*. Academic Press, London.

Ross, A.G.P., Bartley, P.B., Sleigh, A.C., Olds, G.R., Li, Y., Williams, G.M. and McManus, D.P. (2002) Schistosomiasis. *The New England Journal of Medicine* 346, 1212–1220.

Rossi, A., Wippersteg, V., Klinkert, M.Q. and Grevelding, C.G. (2003) Cloning of 5′ and 3′ flanking regions of the *Schistosoma mansoni* calcineurin A gene and their characterization in transiently transformed parasites. *Molecular Biochemical Parasitology* 130, 133–138.

Russel, J.A., Roy, M.K. and Sanford, J.C. (1992) Physical trauma and tungsten toxicity reduce the efficiency of biolistic transformation. *Plant Physiology* 98, 1050–1056.

Sajid, M. and McKerrow, J.H. (2002) Cysteine proteases of parasitic organisms. *Molecular Biochemical Parasitology* 120, 1–21.

Salafsky, B., Fusco, A.C., Whitley, K., Nowicki, D. and Ellenberger, B. (1988) *Schistosoma mansoni*: analysis of cercarial transformation methods. *Experimental Parasitology* 67, 116–127.
Salo, E. and Baguna, J. (2002) Regeneration in planarians and other worms: new findings, new tools, new perspectives. *Journal of Experimental Zoology* 292, 528–539.
Sanchez Alvarado, A. (2003) The freshwater planarian *Schmidtea mediterranea*: embryogenesis, stem cells and regeneration. *Current Opinion in Genetics and Development* 13, 438–444.
Sanchez Alvarado, A. and Newmark, P.A. (1999) Double-stranded RNA specifically disrupts gene expression during planarian regeneration. *Proceedings of the National Academy of Sciences USA* 96, 5049–5054.
Sanford, J.C., Klein, T.M., Wolf, E.D. and Allen, N. (1987) Delivery of substances into cells and tissues using a particle bombardment process. *Particulate Science and Technology* 6, 559–563.
Sanford, J.C., Smith, F.D. and Russel, J.A. (1993) Optimizing the biolistic process for different biological applications. *Methods in Enzymology* 36, 483–513.
Savioli, L., Engels, D., Roungou, J.B., Fenwick, A. and Endo, H. (2004) Schistosomiasis control. *Lancet* 363, 658.
Sbicego, S., Schnaufer, A. and Blum, B. (1998) Transient and stable transfection of *Leishmania* by particle bombardment. *Molecular and Biochemical Parasitology* 94, 123–126.
Scallon, B.J., Bogitsh, B.J. and Clinton, C.E. (1987) Cloning of a *Schistosoma japonicum* gene encoding a major immunogen recognized by hyperinfected rabbits. *Molecular Biochemical Parasitology* 24, 237–245.
Schramm, G., Falcone, F.H., Gronow, A., Haisch, K., Mamat, U., Doenhoff, M.J., Oliveira, G., Galle, J., Dahinden, C.A. and Haas, H. (2003) Molecular characterization of an interleukin-4-inducing factor from *Schistosoma mansoni* eggs. *Journal of Biological Chemistry* 278, 18384–18392.
Schüßler, P., Pötters, E., Winnen, R., Bottke, W. and Kunz, W. (1995) An isoform of ferritin as a component of protein yolk platelets in *Schistosoma mansoni*. *Molecular Reproduction and Development* 41, 325–330.
Siles-Lucas, M. and Hemphill, A. (2002) Cestode parasites: application of *in vivo* and *in vitro* models for studies on the host–parasite relationship. *Advances in Parasitology* 51, 133–230.
Skelly, P.J. and Shoemaker, C.B. (1998) Rapid appearance and asymmetric distribution of glucose transporter SGTP4 at the apical surface of intramammalian-stage *Schistosoma mansoni*. *Proceedings of the National Academy of Sciences USA* 93, 3642–3646.
Skelly, P.J., Da'dara, A. and Harn, D.A. (2003) Suppression of cathepsin B expression in *Schistosoma mansoni* by RNA interference. *International Journal for Parasitology* 33, 363–369.
Smith, J.D. and Clegg, J.A. (1959) Egg shell formation in trematodes and cestodes. *Experimental Parasitology* 8, 286–323.
Spiliotis, M., Kroner, A. and Brehm, K. (2003) Identification, molecular characterization and expression of the gene encoding the epidermal growth factor receptor orthologue from the fox-tapeworm *Echinococcus multilocularis*. *Gene* 323, 57–65.
Spiliotis, M., Tappe, D., Sesterhenn, L. and Brehm, K. (2004) Long-term *in vitro* cultivation of *Echinococcus multilocularis* metacestodes under axenic conditions. *Parasitology Research* 92, 430–432.
Stewart, M.T., Mousley, A., Koubkova, B., Sebelova, S., Marks, N.J. and Halton, D.W. (2003) Development *in vitro* of the neuromusculature of two strigeid trematodes, *Apatemon cobitidis proterorhini* and *Cotylurus erraticus*. *International Journal for Parasitology* 33, 413–424.
Stirewalt, M.A. (1974) *Schistosoma mansoni*: cercariae to schistosomule. *Advances in Parasitology* 12, 115–182.
Sturrock, R.F. (2001) Schistosomiasis epidemiology and control: how did we get here and where should we go? *Memórias do Instituto Oswaldo Cruz* 96, 17–27.
Tan, H.H.C., Thornhill, J.A., Al-Adhami, B.H., Akhkha, A. and Kusel, J.R. (2003) A study of the effect of surface damage on the uptake of Texas Red-BSA by schistosomula of *Schistosoma mansoni*. *Parasitology* 126, 235–240.
Timmons, L., Tabara, H., Mello, C.C. and Fire, A.Z. (2003) Inducible systemic RNA silencing in *Caenorhabditis elegans*. *Molecular Biology of the Cell* 14, 2972–2983.
Trap, C. and Boireau, P. (2000) Proteases in heminthic parasites. *Veterinary Research* 31, 461–471.
Verjovski-Almeida, S., DeMarco, R., Martins, E.A., Guimaraes, P.E., Ojopi, E.P., Paquola, A.C., Piazza, J.P., Nishiyama, M.Y. Jr, Kitajima, J.P., Adamson, R.E., Ashton, P.D., Bonaldo, M.F., Coulson, P.S., Dillon, G.P., Farias, L.P., Gregorio, S.P., Ho, P.L., Leite, R.A., Malaquias, L.C., Marques, R.C., Miyasato, P.A., Nascimento, A.L., Ohlweiler, F.P., Reis, E.M., Ribeiro, M.A., Sa, R.G., Stukart, G.C., Soares, M.B., Gargioni, C., Kawano, T., Rodrigues, V., Madeira, A.M., Wilson, R.A., Menck, C.F., Setubal, J.C., Leite,

L.C. and Dias-Neto, E. (2003) Transcriptome analysis of the acoelomate human parasite *Schistosoma mansoni*. *Nature Genetics* 35, 148–157.

Verjovski-Almeida, S., Leite, L.C., Dias-Neto, E., Menck, C.F. and Wilson, R.A. (2004) Schistosome transcriptome: insights and perspectives for functional genomics. *Trends in Parasitology* 20, 304–308.

Vicogne, J., Pin, J.P., Lardans, V., Capron, M., Noel, C. and Dissous C. (2003) An unusual receptor tyrosine kinase of *Schistosoma mansoni* contains a Venus Flytrap module. *Molecular and Biochemical Parasitology* 126, 51–62.

Webster, P.J. and Mansour, T.E. (1992) Conserved classes of homeodomains in *Schistosoma mansoni*, an early bilateral metazoan. *Mechanisms of Development* 38, 25–32.

Wilm, T., Demel, P., Koop, H.U., Schnabel, H. and Schnabel, R. (1999) Ballistic transformation of *Caenorhabditis elegans*. *Gene* 229, 31–35.

Wippersteg, V., Kapp, K., Kunz, W. and Grevelding, C.G. (2002a) Biolistic gene transfer to generate transgenic schistosomes. *BioRad Technote* 2768, 1–4.

Wippersteg, V., Kapp, K., Kunz, W., Jackstadt, P., Zahner, H. and Grevelding, C.G. (2002b) HSP 70-controlled GFP expression in transiently transformed schistosomes. *Molecular and Biochemical Parasitology* 120, 141–150.

Wippersteg, V., Kapp, K., Kunz, W. and Grevelding, C.G. (2002c) Characterisation of the cysteine protease ER60 in transgenic *Schistosoma mansoni* larvae. *International Journal for Parasitology* 32, 1219–1224.

Wippersteg, V., Ribeiro, F., Liedtke, S., Kusel, J.R. and Grevelding, C.G. (2003) The uptake of Texas red BSA in the excretory system of schistosomes and its colocalisation with ER60 promoter-induced GFP in transiently transformed adult males. *International Journal for Parasitology* 33, 1139–1143.

Wippersteg, V., Liedtke, S. El-Bahay, A.O., Münnich, M., Philipp, C., Ribeiro, F., Kusel, J.R., Rossi, A., Klinkert, M.Q. and Grevelding, C.G. (2004) Analyses of tissue-specific gene activity in transgenic schistosomes. *International Journal of Medical Microbiology* 293 (Suppl. 38), 91.

Wippersteg, V., Sajid, M., Walshe, D., Khiem, D., Salter, J.P., McKerrow, J.H., Grevelding, C.G. and Caffrey, C.R. (2005) Biolistic transformation of *Schistosoma mansoni* with 5′ flanking regions of two peptidase genes promotes tissue-specific expression. *International Journal for Parasitology* 35, 583–589.

Yang, N.S. and Christou, P. (1994) *Particle Bombardment Technology for Gene Transfer*. Oxford University Press, New York.

Yee, J.K., Miyanohara, A., LaPorte, P., Bouic, K., Burns, J.C. and Friedmann, T. (1994) A general method for the generation of high-titer, pantropic retroviral vectors: highly efficient infection of primary hepatocytes. *Proceedings of the National Academy of Sciences USA* 91, 9564–9568.

8 Immunobiology of Schistosomes

S.G. Forrester and E.J. Pearce
Department of Pathobiology, School of Veterinary Medicine, University of Pennsylvania, Philadelphia, PA 19104-6008, USA

Introduction

Flatworm parasites are well recognized for their ability to live for decades in environments where they are in contact with potentially damaging immune factors. This adaptability reflects the fact that these parasites have evolved mechanisms to evade immune effector mechanisms, and more remarkably, to sense and utilize components of the host immune system for their own development. Schistosomes, digenetic trematodes, are undoubtedly the most well-studied parasitic flatworms. These parasites infect over 200 million people in tropical and subtropical zones, and cause severe disease in approximately 5% of those infected. Of immunological interest is the fact that pathology in schistosome-infected individuals is caused largely by the immune response to tissue-trapped parasite eggs. Despite the fact that schistosomes induce strong host immune responses and cause extensive pathology, they can live for years within the host, continuously depositing eggs into the external environment for subsequent infections. Here, we discuss the nature of schistosome-induced immune responses and the mechanisms utilized by these parasites to evade, regulate, and take advantage of, the host immune system. Our primary focus will be on *Schistosoma mansoni*, since this is the most well-studied organism within the genus. We begin this chapter by discussing host–parasite molecular interactions and emerging evidence suggesting that schistosomes are able to respond to host immune factors, and then move into a more detailed discussion of immune evasion mechanisms, immunopathology and immunotherapy; thorough reviews on

 Parasitic Flatworms: Molecular Biology, Biochemistry, Immunology and Physiology (eds A.G. Maule and N.J. Marks)

each of these areas have been published previously (Pearce and Sher, 1987; Brindley *et al.*, 1989; Capron *et al.*, 2002; Davies and McKerrow, 2003; Wynn *et al.*, 2004). Although of great inherent interest, the ultimate goal of studies of the immunobiology of schistosomes has to be the rational development of immunotherapeutics, with an anti-infection vaccine foremost on the list.

Host–Parasite Molecular Interactions

Schistosomes are dioecious parasitic flatworms (having male and female reproductive organs in separate individuals). Like all sexual digeneans, there is an alteration of generations with asexual reproduction occurring in a snail intermediate host and sexual reproduction in the definitive (mammalian) host. To infect the definitive host schistosomes emerge from freshwater snails as free-swimming cercariae. Cercariae locate and burrow into the skin of the mammalian host and transform into schistosomula. The schistosomula migrate through the skin layer into the vasculature *en route* to the hepatic portal vein where they mature into male or female adult parasites. There is no evidence for immune clearance of adult *S. mansoni* worms from the infected mammalian host (Agnew *et al.*, 1993), and available data indicate that if the immune response is to stop infection, it has to act on vulnerable pre-adult stages (Wilson, 1987). Approximately half of the eggs that are released by each female cross the endothelium and basement membrane of the vein, traverse the intervening tissue, basement membrane and epithelium of the intestine, and are subsequently passed to the lumen and out to the external environment, where they hatch to release miracidia which infect the intermediate snail host. Asexual reproduction within the snail gives rise to cercariae.

Importantly for our discussion, successful completion of the mammalian part of the life cycle appears to require immunological components. This is demonstrated clearly by the observations that schistosome fecundity and/or the excretion of parasite eggs is reduced in mice with severe combined immunodeficiency (SCID), and in nude mice and T-cell depleted mice (Harrison and Doenhoff, 1983; Amiri *et al.*, 1992). In part, this appears to be due to the necessity for T-cell mediated inflammation within the intestine to allow the passage of eggs through to the lumen, but a more interesting finding to emerge from these experiments is that reproductive processes in schistosomes are in some way triggered by T cells. Careful examination of the development of schistosomes in RAG–/– mice (which are B-as well as T-cell deficient) revealed a marked delay in growth and development that was linked to reduced fecundity (Davies *et al.*, 2001; Hernandez *et al.*, 2004). The type of lymphocyte that is important for schistosome development was identified as a previously unrecognized population of hepatic CD4+ T cells that, unusually, exists in major histocompatibility complex class II (MHC II)-deficient mice (Davies *et al.*, 2001). How this specific subset of T cells influences schistosome development is unknown. One suggestion based on the fact that the delayed development phenotype had also been reported for schistosomes growing in mice deficient in the cytokine interleukin-7 (IL-7) (Wolowczuk *et al.*, 1999a), is that T hepatic CD4+ cells are an important source of, or are capable of stimulating the production of, this cytokine (Davies *et al.*, 2001). The importance of IL-7 in schistosome development is underscored by the fact that in IL-7–/– mice, fewer parasites make it to the hepatic portal vein compared to in wild-type mice (Wolowczuk *et al.*, 1999b). IL-7, therefore not only influences schistosome development, but may also function as a signal used by parasites to migrate properly through the host. Another cytokine that has been shown to affect parasite biology is tumour necrosis factor-α (TNF-α). TNF-α influences egg production in female parasites (Amiri *et al.*, 1992; Cheever *et al.*, 1999) and helps maintain schistosome viability in the hepatic portal system (Wolowczuk *et al.*, 1999b). Together, these studies suggest that schistosomes receive crucial developmental signals from host cytokines.

The evidence that schistosome biology is modulated by host-derived immune factors has fuelled interest in the actual parasite receptors that can interact with such molecules. The surface of schistosomes consists of a tegument, which is composed of two closely apposed lipid bilayers overlying a syncytial cytoplasmic layer (Hockley, 1973; Hockley and McLaren, 1973). It has been widely speculated that this unusual surface may contain receptors that could bind to host-derived molecules. Our group has discovered two transmembrane receptor serine threonine kinases, SmRK1 (Davies *et al.*, 1998) and SmRK2 (Forrester *et al.*, 2004), which appear to be expressed at the tegumental surface membrane. SmRK1and SmRK2 are members of the transforming growth factor-β (TGF-β) receptor superfamily and experimental studies in which SmRK1 and SmRK2 have been expressed in mammalian cells indicate that these receptors are capable of signalling in response to host TGF-β superfamily cytokines (Beall and Pearce, 2001; S.G. Forrester and E.J. Pearce, unpublished observations). In other systems, TGF-β receptors control a variety of cellular processes including proliferation, recognition, differentiation and apoptosis as well the specification of developmental fate. Using information that has emerged recently from the identification of key components of a schistosome TGF-β signalling system and knowledge of well-established pathways from mammals, we now have good evidence that such a system exists in schistosomes. Aside from the discovery of SmRK1 and SmRK2, other components of this pathway, specifically the transcriptional components Smad1, 2, 4 (Beall *et al.*, 2000; Osman *et al.*, 2004) and 8, and Medea and SARA (Verjovski-Almeida *et al.*, 2003), have been identified. Using heterologous expression systems to characterize these various schistosome signalling proteins, our group has found that SmRK1, when expressed with Smad2 and a mammalian type II TGF-β receptor, can signal in response to human TGF-β (Beall and Pearce, 2001). A recent report revealed that the co-Smad, Smad4, interacts with Smad2 to propagate the downstream signal initiated by TGF-β (Osman *et al.*, 2004). There is currently no information on the function of SmRK2, but based on its localization to the schistosome surface, it could function as a partner for SmRK1 to bind host ligands and initiate signalling. Intriguingly, transcriptome analyses have failed to identify a schistosome TGF-β superfamily ligand (Verjovski-Almeida *et al.*, 2003), indicating that the likely ligand for SmRK1 and SmRK2 is host-derived. These reports therefore demonstrate that schistosomes possess the components required for a receptor-based signalling system to sense immunologic components in the environment.

How Schistosomes Evade the Host Immune System

Once schistosome cercariae penetrate the skin of their host, they can remain in this environment as schistosomula for up to 5 days before entering the vasculature and beginning their migration to the lungs and, subsequently, the hepatic portal vein (Wilson, 1987). During this journey, which may take 3 weeks (Wilson, 1987), and thereafter, schistosome parasites must contend with an assortment of host immune factors. Yet, as is well known, these parasites are able to complete the journey to their final destination and live for years. Research has shown that schistosomes have adapted a variety of mechanisms that enable them to evade a strong and potentially lethal host immune response.

Skin stage parasites

Cercariae are highly adapted to find and rapidly penetrate the skin of the definitive host as well as to evade initial immune responses. Part of the requirement for successful skin penetration and migration into the vasculature is the hydrolysis of host proteins via the release of proteases from the cercarial acetabular glands (McKerrow and Salter, 2002). One such protease of key importance for invasion is elastase, which degrades dermal elastin (Salter *et al.*, 2002), thus allowing migration through the epidermal layer. Upon penetrating the host, cercariae transform into schistosomula by shedding their glycocalyx, a thick layer of carbohydrate-rich material that overlies the tegumental surface membrane, and which presumably plays a role in parasite survival in water. Loss of this layer appears to be achieved by the shedding of the cercarial tegumental membrane and its replacement with a new membrane structure that consists of two closely apposed membranes (Hockley, 1973; Hockley and McLaren, 1973); amongst trematodes, this double lipid bilayer is unique to blood-dwelling parasites (McLaren and Hockley, 1977). Since the glycocalyx is a potent activator of the host complement system, this transformation is the first step towards allowing the parasites to survive within mammalian hosts (Samuelson and Caulfield, 1986; Marikovsky *et al.*, 1990).

In the epidermial layer of the mammalian host, schistosomula encounter a variety of cells whose function it is to eliminate foreign pathogens. Epidermal cells such as keratinocytes and Langerhans cells (LC, a subset of dendritic cells) are able to respond to pathogens by producing a variety of cytokines (Uchi *et al.*, 2000). During the host response to penetrating schistosomula, significant levels of a variety of chemokines/cytokines, including MIP-1α, IL-1β, IL-6, IL-10 and IL-12p40 are produced in the skin within 2 days (Ramaswamy *et al.*, 2000; Hogg *et al.*, 2003; Mountford and Trottein, 2004), indicating that there is a rapid innate response to infection. At least some of these mediators are made by keratinocytes, which are able to secrete cytokines *in vitro* in response to schistosomula excretory/secretory products (Ramaswamy *et al.*, 1995). Macrophages and dendritic cells (DCs), probably LC, also contribute to the early production of cytokines in the skin during infection with schistosomes (Hogg *et al.*, 2003). Moreover, since LC are specialized antigen presenting cells (APC), they would be expected to acquire schistosomula antigens and carry them to the lymph nodes (LN) draining the site of infection, wherein they would meet and activate specific CD4+ T helper (Th) lymphocytes, and in this manner initiate the adaptive immune response.

Cytokines and chemokines made by cells such as keratinocytes, macrophages and LC might be expected to recruit and activate leukocytes, which have the potential to kill schistosomula before the development of any adaptive immunity. *In vitro*, schistosomula have been shown to be susceptible to a wide array of immune effector mechanisms mediated by eosinophils and neutrophils (Butterworth, 1984; Capron and Capron, 1992), macrophages (James and Nacy, 1993) and platelets (Joseph *et al.*, 1983). In the case of eosinophils and neutrophils, cytotoxicity requires targeting of the cells to the parasites by parasite-specific antibodies and/or complement component C3, deposited on the surface of the parasites as a result of alternative or classical complement activation (Butterworth, 1984). Consequently, these effector mechanisms might be expected to work most efficiently in animals that have already responded immunologically to the parasites as a result of earlier exposure. Macrophage-mediated killing of schistosomula is somewhat different however, in that it can work in the absence of either antibody or complement, and is mediated by the toxic gas nitric oxide, NO, which is released from macrophages following activation by IFN-γ and additional stimuli (James and Nacy, 1993).

How schistosomula deal with the immune system has been of key interest in the study of schistosome biology. Available data indicate that these stages are able to manipulate the initiation of immune responses, and that as they develop and migrate to the lungs they additionally acquire the ability to evade recognition by antigen-specific components of the immune response. It has been shown experimentally that schistosomula can produce prostaglandin D2 (PGD2) which interferes with the migration of LC to the draining LN (Angeli *et al.*, 2001b); schistosome glutathione-*S*-transferase (Sm28GST) has been implicated in the production of PGD2 (Herve *et al.*, 2003). In addition to PGD2, schistosomula produce PGE2 and induce the production of PGE2 and IL-10 in both human and mouse keratinocytes (Ramaswamy *et al.*, 1995). An increased level of PGE2 in the skin of infected mice appears to lead to higher production of IL-10, which might be important for immune evasion. In fact, in infected IL-10–/– mice, some reports have indicated that fewer schistosomula make it out of the skin and that a prominent cellular reaction occurs around parasites in this tissue (Ramaswamy *et al.*, 2000). IL-10, in this case, is likely limiting leukocyte infiltration by downregulating the production of pro-inflammatory cytokines and chemokines at the site of infection (Moore *et al.*, 2001). Independently of any effects on IL-10 production, prostaglandins themselves can directly exert potent anti-inflammatory effects that may promote schistosome survival within the skin (Hata and Breyer, 2004; Trottein *et al.*, 2004).

Skin stage schistosomula also secrete molecules that interfere with T cell responses. Chen *et al.* (2002) found evidence that schistosomula secrete a 23-kDa protein that is pro-apoptotic for skin T lymphocytes. This effect would essentially allow the parasite to regulate the host immune response by inactivating the T cells that are important in initiating immune responses to foreign pathogens.

Lung stage parasites

Following infection, it takes 4–7 days for schistosomula to reach the lungs. At this point, the parasite begins to change shape and becomes elongated without increasing in volume, a change presumably associated with a requirement to pass through the capillary beds of the lungs. The evidence suggests that this phase of migration is most strenuous and it is at this point that schistosomula appear to be at greatest risk of immune-mediated damage (Wilson, 1987). Cellular infiltrates around parasites at this stage can disrupt migration and, at the most extreme, cause the schistosomula to take a one-way trip from the vasculature into the alveolar spaces (Crabtree and Wilson, 1987; Coulson and Wilson, 1988). In the lung capillaries, parasites are in close association with endothelial cells. PGE2 secretion by schistosomula has been reported to induce IL-6 production by endothelial cells (Angeli *et al.*, 2001a). It is thought that IL-6 diminishes inflammation in the lungs and favours successful migration of the parasites through this tissue. In the absence of IL-6, eotaxin and IL-5 production in the infected lungs is reportedly increased, and the resultant increase in infiltrating eosinophils has been associated with increased parasite mortality. It seems, therefore, that the IL-6 induced by the parasite may keep the local immune response in check, and help allow the successful migration of schistosomula through the lungs (Angeli *et al.*, 2001a). Thus in schistosomiasis, unlike in many other situations (Kopf *et al.*, 1995), IL-6 may play an anti-, rather than pro-inflammatory role. In addition to inducing IL-6 production, lung stage schistosomula minimize local cellular infiltration by secreting molecules that can downregulate the expression of endothelial adhesion molecules such as E-selectin and VCAM-1, which have been shown to play a role in the binding and transmigration of leukocytes to sites of inflammation (Trottein *et al.*, 1999).

By the time they have reached the lungs, schistosomula are innately resistant to immune effector mechanisms that are capable of killing schistosomula newly transformed from cercariae. In part, this is a reflection of the fact that growing schistosomula develop the ability to avoid activating complement and to evade recognition by antibodies (see below), and therefore are no longer susceptible to antibody or complement-mediated cellular cytotoxicity. However, studies in which these evasion mechanisms have been experimentally bypassed have revealed underlying resistance to immune effector molecules (Moser *et al.*, 1980); the basis of this resistance is unknown.

Adult parasites

The immune evasion mechanisms used by adult schistosomes are perhaps the most intriguing. Adult parasites live for many years in the portal vasculature where they are in constant contact with immune cells, antibodies and complement proteins. Several mechanisms of adult-stage immune evasion have been postulated. Early hypotheses, particularly those by Damian (1967), and Smithers *et al.* (1969), suggested that the parasites protect themselves from immune recognition by coating their surfaces with host antigens, as a form of concealment. These early studies demonstrated that worms grown to maturity in mice and then transferred sequentially to monkeys immunized against mouse tissue were rapidly destroyed by an antibody-mediated response directed against the surface of the worm (Smithers *et al.*, 1969). This finding was of great importance since it indicated that, unlike lung stage schistosomula (Moser *et al.*, 1980), adult parasites are not resistant to immune attack *per se*, but rather avoid immune-mediated damage by evading recognition. The idea that evasion was the result of the acquisition of host molecules has been circumstantially supported by findings that anti-schistosome antisera that bind strongly to surface-exposed antigens on very young schistosomula, fail to bind to parasites that have been recovered from the skin after 24 h or more of infection, or to lung or adult stage parasites, and that this reduction in binding is reciprocally linked to an increased ability of antisera specific for host components to recognize the parasites (McLaren *et al.*, 1975). Whether parasite antigens are directly masked by host molecules remains unclear, and other mechanisms for evasion, such as antigen-shedding, have been postulated (Pearce *et al.*, 1986). Regardless, it is clear that the schistosome surface is highly adapted to allow 'expression without exposure' of molecules such as the glucose transporters (Jiang *et al.*, 1996) that require localization at the surface of the parasite to function appropriately. It is tempting to believe that the unusual second surface membrane plays a pivotal role in this process.

More recent work has suggested that the coating of host molecules to the parasite surface may have more specific immune evasion functions. For example, one of the molecules acquired from the host is decay-accelerating factor (DAF) (Pearce *et al.*, 1990; Horta *et al.*, 1991). DAF is expressed on most mammalian cells and prevents complement-mediated autolysis by accelerating the decay of the C3 convertase that continuously forms on membranes and which, unregulated, catalyses the process that leads to formation of the complement membrane attack complex (MAC) and cell lysis. Parasites that are unable to acquire DAF become susceptible to complement-mediated killing *in vitro*. The importance of evading complement is emphasized by the fact that schistosomes also synthesize at least two complement regulatory proteins. One of these, schistosome complement inhibitory protein (SCIP-1) is capable of inhibiting MAC, probably by binding to C8 and C9 of the complement terminal pathway (Parizade *et al.*, 1994). Later findings revealed that SCIP-1 is actually a surface exposed form of paramyosin (Deng *et al.*, 2003). In addition to SCIP-1, adult schistosomes express a protein called complement C2 receptor inhibitory trispanning (CRIT) (Inal and Sim, 2000; Deng *et al.*, 2003) that binds to the complement protein C2 from human serum and, like paramyosin, in experimental settings is able to inhibit classical pathway-mediated haemolysis of sheep erythrocytes.

Host Immune Response to Schistosome Infection (Th1 versus Th2)

The host immune response during infection with schistosomes ironically both protects the host, by allowing the walling off of tissue-trapped eggs within granulomatous lesions, and thereby preventing potentially lethal egg-derived hepatotoxins from reaching the liver tissue (Doenhoff *et al.*, 1981; Dunne *et al.*, 1991), and simultaneously causes the major pathological changes that lead to severe disease (Cheever and Yap, 1997). In this section, we provide an overview of the host immune response during schistosome infection, with a particular emphasis on the importance of a Th1 (T helper 1) to Th2 immune response shift that occurs during a critical stage of infection and which is of great importance for survival of both the host and parasite. Both Th1 and Th2 cells can mediate granuloma formation, but the associated pathologies that accompany granulomatous inflammation when responses are polarized in either direction differ considerably. Reviews of this area have been published recently (Hoffmann *et al.*, 2002; Pearce and MacDonald, 2002; Wynn *et al.*, 2004). Th1 and Th2 cells are subsets of CD4+ T lymphocytes cells that are defined functionally based on the discrete panels of cytokines that they produce (Mosmann and Coffman, 1989). Th1 cells typically produce IFN-γ and lymphotoxin and usually provide protection against intracellular pathogens. Th2 cells, on the other hand, generally produce IL-4, IL-5, IL-10 and IL-13, and are important for immunity to helminth parasites. Inappropriate Th2 responses are associated with allergies and asthma, and discreet autoimmunities associated with self-reactive Th1 or Th2 responses are well documented.

During infection with *S. mansoni*, there is initially a pro-inflammatory response that in mice is apparent as the production of the Th1 cytokine IFN-γ, and in people experiencing their first infection, as the production of TNF-α, IL-1 and IL-6 (Grzych *et al.*, 1991; de Jesus *et al.*, 2002). However, in humans and mice there is a major shift in the Th response, from pro-inflammatory/Th1 to Th2, at about the time that parasites begin producing eggs (Grzych *et al.*, 1991; Pearce *et al.*, 1991; Wynn *et al.*, 1993; Araujo *et al.*, 1996). It has become clear that this shift can be crucial for host (and therefore parasite) survival. Th2 response development is dependent upon IL-4, a cytokine made by Th2 cells themselves, but also by basophils, eosinophils and additional cell types (Sabin *et al.*, 1996; Min *et al.*, 2004; Voehringer *et al.*, 2004). In mice lacking IL-4, infection proceeds normally until approximately week 6, but after this time an acute lethal disease develops in which the host becomes cachectic and rapidly succumbs (Brunet *et al.*, 1997; Fallon *et al.*, 2000). The function of IL-4 in this setting appears to be to regulate macrophage activation, promoting the expression of arginase rather than inducible nitric oxide synthase by these cells and thereby limiting the production of NO and the downstream products formed when this mediator interacts with products of the respiratory burst (La Flamme *et al.*, 2001a; Herbert *et al.*, 2004).

In the absence of IL-4, the Th1 response established during the period of infection before egg production persists, and continues to promote classical macrophage activation and NO production (Brunet *et al.*, 1999; La Flamme *et al.*, 2001a). This type of immune response is apparently incompatible with the ongoing insult to the liver caused by egg deposition, and in combination these factors lead to severe hepatic damage and death. It is important to note that severe disease in IL-4 deficient mice and in other animals that cannot make Th2 responses, is not associated with an increased parasite burden but rather is the result of immune-mediated damage (Brunet *et al.*, 1997).

IL-10, a cytokine made by Th2 cells but also by macrophages, DCs and regulatory (CD25+) CD4+ T cells also plays an important role in host survival during infection. IL-10 is a potent suppressor of inflammatory responses, serving to inhibit classical macrophage activation and to inhibit the production of the Th1-response promoting cytokine IL-12 by DCs and macrophages. In the absence of IL-10, infected mice fare much less well, developing a disease with some of the hallmarks of that seen in IL-4 deficient mice (Wynn *et al.*, 1998). Not surprisingly perhaps, IL-4/IL-10 double deficient mice are particularly susceptible to infection with *S. mansoni*. These mice develop a highly polarized and enhanced infection-specific Th1 response, a lethal acute wasting condition and increased hepatotoxicity, all more severe than observed in mice deficient in IL-4 alone (Hoffmann *et al.*, 2000).

IL-10 has been heavily implicated in the Th1 to Th2 switch that occurs as egg production begins, and has been shown by some though not all studies to play a significant role in the overall downregulation of T cell responsiveness that occurs during later stages of schistosomiasis (Wynn *et al.*, 1998; Stadecker, 1999; Hoffmann *et al.*, 2002; Sadler *et al.*, 2003). Strikingly, mice lacking IL-10 develop schistosome egg antigen-specific Th1 in addition to Th2 responses during infection (Wynn *et al.*, 1998; McKee and Pearce, 2004). This is directly the result of unregulated IL-12 production, since mice deficient for IL-12 and IL-10 fail to develop Th1 responses but rather develop excessively Th2-skewed responses (Hoffmann *et al.*, 2000). Recent work has focused on identifying the cellular source of the IL-10 that is important in this context, and it seems likely at this juncture that CD25+CD4+ regulatory T cells are playing an important role in the production of this cytokine (Hesse *et al.*, 2004; McKee and Pearce, 2004), although there is additional evidence from these recent studies for IL-10 being produced by macrophages within granulomas themselves (Hesse *et al.*, 2004).

IL-12 and IFN-γ-deficient mice fail to show any unusual susceptibility to schistosomiasis (La Flamme *et al.*, 2001b; Patton *et al.*, 2001), findings that suggest that the Th1 response that develops during the prepatent period of infection serves no essential host-beneficial function. However, as mentioned above, IL-12/IL-10–/– mice develop a severe form of schistosomiasis that is related to the development of excessively strong and polarized Th2 responses (Hoffmann *et al.*, 2000).

These animals display marked mortality during the chronic stages of infection associated with increased granuloma size and hepatic fibrosis. These findings illustrate two points: first, IL-10 plays a role in regulating Th2 as well as Th1 responses, and second, Th2 responses during schistosomiasis have a downside associated with their ability to promote tissue fibrosis. Indeed, there is a clear inverse relationship between Th1 polarization of the egg antigens' specific response during infection and hepatic fibrosis. This was first noted in mice that were immunized with egg antigens plus IL-12 (as a Th1-response inducing adjuvant) before infection. Animals immunized in this way developed significantly less granuloma-associated fibrosis during subsequent infection (Wynn *et al.*, 1995). This effect is primarily linked to the diminution in IL-13 levels that accompanies the suppression of the Th2 response; IL-13 is the major mediator of Th2 cell-driven fibrosis in schistosomiasis and there have been exciting recent advances in the use of soluble IL-13Rα2 as an antagonist of fibrosis during infection (Chiaramonte *et al.*, 1999a,b; Fallon *et al.*, 2000; Wynn, 2003; Wynn *et al.*, 2004). The finding that immunization to alter Th response polarization can have significant effects on the outcome of subsequent infection without altering susceptibility to infection *per se* has raised the interesting possibility of developing anti-pathology vaccines for schistosomiasis (Wynn, 1999).

The relationship between severe (fibrotic) schistosomiasis and Th2 responses is also indicated by studies in human populations living in schistosome-endemic areas. These studies have shown that protection against hepatic fibrosis is controlled by a major genetic locus that is located on 6q23, near the gene encoding the IFN-γ receptor chain, and that mutations that modulate the transcription of the IFN-γ gene are associated with different susceptibility to disease (Dessein *et al.*, 1999, 2004). The presence of periportal fibrosis has been further associated with a poor ability of cells in whole blood cultures from afflicted patients to make IFN-γ in response to schistosome antigens (Booth *et al.*, 2004). The role of IFN-γ is considered to be in the counter-regulation of Th2 responses and the effector functions of IL-13.

The fact that immunization of mice with soluble egg antigens plus IL-12 is able to induce an immunologic state in which, upon infection, mice exhibit antigen-specific Th1 responses and yet survive the acute phase of infection and enter the chronic phase with significantly diminished hepatic fibrosis contrasts dramatically with the situation in IL-4–/– or IL-4/IL-10–/– mice which develop severe lethal disease associated with Th1-related pro-inflammatory cytokine production. At present, the reasons for this difference in outcome is unclear. Interestingly, mice immunized with egg antigens plus complete Freund's adjuvant (CFA) develop an immune response that is superficially similar to that measured in mice immunized with eggs plus IL-12 and yet are acutely sensitive to subsequent infection, dying during the acute phase with signs of severe liver disease (Rutitzky *et al.*, 2001). One possibility to account for these observed differences is that IL-12 but not CFA, promotes the production of sufficiently high levels of IL-10 (Morris *et al.*, 1994; Wynn *et al.*, 1994) to protect against potentially lethal pro-inflammatory mechanisms. The complexity of the relative importance of Th1 versus Th2 responses in disease during infection is further illustrated by the finding that severe liver disease without an excessive fibrotic component has been linked to strong Th1 like responses in patients with schistosomiasis and concurrent malaria (Mwatha *et al.*, 1998, 2003). The immunopathology associated with Th1-polarized responses in schistosomiasis has been reviewed recently (Hoffmann *et al.*, 2002; Stadecker *et al.*, 2004).

Potential Players in Th2 Response Induction during a Schistosome Infection

There is a tight correlation during infection between the deposition of parasite eggs and the development of the Th2 response. This is due to the inherent ability of schistosome eggs to induce Th2 responses, a characteristic that has been demonstrated repeatedly in mice injected via intravenous, subcutaneous or intraperitoneal routes with isolated eggs and/or soluble egg antigens (Pearce and MacDonald,

2002). Egg antigens are highly immunogenic in the absence of added adjuvant and there is reason to believe that this may be due to signature carbohydrate motifs such as the polylactosamine sugar, lacto-*N*-fucopentaose III (LNFPIII), that are represented on multiple schistosome glycoproteins. Deglycosylated eggs fail to induce Th2 response and non-schistosome proteins modified with LNFPII acquire the ability to induce Th2 responses (Okano *et al.*, 1999, 2001; Williams *et al.*, 2001). These studies, therefore, suggest that sugars present on egg antigens may bind specific receptors that are crucial for Th2 induction. Recent work has indicated that the lectin DC-SIGN, which is expressed on DCs can bind LNFPIII, and is thus a good candidate for such a receptor (Van Die *et al.*, 2003). Lipids in eggs also may play key roles in promoting Th2 polarization. A novel lyso-phosphatidylserine from schistosome eggs has been shown to program DCs to induce CD4+ T cells to acquire a regulatory phenotype and produce IL-10 (van der Kleij *et al.*, 2002), which, as discussed above, plays a significant role in suppressing IFN-γ production during infection and thereby allowing polarization of the response in a Th2 direction.

Recent research into Th2 response induction by eggs has focused on the interactions of egg antigens with DCs. DCs are generally found throughout the organs of the body, where, by virtue of the fact that they express panels of pattern recognition receptors, they are able to monitor for the presence of foreign molecules that possess molecular motifs characteristic of pathogenic/non-self organisms (bacterial lipopolysaccharide, LPS, is a typical example of a such a molecular motif) (Mellman and Steinman, 2001; Akira and Takeda, 2004). Using differentially expressed chemokine receptors DCs are able to leave tissue sites and migrate via the lymphatics to draining LN. DCs are specialized for processing and presenting peptide antigens to T cells. Among the cadre of cell types that express major histocompatibility class II (MHCII) molecules (macrophages, B cells and DCs), DCs are most capable of activating naïve T cells that have not previously encountered the peptide for which their antigen receptor is specific (Mellman *et al.*, 1998). This activation step occurs when DCs meet T cells in the LN. Thus DCs are generally credited with initiating T cell responses and moreover of playing a crucial role in dictating the nature, Th1 or Th2, of the ensuing response. DCs were first attributed the latter quality when it was realized that they are potent producers of IL-12, the key cytokine responsible for promoting Th1 response polarization (Trinchieri, 2003). In the schistosome system, it has been found that murine DCs pulsed with soluble egg antigens *in vitro* are able to induce antigen-specific Th2 responses when injected into mice (MacDonald *et al.*, 2001).

It is now well understood that Toll-like receptors (TLR) play an important role in the recognition of viral, bacterial, fungal and protozoal organisms by DCs, and in the ability of DCs to induce Th1 responses to these pathogens (Akira and Takeda, 2004). Ligation of TLRs induces DC activation and maturation, a multifaceted process involving the regulation of expression of hundreds of genes, including those for receptors made by chemokines secreted by LN, and those for IL-12 and a number of surface membrane molecules that are important for DC–T cell interactions. Interestingly, unlike most microbial pathogens, helminth antigens including those from schistosomes do not induce the classical form of DC activation and maturation, and do not induce the production of IL-12 by DC in culture (Whelan *et al.*, 2000; MacDonald *et al.*, 2001). During infection, DCs from the spleen and mesenteric LN appear to be only minimally activated and the available data indicate that this is due to immune system intrinsic maturation signals provided by the ligation of CD40 on DCs (MacDonald *et al.*, 2002a,b; Straw *et al.*, 2003). Moreover, it appears that a major role for IL-10 in allowing Th2 polarization during infection is to prevent excessive DC maturation (McKee and Pearce, 2004).

Current research is focused on characterizing in detail the effect of helminth antigens on DCs in an effort to more clearly understand how these antigens are able to 'program' this important cell type to induce Th2 responses. It is now clear that Th2 response initiation does not require MyD88 (Jankovic *et al.*, 2002; Kaisho *et al.*, 2002), a key component of the major TLR-initiated signal transduction pathway (Akira and Takeda, 2004). However,

TLR-dependent MyD88-independent signalling pathways exist (Akira and Takeda, 2004), and it is feasible that these may be important for the DC response to schistosome egg antigens (Agrawal *et al.*, 2003; Thomas *et al.*, 2003; Thomas and Harn, 2004).

A possibility argued by some is that it is simply the absence of IL-12 production by DCs that have been exposed to helminth antigens that allows them to induce Th2 responses (Jankovic *et al.*, 2001; Barton and Medzhitov, 2002). Recent findings indicate that egg antigens not only fail to induce the production of IL-12 production by DCs, but also inhibit their ability to make IL-12 in response to classical TLR ligands such as LPS (Zaccone *et al.*, 2003; Cervi *et al.*, 2004). The mechanism of this suppression is unknown but may result from the ability of egg antigens in concert with LPS and other TLR-ligands to induce the production of IL-10, a cytokine that attenuates IL-12 production (Kane *et al.*, 2004). However, soluble egg antigens also inhibit IL-12 production in IL-10–/– DCs, indicating that there is an IL-10 independent mechanism of IL-12 suppression (Kane *et al.*, 2004). A candidate for this mechanism is the soluble egg antigen-induced expression of the transcription factor, c-Fos, which has been linked to suppression of IL-12 production (Agrawal *et al.*, 2003). The importance of pathways which suppress Th1 response development and may favour the development of immune responses where T cells make IL-4 and IL-10, which can have potent anti-inflammatory effects, is attested to not only by their crucial role in the survival of schistosome-infected hosts but also in the ability of schistosomes and other helminths and their antigens to ameliorate Th1 response-mediated autoimmune pathologies (La Flamme *et al.*, 2003; Sewell *et al.*, 2003; Summers *et al.*, 2003; David *et al.*, 2004); this is an area of growing interest (Wilson and Maizels, 2004).

Molecules expressed by DC that may be instrumental in the induction of Th2 responses include CD80 and CD86, the ligands for T cell CD28. For instance, mice deficient in both CD80 and CD86 fail to develop a Th2 response when infected, and exhibit impaired granuloma formation (Hernandez *et al.*, 1999). Much of this defect is apparent in mice lacking CD86 but not CD80, suggesting a more important role for CD86 in Th2 development during schistosome infection (Hernandez *et al.*, 1999). Consistent with this, treatment of infected mice with anti-CD86 antibodies reduces the expression of IL-4, IL-5 and IL-13 (Th2 cytokines) whereas anti-CD80 antibodies have no effect on IL-4 or IL-5 expression (Subramanian *et al.*, 1997). It is as well to note that CD86 is also expressed by B lymphocytes, and these cells have also been implicated in Th2 cell development since infected JhD mice, which have a total lack of B cells, fail to develop Th2 responses (Hernandez *et al.*, 1997). Interestingly, μMT mice, which also lack B cells but as the result of a different induced mutation than that in JhD mice, do develop Th2 responses when infected with schistosomes, but exhibit more severe chronic disease than is usual when B cells are present (Jankovic *et al.*, 1998). Severe disease in this model is associated with a failure to modulate granulomatous inflammation, a process that usually occurs over time in infected wild-type mice. Thus more work is needed to clarify the role of B cells in schistosomiasis.

Chemotherapy – An Integral Role for the Immune Response

The only internationally used drug effective for treating schistosomiasis is praziquantel. This drug induces a rapid influx into the worms of surrounding Ca^{2+}, a process that leads to paralysis (Martin, 1997), and changes in the surface membrane architecture that lead to the exposure of worm antigens that are normally cryptic (Brindley and Sher, 1987). Parasites affected in this way become susceptible to antibody-mediated immune attack and are killed as a result of the synergistic actions of chemotherapy and the immune response (Doenhoff *et al.*, 1987; Brindley *et al.*, 1989).

Immunotherapy

As highlighted above, work on the basic immunology of granuloma formation in schistosomiasis has led to the development of experimental anti-pathology vaccines and potent experimental immunotherapies to prevent fibrosis. The latter, based on the neutralization of IL-13, hold great additional promise for the treatment of diseases in addition to schistosomiasis in which fibrosis is a major factor (Wynn, 2003).

The development of a vaccine that prevents infection with schistosomes has long been recognized as the ideal way to control schistosomiasis. Studies of individuals living in endemic areas have revealed that immunity to superinfection, and to reinfection following drug-treatment, can develop over time. This type of naturally developing immunity in people is linked to the production of the Th2-dependent IgE isotype of antibody against parasite antigens (Hagan *et al.*, 1991; Dunne *et al.*, 1992; Rihet *et al.*, 1992), and is under the control of a major genetic locus located near the Th2 cytokine locus on chromosome 5q31–q33 (Marquet *et al.*, 1996; Dessein *et al.*, 2004). Thus there is strong evidence for a Th2-dependent, naturally acquired immunity in people.

Amongst various experimental vaccines, defined and complex, that have been studied (Pearce, 2003), the radiation-attenuated cercarial vaccine has been examined in greatest detail, and has been shown to induce good (although not sterile) immunity through the induction of multiple effector mechanisms that each have more or less significance depending on the number of repeat booster vaccinations an individual mouse receives (Wilson *et al.*, 1999; Wynn and Hoffmann, 2000). The immune response induced by one exposure to this type of vaccine is Th1-like, and the resistance to infection is mediated by IFN-γ-dependent mechanisms including the production of NO by classically activated macrophages; NO is toxic to larval schistosomes (Oswald *et al.*, 1994; Wilson *et al.*, 1996; Jankovic *et al.*, 1999; Street *et al.*, 1999). However, additional boosting leads to an immune response in which cytokines typical of Th1 and Th2 cells are produced and in which anti-schistosome antibody titres rise and antibodies play an important role in protection (Mangold and Dean, 1992; Jankovic *et al.*, 1999). IL-12 and bacterial CpG motifs (which induce IL-12 production by DCs and macrophages that express appropriate TLRs) have been shown to act as adjuvants to boost the Th1 component of irradiated cercariae-induced immunity (Wynn *et al.*, 1996; Mountford *et al.*, 1998; Chiaramonte *et al.*, 2000). It is important to note that vaccinated IL-10–/– mice develop exaggerated immune responses with Th1 and Th2 components, and are almost entirely resistant to challenge infection (Hoffmann *et al.*, 1999). It appears, therefore, that in order for a schistosome vaccine to be successful, it must generate a broad, non-polarized, high-magnitude immune response.

Concluding Remarks

Despite our knowledge of the immunity induced by irradiated cercariae, there is little interest in developing an attenuated vaccine for use in people. Moreover, despite the emergence over the last two decades of several candidate vaccines including glutathione-S-transferase (p28/GST), paramyosin, (Sm97), triose phosphate isomerase (TPI), Sm23, IrV-5 and Sm14, there appears to be little of great promise in terms of a usable defined vaccine against schistosomiasis (Pearce, 2003), and despite the importance and feasibility of developing a vaccine against this important disease (Todd and Colley, 2002), research towards this goal has, for a variety of reasons, reached a nadir (Pearce, 2003). However, with the near completion of the *S. mansoni* genome, the recent publications of the transcriptomes of *S. mansoni* (Verjovski-Almeida *et al.*, 2003) and *S. japonicum* (Hu *et al.*, 2003), and the emergence of techniques for experimentally manipulating gene expression in schistosomes (Boyle *et al.*, 2003; Skelly *et al.*, 2003; Correnti and Pearce, 2004), there is now great potential for understanding the biology of these important parasites in a way that, heretofore, has been impossible. Thus it seems feasible that the near future will see the identification of schistosome genes that are essential for successful parasitism and which encode proteins that are susceptible to immunologic targeting. Hopefully then, vaccine-centred research on schistosomiasis will soon experience a resurgence of interest.

Acknowledgements

The authors wish to acknowledge grant support from The Burroughs Wellcome Fund (EJP is a BWF Scholar in Molecular Parasitology), the National Institutes of Health and the Ellison Medical Foundation.

References

Agnew, A.M., Murare, H.M. and Doenhoff, M.J. (1993) Immune attrition of adult schistosomes. *Parasite Immunology* 15, 261–271.

Agrawal, S., Agrawal, A., Doughty, B., Gerwitz, A., Blenis, J., Van Dyke, T. and Pulendran, B. (2003) Cutting edge: different Toll-like receptor agonists instruct dendritic cells to induce distinct Th responses via differential modulation of extracellular signal-regulated kinase-mitogen-activated protein kinase and c-Fos. *The Journal of Immunology* 171, 4984–4989.

Akira, S. and Takeda, K. (2004) Toll-like receptor signalling. *Nature Reviews in Immunology* 4, 499–511.

Amiri, P., Locksley, R.M., Parslow, T.G., Sadick, M., Rector, E., Ritter, D. and McKerrow, J.H. (1992) Tumour necrosis factor alpha restores granulomas and induces parasite egg-laying in schistosome-infected SCID mice. *Nature* 356, 604–607.

Angeli, V., Faveeuw, C., Delerive, P., Fontaine, J., Barriera, Y., Franchimont, N., Staels, B., Capron, M. and Trottein, F. (2001a) *Schistosoma mansoni* induces the synthesis of IL-6 in pulmonary microvascular endothelial cells: role of IL-6 in the control of lung eosinophilia during infection. *European Journal of Immunology* 31, 2751–2761.

Angeli, V., Faveeuw, C., Roye, O., Fontaine, J., Teissier, E., Capron, A., Wolowczuk, I., Capron, M. and Trottein, F. (2001b) Role of the parasite-derived prostaglandin D2 in the inhibition of epidermal Langerhans cell migration during schistosomiasis infection. *The Journal of Experimental Medicine* 193, 1135–1147.

Araujo, M.I., de Jesus, A.R., Bacellar, O., Sabin, E., Pearce, E. and Carvalho, E.M. (1996) Evidence of a T helper type 2 activation in human schistosomiasis. *European Journal of Immunology* 26, 1399–1403.

Barton, G.M. and Medzhitov, R. (2002) Control of adaptive immune responses by Toll-like receptors. *Current Opinion in Immunology* 14, 380–383.

Beall, M.J. and Pearce, E.J. (2001) Human transforming growth factor-beta activates a receptor serine/threonine kinase from the intravascular parasite *Schistosoma mansoni*. *The Journal of Biological Chemistry* 276, 31613–31619.

Beall, M.J., McGonigle, S. and Pearce, E.J. (2000) Functional conservation of *Schistosoma mansoni* Smads in TGF-beta signaling. *Molecular and Biochemical Parasitology* 111, 131–142.

Booth, M., Mwatha, J.K., Joseph, S., Jones, F.M., Kadzo, H., Ireri, E., Kazibwe, F., Kemijumbi, J., Kariuki, C., Kimani, G., Ouma, J.H., Kabatereine, N.B., Vennervald, B.J. and Dunne, D.W. (2004) Periportal fibrosis in human *Schistosoma mansoni* infection is associated with low IL-10, low IFN-gamma, high TNF-alpha, or low RANTES, depending on age and gender. *The Journal of Immunology* 172, 1295–1303.

Boyle, J.P., Wu, X.J., Shoemaker, C.B. and Yoshino, T.P. (2003) Using RNA interference to manipulate endogenous gene expression in *Schistosoma mansoni* sporocysts. *Molecular and Biochemical Parasitology* 128, 205–215.

Brindley, P.J. and Sher, A. (1987) The chemotherapeutic effect of praziquantel against *Schistosoma mansoni* is dependent on host antibody response. *The Journal of Immunology* 139, 215–220.

Brindley, P.J., Strand, M., Norden, A.P. and Sher, A. (1989) Role of host antibody in the chemotherapeutic action of praziquantel against *Schistosoma mansoni*: identification of target antigens. *Molecular and Biochemical Parasitology* 34, 99–108.

Brunet, L.R., Finkelman, F.D., Cheever, A.W., Kopf, M.A. and Pearce, E.J. (1997) IL-4 protects against TNF-alpha-mediated cachexia and death during acute schistosomiasis. *The Journal of Immunology* 159, 777–785.

Brunet, L.R., Beall, M., Dunne, D.W. and Pearce, E.J. (1999) Nitric oxide and the Th2 response combine to prevent severe hepatic damage during *Schistosoma mansoni* infection. *The Journal of Immunology* 163, 4976–4984.

Butterworth, A.E. (1984) Cell-mediated damage to helminths. *Advances in Parasitology* 23, 143–235.

Capron, M. and Capron, A. (1992) Effector functions of eosinophils in schistosomiasis. *Memorias do Instituto Oswaldo Cruz* 87, 167–170.

Capron, A., Riveau, G.J., Bartley, P.B. and McManus, D.P. (2002) Prospects for a schistosome vaccine. *Current Drug Targets of Immune Endocrine and Metabolic Disorders* 2, 281–290.
Cervi, L., MacDonald, A.S., Kane, C., Dzierszinski, F. and Pearce, E.J. (2004) Cutting edge: dendritic cells copulsed with microbial and helminth antigens undergo modified maturation, segregate the antigens to distinct intracellular compartments and concurrently induce microbe-specific Th1 and helminth-specific Th2 responses. *The Journal of Immunology* 172, 2016–2020.
Cheever, A.W. and Yap, G.S. (1997) Immunologic basis of disease and disease regulation in schistosomiasis. *Chemical Immunology* 66, 159–176.
Cheever, A.W., Poindexter, R.W. and Wynn, T.A. (1999) Egg laying is delayed but worm fecundity is normal in SCID mice infected with *Schistosoma japonicum* and *S. mansoni* with or without recombinant tumor necrosis factor alpha treatment. *Infection and Immunity* 67, 2201–2208.
Chen, L., Rao, K.V., He, Y.X. and Ramaswamy, K. (2002) Skin-stage schistosomula of *Schistosoma mansoni* produce an apoptosis-inducing factor that can cause apoptosis of T cells. *Journal of Biological Chemistry* 277, 34329–34335.
Chiaramonte, M.G., Donaldson, D.D., Cheever, A.W. and Wynn, T.A. (1999a) An IL-13 inhibitor blocks the development of hepatic fibrosis during a T-helper type 2-dominated inflammatory response. *The Journal of Clinical Investigation* 104, 777–785.
Chiaramonte, M.G., Schopf, L.R., Neben, T.Y., Cheever, A.W., Donaldson, D.D. and Wynn, T.A. (1999b) IL-13 is a key regulatory cytokine for Th2 cell-mediated pulmonary granuloma formation and IgE responses induced by *Schistosoma mansoni* eggs. *The Journal of Immunology* 162, 920–930.
Chiaramonte, M.G., Hesse, M., Cheever, A.W. and Wynn, T.A. (2000) CpG oligonucleotides can prophylactically immunize against Th2-mediated schistosome egg-induced pathology by an IL-12-independent mechanism. *The Journal of Immunology* 164, 973–985.
Correnti, J.M. and Pearce, E.J. (2004) Transgene expression in *Schistosoma mansoni*: introduction of RNA into schistosomula by electroporation. *Molecular and Biochemical Parasitology* 137, 75–79.
Coulson, P.S. and Wilson, R.A. (1988) Examination of the mechanisms of pulmonary phase resistance to *Schistosoma mansoni* in vaccinated mice. *American Journal of Tropical Medicine and Hygiene* 38, 529–539.
Crabtree, J.E. and Wilson, R.A. (1987) Pulmonary cellular reactions to *Schistosoma mansoni* schistosomula in normal and vaccinated mice. *Advances in Experimental Medical Biology* 216A, 701–707.
Damian, R.T. (1967) Common antigens between adult *Schistosoma mansoni* and the laboratory mouse. *Journal of Parasitology* 53, 60–64.
David, T., Thomas, C., Zaccone, P., Dunne, D.W. and Cooke, A. (2004) The impact of infection on the incidence of autoimmune disease. *Current Topics in Medicinal Chemistry* 4, 521–529.
Davies, S.J. and McKerrow, J.H. (2003) Developmental plasticity in schistosomes and other helminths. *International Journal for Parasitology* 33, 1277–1284.
Davies, S.J., Shoemaker, C.B. and Pearce, E.J. (1998) A divergent member of the transforming growth factor beta receptor family from *Schistosoma mansoni* is expressed on the parasite surface membrane. *Journal of Biological Chemistry* 273, 11234–11240.
Davies, S.J., Grogan, J.L., Blank, R.B., Lim, K.C., Locksley, R.M. and McKerrow, J.H. (2001) Modulation of blood fluke development in the liver by hepatic CD4+ lymphocytes. *Science* 294, 1358–1361.
de Jesus, A.R., Silva, A., Santana, L.B., Magalhaes, A., de Jesus, A.A., de Almeida, R.P., Rego, M.A., Burattini, M.N., Pearce, E.J. and Carvalho, E.M. (2002) Clinical and immunologic evaluation of 31 patients with acute *Schistosomiasis mansoni*. *Journal of Infectious Diseases* 185, 98–105.
Deng, J., Gold, D., LoVerde, P.T. and Fishelson, Z. (2003) Inhibition of the complement membrane attack complex by *Schistosoma mansoni* paramyosin. *Infection and Immunity* 71, 6402–6410.
Dessein, A.J., Hillaire, D., Elwali, N.E., Marquet, S., Mohamed-Ali, Q., Mirghani, A., Henri, S., Abdelhameed, A.A., Saeed, O.K., Magzoub, M.M. and Abel, L. (1999) Severe hepatic fibrosis in *Schistosoma mansoni* infection is controlled by a major locus that is closely linked to the interferon-gamma receptor gene. *American Journal of Human Genetics* 65, 709–721.
Dessein, A., Kouriba, B., Eboumbou, C., Dessein, H., Argiro, L., Marquet, S., Elwali, N.E., Rodrigues, V., Li, Y., Doumbo, O. and Chevillard, C. (2004) Interleukin-13 in the skin and interferon-gamma in the liver are key players in immune protection in human schistosomiasis. *Immunological Reviews* 201, 180–190.
Doenhoff, M.J., Pearson, S., Dunne, D.W., Bickle, Q., Lucas, S., Bain, J., Musallam, R. and Hassounah, O. (1981) Immunological control of hepatotoxicity and parasite egg excretion in *Schistosoma mansoni* infections: stage specificity of the reactivity of immune serum in T-cell deprived mice. *Transactions of the Royal Society of Tropical Medicine and Hygiene* 75, 41–53.

Doenhoff, M.J., Sabah, A.A., Fletcher, C., Webbe, G. and Bain, J. (1987) Evidence for an immune-dependent action of praziquantel on *Schistosoma mansoni* in mice. *Transactions of the Royal Society Tropical Medicine and Hygiene* 81, 947–951.

Dunne, D.W., Jones, F.M. and Doenhoff, M.J. (1991) The purification, characterization, serological activity and hepatotoxic properties of two cationic glycoproteins (alpha 1 and omega 1) from *Schistosoma mansoni* eggs. *Parasitology* 103, 225–236.

Dunne, D.W., Butterworth, A.E., Fulford, A.J., Ouma, J.H. and Sturrock, R.F. (1992) Human IgE responses to *Schistosoma mansoni* and resistance to reinfection. *Memorias do Instituto Oswaldo Cruz* 87, 99–103.

Fallon, P.G., Richardson, E.J., McKenzie, G.J. and McKenzie, A.N. (2000) Schistosome infection of transgenic mice defines distinct and contrasting pathogenic roles for IL-4 and IL-13: IL-13 is a profibrotic agent. *The Journal of Immunology* 164, 2585–2591.

Forrester, S., Warfel, P.W. and Pearce, E.J. (2004) Tegumental expression of a novel type II receptor serine/threonine kinase (SmRK2) in *Schistosoma mansoni. Molecular and Biochemical Parasitology* 136, 149–156.

Grzych, J.M., Pearce, E., Cheever, A., Caulada, Z.A., Caspar, P., Heiny, S., Lewis, F. and Sher, A. (1991) Egg deposition is the major stimulus for the production of Th2 cytokines in murine *Schistosoma mansoni*. *The Journal of Immunology* 146, 1322–1327.

Hagan, P., Blumenthal, U.J., Dunn, D., Simpson, A.J. and Wilkins, H.A. (1991) Human IgE, IgG4 and resistance to reinfection with *Schistosoma haematobium*. *Nature* 349, 243–245.

Harrison, R.A. and Doenhoff, M.J. (1983) Retarded development of *Schistosoma mansoni* in immunosuppressed mice. *Parasitology* 86, 429–438.

Hata, A.N. and Breyer, R.M. (2004) Pharmacology and signaling of prostaglandin receptors: multiple roles in inflammation and immune modulation. *Pharmacological Therapy* 103, 147–166.

Herbert, D.R., Holscher, C., Mohrs, M., Arendse, B., Schwegmann, A., Radwanska, M., Leeto, M., Kirsch, R., Hall, P., Mossmann, H., Claussen, B., Förster, I. and Bronbacher, F. (2004) Alternative macrophage activation is essential for survival during schistosomiasis and downmodulates T helper 1 responses and immunopathology. *Immunity* 20, 623–635.

Hernandez, H.J., Wang, Y. and Stadecker, M.J. (1997) In infection with *Schistosoma mansoni*, B cells are required for T helper type 2 cell responses but not for granuloma formation. *The Journal of Immunology* 158, 4832–4837.

Hernandez, H.J., Sharpe, A.H. and Stadecker, M.J. (1999) Experimental murine schistosomiasis in the absence of B7 costimulatory molecules: reversal of elicited T cell cytokine profile and partial inhibition of egg granuloma formation. *The Journal of Immunology* 162, 2884–2889.

Hernandez, D.C., Lim, K.C., McKerrow, J.H. and Davies, S.J. (2004) *Schistosoma mansoni*: sex-specific modulation of parasite growth by host immune signals. *Experimental Parasitology* 106, 59–61.

Herve, M., Angeli, V., Pinzar, E., Wintjens, R., Faveeuw, C., Narumiya, S., Capron, A., Urade, Y., Capron, M., Riveau, G. and Trottein, F. (2003) Pivotal roles of the parasite PGD2 synthase and of the host D prostanoid receptor 1 in schistosome immune evasion. *European Journal of Immunology* 33, 2764–2772.

Hesse, M., Piccirillo, C.A., Belkaid, Y., Prufer, J., Mentink-Kane, M., Leusink, M., Cheever, A.W., Shevach, E.M. and Wynn, T.A. (2004) The pathogenesis of schistosomiasis is controlled by cooperating IL-10-producing innate effector and regulatory T cells. *Journal of Immunology* 172, 3157–3166.

Hockley, D.J. (1973) Ultrastructure of the tegument of Schistosoma. *Advances in Parasitology* 11, 233–305.

Hockley, D.J. and McLaren, D.J. (1973) *Schistosoma mansoni*: changes in the outer membrane of the tegument during development from cercaria to adult worm. *International Journal for Parasitology* 3, 13–25.

Hoffmann, K.F., James, S.L., Cheever, A.W. and Wynn, T.A. (1999) Studies with double cytokine-deficient mice reveal that highly polarized Th1- and Th2-type cytokine and antibody responses contribute equally to vaccine-induced immunity to *Schistosoma mansoni*. *The Journal of Immunology* 163, 927–938.

Hoffmann, K.F., Cheever, A.W. and Wynn, T.A. (2000) IL-10 and the dangers of immune polarization: excessive type 1 and type 2 cytokine responses induce distinct forms of lethal immunopathology in murine schistosomiasis. *The Journal of Immunology* 164, 6406–6416.

Hoffmann, K.F., Wynn, T.A. and Dunne, D.W. (2002) Cytokine-mediated host responses during schistosome infections; walking the fine line between immunological control and immunopathology. *Advances in Parasitology* 52, 265–307.

Hogg, K.G., Kumkate, S., Anderson, S. and Mountford, A.P. (2003) Interleukin-12 p40 secretion by cutaneous CD11c+ and F4/80+ cells is a major feature of the innate immune response in mice that develop Th1-mediated protective immunity to *Schistosoma mansoni*. *Infection and Immunity* 71, 3563–3571.

Horta, M.F., Ramalho-Pinto, F.J. and Fatima, M. (1991) Role of human decay-accelerating factor in the evasion of *Schistosoma mansoni* from the complement-mediated killing *in vitro*. *The Journal of Experimental Medicine* 174, 1399–1406.

Hu, W., Yan, Q., Shen, D.K., Liu, F., Zhu, Z.D., Song, H.D., Xu, X.R., Wang, Z.J., Rong, Y.P., Zeng, L.C., Wu, J., Zhang, X., Wang, J.J., Xu, X.N., Wang, S.Y., Fu, G., Zhang, X.L., Wang, Z.Q., Brindley, P.J., McManus, D.P., Xue, C.L., Feng, Z., Chen, Z. and Han, Z.G. (2003) Evolutionary and biomedical implications of a *Schistosoma japonicum* complementary DNA resource. *Nature Genetics* 35, 139–147.

Inal, J.M. and Sim, R.B. (2000) A schistosoma protein, Sh-TOR, is a novel inhibitor of complement which binds human C2. *FEBS Letters* 470, 131–134.

James, S.L. and Nacy, C. (1993) Effector functions of activated macrophages against parasites. *Current Opinion in Immunology* 5, 518–523.

Jankovic, D., Cheever, A.W., Kullberg, M.C., Wynn, T.A., Yap, G., Caspar, P., Lewis, F.A., Clynes, R., Ravetch, J.V. and Sher, A. (1998) CD4+ T cell-mediated granulomatous pathology in schistosomiasis is downregulated by a B cell-dependent mechanism requiring Fc receptor signaling. *The Journal of Experimental Medicine* 187, 619–629.

Jankovic, D., Wynn, T.A., Kullberg, M.C., Hieny, S., Caspar, P., James, S., Cheever, A.W. and Sher, A. (1999) Optimal vaccination against *Schistosoma mansoni* requires the induction of both B cell- and IFN-gamma-dependent effector mechanisms. *The Journal of Immunology* 162, 345–351.

Jankovic, D., Sher, A. and Yap, G. (2001) Th1/Th2 effector choice in parasitic infection: decision making by committee. *Current Opinion in Immunology* 13, 403–409.

Jankovic, D., Kullberg, M.C., Hieny, S., Caspar, P., Collazo, C.M. and Sher, A. (2002) In the absence of IL-12, CD4(+) T cell responses to intracellular pathogens fail to default to a Th2 pattern and are host protective in an IL-10(–/–) setting. *Immunity* 16, 429–439.

Jiang, J., Skelly, P.J., Shoemaker, C.B. and Caulfield, J.P. (1996) *Schistosoma mansoni*: the glucose transport protein SGTP4 is present in tegumental multilamellar bodies, discoid bodies and the surface lipid bilayers. *Experimental Parasitology* 82, 201–210.

Joseph, M., Auriault, C., Capron, A., Vorng, H. and Viens, P. (1983) A new function for platelets: IgE-dependent killing of schistosomes. *Nature* 303, 810–812.

Kaisho, T., Hoshino, K., Iwabe, T., Takeuchi, O., Yasui, T. and Akira, S. (2002) Endotoxin can induce MyD88-deficient dendritic cells to support T(h)2 cell differentiation. *International Immunology* 14, 695–700.

Kane, C.M., Cervi, L., Sun, J., McKee, A.S., Masek, K.S., Shapira, S., Hunter, C.A. and Pearce, E.J. (2004) Helminth antigens modulate TLR-initiated dendritic cell activation. *The Journal of Immunology* 173, 7454–7461.

Kopf, M., LeGros, G., Coyle, A.J., Kosco-Vilbois, M. and Brombacher, F. (1995) Immune responses of IL-4, IL-5, IL-6 deficient mice. *Immunological Reviews* 148, 45–69.

La Flamme, A.C., Patton, E.A., Bauman, B. and Pearce, E.J. (2001a) IL-4 plays a crucial role in regulating oxidative damage in the liver during schistosomiasis. *The Journal of Immunology* 166, 1903–1911.

La Flamme, A.C., Patton, E.A. and Pearce, E.J. (2001b) Role of gamma interferon in the pathogenesis of severe schistosomiasis in interleukin-4-deficient mice. *Infection and Immunity* 69, 7445–7452.

La Flamme, A.C., Ruddenklau, K. and Backstrom, B.T. (2003) Schistosomiasis decreases central nervous system inflammation and alters the progression of experimental autoimmune encephalomyelitis. *Infection and Immunity* 71, 4996–5004.

MacDonald, A.S., Straw, A.D., Bauman, B. and Pearce, E.J. (2001) CD8-dendritic cell activation status plays an integral role in influencing Th2 response development. *The Journal of Immunology* 167, 1982–1988.

MacDonald, A.S., Patton, E.A., La Flamme, A.C., Araujo, M.I., Huxtable, C.R., Bauman, B. and Pearce, E.J. (2002a) Impaired Th2 development and increased mortality during *Schistosoma mansoni* infection in the absence of CD40/CD154 interaction. *The Journal of Immunology* 168, 4643–4649.

MacDonald, A.S., Straw, A.D., Dalton, N.M. and Pearce, E.J. (2002b) Cutting edge: Th2 response induction by dendritic cells: a role for CD40. *Journal of Immunology* 168, 537–540.

Mangold, B.L. and Dean, D.A. (1992) The role of IgG antibodies from irradiated cercaria-immunized rabbits in the passive transfer of immunity to *Schistosoma mansoni*-infected mice. *American Journal of Tropical Medicine and Hygiene* 47, 821–829.

Marikovsky, M., Parizade, M., Arnon, R. and Fishelson, Z. (1990) Complement regulation on the surface of cultured schistosomula and adult worms of *Schistosoma mansoni*. *European Journal of Immunology* 20, 221–227.

Marquet, S., Abel, L., Hillaire, D., Dessein, H., Kalil, J., Feingold, J., Weissenbach, J. and Dessein, A.J. (1996) Genetic localization of a locus controlling the intensity of infection by *Schistosoma mansoni* on chromosome 5q31–q33. *Nature Genetics* 14, 181–184.

Martin, R.J. (1997) Modes of action of antihelmintic drugs. *Veterinary Journal* 154, 11–34.

McKee, A.S. and Pearce, E.J. (2004) CD25+CD4+ cells contribute to Th2 polarization during helminth infection by suppressing Th1 response development. *The Journal of Immunology* 173, 1224–1231.

McKerrow, J.H. and Salter, J. (2002) Invasion of skin by schistosoma cercariae. *Trends in Parasitology* 18, 193–195.

McLaren, D.J. and Hockley, D.J. (1977) Blood flukes have a double outer membrane. *Nature* 269, 147–149.

McLaren, D.J., Clegg, J.A. and Smithers, S.R. (1975) Acquisition of host antigens by young *Schistosoma mansoni* in mice: correlation with failure to bind antibody *in vitro*. *Parasitology* 70, 67–75.

Mellman, I. and Steinman, R.M. (2001) Dendritic cells: specialized and regulated antigen processing machines. *Cell* 106, 255–258.

Mellman, I., Turley, S.J. and Steinman, R.M. (1998) Antigen processing for amateurs and professionals. *Trends in Cell Biology* 8, 231–237.

Min, B., Prout, M., Hu-Li, J., Zhu, J., Jankovic, D., Morgan, E.S., Urban, J.F. Jr, Dvorak, A.M., Finkelman, F.D., LeGros, G. and Paul, W.E. (2004) Basophils produce IL-4 and accumulate in tissues after infection with a Th2-inducing parasite. *The Journal of Experimental Medicine* 200, 507–517.

Moore, K.W., de Waal Malefyt, R., Coffman, R.L. and O'Garra, A. (2001) Interleukin-10 and the interleukin-10 receptor. *Annual Reviews Immunology* 19, 683–765.

Morris, S.C., Madden, K.B., Adamovicz, J.J., Gause, W.C., Hubbard, B.R., Gately, M.K. and Finkelman, F.D. (1994) Effects of IL-12 on *in vivo* cytokine gene expression and Ig isotype selection. *The Journal of Immunology* 152, 1047–1056.

Moser, G., Wassom, D.L. and Sher, A. (1980) Studies of the antibody-dependent killing of schistosomula of *Schistosoma mansoni* employing haptenic target antigens. I. Evidence that the loss in susceptibility to immune damage undergone by developing schistosomula involves a change unrelated to the masking of parasite antigens by host molecules. *The Journal of Experimental Medicine* 152, 41–53.

Mosmann, T.R. and Coffman, R.L. (1989) Th1 and Th2 cells: different patterns of lymphokine secretion lead to different functional properties. *Annual Reviews Immunology* 7, 145–173.

Mountford, A.P. and Trottein, F. (2004) Schistosomes in the skin: a balance between immune priming and regulation. *Trends in Parasitology* 20, 221–226.

Mountford, A.P., Shires, V.L. and Anderson, S. (1998) Interleukin-12 and protective immunity to schistosomes. *Brazilian Journal of Medical Biological Research* 31, 163–169.

Mwatha, J.K., Kimani, G., Kamau, T., Mbugua, G.G., Ouma, J.H., Mumo, J., Fulford, A.J., Jones, F.M., Butterworth, A.E., Roberts, M.B. and Dunne, D.W. (1998) High levels of TNF, soluble TNF receptors, soluble ICAM-1 and IFN-gamma, but low levels of IL-5, are associated with hepatosplenic disease in human schistosomiasis mansoni. *The Journal of Immunology* 160, 1992–1999.

Mwatha, J.K., Jones, F.M., Mohamed, G., Naus, C.W., Riley, E.M., Butterworth, A.E., Kimani, G., Kariuki, C.H., Ouma, J.H., Koech, D. and Dunne, D.W. (2003) Associations between anti-*Schistosoma mansoni* and anti-*Plasmodium falciparum* antibody responses and hepatosplenomegaly, in Kenyan schoolchildren. *The Journal of Infectious Diseases* 187, 1337–1341.

Okano, M., Satoskar, A.R., Nishizaki, K., Abe, M. and Harn, D.A. Jr (1999) Induction of Th2 responses and IgE is largely due to carbohydrates functioning as adjuvants on *Schistosoma mansoni* egg antigens. *The Journal of Immunology* 163, 6712–6717.

Okano, M., Satoskar, A.R., Nishizaki, K. and Harn, D.A. Jr (2001) Lacto-*N*-fucopentaose III found on *Schistosoma mansoni* egg antigens functions as adjuvant for proteins by inducing Th2-type response. *The Journal of Immunology* 167, 442–450.

Osman, A., Niles, E.G. and LoVerde, P.T. (2004) Expression of functional *Schistosoma mansoni* Smad4: role in Erk-mediated transforming growth factor beta (TGF-beta) down-regulation. *Journal of Biological Chemistry* 279, 6474–6486.

Oswald, I.P., Wynn, T.A., Sher, A. and James, S.L. (1994) NO as an effector molecule of parasite killing: modulation of its synthesis by cytokines. *Comparative Biochemistry Physiology Pharmacology Toxicology Endocrinology* 108, 11–18.

Parizade, M., Arnon, R., Lachmann, P.J. and Fishelson, Z. (1994) Functional and antigenic similarities between a 94-kD protein of *Schistosoma mansoni* (SCIP-1) and human CD59. *The Journal of Experimental Medicine* 179, 1625–1636.

Patton, E.A., Brunet, L.R., La Flamme, A.C., Pedras-Vasconcelos, J., Kopf, M. and Pearce, E.J. (2001) Severe schistosomiasis in the absence of interleukin-4 (IL-4) is IL-12 independent. *Infection and Immunity* 69, 589–592.

Pearce, E.J. (2003) Progress towards a vaccine for schistosomiasis. *Acta Tropica* 86, 309–313.

Pearce, E.J. and MacDonald, A.S. (2002) The immunobiology of schistosomiasis. *Nature Reviews in Immunology* 2, 499–511.

Pearce, E.J. and Sher, A. (1987) Mechanisms of immune evasion in schistosomiasis. *Contributions in Microbiology Immunology* 8, 219–232.

Pearce, E.J., Basch, P.F. and Sher, A. (1986) Evidence that the reduced surface antigenicity of developing *Schistosoma mansoni* schistosomula is due to antigen shedding rather than host molecule acquisition. *Parasite Immunology* 8, 79–94.

Pearce, E.J., Hall, B.F. and Sher, A. (1990) Host-specific evasion of the alternative complement pathway by schistosomes correlates with the presence of a phospholipase C-sensitive surface molecule resembling human decay accelerating factor. *The Journal of Immunology* 144, 2751–2756.

Pearce, E.J., Caspar, P., Grzych, J.M., Lewis, F.A. and Sher, A. (1991) Downregulation of Th1 cytokine production accompanies induction of Th2 responses by a parasitic helminth, *Schistosoma mansoni. The Journal of Experimental Medicine* 173, 159–166.

Ramaswamy, K., Salafsky, B., Potluri, S., He, Y.X., Li, J.W. and Shibuya, T. (1995) Secretion of an anti-inflammatory, immunomodulatory factor by Schistosomulae of *Schistosoma mansoni. Journal of Inflammation* 46, 13–22.

Ramaswamy, K., Kumar, P. and He, Y.X. (2000) A role for parasite-induced PGE2 in IL-10-mediated host immunoregulation by skin stage schistosomula of *Schistosoma mansoni. The Journal of Immunology* 165, 4567–4574.

Rihet, P., Demeure, C.E., Dessein, A.J. and Bourgois, A. (1992) Strong serum inhibition of specific IgE correlated to competing IgG4, revealed by a new methodology in subjects from a *S. mansoni* endemic area. *European Journal of Immunology* 22, 2063–2070.

Rutitzky, L.I., Hernandez, H.J. and Stadecker, M.J. (2001) Th1-polarizing immunization with egg antigens correlates with severe exacerbation of immunopathology and death in schistosome infection. *Proceedings of the National Academy of Sciences USA* 98, 13243–13248.

Sabin, E.A., Kopf, M.A. and Pearce, E.J. (1996) *Schistosoma mansoni* egg-induced early IL-4 production is dependent upon IL-5 and eosinophils. *The Journal of Experimental Medicine* 184, 1871–1878.

Sadler, C.H., Rutitzky, L.I., Stadecker, M.J. and Wilson, R.A. (2003) IL-10 is crucial for the transition from acute to chronic disease state during infection of mice with *Schistosoma mansoni. European Journal of Immunology* 33, 880–888.

Salter, J.P., Choe, Y., Albrecht, H., Franklin, C., Lim, K.C., Craik, C.S. and McKerrow, J.H. (2002) Cercarial elastase is encoded by a functionally conserved gene family across multiple species of schistosomes. *Journal of Biological Chemistry* 277, 24618–24624.

Samuelson, J.C. and Caulfield, J.P. (1986) Cercarial glycocalyx of *Schistosoma mansoni* activates human complement. *Infection and Immunity* 51, 181–186.

Sewell, D., Qing, Z., Reinke, E., Elliot, D., Weinstock, J., Sandor, M. and Fabry, Z. (2003) Immunomodulation of experimental autoimmune encephalomyelitis by helminth ova immunization. *International Immunology* 15, 59–69.

Skelly, P.J., Da'dara, A. and Harn, D.A. (2003) Suppression of cathepsin B expression in *Schistosoma mansoni* by RNA interference. *International Journal for Parasitology* 33, 363–369.

Smithers, S.R., Terry, R.J. and Hockley, D.J. (1969) Host antigens in schistosomiasis. *Proceedings of the Royal Society of London Series B: Biological Sciences* 171, 483–494.

Stadecker, M.J. (1999) The regulatory role of the antigen-presenting cell in the development of hepatic immunopathology during infection with *Schistosoma mansoni. Pathobiology* 67, 269–272.

Stadecker, M.J., Asahi, H., Finger, E., Hernandez, H.J., Rutitzky, L.I. and Sun, J. (2004) The immunobiology of Th1 polarization in high-pathology schistosomiasis. *Immunological Reviews* 201, 168–179.

Straw, A.D., MacDonald, A.S., Denkers, E.Y. and Pearce, E.J. (2003) CD154 plays a central role in regulating dendritic cell activation during infections that induce Th1 or Th2 responses. *The Journal of Immunology* 170, 727–734.

Street, M., Coulson, P.S., Sadler, C., Warnock, L.J., McLaughlin, D., Bluethmann, H. and Wilson, R.A. (1999) TNF is essential for the cell-mediated protective immunity induced by the radiation-attenuated schistosome vaccine. *The Journal of Immunology* 163, 4489–4494.

Subramanian, G., Kazura, J.W., Pearlman, E., Jia, X., Malhotra, I. and King, C.L. (1997) B7-2 requirement for helminth-induced granuloma formation and CD4 type 2 T helper cell cytokine expression. *The Journal of Immunology* 158, 5914–5920.

Summers, R.W., Elliott, D.E., Qadir, K., Urban, J.F. Jr, Thompson, R. and Weinstock, J.V. (2003) Trichuris suis seems to be safe and possibly effective in the treatment of inflammatory bowel disease. *The American Journal of Gastroenterology* 98, 2034–2041.

Thomas, P.G. and Harn, D.A. Jr (2004) Immune biasing by helminth glycans. *Cellular Microbiology* 6, 13–22.

Thomas, P.G., Carter, M.R., Atochina, O., Da'Dara, A.A., Piskorska, D., McGuire, E. and Harn, D.A. (2003) Maturation of dendritic cell 2 phenotype by a helminth glycan uses a Toll-like receptor 4-dependent mechanism. *The Journal of Immunology* 171, 5837–5841.

Todd, C.W. and Colley, D.G. (2002) Practical and ethical issues in the development of a vaccine against *Schistosomiasis mansoni*. *American Journal of Tropical Medicine and Hygiene* 66, 348–358.
Trinchieri, G. (2003) Interleukin-12 and the regulation of innate resistance and adaptive immunity. *Nature Reviews in Immunology* 3, 133–146.
Trottein, F., Nutten, S., Angeli, V., Delerive, P., Teissier, E., Capron, A., Staels, B. and Capron, M. (1999) *Schistosoma mansoni* schistosomula reduce E-selectin and VCAM-1 expression in TNF-alpha-stimulated lung microvascular endothelial cells by interfering with the NF-kappaB pathway. *European Journal of Immunology* 29, 3691–3701.
Trottein, F., Faveeuw, C., Gosset, P. and Angeli, V. (2004) Role of the d-prostanoid receptor 1 in the modulation of immune and inflammatory responses. *Critical Reviews in Immunology* 24, 349–362.
Uchi, H., Terao, H., Koga, T. and Furue, M. (2000) Cytokines and chemokines in the epidermis. *Journal of Dermatological Science* 24, S29–S38.
van der Kleij, D., Latz, E., Brouwers, J.F., Kruize, Y.C., Schmitz, M., Kurt-Jones, E.A., Espevik, T., de Jong, E.C., Kapsenberg, M.L., Golenbock, D.T., Tielens, A.G.M. and Yazdanbakhsh, M. (2002) A novel host-parasite lipid cross-talk. Schistosomal lyso-phosphatidylserine activates toll-like receptor 2 and affects immune polarization. *Journal of Biological Chemistry* 277, 48122–48129.
Van Die, I., Van Vliet, S.J., Nyame, A.K., Cummings, R.D., Bank, C.M., Appelmelk, B., Geijtenbeek, T.B. and Van Kooyk, Y. (2003) The dendritic cell-specific C-type lectin DC-SIGN is a receptor for *Schistosoma mansoni* egg antigens and recognizes the glycan antigen Lewis x. *Glycobiology* 13, 471–478.
Verjovski-Almeida, S., DeMarco, R., Martins, E.A., Guimaraes, P.E., Ojopi, E.P., Paquola, A.C., Piazza J.P., Nishiyama, M.Y. Jr, Kitajima, J.P., Adamson, R.E., Ashton, P.D., Bonaldo, M.F., Coulson, P.S., Dillon, G.P., Farias, L.P., Gregoria, S.P., Ho, P.L., Leite, R.A., Malaquias, L.C., Marques, R.C., Miyasato, P.A., Nascimento, A.L., Ohlweiler, F.P., Reis, E.M., Ribeiro, M.A., Sa, R.G., Stukart, G.C., Soares, M.B., Gargioni, C., Kawano, T., Rodrigues, V., Madeira, A.M., Wilson, R.A., Menck, C.F., Setubal, J.C., Leite, L.C. and Dias-Neto, E. (2003) Transcriptome analysis of the acoelomate human parasite *Schistosoma mansoni*. *Nature Genetics* 35, 148–157.
Voehringer, D., Shinkai, K. and Locksley, R.M. (2004) Type 2 immunity reflects orchestrated recruitment of cells committed to IL-4 production. *Immunity* 20, 267–277.
Whelan, M., Harnett, M.M., Houston, K.M., Patel, V., Harnett, W. and Rigley, K.P. (2000) A filarial nematode-secreted product signals dendritic cells to acquire a phenotype that drives development of Th2 cells. *Journal of Immunology* 164, 6453–6460.
Williams, D.L., Asahi, H., Botkin, D.J. and Stadecker, M.J. (2001) Schistosome infection stimulates host CD4(+) T helper cell and B-cell responses against a novel egg antigen, thioredoxin peroxidase. *Infection and Immunity* 69, 1134–1141.
Wilson, R.A. (1987) Cercariae to liver worms: development and migration in the mammalian host. In: Rollinson, D. and Simpson, A.J.G. (eds) *The Biology of Schistosomes*. Academic Press, London, pp. 115–146.
Wilson, M.S. and Maizels, R.M. (2004) Regulation of allergy and autoimmunity in helminth infection. *Clinical Reviews in Allergy and Immunology* 26, 35–50.
Wilson, R.A., Coulson, P.S., Betts, C., Dowling, M.A. and Smythies, L.E. (1996) Impaired immunity and altered pulmonary responses in mice with a disrupted interferon-gamma receptor gene exposed to the irradiated *Schistosoma mansoni* vaccine. *Immunology* 87, 275–282.
Wilson, R.A., Coulson, P.S. and Mountford, A.P. (1999) Immune responses to the radiation-attenuated schistosome vaccine: what can we learn from knock-out mice? *Immunology Letters* 65, 117–123.
Wolowczuk, I., Nutten, S., Roye, O., Delacre, M., Capron, M., Murray, R.M., Trottein, F. and Auriault, C. (1999a) Infection of mice lacking interleukin-7 (IL-7) reveals an unexpected role for IL-7 in the development of the parasite *Schistosoma mansoni*. *Infection and Immunity* 67, 4183–4190.
Wolowczuk, I., Roye, O., Nutten, S., Delacre, M., Trottein, F. and Auriault, C. (1999b) Role of interleukin-7 in the relation between *Schistosoma mansoni* and its definitive vertebrate host. *Microbes and Infection* 1, 545–551.
Wynn, T.A. (1999) Immune deviation as a strategy for schistosomiasis vaccines designed to prevent infection and egg-induced immunopathology. *Microbes and Infection* 1, 525–534.
Wynn, T.A. (2003) IL-13 effector functions. *Annual Reviews in Immunology* 21, 425–456.
Wynn, T.A. and Hoffmann, K.F. (2000) Defining a schistosomiasis vaccination strategy – is it really Th1 versus Th2? *Parasitology Today* 16, 497–501.
Wynn, T.A., Eltoum, I., Cheever, A.W., Lewis, F.A., Gause, W.C. and Sher, A. (1993) Analysis of cytokine mRNA expression during primary granuloma formation induced by eggs of *Schistosoma mansoni*. *The Journal of Immunology* 151, 1430–1440.

Wynn, T.A., Eltoum, I., Oswald, I.P., Cheever, A.W. and Sher, A. (1994) Endogenous interleukin 12 (IL-12) regulates granuloma formation induced by eggs of *Schistosoma mansoni* and exogenous IL-12 both inhibits and prophylactically immunizes against egg pathology. *The Journal of Experimental Medicine* 179, 1551–1561.

Wynn, T.A., Cheever, A.W., Jankovic, D., Poindexter, R.W., Caspar, P., Lewis, F.A. and Sher, A. (1995) An IL-12-based vaccination method for preventing fibrosis induced by schistosome infection. *Nature* 376, 594–596.

Wynn, T.A., Reynolds, A., James, S., Cheever, A.W., Caspar, P., Hieny, S., Jankovic, D., Strand, M. and Sher, A. (1996) IL-12 enhances vaccine-induced immunity to schistosomes by augmenting both humoral and cell-mediated immune responses against the parasite. *The Journal of Immunology* 157, 4068–4078.

Wynn, T.A., Cheever, A.W., Williams, M.E., Hieny, S., Caspar, P., Kuhn, R., Muller, W. and Sher, A. (1998) IL-10 regulates liver pathology in acute murine *Schistosomiasis mansoni* but is not required for immune down-modulation of chronic disease. *The Journal of Immunology* 160, 4473–4480.

Wynn, T.A., Thompson, R.W., Cheever, A.W. and Mentink-Kane, M.M. (2004) Immunopathogenesis of schistosomiasis. *Immunological Reviews* 201, 156–167.

Zaccone, P., Fehervari, Z., Jones, F.M., Sidobre, S., Kronenberg, M., Dunne, D.W. and Cooke, A. (2003) *Schistosoma mansoni* antigens modulate the activity of the innate immune response and prevent onset of type 1 diabetes. *European Journal of Immunology* 33, 1439–1449.

9 Cestode Infection: Immunological Considerations from Host and Tapeworm Perspectives

D.M. McKay[1] and R.A. Webb[2]
[1]*Intestinal Disease Research Programme, McMaster University, Hamilton, Ontario, Canada L8N 3Z5; and* [2]*Department of Biology, York University, Toronto, Ontario, Canada*

Introduction

At the outset of any consideration of the host immune response to cestode infection, the first question is 'why should we be interested in the immune response to cestodes?' The view could be advanced that the availability of effective anthelminthics and appropriate food preparation methods has reduced the impact of cestode infection on humans to almost an historical oddity. Not so. Tapeworm infection of humans and their domestic stock is an age-old problem that continues to exert a significant socio-economic impact by reducing livestock productivity and human well-being. Indeed, tapeworm infection may be a re-emerging threat to human health, particularly in non-Western societies where meat (i.e. tainted meat) is becoming a more frequent addition to traditional agrarian diets. For instance, neurocysticercosis, a consequence of *Taenia solium* infection, is increasingly diagnosed in northern

Indian populations and *Echinococcus* hydatid disease is not uncommon in central and northern Africa (Rajshekhar *et al.*, 2003). Moreover, the expansion of international travel would suggest that cestode infections will increase in Western Europe and North America. So rather than trivializing the impact of cestode infections, a cogent argument can be made for the need for greater awareness of these infections and the development of a comprehensive knowledge of cestode–host interactions, which in the authors' opinion, regrettably lags behind the appreciation of parasitic nematode or trematode (particularly *Schistosome* spp.) interactions with human hosts.

A complete review of cestode taxonomy and the immunological crosstalk between cestodes and their vertebrate, never mind their invertebrate, hosts is beyond the scope of this commentary. Thus, we will consider neither the tetraphyllidea or trypanorhyncha tapeworms nor the interaction of pseudophyllidean worms with their piscine or avian hosts, but rather restrict our review to that of some cyclophyllideans (Taeniids and Hymenolepidids) with their mammalian hosts. The eloquently stated 'Red Queen Hypothesis' (Castrodeza, 1979) draws one's attention to the evolutionary game of 'leapfrog' that parasites play with their hosts. Thus, evaluation of the host response to cestode infection must be balanced by a consideration of how the successful tapeworm evades or subverts either immunological recognition by the host or host events aimed at destroying/eradicating the offending invader.

Cestode Parasites

The cestodes are a large and diverse group of platyhelminths that share two common features: as adults, they have an elongate body and they lack an alimentary canal. Consequently, they must reside in an elongate, nutrient-rich environment. Thus, adult tapeworms are almost invariably found in the definitive hosts' intestine where they absorb nutrients directly across their tegument. The cestode life cycle, with the notable exception of *Hymenolepis nana* (Fig. 9.1), is indirect and involves at least one intermediate host. Unlike adult tapeworms, the juvenile forms are morphologically quite diverse and can be found in almost

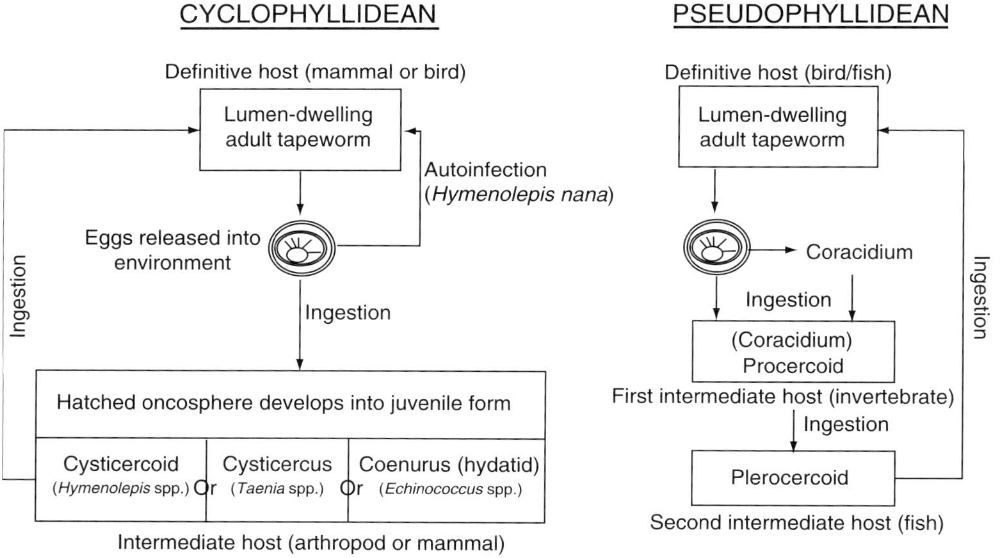

Fig. 9.1. Schemas showing the generalized life cycles of cyclophyllidean cestode parasite that must pass through intermediate and definite hosts and pseudophyllidean cestode parasites whose life cycle involves juvenile stages in two different intermediate hosts (the typical class of host is shown in parentheses for each life-cycle stage).

any organ of their invertebrate or vertebrate hosts (Smyth, 1994): clearly, this presents a number of different immunological challenges. Confining our comments to the Cyclophyllidea, these organisms are characterized by a scolex bearing four suckers (or acetabula) that is typically armed with an array of hooks (an exception being *Hymenolepis diminuta*). Intermediate hosts can be invertebrate (mostly arthropod) or warm-blooded (typically mammals) and the definite host mammalian or avian. The three most common juvenile forms that develop from the hatched oncoshpere are shown in Fig. 9.1, with the less common juvenile stobilocercus form (e.g. *Taenia taeniaeformis*) being omitted from the figure. A number of hymenolepidid species (*H. diminuta, H. nana, Hymenolepis microstoma*) are common laboratory models for assessing tapeworm infection, whereas the taeniids (*Taenia* spp., *Echinococcus* spp.) are the most relevant from a medical and veterinary perspective. With hymenolepidid infections, the ingested larvae pass through the stomach, excyst and attach to, and develop in, the small intestine. Thus, being a lumen-dwelling parasite in the definite host, it is effectively outside of the body (i.e. separated from the mucosa by the enteric epithelial barrier) and so presents a series of immunological challenges in terms of detection. The same is true for adult taeniids in their definitive hosts. In addition, larval forms of taeniids can encyst in the muscle of the intermediate host (e.g. the cysticercus of *T. solium*) or, in the case of hydatid disease, the oncosphere can penetrate the intestinal mucosa and via the draining circulation be deposited in the liver or other viscera where it develops into the hydatid cyst. These modifications to the life cycle represent different antigenic presentations and hence different challenges to the immune system, i.e. how does the mammalian host respond to a tissue dwelling juvenile form of a tapeworm compared to the lumen-dwelling adult form?

Immunology – The Basics

The mammalian immune system is divided into innate and adaptive immune responses. Innate or non-learned events aim to block the entry of potentially noxious antigen and invading pathogens and mobilize a very rapid response if/when the barrier elements are breached. For example, acid production in the stomach, mucus production by the gastrointestinal tract, sloughing of the dead keratinized layers of the skin and the continuous, rapidly renewing epithelial lining of mucosal surfaces are all barrier components of the innate immune system. Cellular elements of the innate immune system, such as dendritic cells, macrophages and neutrophils, have evolved to recognize and respond to conserved structures of bacterial and other pathogens, such as lipopolysaccharide, flagellin and CpG-rich DNA – i.e. pathogen-associated molecular patterns (PAMPs). These molecules bind to pattern recognition receptors (PRRs), such as the Toll-like receptors (TLRs), and evoke generalized immune responses characterized by increased phagocytosis and the production of destructive reactive oxygen and nitrogen metabolites, pro-inflammatory cytokines (e.g. tumour necrosis factor alpha (TNFα)) and chemotactic molecules that will recruit additional immune cells to the site of injury or infection. Moreover, the response to PAMPs up-regulates dendritic cell and macrophage expression of cytokines/chemokines, adhesion molecules, B7 (an important second signal for naïve T cells) and their antigen-presenting function. So while the cellular components of the innate immune system provide a rapid deployment defence they are also critical in driving and optimizing subsequent adaptive immune responses and the generation of antigen-specific immunological memory.

The adaptive arm of the immune response relies on the ability of T and B cells to identify specific antigen via the T cell receptor (TcR) and B cell receptor (or immunoglobulin (Ig)). The T cells only recognize peptide that is processed and presented by antigen presenting cells (APCs) (e.g. dendritic cells) in the context of major histocompatibility class (MHC) I or II antigens. Protein antigen derived from the cytosol or intracellular pathogens is presented in MHC I to $CD8^+$ cytotoxic T cells, whereas extracellular antigen taken in by phagocytosis (e.g. that from helminths) will be presented in MHC II to $CD4^+$ T helper (Th) cells. Cytotoxic $CD8^+$ cells kill infected or abnormal cells by inducing apoptosis via perforin–granzyme release, Fas–Fas ligand interaction or the release of cytotoxic cytokines (e.g. TNFα, lymphotoxin). The primary role of $CD4^+$ T cells is orchestration of immune events and the regulation of humoral immunity. The Th cells have been subdivided into types 1 (Th1) and 2 (Th2) and a variety of regulatory types (i.e. Th3, $Th_{regulatory}$, $CD4^+CD25^+$) (Mittrucker and Kaufmann, 2004). Upon stimulation Th1 cells release interferon-γ (IFNγ), TNFα and interleukin (IL)-2 and will activate phagocytes to enhance cell-mediated immunity. In contrast, IL-4, IL-5, IL-6, IL-10 and IL-13 from Th2 cells are the keys in humoral immune reactions, directing B cell activity and influencing antibody (AB) production and isotype switching from low-affinity IgM to high-affinity IgA, IgE and IgG. Following T cell–B cell interactions, B cells give rise to terminally differentiated plasma cells that produce large amounts of high-affinity Ig that enters the circulation (or gut lumen) and will neutralize antigen, opsonize metazoan pathogens for subsequent attack by macrophages and granulocytes or allow for complement fixation and lysis of the parasite. Importantly, IFNγ can inhibit Th2 events and IL-4 and IL-10 diminish Th1 responses, and so polarization towards either Th1- or Th2-dominated events has the capacity to block the reciprocal Th cell reactions (Fig. 9.2). Thus, the source of the antigen and the nature of the T cell reaction are critical determinants in mobilizing appropriate immune responses: when challenged by a tapeworm a pure $CD8^+$ T cell event will not effectively combat the worm infection.

While the tradition in biomedical science has been to pigeon-hole cells or events into discrete systems, nowhere is the folly of this simplistic approach better illustrated than in the immune system where the vanguard must be connectedness. For example, dendritic cells are key players in innate and adaptive immune responses, humoral immunity involves both

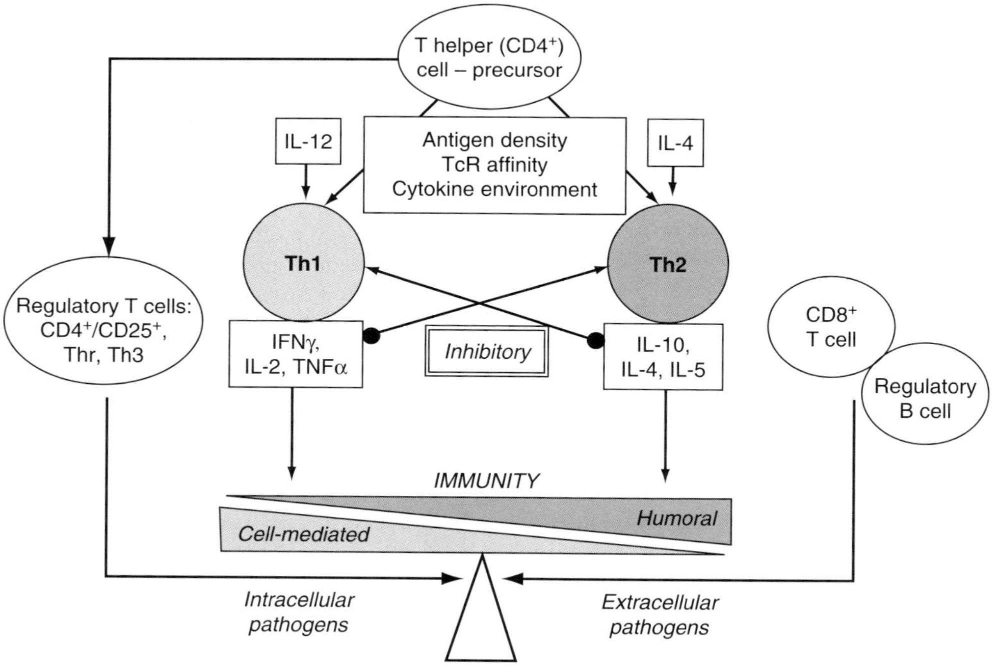

Fig. 9.2. Simplified schema of the development of T helper type 1 (Th1) and 2 (Th2) cells under the influence of interleukin (IL)-12 and IL-4, respectively, and how these two polarized T cell phenotypes produce cross-inhibitory cytokine profiles and promote different aspects of the host immune response (IFNγ, interferon-gamma; TNFα, tumour necrosis factor alpha; Thr, T helper regulatory cell; Th3, T helper type 3 cell).

soluble (i.e. AB, complement) and cellular (e.g. mast cells, eosinophils) elements, $CD8^+$ cells are a source of cytokines as are regulatory B cells, and both $CD8^+$ and $CD4^+$ T cells respond to viral infection. Thus, the efficient immune response that results in pathogen recognition and eradication in the face of minimal immunopathology is dependent on the interaction and coordination of innate and adaptive immune responses. It is also important to appreciate that immune events must be placed in an immunophysiological context where non-classical immune cells such as epithelial cells, stromal cells (i.e. fibroblasts) and nerves have all been shown to have the potential to modulate immune responses (Perdue and McKay, 1994).

The Intestine and Immune Responses

The intestinal lumen is an ideal location for a parasite, affording protection from the external elements and providing a nutrient-rich milieu. Moreover, the lumen is distinct from the body proper and thus with the exception of any abrasive contact points (i.e. hooked scolicies) the cestodes are effectively separated from many of the active immune components by the continuous epithelial cell layer lining the gastrointestinal tract. Indeed, it is imperative that the host maintains the integrity of this epithelial barrier and thus minimizes the entry of antigens and microbes into the mucosa. By selecting the mammalian intestinal lumen as its residence of choice the adult tapeworm avoids contact with the majority of the cellular components of the immune system. Nonetheless, it is equally clear that in non-permissive systems the worms are immunologically eradicated from the host.

The intestinal immune response, or more appropriately the immunophysiological response, to tapeworms can be considered in two categories: (i) creation of an inhospitable environment and (ii) active humoral events aimed at destruction of the tegument, destrobilization and scolex detachment. We provide a brief overview of the response to cestodes, with pertinent species-specific and life-cycle stage-specific events given in the following section.

Create an inhospitable environment

The protective outer layers of the infective larval stages should ensure safe passage through the gastric acid. Once lodged in the small intestine a series of host responses are geared to dislodge and expel the worm: a washer/sweeper event, increased mucus production and a mast-cell response that is the major effector mechanism for the 'leak hypothesis'. Damage to the tegument by gastric acid remains a threat but some hymenolepidids by a circadian migratory pattern move into the duodenum and proximal jejunum shortly after the host commences feeding where the concentrations of acid are highest but potentially destructive intestinal digestive enzymes are inhibited. Increased peristalsis creates a propulsive force to move the worm burden caudally, while increased luminally directed electrogenic Cl^- and Na^+ transport provides the driving force for water movement into the lumen to aid flushing of intestinal contents – the washer/sweeper phenomenon (Wood, 1991). Tapeworm infection is accompanied by goblet cell hyperplasia and the increased mucus production would coat the tegument interfering with worm nutrient absorption, entrap the worms and along with a lubricant function facilitate the caudal movement of dislodged worms.

Mast cell and eosinophil hyperplasia are hallmarks of enteric helminth infection, particularly with nematodes but also cestodes. However, the mast cells and eosinophils may not be activated to secrete their anti-helminth products. When activated, mast cells release a plethora of preformed (e.g. histamine), rapidly synthesized (e.g. prostaglandins (PGs)) and more slowly synthesized (e.g. ILs) mediators. Similarly activated eosinophils release many products including major basic protein and peroxidase. These molecules acting individually or synergistically exert numerous effects including: (i) increasing epithelial electrolyte transport; (ii) increasing endothelial and epithelial permeability to enhance white blood cell extravazation from the vasculature and movement into the surrounding tissue and ease the movement of anti-worm antibodies into the gut lumen; and (iii) increasing inflammatory cell recruitment and modulation of the activity of the resident and recruited immune cells.

Effects of the same spectrum of mediators on nerve and muscle function highlight immunophysiological reactions as a coordinated series of host events in response to infection (Fig. 9.3).

Mast cells can be activated by a variety of stimuli, such as neuropeptides and anaphylotoxin. In the context of helminth infection, mast-cell activation is accomplished by antigen cross-linkage of IgE bound to the high-affinity FcεI receptor. However, antigen-specific IgE production takes 8–10 days after initial antigen exposure and is dependent on T cell–B cell interaction – an adaptive immune response.

Mobilize an active immune response

As extracellular metazoan parasites, cestode attack and expulsion depend on a humoral immune response that will be driven principally by Th2-type cytokines. Following antigen processing and presentation and the interaction of $CD4^+$ T cells with B cells, plasma cells develop that produce large amounts of high-affinity AB: IgG (and subtypes thereof), IgE and in the gut, IgA. The anti-worm AB works in three main ways. First, AB can absorb and neutralize any free antigen; a principal mode of action for the dimeric secretory IgA that enters the gut lumen. Second, by binding to the surface of the cestode (i.e. opsinization) AB (mainly IgG, but also IgE and IgA) allows for attachment of eosinophils and macrophages via receptors for the AB Fc region and subsequent release of lysozymes and other cytotoxic mediators that damage the surface of the parasite – i.e. AB-dependent cell-mediated cytotoxicity (ADCC). Third, AB coating the surface of the tapeworm can fix the initial components of complement resulting in activation of the complement cascade that punches holes in the tegument. In addition, activated complement is chemotactic for inflammatory cells and can also activate mast cells, which in turn, and along with IgE activation, contribute to the inflammatory processes and an immunophysiological response to the worm (Fig. 9.4).

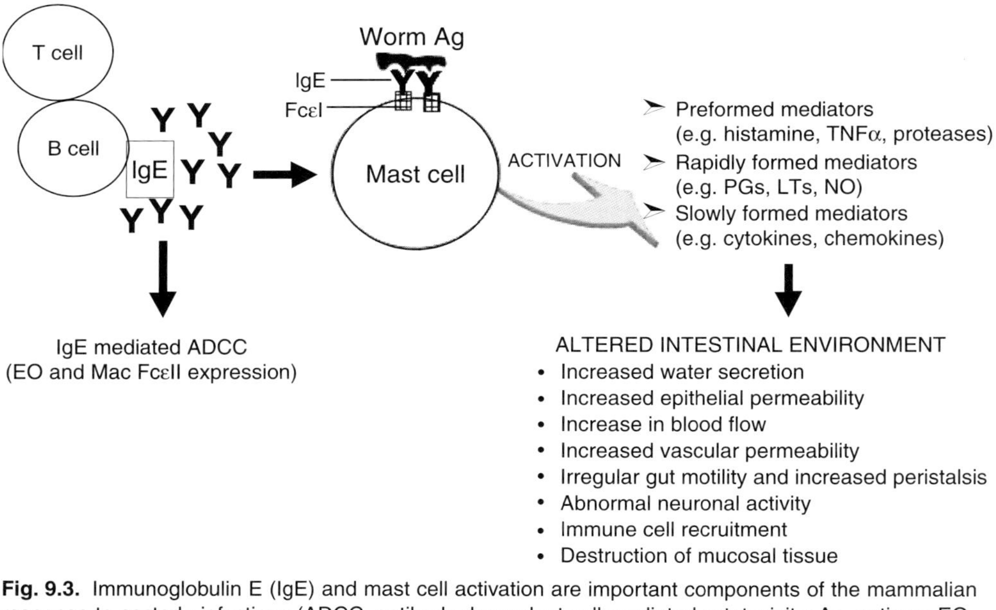

Fig. 9.3. Immunoglobulin E (IgE) and mast cell activation are important components of the mammalian response to cestode infections (ADCC, antibody dependent cell-mediated cytotoxicity; Ag, antigen; EO, eosinophils; FcεI, high affinity receptor for Fc portion of IgE; LTs, leukotrienes; Mac, macrophages; NO, nitric oxide; PGs, prostaglandins; TNFα, tumour necrosis factor alpha).

Specifics of the Host Response to Cestode Infection

Parenteral infections of cestodes

Oncospheres of many cestode species penetrate and develop into metacestodes within cysts (cysticerci, hydatid, multilocular) in the soft tissues of their rodent, ruminant or human hosts. Thus, species such as *Echinococcus granulosus, E. multilocularis, Taenia multiceps, T. ovis, T. saginata* and *T. solium* are of economic and medical importance. Such soft tissue invasion elicits a host immune response to resist the helminths. However, although some cysts may be destroyed as evidenced by involution or calcification, the host response is often 'too little – too late' to eliminate the invaders. The susceptibility of the host to invasion is often due to successful evasive strategies mounted by the tapeworm (see section on 'Evasion and Subversion of Host Response').

Host response to the oncosphere

Oncospheres, released from eggs after passage into the small intestine, become active and penetrate the mucosa to gain access to blood vessels and are swept away to the liver, skeletal muscles, brain or other tissues. However, oncospheres may pass completely through the intestinal wall into the abdominal cavity. Within a few days the oncosphere begins its transformation with the tegument microvilli being replaced by microtriches. During oncosphere migration and transformation, proteins associated with the surface, the excretory and the penetration gland ducts and the hooks stimulate an immune response that can result in protective AB against subsequent invasion

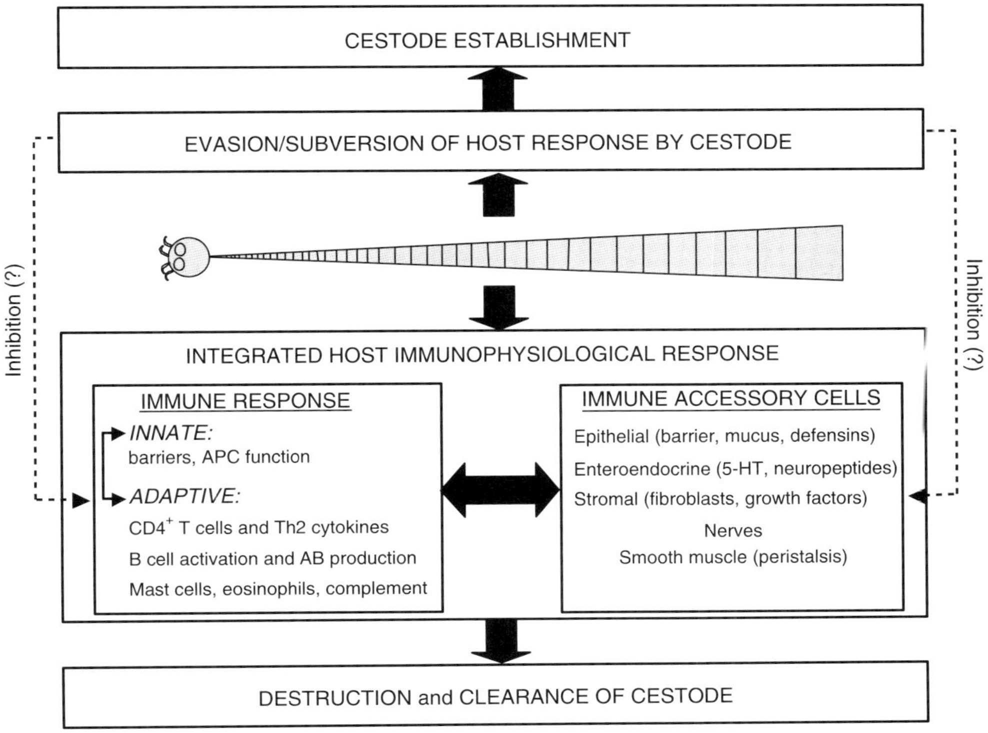

Fig. 9.4. Diagrammatic representation of the major elements in the integrated immunophysiological response to the host aimed at detecting, destroying and eradicating cestode infections (Th2, T helper type 2 cell; AB, antibody; 5-HT, 5-hydroxytryptamine (serotonin); APC, antigen presenting cell).

(Benitez *et al.*, 1996). However, by the time the AB titre rises sufficiently in the host, the oncospheres in the first wave of invasion have undergone transformation to evade this immune response (Rickard and Williams, 1982). Antibodies, primarily of the $IgG_{2\alpha}$ isotype, binding to the surface of the oncosphere activate complement, leading to lysis of the oncospheral membranes. In contrast, AB raised against *T. taeniaeformis* metacestode antigens appears to offer little or no protection against subsequent oncosphere infections (Bogh *et al.*, 1990). Although there has been a suggestion of an intestinal barrier in challenged *T. taeniaeformis* infected mice, Heath and Pavloff (1975) found that oncospheres in passively immunized mice were able to penetrate the gut mucosa and travel to the liver where they were subsequently killed. Additionally, resistance to *T. taeniaeformis* in naïve rats was provided by the adoptive transfer of spleen cells from rats infected with eggs (Asano *et al.*, 1994) where the data suggested that the most vulnerable stages were the oncospheres and the post-oncospheral stages in the liver.

Although sheep are not well protected against *T. ovis* oncospheral invasion by passive transfer of AB in colostrum, mice were protected against *T. taeniaeformis* by the oral administration of colostral IgA and IgG (Lloyd and Soulsby, 1978) or serum IgG (Musoke and Williams, 1975) from infected animals.

In murine *E. granulosus* oncospheral infections, the primary AB response associated with protection against a re-infection was IgG_1 (Zhang *et al.*, 2003). Elevated IgG_1 is associated with Th2 reactions, yet most tissue invasive cestodes seem to produce an initial Th1 or mixed Th1/Th2 response (Cortes *et al.*, 2003; Vuitton, 2003).

Immunization with the monoclonal AB reactive against the 18 kDa major surface protein in the oncosphere of *T. saginata*, HP6, confers protection in cattle (Harrison and Parkhouse, 1986). The HP6-encoding gene of *T. saginata* has been cloned and transfected into normal rat kidney cells. Localization studies revealed HP6 in the endoplasmic reticulum, the Golgi apparatus and at the cell surface (Bonay *et al.*, 2002). The deduced amino acid sequence of HP6 was similar to the mammalian extracellular matrix protein, fibronectin type III. Moreover, the transfected cells are bound to each other and this was blocked by HP6 monoclonal AB (Bonay *et al.*, 2002), suggesting that elaboration of HP6 by the oncosphere could interfere with the binding of immunocompetent cells by binding other host cells, and thereby contribute to evasion of the immune response.

Resistance to parenteral metacestodes

After oncosphere transformation to a metacestode, the parasite evades any initial antigen-specific host immunity. However, the host does mount an immune response to the growing metacestode (which may be undergoing asexual multiplication), as evidenced by the presence of anti-worm AB that specifically binds metacestode antigens, although complement is not activated. There is also a complex sequalae of cellular events that include elaboration of cytokines, chemokines and granuloma formation. Moreover, the former practise of providing liver transplants to humans infected with *E. multilocularis* showed how well the parasite is controlled by the immune response; patients experienced rapid growth of cysts in areas such as the brain due to immunosuppression required for survival of the transplant (Vuitton, 2003).

One parenteral cestode that is widely accepted as a model for cysticercosis caused by *T. solium* in humans is *Taenia crassiceps*, which is found as an adult in foxes and has a rodent intermediate host. Larval *T. crassiceps* reproduce by budding in the peritoneal cavity of mice and can be serially transferred from mouse to mouse. The immune response, which controls larval growth, relies on T cell-mediated immune mechanisms (Lopez-Briones *et al.*, 2001) and treatments resulting in increased delayed type hypersensitivity led to greater resistance, while AB production was unaffected (Bojalil *et al.*, 1993).

The initial response to *T. crassiceps* infection has been studied in BALB/c mice (Toenjes and Kuhn, 2003) where by day 5 post-infection (p.i.) soluble larval antigen preparations (SLAP) induce a cytokine response. *Ex vivo* challenge of spleen and mesenteric lymph node cells (which do not drain the peritoneal cavity) and peritoneal exudate cells with SLAP resulted in increased IL-4 and IL-10 production, and also IFNγ release, indicating a mixed Th1/Th2 response. However, while IL-10 production was sustained at day 7 p.i., this was not the case for IL-4 and IFNγ. SLAP-stimulated splenocyte proliferation was observed at days 5 and 7 p.i. but this response waned in older infections and was eventually replaced by suppression of mitogen-induced T cell proliferation (Spolski *et al.*, 2002a). Generally, in early infections where budding of cysticerci is occurring a mixed Th1/Th2 response dominates (Toenjes and Kuhn, 2003). In older infections, where many of the cysticerci are encased in granulomas and are dying/dead, there is polarization towards a Th2 environment that nonetheless is not effective in controlling parasite growth. Similarly, the increased IFNγ accompanying infections of *E. granulosus* appears to limit parasite growth (Dematteis *et al.*, 2003) and enhanced Th2 responses correlate with progression of *E. multilocularis* lesions (Vuitton, 2003). Brain granulomas in *T. solium* infections are also associated with a mixed Th1/Th2 profile (Restrepo *et al.*, 2001).

One difficulty in interpreting responses in older infections is that many stages of cysticerci are found, from relatively young to those that are dead and are being reabsorbed. Further, the cellular response may be influenced by the stage of each cyst, which is reflected in the make-up of the cells in the granuloma around the cyst. Assessment of cytokine-producing cells in different cyst stages (1–4) of *T. crassiceps* revealed an increase in IL-2 positive cells in stage 2 cysts, diminution in stage 3 and enhancement in stage 4 (Robinson *et al.*, 1997). The IFNγ-positive cells increased from stage 1 to 3 but decreased in stage 4 cysts, while IL-10 producing cells were elevated in cyst stages 2–4. The IL-4-positive cells were markedly increased in stage 4 cysts. These data are consistent with an initial Th1/Th2 response followed by a skewing towards Th2 dominance in involuting cysts. The changes in *T. crassiceps* infections reflect those seen in human neurocysticercosis. Additionally, as an infection of *T. crassiceps* ages in the peritoneal cavity, unusual parasite-derived mononucleated and multinucleated cells appear. These 'cells' rapidly increase in the first 5 weeks p.i. and then diminish, and probably contribute to the Th2 polarization (Padilla *et al.*, 2001).

The role of the Th1/Th2 response is becoming clearer. Infections of *T. crassiceps* in signal transducer and activator of transcription (STAT)-6 knockout (KO) or CD40 KO mice result in a highly effective resistance to growth of cysts compared to controls (Rodriguez-Sosa *et al.*, 2003a). Immune events that paralleled the inhibition of cyst growth in the KO mice included: (i) enhanced IL-12 and nitric oxide activity in LPS-stimulated macrophages; (ii) reduced IgE and IgG_1; and (iii) diminution of IL-5 and IL-13 production by antigen-stimulated splenocytes. Moreover, anti-CD3 activation of splenocytes from CD40 KO resulted in less IL-4, IL-5 and IL-13, but greater IFNγ production compared to control (Rodriguez-Sosa *et al.*, 2003a): evidence supporting resistance as a consequence of a Th1 response.

The cellular response was investigated to determine the involvement of lymphocytes and cytokines regulating the site-specific response to *T. crassiceps* in BALB/c mice. The percentage of γ/δ^+ TcR and $CD8^+$ T cells increased and remained high for the first 40 days p.i., while $CD4^+$ T cells and $B220^+/CD5^-$ B cells were increased by 10 days p.i. and declined thereafter. In contrast $B220^+/CD5^+$ B cells decreased threefold between 10 and 50 days p.i. (Toenjes *et al.*, 1999a). However, γ/δ^+ TcR T cell KO mice are no more susceptible to *T. crassiceps* than normal mice, and the best correlation to susceptibility was linked to IL-10 levels (Toenjes *et al.*, 1999b). Further CD8 KO mice had no increased parasite load compared to controls (Lopez-Briones *et al.*, 2001), although $CD8^+$ T cells form the predominant cell type in the pericystic adventitia of larger mature cysts of *E. granulosus* (Sakamoto and Cabrera, 2003). However, CD4, TcRα, β or γ chain and RAG1-deficient mice developed more cysticerci of *T. crassiceps* and had a higher mortality rate (Lopez-Briones *et al.*, 2001). Noteworthy was the observation of undiminished IFNγ release from antigen-stimulated spleen cells from CD4 KO animals that were highly susceptible to *T. crassiceps* infection. Given that specific proliferative spleen responses, Th2 cytokine production or AB responses were not observed in the highly susceptible strains (CD4, TcRα, TcRβ, or RAG1

KO), the data imply a pivotal role for CD4⁺TcRα/β⁺ T cells in resistance to *T. crassiceps* infection. There is also an increase in eosinophil number throughout parenteral *T. crassiceps* and *T. solium* infections (Londono *et al.*, 2002; Cortes *et al.*, 2003), but neither eosinophil numbers nor IgE levels correlate with susceptibility (Rodriguez-Sosa *et al.*, 2003a).

The PGE_2 is produced by *T. taeniaeformis* infection, although whether or not it is of parasite origin is not clear. Its role has been investigated in mice infected with *T. crassiceps* infections of long duration (2–4 months) (Terrazas *et al.*, 1999) where cyclooxygenase inhibition (and hence PG synthesis) by indomethacin resulted in fewer parasites, while PGE_2 treatment resulted in an increased cyst yield. Indomethacin also enhanced splenocyte-stimulated release of IL-2 and IFNγ, and down-regulated IL-4, IL-6 and IL-10 synthesis. It is feasible that PGE_2, IL-4 and IL-10 from peritoneal macrophages regulate the production of IL-2 and IFNγ by T cells as well as IL-12 and TNFα from activated macrophages.

Understanding the role of antigen-presenting peritoneal exudate cells (predominantly macrophages) in cestode infections has developed rapidly. Macrophages generating high levels of nitric oxide are associated with resistance to *T. crassiceps*. Macrophages from acute infections produce high levels of IL-12, nitric oxide, low amounts of IL-6 and PGE_2, and promote a vigorous T cell response to non-worm antigen. In contrast, macrophages from chronic infections (8–12 weeks p.i.) display opposite qualities and influence $CD4^+$ T cells to produce IL-4 (Rodriguez-Sosa *et al.*, 2002). This switch in macrophage phenotype occurs around the time that parasite numbers increase (~4 weeks p.i.). The alternatively activated macrophages from chronic infections express high levels of the low-affinity IgE receptor (FcεII or CD23) and the chemokine CCR5 (CD195 binds, among others, macrophage inflammatory protein (MIP)-1α and MIP-1β) receptor, which is not true of macrophages from acute infections or infected STAT-6 KO mice where the numbers of parasites are effectively controlled. Also, macrophage migration inhibitory factor (MIF) seems to play a role in *T. crassiceps* resistance because MIF KO mice harbour greater parasite loads than control mice (Rodriguez Sosa *et al.*, 2003b). Neither IgG_1 titres nor IL-4 levels were significantly different between MIF KO and normal mice, although IL-13 levels increased in the KO mice. Levels of nitric oxide were very low in MIF KO mice, correlating with reduced macrophage infiltration. Collectively these data support the hypothesis that murine susceptibility to *T. crassiceps* is due to diminished macrophage activity rather than a reduced Th1/IFNγ activity.

Lumenal Infections: The Immune Response to Hymenolepidid Cestodes

There is surprisingly little information on the immune response to gut–lumen-dwelling adult stages of taenids and echinococcid cestodes (Moreno *et al.*, 2004). Knowledge of the immune response to hymenolepidids has been mostly obtained from *H. diminuta* or *H. nana* infection of rodents. The former model is particularly useful because *H. diminuta* is not auto-infective and responses can be compared in the non-permissive murine and permissive rat hosts.

Primary infections of *H. diminuta* develop normally in the mouse until ~5–6 days p.i., when progressively more severe tegumental and parenchymal damage is observed until ~9–12 days p.i. when the worms are expelled. There is already massive damage to the tegument 1 day after a secondary infection (McCaigue and Halton, 1987). The rapidity of this response indicates immunological memory: a hallmark of adaptive immunity. Indeed participation of elements of the adaptive immune system in murine spontaneous eradication of *H. diminuta* is supported by the fact that the worms will develop to maturity in mice immunosuppressed by daily hydrocortisone treatments (McKay *et al.*, 1990b) or in STAT-6 KO mice (McKay and Khan, 2003), a key signalling molecule in IL-4 and IL-13 driven events. The latter observation is consistent with *in vitro* analysis of mesenteric lymph node cells from *H. diminuta*-infected mice that revealed enhanced IL-3, IL-4 and IL-5 production and significantly less IFNγ upon mitogenic stimulation (Palmas *et al.*, 1997). Similarly isolated lamina propria lymphocytes proliferate when challenged with worm antigen, indicating the presence of effector or memory cells in the previously infected mice. The data are not extensive but are compatible with Th2-dominated events underlying worm expulsion. Likewise the literature suggests that the actual effector mechanisms are likely: (i) AB- and complement-mediated damage of the tegument (Andreassen *et al.*, 1990); (ii) targeted attack of the worm by eosinophils and macrophages and release of their granule-stored mediators/synthesis of reactive oxygen species (ROS); and (iii) mast cell activation and goblet cell hyperplasia that accompanies both primary and secondary infections (McKay *et al.*, 1990b). Thus, the response to *H. diminuta* fits the generalized pattern of immune events following infection with a lumen-dwelling helminth.

Additionally, we demonstrated increased substance P and reduced vasoactive intestinal polypeptide levels, and increased numbers of serotonin-positive enteroendocrine cells that correlated exactly with *H. diminuta* expulsion following primary and secondary infections (McKay *et al.*, 1990a, 1991): whether these neurohormonal changes are a cause or an effect of worm rejection remains to be clarified. Indeed, the role of the enteric nervous system needs to be incorporated into the consideration of the enteric response to cestode infection (McKay and Fairweather, 1997).

H. nana infection has the complication that ingested gravid eggs are infective and so the worm presents an additional threat to the host. In general the murine response to *H. nana* parallels that of *H. diminuta* – there is a Th2 cytokine response and evidence of participation by mast cells and eosinophils, such that mice that lack either cell type display significantly slower rates of worm loss, yet importantly the rejection response is not absent (Watanabe *et al.*, 1994). Two additional points are noteworthy. First, gut tissues from *H. nana*-infected mice have increased ROS levels and the eosinophils produce more ROS upon *in vitro* stimulation (Niwa and Miyazato, 1996). Second, the cytokine response to egg ingestion is characterized by increased IFNγ (Asano and Muramatsu, 1997), which would be expected to mobilize macrophages to engulf and destroy the eggs. The authors speculated that the tissue phase (i.e. eggs) of *H. nana* infection is combated by Th1 events, while the luminal phase (strobilated worm) evoked Th2 cytokine production.

Thus, for the Hymenolepidids the data can be interpreted to fit the Th2 paradigm with the involvement of the predicted immune/physiological effector mechanisms: AB, complement, mast cells, eosinophils and goblet cells.

Finally, a similar spectrum of immune events has been examined in *H. diminuta*-infected rats and the situation is not unlike that in mice. Infected rats, at least initially, can develop mast cell, eosinophil, goblet cell and IgG responses along with irregular peristalsis (McKay *et al.*, 1990b, 1991; Dwinell *et al.*, 1997). Despite these changes the worm persists and the host, in terms of behaviour and weight gain, seems oblivious of its companion. However, there is a lack of consistency in tissue mRNA cytokine levels that accompany infection with 5, 10 or 50 cysticercoids (personal observation). One is left with the unsatisfactory realization that the reason *H. diminuta* persists in the rat lies in the vagaries of a qualitatively or quantitatively different immune response to the worm, and/or an ability of the worm to block or overcome the immune response.

Cestode Infections and the Immune Response – Where Do We Go from Here?

In comparing the response of the host to a parenteral or luminal infection, one is struck by the effect of STAT-6 manipulation. STAT-6 KO mice are Th2 cell-deficient, producing primarily a Th1 response, lack trefoil factor-2 (TFF2), a peptide involved in epithelial restitution and mucosal secretion in the gastrointestinal tract (Nikolaidis *et al.*, 2003) and exhibit a reduced goblet cell response to *H. diminuta* infection (McKay and Khan, 2003). Thus, resistance to lumen-dwelling helminths aligns with Th2 responses and resistance to parenteral infections is more likely dependent on Th1 responses.

However, STAT-4 KO mice, which produce primarily a Th2 response (because they have reduced IL-12 signalling), have not been used in studies on cestode establishment and growth. In addition, the use of either STAT-4 or STAT-6 KO mice with adoptive transfer of specific cell types producing particular cytokines will aid delineation of the host immune response.

The use of spleen cells in analysing the status of a parenteral cestode infection has not produced consistent results, suggesting that greater commonality might be observed if local rather than systemic responses are examined (e.g. the spleen does not directly drain the peritoneal cavity). Moreover, because many cells of cestode origin are located in the peritoneal cavity care needs to be taken when isolating these cells to prevent contamination with cestode tissue. However, *ex vivo* studies on parenteral cestodes may allow more precise manipulation of host cell types and cytokines to establish their role in retarding, or enhancing, different stages of cyst growth. In addition, there are many opportunities to examine some of the features underlying the ability of the cestode to evade the host immune response, including mechanisms by which *T. crassiceps* alters the endocrine milieu of the host to favour cyst growth (Morales-Montor *et al.*, 2002).

Evasion and Subversion of the Host Response

Even a cursory review of the pertinent literature leaves the reader in little doubt that cestodes can subvert host immune responses by mechanisms that vary from membrane fluidity (i.e. membrane turnover) (Taylor *et al.*, 1997) to active immunomodulation that favours development of immunosuppressive/immunoregulatory cells (Dai and Gottstein, 1999) and/or an immune-suppressed environment. In this respect the cestode is no different than other helminth parasites. However, the data are fragmented, often quite vague and confused by the specificity of the host–parasite interaction. Therefore we have opted to provide the reader with tabulated data on some of the main findings illustrating cestode manipulation of their mammalian host immune systems (Table 9.1). Awareness of how tapeworms modulate the immune system is more than an academic pursuit, because it has the potential to lead to the isolation and characterization of novel molecules that could be the basis of new medicines.

The Beneficial Tapeworm

The immune system has evolved to combat the threat of infectious organisms, mostly microbial but also metazoans. The hygiene hypothesis has recently been postulated to explain the rise in allergic/atopic and autoimmune type disorders in the latter half of the 20th century in Westernized societies (Yazdanbakhsh *et al.*, 2002). Essentially, it is reasoned that the reduced threat from microbial pathogens (due to improved sanitation and hygiene, and probably widespread antibiotic use and vaccinations also) increases the chance of T cells developing that are auto-reactive or responsive to allergens. Thus, in a paradoxical manner low-grade pathogens may in fact offer some benefit to the host by protecting against autoimmune disease. Extrapolating from this, an appreciation of how helminths impact host immunity would, theoretically, allow use of helminths to combat other diseases; the caveat being that the consequences of the helminth infection should not be more deleterious than the disease to be treated.

A number of investigators are pursuing this strategy of immune distraction as a potential therapeutic modality. Conceptually, because helminth infections typically result in a host Th2 response, they could prevent or reduce the severity of disorders that are characterized by Th1 cytokines (noting that a helminth-driven Th2 response is clear in mice but the evidence is less convincing in humans). The potential impact of helminth infection on human Crohn's disease (CD; a major type of human inflammatory bowel disease) is an intriguing example of this, which we reviewed in detail elsewhere (Hunter and McKay, 2004). CD is associated with Th1 cytokines and the prevalence of CD is highest in North America, western Europe and Japan, with much lower incidence in developing areas of the globe where helminth infections are pandemic.

Table 9.1. Cestode manipulation of some host immune/physiological responses.

Taenia cassiceps
Promotion of the development of macrophages with altered antigen presenting function (Rodriguez-Sosa *et al*., 2002)
Induction of apoptosis in CD4$^+$/CD19$^+$ splenocytes (Lopez-Briones *et al*., 2003)
Production of an IFNγ analogue (Spolski *et al*., 2002b)
Taenia solium
Soluble factor blocks ConA-induced IL-2, IL-4 and IL-8 production and macrophage-derived TNFα (Arechavaleta *et al*., 1998)
Taenia taeniaeformis
Blocks gastric acid production (Oku *et al*., 2002)
Echinococcus granulosus
E4 carbohydrate antigen blocks immune cell proliferation and IL-10 production (Dematteis *et al*., 2001)
Antigen B stimulation of PBMC IL-4 and IL-13 production and inhibition of IFNγ (Rigano *et al*., 2001)
Echinococcus multilocularis
Isolated cells from infected mice are less proliferative upon *in-vitro* activation and increase inducible nitric oxide synthase expression in peritoneal cells, yet splenocytes have normal cytokine responses to ConA (Dai and Gottstein, 1999)
Neutral glycosphingolipids block immune cell proliferation and IL-2 production evoked by exposure to pokeweed (a T cell mitogen) (Persat *et al*., 1996)
Protoscolices block immune cell proliferation via generation of a CD8$^+$ 'suppressive' T cell (Kizaki *et al*., 1993)
Macrophages from infected mice secrete products that block ConA activation of lymphocytes (Rakha *et al*., 1991)
Hymenolepis diminuta
Significant membrane fluidity and turnover (Taylor *et al*., 1997)
Antigen inhibition of murine and human splenocyte and PBMC–ConA-induced proliferation and IL-2 production (Wang and McKay, 2005)
Production of an IL-12 p40 subunit analogue (Wang and McKay, 2005)

Abbreviations: CD, cluster designation (surface markers); ConA, concanavalin A (a T cell mitogen); IFNγ, interferon-gamma; IL, interleukin; PBMC, peripheral blood mononuclear cells; TNFα, tumour necrosis factor-alpha.

These epidemiological data support the paradigm. Moreover, the disease severity and symptomatology in murine models of Th1-dominated colitis is significantly reduced if the mice have been infected with either non-permissive nematodes (Khan *et al*., 2002) or cestodes (i.e. *H. diminuta*) (Reardon *et al*., 2001), and preliminary data have been presented showing that CD patients can experience a therapeutic benefit following consumption of viable *Trichuris suis* eggs (Summers *et al*., 2003). The mechanism underlying these observations may well stem from Th2 antagonism of Th1 events. However, the potential involvement of regulatory T cells (or mediators) or active participation of helminth-derived factors in the amelioration of Th1-type colitis should not be overlooked and need to be the focus of future investigations.

We do not suggest that parasitic infection per se is not a malevolent condition, but the notion of a 'beneficial or harmonious parasite' is not unprecedented (Desowitz, 1981) and should not be surprising in the light of the ability of parasites to manipulate host immune reactions. Rather we suggest that there may be a role for the 'therapeutic helminth' and that a more tolerant view of analyses of host–helminth interactions has the potential to reveal additional facets of the host immune response that may lead to the development of novel ways to treat both infectious and allergic/autoimmune disorders.

Concluding Remarks

We hope to have convinced the reader of the need for, and value of, continued analysis of cestode–host interactions. While the cestodes seem to lack the cachet of their trematode and nematode cousins, the host response to cestode infection is for the most part consistent with both Th1 (mostly with parenteral infections) and Th2-dominated (lumen-dwelling stages) humoral responses and the creation of an inhospitable environment. Certainly some exceptions exist and one must resist the temptation to generalize particularly when considering responses to different stages of the worms' life cycle. Moreover, this view of the immune response is a broad spectrum picture, encompassing a role for many cells and mediators – and appropriately so, because this is undoubtedly the *in vivo* reality. Recent years have seen the introduction of new immunological tools, which can be used to provide exquisite detailed knowledge on the host immune response to cestodes. For example, the variety of cytokine and signalling molecule gene KO and transgenic mice allows for a role for any particular mediator in the host rejection response to be unequivocally ruled in or out. Also, chimeric or conditional KO mice will permit assessment of the relative importance of different cellular sources of key mediators. Knowledge of T cell subtypes and the variety of cytokines/chemokines has increased and can be used as the platform from which future studies on the cestode–host immune interactions can be launched. We suggest that the stage is set for concerted efforts to provide a comprehensive picture of the cestode–host interaction.

Acknowledgements

Studies conducted in the authors' laboratories were supported by operating grants from the Canadian Institutes of Health Research and the Canadian Crohn's and Colitis Foundation to DMM and the Natural Sciences and Engineering Research Council to RAW.

References

Andreassen, J., Hoole, D. and Befus, D. (1990) Complement-mediated leucocyte adherence of newly excysted and 4-day-old *Hymenolepis diminuta*. *International Journal for Parasitology* 20, 905–911.

Arechavaleta, F., Molinari, J.L. and Tato, P. (1998) A *Taenia solium* metacestode factor non-specifically inhibits cytokine production. *Parasitology Research* 84, 117–122.

Asano, K. and Muramatsu, K. (1997) Importance of interferon-γ in protective immunity against *Hymenolepis nana* cysticercoids derived from challenge infection with eggs in Balb/c mice. *International Journal for Parasitology* 27, 1437–1443.

Asano, K., Ito, A., Ikeda, K. and Okamoto, K. (1994) Adoptive transfer of protection to *Taenia taeniaeformis* infection in rats with immune spleen cells. *International Journal for Parasitology* 24, 413–415.

Benitez, L., Garate, T., Harrison, L.J., Kirkham, P., Brookes, S.M. and Parkhouse, R.M. (1996) Cloning and sequencing of the gene encoding the principal 18-kDa secreted antigen of activated oncospheres of *Taenia saginata*. *Molecular and Biochemical Parasitology* 78, 265–268.

Bogh, H.O., Lightowlers, M.W, Sullivan, N.D., Mitchell, G.F. and Rickard, M.D. (1990) Stage-specific immunity to *Taenia taeniaeformis* infection in mice. A histological study of the course of infection in mice vaccinated with either oncosphere or metacestode antigens. *Parasite Immunology* 12, 153–162.

Bojalil, R., Terrazas, L.I., Govezensky, T., Sciutto, E. and Larralde, C. (1993) Thymus-related cellular immune mechanisms in sex-associated resistance to experimental murine cysticercosis (*Taenia crassiceps*). *Journal of Parasitology* 79, 384–389.

Bonay, P., Gonzalez, L.M., Benitez, L., Foster, M., Harrison, L.J., Parkhouse, R.M. and Garate, T. (2002) Genomic and functional characterisation of a secreted antigen of *Taenia saginata* oncospheres. *Molecular and Biochemical Parasitology* 121, 269–273.

Castrodeza, C. (1979) Non-progressive evolution, the Red Queen hypothesis, and the balance of nature. *Acta Biotheories* 28, 11–18.

Cortes, I., Molinari, J.L., Solano, S., Hernandez-Mendoza, L., Ramirez, A. and Tato, P. (2003) *Taenia solium* metacestode antigens which are protective for pigs induce Th1/Th2 mixed responses in mice. *Parasitology Research* 90, 273–279.

Dai, W.J. and Gottstein, B. (1999) Nitric oxide-mediated immunosuppression following murine *Echinococcus multilocularis* infection. *Immunology* 97, 107–116.

Dematteis, S., Pirotto, F. Marques, J., Nieto, A., Oru, A. and Baz, A. (2001) Modulation of the cellular response by a carbohydrate rich fraction of *Echinococcus granulosus*. Parasite Immunology 23, 1–9.

Dematteis, S., Rottenberg, M. and Baz, A. (2003) Cytokine response and outcome of infection depends on the infective dose of parasites in experimental infection by *Echinococcus granulosus*. *Parasite Immunology* 25, 189–197.

Desowitz, M. (1981) Harmonious Parasites. In: Desowitz, M. (ed.) *New Guinea Tapeworms and Jewish Grandmothers*. Pickwick, London, pp. 179–205.

Dwinell, M.B., Bass, P., Schaefer, D.M. and Oaks, J.A. (1997) Tapeworm infection decreases intestinal transit and enteric aerobic bacterial populations. *American Journal of Physiology* 273, G480–G485.

Harrison, L.J. and Parkhouse, R.M. (1986) Passive protection against *Taenia saginata* infection in cattle by a mouse monoclonal antibody reactive with the surface of the invasive oncosphere. *Parasite Immunology* 8, 319–332.

Heath, D.D. and Pavloff, P. (1975) The fate of *Taenia taeniaeformis* oncospheres in normal and passively protected rats. *International Journal for Parasitology* 5, 83–88.

Hunter, M.M. and McKay, D.M. (2004) Review article: helminths as therapeutic agents for inflammatory bowel disease. *Alimentary Pharmacology and Therapeutics* 19, 167–177.

Khan, W.I., Blennerhassett, P.A., Varghese, A.K., Chowdhury, S.K., Omsted, P., Deng, Y. and Collins, S.M. (2002) Intestinal nematode infection ameliorates experimental colitis in mice. *Infection and Immunity* 71, 5913–5917.

Kizaki, T., Ishige, M., Bingyan, W., Day, N.K., Good, R.A. and Onoe, K. (1993) Generation of $CD8^+$ suppressor T cells by protoscoleces of *Echinococcus multilocularis in vitro*. *Immunology* 79, 412–417.

Lloyd, S. and Soulsby, E.J. (1978) The role of IgA immunoglobulins in the passive transfer of protection to *Taenia taeniaeformis* in the mouse. *Immunology* 34, 939–945.

Londono, D.P., Alvarez, J.I., Trujillo, J., Jaramillo, M.M. and Restrepo, B.I. (2002) The inflammatory cell infiltrates in porcine cysticercosis: immunohistochemical analysis during various stages of infection. *Veterinary Parasitology* 109, 249–259.

Lopez-Briones, S., Soloski, M.J., Bojalil, R., Fragoso, G. and Sciutto, E. (2001) $CD4^+$ TCRαβ T cells are critically involved in the control of experimental murine cysticercosis in C57BL/6J mice. *Parasitology Research* 87, 826–832.

Lopez-Briones, S., Sciutto, E., Ventira, J.L., Zentella, A. and Fragoso, G. (2003) $CD4^+$ and $CD19^+$ splenocytes undergo apoptosis during an experimental infection with *Taenia crassiceps*. *Parasitology Research* 90, 157–163.

McCaigue, M.D. and Halton, D.W. (1987) Immunological damage to *Hymenolepis diminuta* following a challenge infection in C57 mice. *International Journal for Parasitology* 17, 795–803.

McKay, D.M. and Fairweather, I. (1997) A role for the enteric nervous system in the response to helminthic infection. *Parasitology Today* 13, 63–69.

McKay, D.M. and Khan, W.I. (2003) STAT-6 is an absolute requirement for murine expulsion of the tapeworm, *Hymenolepis diminuta*. *Journal of Parasitology* 89, 188–189.

McKay, D.M., Halton, D.W., Johnston, C.F., Fairweather, I. and Shaw, C. (1990a) *Hymenolepis diminuta*: changes in intestinal morphology and the enterochromaffin cell population associated with infection in male C57 mice. *Parasitology* 101, 107–113.

McKay, D.M., Halton, D.W., McCaigue, M.D., Johnston, C.F., Fairweather, I. and Shaw, C. (1990b) *Hymenolepis diminuta*: intestinal goblet cell response in male C57 mice. *Experimental Parasitology* 71, 9–20.

McKay, D.M., Halton, D.W., Shaw, C., Johnston, C.F., Fairweather, I. and Buchanan, K.D. (1991) *Hymenolepis diminuta*: changes in the levels of certain intestinal regulatory peptides in infected C57 mice. *Experimental Parasitology* 73, 15–26.

Mittrucker, H.W. and Kaufmann, S.H. (2004) Mini-review. Regulatory T cells and infection: suppression revisited. *European Journal of Immunology* 34, 306–312.

Morales-Montor, J., Hallal-Calleros, C., Romano, M.C. and Damian, R.T. (2002) Inhibition of P-450 aromatase prevents feminisation and induces protection during cysticercosis. *International Journal for Parasitology* 32, 1379–1387.

Moreno, M., Benavidez, U., Carol, H., Rosenkranz, C., Welle, M., Carmona, C., Nieto, A. and Chabalgoity, J.A. (2004) Local and systemic immune responses to *Echinococcus granulosus* in experimentally infected dogs. *Veterinary Parasitology* 119, 37–50.

Musoke, A.J. and Williams, J.F. (1975) The immunological response of the rat to infection with *Taenia taeniaeformis*. V. Sequence of appearance of protective immunoglobulins and the mechanism of action of 7Sγ2α antibodies. *Immunology* 29, 855–866.

Nikolaidis, N.M., Zimmermann, N., King, N.E., Mishra, A., Pope, S.M., Finkelman, F.D. and Rothenberg, M.E. (2003) Trefoil factor-2 is an allergen-induced gene regulated by Th2 cytokines and STAT-6 in the lung. *American Journal of Respiratory Cell and Molecular Biology* 29, 458–464.

Niwa, A. and Miyazato, T. (1996) Reactive oxygen intermediates from eosinophils in mice infected with *Hymenolepis nana*. *Parasite Immunology* 18, 285–295.

Oku, Y., Yamanouchi, T., Matsuda, K., Abelaa, J.A., Ooi, H.K., Ohtsubo, R., Goto, Y, and Kamiya, M. (2002) Retarded gastric acid secretion in rats infected with larval *Taenia taeniaeformis*. *Parasitology Research* 88, 872–873.

Padilla, A., Govezensky, T., Sciutto, E., Jimenez-Garcia, L.F., Gonsebatt, M.E., Ramirez, P. and Larralde, C. (2001) Kinetics and characterization of cellular responses in the peritoneal cavity of mice infected with *Taenia crassiceps*. *Journal of Parasitology* 87, 591–599.

Palmas, C., Bortoletti, G., Gabriele, F., Wakelin, D. and Conchedda, M. (1997) Cytokine production during infection with *Hymenolepis diminuta*. *International Journal for Parasitology* 27, 855–859.

Perdue, M.H. and McKay, D.M. (1994) Integrative immunophysiology in the intestinal mucosa. *American Journal of Physiology* 267, G151–G165.

Persat, F., Vincent, C., Schmitt, D., and Mojon, M. (1996) Inhibition of human peripheral blood mononuclear cell proliferative response by glycosphingolipids from metacestodes of *Echinococcus multilocularis*. *Infection and Immunity* 64, 3682–3687.

Rajshekhar, V., Joshi, D.D., Doanh, N.Q., van De, N. and Xiaonong, Z. (2003) *Taenia solium*/cysticercosis in Asia: epidemiology, impact and issues. *Acta Tropica* 87, 53–60.

Rakha, N.K., Dixon, J.B., Carter, S.D., Craig, P.S., Jenkins, P. and Folkard, S. (1991) *Echinococcus multilocularis* antigens modify accessory cell function of macrophages. *Immunology* 74, 652–656.

Reardon, C., Sanchez, A., Hogaboam, C.M. and McKay, D.M. (2001). Tapeworm infection reduces the ion transport abnormalities induced by dextran sulphate sodium (DSS) colitis. *Infection and Immunity* 69, 4417–4423.

Restrepo, B.I., Alvarez, J.I., Castano, J.A., Arias, L.F., Restrepo, M., Trujillo, J., Colegial, C.H. and Teale, J.M. (2001) Brain granulomas in neurocysticercosis patients are associated with a Th1 and Th2 profile. *Infection and Immunity* 69, 4554–4560.

Rickard, M.D. and Williams, J.F. (1982) Hydatidosis/cysticercosis: immune mechanisms and immunization against infection. *Advances in Parasitology* 21, 229–296.

Rigano, R., Profumo, E., Bruschi, F., Carulli, G., Azzarra, A., Ioppolo, S., Buttan, B., Ortona, E., Margutti, P. Teggi, A. and Siracusano, A. (2001) Modulation of human immune response by *Echinococcus granulosus* antigen B and its possible role in evading host defences. *Infection and Immunity* 69, 288–296.

Robinson, P., Atmar, R.L., Lewis, D.E. and White, A.C. Jr (1997) Granuloma cytokines in murine cysticercosis. *Infection and Immunity* 65, 2925–2931.

Rodriguez-Sosa, M., Satoskar, A.R., Calderon, R., Gomez-Garcia, L., Saavedra, R., Bojalil, R. and Terrazas, L.I. (2002) Chronic helminth infection induces alternatively activated macrophages expressing high levels of CCR5 with low interleukin-12 production and Th2-biasing ability. *Infection and Immunity* 70, 3656–3664.

Rodriguez-Sosa, M., Satoskar, A.R., David, J.R. and Terrazas, L.I. (2003a) Altered T helper responses in CD40 and interleukin-12 deficient mice reveal a critical role for Th1 responses in eliminating the helminth parasite *Taenia crassiceps*. *International Journal for Parasitology* 33, 703–711.

Rodriguez-Sosa, M., Rosas, L.E., David, J.R., Bojalil, R., Satoskar, A.R. and Terrazas, L.I. (2003b) Macrophage migration inhibitory factor plays a critical role in mediating protection against the helminth parasite *Taenia crassiceps*. *Infection and Immunity* 71, 1247–1254.

Sakamoto, T. and Cabrera, P.A. (2003) Immunohistochemical observations on cellular response in unilocular hydatid lesions and lymph nodes of cattle. *Acta Tropica* 85, 271–279.

Smyth, J.D. (1994) *Introduction to Animal Parasitology*, 3rd edn, chap. 23. Cambridge University Press, Cambridge, UK, pp. 321–345.

Spolski, R.J., Alexander-Miller, M.A. and Kuhn, R.E. (2002a) Suppressed cytotoxic T lymphocyte responses in experimental cysticercosis. *Veterinary Parasitology* 106, 325–330.

Spolski, R.J., Thomas, P.G., See, E.J., Mooney, K.A. and Kuhn, R.E. (2002b) Larval *Taenia crassiceps* secretes a protein with characteristics of murine interferon-γ. *Parasitology Research* 88, 431–438.

Summers, R.W., Elliott, D.E., Qadir, K., Urban, J.F., Thompson, R. and Weinstock, J.V. (2003) *Trichuris suis* seems to be safe and possibly effective in the treatment of inflammatory bowel disease. *American Journal of Gastroenterology* 98, 2034–2041.

Taylor, K., Kusel, J.R. and Hoole, D. (1997) Membrane fluidity on the surface of *Hymenolepis diminuta* (Cestoda): immunological implications. *Parasitology* 115, 667–671.

Terrazas, L.I., Bojalil, R., Rodriguez-Sosa, M., Govezensky, T. and Larralde, C. (1999) *Taenia crassiceps* cysticercosis: a role for prostaglandin E_2 in susceptibility. *Parasitology Research* 85, 1025–1031.

Toenjes, S.A. and Kuhn, R.E. (2003) The initial immune response during experimental cysticercosis is of the mixed Th1/Th2 type. *Parasitology Research* 89, 407–413.

Toenjes, S.A., Spolski, R.J., Mooney, K.A. and Kuhn, R.E. (1999a) The specific immune response of BLAB/c mice infected with larval *Taenia crassiceps* is a mixed Th1/Th2-type response. *Parasitology* 118, 623–633.

Toenjes, S.A., Spolski, R.J., Mooney, K.A. and Kuhn, R.E. (1999b) γδT cells do not play a major role in controlling infection in experimental cysticercosis. *Parasitology* 119, 413–418.

Vuitton, D.A. (2003) The ambiguous role of immunity in echinococcosis: protection of the host or of the parasite? *Acta Tropica* 85,119–132.

Wang, A. and McKay, D.M. (2005) Immune modulation by a high molecular weight fraction from the rat tapeworm *Hymenolepis diminuta*. *Parasitology* 130, 575–585.

Watanabe, N., Nawa, Y., Okamoto, K. and Kobayashi, A. (1994) Expulsion of *Hymenolepis nana* from mice with congenital deficiencies of IgE production or of mast cell development. *Parasite Immunology* 16, 137–144.

Wood, J.D. (1991) Communication between minibrain in gut and enteric nervous system. *News in Physiological Sciences* 6, 64–69.

Yazdanbakhsh, M., Kremsner, P.G. and van Ree, R. (2002) Allergy, parasites and the hygiene hypothesis. *Science* 296, 490–494.

Zhang, W., You, H., Li, J., Zhang, Z., Turson, G., Aili, H., Wang, J. and McManus, D.P. (2003) Immunoglobulin profiles in a murine intermediate host model of resistance for *Echinococcus granulosus* infection. *Parasite Immunology* 25, 161–168.

10 Signal Transduction at the Host–Parasite Interface

T.P. Yoshino, J.J. Vermeire and J.E. Humphries
Department of Pathobiological Sciences, University of Wisconsin-Madison, School of Veterinary Medicine, 2015 Linden Drive, Madison, WI 53706, USA

Introduction

Parasitic flatworms of the Phylum Platyhelminthes represent an ancient and diverse group of metazoan organisms comprised of two major classes: Trematoda (aspidobothrids and flukes) and Cercoeromorphae (monogeneans and tapeworms) (Marquardt *et al.*, 2000). Although differing in specific details of body form, developmental stages, reproductive structures and strategies, life histories and the like, parasitic flatworms share in common many basic 'necessities of life' to ensure their continued survival. Essential among these necessities is their ability to communicate with their external environment by reception of chemical cues or signals from the host, the host's environment or other parasites (conspecifics) co-inhabiting a given host. The major interface at which much of the initial chemical communication between these parasites and their hosts takes place is at the surface of the outer body wall or tegument; the site not only for acquisition of simple nutrients such as sugars, amino acids or lipids, but for binding external 'signalling' molecules capable of initiating internal cellular, biochemical or molecular events involved in numerous parasite functions from host finding and tissue tropism to triggering of developmental or metabolic processes.

The tegument of parasitic platyhelminths has been studied extensively and much is known regarding its cell structure and role in transporting a variety of host molecules. This organ is comprised of a single cellular syncytium that encompasses the entire parasite's body. It overlies a basement membrane through which subtending nucleated cytons (perikarya) connect with the tegumental syncytium via thin cytoplasmic bridges or trabeculae (Hockley, 1973; Holy *et al.*, 1991). In terms of function, the molecules expressed at the surface of the tegument serve in multiple capacities including nutrient acquisition, especially the transport of simple sugars, amino acids and lipids through the tegumental membrane (reviewed by Dalton *et al.*, 2004), protection from attack by the host's immune system or entry into harsh chemical environments and in the regulation of cellular differentiation and development. This latter area of receptor involvement in platyhelminth development remains poorly understood, although it is believed that signals received from the host environment (or other co-infecting parasites) at the tegumental surface and transmitted through this syncytial barrier are critically involved in this process. However, to date, we are only beginning to identify the tegumental receptors with putative roles in regulating tissue differentiation/development and, as a consequence, have only a rudimentary understanding of the molecular signalling pathways by which 'information' is transmitted from surface receptors, through the tegumental syncytium and ultimately to target tissues within the parasite's body (Fig. 10.1).

The aim of this chapter is to review what is known about the signal-transducing tegumental receptors and their pathways potentially mediating chemical communication at the host–parasite interface in the Platyhelminthes. As other chapters in this book include discussions of the role of tegumental molecules in the context of immunologic responses, neuronal/sensory function and nutritional or ion transport, we have focused our attention on the cell signalling cascades initiated by tegumental receptor–ligand interactions, potentially those associated with parasite growth and development. In this chapter, discussion of molecular signalling in parasitic flatworms will be divided into three areas including: (i) tegumental involvement in transducing external signals; (ii) molecular conservation of putative signal transduction molecules in parasitic flatworms; and (iii) signal-transducing receptors of platyhelminths and their signalling cascades. It should be noted from the outset that we will be focusing on those systems in which sufficient information is available to begin to build a more comprehensive picture of the molecules involved in signal reception and transduction through the tegumental syncytium. In this regard we have restricted our discussion mainly to the coverage of the digenean trematodes and cestodes, where the greatest accumulation of information on specific receptors and signalling molecules is currently available.

Tegumental Involvement in Signal Transduction: Perception and Reality

The most extensive and intimate contact between parasitic platyhelminths and their host is at the tegumental surface, and this provides the greatest potential for molecular communication between these organisms. However, the molecular make-up of the tegument of parasitic flatworms is extremely complex, due in part to a diversity of surface glycolipids and glycoproteins. Therefore, relatively few surface molecules known to be associated with cellular signalling systems, in particular those that may be serving parasite development- or growth-regulatory functions, have been identified and characterized. Maturation/growth-promoting effects of exposure to host hormones or cytokines in the schistosome blood flukes (reviewed by de Mendonca *et al.*, 2000; Davies and McKerrow, 2003), and estrogen exposure in *Taenia* tapeworms (Terrazas *et al.*, 1994) are suggestive of ligand–receptor-mediated effects. However, because of the potential complex interactions between endocrine and immune constituents, it remains unclear whether these host factors are acting directly, through binding to parasite receptors, or indirectly, by initiating specific alterations in immune status or metabolic state/pathways of the host, to favour parasite survival (de Mendonca *et al.*, 2000; Saule *et al.*, 2002; Davies and McKerrow, 2003). The role of the adaptive immune system in regulating development of *Schistosoma mansoni* in mice serves to illustrate this point.

Amiri *et al.* (1992) showed that exposure of female *S. mansoni in vivo* (SCID mice) and

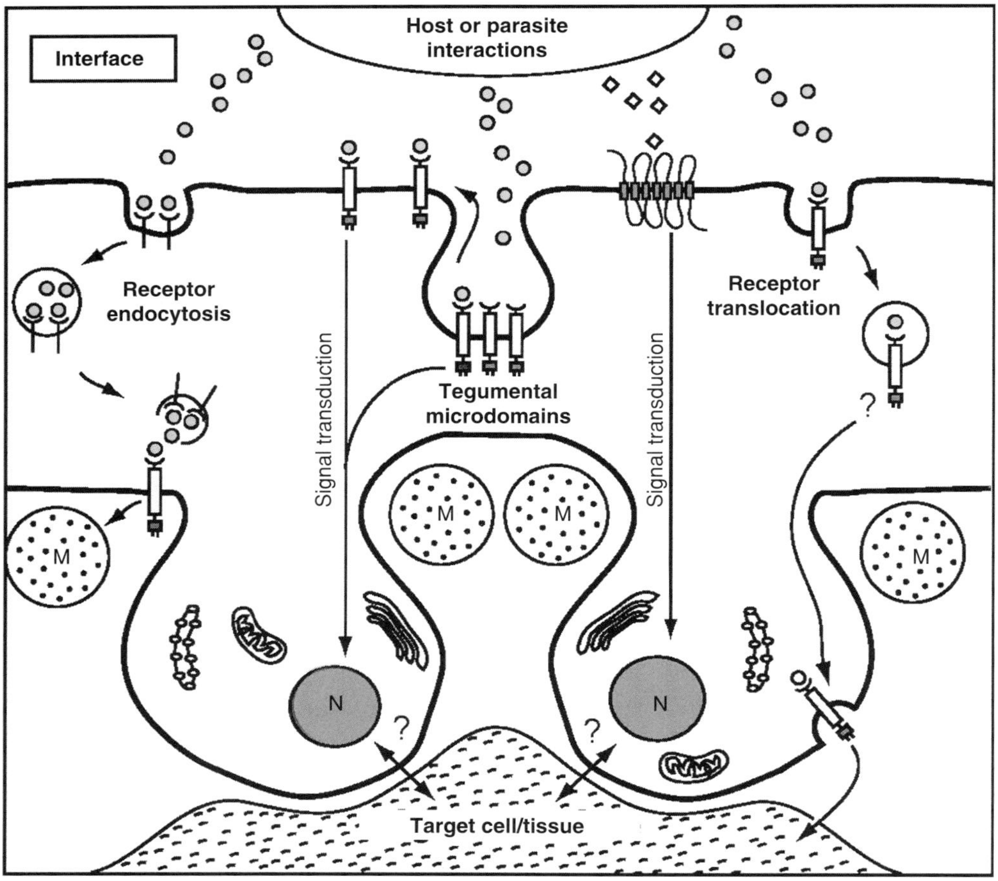

Fig. 10.1. Hypothetical scheme of platyhelminth tegumental signal transduction. Signals emanating from the host–parasite interface are transduced through the tegument by receptors at the surface and relayed to sub-tegumental cytons or to nearby non-tegumental cells via their associated signal transduction pathways. Surface receptors may exist as 'free' receptor–ligand complexes or as clusters within tegumental membrane microdomains, where upon ligand binding they may redistribute in the surrounding tegument surface to form receptor complexes. Receptor–ligand complexes may then be internalized via receptor endocytosis and recycled, or the activated receptor may be translocated through the tegument in order to signal to other target cells or tissues. N, nucleus; M, muscle.

in vitro to TNF-alpha induces egg production in mature female worms, suggesting that direct interaction of TNF may be responsible for changes in female worm fecundity. However, follow-up experiments, using $TNF^{-/-}$ knockout mice showed that female worm development and egg production were not linked directly to TNF or TNF receptor deficiencies (data cited in Davies *et al.*, 2004). These authors reconcile these apparently disparate observations by hypothesizing that TNF may be exerting its influence only after female worms have reached maturity and not during the prepatent developmental period (Davies *et al.*, 2004). Furthermore, they present evidence that an intact adaptive immune system is required for normal prepatent schistosome development (Davies *et al.*, 2001), especially of the male worm (Hernandez *et al.*, 2004). As mature male worms have been suspected of providing as yet unidentified molecular signals to stimulate female sexual maturation at pairing (Basch, 1988, 1990), specifically vitellogenesis (Popiel and Basch, 1984), Hernandez *et al.* (2004)

hypothesize that the signals provided by the adaptive immune system promote normal male worm development, which in turn provides essential molecular cues to trigger egg production in females. Clearly, the molecular interactions involved in this parasite–host relationship are complex and little definitive is known as to the identity of the host-signalling molecules or parasite receptors responsible for mediating the observed effects. Moreover, the specific male ligand(s) and female 'receptors' involved in stimulating vitellogenesis in female worms, and whether these receptor interactions are coupled to tegumental signal transduction systems, have yet to be explored. The molecular interactions between male and female *S. mansoni* recently have been reviewed by LoVerde *et al.* (2004) and Fitzpatrick *et al.* (2005).

Conspecific parasite–parasite signalling, similar to that proposed in blood flukes, has also been speculated to occur in cestodes. *Hymenolepis* developing under crowded conditions within the gut of the rat host typically are smaller (stunted) than lighter intensity infections, prompting speculation that 'crowding factors' elaborated from conspecifics may be regulating worm growth (Roberts, 1961). Subsequent experiments have shown that cGMP, produced and secreted by 'crowded' worms, exerted an *in vitro* mitosis-inhibiting effect in cells comprising the neck region (proglottid generating region) of normal adult worms (Zavras and Roberts, 1985), implicating cGMP as a potential component of the crowding effect. How the exogenous cGMP mediates its anti-mitotic signal across the tegument is not known, although a process involving cGMP receptor binding and/or vesicular transport cannot be ruled out (Threadgold and Hopkins, 1981; J.A. Oaks, University of Wisconsin-Madison, 2004, personal communication).

Many other examples of host–parasite or parasite–parasite molecular interactions may be cited. However, the central message carried by the above examples is that parasitic platyhelminths are constantly receiving a diversity of chemical cues at the tegumental surface, many of which require 'transmission' of specific signals through this syncytial layer to other cellular systems comprising these metazoan organisms. As stated earlier, we are only beginning to identify and characterize receptors (both tegumental and non-tegumental) that are known to be coupled to signalling pathways. In the following sections we summarize the conserved elements comprising known signal transduction pathways in parasitic flatworms and provide an update of known signal-coupled receptors and their associated pathways in this group of helminths.

Elements of Platyhelminth Signal Transduction Systems

Like all living organisms, parasites depend on well-developed signalling systems, at both the organismal level, sensing the surrounding host milieu and also at the cellular level, assessing the microenvironment encompassing each cell. This requires numerous transmembrane receptors capable of binding extracellular ligands, which in turn are coupled to a diversity of intracellular signalling molecules through which the signal is transmitted. In a generalized signal-transduction pathway, an extracellular signal traverses the membrane via a receptor, and inside, the 'message' is relayed through a signalling cascade by means of the sequential activation and deactivation of signalling proteins, most commonly involving phosphorylating (kinases) and dephosphorylating (phosphatases) enyzmes. Ultimately, successful signal transduction results in modulation of gene transcription and/or cell activity.

Molecular and EST data

Signal transduction in platyhelminths is still a relatively unexplored area, encompassing only a limited number of genera of parasitic flatworms, such as *Schistosoma*, *Fasciola*, *Hymenolepis* and *Echinococcus*. If little is known regarding signal transduction within these more intensely studied platyhelminths, one can appreciate the tremendous knowledge gap that currently exists in our understanding of the role of cellular communication in platyhelminth biology.

Until recently, flatworm signal transduction studies were mostly restricted to those of a functional nature, using inhibitory compounds and ligand-binding assays, and as a consequence, knowledge at the molecular level has been sparse. However, advances in biotechnology, especially in the areas of genomics and proteomics, have dramatically transformed this situation, allowing for platyhelminth signalling components to be assessed at the nucleotide and protein levels. Recent EST projects have inundated the databases with cDNA sequences, some of which display homology to signal transduction molecules of other organisms. To illustrate, a recent *S. mansoni* transcriptome project generated a total of approximately 31,000 assembled ESTs, of which at least 900 sequences are categorized as 'cell communication' molecules according to Gene Ontology (GO) classifications (Verjovski-Almeida *et al.*, 2003). A large number of assembled ESTs (approximately 13,000 non-redundant assembled sequences) also have been generated for *S. japonicum* (Hu *et al.*, 2003). It is estimated that 90% and 92% of the transcribed genes in *Schistosoma japonicum* and *S. mansoni*, respectively, have been sequenced, which presumably extrapolates to approximately 90% of their signalling molecules as well. Therefore the databases represent a rich and accessible resource, containing the cDNA sequences to encode the majority of schistosome signal transduction molecules.

Platyhelminth signalling proteins

One might expect the platyhelminths, as an early emerging group within the metazoans, to have an elementary set of signalling molecules from which the mammalian families evolved and expanded. However, the available data suggest flatworms do not possess such a primitive, ancestral version, and indeed, many of the major signal transduction families known in mammals appear to be represented within the phylum Platyhelminthes (Table 10.1). In fact, of the vertebrate signalling pathways that have been thoroughly examined to the point of their becoming exemplary or 'classical pathways' (Heldin and Purton, 1997), various components comprising these classical pathways can be found amongst the known flatworm signalling proteins. Consequently it is proposed that the same or similar pathways should be represented in platyhelminth-signalling networks. For example, in vertebrate systems, many signal transduction proteins have diverged, creating multiple types or isoforms, and it appears that this molecular variation is mirrored in flatworms. To illustrate, schistosome gene sequences similar to vertebrate protein kinase C (PKC) α, β, γ and ε isozymes (GenBank Accession Numbers AI067349, AI068067, AI975148, BU724154, BU798216, CD062663), both Type I and Type II receptors for activated PKC (RACK) (GenBank Accession Numbers AF422164, BU722968, CD098497, CD088499) and multiple forms of Gα subunits $G\alpha_i$, $G\alpha_o$ and $G\alpha_s$ (GenBank Accession numbers AAN59790, AI723358, CD178311, AAN59791; Iltzsch *et al.*1992; Davis *et al.*, 1995) are present (Table 10.1). This indicates complexity in the flatworm-signalling repertoire, suggesting that the divergence of many ancestral signalling proteins occurred before the emergence of the Platyhelminthes.

As pointed out, both receptors and intracellular signalling proteins exist in parasitic flatworms. However, in comparison with other organisms, fewer transmembrane receptors have been identified and annotated than one might expect, with just five G protein-coupled receptors (GPCRs), four receptor tyrosine kinases (RTKs) and two receptor serine/threonine kinases (RSTKs) (based on the NCBI EST database; Table 10.1). Taking the schistosome GPCR superfamily as an example, it includes receptors for a wide range of ligands, such as photons, lipids, peptides and biogenic amines, and is presumed to be involved in regulating a multitude of cellular activities. Yet only three schistosome GPCR genes (GenBank Accession Number AY354457; Hoffmann *et al.*, 2001; Hamdan *et al.*, 2002) have been identified to date. By comparison, at least 200 GPCRs have been identified in *Drosophila* (Brody and Cravchik, 2000), while *Caenorhabditis elegans* (Bargmann, 1998) and humans possess extensive GPCR superfamilies, numbering more than 1000 (Bockaert and Pin, 1999). This paucity of GPCRs might not be unexpected if very little schistosome genomic information was available, but as previously mentioned, recent EST projects are thought to have covered

Table 10.1. Summary of platyhelminth signalling proteins for which molecular evidence is available.

	Trematoda	Cercoero-morphae	Turbellaria	References
GPCRs				
Histamine	C[a]	–	–	Hamdan *et al.* (2002)
Peptide	C	E	C	AY354457, AF329279
Serotonin	E[b]	–	C	CV753613; Saitoh *et al.* (1997)
Rhodopsin	C	–	–	Hoffmann *et al.* (2001)
Other	E	–	–	CD079649.1; CN652751
RSTKs				
TGF-β	C	C	–	Davies *et al.* (1998); Forrester *et al.* (2004); AJ841786
RTK				
EGF	C/E	C	–	Shoemaker *et al.* (1992); BU796257; Spiliotis *et al.* (2003)
FGF	C	C	C	Cebria *et al.* (2002)
Other	–	–	E	Vicogne *et al.* (2003); Konrad *et al.* (2003); BU275976; AY066388
Cell adhesion receptors	E	–	–	CD064430.1
Orphan receptor	C	–	–	Inal (1999)
Carbohydrate/lipid receptors	E	E	E	CD115089.1; CD119026.1; CD178551.1; BQ173114; AY066690; AY068249
Nuclear receptors				
RXR	C	–	–	Freebern *et al.* (1999a,b)
Serine/threonine kinases				
PKC related	C/E	E	E	AY337620; BU798216; BQ173746; AY067126; AY067127
MAPK related	C/E	C/E	E	AY594257; BU804196; Spiliotis *et al.* (2005); CN653216; AY066128;
Other	E	C	E	BU802566; AJ785001; AY067132; AY067150
Tyrosine kinases	C/E	–	E	Kapp *et al.* (2001); Knobloch *et al.* (2002); CD167082.1; AY067121; AY066962
Trimeric G-proteins	C/E	–	–	Iltzsch *et al.* (1992); Davis *et al.* (1995); AF540396; AF540395; AF540394; BU800660
G-proteins				
Ras related	C/E	C/E		Kampkotter *et al.* (1999); CD080686.1; CD081865.1; Spiliotis and Brehm (2004); Spiliotis *et al.* (2005); CN649575
Rho related	C/E			Santos *et al.* (2002); Vermeire *et al.* (2003)
Adaptor proteins				
14-3-3	C			Schechtman *et al.* (1995); Zhang *et al.* (2000); Siles-Lucas *et al.* (1998); Nunes *et al.* (2004)
RACK	C/E	C		AF422164; BU804503

continued

Table 10.1. Summary of platyhelminth signalling proteins for which molecular evidence is available. (*cont'd*)

	Trematoda	Cercoero-morphae	Turbellaria	References
SH2/SH3 domain-containing proteins	C			McGonigle *et al.* (2001b)
Smads	C	C		Beall *et al.* (2000); Zavala-Gongora *et al.* (2003); Osman *et al.* (2004)

[a]Represents complete coding sequence (cds).
[b]Represents partial coding sequence and/or EST.
Note: Unpublished sequences (C,E) are referenced by accession number. EST sequences may have been listed as examples of multiple ESTs in the database.
Abbreviations: GPCR = G-protein coupled receptor; RSTK = receptor serine/threonine kinase; TGF-β = transforming growth factor; RTK = receptor tyrosine kinase; EGF = epidermal growth factor; FGF = fibroblast growth factor; RXR = retinoid X receptor; PKC = protein kinase C; MAPK = mitogen-activated protein kinase; G-protein = guanine nucleotide-binding protein; RACK = receptor for activated C kinase; SH2 = src homology.

greater than 90% of expressed genes in *S. japonicum* and *S. mansoni*. In addition, GPCRs, by definition, couple with trimeric G-proteins and nearly a full complement of these subunits has already been identified in schistosomes (Table 10.1), suggesting an intact and fully functional GPCR signalling system. Why, therefore, have so few schistosome GPCRs been identified? One explanation might lie in the methods used to classify schistosome ESTs, in that they are categorized based on similarity with characterized proteins of other species. On comparison with mammalian receptors, annotated platyhelminth receptors share 20–25% identity at the amino acid level (Table 10.2), so perhaps more GPCRs and receptors in general are present within the schistosome transcriptomes but have not yet been classified as such due to their low homology with receptors of other species. This notion is supported by a similarly low percentage of amino acid identity for flatworm homologues of the human transforming growth factor (TGF)-β (TGF-β) RTK and epidermal growth factor (EGF) receptor families.

In contrast to receptors, intracellular signalling molecules have greater homology between flatworms and humans, with identities ranging from 30% to 77%, suggesting that these proteins have been highly conserved throughout evolution (Table 10.2). One explanation for the difference in sequence conservation between receptors vs intracellular signalling molecules may be that functional specificity (i.e. triggering of a specific cellular response) is dictated by an initial receptor–ligand interaction and, therefore, genes encoding such receptors might be predicted to be more dissimilar between different species (i.e. more species specific). Whereas, once a receptor has been appropriately bound by its specific ligand (activated), the signal is transduced through a common set of conserved signalling molecules.

As summarized in Table 10.1, parasitic flatworms possess a substantial repertoire of genes representing intracellular signal-transducing molecules and their associated membrane receptors; evidence of an extensive platyhelminth-signalling system. Flatworm-signalling proteins most likely regulate a plethora of cellular functions involved in development, reproduction, metabolism and behaviour. An important and intriguing question, however, is what role signal-transducing mechanisms might play at the host–parasite interface. It is assumed that the tegument is responsible, at least in part, for communicating external chemical signals to sub-tegumental tissues or cells of the worm, although little is known of what signalling proteins and pathways may be involved.

Table 10.2. Comparison of platyhelminth and human signalling proteins based on tBLASTn analyses and alignment against the human homologue with greatest similarity.

Platyhelminth protein	Accession number	Human homologue	Accession number	% Identity	% Similarity
Class Turbellaria					
(**R**) *Girardia*[a] GPCR	AF329279	NPY receptor	AY236540	26.6	37.5
(**R**) *Girardia*[a] GPCR	AB004540	5-HT receptor	AF498978	27.9	38.5
(**R**) *Girardia*[a] RTK (DjFGFR2)	AB074426	FGF receptor	NM_022969	22.4	36.6
Class Trematoda (flukes)					
(**R**) *Schistosoma* GPCR	AF031196	Adrenergic receptor α	AY548167	22.3	33.5
(**R**) *Schistosoma* GPCR (RHO)	AF155134	Melanopsin	NM_033282	24.8	39
(**R**) *Schistosoma* GPCR (SmNPR1)	AY354457	Neuropeptide Y2 receptor	U42766	21.3	35.9
(**R**) *Schistosoma* RSTK (SmRK1)	AF031557	TGF β receptor	NM_004612	28.4	38.2
(**R**) *Schistosoma* RTK	AF101194	RTK	M10051	13.4	24.6
(**R**) *Schistosoma* RTK (SER)	M86396	EGF receptor	NM_005228	18.8	27.1
Schistosoma RXR	AF094759	RXR α	BC063827	3.2	7.2
Schistosoma RXR (RXR-2)	AF129816	RXR α	BC063827	18.2	25.2
Schistosoma MAPK	AY594257	MAPK	BC013992	61.1	71.8
Schistosoma RACK	AF422164	RACK1	BC000214	69.9	79.9
Schistosoma Rho	AF140785	Rho (rac)	NM175744	52.1	61.9
Schistosoma Ras	U53177	H-Ras	AF493916.1	75.7	83.1
Schistosoma 14-3-3 ε	AF375996	14-3-3 ε	U28936	61.1	71.2
Schistosoma 14-3-3 (Sm14-3-3)	U24281	14-3-3	U28964	58.6	67.6
Class Cercoeromorphae (tapeworms)					
(**R**) *Echinococcus* RTK (EmER)	AJ515524	EGF receptor	NM_005228	21.1	29.3
Echinococcus Ras (EmRas)	AJ785000	H-Ras	AF493916.1	77.2	83.6
Echinococcus Raf (EmRaf)	AJ785001	Raf	NM_002880.2	30.9	42.3
Echinococcus 14-3-3 ε	AF207904	14-3-3	NM_145690	54.1	63.1
Echinococcus 14-3-3	AF529418	14-3-3	U28964	56.6	65.1

[a]*Girardia = Dugesia.*

Abbreviations: GPCR = G protein coupled receptor; RXR = retinoid X receptor; RTK = receptor tyrosine kinase; RSTK = receptor serine/threonine kinase; RACK = receptor for activated protein kinase C, MAPK = mitogen-activated protein kinase.

Note: Transmembrane receptors are preceded by (**R**).

Platyhelminth Signal Transduction Pathways: Tegument and Beyond

The tegument of parasitic platyhelminths, although structurally similar in their syncytial organization, is not homogenous in surface architecture or protein composition. Adult schistosomes, for example, have a double lipid bilayer membrane that constantly undergoes turnover by shedding of its surface membrane. Cestodes, by contrast, possess a single bilayered or unit membrane surface with apparently little turnover. Despite these differences, both of these flatworm groups face a similar task of transducing signals received via surface receptors through the tegumental syncytial cytoplasm either to nuclei of the perikarya (sub-tegumental cytons) or to other nearby non-tegumental cells. Whether signalling-coupled receptors occur randomly in the surface tegumental membrane or within specialized structures still remains an open question, although both represent equally likely possibilities. In schistosomes, for example, it has been hypothesized that receptors and functional surface molecules exist in tegumental microdomains (Kunz, 2001), yet this remains relatively unexplored. Using indirect immunofluorescence microscopy the SmRK-1 receptor has been localized to anterior dorsal tubercles of adult male worms (Davies *et al.*, 1998). Moreover, caveolae-like structures in these tubercles of the tegument of adult male *S. mansoni* have been speculated to be foci of signal-transducing molecules (Rocoonin *et al.*, 1999). These specialized regions on the tegumental surface could maintain a complement of receptors and other signal-transducing molecules utilized by the parasite to receive stimuli from the surrounding environment. Electron microscopic studies have put putative photoreceptors in close physical proximity to structures located on the surface of the tegument (Short and Gagne, 1975) as well as in subtegumentary 'cilium-containing chambers' (Hoffmann *et al.*, 2001). In this case, a tegumental surface receptor may not be required as rhodopsin is presumed to be directly activated by light.

As alluded to above, because the eventual consequence of tegumental signal reception and transduction will likely be the triggering of molecular or biochemical pathways in non-tegumentary cells or tissues, the issue of intercellular transfer of transducing signals must be addressed. As Kunz (2001) points out, platyhelminths, being acoelomates, lack a fluid-filled (pseudo)coelomic cavity, precluding the transfer of chemical messages between cell/organs via the (pseudo)coelomic fluid. However, the syncytial nature not only of the tegument, but other tissues of flatwoms, provides the opportunity for different tissue types to be in intimate proximity to the proximal (inner) tegumental membrane or that of the subtegumentary perikarya, thus providing for cell–cell contact and potential molecular interaction (Kunz, 2001; Dalton *et al.*, 2004; LoVerde *et al.*, 2004). One might imagine that intercellular signalling through vesicular exocytosis of second messengers, membrane electrocoupling or perhaps gaseous signalling (nitric oxide), may then serve in the transfer of chemical signals to the eventual target tissues (Fig. 10.1).

What do we know about receptor-mediated signalling pathways in parasitic platyhelminths? Unfortunately, as a larger taxonomic group, we know very little. In-depth studies of the human blood fluke, *S. mansoni*, have provided some of the most complete information on receptor-mediated signal transduction, and are summarized in the section on 'Transforming Growth Factor Receptor (TGFR) Pathway'.

Transforming growth factor receptor (TGFR) pathway

The TGF-β superfamily of cytokines consists of a wide range of proteins that regulate various physiological processes, including embryonic development, homeostasis, wound healing, chemotaxis and cell cycle control (Leask and Abraham, 2004). In mammalian cells, the classical TGFR pathway is initiated by ligand binding at the cell surface to the TGF-β type II receptor, which in turn triggers the recruitment of the TGF-β type I RSTK (Fig. 10.2B). The type I receptor kinase is subsequently activated via phosphorylation within its GS domain by the ligand-type II receptor complex. This activation of the type I receptor kinase allows for the transduction of the signal into the cell via recruitment and phosphorylation of receptor-regulated Smad (R-Smad) proteins (Leask and Abraham,

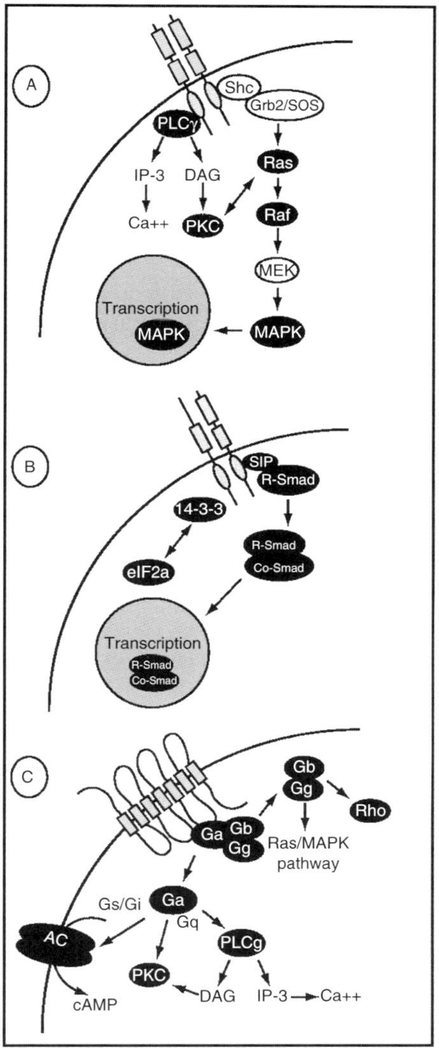

Fig. 10.2. Diagrammatic representation of generalized epidermal growth factor (EGF) (A), transforming growth factor-β (B) and G protein-coupled receptor (C) signal transduction pathways. Those components denoted by shading have been identified in platyhelminths as complete coding sequences. AC, adenylate cyclase; cAMP, cyclic 3′,5′ adenosine monophosphate; Co-Smad, co-regulatory smad protein; R-Smad, receptor-regulated Smad protein; SIP, SmRK1 interacting protein; DAG, diacylglycerol; eIF2α, eukaryotic initiation factor subunit 2α; Gα, Gα subunit; Gβ, Gβ subunit; Gγ, Gγ subunit; IP-3, inositol 1,4, 5-triphosphate; MAPK, mitogen-activated protein kinase; PLCγ, phospholipase Cγ; PKC, protein kinase C; MEK, MAPK kinase; SOS, Son of Sevenless (guanine nucleotide exchange).

2004). The R-Smad proteins interact with type I receptor and co-regulatory Smad (co-Smad) proteins in the signal cascade by virtue of conserved MH1 and MH2 domains at their N- and C-termini, which are separated by a variable, non-conserved region. These domains possess DNA binding and transcriptional activation functions, respectively. The inactive MH1 domain interacts with the MH2 domain in an auto-inhibitory manner. However, once C-terminal phosphorylation by the type I receptor occurs, this inhibitory association is alleviated and the R-Smad is then able to interact with its appropriate co-Smad. The phosphorylated R-Smad/co-Smad complex is capable of translocation to the cell nucleus where, in coordination with nuclear factors, it is able to exert its transcriptional activation function.

TGF-β receptor signalling in schistosomes

The TGF-β signalling pathway in *S. mansoni* has been studied in detail by incorporating a combination of molecular and functional assays, and at present it is the best characterized of signalling pathways among the platyhelminths. Davies *et al.* (1998) were the first to identify and characterize the TGF-β type I receptor (SmRK1) and subsequently found that it was capable of interacting with mammalian TGF-β type II receptor in transfected COS7 cells when bound by ligand (TGF-β) (Beall and Pearce, 2001). In this heterologous expression system, TGF-β ligand-binding to dimerized type I and II receptors also was capable of complexing with schistosome Smad2, a downstream signalling protein. This demonstrates not only a structural conservation of schistosome type I TGF-β receptor with its mammalian counterpart, but a functional conservation, with its ability to bind components of the downstream signalling cascade. Further studies identified a type II TGF-β receptor homologue (SmRK2; Forrester *et al.*, 2004), providing additional support for a fully functional TGF-β receptor in the schistosome tegument.

In addition, a number of SmRK1-interacting proteins associated with activated TGF-β receptor signalling have recently been identified in *S. mansoni* (Fig. 10.2B). These include Smad (Smad1, Smad2; Beall *et al.*, 2000; Osman *et al.*, 2001) and co-Smad (Smad4;

Osman *et al.*, 2004) homologues, which have been localized in the schistosome nucleus. Furthermore, using yeast two-hybrid assays employing human type I TGF-β receptor as bait, the adaptor protein 14-3-3ε (McGonigle *et al.*, 2001a), SmRK1 interacting protein (SIP) (McGonigle *et al.*, 2001b) and eukaryotic initiation factor subunit 2 α (eIF2α) (McGonigle *et al.*, 2002) also have been found. In functional assays, 14-3-3ε and eIF2α were shown to interact both with each other and also with SmRK1 (McGonigle *et al.*, 2002). Perhaps one of the most enlightening patterns to emerge from the above studies is that schistosome receptors and signalling molecules associated with the TGF-β receptor family can functionally interact with those of mammalian systems, again demonstrating a high level of functional conservation for this cellular signal transducing system. Recent cloning of Smad homologues in *Echinococcus* (Zavala-Gongora *et al.*, 2003) further suggests that similar TGF-β signalling pathways likely occur throughout the Platyhelminthes.

Given the extensive body of accumulated findings, there is strong evidence for a TGF-β signalling network in the tegument of adult schistosomes. A major question still exists as to its specific function in the biology of these parasites. Receptor localization at the tegumental surface would imply a potentially important role for TGF-β signalling in establishing and/or maintaining a compatible host–parasite relationship. Alternatively, as Osman *et al.* (2004) have speculated, TGF-β-like molecules and their receptors may also be involved in male worm-induced maturation of females during pairing (reviewed by LoVerde *et al.*, 2004). Clearly, additional work that incorporates rapidly developing transgenic technologies (Davis *et al.*, 1999; Boyle and Yoshino, 2003; Correnti and Pearce, 2004) is needed to begin addressing questions of receptor functionality in parasitic flatworms.

Epidermal growth factor receptor (EGFR) pathway

As shown in Fig. 10.2A, the classical EGFR pathway in mammalian cells, as reviewed by Wong and Guillard (2004), is initiated upon ligand binding of EGFR resulting in dimerization and activation of its own tyrosine kinase domain by auto-phosphorylation of specific tyrosine residues in its cytoplasmic tail. These residues are targets of PKC phosphorylation and allow for the binding of src homology 2 (SH2) domain-containing proteins including among others: phospholipase Cγ (PLCγ), Ras-GTPase-activating protein (Ras-GAP), Grb2 and Shc. Adaptor proteins like Grb2 and Shc serve as intermediates between receptor tyrosine kinases (RTKs) and downstream signal cascades by linking activated EGFR to the Ras signalling pathway. At the plasma membrane, Ras, activated via GDP to GTP exchange, induces a kinase cascade including Raf (Mek kinase), mitogen-activated protein (MAP) kinase kinase (MAPKK or Mek) and MAP kinase (MAPK). Activated MAPK typically translocates to the nucleus, where it can phosphorylate and activate transcription factors such as Elk-1 and c-Myc (Wong and Guillard, 2004). PLCγ activation also may occur when it is phosphorylated on specific tyrosine residues upon binding to an activated EGFR. PLCγ hydrolysis of the membrane phospholipid phosphatidylinositol-4,5-bisphosphate generates the second messengers diacylglycerol (DAG) and inositol 1,4,5-trisphosphate (IP-3). At the membrane DAG activates PKC, a serine/threonine protein kinase involved in a diverse array of cellular processes (Newton, 1995). IP-3 induces Ca^{2+} release from intracellular stores, affecting a variety of Ca^{2+}-dependent chemical reactions in the cell.

EGFR signalling: roles in the host–parasite relationship

The RTK genes with homology to members of the EGFR family have been identified in the cestode *Echinococcus multilocularis* (EmER; Spiliotis *et al.*, 2003) as well as in the blood flukes *S. mansoni* (SER; Shoemaker *et al.*, 1992) and *S. japonicum* (Hu *et al.*, 2003) (Fig. 10.2A). The *E. multilocularis* EGFR homologue, EmER, was the first molecule of its kind to be identified in a cestode. This receptor was shown to be expressed in the metacestode and to a greater extent, the protoscolex larval stages of *E. multilocularis*. EmER showed greatest overall sequence homology with the *S. mansoni*

EGFR homologue, SER (42% identity, 55% similarity). The amino acid sequence encoded by the EmER open reading frame contains the prototypical 'L1–C–L2' extracellular ligand-binding domain, a putative transmembrane domain and a hydrophobic, kinase-containing intracellular domain (Spiliotis *et al.*, 2003). The tyrosine–kinase domain is the most highly conserved region between EmER and SER (61% identity, 75% similarity), as is common among members of the EGFR family.

This family of receptor kinases is thought to signal through multiple pathways, involving activation of PKC and following the Ras/MAPK pathway, of which a number of members have been described in several *Schistosoma* spp. (Table 10.2; Schussler *et al.*, 1997; Kampkotter *et al.*, 1999) and the cestode *E. multilocularis* (Table 10.2; Spiliotis *et al.*, 2005). Other studies have shown that stimulation of putative guanine nucleotide-binding regulatory proteins activates endogenous schistosome phospholipase C, resulting in the production of inositol phosphate second messengers (Wiest *et al.*, 1992). These findings provide evidence supporting a functional signalling network involving both the Ras-MAPK and PLC-type pathways. Signalling through EGF RTKs is known to regulate a wide range of biological processes in mammalian cells, including cellular proliferation and differentiation, and it is presumed that flatworm RTKs binding exogenous EGF or EGF-like ligands from the host also serve as potential triggers of physiological events crucial to parasite development.

Despite the structural similarities between *S. mansoni* SER and *E. multilocularis* EmER, an interesting dissimilarity has been noted between these parasite genes in their mode of expression. Northern blot analyses of EmER identified only full-length mRNAs encoding transcripts (Spiliotis *et al.*, 2003), while in schistosomes, along with full-length SER mRNAs, three additional variant classes (2, 4, 5) of transcripts encoding truncated forms of SER have been identified (Shoemaker *et al.*, 1992). These truncated splice variants include the secretory leader sequence and only negligible segments of the ligand-binding domain. Possible roles in the host–parasite relationship have been proposed for these putatively secreted, or membrane-bound truncated variants in immune avoidance, including acting as 'dummy-epitopes' for blocking antibodies or as ligands for other receptors, be they host or parasite (Shoemaker *et al.*, 1992).

The theory that EGFR splice variants may be involved in antigenic masking or antigenic decoy suggests a localization of receptors at the tegumental surface. Moreover, if exogenous host EGF is serving as the natural ligand, a surface receptor would most likely serve as the initial site for molecular interaction. However, immunohistochemical studies by Ramachandran *et al.* (1996) have localized SER primarily to the muscle layer of adult male and female worms. One possible explanation for SER not being seen in the tegument may be that the variant SERs are in fact present at the tegumental surface, but are not recognized by the antibody used in their studies. Alternatively, as the authors suggest, the resolving power for precisely localizing immunoreactive antigens by fluorescence microscopy does not exclude the possible expression of SER at the worm's tegumental surface.

Expression of SER was confirmed in *in vitro*-transformed *S. mansoni* mother sporocysts (Vermeire *et al.*, 2004), suggesting a potential functional pathway in larval schistosome stages. Moreover, a novel EGF-like homologue (L-EFG) with neurotrophic activity has been described in the pond snail, *Lymnaea stagnalis* (Wildering *et al.*, 2001). L-EGF is unique in that its structure is that of a secreted peptide, the first of its type to be described in an invertebrate system. A role in snail reproduction has been proposed for this molecule given that it is expressed in the albumen gland, an organ involved in the secretion of nutritive perivitelline fluid on to fertilized oocytes before they are packaged into egg masses. Expression of a similar molecule in *Biomphalaria* spp., the intermediate host of *S. mansoni*, is plausible, although, as is the case with SER expression in adult worms, the functional significance of snail EGF-like ligand binding to larval schistosomes is unknown. With the potential link between snail L-EGF expression and reproduction, it might be speculated that larval trematodes may be competing for snail EGF, thus leading to host ligand depletion and disruption of snail reproduction (parasitic castration; De Jong-Brink, 1995). Thus, expressing a parasite

EGF receptor could afford a survival advantage to developing schistosome larvae.

G Protein-coupled receptors (GPCRs)

These receptors bind an array of signal-mediating ligands including hormones, neurotransmitters and chemokines in order to transduce external stimuli to downstream effectors. In mammalian systems, the seven transmembrane GPCRs accomplish this via heterotrimeric G-proteins composed of Gα, Gβ and Gγ subunits (Hamm, 1998) (Fig. 10.2C). The four major families of heterotrimeric Gα-proteins have been assigned putative functions: (i) potentiation of adenylate cyclase ($G\alpha_s$); (ii) inhibition of adenylate cyclase ($G\alpha_i$, $G\alpha_o$); (iii) activation of phospholipase C ($G\alpha_q$); and (iv) a group ($G\alpha_{12}$–$G\alpha_{13}$) with unknown function. Receptor activation by extracellular ligand binding induces conformational changes in the transmembrane helices, which in turn expose cryptic G-protein binding sites on intracellular loops. An activated receptor facilitates an exchange of GDP for GTP, yielding an activated G-protein. The active conformation caused by GTP binding leads to the dissociation of the Gα and Gβγ dimer. Both Gα and the Gβγ dimer act in highly specific manner to modulate a multitude of target enzymes including adenylate cyclase, PLC or photoreceptor cGMP phosphodiesterase (Hamm, 1998). Among a plethora of other target proteins, Gβγ has been demonstrated as an activator of a pheromone-stimulated MAPK cascade (Inouye *et al.*, 1997), and members of the Rho family of small GTPases including Rho, Rac and Cdc42 (Aspenstrom, 1999), involved in the control of gene transcription, cell cycle regulation, apoptosis and cytoskeletal responses to extracellular signalling (Mackay and Hall, 1998).

GPCRs in parasitic platyhelminths

Several GPCRs, their associated heterotrimeric G protein subunits and a family of Rho GTPases have been identified and characterized in trematodes (Fig. 10.2C). Mansour (1984) was the first to study, at the biochemical level, tegumental receptors of *S. mansoni* and the liver fluke, *Fasciola hepatica*, that were capable of binding serotonin (5-hydroxytryptamine, 5-HT) and subsequently activating adenylate cyclase and production of the second messenger cyclic 3′,5′ adenosine monophosphate (AMP) (cAMP); they designated this novel class of trematode membrane receptors as serotonin receptors. Further studies on these receptors resulted in the identification of $G\alpha_s$, $G\alpha_o$ and $G\alpha_i$ homologues in *S. mansoni* and *F. hepatica* (Mansour and Mansour, 1989) and the cloning and characterization of $G\alpha_s$ (Iltzsch *et al.*, 1992) and $G\alpha_o$ (Genbank Accession Number AAN59790) subunits from *S. mansoni*. More recently several Rho family members including Rho, Rac and Cdc42 have been identified in these blood flukes (Santos *et al.*, 2002; Vermeire *et al.*, 2003), suggesting the presence of an intricate system of signal transduction associated with trematode GPCRs and their potential involvement in biochemical pathways such as energy metabolism.

Hamdan *et al.* (2002) describe another biogenic amine-responsive GPCR in the tegument of *S. mansoni*, which specifically binds to and is activated by histamine. This cloned receptor shares approximately 30% amino acid sequence homology with other major amine GPCRs, and, when transfected into mammalian cells exhibited the following characteristics:

- was expressed as a plasma membrane protein;
- possessed strong binding affinity to histamine and
- in response to histamine binding, resulted in elevated intracellular Ca^{2+} and cAMP.

Clearly, following heterologous expression in mammalian cells, the schistosome histamine GPCR is capable of initiating a signalling cascade that generates second messengers that are characteristic of many other GPCR-primed pathways. It is unknown if a similar signalling pathway is associated with ligand activation of this candidate schistosome histamine receptor *in vivo*.

The putative function of trematode sensory GPCRs in the free-swimming cercariae and miracidia may be rooted in the molecular events underlying the host-finding behaviour. The miracidial and cercarial stages of *S. mansoni* are known to exhibit positive phototropic behaviour that is presumed to aid in the location of an appropriate intermediate or definitive host, respectively. Recently, an identified rhodopsin-like GPCR gene was found to display developmentally regulated expression, in that its transcript levels were higher in both cercariae and miracidia than in adult worms (Hoffmann *et al.*, 2001) or mother sporocysts (Vermeire *et al.*, 2004). Because this rhodopsin-like GPCR has been ascribed a putative function as part of a light-sensing organ system and has been localized to subtegumental structures found near the anterior end of cercariae (Hoffmann *et al.*, 2001), it is speculated that its up-regulation in the free-living larval stages is associated with photoreception and photoreactive behaviour. Although the broad function of this GPCR seems apparent, little is known about the downstream signalling molecules involved in rhodopsin activation. It is likely that other larval trematode behaviours, such as chemotactic responses to host molecular signals (e.g. arginine gradients; Grabe and Haas, 2004) are mediated through similar sensory GPCRs (Speca *et al.*, 1999).

Finally, the possibility that GPCR signalling systems are present in cestodes is supported by pharmacological studies on *Hymenolepis*, in which a nitroimipramine-inhibitable, high-affinity serotonin receptor was found to be coupled with membrane adenylate cyclase activation, the production of cAMP (Ribeiro and Webb, 1987) and the stimulation of cAMP-dependent protein phosphorylation (Ribeiro and Webb, 1991). More recently, the activity of a GPCR in neuropeptide F (NPF) activity in *Mesocestoides* was supported by the selective effects of guanosine trilithium salts on NPF-mediated larval motility (Hrckova *et al.*, 2004). Although these findings demonstrate the occurrence of ligand-activated GPCRs in tapeworm muscle, they provide further evidence for the widespread occurrence of GPCRs and their signalling pathways in cestodes.

Concluding Remarks

The ability of a parasite to communicate with its external and internal environments represents a critical adaption for survival. Current data suggest that parasitic flatworms possess tegumental receptors coupled to and capable of triggering signal-transducing systems involved in the initiation of effector reactions both within and beyond the syncytial border. In the human blood flukes, the most thoroughly studied of the parasitic platyhelminths and best characterized at the molecular level, several signal-transducing receptors have been identified including homologues of the mammalian TGF-β and EGF receptor families as well as several GPCRs. Importantly, molecules associated with 'classical' signal transduction cascades of these receptors also have been identified including the enzymes PKC, adenylate cyclase, small GTPases (Ras, Rho) as well as a number of adaptor molecules or cofactors like 14-3-3, eIF2α and several Smads/co-Smads. Heterologous mammalian expression systems have proved effective for evaluating the functionality of these receptors and putative signalling molecules. The fact that appropriate signalling function can be demonstrated in this fashion speaks to the high degree of structural and functional conservation exhibited by both receptors and their signalling cascades. However, although signalling-coupled receptors and a variety of cell signalling molecules have been identified in both trematodes and cestodes, very little is known about the nuclear or cytoplasmic factors targeted by specific signalling cascades or their functional linkage to regulating parasite behaviours, growth or development. The steady increase in genomic, transcriptomic and proteomic information being generated for various parasitic flatworms, including human blood flukes, will undoubtedly provide excellent research opportunities for expanding our understanding of how molecular signals are received and what signalling systems are involved in the transfer of chemical messages both within and between cells comprising these complex organisms.

Acknowledgements

J.J.V. and J.E.H. both contributed equally to the writing of this chapter. Supported in part by NIH grants AI15503 and AI061436 to T.P.Y.

References

Amiri, P., Locksley, R.M., Parslow, T.G., Sadick, M., Rector, E., Ritter, D. and McKerrow, J.H. (1992) Tumour necrosis factor alpha restores granulomas and induces parasite egg-laying in schistosome-infected SCID mice. *Nature* 356, 604–607.

Aspenstrom, P. (1999) Effectors of the Rho GTPases. *Current Opinion in Cell Biology* 11, 95–102.

Bargmann, C. (1998) Neurobiology of the *Caenorhabditis elegans* genome. *Science* 282, 2028–2033.

Basch, P.F. (1988) *Schistosoma mansoni*: nucleic acid synthesis in immature females from single-sex infections, paired *in vitro* intact males and male segments. *Comparative Biochemistry and Physiology B* 90, 389–392.

Basch, P.F. (1990) Single sex schistosomes and chemical messages. Reply. *Parasitology Today* 6, 298.

Beall, M.J. and Pearce, E.J. (2001) Human transforming growth factor-β activates a receptor serine/threonine kinase from the intravascular parasite *Schistosoma mansoni*. *Journal of Biological Chemistry* 276, 31613–31619.

Beall, M.J., McGonigle, S. and Pearce, E.J. (2000) Functional conservation of *Schistosoma mansoni* Smads in TGF-β signaling. *Molecular and Biochemical Parasitology* 111, 131–142.

Bockaert, J. and Pin, J.P. (1999) Molecular tinkering of G protein-coupled receptors: an evolutionary success. *European Molecular Biology Organization Journal* 18, 1723–1729.

Boyle, J.P. and Yoshino, T.P. (2003) Gene manipulation in parasitic helminths. *International Journal for Parasitology* 33, 1259–1268.

Brody, T. and Cravchik, A. (2000) *Drosophila melanogaster* G protein-coupled receptors. *Journal of Cell Biology* 150, F83–F88.

Cebria, F., Kobayshi, C., Umesono, Y., Nakazawa, M., Mineta, K., Ikeo, K., Gojobori, T., Itoh, M., Tairi, M., Alvarado, A.S. and Agata, K. (2002) FGFR-related gene *nou-darake* restricts brain tissues to the head region of planarians. *Nature* 419, 620–624.

Correnti, J.M. and Pearce, E.J. (2004) Transgene expression in *Schistosoma mansoni*: introduction of RNA into schistosomula by electroporation. *Molecular and Biochemical Parasitology* 137, 75–79.

Dalton, J.P., Skelly, P. and Halton, D.W. (2004) Role of the tegument and gut in nutrient uptake by parasitic platyhelminths. *Canadian Journal of Zoology* 82, 211–232.

Davies, S.J. and McKerrow, J.H. (2003) Developmental plasticity in schistosomes and other helminthes. *International Journal for Parasitology* 33, 1277–1284.

Davies, S.J., Shoemaker, C.B. and Pearce, E.J. (1998) A divergent member of the transforming growth factor-β receptor family from *Schistosoma mansoni* is expressed on the parasite surface membrane. *The Journal of Biological Chemistry* 273, 11234–11240.

Davies, S.J., Grogan, J.L., Blank, R.B., Lim, K.C., Locksley, R.M. and McKerrow, J.H. (2001) Modulation of blood fluke development in the liver by hepatic CD4+ lymphocytes. *Science* 294, 1358–1361.

Davies, S.J., Lim, K.C., Blank, R.B., Kim, J.H., Lucas, K.D., Hernandez, D.C., Sedgwick, J.D. and McKerrow, J.H. (2004) Involvement of TNF in limiting liver pathology and promoting parasite survival during schistosome infection. *International Journal for Parasitology* 34, 27–36.

Davis, R.E., Hardwick, C., Taviernier, P., Hodgson, S. and Singh, H. (1995) RNA trans-splicing in flatworms. Analysis of transpliced mRNAs and genes in the human parasite, *Schistosoma mansoni*. *Journal of Biological Chemistry* 270, 21813–21819.

Davis, R.E., Parra, A., LoVerde, P.T., Ribeiro, E., Glorioso, G. and Hodgson, S. (1999) Transient expression of DNA and RNA in parasitic helminths by using particle bombardment. *Proceedings of the National Academy of Sciences USA* 96, 8687–8692.

De Jong-Brink, M. (1995) How schistosomes profit from the stress responses they elicit in their hosts. *Advances in Parasitology* 35, 177–256.

Fitzpatrick, J.M., Johnson, D.A., Williams, G.W., Williams, D.J., Freeman, T.C., Dunne, D.W. and Hoffman, K.F. (2005) An olignucleotide microarray for transcriptome analysis of *Schistoma mansoni* and its application/use to investigate gender-associated gene expression. *Molecular and Biochemical Parasitology* 141, 1–13.

Forrester, S.G., Warfel, P.W. and Pearce, E.J. (2004) Tegumental expression of a novel type II receptor serine/threonine kinase (SmRK2) in *Schistosoma mansoni*. *Molecular and Biochemical Parasitology* 136, 149–156.

Freebern, W.J., Osman, A., Niles, E.G., Christen, L. and LoVerde, P.T. (1999a) Identification of a cDNA encoding a retinoid X receptor homologue from *Schistosoma mansoni*. Evidence for a role in female-specific gene expression. *Journal of Biological Chemistry* 274, 4577–4585.

Freebern, W.J., Niles, E.G. and LoVerde, P.T. (1999b) RXR-2, a member of the retinoid x receptor family in *Schistosoma mansoni*. *Gene* 233, 33–38.
Grabe, K. and Haas, W. (2004) Navigation within host tissues: *Schistosoma mansoni* and *Trichobilharzia ocellata* schistosomula response to chemical gradients. *International Journal for Parasitology* 34, 927–934.
Hamdan, F.F., Abramovitz, M., Mousa, A., Xie, J., Durocher, Y. and Ribeiro, P. (2002) A novel *Schistosoma mansoni* G protein-coupled receptor is responsive to histamine. *Molecular and Biochemical Parasitology* 119, 75–86.
Hamm, H.E. (1998) The many faces of G protein signaling. *Journal of Biological Chemistry* 273, 669–672.
Heldin, C.H. and Purton, M. (1997) *Signal Transduction*, 2nd edn. Chapman & Hall, London.
Hernandez, D.C., Lim, K.C., McKerrow, J.H. and Davies, S.J. (2004) *Schistosoma mansoni:* sex-specific modulation of parasite growth by host immune signals. *Experimental Parasitology* 106, 59–61.
Hockley, D.J. (1973) Ultrastructure of the tegument of *Schistosoma*. *Advances in Parasitology* 11, 233–305.
Hoffmann, K.F., Davis, E.M., Fischer, E.R. and Wynn, T.A. (2001) The guanine protein coupled receptor rhodopsin is developmentally regulated in the free-living stages of *Schistosoma mansoni*. *Molecular and Biochemical Parasitology* 112, 113–123.
Holy, J.M., Oaks, J.A., Mika-Grieve, M. and Grieve, R. (1991) Development and dynamics of regional specialization within the syncytial epidermis of the rat tapeworm, *Hymenolepis diminuta*. *Parasitology Research* 77, 161–172.
Hrckova, G., Velebny, S., Halton, D.W., Day, T.A. and Maule, A.G. (2004) Pharmacological characterization of neuropeptide F (NPF)-induced effects on the motility of *Mesocestoides corti* (syn. *Mesocestoides vogae*) larvae. *International Journal for Parasitology* 34, 83–93.
Hu, W., Yan, Q., Shen, D.K., Liu, F., Zhu, Z.D., Song, H.D., Xu, X.R., Wang, Z.J., Rong, Y.P., Zeng, L.C., Wu, J., Zhang, X., Wang, J.J., Xu, X.N., Wang, S.Y., Fu, G., Zhang, X.L., Wang, Z.Q., Brindley, P.J., McManus, D.P., Xue, C.L., Feng, Z., Chen, Z. and Han, Z.G. (2003) Evolutionary and biomedical implications of a *Schistosoma japonicum* complementary DNA resource. *Nature Genetics* 35, 139–147.
Iltzsch, M.H., Bieber, D., Vijayasarathy, S., Webster, P., Zurita, M., Ding, J. and Mansour, T.E. (1992) Cloning and characterization of a cDNA coding for the α-subunit of a stimulatory G protein from *Schistosoma mansoni*. *Journal of Biological Chemistry* 267, 14504–14508.
Inal, J.M. (1999) *Schistosoma* TOR (trispanning orphan receptor), a novel, antigenic surface receptor of the blood-dwelling *Schistosoma* parasite. *Biochimica et Biophysica Acta* 1445, 283–298.
Inouye, C., Dhillon, N. and Thorner, J. (1997) Ste5 RING-H2 domain: role in Ste4-promoted oligomerization for yeast pheromone signalling. *Science* 278, 103–106.
Kampkotter, A., Ridgers, I., Johnston, D.A., Rollinson, D., Kunz, W. and Grevelding, C.G. (1999) *Schistosoma mansoni*: cloning and characterization of the Ras homologue. *Experimental Parasitology* 91, 280–283.
Kapp, K., Schubler, P., Kunz, W. and Grevelding, C.G. (2001) Identification, isolation and characterization of a Fyn-like tyrosine kinase from *Schistosoma mansoni*. *Parasitology* 122, 317–327.
Knobloch, J., Winnen, R., Quack, M., Kunz, W. and Grevelding, C.G. (2002) A novel Syk-family tyrosine kinase from *Schistsoma mansoni* which is preferentially transcribed in reproductive organs. *Gene* 294, 87–97.
Konrad, C., Kroner, A., Spiliotis, M., Zavala-Gongora, R. and Brehm, K. (2003) Identification and molecular characterisation of a gene encoding a member of the insulin receptor family in *Echinococcus multilocularis*. *International Journal for Parasitology* 33, 301–312.
Kunz, W. (2001) Schistosome male–female interactions: induction of germ-cell differentiation. *Trends in Parasitology* 17, 227–231.
Leask, A. and Abraham, D.J. (2004) TGF-β signaling and the fibrotic response. *FASEB Journal* 18, 816–827.
LoVerde, P.T., Niles, E.G. Osman, A. and Wu, W. (2004) *Schistosoma mansoni* male–female interactions. *Canadian Journal of Zoology* 82, 357–374.
Mackay, D.J.G. and Hall, A. (1998) Rho GTPases. *Journal of Biological Chemistry* 273, 20685–20688.
Mansour, T.E. (1984) Serotonin receptors in parasitic worms. *Advances in Parsitology* 23, 1–36.
Mansour, J.M. and Mansour, T.E. (1989) Identification of GTP-binding proteins in *Fasciola hepatica* and *Schistosoma mansoni* by immunoblotting. *Molecular and Biochemical Parasitology* 36, 11–18.
Marquardt, W.C., Demaree, R.S. and Grieve, R.B. (2000) *Parasitology and Vector Biology*, 2nd edn. Harcourt/Academic Press, San Diego, California, 702 pp.
McGonigle, S., Beall, M.J., Feeney, E.L. and Pearce, E.J. (2001a) Conserved role for 14-3-3ε downstream of type I TGF-β receptors. *FEBS Letters* 490, 65–69.
McGonigle, S., Feeney, E.L., Beall, M.J. and Pearce, E.J. (2001b) SIP, a novel SH3 domain-containing protein, interacts with *Schistosoma mansoni* receptor kinase 1. *Molecular and Biochemical Parasitology* 114, 119–123.

McGonigle, S., Beall, M.J. and Pearce, E.J. (2002) Eukaryotic initiation factor 2α associates with TGF-β receptors and 14-3-3? and acts as a modulator of the TGF-β response. *Biochemistry* 41, 579–587.

de Mendonca, R.L., Escriva, H., Bouton, D., Laudet, V. and Pierce, R.J. (2000) Hormones and nuclear receptors in schistosome development. *Parasitology Today* 16, 233–240.

Newton, A.C. (1995) Protein kinase C: structure, function, and regulation. *Journal of Biological Chemistry* 270, 28495–28498.

Nunes, C.P., Zaha, A., Gottstein, B., Muller, N. and Siles-Lucas, M.del M. (2004) 14-3-3 gene characterization and description of a second 14-3-3 isoform in both *Echinococcus granulosus* and *E. multilocularis*. *Parasitology Research* 93, 403–409.

Osman, A., Niles, E.G. and LoVerde, P.T. (2001) Identification and characterization of a Smad2 homologue from *Schistosoma mansoni*, a transforming growth factor-β signal transducer. *Journal of Biological Chemistry* 276, 10072–10082.

Osman, A., Niles, E.G. and LoVerde, P.T. (2004) Expression of functional *Schistosoma mansoni* Smad4: role in Erk-mediated transforming growth factor β (TGF-β) down-regulation. *The Journal of Biological Chemistry* 279, 6474–6486.

Popiel, R. and Basch, P.F. (1984) Reproductive development of female *Schistosoma mansoni* (Digenea: Schistosomatidae) following bisexual pairing of worms and worm segments. *Journal of Experimental Zoology* 232, 141–150.

Ramachandran, H., Skelly, P.J. and Shoemaker, C.B. (1996) The *Schistosoma mansoni* epidermal growth factor receptor homologue, SER, has tyrosine kinase activity and is localized in adult muscle. *Molecular and Biochemical Parasitology* 83, 1–10.

Ribeiro, P. and Webb, R.A. (1987) Characterization of a serotonin transporter and an adenylate cyclase-linked serotonin receptor in the cestode *Hymenolepis diminuta*. *Life Sciences* 40, 755–768.

Ribeiro, P. and Webb, R.A. (1991) Serotonin stimulates protein phosphorylation in the cestode *Hymenolepis diminuta*. *Comparative Biochemistry and Physiology C* 100, 483–489.

Roberts, L.S. (1961) The influence of population density on patterns and physiology of growth in *Hymenolepis diminuta* (Cestoda: Cyclophyllidea) in the definitive host. *Experimental Parasitology* 11, 332–337.

Rocoonin, E.L., Davies, S.J. and Pearce, E.J. (1999) Caveolae-like structures in the surface membrane of *Schistosoma mansoni*. *Molecular and Biochemical Parasitology* 104, 285–297.

Saitoh, O., Yuruzume, E., Watanabe, K. and Nakata, H. (1997) Molecular identification of a G protein-coupled receptor family, which is expressed in planarians. *Gene* 195, 55–61.

Santos, T.M., Machado, C.R., Franco, G.R. and Pena, S.D.J. (2002) Characterization and comparative functional analysis in yeast of a *Schistosoma mansoni* Rho1 GTPase gene. *Molecular and Biochemical Parasitology* 125, 103–112.

Saule, P., Adriaenssens, E., Delacre, M., Chassande, O., Bossu, M. Auriault, C. and Wolowczuk, I. (2002) Early variations of host thyroxine and interleukin-7 favor *Schistosoma mansoni* development. *Journal of Parasitology* 88, 849–855.

Schechtman, D., Ram, D., Tarrab-Hazdai, R., Arnon, R. and Schechter, I. (1995) Stage-specific expression of the mRNA encoding a 14-3-3 protein during the life cycle of *Schistosoma mansoni*. *Molecular and Biochemical Parasitology* 73, 275–278.

Schussler, P., Grevelding, C.G. and Kunz, W. (1997) Identification of Ras, MAP kinases, and a GAP protein in *Schistosoma mansoni* by immunoblotting and their putative involvement in male–female interaction. *Parasitology* 115, 629–634.

Shoemaker, C.B., Ramachandran, H., Landa, A., Mitermayer, G.R. and Stein, L.D. (1992) Alternative splicing of the *Schistosoma mansoni* gene encoding a homologue of epidermal growth factor receptor. *Molecular and Biochemical Parasitology* 53, 17–32.

Short, R.B. and Gagne, H.T. (1975) Fine structure of possible photoreceptor in cercariae of *Schistosoma mansoni*. *Journal of Parasitology* 61, 69–74.

Siles-Lucas, M., Felleisen, R.S., Hemphill, A., Wilson, W. and Gottstein, B. (1998) Stage-specific expression of the 14-3-3 gene in *Echinococcus multilocularis*. *Molecular and Biochemical Parasitology* 91, 281–293.

Speca, D.J., Lin, D.M., Sorensen, P.W., Isacoff, E.Y., Ngai, J. and Dittman, A.H. (1999) Functional identification of a goldfish odorant receptor. *Neuron* 23, 487–498.

Spiliotis, M. and Brehm, K. (2004) *Echinococcus multilocularis*: identification and molecular characterization of a Ral-like small GTP-binding protein. *Experimental Parasitology* 107, 163–172.

Spiliotis, M., Kroner, A. and Brehm, K. (2003) Identification, molecular characterization and expression of the gene encoding epidermal growth factor receptor orthologues from the fox-tapeworm *Echinococcus multilocularis*. *Gene* 323, 57–65.

Spiliotis, M., Tappe, D., Bruckner, S., Mosch, H.U. and Brehm, K. (2005) Molecular cloning and characterization of ras- and raf-homologues from the fox-tapeworm *Echinococcus multilocularis*. *Molecular and Biochemical Parasitology* 139, 225–237.

Terrazas, L.I., Bojalil, R., Govezensky, T. and Larraide, C. (1994) A role for 17-β-estradiol in immunoendocrine regulation of murine cysticercosis (*Taenia crassiceps*). *Journal of Parasitology* 80, 563–568.

Threadgold, L.T. and Hopkins, C.A. (1981) *Schistocephalus solidus* and *Ligula intestinalis*: pinocytosis by the tegument. *Experimental Parasitology* 51, 444–456.

Verjovski-Almeida, S., DeMarco, R., Martins, E.A., Guimaraes, P.E., Ojopi, E.P., Paquola, A.C., Piazza, J.P., Nishiyama, M.Y. Jr, Kitajima, J.P., Adamson, R.E., Ashton, P.D., Bonaldo, M.F., Coulson, P.S., Dillon, G.P., Farias, L.P., Gregorio, S.P., Ho, P.L., Leite, R.A., Malaquias, L.C., Marques, R.C., Miyasato, P.A., Nascimento, A.L., Ohlweiler, F.P., Reis, E.M., Ribeiro, M.A., Sa, R.G., Stukart, G.C., Soares, M.B., Gargioni, C., Kawano, T., Rodrigues, V., Madeira, A.M., Wilson, R.A., Menck, C.F., Setubal, J.C., Leite, L.C. and Dias-Neto, E. (2003) Transcriptome analysis of the acoelomate human parasite *Schistosoma mansoni*. *Nature Genetics* 35, 148–157.

Vermeire, J.J., Osman, A., LoVerde, P.T. and Williams, D.L. (2003) Characterisation of a Rho homologue of *Schistosoma mansoni*. *International Journal for Parasitology* 33, 721–731.

Vermeire, J.J., Boyle, J.P. and Yoshino, T.P. (2004) Differential gene expression and the effects of *Biomphalaria glabrata* embryonic (Bge) cell factors during larval *Schistosoma mansoni* development. *Molecular and Biochemical Parasitology* 135, 153–157.

Vicogne, J., Pin, J.P., Lardans, V., Capron, M., Neol, C. and Dissous, C. (2003) An unusual receptor tyrosine kinase of *Schistosoma mansoni* contains a Venus Flytrap module. *Molecular and Biochemical Parasitology* 126, 51–62.

Wiest, P.M., Li, Y.N., Burnham, D.C., Olds, G.R. and Bowen, W.D. (1992) *Schistosoma mansoni*: characterization of phosphoinositide response. *Experimental Parasitology* 74, 38–45.

Wildering, W.C., Hermann, P.M. and Bulloch, A.G.M. (2001) *Lymnaea* epidermal growth factor promotes axonal regeneration in CNS organ culture. *Journal of Neuroscience* 21, 9345–9354.

Wong, R.W.C. and Guillard, L. (2004) The role of epidermal growth factor and its receptors in mammalian CNS. *Cytokine and Growth Factor Reviews* 15, 147–156.

Zavala-Gongora, R., Kroner, A., Wittek, B., Knaus, P. and Brehm, K. (2003) Identification and characterisation of two distinct Smad proteins from the fox-tapeworm *Echinococcus multilocularis*. *International Journal for Parasitology* 33, 1665–1677.

Zavras, E.T. and Roberts, L.S. (1985) Developmental physiology of cestodes: cyclic nucleotides and the identity of putative crowding factors in *Hymenolepis diminuta*. *Journal of Parasitology* 71, 96–105.

Zhang, Y., Bickle, Q.D. and Taylor, M.G. (2000) Cloning of *Schistosoma japonicum* 14-3-3ε (Sj14-3-3 ε), a new member of the 14-3-3 family of proteins from schistosomes. *International Journal for Parasitology* 30, 991–994.

11 Parasite Effects on the Snail Host Transcriptome

M. Knight and N. Raghavan
Biomedical Research Institute, 12111 Parklawn Drive, Rockville, MD 20852, USA

Introduction

Parasitic platyhelminth infections continue to affect humans in many countries of the Western Hemisphere, the Pacific, Middle East, Asia and Africa. To limit the spread of diseases caused by these parasites, it has long been recognized that control strategies aimed at long-term reduction in parasite transmission cannot ignore the role that invertebrate hosts play in these diseases. In the case of schistosomiasis, a disease that according to the World Health Organization (WHO) is still found in 74 countries worldwide (WHO Fact Sheet No. 115, 1996; Chitsulo *et al.*, 2000), reduction in transmission has been attempted in numerous ways, including the application of molluscicides to eliminate and curtail the expansion of snail host populations. Mass chemotherapy, especially in school-aged children, has also been an effective control tool (Doenhoff *et al.*, 2000; Sturrock, 2001). However, for several reasons, our ability to stop the spread of schistosomiasis over the long term has been difficult to sustain. Reasons for this include the snails' propensity for re-colonization, the lack of surveillance of treated sites due to civil strife, and the growing expansion of irrigation projects. Also in these afflicted countries other prevailing diseases (e.g. malaria, tuberculosis and HIV-AIDS) receive more attention, thereby severely limiting eradication efforts of snail-transmitted diseases.

The development of a vaccine would be a significant advance towards reducing the burden of schistosomiasis. However, efforts to develop a molecular-based vaccine for this disease have proven to be a challenging task. Furthermore, the intensity of research towards this worthwhile goal appears to have waned, compared to the enthusiasm that characterized this field of research in the 1980s (Pearce, 2003). Until a vaccine becomes available, alternative methods to control schistosomiasis have been suggested (Jordan *et al.*, 1980). One of these involves developing novel tools that will disrupt transmission at the level of the snail host. An earlier suggestion was to use non-susceptible/resistant snails to replace existing susceptible ones in an endemic area as a form of biological control (Hubendick, 1958). Indeed, several studies have used other species of non-schistosome transmitting snails as an integral part of early schistosomiasis control efforts (Jordan, 1972). Whether resistant snails (including those genetically transformed) can be used to displace a compatible population, with all the known complexities of the nature of the snail–parasite association, is debatable. However, to achieve some degree of success with this approach, a better understanding of the host–parasite relationship, especially at the molecular level, is needed. Consequently, the search for genes that are modulated (up/down-regulated) in the snail, leading either to parasite development (susceptible) or destruction (resistant), is now a major focus of research in several laboratories, including ours.

Most studies on this subject have been directed towards *Schistosoma mansoni* and its New World host *Biomphalaria glabrata*, mainly because representative snail isolates and parasite strains displaying different patterns of compatibility were available and easily maintained in the laboratory. Furthermore, there exists a cell-line derived from *B. glabrata,* the embryonic cell line, Bge (Hansen, 1976), which is allowing exploration of the molecular signalling events underlying the host–parasite interaction in an *in vitro* system (Ivanchenko *et al.*, 1999; Humphries *et al.*, 2001). In the case of other schistosomes that are also of significant medical importance, namely *S. hematobium* and *S. japonicum,* comparable studies on how these parasites alter gene expression in their respective snail hosts (*Bulinus* and *Oncomelania* sp.) are not yet available. This apparent disparity in attention is not due to the lack of interest in these parasites and their snail hosts, but rather to the difficulty of maintaining these organisms outside their natural environments.

Aside from a *B. glabrata* field isolate (Salvadore strain or BS-90) that was found to occur naturally as resistant to infection (Paraense and Correa, 1963), most snails of this species that have been studied for parasite resistance have been derived in the laboratory. The development of pedigree snail lines, breeding true for either resistance or susceptibility to parasite infection (Newton, 1955; Richards and Merritt, 1972) has made it possible for us to begin to undertake studies to determine the molecular basis of this interaction.

As early as the 1950s, it was known that the genetic make-up of both the snail host and parasite affects the outcome of infection in the invertebrate host. By exhaustive genetic crossing experiments using snails of different susceptibility phenotypes (resistant/susceptible), exposed either as adults or juveniles to different strains of *S. mansoni,* it was shown that the genetics of snail susceptibility to parasite infection was complex and involved multiple genes (Richards and Shade, 1987). The outcome of the snail–parasite infection was found to fall into four different categories, depending on the age of the snail when infected (reviewed in Lewis *et al.*, 2001). In adult snails, resistance was shown to be a single gene trait that followed a simple Mendelian pattern of inheritance, with resistance being dominant (Richards, 1970). In juvenile snails on the other hand, resistance was shown to be a quantitative trait, as was susceptibility to infection of snails at this age.

About a decade ago, investigators engaged in work towards understanding the genetic basis of schistosome infections in the snail host and mechanisms involved in the snail's internal defence system (IDS) began bringing a molecular dimension to their investigations. Results from these studies, though rudimentary when compared to other parasite–vector systems (e.g. malaria and its mosquito vector), are leading to the identification and characterization of genes that appear to be modulated in the snail host following exposure to the parasite. It is hoped that finding the genes involved in the snail's ability to resist invading parasites will allow genetic transformation of susceptible to resistant snails to become a reality.

In this chapter, we will attempt to summarize progress that has been made in this small, but growing, field of research. Though most of this review will cover what is currently known regarding changes in the *B. glabrata* snail host transcriptome accompanying *S. mansoni* exposure, comparable studies in other platyhelminth–invertebrate host relationships, where available, will be discussed in an effort to bring more attention to this important, but underfunded line of research.

Effect of Parasite Infection on the Snail Host: Towards a Molecular Understanding

It has been suggested that variations in snail host compatibility and parasite infectivity have helped to shape the genomes of both these organisms (Cribb *et al.*, 2001).

Interest in the dynamic interaction between the snail host and parasite, represented by the parasite's ability or inability to survive within a hostile environment and the snail's challenge to infection focused in the early 1990s on searching for genes associated with this intricate and highly specific relationship. Upon infection, there are two major outcomes in the snail host: dynamic equilibrium or destruction of the parasite. Lively (2001) showed that central to the host–parasite interaction are events that shape this relationship: control of host population density by parasites, co-evolution and co-speciation of host and parasite. The evolutionary history of the molluscs can be traced better due to fossil records, while those of the trematodes have been sadly lacking. From this growing interest in the snail host–trematode interaction and co-evolution, progress continues to be made on several fronts (innate immunity, biochemistry and molecular biology) towards the identification of genes that are modulated in the snail host in response to parasite infection.

Innate immunity

Mechanisms of snail innate immunity to trematode infection are known to involve genetic factors associated with both the humoral (haemolymph) and cellular (haemocyte) components of the snail's IDS (Adema and Loker, 1997; Matricon-Gondran and Letocart, 1999a,b,c). Although we are aware of the involvement of these components in the snail's active defence repertoire, scant information exists on the nature of gene products that are involved in this process. Earlier studies have shown that in *B. glabrata* snails, enlargement of the haemapoietic tissue, also known as the amoebocyte-producing organ (APO), can be observed in response to invading trematodes (schistosomes and echinostomes) (Lie and Heyneman, 1976; Lie *et al.*, 1976). Also in response to infection is the recruitment of haemocytes into the snail's open circulatory system. In a resistant snail, these engulf the invading miracidia in a cellular response leading to the destruction of the invading parasite (Adema and Loker, 1997; Bayne *et al.*, 2001; Yoshino *et al.*, 2001).

In *B. glabrata* snails that display susceptibility to a given strain of *S. mansoni*, although the haemocytes adhere to the parasite these cells are not cytotoxic unlike those in resistant snails, the miracidia are not destroyed and survive to develop into sporocysts that give rise to the infective, cercarial form of the parasite (Coustau and Yoshino, 1994). This disparity in haemocyte behaviour has been helpful in designing experiments to identify cellular and biochemical factors that account for the cytotoxicity of these cells towards metazoan parasites. These experiments showed that whereas haemocytes taken from parasite-resistant strains of *B. glabrata* can destroy schistosome sporocysts *in vitro*, cells from susceptible snails are ineffective in similar experiments (Bayne *et al.*, 1980a,b). Additional evidence for the direct involvement of haemocytes in parasite destruction comes from *in vivo* studies showing that adoptive transfer of the APO from a resistant into a susceptible snail renders the latter less susceptible to infection (Sullivan and Spence, 1994, 1999). Similar comparative studies have shown differences in surface proteins that may act as receptors for attachment of parasite surface molecules, ES products or snail haemolymph ligands (Bayne *et al.*, 2001; Yoshino *et al.*, 2001).

The role that the snail haemolymph plays as part of the innate defence arsenal against schistosomes was demonstrated by passive transfer studies where haemolymph from resistant but not susceptible snails was shown to possess a heat-labile factor that specifically affected haemocyte function (Granath and

Yoshino, 1984). Evidence also exists for a variety of soluble factors that may contribute to resistance of the snail to infection, e.g. plasma factors (Bayne *et al.*, 1985), fibrinogen-related proteins (FREPs) (Adema and Loker, 1997) and cytokine-like components (Granath *et al.*, 1994).

Mechanism of cellular signalling of haemocytes, triggered by the adherence of parasite and/or snail ligands is a growing area of study (refer to the section on 'Immunobiology and Host–Parasite Interaction' in this book by T.P. Yoshino). The discovery of genes encoding integrin and protein kinase C, both molecules with known roles in cell-adherence and signalling in *B. glabrata* haemocytes, have helped to fuel this growing interest (Lardans *et al.*, 1998; Davids *et al.*, 1999). Although the nature of haemocyte receptors involved in recognition and signalling (activated in response to miracidial invasion) are not yet known, surface-labelling experiments with biotin identified a 66-kDa band that was upregulated on the surface of susceptible snail haemocytes. A high molecular weight surface protein (210 kDa) that was found in both resistant and susceptible snails was found to be absent or reduced in non-adherent haemocytes (Coustau and Yoshino, 1994).

Soluble ES products released from the parasite may provide the trigger that changes the gene expression profile of haemocytes (Lodes *et al.*, 1991; Loker *et al.*, 1992; Coustau *et al.*, 2003). These changes, depending on the nature of the parasite ligand, can either stimulate haemocyte activity or cause their suppression. The nature of soluble trematode-related factors that up- or downregulate haemocyte gene activity are far from clear. Lately however, in light of recent findings that a group of diverse FREPs, are expressed from haemocytes, more attention is being drawn towards the involvement of carbohydrate-binding proteins in snail defence. Lectins, either on the surface of haemocytes or as haemolymph components (such as agglutinins), are known to play a significant role in the snail's IDS (Loker, 1994; Bayne *et al.*, 2001). Aside from FREPs that may be capable of precipitating soluble parasite products, other snail lectins may have a direct or indirect role in rendering haemocytes cytotoxic. Recently, the gene encoding another haemocyte lectin (selectin) was identified. The expression of this gene has also been detected in the snail embryonic cell line, Bge (Dulcermortier *et al.*, 1999; Table 11.1). It is hoped that because of their similarity in behaviour towards miracidia (attachment) Bge cells will become a useful model system for studying haemocyte function (Yoshino *et al.*, 2001).

Biochemical involvement

Certain enzymes, especially those with a role in generating reactive oxygen species (ROS), are believed to play a role in haemocyte-mediated killing of parasitic trematodes (Bayne *et al.*, 2001). In particular, hydrogen peroxide (H_2O_2), the end product of reactive oxygen (O_2^-) resulting from activation of an NADPH membrane-bound complex, was found to be highly cytotoxic against sporocysts, as was another ROS, nitric oxide (NO), which is generated by an inducible nitric oxide synthase (iNOS). It has been hypothesized that differences between enzymes in biochemical pathways producing these two sporocyst-killing reagents may offer an insight into whether or not snails are resistant or susceptible to parasite infection. To date, however, no modulation of genes coding for either NADPH oxidase or NOS has been identified in snails displaying different susceptibility phenotypes. Likewise, the gene encoding catalase, the enzyme which converts H_2O_2 to water (H_2O) and oxygen (O_2), that hypothetically could be upregulated in susceptible snails by the parasite to counteract the parasite-killing effect of H_2O_2, has not yet been identified in haemocytes. Catalase activity with high substrate affinity has been demonstrated in *B. alexandrina* and *B. truncatus* snails (Nabih *et al.*, 1968; Nabih and El Ansary, 1993) and with the recent identification from Bge cells of a transcript that appears to be the snail orthologue of this enzyme, we may now be in the position to evaluate the modulation of this gene in parasite infection (N. Raghavan and M. Knight, Biomedical Research Institute, Rockville, MD, USA, 2002, unpublished).

Other enzymes that have been implicated in playing a role in the phagocytotic capacity

Table 11.1. EST transcripts identified from the *B. glabrata* embryonic cell line Bge.

Gene	Number or transcripts	% Total
No hits	1070	69.98
Ribosomal protein	114	7.46
Cytochrome *c* oxidase	87	5.69
ATP synthase	69	4.51
NADH dehydrogenase	39	2.55
Cytochrome *b*	36	2.35
Lymphocyte antigen 6 complex	18	1.18
Soma ferritin	14	0.92
Guanine nucleotide binding protein	10	0.65
Hypothetical protein	8	0.52
Ubiquitin	7	0.46
Valosin-containing protein	7	0.46
Tubulin	5	0.33
NICE-4	4	0.32
Cytochrome p450	3	0.20
Other low copy genes (see below)	**32**	**2.09**
FREP/selectin	1	0.06
Haemolysin	1	0.06
Actin	1	0.06
Pyruvate kinase	1	0.06
Total	**1529**	

GenBank accession numbers for the representative from each group are given in parenthesis. Ribosomal protein (CO501265), cytochrome *c* oxidase (CO501267), ATP synthase (CO501268), NADH dehydrogenase (CO501269), cytochrome *b* (CO501270), lymphoctye antigen 6 complex (CO501271), soma ferritin (CO501272), guanine nucleotide binding protein (CO501273), hypothetical protein (CO501274), ubiquitin (CO501266), valosin-containing protein (CO501276), tubulin (CO501277), NICE-4 (CO501278), cytochrome p450 (CO501279), FREP/selectin (CO501280), haemolysin (CO501281), actin (CO501282), pyruvate kinase (CO501283).

of *B. glabrata* haemocytes are cyclase adenyl synthase and its respective phosphodiesterases (Bezerra *et al.*, 1999). From studying the role of cyclic nucleotides in the phagocytotic capacity of *B. glabrata* haemocytes, it was observed that cyclic adenylate monophosphate (cAMP) and not cyclic guanylate monophosphate (cGMP), regulates snail haemocyte function. These results, though implicating a direct role of enzymes involved in this biochemical pathway in haemocyte function, have not yet led to the demonstration that the genes encoding these enzymes are regulated in these cells in response to parasite infection.

In contrast to parasitic helminths, very few proteases of the snail hosts have been characterized at the functional and molecular levels (Sajid and McKerrow, 2002). Proteolytic enzymes have been implicated in playing a role in parasite destruction or survival in the snail host. Elevation of lysosomal enzyme activity in response to schistosome or snail infection has also been shown (Cheng and Garrabrant, 1977; Kassim and Richards, 1978a,b). Recently, serine protease and phenol oxidase activities, the latter with a known role in melanization leading to killing of invertebrate parasites, were detected in *B. glabrata* haemocytes (Bahgat *et al.*, 2002). *B. glabrata* protease inhibitor, α-macroglobulin from haemolymph, has been shown to inhibit cysteine protease activity of schistosome sporocysts (Fryer *et al.*, 1996). The role that proteolytic enzymes play in the haemolymph-mediated destruction of metazoan parasites is far from clear. However, since the haemolymph rather than haemocytes have been shown to play a more significant role in the *B. glabrata*/echinostome resistance, this host–parasite model may be useful in the molecular characterization of proteolytic-related trematode-killing factors (Adema and Loker, 1997; Ataev and Coustau, 1999). In our laboratory, gelatin zymology showed the presence of this cysteine protease in the hepatopancreas and ovotestis. Although this protease activity was detected in both resistant and susceptible snails, a higher basal level was found in extracts prepared from resistant compared to susceptible snails (M. Knight, Biomedical Research Institute, Rockville, MD, USA, 2004, unpublished).

In a *B. glabrata* resistant snail line (LAC) that was derived from a previously susceptible snail stock NMRI (Cooper *et al.*, 1994), histological studies indicated that they possessed abnormal albumen glands compared to their susceptible counterparts. Concurrent with these results was the observation that the LAC snail displayed a reduced fecundity, manifested by the presence of abnormal egg clutches. There is as yet, no evidence that the reduced reproductive capacity seen in these snails is directly related to their parasite refractory phenotype.

Differential screening of albumen gland cDNA libraries made from the LAC and NMRI snails, using antibodies made against soluble albumen gland proteins from the resistant and susceptible phenotypes, identified several clones that were uniquely expressed in the resistant snail. Nucleotide sequence analysis of these clones indicated that although most showed no similarity to sequences in the public databases, one clone (BgTPx) contained an open reading frame (ORF) with significant sequence similarity to a known class of human thiol-specific antioxidant enzyme, thioredoxin peroxidase 1 (TPx 1) (also known as peroxiredoxin 2 [Prx 2]; C. Cousin, Washington, DC, 2001, unpublished). Thioredoxin peroxidases are a family of antioxidant enzymes that also control cytokine-induced peroxide levels. Multiple sequence alignment of the deduced amino acid sequence of the snail enzyme (BgTPx) to that of the parasites *S. mansoni*, *S. japonicum* and the vertebrate hosts, human, mouse and rat is shown in Fig. 11.1. Interestingly, the thioredoxin peroxidases seem to be highly conserved among the snail (*B. glabrata)*, human (*H. sapiens*) and the parasites (*S. mansoni* and *S. japonicum*). The alignment from Fig. 11.1 shows the highly conserved peroxidatic cysteine indicated by a '**C**' (generally near residue 50 in the human Prx 2) and the resolving cysteine (near residue 170 in the human Prx 2), indicating these sequences to be of the 'typical 2-Cys peroxiredoxins' (Wood *et al.*, 2003). In addition, they show the invariant active site residues (*) and the conserved domains required for decamer formation. Likewise, an unrooted, phylogenetic analysis of the snail BgTPx sequence revealed a closer evolutionary relationship between the vertebrate orthologues and the snail (Fig. 11.2) in contrast to the parasite. One obvious interpretation of this finding is that this may represent an example of molecular co-evolution at work for an enzyme that may play a protective role against pathogens in both the definitive and intermediate hosts of this parasite.

Characterization of the enzymatic activity of the parasite and snail recombinant TPx are in progress, and the recombinant snail enzyme BgTPx has been found to be enzymatically active in an *in vitro* assay (Sayed and Williams, 2004). It remains to be seen whether the results obtained showing the transcript corresponding to BgTPx to be upregulated in parasite-exposed snails, compared to uninfected controls, correlates with enhanced thioredoxin peroxidase activity in infected snails.

Parasite-induced changes in gene expression

Efforts to identify genes that are regulated (up or down) in the intermediate snail host in response to trematode infection have involved the utilization of several gene discovery techniques. Of these, the most widely applied to date has been the differential display-reverse transcription-polymerase chain reaction (DD-RT-PCR) technique which can conveniently decipher, from small amounts of RNA, changes in gene expression profiles in response to any given factor (Liang and Pardee, 1992). With this technique, Lockyer *et al.* (2000) showed in adult *B. glabrata* snails that the transcript encoding cytochrome p450 was upregulated in a partially resistant snail line in response to *S. mansoni* infection. In other studies comparing gene transcription profiles in resistant and susceptible, infected vs uninfected snails, genes encoding enzymes of mobile elements

```
               1                                                          60
B. glabrata    NSARAGGQVYPQETKRTIGHTIQWSQAVISKPAPDWNGTAVI-KGEFKDIKLSDFKGKYL
S. mansoni     MLLQGFIKGALRYNRSSVSNLCRHYAAHVQRPAPDFCGTAVV-DGQFKEIKLRDFAGKYL
S. japonicum   ------------------------MVLIPNKPAPEFHGCAVI-DGDFKEINLKDYSGKYV
H. sapiens     ---------------------MASGNARIGKPAPDFKATAVV-DGAFKEVKLSDYKGKYV
M. musculus    ---------------------MSSGNAKIGYPAPNFKATGVMPDGQFKDTSLSEYKGKYV
R. norvegicus  ---------------------MSSGNAKTGHPAPNFKATAVMPDGQFKNISLSDXKGKYV

                    Region I  Cp-loop                        Region II
               61    *   *| C                                               120
B. glabrata    VFFFYPLDFTFVCPTEIIAFSDRIDEFQKINTKVVGCSVDSQFTHLAWTNVPRSQGGLGK
S. mansoni     VLFFYPLDFTFVCPTELTAFSDRIDEFKNEGVEVVGVSTDSHFSHLAWINTPRKEGGLGG
S. japonicum   VLFFYPADFTFVCPTEIIAFSDEVDQFKSRNCQVIACSTDSKYSHLAWTKQDRKSGGLGD
H. sapiens     VLFFYPLDFTFVCPTEIIAFTTVKRTSAKLGCEVLGVSVDSQFTHLAWINTPRKEGGLGP
M. musculus    VFFFYPLDFTFVCPTEIIAFRDRADEFKKLTCQVIGVSVDSHFCHLPWINTPKKQGGLGP
R. norvegicus  VFFFYPLDFTFVCPTEIIAFSDRAEEFKKLNCQVIGASVNSHFCHLAXINTPKKQGGSGP
                    PXDFTFVCPTE                       XDSXFXHXXWXXTX
                                                        T Y       S
                                                                  V

               121                         *                                180
B. glabrata    INYPLLSDITHEISKAYGVYLQDLGHSLRGLFIIDPKGTLRQITMNDLPVGRSVDETLRL
S. mansoni     LRYPLLADYQKQVTRDYGVLHEELGVALRGLFIISADGIIRQITINDLPAGRSVDEVLRL
S. japonicum   MRIPLLADPTKSIARAYGVLDEEEGNAFRGLFIIDPKGILRQITVNDKPVGRSVDETLRL
H. sapiens     LNIPLLADVTRRLSEDYGVLKNDEGIAYRGLFIIDGKGVLRQITVNDLPVGRSVDEALRL
M. musculus    MNIPLISDPKRTIAQDYGVLKADEGISFRGLFIIDDKGILQQITINDLPVAVSVDEIIRL
R. norvegicus  MSIPLVSDPKPTIAQDDGVLKADEGFSFRGLFLIDEKGSLRQMTINDLPVGRSVDEILRL

               181           C                             224
B. glabrata    VQAFQYTDKHGEVCPAGWKPGSATIIPDPKKSKEYFKQQSNDEL
S. mansoni     VRAFQYTDKYGEVCPADWQPKGPTIKPDLKKYKEYFHKVN----
S. japonicum   LDAFQFVEKYGEVCPVNWKRGQHGIKVNH---------------
H. sapiens     VQAFQYTDEHGEVCPAAWKPGRDTIKPNVDDSKEYFSKHN----
M. musculus    VQAFQFTDKHGEVCPAGWKPSSDTIKPDVNKSKEYFSKQK----
R. norvegicus  VQAFQFTDKHGEVCPAGWKPGSDTIKPDVNKSKEYFSKQK----
```

Fig. 11.1. Multiple sequence alignment of the deduced amino acid sequence of thioredoxin peroxidase 1 (Peroxiredoxin 2, TSA, natural killer cell enhancing factor B) from the snail host *B. glabrata* clone BgTPx (GenBank Accession # AAK26236), the parasites *S. mansoni* (GenBank Accession # AAG15506) and *S. japonicum* (GenBank Accession # BAD01572), and the vertebrate hosts *Homo sapiens* (Genbank Accession # CAA80269), *Mus musculus* (GenBank Accession # XP_147236) and *Rattus norvegicus* (GenBank Accession # XP_212921). The computer program ClustalW (version 1.82) was used to generate this alignment. The numbering system used in the alignment represents the amino acid position in *B. glabrata*. Thioredoxin peroxidases are a family of antioxidant enzymes that also control cytokine-induced peroxide levels and seem to be highly conserved among snail (*B. glabrata*), human (*H. sapiens*) and flatworm parasites (*S. mansoni* and *S. japonicum*). '**C**' represents the two conserved, redox-active cysteines. The conserved active site residues are indicated by a '*'. The upper case bold lettering shown below the alignment (residues 61–120) represents the consensus sequence for all classes of peroxiredoxins. Also identified in this alignment are conserved regions I (loop-helix active-site motif) and II, which together play a role in decamer formation. One obvious interpretation of this finding (along with Fig. 11.2) is that this may represent an example of molecular co-evolution at work for an enzyme that may play a protective role against pathogens in both the definitive and intermediate hosts of this parasite.

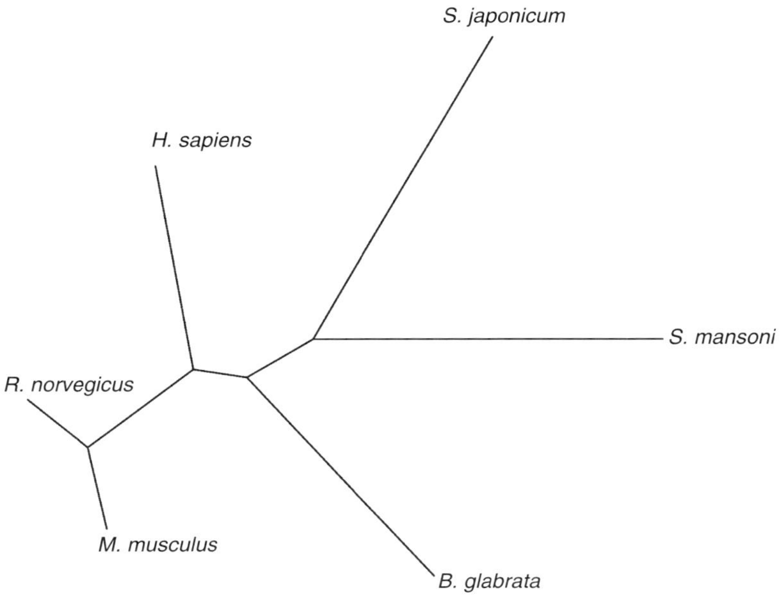

Fig. 11.2. Neighbour-joining phylogenetic tree of the predicted peroxiredoxins shown in Fig. 11.1. This unrooted tree was generated using the computer program PHYLIP (Felsenstein, 1989) version 3.5c. GenBank accession #s of the sequences used in this analysis are shown in Fig. 11.1 legend.

were found to be downregulated in exposed resistant snails (Ongele, 2000; Miller *et al.*, 2001). In *B. glabrata* snails that display a transient age-dependent expression of the resistance phenotype (strain 93375), an upregulation of the gene encoding a zinc-dependent metalloprotease was observed in adult-resistant snails by DD-RT-PCR, compared to susceptible juvenile snails (Ongele, 2000). To detect the differences between the resistant and susceptible snail phenotypes, especially with regard to the haemocytes, Schneider and Zelck (2001) used DD-RT-PCR on RNA extracted from isolated haemocytes. A number of differentially expressed fragments from resistant and susceptible snails were detected by this technique and were further confirmed using single-strand conformation polymorphism. The most interesting phenotype-specific fragments detected showed sequence homologies to an adhesion molecule, defensin, serine/threonine kinases, peroxidases and glycosidases.

Other RNA fingerprinting tools are being employed to examine the parasite-mediated alteration of gene expression in the snail host. For example, using RNA arbitrarily primed PCR (RAP-PCR), an upregulation of several novel transcripts, including that encoding cytomyoglobin, has recently been discovered (Lockyer *et al.*, 2004). A major factor hampering progress in all these gene discovery studies is the paucity of sequences that have been described from this snail host and other molluscs. This reality makes it difficult to predict the identity of gene sequences that are generated from this snail host when there are no related sequences with which to compare. This factor notwithstanding, other approaches towards gene discovery in the snail are underway. Because of the known bias of the DD-RT-PCR towards abundant transcripts, others have chosen to use another RNA profiling tool, ORF expressed sequence tags (ORESTES), which has been shown to target internal regions of rarer transcripts (Dias Neto *et al.*, 2000). Also, with the same technique, Lockyer *et al.* (2004) showed the upregulation of several transcripts upon infection of resistant snails. By comparing transcriptomes generated by ORESTES from normal Bge cells to those co-cultured with miracidia, several transcripts were found to be upregulated in these cells in response to parasite exposure (ethylene receptor, presynaptic cytomatrix protein, formate dehydrogenase, casein kinase, focal adhesion kinase, tyrosine kinase, catalase, etc.; N. Raghavan and M. Knight, Biomedical Research Institute, Rockville, MD, USA, 2002, unpublished). Additionally, conventional EST analysis of these cells revealed an interesting repertoire of enzymes, including housekeeping and defence-related genes (Table 11.1; N. Raghavan and M. Knight, Biomedical Research Institute, Rockville, MD, USA, 2002, unpublished). Since pyruvate kinase deficiency in mice has been shown to protect this host against malaria, it will be interesting to determine whether or not differences in the level of this enzyme have an effect on the outcome of the schistosome/snail host infection (Min-Oo *et al.*, 2003).

Differential gene expression in the snail that accompanies parasite exposure has also been examined by subtractive hybridization studies. In an earlier attempt to isolate transcripts uniquely expressed in parasite-exposed resistant snails, Miller *et al.* (1996) identified two transcripts that were highly expressed in parasite-exposed resistant snails compared to unexposed controls. Since both transcripts showed no significant resemblance to any sequences in the public databases, further characterization of these genes would depend on the use of other technologies such as RNA interference (RNAi) that can be used to disrupt specific gene activity in an effort to determine their biological function. Recent reports showing that this method can be used to interfere with gene activity in the parasite are encouraging (Boyle *et al.*, 2003a; Skelly *et al.*, 2003). It is possible that in the not too distant future, the same approach may prove helpful in unravelling some of the complex molecular events underlying the snail–parasite relationship.

Suppression-subtractive hybridization (SSH) is another hybridization strategy that is being pursued to examine differential gene expression following parasite infection in the snail host. At least two major laboratories are currently engaged in using this approach towards the isolation of genes that are specifically expressed in either resistant or susceptible snails in response to infection. Initial results from these studies showed differential expression of several transcripts in a resistant (BS-90) and susceptible snail (M-line) in response to *S. mansoni* infection (Nowak *et al.*, 2004). It is likely that in the coming year, we will see a dramatic increase in the number of snail genes that are identified using this approach.

As was done for parasite transcripts (Hoffman and Dunne, 2003), another useful method that is currently being applied to identify genes that are differentially expressed in the snail host in relation to changing conditions induced by parasite infection is to examine modulation of transcripts by hybridization to snail microarrays. The creation of snail gene microarrays to examine changes in the gene expression profiles is in a very early stage of development (A.E. Lockyer, London, 2004, personal communication). A major prerequisite for the development of gene arrays is the generation of expressed sequence tags (ESTs) from a variety of snail tissues. In this regard, 8207 *B. glabrata* ESTs are currently available (October 2006) from GenBank. A large number of sequences have been generated from haemocytes isolated from resistant snails with and without parasite exposure (Fig. 11.3; Raghavan *et al.*, 2003). This method of gene discovery has been extremely useful in the identification of transcripts that may be modulated upon exposure to parasite.

An intriguing result from the generation of haemocyte ESTs from parasite exposed and unexposed snails was the identification of abundant sequences that correspond to enzymes of mobile elements, particularly, reverse transcriptase (RT). Differences in basal levels of active RT were found to exist between resistant and susceptible snails, with resistant snails showing higher levels of enzyme activity compared to susceptible snails. The role of mobile genetic elements (MGE) in differential gene regulation (if any) in this snail host remains to be determined. Work is underway to sequence bacterial artificial chromosome (BAC) clones corresponding to several *B. glabrata* retrotransposable elements, including the *B. glabrata* reverse transcriptase (BgRT) and a LINE-1-related element (Knight *et al.*, 1992). Characterization of these transposable elements may lead to the development of delivery vectors for genetic transformation studies in either the parasite or the snail host. In light of a recent report showing that L1 retrotransposons may play a role in transcription regulation, it is likely that these related mobile sequences in

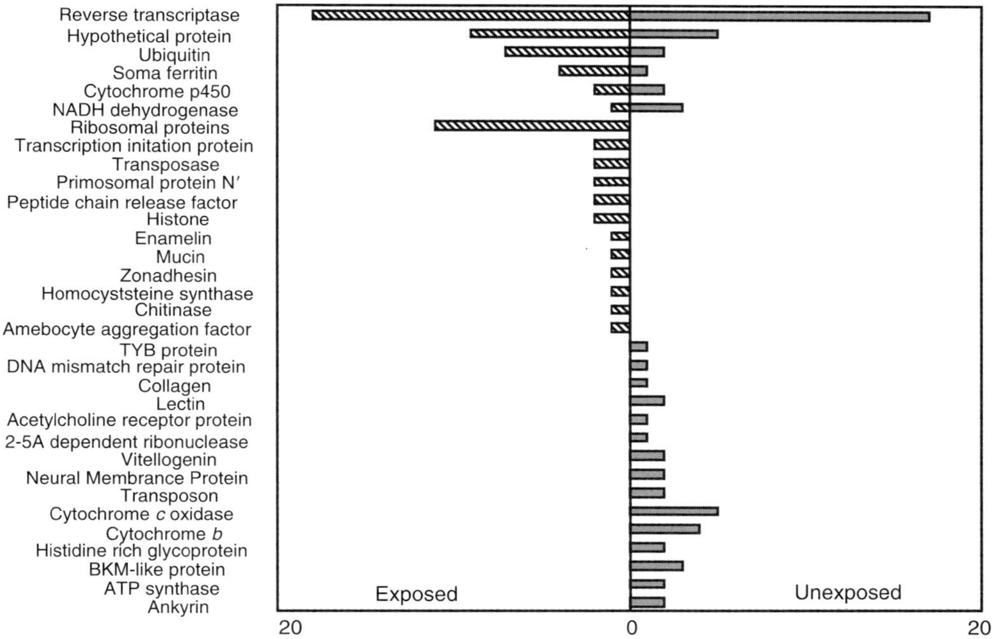

Fig. 11.3. Comparison of the differential distribution of haemocyte EST transcripts identified from 5 h exposed and unexposed haemocyte cDNA libraries of the *S. mansoni* resistant strain (BS-90) of *B. glabrata*. The scale on the *X*-axis represents the actual number of specific transcripts identified in the current cohort. (Reproduced from Raghavan *et al.*, 2003, *Molecular and Biochemical Parasitology* 126, 181–191.)

the snail may also be involved in a similar biological role (Han *et al.*, 2004).

Parasite-induced changes in snail neurobiology

Previous results have shown that parasite infection of the snail host affects the neuroendocrine system of the snail in a way that results in reduced reproduction of these snails: a phenomenon referred to as 'parasitic castration' (Crews and Yoshino 1989; Sorensen and Minchella, 2001). The *Lymnaea stagnalis–Trichobilharzia ocellata* host–parasite combination has been the model system adopted for most of these studies (de Jong-Brink *et al.*, 2001). However, comparable studies as to how parasite infection affects the snail's central nervous system (CNS) in releasing neuropeptides and bioamines that modulate snail reproduction and the internal defence system of *B. glabrata* are gaining attention. From these studies, the existence of several bioamines were reported from the haemolymph and extracts of various snail tissues (albumen gland, ovotestis and brain) (Manger *et al.*, 1996; Santhanagopalan and Yoshino, 2000; Boyle and Yoshino, 2003). For example, serotonin (5-hydroxytryptamine, 5-HT) has been reported to stimulate snail reproduction (egg-laying) and levels of this bioamine and dopamine were found to be depressed in *B. glabrata* snails following parasite exposure (Manger *et al.*, 1996). Furthermore, a reduction in reproduction was seen after treatment of snails with low doses of the serotonin receptor antagonist, methiothepin, directly implicating a role of this neurotransmitter in *B. glabrata* reproductive behaviour (Muschamp and Fong, 200´). The uptake of serotonin via a surface-exposed serotonin transporter (SERT) on the developing intramolluscan stage (sporocysts) of the parasite has also been shown (Boyle *et al.*, 2003b). The gene encoding this transporter (SERT) has recently been cloned from adult *S. mansoni* worms (O.V. Mortensen and M.S. Sonders, Albert Einstein College of Medicine, New York, USA, 2004, personal communication), thus making it possible to determine the stage and sex-specificity of its expression in relation to parasite development in both the vertebrate and snail host. Furthermore, investigating transcriptional regulation of this gene in the developing sporocyst may help in studying how the level of expression of the SERT-encoding gene, by the parasite, alters serotonin levels in the infected snail host. Also, cloning of the gene encoding the snail serotonin receptor may be important towards finding an explanation for the depletion of serotonin in infected snails. The gene encoding this receptor from *L. stagnalis* has been cloned (Sugamori *et al.*, 1993), perhaps offering a chance at manipulating its function gene by modern technology such as RNAi to directly study the role of serotonin in relation to snail host reproduction and parasite development.

Differential gene regulation by parasite infection has also been shown for several snail neuropeptides that play a role in snail defence (de Jong-Brink *et al.*, 2001). Of all the genes so far shown as being either down- (molluscan defence molecule, MDM) or upregulated (granularin, FMRFamide) in infected snails, all have been identified from studies conducted on the *L. stagnalis/T. ocellata* snail host–parasite model system. However, from recent gene discovery studies, several ESTs identified from haemocytes and brain cDNA libraries of *B. glabrata* have shown significant sequence identities to invertebrate neuropeptides (Knight *et al.*, 1998; Raghavan *et al.*, 2003). The modulation of these neuropeptides in this snail host by parasite infection remains to be determined.

Concluding Remarks

It is clear that significant progress has been made towards the identification of molecular determinants whose modulation may affect the outcome of this ancient and complex snail–metazoan parasite relationship. Continued interest in this line of research is, however, essential if discoveries made from the above studies are to be translated into novel methods to reduce the spread of snail-transmitted diseases. In this regard, at least for human schistosomiasis, it is possible that once the genome sequence of the parasite (*S. mansoni*) becomes available (N.M. El-Sayed and P. LoVerde, SUNY at Buffalo, New York, USA, 2004, personal communication), comparative genome analysis of the parasite to the already available human genome sequence will help in the identification of new genes that can be manipulated for controlling this disease. Of the two natural hosts that are central to the transmission of schistosomiasis it has been suggested that the snail, rather than the human, has played the larger role in influencing parasite genetic modification. The relative ease at which any given parasite strain, regardless of geographical origin, can successfully infect the mammalian host, but not all snail stocks, would argue for a more dynamic genetic interplay with the intermediate snail host, leading to a wide range of compatibilities over time (reviewed by Lewis *et al.*, 2001). Thus, a snail genome project that is currently underway might be the best approach to dete mine whether or not molecular co-evolution has allowed these two organisms (parasite and snail) to adapt to each other. While the large genome size of the snail (930 Mb; Gregory, 2003) might be a daunting factor in considering a whole genome-sequencing project, it is not insurmountable. Such an approach might have long-range benefits for controlling schistosomiasis, and there is now a move towards fulfilling this objective. The status and progress of the snail genome project can be obtained from the Genome Sequencing Center at Washington University at the URL: http://genome.wustl.edu/ISAgenome.cgi?GENOME=Biomphalaria%20glabrata.

Acknowledgements

We thank Dr Fred Lewis, Dr Peter FitzGerald and Andre Miller for their help in discussing and reviewing this manuscript. We also thank Dr Charles Richards for his years of dedication to working with the *B. glabrata* snail host that makes our research possible. This work was supported in part by NIH-NIAID Grant AI-27777, UNDP/World Bank/WHO Special Programme for Research and Training in Tropical Diseases (TDR) and The Sandler Foundation.

References

Adema, C.M. and Loker, E.S. (1997) Specificity and immunobiology of larval digenean-snail associations. In: Fried, B. and Graczyk, T.K. (ed.) *Advances in Tremtode Biology*. CRC Press, Boca Raton, New York, pp. 229–263.

Ataev, G.L. and Coustau, C. (1999) Cellular response to *Echinostoma caproni* infection in *Biomphalaria glabrata* strains selected for susceptibility/resistance. *Developmental and Comparative Immunology* 23, 187–198.

Bahgat, M., Doenhoff, M., Kirschfink, M. and Ruppel, A. (2002) Serine protease and phenoloxidase activities in hemocytes of *Biomphalaria glabrata* snails with varying susceptibility to infection with the parasite *Schistosoma mansoni*. *Parasitology Research* 88, 489–494.

Bayne, C.J., Buckley, P.M. and DeWan, P.C. (1980a) Macrophage-like hemocytes of resistant *Biomphalaria glabrata* are cytotoxic for sporocysts of *Schistosoma mansoni in vitro*. *Journal of Parasitology* 66, 413–419.

Bayne, C.J., Buckley, P.M. and DeWan, P.C. (1980b) *Schistosoma mansoni*: cytotoxicity of hemocytes from susceptible snail hosts for sporocysts in plasma from resistant *Biomphalaria glabrata*. *Experimental Parasitology* 50, 409–416.

Bayne, C.J., Boswell, C.A., Loker, E.S. and Yui, M.A. (1985) Plasma components which mediate cellular defences in the gastropod mollusc *Biomphalaria glabrata*. *Developmental and Comparative Immunology* 9, 523–530.

Bayne, C.J., Hahn, U.K. and Bender, R.C. (2001) Mechanisms of molluscan host resistance and of parasite strategies for survival. *Parasitology* 123, S159–S167.

Bezerra, F.S., Coelho, P.M., Chaves, M.M., Martins, R.L. and Nogueira-Machado, J.A. (1999) Comparative study on the effect of cyclic nucleotides related to the function of *Biomphalaria glabrata* hemocytes and human granulocytes. *Acta Tropica* 72, 275–280.

Boyle, J.P. and Yoshino, T.P. (2003) Gene manipulation in parasitic helminths. *International Journal of Parasitology* 33, 259–268.

Boyle, J.P., Wu, X.J., Shoemaker, C.B. and Yoshino, T.P. (2003a) Using RNA interference to manipulate endogenous gene expression in *Schistosoma mansoni* sporocysts. *Molecular and Biochemical Parasitology* 128, 205–215.

Boyle, J.P., Hillyer, J.F. and Yoshino, T.P. (2003b) Pharmacological and autoradiographical characterization of serotonin transporter-like activity in sporocysts of the human blood fluke, *Schistosoma mansoni*. *Journal of Comparative Physiology A: Sensory, Neural, and Behavioral Physiology* 189, 631–641.

Cheng, T.C. and Garrabrant, T.A. (1977) Acid phosphatase in granulocytic capsules formed in strains of *Biomphalaria glabrata* totally and partially resistant to *Schistosoma mansoni*. *International Journal of Parasitology* 7, 467–472.

Chitsulo, L., Engels, D., Montresor, A. and Savioli, L. (2000) The global status of schistosomiasis and its control. *Acta Tropica* 77, 41–51.

Cooper, L.A., Richards, C.S., Lewis, F.A. and Minchella, D.J. (1994) *Schistosoma mansoni*: relationship between low fecundity and reduced susceptibility to parasite infection in the snail *Biomphalaria glabrata*. *Experimental Parasitology* 79, 21–28.

Coustau, C. and Yoshino, T.P. (1994) Surface membrane polypeptides associated with hemocytes from *Schistosoma mansoni*-susceptible and -resistant strains of *Biomphalaria glabrata* (Gastropoda). *Journal of Invertebrate Pathology* 63, 82–89.

Coustau, C., Mitta, G., Dissous, C., Guillou, F., Galinier, R., Allienne, J.F. and Modat, S. (2003) *Schistosoma mansoni* and *Echinostoma caproni* excretory–secretory products differentially affect gene expression in *Biomphalaria glabrata* embryonic cells. *Parasitology* 127, 533–542.

Crews, A.E. and Yoshino, T.P. (1989) *Schistosoma mansoni*: effect of infection on reproduction and gonadal growth in *Biomphalaria glabrata*. *Experimental Parasitology* 68, 326–334.

Cribb, T.H., Bray, R.A. and Littlewood, D.T. (2001) The nature and evolution of the association among digeneans, molluscs and fishes. *International Journal of Parasitology* 31, 997–1011.

Davids, B.J., Wu, X.J. and Yoshino, T.P. (1999) Cloning of a beta integrin subunit cDNA from an embryonic cell line derived from the freshwater mollusc, *Biomphalaria glabrata*. *Gene* 228, 213–223.

de Jong-Brink, M., Bergamin-Sassen, M. and Solis Soto, M. (2001) Multiple strategies of schistosomes to meet their requirements in the intermediate snail host. *Parasitology* 123(Suppl.), S129–S141.

Dias Neto, E., Correa, R.G., Verjovski-Almeida, S., Briones, M.R., Nagai, M.A., Da Silva, W. Jr, Zago, M.A., Bordin, S., Costa, F.F., Goldman, G.H., Carvalho, A.F., Matsukuma, A., Baia, G.S., Simpson, D.H., Brunstein, A., De Oliveira, P.S., Bucher, P., Jongeneel, C.V., O'Hare, M.J., Soares, F., Brentani, R.R., Reis, L.F., De Souza, S.J. and Simpson, A.J. (2000) Shotgun sequencing of the human transcriptome with ORF expressed sequence tags. *Proceedings of the National Academy of Sciences USA* 97, 3491–3496.

Doenhoff, M.J., Kimani, G. and Cioli, D. (2000) Praziquantel and the control of schistosomiasis. *Parasitology Today* 16, 364–366.

Duclermortier, P., Lardans, V., Serra, E., Trottein, F. and Dissous, C. (1999) *Biomphalaria glabrata* embryonic cells express a protein with a domain homologous to the lectin domain of mammalian selectins. *Parasitology Research* 85, 481–486.

Felsenstein, J. (1989) PHYLIP – Phylogeny Inference Package (Version 3.2). *Cladistics* 5, 164–166.

Fryer, S.E., Bender, R.C. and Bayne, C.J. (1996) Inhibition of cysteine proteinase from *Schistosoma mansoni* larvae by alpha-macroglobulin from the plasma of *Biomphalaria glabrata*. *Journal of Parasitology* 82, 343–347.

Granath, W.O. Jr and Yoshino, T.P. (1984) *Schistosoma mansoni*: passive transfer of resistance by serum in the vector snail, *Biomphalaria glabrata*. *Experimental Parasitology* 58, 188–193.

Granath, W.O. Jr, Connors, V.A. and Tarleton, R.L. (1994) Interleukin 1 activity in haemolymph from strains of the snail *Biomphalaria glabrata* varying in susceptibility to the human blood fluke, *Schistosoma mansoni*: presence, differential expression, and biological function. *Cytokine* 6, 21–27.

Gregory, T.R. (2003) Genome size estimates for two important freshwater molluscs, the zebra mussel (*Dreissena polymorpha*) and the schistosomiasis vector snail (*Biomphalaria glabrata*). *Genome* 46, 841–844.

Han, J.S., Szak, S.T. and Boeke, J.D. (2004) Transcriptional disruption by the L1 retrotransposon and implications for mammalian transcriptomes. *Nature* 429, 268–274.

Hansen, E.L. (1976) A cell line from embryos of *Biomphalaria glabrata* (Pulmonata): establishment and characteristics. In: Maramorsch, K. (ed.) *Invertebrate Tissue Culture*. Academic Press, New York, pp. 75–99.

Hoffmann, K.F. and Dunne, D.W. (2003) Characterization of the schistosoma transcriptome opens up the world of helminth genomics. *Genome Biology* 5, 203.

Hubendick, B. (1958) A possible method of schistosome-vector control by competition between resistant and susceptible strains. *Bulletin of the World Health Organization* 18, 1113–1116.

Humphries, J.E., Elizondo, L. and Yoshino, T.P. (2001) Protein kinase C regulation of cell spreading in the molluscan *Biomphalaria glabrata* embryonic (Bge) cell line. *Biochimica et Biophysica Acta* 1540, 243–252.

Ivanchenko, M.G., Lerner, J.P., McCormick, R.S., Toumadje, A., Allen, B., Fischer, K., Hedstrom, O., Helmrich, A., Barnes, D.W. and Bayne, C.J. (1999) Continuous *in vitro* propagation and differentiation of cultures of the intramolluscan stages of the human parasite *Schistosoma mansoni*. *Proceedings of the National Academy of Sciences USA* 96, 4965–4970.

Jordan, P. (1972) Epidemiology and control of schistosomiasis. *Brazilian Medical Bulletin* 28, 55–59.

Jordan, P., Christie, J.D. and Unrau, G.O. (1980) Schistosomiasis transmission with particular reference to possible ecological and biological methods of control. *Acta Tropica* 37, 95–135.

Kassim, O.O. and Richards, C.S. (1978a) *Schistosoma mansoni*: lysozyme activity in *Biomphalaria glabrata* during infection with two strains. *Experimental Parasitology* 46, 213–217.

Kassim, O.O. and Richards, C.S. (1978b) *Biomphalaria glabrata*: lysozyme activities in the hemolymph, digestive gland, and headfoot of the intermediate host of *Schistosoma mansoni*. *Experimental Parasitology* 46, 218–224.

Knight, M., Miller, A., Raghavan, N., Richards, C. and Lewis, F. (1992) Identification of a repetitive element in the snail *Biomphalaria glabrata*: relationship to the reverse transcriptase-encoding sequence in LINE-1 transposons. *Gene* 118, 181–187.

Knight, M., Miller, A.N., Geoghagen, N.S.M., Lewis, F.A. and Kerlavage, A.R. (1998) Expressed sequence tags (ESTs) of *Biomphalaria glabrata*. An intermediate snail host of *Schistosoma mansoni*: use in the identification of RFLP markers. *Malacologia* 39, 175–182.

Lardans, V., Serra, E., Capron, A. and Dissous, C. (1998) Characterization of an intracellular receptor for activated protein kinase C (RACK) from the mollusc *Biomphalaria glabrata*, the intermediate host for *Schistosoma mansoni*. *Experimental Parasitology* 88, 194–199.

Lewis, F.A., Patterson, C.N., Knight, M. and Richards, C.S. (2001) The relationship between *Schistosoma mansoni* and *Biomphalaria glabrata*: genetic and molecular approaches. *Parasitology* 123, S169–S179.

Liang, P. and Pardee, A.B. (1992) Differential display of eukaryotic messenger RNA by means of the polymerase chain reaction. *Science* 257, 967–971.

Lie, K.J. and Heyneman, D. (1976) Studies on resistance in snails. 5. Tissue reactions to *Echinostoma lindoense* in naturally resistant *Biomphalaria glabrata*. *Journal of Parasitology* 62, 292–297.

Lie, J.K., Heyneman, D. and Jeong, K.H. (1976) Studies on resistance in snails. 4. Induction of ventricular capsules and changes in the amebocyte-producing organ during sensitization of *Biomphalaria glabrata* snails. *Journal of Parasitology* 62, 286–291.

Lively, C.M. (2001) Trematode infection and the distribution and dynamics of parthenogenetic snail populations. *Parasitology* 123(Suppl.), S19–S26.
Lockyer, A.E., Jones, C.S., Noble, L.R. and Rollinson, D. (2000) Use of differential display to detect changes in gene expression in the intermediate snail host *Biomphalaria glabrata* upon infection with *Schistosoma mansoni*. *Parasitology* 120, 399–407.
Lockyer, A.E., Noble, L.R., Rollinson, D. and Jones, C.S. (2004) *Schistosoma mansoni*: resistant specific infection-induced gene expression in *Biomphalaria glabrata* identified by fluorescent-based differential display. *Experimental Parasitology* 107, 97–104.
Lodes, M.J., Connors, V.A. and Yoshino, T.P. (1991) Isolation and functional characterization of snail hemocyte-modulating polypeptide from primary sporocysts of *Schistosoma mansoni*. *Molecular and Biochemical Parasitology* 49, 1–10.
Loker, E.S. (1994) On being a parasite in an invertebrate host: a short survival course. *Journal of Parasitology* 80, 728–747.
Loker, E.S., Cimino, D.F. and Hertel, L.A. (1992) Excretory–secretory products of *Echinostoma paraensei* sporocysts mediate interference with *Biomphalaria glabrata* hemocyte functions. *Journal of Parasitology* 78, 104–115.
Manger, P., Li, J., Christensen, B.M. and Yoshino, T.P. (1996) Biogenic monoamines in the freshwater snail, *Biomphalaria glabrata*: influence of infection by the human blood fluke, *Schistosoma mansoni*. *Comparative Biochemistry and Physiology A – Physiology* 114, 227–234.
Matricon-Gondran, M. and Letocart, M. (1999a) Internal defences of the snail *Biomphalaria glabrata*. *Journal of Invertebrate Pathology* 74, 224–234.
Matricon-Gondran, M. and Letocart, M. (1999b) Internal defences of the snail *Biomphalaria glabrata*. *Journal of Invertebrate Pathology* 74, 235–247.
Matricon-Gondran, M. and Letocart, M. (1999c) Internal defences of the snail *Biomphalaria glabrata*. *Journal of Invertebrate Pathology* 74, 248–254.
Miller, A.N., Ofori, K., Lewis, F. and Knight, M. (1996) *Schistosoma mansoni*: use of a subtractive cloning strategy to search for RFLPs in parasite-resistant *Biomphalaria glabrata*. *Experimental Parasitology* 84, 420–428.
Miller, A.N., Raghavan, N., FitzGerald, P.C., Lewis, F.A. and Knight, M. (2001) Differential gene expression in haemocytes of the snail *Biomphalaria glabrata*: effects of *Schistosoma mansoni* infection. *International Journal of Parasitology* 31, 687–696.
Min-Oo, G., Fortin, A., Tam, M.F., Nantel, A., Stevenson, M.M. and Gros, P. (2003) Pyruvate kinase deficiency in mice protects against malaria. *Nature Genetics* 35, 357–362.
Muschamp, J.W. and Fong, P.P. (2001) Effects of the serotonin receptor ligand methiothepin on reproductive behavior of the freshwater snail *Biomphalaria glabrata*: reduction of egg laying and induction of penile erection. *Journal of Experimental Zoology* 289, 202–207.
Nabih, I. and El Ansary, A. (1993) Kinetic potentials of certain scavenger enzymes in fresh water snails susceptible and non-susceptible to schistosoma infection. *Cell and Molecular Biology* 39, 449–454.
Nabih, I., El-Wassimi, M.T. and Metri, J. (1968) Structure and activity in molluscicides II. Activity assay of catalase and peroxidase in snails. *Journal of Pharmaceutical Sciences* 57, 1609–1611.
Newton, W.L. (1955) The establishment of a strain of *Australorbis glabratus* which combines albinism and high susceptibility to infection with *Schistosoma mansoni*. *Journal of Parasitology* 41, 526–528.
Nowak, T.S., Woodards, A.C., Jung, Y., Adema, C.M. and Loker, E.S. (2004) Identification of transcripts generated during the response of resistant *Biomphalaria glabrata* to *Schistosoma mansoni* infection using suppression subtractive hybridization. *Journal of Parasitology* 90, 1034–1040.
Ongele, E. (2000) Differential display analysis of developmentally expressed genes modulated by *Schistosoma mansoni* infection in *Biomphalaria glabrata*. PhD thesis. Howard University, Washington, DC.
Paraense, W. and Correa, L. (1963) Variation in susceptibility of populations of *Australorbis glabratus* to a strain of *Schistosoma mansoni*. *Revista do Instituto de Medicina Tropical de São Paulo* 5, 15–22.
Pearce, E.J. (2003) Progress towards a vaccine for schistosomiasis. *Acta Tropica* 86, 309–313.
Raghavan, N., Miller, A.N., Gardner, M., FitzGerald, P.C., Kerlavage, A.R., Johnston, D.A., Lewis, F.A. and Knight, M. (2003) Comparative gene analysis of *Biomphalaria glabrata* hemocytes pre- and post-exposure to miracidia of *Schistosoma mansoni*. *Molecular and Biochemical Parasitology* 126, 181–191.
Richards, C.S. (1970) Genetics of a molluscan vector of schistosomiasis. *Nature* 227, 806–810.
Richards, C.S. and Merritt, J.W. Jr (1972) Genetic factors in the susceptibility of juvenile *Biomphalaria glabrata* to *Schistosoma mansoni* infection. *American Journal of Tropical Medicine and Hygiene* 21, 425–434.

Richards, C.S. and Shade, P.C. (1987) The genetic variation of compatibility in *Biomphalaria glabrata* and *Schistosoma mansoni*. *Journal of Parasitology* 73, 1146–1151.
Sajid, M. and McKerrow, J.H. (2002) Cysteine proteases of parasitic organisms. *Molecular and Biochemical Parasitology* 120, 1–21. *Erratum* in: *Molecular and Biochemical Parasitology* 121, 159.
Santhanagopalan, V. and Yoshino, T.P. (2000) Monoamines and their metabolites in the freshwater snail *Biomphalaria glabrata*. *Comparative Biochemistry and Physiology – Part A: Molecular and Integrative Physiology* 125, 469–478.
Sayed, A.A. and Williams, D.L. (2004) Biochemical characterization of 2-Cys peroxiredoxins from *Schistosoma mansoni*. *Journal of Biological Chemistry* 279, 26159–26166.
Schneider, O. and Zelck, U.E. (2001) Differential display analysis of hemocytes from schistosome-resistant and schistosome-susceptible intermediate hosts. *Parasitology Research* 87, 489–491.
Skelly, P.J., Da'dara, A. and Harn, D.A. (2003) Suppression of cathepsin B expression in *Schistosoma mansoni* by RNA interference. *International Journal of Parasitology* 33, 363–369.
Sorensen, R.E. and Minchella, D.J. (2001) Snail-trematode life history interactions: past trends and future directions. *Parasitology* 123, S3–S18.
Sturrock, R.F. (2001) Schistosomiasis epidemiology and control: how did we get here and where should we go? *Memórias do Instituto Oswaldo Cruz* 96, S17–S27.
Sugamori, K.S., Sunahara, R.K., Guan, H.C., Bulloch, A.G., Tensen, C.P., Seeman, P., Niznik, H.B. and Van Tol, H.H. (1993) Serotonin receptor cDNA cloned from *Lymnaea stagnalis*. *Proceedings of the National Academy of Sciences USA* 90, 11–15.
Sullivan, J.T. and Spence, J.V. (1994) Transfer of resistance to *Schistosoma mansoni* in *Biomphalaria glabrata* by allografts of amoebocyte-producing organ. *Journal of Parasitology* 80, 449–453.
Sullivan, J.T. and Spence, J.V. (1999) Factors affecting adoptive transfer of resistance to *Schistosoma mansoni* in the snail intermediate host, *Biomphalaria glabrata*. *Journal of Parasitology* 85, 1065–1071.
Wood, Z.A., Schroder, E., Robin Harris, J. and Poole, L.B. (2003) Structure, mechanism and regulation of peroxiredoxins. *Trends in Biochemical Sciences* 28, 32–40.
Yoshino, T.P., Boyle, J.P. and Humphries, J.E. (2001) Receptor–ligand interactions and cellular signalling at the host–parasite interface. *Parasitology* 123, S143–S157.

12 Developments in the Chemotherapy of Parasitic Flatworms

G.C. Coles
Department of Clinical Veterinary Science, University of Bristol, Langford House, BS40 5DU, UK

Introduction

The end of patent cover for most of the modern anthelmintics means that market prices have fallen and therefore funds are lower for investment in research into animal health by pharmaceutical companies. As it is the animal health market that drives any search for anti-parasitic drugs by the pharmaceutical companies, few companies are now looking for modern anti-nematodal drugs and the market size for drugs that kill trematodes and cestodes are such that a search for new drugs cannot be financially justified. Unless a novel anthelmintic that kills nematodes happens to have activity against trematodes or cestodes, it is unlikely that new drugs will become available from industry. The macrocyclic lactones (abamectin, doramectin, eprinomectin, ivermectin and moxidectin) are such good drugs that the only reason for the development of novel anthelmintics is the need to control drug-resistant nematodes, which are becoming an increasing problem, not only in sheep and goats but also in cattle (Coles, 2002a, 2005). As there have been few new drugs in the last two decades, the most significant development in the chemotherapy of trematodes and cestodes has been the development of drug resistance in mongeneans, digeneans and cestodes. The one exception to the lack of new drugs is nitazoxanide, a broad-spectrum drug with activity against some bacteria, protozoa and helminths that was approved for use in the USA in 2003 (Hussar, 2004).

Parasites of Animals

In farm animals, the major uses of drugs to kill trematodes and cestodes are for the treatment of *Fasciola hepatica* in grazing animals in temperate regions and *Fasciola gigantica* in the tropics. Both are serious infections that reduce growth and heavy infections, particularly in sheep, can lead to death. The tapeworms of importance are primarily *Anoplocephala perfoliata* in horses, believed to be a cause of colic (Proudman *et al.*, 1998), and *Moniezia expansa* in sheep. Although the significance of *M. expansa* in sheep is still a matter of debate (Elliott, 1986), in southern Africa and north-east Asia, the tapeworms are considered to be highly pathogenic. This might be due to the shortage of nutrition for young animals in these more arid regions. Financially the most important market has been for the treatment of tapeworms in dogs and cats, especially those seen by the owners (i.e. *Dipylidium caninum* and *Taenia* species). Praziquantel has played a special role in attempts to eradicate *Echinococcus granulosus* and to control *Echinococcus multilocularis*. Both species are very serious human pathogens as, if people become infected by ingesting tapeworm eggs, cysts develop.

Fasciolicides sold in the UK are listed in Table 12.1. The availability of drugs differs between countries and trade names are often not the same. The most important fasciolicide is triclabendazole because of its excellent activity against young flukes, enabling one dose to protect animals for a prolonged period. However, where resistance to triclabendazole has become important (especially in Ireland), this is now being replaced by closantel. *Dicrocoelium dendriticum*, the 'small liver fluke', is usually less important than *Fasciola* but can be a significant pathogen causing sizeable production loss. A range of benzimidazoles and probenzimidazoles are effective at elevated doses, but triclabendazole is not effective (Otrano and Traversa, 2002). As increased doses of albendazole kill liver flukes and cestodes, a dose of 20 mg/kg albendazole (Himonas and Liakos, 1980) should kill trematodes, cestodes and nematodes unless benzimidazole-resistant nematodes are present. Paramphistomes in cattle respond to oxyclozanide (16.6 mg/kg) (Spence *et al.*, 1996). As expected, given its activity against schistosomes in humans, *Schistosoma bovis* is killed in goats by 60 mg/kg praziquantel (96% efficiacy) (Johansen *et al.*, 1996), and cattle by 2 × 25 mg/kg (Bushara *et al.*, 1983). In water buffaloes infected with *Schistosoma japonicum*, a single dose of 25 mg/kg cured 95% with the remainder being cured with a repeat dose (Kenworthy *et al.*, 2003).

Praziquantel given at 23 mg/kg thrice a day for 3 days was effective against *Paragonimus kellicotti* in artificially infected cats (Bowman *et al.*, 1991) and fenbendazole given twice a day for 10–14 days killed the fluke in dogs (Dubey *et al.*, 1979). Triclabendazole at 100 mg/kg for 2 days was effective against *Paragonimus westermani* infection in dogs (Liu *et al.*, 1999).

Until the introduction of praziquantel for use in horses (Lyons *et al.*, 1994; Coles *et al.*, 2003), control of *A. perfoliata* was based on use of a double dose of pyrantel (38 mg/kg pyrantel

Table 12.1. Fasciolicides available in the UK.

Chemical name	Action group	Dose (mg/kg)	Effective against immature *Fasciola*
Closantel	Uncouplers	10	Yes
Nitroxynil		10	Yes
Oxyclozanide		15	No
Albendazole	Tubulin inhibitors	7.5	No
Ricobendazole		7.5	No
Triclabendazole	Unknown	10 sheep, 12 cattle	Yes
Clorsulon	Glycolysis inhibitor	2	No

embonate) (Slocombe, 2004). If used at half this dose, the dose used for nematode control, the percentage removal of tapeworms was only 88% (Lyons *et al.*, 1989). Benzimidazoles effective against *M. expansa* in sheep did not remove *A. perfoliata* in horses. In sheep the major drug for control of *Moniezia* has been the modern benzimidazoles, but in some countries niclosamide is available and praziquantel is being increasingly licensed. In addition nitroscanate is used in dogs, although, where there is a risk of infection with *Echinococcus* species, the only drug with proven efficacy is praziquantel. A list of chemicals for control of tapeworms in dogs and cats is given in Table 12.2.

Monogeneans can be important causes of loss in farmed fish. Benzimidazoles have been used to control *Pseudodactylogyrus* on eels and a range of drugs including toltrazuril (Schmahl and Mehlhorn 1988), praziquantel (Schmahl and Mehlhorn, 1985), niclosamide, levamisole and metrifonate (Schmahl and Taraschewski, 1987) can kill some or all of the genera *Gyrodacytylus*, *Dactylogyrus* and *Diplozoon*, although therapeutic indexes may be low for niclosamide and levamisole.

Infections of Humans

The drug of choice for all species of *Schistosoma*, the most important trematode of humans, is praziquantel (Chapters 13 and 14, this volume). As would be expected this is also the drug of choice for the majority other species of fluke infecting humans (summarized in Table 12.3). A detailed review of the different

Table 12.2. Anthelmintics used for control of tapeworms in dogs and cats.

Anthelmintic	Dose (mg/kg)	Species of tapeworms for which claims are made
Fenbendazole[a]	100	*Taenia* sp.
Flubendazole[a]	22	*Taenia pisiformis*
Nitroscanate[a,b]	50	*Taenia* sp. and *Dipylidium caninum*
Praziquantel	5	*Taenia* sp., *D. caninum*, *Echinococcus* sp.

[a]Also kills nematodes.
[b]Not used in cats.

Table 12.3. Treatments for trematodes, other than *Schistosoma* sp., in humans.

Fluke	Drug	Dose (mg/kg)	Efficacy	Reference
Clonorchis	Praziquantel	25 × 3	100%	Bunnag and Harinasuta (1980)
Clonorchis	Albendazole	10 for 7 days	100% cure	Liu *et al.* (1991)
Fasciolopsis	Praziquantel	15	100% cure	Harinasuta *et al.* (1984)
Metagonimus	Praziquantel	20 × 1	88% cure	Rim *et al.* (1978)
Opisthorcis	Praziquantel	25 × 3	100%	Bunnag and Harinasuta (1980)
Opisthorchis, *Haplorchis* and *Echinostome* sp.	Praziquantel	40	100%	Pungpak *et al.* (1998)
Paragonimus	Praziquantel	3 × 25 over 2 days	86% cure	Rim *et al.* (1981)
Paragonimus	Praziquantel	27 over 2 days	100%	Udonsi (1989)
Paragonimus	Triclabendazole	2 × 10	100%	Gao *et al.* (2003)
Paragonimus	Triclabendazole	2 × 10, when repeated	91%, 100%	Calvopina *et al.* (1998, 2003)

drugs used against *Paragonimus* has been given by Choi (1990) who emphasized the activity of praziquantel. Although praziquantel is usually considered to be effective against *Clonorchis sinensis* (Mairiang and Mairiang, 2003), its lack of activity has been reported in North Vietnam (Tinga *et al.*, 1999). As in cattle and sheep, the preferred drug for treatment of *F. hepatica* in humans is triclabendazole (el-Karaksy *et al.*, 1999).

Adult tapeworms

Adult tapeworm infections in humans are usually treated with praziquantel (Table 12.4), but with adult *Taenia solium* there may be neurological side effects if patients are also infected with cysts, so niclosamide, that has no effect on the cysts, is the preferred treatment (Garcia *et al.*, 2003).

Larval tapeworms

Ingestion of eggs from the final host (humans, *T. solium* or canines, *E. granulosus* and *E. multilocularis*) can result in serious disease. Cysts of *T. solium*, the human pork tapeworm can develop in the brain and thus may cause seizures, epilepsy, hydrocephalus and dementia (Carpio, 2002). *E. granulosus* cysts, which most commonly settle in the liver, can resemble a slow growing tumour but anaphylactic shock and even death may result from rupture of a cyst. If people become infected by ingestion of eggs of *E. multilocularis* the parasite grows in the liver where it resembles a diffuse slow growing tumour. Left untreated the infection is often fatal.

Taenia solium

A dose of 10 mg/kg praziquantel is recommended for treatment of *T. solium* in both humans and pigs (Sarti *et al.*, 2000) but prevalence has also been reduced by the use of 2 g/patient of niclosamide (Allan *et al.*, 1997). Death of all cysts by 12 weeks was found following treatment of pigs with a single oral dose of 30 mg/kg oxfendazole (Gonzalez *et al.*, 1998) but single lower doses are not so effective (Gonzalez *et al.*, 1997). A definitive dose for treatment of people with neurocysticercosis with either albendazole or praziquantel has not been established (Carpio, 2002) so optimal doses and the length of treatment for reliable cure are not known.

Echincoccus granulosus

Benzimidazoles (mebendazole or albendazole) have been the most used treatment, but a combination of albendazole plus praziquantel may be more effective than a benzimidazole on its own (Mohamed *et al.*, 1998; Ayles *et al.*, 2002; El-On, 2003). As cimetidine increases benzimidazole serum concentrations it may enhance their activity (Bekhti and Pirotte, 1987; Wen *et al.*, 1994). Whilst for some patients chemotherapy is the sole treatment, even where there will be surgical intervention, treatment with benzimidazoles is recommended before surgery (Keshmiri *et al.*, 2001). A range of doses have been used for treatments of patients but 12–15 mg/kg/day over weeks or months may be best, although an emulsion

Table 12.4. Therapy of adult tapeworms in humans.

Tapeworm	Anthelmintic	Dose (mg/kg)	Efficacy	Reference
Diphyllobothrium latum	Praziquantel	15	100%	Chung *et al.* (1997)
Hymenolepis nana	Albendazole	200	66%	Pamba *et al.* (1989)
Taenia saginata and *H. nana*	Praziquantel	10 and 20	100%	Bouree (1991)
H. nana	Praziquantel	15	84% cure	Mason and Patterson (1994)
T. saginata	Praziquantel	10	100%	Ruiz Perez *et al.* (1989)
Taenia solium	Niclosamide	2 g/patient	–	Pearson and Hewlett (1985)
T. solium	Praziquantel	5	–	Flisser *et al.* (1993)

formulation may increase activity. An efficacy of 98% and a cure rate of 75% have been found following administration from 3 months to more than 1 year (Chai *et al.*, 2004).

Echinococcus multilocularis

Both mebendazole and albendazole are used in the long-term treatment of infections and an overall success rate of 97% has been obtained. Albendazole may be the preferred drug as it is cheaper and easier to take than mebendazole (Reuter *et al.*, 2000). Using nitazoxanide with albendazole in a murine model suggests that there may be improved anti-parasitic activity (Stettler *et al.*, 2004).

New Anti-parasitics: Nitazoxanide

The activity of nitazoxanide, 2-acetolyloxy-*N*-(5-nitro-2-thiazolyl)benzamide, was first described against tapeworms in 1984 (Rossignol and Maisonneuve, 1984) and has subsequently been shown to be a broad-spectrum drug with activity against some bacteria and a range of protozoa including cryptosporidium (White, 2001; Gilles and Hoffman, 2002). In bacteria and protozoa, it inhibits pyruvate ferredoxin oxidoreductase but its mode of action against cestodes is not known.

Initial reports of activity against *F. hepatica* (Rossignol *et al.*, 1998) were confirmed in larger trials in Peru. Using 500 mg/kg/day for 7 days, 60% of adults were cured based on faecal egg counts (Favennec *et al.*, 2003). Results from treatment of cestodes showed activity against *Hymenolepis nana* (Romero-Cabello *et al.*, 1997; Diaz *et al.*, 2003) and *in vitro* effects against *E. multilocularis* metacestodes (Stettler *et al.*, 2003) and protoscoles and metacestodes of *E. granulosus* (Walker *et al.*, 2004). A combination of nitazoxanide and albendazole was more effective than either drug on its own against alveolar echinococcus in mice (Stettler *et al.*, 2004). The relatively low activity against flukes and tapeworms combined with the need for a week's course of drug suggests that nitazoxanide is unlikely to be used for treating these infections unless serious cases of drug resistance are encountered. Rather the activity against flukes, cestodes and nematodes (Juan *et al.*, 2002) will be an added advantage if the drug is being used for the control of protozoal infections, and patients also have helminth infections.

Mode of Action of Fasciolicides and Cestodicides

The detailed mechanism of resistance of nematodes to two of the three classes of anthelmintics, the imidazothiazoles (levamisole) and the macrocyclic lactones, is not known. It is, therefore, not surprising that the molecular mechanism of action of chemicals killing trematodes and cestodes remains unknown. Based on detailed studies in *Schistosoma* it seems likely that praziquantel will act on β-subunits of voltage-gated chloride channels (Chapter 14, this volume). If these β-units were co-expressed in mammalian cells, the cells became sensitive to praziquantel (Kohn *et al.*, 2001). It appears that the lack of two conserved serines, which form a site for kinase C phosphorylation, may explain why schistosome cells are sensitive to praziquantel (Kohn *et al.*, 2003). Probably this will be a common mechanism of action against trematodes and cestodes. The result of the effect on the calcium channel is muscular contraction and blebbing of the surface of the worms, which allows attack by the host immune system. As the dose used for cestodes is usually lower than for schistosomes it seems probable that the receptors in cestodes are more sensitive to praziquantel, but the molecular differences in the cestodes receptors have not been investigated. The reasons for differing results of trials with praziquantel against *F. hepatica* in patients (Farag *et al.*, 1986; Moreau *et al.*, 1995) are not known. The changes in the tegument of *Schistosoma* when they are incubated in praziquantel *in vitro* are not found in *Fasciola* (Becker *et al.*, 1980). As the action of praziquantel on miracidia of *Schistosoma* and *Fasciola* is similar (G.C. Coles, Bristol, 2005; unpublished data) it seems likely that the lack of activity against *F. hepatica* has to do with the differences in structure of the worms' surfaces rather than the receptors per se.

Where benzimidazoles kill tapeworms or flukes, it seems reasonable to conclude that this will be by binding to β-tubulin as occurs with nematodes, but the exact details remain to be established. Although the effects of triclabendazole on *F. hepatica* are compatible with an action on tubulin, as illustrated with disruption of microtubule function (Stitt and Fairweather, 1992), albendazole kills adult triclabendazole-resistant *F. hepatica* (Coles and Stafford, 2001). This suggests that the two benzimidazoles do not act at identical sites on the tubulin molecule. Closantel, nitroxynil and oxyclozanide are uncouplers in mammalian mitochondria so that it is tempting to assume that they act in this way against flukes. Effects on fluke mitochondria are suggested by studies with salicylanilides (Fairweather and Boray, 1999a,b). However, energy production in *Fasciola* is primarily anaerobic, so the detailed mode of action remains to be established. It has been suggested that clorsulon acts by inhibition of glycolysis in flukes (Schulman and Valentino, 1980), but as pointed out by Fairweather and Boray (1999a) the concentrations of drug used were very high, so it will be of interest to see whether there is a change in the target enzyme if resistance develops to this drug. A much more detailed discussion of the action of fasciolicides is given by Fairweather and Boray (1999a,b).

Resistance

The definition of resistance provided by Prichard *et al.* (1980) is suitable for trematodes and cestodes. 'Resistance is present when there is a greater frequency of individuals within a population able to tolerate doses of compound than in a normal population of the same species and is heritable'. The definition has been refined for *Schistosoma* sp. to stress the difference between tolerance and resistance. Tolerance is when a drug has never worked against a parasite, e.g. oxamniquine and *S. japonicum*. Genes for resistance are present if an isolate of a parasite is significantly less susceptible than the most susceptible isolate of the same species (Coles and Kinoti, 1997).

Resistance develops when naturally occurring worms have a selective advantage over susceptible worms in the presence of drugs. The evidence from studies with *Schistosoma mansoni* and oxamniquine suggests that there can be regional differences in the percentage of parasites naturally resistant to drugs (Coles *et al.*, 1987). Due to lack of suitable probes, there are as yet no figures on the variation of susceptibility within any species of nematode, trematode or cestode, except for cyathostomins in horses where it has been suggested that in wild (unselected) populations the gene frequency for benzimidazole resistance could be as high as 3% (Pape *et al.*, 2003).

The key issue in the development of resistance in parasitic helminths is the percentage contribution that worms surviving treatment make to the next generation. This is high where either there are few free-living stages (usually caused by drought or treatment of animals followed by movement to clean pasture) or where all animals or humans are treated. The stages not exposed to drugs are technically in refugia as they avoid the action of the drug. If refugia are maintained, resistance should be slow to develop as the bulk of the next generation will be coming from unselected organisms. This is perhaps best illustrated with bovine nematodes where in the UK there have only been two documented cases of anthelmintic resistance (Stafford and Coles, 1999; Coles *et al.*, 2001). This is almost certainly due to the past habit of only treating calves and not adult cattle so that the bulk of nematode eggs contaminating the pasture come from untreated adult animals. The lack of surveys may provide an optimistic view suggesting the absence of resistance in bovine nematodes, but this also applies to resistance in trematodes and cestodes. Funding of surveys by public bodies may reveal problems on which money has to be spent, so not looking for problems saves money! Recent articles on refugia and its practical application to the control of nematodes should be consulted for further details (Van Wyk, 2001; Coles, 2002b, 2003). Only where there has been mass treatment is resistance likely to develop, so that there have been no reports of resistance developing in human infections with trematodes and cestodes, except in *S. mansoni*. Documented resistance in other flatworms has been confined to *F. hepatica*, the sheep tapeworm *M. expansa* and the monogenean *Pseudodactylogyrus*.

Resistance in *Fasciola*

Resistance in the field and laboratory was first discussed by Boray (1990). Resistance to rafoxanide was present on the majority of Australian properties surveyed. This resistance extended to closantel and nitroxynil but probably not to oxyclozanide. By selection Boray (1990) produced an isolate resistant to triclabendazole. Subsequently, triclabendazole resistance was found in the field in Australia (Overend and Bowen, 1995). From discussions with veterinarians, triclabendazole resistance now appears to be widespread in Ireland and in Northern Ireland with what is now a large foci in Scotland (Mitchell *et al.*, 1998) and smaller foci in Wales (Thomas *et al.*, 2000). There is a well-documented outbreak in the Netherlands (Moll *et al.*, 2000). The true extent of the problem is not known due to both the lack of simple rapid tests for resistance and to the lack of surveys for prevalence. In Ireland, control is switching to closantel, but how long before resistance to this drug will be described is not known. When this happens there will be no fasciolicides remaining that kill immature flukes making control more difficult. Currently the other available fasciolicides kill triclabendazole resistant flukes (Coles *et al.*, 2000; Coles and Stafford, 2001). The mechanism of resistance of flukes to fasciolicides is not known and no validated tests for resistance exist.

The management of resistance is problematic. Rabbits and deer are good hosts for *Fasciola* and can easily move resistant flukes from farm to farm, so treatment and quarantining of moved sheep and cattle to prevent the spread of resistance may not be as effective as desired. Combinations of some fasciolicides show synergy (Boray, 1997; Fairweather and Boray, 1999a,b) and in theory this could be very useful. However, the difficulty and expense of registering combinations in the European Union (EU) combined with the small market size suggests that combinations of fasciolicides will not be made available. Combinations of anthelmintics that are widely used in some countries in the southern hemisphere are also unlikely to be commercialized in Europe.

Resistance in monogeneans

In commercial eel farming, the major parasite is the monogenean *Pseudodactylogyrus* sp., which has been controlled by addition of 1 ppm mebendazole. Typically this may have been added 6–12 times/year. As tanks are recirculating there will have been gradually decreasing concentration of drug, which is likely to have encouraged the selection for resistance. Waller and Buchman (2001) reported that 1–2 ppm mebendazole had no effect on the population of flukes from eels in Denmark.

Resistance in cestodes

There has been one report of resistance to benzimidazoles in cestodes of *E. granulosus* in experimental animals (Morris and Taylor, 1990). Resistance is otherwise confined to the failure of benzimidazoles to control *M. expansa* in sheep (Southworth *et al.*, 1996) and the failure of niclosamide in sheep in South Africa (summarized by Coles, 2002b). Recent evidence suggests that there is reduced activity of praziquantel in sheep in New Zealand (Mason *et al.*, 2002) but further research is required to confirm this. The major tapeworm of the horse is *A. perfoliata* that normally attaches at the ileocaecal junction. Recently, many or most of the tapeworms have been found attached in the caecum rather than at the junction (Williamson *et al.*, 1997; Yue *et al.*, 2003). Pyrantel is used at double doses to kill *A. perfoliata*, but as it is an anthelmintic it may be commonly given at a dose that is sub-therapeutic for the tapeworm. *In vitro* tests with the tapeworms failed to find a difference in response between worms attached at the ileocaecal junction and those lower down in the caecum. As it appears likely that the concentration of pyrantel, which is not absorbed by the host, will be higher at the ileocaecal junction than in the caecum, Yue *et al.* (2003) suggested that this might be a behavioural response of the worm to avoid the action of the drug.

Detection of resistance

In theory four types of tests can be used to detect resistance in parasitic helminths. The standard way to confirm any suspected novel case of resistance in farm animals will be the treatment and post-mortem of animals, but the cost involved excludes its routine use. Naturally or artificially infected animals are weighed and part of the group treated by injection or by mouth with the correct dose of drug under investigation. After a suitable period, the animals are killed and the numbers of worms present counted. The time of treatment will be based on the developmental stage of the parasite under investigation and the gap between treatment and post-mortem will be affected by the rate at which the drug works and parasites are passed out of the host. Post-mortem trials have been used to investigate the efficacy of alternative anthelmintics to triclabendazole-resistant *F. hepatica* (Coles *et al.*, 2000; Moll *et al.*, 2000; Coles and Stafford, 2001) and to confirm praziquantel resistance in *S. mansoni* using infected mice.

Once a new case of resistance has been confirmed, detection will then most likely rely on the reduction of egg counts after treatment (faecal egg count reduction test (FECRT)). Allowance must be made for temporary sterilization of the worms, or in the case of tapeworms, destrobilation. Interpretation of results may be further complicated by development of immature forms in cases where drugs only kill adult worms. No such tests have been adequately investigated for detecting resistance in cestodes or trematodes.

Incubation of free-living stages of the parasite in drug can be very useful, the classical example being the egg hatch test for benzimidazole resistance in nematodes (see Coles *et al.*, 1992). As eggs of *Fasciola* are killed by incubation in solutions of benzimidazoles (Coles and Briscoe, 1978) it is probable this test could be used to pick up albendazole resistance in *F. hepatica*. There is a difference in the response of susceptible and praziquantel-resistant free-living stages of *S. mansoni* to praziquantel (Liang *et al.*, 2001) but further research is required to make this test suitable for routine use in the clinic.

It should be possible to develop either biochemical- or molecular-based tests for resistance in flatworms, as has been successfully used to detect insecticide resistance in insects. However, the necessary information is not available on the mechanism of resistance to be able to design tests of this type for use with trematodes or cestodes. When developed they will hopefully be based on real-time polymerase chain reaction (PCR) or pyrosequencing.

Concluding Remarks

The only parasite for which new fasciolicides or cestodicides are really needed at present is *F. hepatica*. If closantel/nitroxynil resistance develops where there is already triclabendazole resistance, farmers will be left without any drug with significant activity against immature flukes, making control in endemic areas more difficult. There remains the challenge of producing standardized tests for the *in vivo* and *in vitro* detection of resistance and the production of sensitive molecular-based tests that can be used to investigate the effects of management on the development of resistance. It is disappointing that so little progress has been made over the last few years to understand how existing drugs work and develop new drugs either through serendipity or the use of gene sequencing to identify novel targets for drugs.

References

Allan, J.C., Velasquez-Tohom, M., Fletes, C., Torres-Alvarez, R., Lopez-Virula, G., Yurrita, P., Soto de Alfero, H., Rivera, A. and Garcia-Noval, J. (1997) Mass chemotherapy for intestinal *Taenia solium* infection. *Transactions of the Royal Society of Tropical Medicine and Hygiene* 91, 595–598.

Ayles, H.M., Corbett, E.L., Taylor, I., Cowie, A.G., Bligh, J., Walmsley, K. and Bryceson, A.D. (2002) A combined medical and surgical approach to hydatid disease: 12 years' experience at the hospital for tropical diseases. *Annals of the Royal College of Surgery, England* 84, 100–105.

Becker, B., Mehlhorn, H., Andrews, P., Thomas, H. and Eckert, J. (1980) Light and electron microscope studies on the effect of praziquantel in *Schistosoma mansoni, Dicrocoelium dendriticum,* and *Fasciola hepatica* (Trematoda) *in vitro. Zeitschrift fur Parasitenkunde* 63, 113–128.

Bekhti, A. and Pirotte, J. (1987) Cimetidine increases serum mebendazole concentrations. Implications for treatment of hepatic hydatid disease. *British Journal of Clinical Pharmacology* 24, 390–392.

Boray, J.C. (1990) Drug resistance in *Fasciola hepatica*. In: Boray, J.C., Martin, P.J. and Roush, R.T. (eds) *Resistance of Parasites to Antiparasitic Drugs*. MSD AGVET, Rahway, New Jersey, pp. 51–60.

Boray, J.C. (1997) Chemotherapy of infections with Fasciolidae. In: Boray, J.C. (ed.) *Immunology, Pathobiology and Control of Fasciolosis*. MSD AGVET, Rahway, New Jersey, pp. 83–97.

Bouree, P. (1991) Successful treatment of *Taenia saginata* and *Hymenolepis nana* by single oral dose of praziquantel. *Journal of the Egyptian Society for Parasitology* 21, 303–307.

Bowman, D.D., Frongillo, M.K., Johnson, R.C., Beck, K.A., Hornbuckle, W.F. and Blue, J.T. (1991) Evaluation of praziquantel for treatment of experimentally induced paragonimiasis in dogs and cats. *American Journal of Veterinary Research* 52, 68–71.

Bunnag, D., and Harinasuta, T. (1980) Studies on the chemotherapy of human Opisthociasis in Thailand. 1. Clinical trail of praziquantel. *Southeast Asian Journal of Tropical Medicine and Public Health* 11, 528–530.

Bushara, H.O., Majid, B.Y., Majid, A.A., Khitma, I., Gameel, A.A., Karib, E.A., Hussein, M.F. and Taylor, M.G. (1983) Observations on cattle schistosomiasis in the Sudan, a study in comparative medicine. V. The effect of praziquantel therapy on naturally acquired resistance to *Schistosoma bovis. American Journal of Tropical Medicine and Hygiene* 32, 1370–1374.

Calvopina, M., Guderian, R.H., Paredes, W., Chico, M. and Cooper, P.J. (1998) Treatment of human paragonimiasis with triclabendazole: clinical tolerance and drug efficacy. *Transactions of the Royal Society of Tropical Medicine and Hygiene* 92, 566–569.

Calvopina, M., Guderian, R.H., Paredes, W.Y. and Cooper, P.J. (2003) Comparison of two single-day regimes of triclabendazole for the treatment of human pulmonary paragonimiasis. *Transactions of the Royal Society of Tropical Medicine and Hygiene* 97, 451–454.

Carpio, A. (2002) Neurocyticercosis: an update. *Lancet Infectious Diseases* 2, 751–762.

Chai, J., Menghebat, W.J., Deyu, S., Bin, L., Jincao, S., Chen, F., Xiong, L., Yiding, M., Xiuling, W., Dolikun, G., Yanchun, W., Fanghua, G. and Shuhua, X. (2004) Observations on clinical efficacy of albendazole emulsion in 264 cases of hepatic cystic echinococcosis. *Parasitology International* 53, 3–10.

Choi, D.W. (1990) *Paragonimus* and paragonimiasis. *Korean Journal of Parasitology* 28 (suppl.), 79–102.

Chung, P.R., Sohn, W.M., Jung, Y., Pai, S.H. and Nam, M.S. (1997) Five human cases of *Diphylobothrium latum* infection through eating raw flesh of redlip mullet, *Liza haematocheila. Korean Journal of Parasitology* 35, 283–285.

Coles, G.C. (2002a) Cattle resistant to anthelmintics: why so few cases? *Veterinary Research* 33, 481–489.

Coles, G.C. (2002b) Sustainable use of anthelmintics in grazing animals. *Veterinary Record* 151, 165–169.

Coles, G.C. (2003) Strategies to minimise anthelmintic resistance in large animal practice. *In Practice* 25, 494–499.

Coles, G.C. (2005) Anthelmintic resistance – looking to the future: a UK perspective. *Research in Veterinary Science* 78, 99–108.

Coles, G.C. and Briscoe, M.G. (1978) Benzimidazoles and fluke eggs. *Veterinary Record* 103, 360–361.

Coles, G.C. and Kinoti, G.K. (1997) Defining resistance in *Schistosoma. Parasitology Today* 13, 157–158.

Coles, G.C. and Stafford, K.A. (2001) Activity of oxyclozanide, nitroxynil, clorsulon and albendazole against adult triclabendazole-resistant *Fasciola hepatica. Veterinary Record* 148, 723–724.

Coles, G.C., Mutahi, W.T., Kinoti, G.K., Bruce, J.I. and Katz, N. (1987) Tolerance of Kenyan *Schistosoma mansoni* to oxamniquine. *Transactions of the Royal Society of Tropical Medicine and Hygiene* 81, 782–785.

Coles, G.C., Bauer, C., Borgsteede, F.H.M., Geerts, S., Klei, T.R., Taylor, M.A. and Waller, P.J. (1992) World Association for the Advancement of Veterinary Parasitology (W.A.A.V.P.) methods for the detection of anthelmintic resistance in nematodes of veterinary importance. *Veterinary Parasitology* 44, 35–44.

Coles, G.C., Rhodes, A.C and Stafford, K.A. (2000) The activity of closantel against adult triclabendazole-resistant *Fasciola hepatica*. *Veterinary Record* 146, 504.
Coles, G.C., Watson, C.L. and Anziani, O.S. (2001) Ivermectin-resistant *Cooperia* in cattle. *Veterinary Record* 148, 283.
Coles, G.C., Hillyer, M.H., Taylor, F.G.R. and Villard, I. (2003) Efficacy of an ivermectin–praziquantel combination in equids. *Veterinary Record* 152, 178–179.
Diaz, E., Mondragon., J., Ramirez, E. and Bernal, R. (2003) Epidemiology and control of intestinal parasites with nitazoxanide in children in Mexico. *American Journal of Tropical Medicine and Hygiene* 68, 384–385.
Dubey, J.P., Miller, T.B. and Sharma, S.P. (1979) Fenbendazole for the treatment of *Paragonimus kellicotti* infection in dogs. *Journal of the American Veterinary Medical Association* 174, 835–837.
el-Karaksy, H., Hassanein, B., Okasha, S., Behairy, B. and Gadallah, I. (1999) Human fasciolosis in Egyptian children: successful treatment with triclabendazole. *Journal of Tropical Pediatrics* 45, 135–138.
Elliott, D.C.V. (1986) Tapeworm (*Moniezia expansa*) and its effect on sheep production: the evidence reviewed. *New Zealand Veterinary Journal* 34, 61–65.
El-On, J. (2003) Benzimidazole treatment of cystic echinococcis. *Acta Tropica* 85, 243–252.
Fairweather, I. and Boray, J.C. (1999a) Fasciolicides: efficacy, actions, resistance and its management. *Veterinary Journal* 158, 81–112.
Fairweather, I. and Boray, J.C. (1999b) Mechanisms of fasciolicide action and drug resistance in *Fasciola hepatica*. In: Dalton, J.P. (ed.) *Fasciolosis*. CAB International, Wallingford, UK, pp. 225–276.
Farag, H.F., Ragab, M., Salem, A. and Sadek, N. (1986) A short note on praziquantel in human fascioliasis. *Journal of Tropical Medicine and Hygiene* 89, 79–80.
Favennec, L., Ortiz, J.J., Gargala, G., Lopez Chegne, N., Ayoub, A. and Rossignol, J.F. (2003) Double-blind, randomized, placebo-controlled study of nitazoxanide in the treatment of fascioliasis in adults and children from northern Peru. *Alimentary Pharmacology and Therapeutics* 17, 265–270.
Flisser, A., Madrazo, I., Plancarte, A., Schantz, P., Allan, J., Craig, P. and Sarti, E. (1993) Neurological symptoms in occult neurocysticercosis after single taeniacidal dose of praziquantel. *Lancet* 342, 748.
Gao, J., Liu, Y., Wang, X. and Hu, P. (2003) Triclabendazole in the treatment of *Paragonimus skrjabini*. *Chinese Medical Journal* 116, 1683–1686.
Garcia, H.H., Gonzales, A.E., Evans, A.C, Gilman, R.H. and Cysticercosis working group in Peru (2003) *Taenia solium* cysticercosis. *Lancet* 362, 547–556.
Gilles, H.M. and Hoffman, P.S. (2002) Treatment of intestinal parasitic infections: a review of nitazoxanide. *Trends in Parasitology* 18, 95–97.
Gonzalez, A.E.R., Falcon, N., Gavidia, C., Garcia, H.H., Tsang, V.C., Bernal, T., Romero, M. and Gilman, R.H. (1997) Treatment of porcine cyticercosis with oxfendazole: a dose-response trial. *Veterinary Record* 141, 420–422.
Gonzalez, A.E., Falcon, N., Gavidia, C., Garvia, H.H., Tsang, V.C., Bernal, T., Romero, M. and Gilman, R.H. (1998) Time-response curve of oxfendazole in the treatment of swine cysticercosis. *American Journal of Tropical Medicine and Hygiene* 59, 832–836.
Harinasuta, T., Bunnag, D. and Radomyos, P. (1984) Efficacy of praziquantel on fasciolopsiasis. *Arzneimittelforschung* 34, 1214–1215.
Himonas, C.A. and Liakos, V. (1980) Efficacy of albendazole against *Dicrocoelium dendriticum* in sheep. *Veterinary Record* 107, 288–289.
Hussar, D.A. (2004) New drugs of 2003. *Journal of the American Pharmaceutical Association* 44, 168–206.
Johansen, M.V., Monrad, J. and Christensen, N.O. (1996) Effects of praziquantel on experimental *Schistosoma bovis* infection in goats. *Veterinary Parasitology* 62, 83–89.
Juan, J.O., Lopez Chegne, N., Gargala, G. and Favennec, L. (2002) Comparative clinical studies of nitazoxanide, albendazole and praziquantel in the treatment of ascariasis, trichuriasis and hymenolepiasis in children from Peru. *Transactions of the Royal Society of Tropical Medicine and Hygiene* 96, 193–196.
Kenworthy, J.D., Ye, P., Wu, G.C., Yu, H., Shi, Y.J., Li, H. and Coles, G.C. (2003) Field evaluation of a test for praziquantel resistance in *Schistosoma* sp. *Veterinary Parasitology* 113, 83–87.
Keshmiri, M., Baharvahdat, H., Fattahi, S.H., Davachi, B., Dabiri, R.H., Baradaran, H. and Rajabzadeh, F. (2001) Albendazole versus placebo in treatment of echinococcosis. *Transactions of the Royal Society of Tropical Medicine and Hygiene* 95, 190–194.
Kohn, A.B., Anderson, P.A., Roberts-Misterly, J.M. and Greenberg, R.M. (2001) Schistosome calcium channel beta subunits: unusual modulatory effects and potential role in the action of the antischistosomal drug praziquantel. *Journal of Biological Chemistry* 276, 36873–36876.

Kohn, A.B., Roberts-Misterly, J.M., Anderson, P.A. and Greenberg, R.M. (2003) Creation by mutagenesis of a mammalian Ca(2+) channel beta subunit that confers praziquantel sensitivity to a mammalian Ca(2+) channel. *International Journal for Parasitology* 33, 1303–1308.

Liang, Y.S., Coles, G.C., Doenhoff, M.J. and Southgate, V. (2001) *In vitro* responses of praziquantel-resistant and -susceptible *Schistosoma mansoni* to praziquantel. *International Journal for Parasitology* 31, 1227–1235.

Liu, Y.H., Wang, X.G., Gao, P. and Qian, M.X. (1991) Experimental trial of albendazole in the treatment of *Clonorchis sinensis*. *Chinese Medical Journal* 104, 27–31.

Liu, Y., Gao, J., Wang, X., Yu, D. and Su, Q. (1999) Experimental observation of effects of triclabendazole on *Paragonimus westermani* infections in dogs. *Chinese Medical Journal* 112, 345–348.

Lyons, E.T., Drudge, J.H., Tolliver, S.C., Swerczek, T.W. and Collins, S.S. (1989) Determination of the efficacy of pyrantel pamoate at the therapeutic dose rate against the tapeworm *Anoplocephala perfoliata* in equids using a modification of the critical test method. *Veterinary Parasitology* 31, 13–18.

Lyons, E.T., Tolliver, S.C., Drudge, J.H., Granstrom, D.E. and Stamper, S. (1994) Activity of praziquantel against *Anoplocephala perfoliata* (Cestoda) in horses. *Journal of the Helminthological Society of Washington* 59, 1–4.

Mairiang, E. and Mairiang, P. (2003) Clinical manifestation of opisthorchiasis and treatment. *Acta Tropica* 88, 221–227.

Mason, P.R. and Patterson, B.A. (1994) Epidemiology of *Hymenolepis nana* infections in primary school children in urban and rural communities in Zimbabwe. *Journal of Parasitology* 80, 245–250.

Mason, P., Moffat, J. and Cole, D. (2002) Tapeworm in sheep revisited. *Proceedings of the 32nd Seminar of The Society of Sheep and Beef Cattle Veterinarians*, 22–24 May 2002, NZVA, Wellington, pp. 147–151.

Mitchell, G.B.B., Maris, L. and Bonniwell, M.A. (1998) Triclabendazole-resistant liver fluke in Scottish sheep. *Veterinary Record* 143, 399.

Mohamed, A.E., Yasawy, M.I. and Al Karawi, M.A. (1998) Combined albendazole and praziquantel versus albendazole alone in the treatment of hydatid disease. *Hepatogastroenterology* 45, 1690–1694.

Moll, L., Gaasenbeek, C.P., Vellema, P. and Borgsteede, F.H. (2000) Resistance of *Fasciola hepatica* against triclabendazole in cattle and sheep in the Netherlands. *Veterinary Parasitology* 91, 153–158.

Moreau, J.A., Fernandez, J., Recco, P., Seguela, J.P. and Frexinos, J. (1995) Efficacy and tolerance of praziquantel (Biltricide) in the treatment of distomatosis caused by *Fasciola hepatica*. *Gastroenterologie Clinical et Biologie* 19, 514–519.

Morris, D.L. and Taylor, D.H. (1990) *Echinococcus granulosus*: development of resistance to albendazole in an animal model. *Journal of Helminthology* 164, 171–174.

Otrano, D. and Traversa, D. (2002) A review of dicrocoeliosis of ruminants including recent advances in the diagnosis and treatment. *Veterinary Parasitology* 107, 317–335.

Overend, D.J. and Bowen, F.L. (1995) Resistance of *Fasciola heptica* to triclabendazole. *Australian Veterinary Journal* 72, 275–276.

Pamba, H.O., Bwibo, N.O., Chunge, C.N. and Estambale, B.B. (1989) A study of the efficacy and safety of albendazole (Zentel) in the treatment of intestinal helminthiasis in Kenyan children less than 2 years of age. *East African Medical Journal* 66, 197–202.

Pape, M., Posedi, J., Failing, K., Schnieder, T. and von Samson-Himmelstjerna, G. (2003) Analysis of the beta-tubulin codon 200 genotype distribution in a benzimidazole-susceptible and -resistant cythostome population. *Parasitology* 127, 53–59.

Pearson, R.D. and Hewlett, E.L. (1985) Niclosamide therapy for tapeworm infections. *Annals of Internal Medicine* 102, 550–551.

Prichard, R.K., Hall, C.A., Kelly, J.D., Martin, I.C.A. and Donald, A.D. (1980) The problem of anthelmintic resistance in nematodes. *Australian Veterinary Journal* 56, 239–250.

Proudman, C.J., French, N.J. and Trees, A.J. (1998) Tapeworm infection is a significant risk factor for spastic colic and ileal impaction colic in the horse. *Equine Veterinary Journal* 30, 194–199.

Pungpak, S., Radomyos, P., Radomyos, B.E., Schelp, F.P., Jongsuksuntigul, P. and Bunnag, D. (1998) Treatment of *Opisthorchis viverrini* and intestinal fluke infections with praziquantel. *Southeast Asian Journal of Tropical Medicine and Public Health* 29, 246–249.

Reuter, S., Jensen, B., Buttenschoen, K., Wolfgang, K. and Kern, P. (2000) Benzimidazoles in the treatment of alveolar echinococcosis: a comparative study and review of the literature. *Journal of Antimicrobial Chemotherapy* 46, 451–456.

Rim, H.J., Chu, D.S., Lee, J.S., Joo, K.H. and Won, C.Y. (1978) Anthelmintic effects of various drugs against metagonimiasis. *Korean Journal of Parasitology* 16, 117–122.

Rim, H.J., Chang, Y.S., Lee, J.S., Joo, K.H., Suh, W.H. and Tsuji, M. (1981) Clinical evaluation of praziquantel (Embay 8440: Biltricide) in the treatment of *Paragonimus westermani*. *Korean Journal of Parasitology* 19, 27–37.

Romero-Cabello, R., Guerrero, K.R., Munoz-Garcia, M.R. and Geyne Cruz, A. (1997) Nitazoxanide for the treatment of intestinal protozoan and helminthic infections in Mexico. *Transactions of the Royal Society of Tropical Medicine and Hygiene* 91, 701–703.

Rossignol, J.F. and Maisonneuve, H. (1984) Nitazoxanide in the treatment of *Taenia saginata* and *Hymenolepis nana* infections. *American Journal of Tropical Medicine and Hygiene* 33, 511–512.

Rossignol, J.F., Abaza, H. and Friedman, H. (1998) Successful treatment of human fascioliasis with nitazoxanide. *Transactions of the Royal Society of Tropical Medicine and Hygiene* 92, 103–104.

Ruiz Perez, A., Diaz Hernandez, A., Perez Avila, J. and Diaz Jidy, M. (1989) Our experience in the treatment of *Taenia saginata* (Goenze, 1762) with praziquantel. *Revista Cubana Medicale Tropicale* 41, 49–55.

Sarti, E., Schantz, P.M., Avila, G., Ambrosio, J., Medina-Santillan, R. and Flisser, A. (2000) Mass treatment against human taeniasis for the control of cysticercosis: a population-based intervention study. *Transactions of the Royal Society of Tropical Medicine and Hygiene* 94, 85–89.

Schmahl, G. and Mehlhorn, H. (1985) Treatment of fish parasites. 1. Praziquantel effective against Monogenea (*Dactylogyrus vastator, Dactylogyrus extensus, Diplozoon paradoxum*). *Zeitschrift fur Parasitenkunde* 71, 727–737.

Schmahl, G. and Mehlhorn, H. (1988) Treatment of fish parasites. 4. Effects of sym. triazinone (toltrazuril) on monogenea. *Parasitology Research* 75, 132–143.

Schmahl, G. and Taraschewski, H. (1987) Treatment of fish parasites. 2. Effects of praziquantel, niclosamide, levamisole-HCl and metrifonate on monogenea (*Gyrodactylus aculeati, Diplozoon paradoxum*). *Parasitology Research* 73, 341–351.

Schulman, M.D. and Valentino, D. (1980) *Fasciola hepatica*: effect of 4-amino-6-trichloroethenyl-1, 3-benzenedisulphonamide on glycolysis *in vitro*. *Experimental Parasitology* 49, 206–215.

Slocombe, J.O. (2004) A modified critical test for the efficacy of pyrantel pamoate for *Anoplocephala perfoliata* in equids. *Canadian Journal of Veterinary Research* 68, 112–117.

Southworth, J., Harvey, C. and Larson, S. (1996) Use of praziquantel for the control of *Moniezia expansa* in lambs. *New Zealand Veterinary Journal* 44, 112–115.

Spence, S.A., Fraser, G.C. and Chang, S. (1996) Responses in milk production to control of gastrointestinal nematode and paraphistome parasites in dairy cattle. *Australian Veterinary Journal* 74, 456–459.

Stafford, K. and Coles, G.C. (1999) Nematode control practices and anthelmintic resistance in dairy calves in the south west of England. *Veterinary Record* 144, 659–661.

Stettler, M., Fink, R., Walker, M., Gottstein, B., Geary, T.G., Rossignol, J.F. and Hemphill, A. (2003) *In vitro* parasiticidal effect of nitazoxanide against *Echinococcus multilocularis* metacestodes. *Antimicrobial Agents and Chemotherapy* 47, 467–474.

Stettler, M., Rossignol, F., Fink, R., Walker, M., Gottstein, B., Merli, M., Theurillat, R., Thorman, W., Dricot, E., Segers, R. and Hemphill, A. (2004) Secondary and primary murine alveolar echinococcosis: combined albendazole/nitazoxanide chemotherapy exhibits profound anti-parasitic activity. *International Journal for Parasitology* 34, 615–624.

Stitt, A.W and Fairweather, I. (1992) Spermatogenesis in *Fasciola hepatica*: an ultrastructrual comparison of the effects of the anthelmintic, triclabendazole ('Fasinex') and the microtubule inhibitor, tubulozole. *Invertebrate Reproduction and Development* 22, 139–150.

Thomas, I., Coles, G.C. and Duffus, K. (2000) Triclabendazole-resistant *Fasciola hepatica* in south-west Wales. *Veterinary Record* 146, 200.

Tinga, N., De, N., Vien, H.V., Chasu, L., Toan, N.D., Kager, P.A. and Vries, P.J. (1999) Little effect of praziquantel or artemesian on clonorchiasis in Northern Vietnam. A pilot study. *Tropical Medicine and International Health* 4, 814–818.

Udonsi, J.K. (1989) Clinical field trials of praziquantel in pulmonary paragonimiasis due to *Paragonimus uterobilateralis* in endemic populations of the Igwun Basin, Nigeria. *Tropical Medicine and Parasitology* 40, 65–68.

Van Wyk, J.A. (2001) Refugia – overlooked as perhaps the most potent factor concerning development of anthelmintic resistance. *Ondesterpoort Journal of Veterinary Research* 68, 55–67.

Walker, M., Rossignol, J.F.G., Torgerson, P. and Hemphill, A. (2004) *In vitro* effects of nitazoxanide on *Echinococcus granulosus* protoscolices and metacestodes. *Journal of Antimicrobial Chemotherapy* (advanced publication) 54, 609–616.

Waller, P.J. and Buchmann, K. (2001) Anthelmintic resistance and parasite control in commercial eel farms: consequences for producers. *Veterinary Record* 148, 783–784.

Wen, H., Zhang, H.W., Muhmut, M., Zou, P.F., New, R.R. and Craig, P.S. (1994) Initial observation on albendazole in combination with cimetidine for the treatment of human cystic echinococcosis. *Annals of Tropical Medicine and Parasitology* 88, 49–52.

White, A.C. Jr (2001) Nitazoxanide: an important advance in anti-parasitic therapy. *American Journal of Tropical Medicine and Hygiene* 68, 382–383.

Williamson, R.M.C., Gasser, R.B., Middleton, D. and Beveridge, I. (1997) The distribution of *Anoplocephala perfoliata* in the intestine of the horse and associated pathological changes. *Veterinary Parasitology* 73, 225–241.

Yue, C., Lawrence, S. and Coles, G.C. (2003) Changing behaviour in the common horse tapeworm. *Veterinary Record* 153, 663–664.

13 Drug Resistance in Schistosomes

T.A. Day[1] and S. Botros[2]
[1]*Department of Biomedical Sciences, Veterinary Medicine Building, Iowa State University, Ames, IA 50011, USA;* [2]*Theodor Bilharz Research Institute, Cairo, Warrak El-Hadar, Imbaba, PO Box 30, Giza, Egypt 12411*

Introduction

Schistosomiasis afflicts more than 200 million people throughout the world, making it by far the most significant flatworm disease and one of the most important infectious diseases in the world. A great majority of those infected are in Africa (Chitsulo *et al.*, 2000, 2004), but the disease is present in Asia and South America as well as Pacific and Caribbean Islands.

People become infected when the skin is penetrated by free-swimming schistosome cercariae which emerge from snail intermediate hosts. After a somewhat circuitous route through the human body, the adult schistosomes reside in the vasculature supplying the intestines, liver or urinary bladder. Female worms are prodigious egg producers, expelling as many as 3000 eggs per day. These eggs are the primary cause of most schistosomiasis-related pathology. Some eggs pass out of the body via the faeces or urine and, if they reach fresh water, hatch into a free-swimming larval stage called a miracidium. Miracidia have a few hours to find and penetrate an appropriate species of freshwater snail. Successful parasitism of this intermediate host then results in the development of mother sporocysts which will produce

 Parasitic Flatworms: Molecular Biology, Biochemistry, Immunology and Physiology (eds A.G. Maule and N.J. Marks)

the free-swimming, infective cercariae. The complex life cycle requires the mutual presence of appropriate snail species and the untreated waste of infected humans.

Although half a dozen species have been found to infect humans, the three most important species are *Schistosoma mansoni*, *Schistosoma japonicum* and *Schistosoma haematobium*. The first two species are the agents of intestinal schistosomiasis, where eggs are passed in the faeces, and the third species causes urinary schistosomiasis.

Chemotherapy is still essentially the only means of controlling the disease, and it has dramatically reduced the impact of the disease in many countries. This chapter will provide background regarding schistosomiasis chemotherapy, then it will detail some of the specific challenges to detecting and monitoring resistance to anti-schistosomals, and finally it will briefly review cases of suspected drug resistance among schistosomes.

Schistosomiasis Chemotherapy

Praziquantel was developed in the late 1970s (Gonnert and Andrews, 1977) and tested in humans soon after (Katz *et al.*, 1979; Santos *et al.*, 1979). It has been identified by the World Health Organization (WHO) as the drug of choice for the treatment of schistosomiasis caused by *S. mansoni*, *S. japonicum* or *S. haematobium* since 1982 (Andrews *et al.*, 1983). The mechanism of praziquantel's anti-schistosomal activity remains unclear, but recent insights are reviewed in Chapter 14. More information about other uses for praziquantel for other flatworm diseases can be found in Chapter 12.

Predecessors to praziquantel as the preferred treatment for schistosomiasis include oxamniquine, and the related compound hycanthone (Cioli *et al.*, 1995). These drugs act by inhibiting nucleic acid synthesis, but only after parasite-mediated biotransformation of the drug by a schistosome enzyme (Cioli *et al.*, 1993, 1985). Until relatively recently, oxamniquine continued to be the primary drug used to treat schistosomiasis in Brazil.

Negative effects are extremely rare with praziquantel and it provides some significant advantages over oxamniquine. Firstly, it is effective against all major species of schistosomes infecting humans, whereas oxamniquine is only effective against *S. mansoni*, and in many locations mixed infection is not uncommon. Further, praziquantel is effective in a single oral dose, whereas oxamniquine requires more than one dose and compliance can be an issue. Because of spectrum, ease of administration and wide safety margin, praziquantel has become the centrepiece of successful control efforts in several countries. Praziquantel is presently the only anti-schistosomal drug commercially available in most parts of the world (Fenwick *et al.*, 2003).

Successful anti-schistosomal chemotherapy consists of two different endpoints. Firstly, the treatment can cure the infection – completely eradicating schistosomes from the host. Alternatively, the treatment can dramatically reduce the infection, thus decreasing egg production and associated morbidity. Even when patients are not cured by anti-schistosomal application, there is almost always a dramatic reduction in the rate of egg production and, therefore, decreased schistosomiasis-related morbidity. Since the goal of most chemotherapy campaigns is to control morbidity, either of these endpoints is a positive outcome.

For example, oxamniquine typically produces cure rates of 80–92%, and in the patients not cured there is a 75–97% reduction in egg excretion. The recommended dose of praziquantel results in similar cure rates of 70–95% and egg reduction rates of up to 95% in patients not cured (Gryseels *et al.*, 1987; Behbehani and Savioli, 1998; Barakat *et al.*, 2000; El-Khoby *et al.*, 2000; Degu *et al.*, 2002). A patient with a dramatic reduction in egg excretion is a success in terms of morbidity reduction, but represents an important risk in that the patient now harbours worms that have survived exposure to the drug.

Impact of Praziquantel

In order to grasp the significance of the issue of drug resistance among schistosomes, it is useful to consider the positive impact that chemotherapy has had on public health in many countries. Egypt is an apt example.

The Egyptian Ministry of Health in conjunction with the Schistosomiasis Research

Project initiated a selective chemotherapy programme in 1988 that provided praziquantel free of charge to patients throughout Egypt. The scope of this effort increased in 1992 when the World Bank loaned the Egyptian Government US$27 million (over 6 years) to extend the schistosomiasis control programme – almost US$13 million of that went directly to the purchase of praziquantel. Further, after years of effective and safe use of the drug, the Ministry of Health reversed a prior rule that treatment could only be given after a positive diagnosis. This opened the way for mass treatments in highly endemic villages and large-scale treatments in school children (El-Khoby *et al.*, 1998).

The praziquantel-based programmes were very successful. The Ministry of Health estimated that throughout Egypt there was a decrease in *S. mansoni* prevalence from 39% in 1983 (Scott, 1937; Cline *et al.*, 1989) to 12% in 1996 (El-Khoby *et al.*, 1998). At present, the situation is even better in some historically endemic villages where prevalence rates are below 4% (El-Khoby *et al.*, 2000). It is certain that praziquantel has dramatically reduced the human, social and economic impact of schistosomiasis in Egypt and other countries. As a testament to the impact of the drug, it is the cornerstone of a US$30 million Schistosomiasis Control Initiative targeting schistosomiasis-associated morbidity in sub-Saharan Africa (www.schisto.org).

Challenges to Detecting and Monitoring Anti-schistosomal Resistance

There is a high background of drug failure

A critical issue is that true drug resistance among schistosomes in the field is almost always masked by an expected failure rate of 5–30%. As an illustration, consider a village in which the schistosomes have a normal susceptibility to praziquantel. Treatment might cure 85% of the infected patients (and reduce the egg burden in the other 15%). But, this could be quantified as a 15% drug failure rate. In a parallel village where 2% of the worm population is resistant to the recommended dose of praziquantel, that failure rate would be 17%. Attributing significance to the difference between 15% and 17% failure is thorny, since both are well within the expected range of treatment failure which is 5–30% (Behbehani and Savioli, 1998).

Even if it is suspected that a high failure rate might be signalling underlying drug resistance, there are a large number of factors other than the worms which could account for the lower-than-expected cure rates. So, it is very difficult to attribute even significantly lower-than-expected cure rates to schistosome drug resistance. This will be illustrated in the anecdotes from Senegal.

Methods for detecting infection are variable

Another factor that obfuscates the detection of any potential drug resistance is the large degree of variability in the methods that are used to test for infection. The most common approach is parasitological – examining the urine and faeces of subjects for schistosome eggs. Although standards have been suggested, there is no uniform protocol for the number of stools or slides that are examined for uniform diagnosis. For that matter, even when uniformity of protocol is in place, the technicians examining the slides introduce a level of variability. There are serological tests that might prove to be more consistent, but since their use is quite sporadic, the problem persists.

For the most part, these parasitological approaches to detecting schistosomiasis are sufficiently consistent and vigorous to meet the population-based public health interests, but they are not sufficiently consistent or vigorous to detect with significance a 2% or 3% increase in drug failure. The variability of detection simply swamps the types of increases in failures that could signal an increase in drug resistance.

Factors other than resistance can produce drug failure

Even if the high background of failure and the variable detection methods are accounted for, a number of factors other than resistance can alter the apparent efficacy of the drug. Firstly, with almost every anti-schistosomal, there is variability in vulnerability based on the maturity of the worms; immature worms are known to be quite recalcitrant to praziquantel therapy (Fallon *et al.*, 1997). Secondly, as with most chemotherapeutics, the ultimate efficacy of the drug depends (at least in part) on a contribution by the host's immune system, such that less-than-expected efficacy can be observed in immunocompromised patients (Sabah *et al.*, 1985; Brindley and Sher, 1987; Brindley *et al.*, 1989). Underperformance of the drug has also been observed in undernourished patients where drug absorption can be severely impaired (A. Fenwick, Imperial College, 2005, personal communication). Lower-than-expected cure rates could be observed in populations that inordinately metabolize a drug – for example, the induction of cytochrome P450 speeds the metabolism of praziquantel. Another issue that can limit the efficacy is vomiting after consumption of the drug; when one child vomits, it can induce serial vomiting in the dosing group. All of these factors can lead to apparent underperformance of the drug, but are not attributable to either tolerance or resistance of the parasites.

It is difficult to substantiate a decreased response in the worms

Even with all other factors accounted for, it remains very difficult to conclusively attribute lower-than-expected cure rates to the worms. The only sure way to do this is to collect eggs from patients not responding to therapy and use the resultant miracidia to infect snails. The cercariae emerging from those snails can then be used to generate murine infections, and if these murine infections are significantly more difficult than control infections to cure with praziquantel, there is some indication that the worms contributed to the lack of cure. But, this is a rather long and arduous process.

Alternatively, other means of assessing the response of worms to praziquantel have been used. For example, one of the most prominent effects of the drug is an abrupt contraction of the somatic musculature of the adult worms. These contractions can be quantified and there is a good correlation between the magnitude of contraction elicited by the drug *in vitro* and the sensitivity of those worms to praziquantel *in vivo* (Ismail *et al.*, 1996, 1999). However, there is some question as to whether or not this is a solid indicator of praziquantel's *in vivo* activity, and it is by no means less arduous than treating murine infections as described above. What these studies have brought to the table is an unequivocal demonstration that some worms derived from the field do not respond as robustly to praziquantel, either *in vivo* or *in vitro*.

Another potential assay is the observation of miracidial responses to praziquantel *in vitro*. In some cases, it has been shown that the *in vitro* miracidial response can correlate with the sensitivity of the related worms *in vivo* (Liang *et al.*, 2001; Kenworthy *et al.*, 2003). However, this assay does not seem to work in every case (S. Botros, personal communication). Unequivocally, a predictive assay utilizing miracidia (which can be directly derived from the eggs passing from uncured patients) would be a significant asset for investigators endeavouring to monitor for emergence of praziquantel-resistant schistosomes.

Praziquantel's mechanism of action is not known

Despite its widespread use since the late 1970s, the molecular basis for praziquantel's anti-schistosomal action is still not known (Andrews, 1985; Day *et al.*, 1992). There is new, compelling information on this front that is reviewed within this volume (Chapter 14).

Lacking this information, there is no clear place to look for substantial molecular differences that could account for decreased response to the drug, so there is no promising development for rapid molecular tests for genetic profiles correlating with resistance. It is not clear if resistance can be attributed to a single gene or, as presently appears more likely, is a polygenic trait. The most likely scenario is that liberal drug application is simply selecting for worms from the extremes of the normal phenotypic range. But, the primary point in the present context is that the dearth of knowledge regarding the molecular basis of praziquantel's action leaves investigators without sensitive tools for the discovery and detection of resistant worms.

Rather than knowledge of the molecular mechanism of action leading us to uncover resistant strains of worms, it is much more likely that the resistant strains of worms could lead us to knowledge of the molecular mechanism of action. Genetic or proteomic analysis of worms with diminished response could point to the molecules critical to praziquantel function.

Defining resistance

As described elsewhere in this volume (Chapter 12), there has been a lack of clarity in the use of the term resistance with respect to schistosomes (and other flatworms). *Resistance* has been defined as a heritable change in the susceptibility of a population in response to exposure to the drug (Prichard *et al.*, 1980). In contrast, *tolerance* implies a native lack of susceptibility that is not in response to exposure to the drug (Coles and Kinoti, 1997). So, resistance is a phenomenon of population genetics under the selective pressure of the drug, whereas tolerance is simply innate in the population. With schistosomes, it is often not possible to make this distinction, because in most cases of liberal anti-schistosomal drug use, the endogenous susceptibility of a given population is not known and such a determination is not a priority of the control programmes distributing the drug.

For example, if praziquantel was distributed to a particular village in Egypt as a part of the national control programme, it is very possible that efficacy of the drug would not be monitored in any detailed fashion. If 5 years later, the drug produces lower-than-expected cure rates in that village, it is not possible to determine if that is a result of the previous drug pressure, or simply the native state of that population of worms. At that point, it is not possible to distinguish between resistance and tolerance.

Cases of Drug Resistance in Schistosomes

Here, we will briefly review the cases of suspected drug resistance in schistosomes. In doing so, some of the difficulties discussed above will be illustrated. The data related to oxamniquine and hycanthone have been reviewed very well elsewhere (Cioli *et al.*, 1995) and will only be summarized here. More detail will be provided regarding data from Senegal and Egypt regarding praziquantel.

Resistance to hycanthone and oxamniquine

Schistosome resistance to hycanthone was found in the laboratory in the early 1970s, and the progeny of worms not cured by the drug proved quite insensitive (Rogers and Bueding, 1971). Hycanthone resistance was also found in Brazil, where two patients could not be cured by the drug. When eggs derived from these patients were used to establish infections in laboratory mice, the murine infections were also resistant to hycanthone (Katz, 1975). Schistosome resistance to oxamniquine was also generated in the laboratory (Cioli *et al.*, 1992; Pica-Mattoccia *et al.*, 1992; Drescher *et al.*, 1993). Subsequently, there have been a number of documented cases of schistosomes which did not respond to hycanthone or oxamniquine, and resistance to one of the drugs translated into resistance to the other, substantiating their shared mechanism of action (Campos *et al.*, 1976; Katz *et al.*, 1977; Dias *et al.*, 1978, 1982; Guimaraes *et al.*, 1979; Bruce *et al.*, 1987; Coles *et al.*, 1987; Kinoti, 1987; Bonesso-Sabadini and Dias, 2002).

These resistant worms proved to be a valuable tool in uncovering the anti-schistosomal mechanism for these drugs. In a series of extremely demanding experiments involving genetic crosses, Cioli found that oxamniquine resistance was controlled by a single, autosomal recessive gene (Cioli and Pica Mattoccia, 1984; Cioli *et al.*, 1992, 1993). This opened the door to further studies showing that the action of oxamniquine and hycanthone depended on the biotransformation of the parent drug into a molecule capable of interfering with nucleic acid synthesis. Oxamniquine resistance is based on a lack of the schistosome enzyme responsible for the activating biotransformation; worms without the functional enzyme activity were not susceptible to the drug (Pica-

Mattoccia *et al.*, 1992). This also led to the elucidation of why oxamniquine is relatively ineffective against *S. haematobium*; the native enzyme in this species cannot biotransform oxamniquine (Pica-Mattoccia *et al.*, 1997).

Despite the clear demonstration that schistosomes are capable of developing significant resistance to the action of these drugs and the discovery of resistant schistosomes in the field, the problem of resistance never dramatically hindered the clinical usefulness of oxamniquine, even in Brazil where the drug was most aggressively applied. This has been attributed to a decreased reproductive capacity associated with worms lacking the relevant enzyme activity (Cioli *et al.*, 1993), such that the resistant worms were selected against due to other biological costs associated with oxamniquine resistance.

Resistance to praziquantel

As with oxamniquine, scientists have been able to induce resistance to praziquantel in the laboratory (Fallon and Doenhoff, 1994). Schistosome-infected mice were given sub-curative doses of praziquantel for seven passages through the life cycle, and with successive passages the dosage was increased. By the seventh passage, three doses of 300 mg/kg eliminated only 7% of the worms, whereas this dose eliminated at least 80% of the worms in praziquantel naïve infections. Praziquantel pressure produced a population of worms that was significantly less responsive to the drug in only seven generations.

These studies demonstrate the genetic capacity of a population of schistosomes to shift toward less responsive phenotypes under drug pressure. In light of the aggressive praziquantel pressure applied to the populations of schistosomes in the field (for example, Egypt), these results raise legitimate concerns regarding the possibility of praziquantel resistance emerging in the field.

Senegal

One focus of potential resistance to praziquantel has been Northern Senegal, where concern was raised by a 1994 study which found cure rates of only 18% (Gryseels *et al.*, 1994; Stelma *et al.*, 1995). Although different protocols for follow-up found higher cure rates, even at best the cure rate was half that expected (Stelma *et al.*, 1997). Further studies continued to find lower-than-expected cure rates in Senegal (Picquet *et al.*, 1998; De Clercq *et al.*, 1999; Ernould *et al.*, 1999).

The failure of praziquantel to produce expected efficacy prompted a great deal of discussion and attentive study of these foci in Senegal. Patent snails collected from these areas in Senegal were used to generate miracidia for infection of laboratory mice. In these murine infections, praziquantel proved less effective against the Senegalese isolate as compared to other laboratory isolates (Fallon *et al.*, 1995). This demonstrated that at least some of the lower-than-expected efficacy of the drug could be attributed to the worms (rather than host factors), raising the alarming prospect that there were worms in Senegal which could provide the substrate for a growing population of schistosomes that would not respond to praziquantel. The pursuant studies in Senegal and ongoing evaluation of the significance of the results have helped to clarify some of the issues that can contribute to lower-than-expected cure rates.

Firstly, low cure rates can be a function of very high intensity infections. Indeed the intensity of infection in Northern Senegal was extremely high compared to a great majority of schistosomiasis cases (Stelma *et al.*, 1993; Gryseels *et al.*, 2001). One important study assembled and standardized data from a number of reports of praziquantel efficacy in the field and compared these to the troubling data emerging from Senegal (Danso-Appiah and De Vlas, 2002). Analysis suggested that the expected cure rate of 80% is indeed reasonable for lower intensity infections, but a lower cure rate should be expected with higher intensity infections. Thus, the high intensity of infection accounted for some, but not all of the lower-than-expected cure rates in Senegal.

Studies of schistosomiasis in Senegal also reinforced in fact what many scientists knew in principle: the time between treatment and disease testing will have an effect on the apparent cure rate. Even in the initial study, the investigators found a shorter follow-up produced a better apparent cure rate, 18% at 12 weeks and 38% at 8 weeks (Gryseels *et al.*, 1994; van Lieshout *et al.*, 1999). Further examination reinforced this principle (Danso-Appiah and De Vlas, 2002). Studies in Senegal also attributed some of the lower-than-expected cure rates to more sensitive diagnosis, pointing to the variability of apparent cure rates that can be generated by different approaches to detecting and defining 'cure'.

Even after considering the high burden of infection, the short follow-up times and the sensitive diagnosis, the cure rates from Senegal were lower than expected. Since worms derived from these regions produced less responsive murine infections (Fallon *et al.*, 1995), one explanation was that the schistosomes were praziquantel-tolerant, but other possibilities remained. Specifically, since transmission of schistosomiasis was very intense in Senegal at the time, it was possible that rapid reinfection was playing a role in the high rate of apparent drug failure. Supporting this explanation, patients from the problematic foci who were relocated to urban areas (without ongoing transmission) experienced normal cure rates.

A clear lesson from this series of investigations is that praziquantel produces the expected cure rates of 70–95% in 'normal' endemic situations, but lower cure rates should not be startling when the intensity of infection is high or the levels of transmission are elevated (Gryseels *et al.*, 2001). Although these factors might not account for all of the drug failure that occurred in Senegal (Danso-Appiah and De Vlas, 2002), there has been no compelling evidence for the emergence of a praziquantel-resistant strain of schistosomes.

All of this emphasizes the extreme difficulty in attributing lower-than-expected cure rates to schistosome resistance. Even in this case where the initial report was an 82% failure rate, closer examination found that this high failure rate was most likely not a function of drug resistance.

Egypt

Whereas the drug failures in northern Senegal elicited the most focused field-based studies of praziquantel resistance, a group of worms isolated from patients not cured by praziquantel in Egypt have been the substrate of the most in-depth laboratory study of potentially resistant worms.

A 1994 study identified a number of patients not cured after receiving three successive doses of praziquantel in the Nile Delta region of Egypt (Ismail *et al.*, 1996). The study included eight villages with a 25% prevalence of *S. mansoni*, and the infections were much less intense than those in the Senegal focus described above. After a single dose of 40 mg/kg praziquantel, 21% of these patients continued to pass viable eggs, well within the normal range of expectations. These patients were treated again, and 28% of these were still passing viable eggs after a second dose of 40 mg/kg. For these patients, a third dose of 60 mg/kg was administered with a 27% failure rate. In each case, the cure rate was not abnormally low, but the study identified 24 patients (1.6% of the original patients) who continued to pass viable eggs after these three doses of praziquantel.

As has already been pointed out, this failure rate is really well within the expected cure rates for praziquantel. Using the range of an expected 70–95% cure rate mentioned above, one could expect a failure rate ranging from 0.001% to 2.7% after three treatments, and the 1.6% failure rate reported here is squarely within that expectation. So, these failures in and of themselves are not an indication of inordinate drug failure or resistance.

However, the results raised the question as to whether this type of treatment is selecting for worms less responsive to praziquantel, leaving only less responsive worms to reproduce in that given area. The only clear way to determine if the failures are attributable to worms not responding to praziquantel is the rather arduous protocol of establishing laboratory murine infections from the eggs expelled by the patients not being cured.

Three-fourths of the isolates derived from patients not cured by praziquantel produced murine infections that were significantly more difficult to cure than control infections. This indicated that these worms did, in fact, have a decreased sensitivity. The differences were significant, but not stunning in magnitude. For one example, the ER-6 isolate required 680 mg/kg to clear half the worms in murine infections, compared to 96 mg/kg in field-derived controls (Ismail *et al.*, 1996). In general, the elevated doses for the isolates from non-responding patients were between three and eight times those of controls. This suggested that a great majority of these patients not cured by praziquantel were harbouring worms less sensitive than naïve worms to praziquantel.

In order to confirm this, the isolates were examined with respect to their responses to praziquantel *in vitro*. Two of the hallmark effects are contraction of the worm's musculature (Pax *et al.*, 1978) and a disruption of the worm's tegument (Bricker *et al.*, 1983). The Egyptian isolates that were significantly more difficult to cure in mice were also significantly less responsive to the action of praziquantel *in vitro*, both in terms of muscle contraction (Ismail *et al.*, 1999) and tegumental disruption (William *et al.*, 2001a).

These results demonstrated that Egyptian patients that were not cured by the recommended doses of praziquantel-harboured worms that were demonstrably less responsive to the drug both *in vivo* and *in vitro*. This fact produced concern that these worms could result in a large population of schistosomes that would not be cured by praziquantel. However, further study of these worms found two issues that would limit the spread of these less responsive worms in the field. Firstly, some of the isolates reverted to a praziquantel-sensitive state with the removal of praziquantel pressure. Secondly, all of the isolates that did not revert to a praziquantel sensitivity suffered from decreased biological fitness.

After multiple passages through the life cycle in the absence of drug pressure, one-third of the experimental isolates reverted to sensitivity not significantly different from controls (William *et al.*, 2001b). For example, the ER-6 isolate mentioned above initially required 680 mg/kg praziquantel to effect a 50% reduction in worm load in murine infections, but after only six passages through the life cycle over 5 years this was reduced to 113 mg/kg, not significantly different from control infections. These results are significant in that they demonstrate the ability of at least some of the less responsive isolates to revert to a sensitive state. Based on other cases of helminth drug resistance, the reversion from a drug-resistant state back to susceptibility has been infrequent (Conder and Campbell, 1995; Anderson *et al.*, 1998; Geerts and Gryseels, 2000). It is important to point out that the schistosome isolates studied here were not genetically homogeneous, but rather the product of the genetically heterogeneous pool of eggs surviving three doses of praziquantel. This is probably a good approximation of the situation that would occur in the field, and so the capacity of some of the isolates to revert to susceptibility bodes well.

All of the experimental isolates that retained their decreased sensitivity to praziquantel showed compromised reproductive fitness in the laboratory, expressed most frequently as a decreased cercarial production from snails infected with those isolates compared to controls. For example, the tota cercarial production of snails infected with one isolate was only 57% of control isolates (William *et al.*, 2001b). So, despite the presence of some worms more capable of surviving exposure to the recommended therapeutic levels of praziquantel, it is not clear that these worms will provide the genetic stock for an epidemiologically significant core of resistant schistosomes in any of these villages.

Subsequent studies in the same villages have examined this very issue. Since the 1994 studies, praziquantel has been used frequently and widely throughout Egypt, including in these villages in the Delta region. Praziquantel treatment in these villages has come in at least two forms: (1) biannual treatment of school children and (2) at least two mass treatment campaigns involving mass distribution of the drug by the Ministry of Health and Population authorities (WHO, 2000).

The 2004 studies showed that the prevalence of infection had significantly decreased in the villages since the original study in 1994, dropping from 25% to below 11% (Botros *et al.*, 2005). In addition, the egg burden in those still-infected was small, with the vast majority shedding fewer than 100 eggs per gram of faeces. This is more anecdotal testimony to the efficacy of praziquantel-based control programmes, and clear indication that the drug has indeed been aggressively applied in the intervening years since the original discovery of the less responsive isolates.

The encouraging result was that the 2004 follow-up study found no patients not cured by the same dosing protocol used in the 1994 study (which had found a 1.6% failure rate after three doses). Because the sample size in the follow-up study was not as large as the original study, it could not conclude that there had actually been a decrease in the failure rate, but the data did show that there was not a dramatic increase in the failure rate in these villages. For example, if the real frequency of drug failure had doubled from 1.6% to 3.2%, the chances of the follow-up study not finding a single patient with treatment failure would have been less than 1 in 1000 (Botros *et al.*, 2005). In the very villages where patients were found to harbour unresponsive isolates before 10 years, and where praziquantel had been applied in the intervening years, no significant increase in praziquantel failure rate could be found.

These results do not support any rapid emergence of resistance to praziquantel in these villages in Egypt. Nor is there any other evidence of resistance to the drug limiting the efficacy of chemotherapy-based control programmes throughout Egypt – the country which has generated the most sustained and intense use of praziquantel. The infections remaining after chemotherapy campaigns tend to be light, and reproduction is potentially further slowed by compromises in reproductive capacity of the surviving worms.

Concluding Remarks

There is no substantive report of praziquantel resistance producing a dramatic public health impact at the time of this writing. The two most studied locations where praziquantel resistance has been suspected have been Senegal and Egypt. In Senegal, a great deal of the original drug failure could be explained by the heavy infections and rapid reinfection, and there has not been a documented surge in praziquantel-resistant schistosomiasis. In Egypt, even though some patients have been found to harbour worms that are not responsive to the recommended doses of the drug, there has not been any evidence of an increasing population of praziquantel-resistant schistosomes. On the contrary, the very villages that produced the praziquantel-resistant worms a decade ago have not experienced an increase in drug failure.

However, there is very clear data showing that schistosomes have the biological and genetic capacity to develop resistance to chemotherapeutics. This has been demonstrated with both oxamniquine and praziquantel in the laboratory, where exposure to sub-curative doses of the drug over generations will result in a population of significantly less susceptible worms. Further, it is known that the recommended dose of praziquantel leaves a significant number of worms alive and producing eggs after exposure to the drug. There are documented cases of worms found in the field that are demonstrably less sensitive to the drugs than would be expected.

Given these facts, even though there has not yet been a substantive report of praziquantel resistance of public health significance, it seems warranted to keep a watchful eye for potential resistance as the drug is administered aggressively to combat schistosomiasis-related morbidity. There remain, however, a great number of challenges that make it unclear as to exactly how this can be done most effectively. It remains very difficult to identify a nascent pocket of resistance in the face of an expected drug failure rate of 5–30%.

Ongoing epidemiological and laboratory studies continue to elucidate a number of factors that can contribute to lower-than-expected cure rates for praziquantel. As these factors become more clearly and precisely understood, variability in the field results with the drug will be less ambiguous and significant drug failures may be identified more readily.

Laboratory studies continue to explore the molecular mechanisms of praziquantel's anti-schistosomal action. The identification of the critical target molecules within the parasite would produce hope for finding the molecular basis underlying worms capable of surviving exposure to the drug. This could lead to a simple molecular monitoring assay. Likewise, efforts to identify other non-molecular, field-amenable assays for unresponsive worms are ongoing. Any means of clearly attributing drug failure to resistant worms would be of great value.

Given the immense value of praziquantel, the clear demonstration that schistosomes are capable of developing resistance to drugs, and the widespread applications of praziquantel that are planned for the future, diligence in monitoring for the potential emergence of resistance is in order.

References

Anderson, T.J., Blouin, M.S. and Beech, R.N. (1998) Population biology of parasitic nematodes: applications of genetic markers. *Advances in Parasitology* 41, 219–283.

Andrews, P. (1985) Praziquantel: mechanisms of anti-schistosomal activity. *Pharmacology and Therapeutics* 29, 129–156.

Andrews, P., Thomas, H., Pohlke, R. and Seubert, J. (1983) Praziquantel. *Medicinal Research Reviews* 3, 147–200.

Barakat, R., Farghaly, A., El Masry, A.G., El-Sayed, M.K. and Hussein, M.H. (2000) The epidemiology of schistosomiasis in Egypt: patterns of *Schistosoma mansoni* infection and morbidity in Kafer El-Sheikh. *American Journal of Tropical Medicine and Hygiene* 62, 21–27.

Behbehani, M. and Savioli, L. (1998) Report of the WHO informal consultation on monitoring of drug efficacy in the control of schistosomiasis and intestinal nematodes. World Health Organization, Geneva.

Bonesso-Sabadini, P.I.P. and Dias, L.C. (2002) Altered response of strain of *Schistosoma mansoni* to oxamniquine and praziquantel. *Memorias do Instituto Oswaldo Cruz* 97, 381–385.

Botros, S., Sayed, H., Amer, N., El-Ghannam, M., Bennett, J.L. and Day, T.A. (2005) Current status of sensitivity to praziquantel in a focus of potential drug resistance in Egypt. *International Journal for Parasitology* 35, 787–791.

Bricker, C.S., Depenbusch, J.W., Bennett, J.L. and Thompson, D.P. (1983) The relationship between tegumental disruption and muscle contraction in *Schistosoma mansoni* exposed to various compounds. *Zeitschrift für Parasitenkund* 69, 61–71.

Brindley, P.J. and Sher, A. (1987) The chemotherapeutic effect of praziquantel against *Schistosoma mansoni* is dependent on host antibody response. *Journal of Immunology* 139, 215–220.

Brindley, P.J., Strand, M., Norden, A.P. and Sher, A. (1989) Role of host antibody in the chemotherapeutic action of praziquantel against *Schistosoma mansoni*: identification of target antigens. *Molecular and Biochemical Parasitology* 34, 99–108.

Bruce, J.I., Dias, L.C., Liang, Y.S. and Coles, G.C. (1987) Drug resistance in schistosomiasis: a review. *Memorias do Instituto Oswaldo Cruz* 82(Suppl. 4), 143–150.

Campos, R., Moreira, A.A., Sette, H., Chamone, D.A. and Da Silva, L.C. (1976) Hycanthone resistance in a human strain of *Schistosoma mansoni*. *Transactions of the Royal Society of Tropical Medicine and Hygiene* 70, 261–262.

Chitsulo, L., Engels, D., Montresor, A. and Savioli, L. (2000) The global status of schistosomiasis and its control. *Acta Tropica* 77, 41–51.

Chitsulo, L., Loverde, P. and Engels, D. (2004) Schistosomiasis. *Nature Reviews Microbiology* 2, 12–13.

Cioli, D. and Pica Mattoccia, L. (1984) Genetic analysis of hycanthone resistance in *Schistosoma mansoni*. *American Journal of Tropical Medicine and Hygiene* 33, 80–88.

Cioli, D., Pica-Mattoccia, L., Rosenberg, S. and Archer, S. (1985) Evidence for the mode of antischistosomal action of hycanthone. *Life Sciences* 37, 161–167.

Cioli, D., Pica-Mattoccia, L. and Moroni, R. (1992) *Schistosoma mansoni*: hycanthone/oxamniquine resistance is controlled by a single autosomal recessive gene. *Experimental Parasitology* 75, 425–432.

Cioli, D., Pica-Mattoccia, L. and Archer, S. (1993) Drug resistance in schistosomes. *Parasitology Today* 9, 162–166.

Cioli, D., Pica-Mattoccia, L. and Archer, S. (1995) Antischistosomal drugs: past, present ... and future? *Pharmacology and Therapeutics* 68, 35–85.

Cline, B.L., Richards, F.O., El Alamy, M.A., El Hak, S., Ruiz-Tiben, E., Hughes, J.M. and McNeeley, D.F. (1989) 1983 Nile Delta schistosomiasis survey: 48 years after Scott. *American Journal of Tropical Medicine and Hygiene* 41, 56–62.

Coles, G.C. and Kinoti, G. (1997) Defining resistance in schistosoma. *Parasitology Today* 13, 157–158.

Coles, G.C., Mutahi, W.T., Kinoti, G.K., Bruce, J.I. and Katz, N. (1987) Tolerance of Kenyan *Schistosoma mansoni* to oxamniquine. *Transactions of the Royal Society of Tropical Medicine and Hygiene* 81, 782–785.

Conder, G.A. and Campbell, W.C. (1995) Chemotherapy of nematode infections of veterinary importance, with special reference to drug resistance. *Advances in Parasitology* 35, 1–84.
Danso-Appiah, A. and De Vlas, S.J. (2002) Interpreting low praziquantel cure rates of *Schistosoma mansoni* infections in Senegal. *Trends in Parasitology* 18, 125–129.
Day, T.A., Bennett, J.L. and Pax, R.A. (1992) Praziquantel: the enigmatic antiparasitic. *Parasitology Today* 8, 342–344.
De Clercq, D., Vercruysse, J., Picquet, M., Shaw, D.J., Diop, M., Ly, A. and Gryseels, B. (1999) The epidemiology of a recent focus of mixed *Schistosoma haematobium* and *Schistosoma mansoni* infections around the 'Lac de Guiers' in the Senegal River Basin, Senegal. *Tropical Medicine and International Health* 4, 544–550.
Degu, G., Mengistu, G. and Jones, J. (2002) Praziquantel efficacy against schistosomiasis mansoni in schoolchildren in north-west Ethiopia. *Transactions of the Royal Society of Tropical Medicine and Hygiene* 96, 444–445.
Dias, L.C., Pedro, R.J., Rigo, E., Goto, M.M. and Mafra, G.L. (1978) A human strain of *Schistosoma mansoni* resistant to schistosomicides. *Revista de Saude Publica* 12, 110.
Dias, L.C., Pedro, R.J. and Debelardini, E.R. (1982) Use of praziquantel in patients with schistosomiasis mansoni previously treated with oxamniquine and/or hycanthone: resistance of *Schistosoma mansoni* to schistosomicidal agents. *Transactions of the Royal Society of Tropical Medicine and Hygiene* 76, 652–659.
Drescher, K.M., Rogers, E.J., Bruce, J.I., Katz, N., Dias, L.C. and Coles, G.C. (1993) Response of drug resistant isolates of *Schistosoma mansoni* to antischistosomal agents. *Memorias do Instituto Oswaldo Cruz* 88, 89–95.
El-Khoby, T., Galal, N. and Fenwick, A. (1998) The USAID/Government of Egypt's Schistosomiasis Research Project (SRP). *Parasitology Today* 14, 92–96.
El-Khoby, T., Galal, N., Fenwick, A., Barakat, R., El-Hawey, A., Nooman, Z., Habib, M., Abdel-Wahab, F., Gabr, N.S., Hammam, H.M., Hussein, M.H., Mikhail, N.N., Cline, B.L. and Strickland, G.T. (2000) The epidemiology of schistosomiasis in Egypt: summary findings in nine governorates. *American Journal of Tropical Medicine and Hygiene* 62, 88–99.
Ernould, J.C., Ba, K. and Sellin, B. (1999) Increase of intestinal schistosomiasis after praziquantel treatment in a *Schistosoma haematobium* and *Schistosoma mansoni* mixed focus. *Acta Tropica* 73, 143–152.
Fallon, P.G. and Doenhoff, M.J. (1994) Drug-resistant schistosomiasis: resistance to praziquantel and oxamniquine induced in *Schistosoma mansoni* in mice is drug specific. *American Journal of Tropical Medicine and Hygiene* 51, 83–88.
Fallon, P.G., Sturrock, R.F., Niang, A.C. and Doenhoff, M.J. (1995) Short report: diminished susceptibility to praziquantel in a Senegal isolate of *Schistosoma mansoni*. *American Journal of Tropical Medicine and Hygiene* 53, 61–62.
Fallon, P.G., Mubarak, J.S., Fookes, R.E., Niang, M., Butterworth, A.E., Sturrock, R.F. and Doenhoff, M.J. (1997) *Schistosoma mansoni*: maturation rate and drug susceptibility of different geographic isolates. *Experimental Parasitology* 86, 29–36.
Fenwick, A., Savioli, L., Engels, D., Bergquist, R.N. and Todd, M.H. (2003) Drugs for the control of parasitic diseases: current status and development in schistosomiasis. *Trends in Parasitology* 19, 509–515.
Geerts, S. and Gryseels, B. (2000) Drug resistance in human helminths: current situation and lessons from livestock. *Clinical Microbiology Reviews* 13, 207–222.
Gonnert, R. and Andrews, P. (1977) Praziquantel, a new board-spectrum antischistosomal agent. *Zeitschrift für Parasitenkund* 52, 129–150.
Gryseels, B., Nkulikyinka, L. and Coosemans, M.H. (1987) Field trials of praziquantel and oxamniquine for the treatment of schistosomiasis mansoni in Burundi. *Transactions of the Royal Society of Tropical Medicine and Hygiene* 81, 641–644.
Gryseels, B., Stelma, F.F., Talla, I., van Dam, G.J., Polman, K., Sow, S., Diaw, M., Sturrock, R.F., Doehring-Schwerdtfeger, E. and Kardorff, R. (1994) Epidemiology, immunology and chemotherapy of *Schistosoma mansoni* infections in a recently exposed community in Senegal. *Tropical and Geograpical Medicine* 46, 209–219.
Gryseels, B., Mbaye, A., De Vlas, S.J., Stelma, F.F., Guisse, F., Van Lieshout, L., Faye, D., Diop, M., Ly, A., Tchuem-Tchuente, L.A., Engels, D. and Polman, K. (2001) Are poor responses to praziquantel for the treatment of *Schistosoma mansoni* infections in Senegal due to resistance? An overview of the evidence. *Tropical Medicine and International Health* 6, 864–873.

Guimaraes, R.X., Tchakerian, A., Dias, L.C., de Almeida, F.M., Vilela, M.P., Cabeca, M. and Takeda, A.K. (1979) Resistance to hycanthone and oxamniquine in patients with a clinical hepato-intestinal form of schistosomiasis. *AMB; Revista da Associaco Medica Brasileira* 25, 48–50.
Ismail, M., Metwally, A., Farghaly, A., Bruce, J., Tao, L.F. and Bennett, J.L. (1996) Characterization of isolates of *Schistosoma mansoni* from Egyptian villagers that tolerate high doses of praziquantel. *American Journal of Tropical Medicine and Hygiene* 55, 214–218.
Ismail, M., Botros, S., Metwally, A., William, S., Farghally, A., Tao, L.F., Day, T.A. and Bennett, J.L. (1999) Resistance to praziquantel: direct evidence from *Schistosoma mansoni* isolated from Egyptian villagers. *American Journal of Tropical Medicine and Hygiene* 60, 932–935.
Katz, N. (1975) Clinical evaluation of niridazole and hycanthone in schistosomiasis mansoni endemic areas. *Journal of Toxicology and Environmental Health* 1, 203–209.
Katz, N., Zicker, F. and Pereira, J.P. (1977) Field trials with oxamniquine in a schistosomiasis mansoni-endemic area. *American Journal of Tropical Medicine and Hygiene* 26, 234–237.
Katz, N., Rocha, R.S. and Chaves, A. (1979) Preliminary trials with praziquantel in human infections due to *Schistosoma mansoni*. *Bulletin of the World Health Organisation* 57, 781–785.
Kenworthy, J.D., Ye, P., Wu, G.C., Yu, H., Shi, Y.J., Li, H. and Coles, G.C. (2003) Field evaluation of a test for praziquantel resistance in *Schistosoma* sp. *Veterinary Parasitology* 113, 83–87.
Kinoti, G. (1987) The significance of variation in susceptibility of *Schistosoma mansoni* to the antischistosomal drug oxamniquine. *Memorias do Instituto Oswaldo Cruz* 82, 151–156.
Liang, Y.S., Coles, G.C., Doenhoff, M.J. and Southgate, V.R. (2001) *In vitro* responses of praziquantel-resistant and -susceptible *Schistosoma mansoni* to praziquantel. *International Journal for Parasitology* 31, 1227–1235.
Pax, R., Bennett, J.L. and Fetterer, R. (1978) A benzodiazepine derivative and praziquantel: effects on musculature of *Schistosoma mansoni* and *Schistosoma japonicum*. *Naunyn Schmiedebergs Archives of Pharmacology* 304, 309–315.
Pica-Mattoccia, L., Archer, S. and Cioli, D. (1992) Hycanthone resistance in schistosomes correlates with the lack of an enzymatic activity which produces the covalent binding of hycanthone to parasite macromolecules. *Molecular and Biochemical Parasitology* 55, 167–175.
Pica-Mattoccia, L., Novi, A. and Cioli, D. (1997) Enzymatic basis for the lack of oxamniquine activity in *Schistosoma haematobium* infections. *Parasitology Research* 83, 687–689.
Picquet, M., Vercruysse, J., Shaw, D.J., Diop, M. and Ly, A. (1998) Efficacy of praziquantel against *Schistosoma mansoni* in northern Senegal. *Transactions of the Royal Society of Tropical Medicine and Hygiene* 92, 90–93.
Prichard, R.K., Hall, C.A., Kelly, J.D., Martin, I.C. and Donald, A.D. (1980) The problem of anthelmintic resistance in nematodes. *Australian Veterinary Journal* 56, 239–251.
Rogers, S.H. and Bueding, E. (1971) Hycanthone resistance: development in *Schistosoma mansoni*. *Science* 172, 1057–1058.
Sabah, A.A., Fletcher, C., Webbe, G. and Doenhoff, M.J. (1985) *Schistosoma mansoni*: reduced efficacy of chemotherapy in infected T-cell-deprived mice. *Experimental Parasitology* 60, 348–354.
Santos, A.T., Blas, B.L., Nosenas, J.S., Portillo, G.P., Ortega, O.M., Hayashi, M. and Boehme, K. (1979) Preliminary clinical trials with praziquantel in *Schistosoma japonicum* infections in the Philippines. *Bulletin of the World Health Organisation* 57, 793–799.
Scott, J. (1937) The incidence and distribution of human schistosomiasis in Egypt. *American Journal of Hygiene* 25, 566–614.
Stelma, F.F., Talla, I., Polman, K., Niang, M., Sturrock, R.F., Deelder, A.M. and Gryseels, B. (1993) Epidemiology of *Schistosoma mansoni* infection in a recently exposed community in northern Senegal. *American Journal of Tropical Medicine and Hygiene* 49, 701–706.
Stelma, F.F., Talla, I., Sow, S., Kongs, A., Niang, M., Polman, K., Deelder, A.M. and Gryseels, B. (1995) Efficacy and side effects of praziquantel in an epidemic focus of *Schistosoma mansoni*. *American Journal of Tropical Medicine and Hygiene* 53, 167–170.
Stelma, F.F., Sall, S., Daff, B., Sow, S., Niang, M. and Gryseels, B. (1997) Oxamniquine cures *Schistosoma mansoni* infection in a focus in which cure rates with praziquantel are unusually low. *Journal of Infectious Disease* 176, 304–307.
van Lieshout, L., Stelma, F.F., Guisse, F., Falcao Ferreira, S.T., Polman, K., van Dam, G.J., Diakhate, M., Sow, S., Deelder, A. and Gryseels, B. (1999) The contribution of host-related factors to low cure rates of praziquantel for the treatment of *Schistosoma mansoni* in Senegal. *American Journal of Tropical Medicine and Hygiene* 61, 760–765.

WHO (2000) Report of the WHO informal consultation on schistosomiasis in low transmission areas: control strategies and criteria for elimination. World Health Organization, London.

William, S., Botros, S., Ismail, M., Farghally, A., Day, T.A. and Bennett, J.L. (2001a) Praziquantel-induced tegumental damage *in vitro* is diminished in schistosomes derived from praziquantel-resistant infections. *Parasitology* 122, 63–66.

William, S., Sabra, A., Ramzy, F., Mousa, M., Demerdash, Z., Bennett, J.L., Day, T.A. and Botros, S. (2001b) Stability and reproductive fitness of *Schistosoma mansoni* isolates with decreased sensitivity to praziquantel. *International Journal for Parasitology* 31, 1093–1000.

14 Praziquantel: Mechanism of Action

R.M. Greenberg
Marine Biological Laboratory, 7 MBL Street, Woods Hole, MA 02543, USA

Introduction

Schistosomiasis affects approximately 200 million people worldwide, an estimated 85% of whom live in Africa (Engels *et al.*, 2002), with as many as 280,000 deaths per year attributed to the disease (van der Werf *et al.*, 2003). The causative agents of schistosomiasis are trematode flatworms of the genus *Schistosoma*. Investigations into vaccine strategies against schistosomiasis continue (see reviews by Bergquist *et al.*, 2002; Capron *et al.*, 2002; Hagan and Sharaf, 2003; Wilson and Coulson, this volume), and public health measures are clearly important for controlling the disease. Currently, however, treatment of schistosomiasis remains dependent on chemotherapeutic intervention, particularly with praziquantel.

Praziquantel (Fig. 14.1) is a pyrazinoisoquinoline anthelmintic that was discovered in the 1970s and subsequently introduced for the treatment of schistosomiasis (reviewed by Andrews *et al.*, 1983). It is an asymmetric molecule, and standard preparations are composed of equal proportions of levo (–) and dextro (+) optical isomers. The (–) enantiomer is the active form, based on experiments both *in vivo* (Andrews *et al.*, 1983; Liu *et al.*, 1986; Tanaka *et al.*, 1989; Wu *et al.*, 1991; Xiao *et al.*, 1999) and *in vitro* (Xiao and Catto, 1989; Staudt *et al.*, 1992).

Praziquantel is the current drug of choice against schistosomiasis. Indeed, in recent years, praziquantel has effectively become the sole anti-schistosomal agent that is commercially available (Fenwick *et al.*, 2003; Hagan *et al.*, 2004). Praziquantel is effective against all species of schistosomes and shows minimal side effects. It is also active against other trematode and cestode infections, though not against nematodes (reviewed by Andrews, 1985). However, schistosomes do show stage- and sex-dependent differences in susceptibility to praziquantel (Xiao *et al.*, 1985; Sabah *et al.*, 1986; Pica-Mattoccia and Cioli, 2004).

The effectiveness of praziquantel has been demonstrated repeatedly in large-scale schistosomiasis control efforts in various regions, but it has been severely underutilized in areas such as sub-Saharan Africa (reviewed by Fenwick

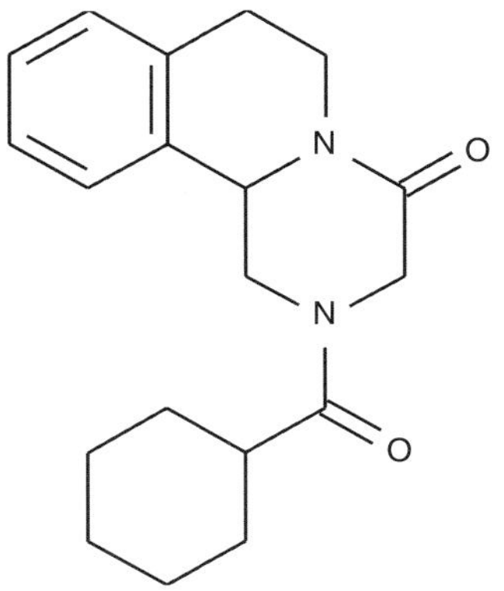

Fig. 14.1. Chemical structure of praziquantel.

et al., 2003; Hagan *et al.*, 2004). An attempt to remedy this situation can be found in the recent launch of the Schistosomiasis Control Initiative (www.schisto.org), funded in large measure by the Bill and Melinda Gates Foundation, which has been initiated in part to develop sustainable schistosomiasis control programmes in this neglected region.

With praziquantel serving in effect as the only anti-schistosomal treatment in widespread use, the possibility of emerging drug resistance has alarming implications. Recent discussions of this problem can be found elsewhere (see e.g. Cioli, 2000; Doenhoff *et al.*, 2002), as well as in this volume (Day and Botros, this volume), and will not be explored further here. However, the potential for the emergence of praziquantel-resistant strains of schistosomes is of special concern considering that the mechanism of action of the drug remains undefined.

Praziquantel Mode of Action

Despite widespread use and nearly 30 years of experimental inquiry, the precise manner in which praziquantel acts continues to be largely unresolved (reviewed by Day *et al.*, 1992; Redman *et al.*, 1996; Cioli and Pica-Mattoccia, 2003). This chapter will focus on recent advances in identifying the molecular target of praziquantel, spotlighting on those experiments indicating a critical role for voltage-gated calcium (Ca^{2+}) channel proteins.

Praziquantel is effective against several different platyhelminths, but typically shows minimal activity against other organisms, including other helminths such as nematodes (Andrews *et al.*, 1983). Thus, this compound appears to be highly selective against members of the phylum Platyhelminthes. Based on this apparent specificity, one might predict that the molecular target of praziquantel should be present only in members of this phylum. Accordingly, the drug's target (or targets) could be encoded by a novel gene found exclusively in the flatworms. Recent examination of schistosome genomes and transcriptomes has identified several sequences that show no apparent homology to genes from any other phyla (reviewed by Hu *et al.*, 2004; LoVerde *et al.*, 2004; McManus *et al.*, 2004; Verjovski-Almeida *et al.*, 2004). Alternatively, the target for praziquantel could be encoded by a member of a gene family found in other phyla as well as in the platyhelminths, but with particular structural signatures that interact specifically with the drug. These signatures could represent minor differences in the primary sequence of the protein that have major consequences. Changes of single amino acid residues in critical domains can produce dramatic modifications in the functional and pharmacological properties of typical receptors and channels (see e.g. Heinemann *et al.*, 1992; Satin *et al.*, 1992).

The questions surrounding the mechanism of praziquantel action can be illustrated by revisiting a modified version of a figure initially published over a decade ago (Fig. 14.2; Day *et al.*, 1992; Harder, 2002). As will be discussed in more detail below, praziquantel produces a clear, well-documented effect on intracellular Ca^{2+} levels in schistosomes. This disruption of Ca^{2+} homeostasis triggers a cascade of events that eventually leads to the elimination of adult parasites from the host. However, as shown in Fig. 14.2, the entities which interact with praziquantel and mediate these events in effect represent a black box, with the molecular target of the drug and the location of that target remaining undefined.

Though elucidating the mode of action of praziquantel has proved a daunting task, the effects of the drug on adult schistosomes do provide clues to potential targets for the drug. Application of praziquantel results in a

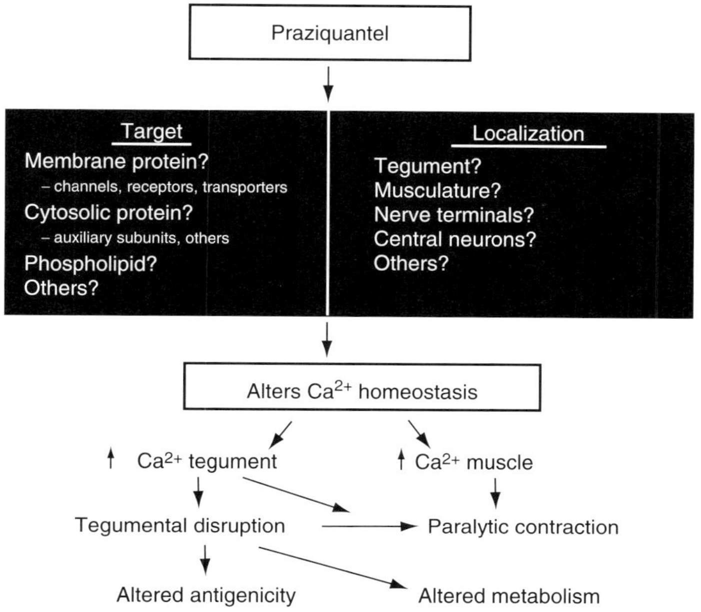

Fig. 14.2. Questions regarding mechanism of action of praziquantel. Praziquantel affects Ca^{2+} homeostasis in the adult worm, which results in a cascade of events that ultimately lead to elimination of parasites from the host. Entities mediating these effects and their location are not currently defined and are presented within a 'black box'. Possible sites of action are shown. Figure updated from Day *et al.* (1992) and Harder (2002).

disruption of Ca^{2+} homeostasis in the adult parasite (reviewed by Andrews, 1985; Day *et al.*, 1992; Redman *et al.*, 1996). Within seconds of exposure of adult schistosomes to the drug, there is a rapid, sustained contraction of the worm's musculature (Fetterer *et al.*, 1980). Praziquantel also produces vacuolization and disruption of the parasite tegument (Becker *et al.*, 1980; Mehlhorn *et al.*, 1981), which is associated with the subsequent exposure of parasite antigens on the surface of the worm (Harnett and Kusel, 1986). Both the contraction of the musculature and the effects on the tegument are considered to be linked to the praziquantel-dependent disruption of Ca^{2+} homeostasis (reviewed by Day *et al.*, 1992; Redman *et al.*, 1996).

Application of praziquantel results in a rapid uptake of $^{45}Ca^{2+}$ in adult male schistosomes (and a much slower influx of Na^{+}; Pax *et al.*, 1978). The effects of praziquantel on both contraction of the worm's musculature and disruption of the parasite tegument are both Ca^{2+}-dependent processes. Thus, removal of Ca^{2+} from the extraworm bathing medium blocks each of these responses (Pax *et al.*, 1978; Wolde Mussie *et al.*, 1982; Xiao *et al.*, 1984), but in both cases, there is a delay before these inhibitory effects are seen. For example, the inhibitory effects of Ca^{2+}-free medium on the praziquantel-dependent contraction of the musculature require at least 10 min to become apparent, a delay thought to be indicative of the time required for the depletion of sequestered intracellular Ca^{2+} stores. These results indicate that extracellular Ca^{2+} is not required for the initiation of praziquantel-dependent effects, but is required for full and sustained responses.

Both the tegument and the sarcolemma appear to contain praziquantel-sensitive sites (Blair *et al.*, 1992). When intact worms are bathed in high magnesium, they exhibit a praziquantel-dependent biphasic muscle contraction instead of the tonic contraction that normally occurs. However, although worms with the tegument removed continue to respond to praziquantel, they show only a single, prominent phasic contraction, indicating involvement of a tegumental site for the full response. Furthermore, application of praziquantel to detegumented parasites incubated in Ca^{2+}-free medium does not produce the muscular contraction. Interestingly, micromolar (1–2 μM) concentrations of praziquantel apparently interact with both sarcolemmal and intracellular sites to produce a sustained Ca^{2+}-dependent contraction of the penile retractor muscle from the mollusc *Lymnaea stagnalis* (Gardner and Brezden, 1984).

Several cellular factors are involved in regulating intracellular levels of Ca^{2+} and have been discussed as plausible targets mediating the effects of praziquantel on Ca^{2+} homeostasis (reviewed by Redman *et al.*, 1996). They included various channels that carry Ca^{2+} such as voltage-, ligand- and second messenger-gated Ca^{2+} channels and intracellular Ca^{2+} release channels. Also considered were pumps and transporters that regulate intracellular Ca^{2+} concentrations, as well as intracellular Ca^{2+} buffers. Early experiments indicated that praziquantel is not acting as an ionophore (Pax *et al.*, 1978). In addition to these factors, there are reports of praziquantel altering the structure of membrane bilayer phospholipids or membrane fluidity (Harder *et al.*, 1988; Lima *et al.*, 1994), effects which could result in changes in ion permeability of membranes or to indirect effects on membrane receptors and channels. However, there have been few lines of direct evidence connecting these potential targets to praziquantel action. For example, it has been reported that even high concentrations of praziquantel (up to 100 μM) have no effect on schistosome (Na^+–K^+)-ATPase or (Ca^{2+}–Mg^{2+})-ATPase activities (Cunha and Noel, 1997).

Recent reports have indicated that voltage-gated Ca^{2+} channels may be targets of praziquantel action (Kohn *et al.*, 2001a, 2003a,b). Voltage-gated Ca^{2+} channels are important entry sites for extracellular Ca^{2+}, and are therefore critical in regulating levels of intracellular Ca^{2+}. Until recently, the question of whether voltage-gated Ca^{2+} channels might play a role in praziquantel action had not been tested directly, largely because Ca^{2+} currents had never been recorded from schistosome cells (see below). Nevertheless, based on their pharmacological studies on praziquantel-induced contraction in both intact and detegumented worms, Blair *et al.* (1992) suggested that Ca^{2+} channels might be involved in the action of the drug. In this regard, praziquantel has been shown to prolong the Ca^{2+}-dependent plateau phase of the cardiac action potential in rats, which is carried by voltage-gated Ca^{2+} channels; however, high concentrations (50 μM) of the drug are required for this effect (Chubb *et al.*, 1978). Evidence against the involvement of voltage-gated Ca^{2+} channels includes the observation that methoxyverapamil (D-600), an inhibitor of one class of mammalian Ca^{2+} channels (L-type), does not block the praziquantel-dependent Ca^{2+} influx in schistosomes, but does block the tonic contraction resulting from increased K^+ concentrations (Fetterer *et al.*, 1980). However, recent results from cloning and expression of Ca^{2+} channel proteins point to a critical role for voltage-gated Ca^{2+} channels in praziquantel action.

Voltage-gated Ca^{2+} Channels

Voltage-gated ion channels are membrane protein complexes that underlie electrical excitability in cells. They include the voltage-gated potassium, sodium and Ca^{2+} channels. These channels act as ion-selective pores that are gated by changes in membrane potential. When activated, they allow ions to flow down their electrochemical gradient across the cell membrane. In addition to their role in contributing to impulse propagation, voltage-gated Ca^{2+} channels are important regulators of intracellular Ca^{2+} levels. Ca^{2+} channels provide the pathway for the Ca^{2+} influxes that mediate excitation–contraction coupling, excitation–secretion coupling and other Ca^{2+}-dependent processes in muscles, nerves and other excitable cells (reviewed by Hofmann *et al.*, 1999; Catterall 2000). Voltage-gated Ca^{2+} channels are therefore critical to behaviour and survival of the animal. Indeed, organisms as diverse as cone snails and spiders produce venoms that contain toxins targeted against specific subtypes of Ca^{2+} channels.

Voltage-gated Ca^{2+} channels are comprised of a pore-forming α_1 subunit that is associated with auxiliary subunits that modulate the properties of the channel (see Fig. 14.2). The α_1 subunit contains four homologous domains, each containing six transmembrane regions (S1–S6) that are linked into a single molecule. The α_1 subunit has a predicted structure that places it within the ion channel superfamily (Doyle *et al.*, 1998). The simplest ion channels are tetrameric structures that contain the fifth and sixth transmembrane regions and the P loop, a region between S5 and S6 that acts as the selectivity filter of the pore. It is in this region that the residues which define a channel's ionic selectivity and much of its pharmacology reside. The fourth transmembrane segment in each domain (S4) is thought to serve as the voltage sensor of the channel and contains a positively charged amino acid (lysine or arginine) at every third residue.

There are two major classes of Ca^{2+} currents that have been described in both vertebrate and invertebrate cells. These are low voltage-activated (LVA; T-type) and high-voltage-activated (HVA). HVA currents can be further divided into L-type, which, in vertebrates, are sensitive to the dihydropyridine class of Ca^{2+} channel blockers (nifedipine, nimodipine, etc.); and a variety of currents collectively known as non-L-type, which are not sensitive to the dihydropyridines. Experiments using heterologous expression of cloned Ca^{2+} channel subunits have shown a correspondence between these different currents and the different subtypes of cloned α_1 subunits. LVA currents are gated by the Ca_v3 class of α_1 subunits (reviewed by Perez-Reyes, 2003), while HVA L-type currents and non-L-type currents are gated by Ca_v1 α_1 and Ca_v2 α_1 subunits, respectively. From similar studies, it has also become clear that the pharmacological distinction between L-type and non-L-type channels breaks down in the invertebrates (reviewed by Jeziorski *et al.*, 2000a). Thus, heterologous expression of invertebrate α_1 subunits that are structurally members of the L-type family, show little if any sensitivity to dihydropyridines and other agents that modulate vertebrate L-type channels with high potency.

The α_1 subunits of HVA channels are associated with auxiliary proteins that have important modulatory effects on α_1 subunits (reviewed by Hofmann *et al.*, 1999; Catterall, 2000; Arikkath and Campbell, 2003). These include the α_2/δ and β subunits (Fig. 14.3), and also γ subunits (not shown). Ca^{2+} channel β subunits ($Ca_v\beta s$) are intracellular proteins that have been studied extensively and are particularly important components of Ca^{2+} channel complexes. Co-expression of β subunits with α_1 subunits results in increases in current density and ligand binding. $Ca_v\beta s$ appear to participate in membrane trafficking of the α_1 subunit, and it has been proposed that they do this in part by masking an endoplasmic reticulum retention site on the α_1 subunit (Bichet *et al.*, 2000). β subunits also have important effects on various biophysical properties of Ca^{2+} channels, including the voltage-dependence of channel activation and steady-state inactivation, rates of inactivation (reviewed by Birnbaumer *et al.*, 1998; Walker and DeWaard, 1998; Hanlon and Wallace, 2002; Dolphin, 2003), and the rate of recovery from inactivation (Jeziorski *et al.*, 2000b).

Several lines of evidence show that there is a specific region in the intracellular loop between domains I and II of the α_1 subunit that serves as the primary site for binding of β subunits. This site has been named the alpha interaction domain (AID; Pragnell *et al.*, 1994). $Ca_v\beta s$ contain a beta interaction domain (BID) that, until very recently, was thought to serve as the primary site for interaction with the α_1 subunit (DeWaard *et al.*, 1994). However, resolution of the structure of Ca^{2+} channel β subunits has indicated that such a view may be oversimplified.

Based on homology modelling, $Ca_v\beta s$ were proposed to be members of the membrane-associated guanylate kinase (MAGUK) family of proteins (Hanlon *et al.*, 1999). MAGUKs act as scaffolding proteins, often concentrated at synapses and playing important roles in clustering ion channels and neurotransmitter receptors (reviewed by Dimitratos *et al.*, 1999). MAGUKs typically contain one or more PDZ domains that are located N-terminal to an Src-homology 3 (SH3) domain, a bridging region (the HOOK domain) and a guanylate kinase (GK)-like domain. Recently, the crystal structure of the conserved core of β subunits has been

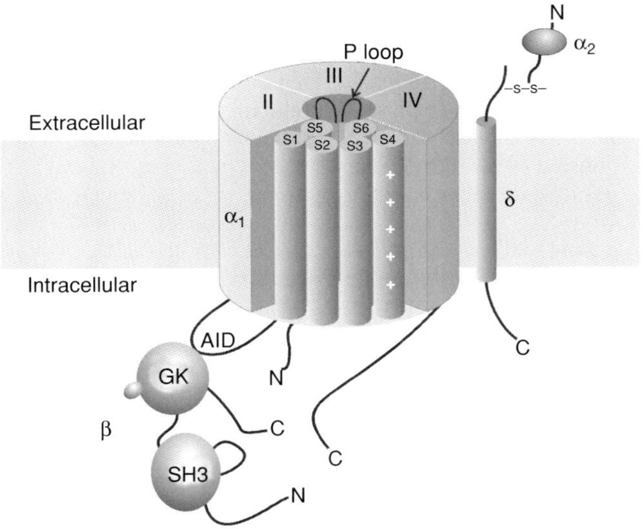

Fig. 14.3. Simplified structure of voltage-gated Ca^{2+} channels. The pore-forming α_1 subunit consists of four homologous domains surrounding a central pore. Each of the domains is comprised of six transmembrane regions (S1–S6). The P loop, which dips into the membrane between S5 and S6, forms the selectivity filter of the channel. The S4 transmembrane region contains a series of regularly spaced, positively charged residues (+) and is thought to form the voltage sensor of the channel. Also shown are auxiliary β and $\alpha_2\delta$, subunits, which modulate α_1 function. The γ subunit has been omitted. Domains of the β subunit defined by recent homology modelling and high-resolution structural analysis (see text) are shown. The β subunit interacts via its guanylate kinase (GK) domain with the alpha interaction domain (AID) on the I–II loop of the α_1 subunit. Size of subunits and domains are not necessarily to scale.

resolved, both alone and in complex with the AID of the α_1 subunit (Chen *et al.*, 2004; Opatowski *et al.*, 2004; Petegem *et al.*, 2004; commentary by Yue, 2004). These structures are consistent with $Ca_v\beta$s indeed being members of the MAGUK family, though with some unusual characteristics. For example, $Ca_v\beta$s appear to lack the PDZ domains typically found in other MAGUKs, and the orientation of the SH3 and GK domains has been modified. Furthermore, the GK domain in $Ca_v\beta$s does not have a nucleotide binding site.

The modified GK domain of the β subunit forms a hydrophobic binding pocket with which the AID of the α_1 subunit interacts. Surprisingly, at least in the conformation that was crystallized, the BID region is found buried within the β subunit protein. It is therefore unlikely to be involved directly in such protein–protein interactions as binding to the AID. However, as Chen *et al.* (2004) have discussed, the BID still appears to play an essential structural role in β subunits, spanning the SH3 and GK domains and their connecting HOOK region, and containing two β strands that are integral parts of the SH3 and GK domains. Indeed, the BID and surrounding areas are the most highly conserved regions of β subunit proteins.

Schistosome and Flatworm Ca^{2+} Currents

Clearly, one reason for studying voltage-gated Ca^{2+} channels in schistosomes and other flatworms is to understand their possible role in praziquantel action. Another important incentive is to better understand the physiology of excitable cells in these organisms, and also to garner clues about the evolution of ion channels. Furthermore, structural and functional characterization of these channels might provide targets for new anti-parasitic agents.

To date, no native Ca^{2+} currents from schistosome cells have been published. In fact, voltage-clamp analyses of muscle cells isolated from adult *Schistosoma mansoni* have revealed no inward currents at all, apparently due to technical limitations (Day *et al.*, 1993). However, isolated muscle cells from *S. mansoni* do show Ca^{2+}-dependent contractility (Day *et al.*, 1994).

However, cells from other flatworms do show evidence of voltage-gated Ca^{2+} currents. For example, neurons from the polyclad flatworm *Notoplana acticola* contain a normal complement of ionic currents, including cadmium-sensitive Ca^{2+} currents, which are implicated in the generation of action potentials (Keenan and Koopowitz, 1984). Voltage-gated Ca^{2+} currents from nerve and muscle cells have been characterized using voltage-clamp in *Bdelloura candida*, a triclad ectoparasitic flatworm that resides on the legs and gills of horseshoe crabs (*Limulus polyphemus*; Blair and Anderson, 1993, 1994). Both of these cell types had Ca^{2+} currents that activated at –30 mV, reached peak amplitude in approximately 5 ms, and inactivated slowly. The Ca^{2+} current from *Bdelloura* neurons was comparatively insensitive to organic Ca^{2+} channel blockers such as nifedipine and verapamil, as well as to the cone snail toxin ω-conotoxin GVIA. Furthermore, the neuronal Ca^{2+} current did not show sensitivity to 10 μM PZQ (Blair and Anderson, 1996). The Ca^{2+} current from muscle cells was not stable enough for pharmacological analysis. More recently, Ca^{2+} currents from muscle cells of the triclad turbellarian *Dugesia tigrina* have been described (Cobbett and Day, 2003). These currents were also not stable enough for thorough pharmacological analysis.

Clearly, elucidation of the properties of native Ca^{2+} currents in schistosomes and other flatworms has presented major challenges. A molecular approach offers an alternative for studying the physiological and pharmacological properties of cloned Ca^{2+} channel subunits expressed in heterologous systems. Ultimately, however, the characterization of native currents within schistosome cells will be essential for understanding the properties of these channels and the physiological roles they play within the parasite.

Schistosome Ca^{2+} Channel Subunits

The structure and function of schistosome Ca^{2+} channel subunits have recently begun to be elucidated (Kohn *et al.*, 2001a,b). Schistosome adults express at least three subtypes of HVA Ca^{2+} channel α_1 subunits. Two of these subtypes are most similar to α_1 subunits from the non-L-type class, while one is most closely related to the L-type class of α_1 subunits. Schistosomes (and presumably other platyhelminths) may be unique among the invertebrates in that they have genes encoding two subtypes of non-L-type α_1 subunits, while other invertebrates that have been examined contain only a single representative for each of these two classes (Jeziorski *et al.*, 2000a; Littleton and Ganetzky, 2000).

Schistosomes express at least two subtypes of $Ca_v\beta$s (Kohn *et al.*, 2001a, 2003b). This finding is also unique among the invertebrates; to date, only a single $Ca_v\beta$ gene has been identified in the genomes of other invertebrate species. More remarkably, one of these schistosome β subunit subtypes exhibits entirely unique structural and functional properties. To date, there have been no representatives of this variant subtype identified in any vertebrate or invertebrate phyla other than the platyhelminths (see Fig. 14.4).

These variant $Ca_v\beta$s are clearly members of the β subunit family, but they have quite distinct structural features. For example, the variant *S. mansoni* $Ca_v\beta$ is approximately 25% larger than the conventional schistosome $Ca_v\beta$, and is as much as 50% larger than β subunits from other species. However, the feature of these variant $Ca_v\beta$s that is most striking is found in the BID, where two conserved serine residues that represent consensus protein kinase C (PKC) phosphorylation sites are replaced by other residues (cysteine, alanine).

The unique structural characteristics of these variant schistosome β subunits appear to be further reflected in their unusual functional properties (Kohn *et al.*, 2001a). As discussed, conventional β subunits enhance currents through α_1 subunits. In contrast, co-expression of these variant β subunits in *Xenopus* oocytes with either a jellyfish (*Cy*Ca_v1) or human (Ca_v2.3) α_1 subunit results in a dramatic reduction in current. Other than this anomalous effect

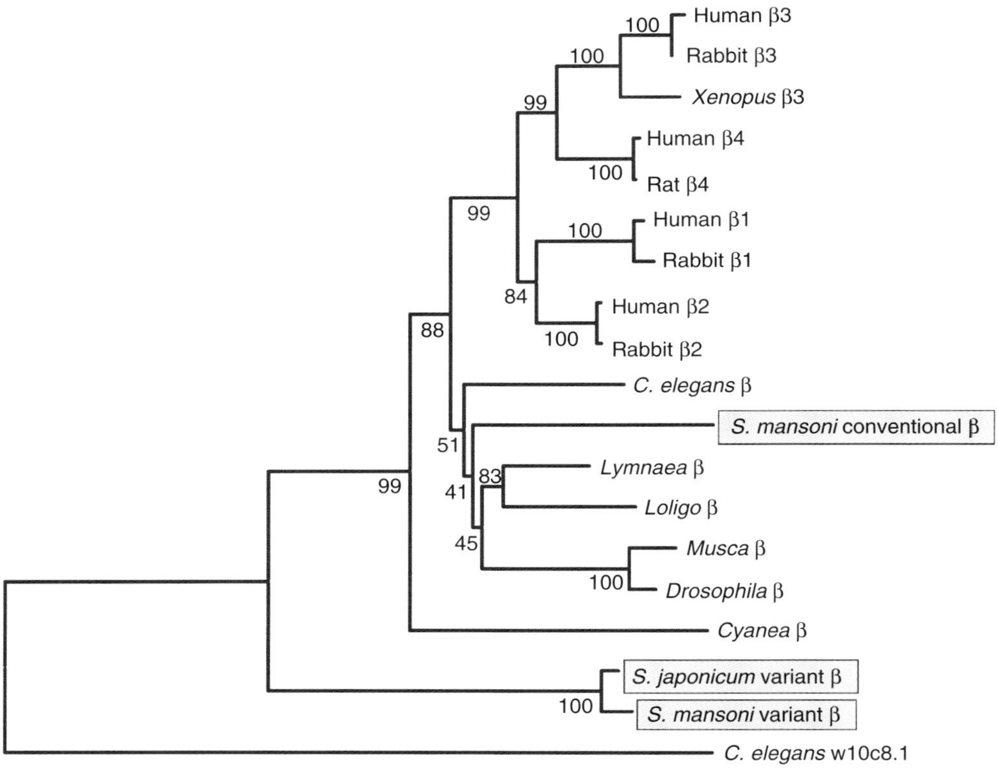

Fig. 14.4. Phylogenetic tree of Ca^{2+} channel β subunits. Amino acid sequences were aligned using Clustal X (Thompson *et al.*, 1997), and a tree constructed using the neighbour-joining method, as implemented in MEGA 2.1 (Kumar *et al.*, 2001). The schistosome conventional and variant β subunits are boxed. Numbers represent bootstrap values (1000 replications). Sequences and NCBI accession numbers are: Human β1, NP_954856; Rabbit β1, AAA31180; Human β2, NP_000715; Rabbit β2, CAA45576; Human β3, NP_000716; Rabbit β3, CAA45578; *Xenopus laevis* (toad) β3, AAA75519; Human β4, NP_000717; Rat β4, A45982; *Caenorhabditis elegans* (nematode) β, AAB53056; *Schistosoma mansoni* conventional β, AY033599; *Lymnaea stagnalis* (snail) β, AAO83844; *Loligo bleekeri* (squid) β, BAB88219; *Musca domestica* (housefly) β, A54844; *Drosophila melanogaster* (fruitfly) β, AAF21096; *Cyanea capillata* (jellyfish) β, AAB87751; *S. japonicum* variant β, AAK51116; *S. mansoni* variant β, AAK51117; *C. elegans* β-like sequence (w10c8.1), AAK21500.

on current levels, however, the variant schistosome Ca_vβs modulate α_1 subunits in a conventional manner. For instance, co-expression of the variant β subunits results in a hyperpolarizing shift in the current–voltage relationship of α_1 subunits, an effect also found with other β subunits. Thus, although the variant Ca_vβs have unusual properties, these proteins appear to be behaving as authentic, functional β subunits.

Of particular interest for this discussion, the variant schistosome Ca_vβs confer praziquantel sensitivity to an otherwise praziquantel-insensitive mammalian α_1 subunit (Kohn *et al.*, 2001a). Thus, when the mammalian $Ca_v2.3$ α_1 subunit is expressed in *Xenopus* oocytes, it shows no sensitivity to 100 nM praziquantel. However, when one of the variant schistosome Ca_vβs is co-expressed in *Xenopus* oocytes with $Ca_v2.3$, peak currents are increased 1.5- to 2-fold in the presence of the drug. This increase in Ca^{2+} influx through channels containing these variant β subunits is consistent with the effects of praziquantel on Ca^{2+} homeostasis in

Variant schistosome $Ca_v\beta$s: PPYEIVPCMRPVVFVGPALKGYEVTDMMQKAIFD
$Ca_v\beta$ consensus: PPYDVVPSMRPVVLVGPSLKGYEVTDMMQKALFD

Fig. 14.5. Comparison of the amino acid sequence of the BID from the variant schistosome β subunits with a consensus BID sequence. The cysteine and alanine residues which substitute in the variant BIDs for the conserved serines are shaded. The two consensus PKC phosphorylation sites conserved in the consensus BID sequence are boxed with a solid line, and those missing consensus sites in the variant β subunits are boxed with a dashed line.

the parasite. Other $Ca_v\beta$s, including the conventional schistosome β subunit, do not confer praziquantel sensitivity to the mammalian α_1 subunit (Kohn *et al.*, 2003b). These observations implicate the variant schistosome $Ca_v\beta$s in the mechanism of action of praziquantel.

Site-directed mutagenesis has been used to localize the specific amino acid residues that might be critical for determining the unusual functional properties and pharmacological sensitivities of the variant β subunits. Both the reduction in current levels and the ability of the variant $Ca_v\beta$s to confer praziquantel sensitivity have been mapped to the two conserved consensus PKC sites in the BID that are not present in these subunits (Kohn *et al.*, 2003a,b). Incorporation of serine residues at either or both of these sites in the variant $Ca_v\beta$ from *S. mansoni* restores the consensus PKC sites (see Fig. 14.5) and results in a β subunit which now enhances currents through α_1 subunits and no longer confers sensitivity to praziquantel. A double mutation, which incorporates a serine but also eliminates the consensus PKC sequence at a second site, behaves like the wild-type schistosome subunit; thus, the novel effects of this subunit appear to result from the absence of PKC sites in the BID region, rather than simply the presence of residues other than serines. Finally, elimination of both PKC sites in the BID of a conventional β subunit creates a β subunit that now confers praziquantel sensitivity to the mammalian α_1 subunit.

Thus, the absence of consensus PKC sites in the β subunit BID is associated with the capability of a β subunit to confer susceptibility to praziquantel. The presence of a single consensus PKC site in the BID is sufficient to transform a variant $Ca_v\beta$ into a β subunit with characteristics typically found for conventional β subunits. Based on these results, we have hypothesized that the unique modulatory properties and pharmacological sensitivities of the variant $Ca_v\beta$s from schistosomes are dependent on the absence of the consensus PKC phosphorylation sites found in the BIDs of other β subunits.

Phosphorylation of voltage-gated Ca^{2+} channel subunits by protein kinases such as PKC plays an important role in regulating channel properties (reviewed by Rossie, 1999; Kamp and Hell, 2000; Keef *et al.*, 2001). The mammalian β2a subunit has been shown to be phosphorylated by PKC *in vitro*, with a stoichiometry estimated to be 1–2 moles of phosphate per mole of β2a protein (Puri *et al.*, 1997). Several consensus PKC sites are present in β subunits; however, the specific sites that are phosphorylated by PKC have not yet been defined either *in vitro* or *in vivo*.

Concluding Remarks

Assuming that the variant $Ca_v\beta$s found in schistosomes are indeed involved in praziquantel action, and that creation of a PKC site in the BID can eliminate susceptibility to the drug, then one mechanism for acquiring praziquantel resistance might involve acquisition of one or both of these crucial PKC sites by mutating residues to serine. Schistosome isolates reported to have reduced praziquantel sensitivity have been tested for the presence of these (or other) mutations (Valle *et al.*, 2003). Worms with reduced susceptibility to praziquantel showed neither changes in primary structure of schistosome β subunits, nor changes in expression levels of those subunits. Thus, a reduction in praziquantel susceptibility in these strains is apparently not based on altered $Ca_v\beta$ structures or expression levels. However, various Egyptian isolates with reduced praziquantel susceptibility show differential stability (William *et al.*, 2001), which may indicate that there are alternative pathways for acquiring reduced praziquantel susceptibility. Thus, different strains of parasites may be acquiring resistance to praziquantel via different mechanisms.

Several issues regarding schistosome Ca^{2+} channels and their role in praziquantel action remain unresolved. For example, the properties of the schistosome α_1 subunits, either alone or in combination with the two schistosome $Ca_v\beta$ subtypes, have not yet been described. Are all schistosome α_1/β combinations able to form functional channels? Is there only a particular α_1/β combination that generates praziquantel-sensitive channels, or can any schistosome channel that contains a variant β subunit show sensitivity to the drug? In which cells are these combinations expressed? How exactly does praziquantel act on these channels? Is the interaction directly with the variant β subunit, does praziquantel instead interfere with α_1/β subunit interaction, or is praziquantel acting somehow indirectly on Ca^{2+} channels that contain these variant $Ca_v\beta$s? Of particular interest is the specific biological role or roles that these variant schistosome $Ca_v\beta$s might play in schistosomes. Unlike other β subunits, which enhance Ca^{2+} currents, these β subunits appear to inhibit currents. What special characteristics of schistosome α_1 subunits would require the development of this unusual type of modulation? Are there any other organisms that use a similar strategy? Some answers to these questions may come from analysis of sequences and expression studies in heterologous systems. Some answers may also arise from analyses of the schistosome genome and transcriptome. Ultimately, elucidation of the properties of native Ca^{2+} (and other) currents will be required to obtain a full understanding of the role they play in schistosome physiology and in praziquantel action.

Acknowledgements

I thank Lynn Milstead for help with manuscript preparation. R.M.G. is supported by NIH grant #AI 40522 and by the Neal Cornell Research Fund at the Marine Biological Laboratory.

References

Andrews, P. (1985) Praziquantel: mechanisms of anti-schistosomal activity. *Pharmacology and Therapeutics* 29, 129–156.

Andrews, P., Thomas, H., Pohlke, R. and Seubert, J. (1983) Praziquantel. *Medicinal Research Reviews* 3, 147–200.

Arikkath, J. and Campbell, K.P. (2003) Auxiliary subunits: essential components of the voltage-gated calcium channel complex. *Current Opinion in Neurobiology* 13, 298–307.

Becker, B., Mehlhorn, H., Andrews, P., Thomas, H. and Eckert, J. (1980) Light and electron microscopic studies on the effect of praziquantel on *Schistosoma mansoni, Dicrocoelium dendriticum*, and *Fasciola hepatica* (trematoda) *in vitro*. *Zeitschrift fur Parasitenkunde* 63, 113–128.

Bergquist, R., Al-Sherbiny, M., Barakat, R. and Olds, R. (2002) Blueprint for schistosomiasis vaccine development. *Acta Tropica* 82, 183–192.

Bichet, D., Cornet, V., Geib, S., Carlier, E., Volsen, S., Hoshi, T., Mori, Y. and De Waard, M. (2000) The I–II loop of the Ca^{2+} channel α_1 subunit contains an endoplasmic reticulum retention signal antagonized by the beta subunit. *Neuron* 25, 177–190.

Birnbaumer, L., Qin, N., Olcese, R., Tareilus, E., Platano, D., Costantin, J. and Stefani, E. (1998) Structures and functions of calcium channel β subunits. *Journal of Bioenergetics and Biomembranes* 30, 357–375.

Blair, K.L. and Anderson, P.A.V. (1993) Properties of voltage-activated ionic currents in cells from the brains of the triclad flatworm *Bdelloura candida*. *Journal of Experimental Biology* 185, 267–286.

Blair, K.L. and Anderson, P.A.V. (1994) Physiological and pharmacological properties of muscle cells isolated from the flatworm *Bdelloura candida* (triclada). *Parasitology* 109, 325–335.

Blair, K.L. and Anderson, P.A.V. (1996) Physiology and pharmacology of turbellarian neuromuscular systems. *Parasitology* 113, S73–S82.

Blair, K.L., Bennett, J.L. and Pax, R.A. (1992) Praziquantel: physiological evidence for its site(s) of action in magnesium-paralysed *Schistosoma mansoni*. *Parasitology* 104, 59–66.

Capron, A., Riveau, G.J., Bartley, P.B. and McManus, D.P. (2002) Prospects for a schistosome vaccine. *Current Drug Targets. Immune, Endocrine and Metabolic Disorders* 2, 281–290.

Catterall, W.A. (2000) Structure and regulation of voltage-gated Ca^{2+} channels. *Annual Review of Cell and Developmental Biology* 16, 521–555.

Chen, Y.-H., Li, M.-H., Zhang, Y., He, L.-L., Yamada, Y., Fitzmaurice, A., Shen, Y., Zhang, H., Tong, L. and Yang, J. (2004) Structural basis of the α_1–β subunit interaction of voltage-gated Ca^{2+} channels. *Nature* 429, 675–680.

Chubb, J.M., Bennett, J.L., Akera, T. and Brody, T.M. (1978) Effects of praziquantel, a new anthelmintic, on electromechanical properties of isolated rat atria. *Journal of Pharmacology and Experimental Therapeutics* 207, 284–293.

Cioli, D. (2000) Praziquantel: is there real resistance and are there alternatives? *Current Opinion in Infectious Diseases* 13, 659–663.

Cioli, D. and Pica-Mattoccia, L. (2003) Praziquantel. *Parasitological Research* 90, S3–S9.

Cobbett, P. and Day, T.A. (2003) Functional voltage-gated Ca^{2+} channels in muscle fibers of the platyhelminth *Dugesia tigrina. Comparative Biochemistry and Physiology A: Molecular and Integrative Physiology* 134, 593–605.

Cunha, V.M. and Noel, F. (1997) Praziquantel has no direct effect on (Na^+–K^+)-ATPases and (Ca^{2+}–Mg^{2+})-ATPases of *Schistosoma mansoni. Life Sciences* 60, 289–294.

Day, T.A., Bennett, J.L. and Pax, R.A. (1992) Praziquantel: the enigmatic antiparasitic. *Parasitology Today* 8, 342–344.

Day, T.A., Orr, N., Bennett, J.L. and Pax, R.A. (1993) Voltage-gated currents in muscle cells of *Schisotosoma mansoni. Parasitology* 106, 471–477.

Day, T.A., Bennett, J.L. and Pax, R.A. (1994) Serotonin and its requirement for maintenance of contractility in muscle fibres isolated from *Schistosoma mansoni. Parasitology* 108, 425–432.

De Waard, M.D., Pragnell, M. and Campbell, K.P. (1994) Ca^{2+} channel regulation by a conserved β subunit domain. *Neuron* 13, 495–503.

Dimitratos, S.D., Woods, D.F., Stathakis, D.G. and Bryant, P.J. (1999) Signaling pathways are focused at specialized regions of the plasma membrane by scaffolding proteins of the MAGUK family. *Bioessays* 21, 912–921.

Doenhoff, M.J., Kusel, J.R., Coles, G.C. and Cioli, D. (2002) Resistance of *Schistosoma mansoni* to praziquantel: is there a problem? *Transactions of the Royal Society of Tropical Medicine and Hygiene* 96, 465–469.

Dolphin, A.C. (2003) β subunits of voltage-gated calcium channels. *Journal of Bioenergetics and Biomembranes* 35, 599–620.

Doyle, D.A., Cabral, J.M., Pfuetzner, R.A., Kuo, A., Gulbis, J.M., Dohen, S.L., Chair, B.T. and MacKinnon, R. (1998) The structure of the potassium channel: molecular basis of K^+ conduction and selectivity. *Science* 280, 69–77.

Engels, D., Chitsulo, L., Montresor, A. and Savioli, L. (2002) The global epidemiological situation of schistosomiasis and new approaches to control and research. *Acta Tropica* 82, 139–146.

Fenwick, A., Savioli, L., Engels, D., Bergquist, N.R. and Todd, M.H. (2003) Drugs for the control of parasitic diseases: current status and development in schistosomiasis. *Trends in Parasitology* 19, 509–515.

Fetterer, R.H., Pax, R.A. and Bennett, J.L. (1980) Praziquantel, potassium and 2,4-dinitrophenol: analysis of their action on the musculature of *Schistosoma mansoni. European Journal of Pharmacology* 64, 31–38.

Gardner, D.R. and Brezden, B.L. (1984) The sites of action of praziquantel in smooth muscle of *Lymnaea stagnalis. Canadian Journal of Physiology and Pharmacology* 62, 282–287.

Hagan, P. and Sharaf, O. (2003) Schistosomiasis vaccines. *Expert Opinion in Biological Therapeutics* 3, 1271–1278.

Hagan, P., Appleton, C.C., Coles, G.C., Kusel, J.R. and Tchuem-Tchuente, L.A. (2004) Schistosomiasis control: keep taking the tablets. *Trends in Parasitology* 20, 92–97.

Hanlon, M.R. and Wallace, B.A. (2002) Structure and function of voltage-dependent ion channel regulatory β subunits. *Biochemistry* 41, 2886–2894.

Hanlon, M.R., Berrow, N.S., Dolphin, A.C. and Wallace, B.A. (1999) Modelling of a voltage-dependent Ca^{2+} channel β subunit as a basis for understanding its functional properties. *FEBS Letters* 445, 366–370.

Harder, A. (2002) Chemotherapeutic approaches to schistosomes: current knowledge and outlook. *Parasitology Research* 88, 395–397.

Harder, A., Goossens, J. and Andrews, P. (1988) Influence of praziquantel and Ca^{2+} on the bilayer–isotropic–hexagonal transition of model membranes. *Molecular and Biochemical Parasitology* 29, 55–60.

Harnett, W. and Kusel, J.R. (1986) Increased exposure of parasite antigens at the surface of adult male *Schistosoma mansoni* exposed to praziquantel *in vitro*. *Parasitology* 93, 401–405.

Heinemann, S.H., Terlau, H. and Imoto, K. (1992) Molecular basis for pharmacological differences between brain and cardiac sodium channels. *Pflugers Archives* 422, 90–92.

Hofmann, F., Lacinova, L. and Klugbauer, N. (1999) Voltage-dependent calcium channels: from structure to function. *Reviews of Physiology, Biochemistry, and Pharmacology* 139, 33–87.

Hu, W., Brindley, P.J., McManus, D.P., Feng, Z. and Han, Z.-G. (2004) Schistosome transcriptomes: new insights into the parasite and schistosomiasis. *Trends in Molecular Medicine* 10, 217–225.

Jeziorski, M.C., Greenberg, R.M. and Anderson, P.A.V (2000a) The molecular biology of invertebrate voltage-gated Ca^{2+} channels. *Journal of Experimental Biology* 203, 841–856.

Jeziorski, M.C., Greenberg, R.M. and Anderson, P.A.V. (2000b) Calcium channel β subunits differentially modulate recovery of the channel from inactivation. *FEBS Letters* 483, 125–130.

Kamp, T.J. and Hell, J.W. (2000) Regulation of cardiac L-type calcium channels by protein kinase A and protein kinase C. *Circulation Research* 87, 1094–1102.

Keef, K.D., Hume, J.R. and Zhong, J. (2001) Regulation of cardiac and smooth muscle Ca^{2+} channels ($Ca_v1.2a,b$) by protein kinases. *American Journal of Physiology: Cell Physiology* 281, C1743–C1755.

Keenan, L. and Koopowitz, H. (1984) The ionic bases of action-potentials in identified flatworm neurons. *Journal of Comparative Physiology* 155, 197–208.

Kohn, A.B., Anderson, P.A.V., Roberts-Misterly, J.M. and Greenberg, R.M. (2001a) Schistosome calcium channel β subunits: unusual modulatory effects and potential role in the action of the antischistosomal drug praziquantel. *Journal of Biological Chemistry* 276, 36873–36876.

Kohn, A.B., Lea, J.M., Roberts-Misterly, J.M., Anderson, P.A.V. and Greenberg, R.M. (2001b) Structure of three high voltage-activated Ca^{2+} channel α_1 subunits from *Schistosoma mansoni*. *Parasitology* 123, 489–497.

Kohn, A.B., Roberts-Misterly, J.M., Anderson, P.A.V. and Greenberg, R.M. (2003a) Creation by mutagenesis of a mammalian Ca^{2+} channel β subunit that confers praziquantel sensitivity to a mammalian Ca^{2+} channel. *International Journal for Parasitology* 33, 1303–1308.

Kohn, A.B., Roberts-Misterly, J.M., Anderson, P.A.V., Khan, N. and Greenberg, R.M. (2003b) Specific residues in the beta interaction domain of a schistosome Ca^{2+} channel β subunit are key to its role in sensitivity to the antischistosomal drug praziquantel. *Parasitology* 127, 349–356.

Kumar, S., Tamura, K., Jakobsen, I.B. and Nei, M. (2001) MEGA2: molecular evolutionary genetics analysis software. *Bioinformatics* 17, 1244–1245.

Lima, S.F., Vieira, L.Q., Harder, A. and Kusel, J. (1994) Effects of culture and praziquantel on membrane fluidity parameters of adult *Schistosoma mansoni*. *Parasitology* 109, 57–64.

Littleton, J.T. and Ganetzky, B. (2000) Ion channels and synaptic organization: analysis of the *Drosophila* genome. *Neuron* 26, 35–43.

Liu, Y.H., Qian, M.X., Wang, X.G., Jia, J., Wang, Q.N., Jiang, Y.F., Wang, R.Q., Yan, S.W., Chen, B.Y., Li, J.S., Qiu, Z.Y. and Shen, J.K. (1986) Comparative efficacy of praziquantel and its optic isomers in experimental therapy of schistosomiasis japonica in rabbits. *Chinese Medical Journal* 99, 935–940.

LoVerde, P.T., Hirai, H., Merrick, J.M., Lee, N.H. and El-Sayed, N. (2004) *Schistosoma mansoni* genome project: an update. *Parasitology International* 53, 183–192.

McManus, D.P., Hu, W., Brindley, P.J., Feng, Z. and Han, Z.-G. (2004) Schistosome transcriptome analysis at the cutting edge. *Trends in Parasitology* 20, 301–304.

Mehlhorn, H., Becker, B., Andrews, P., Thomas, H. and Frenkel, J.K. (1981) *In vivo* and *in vitro* experiments on the effects of praziquantel on *Schistosoma mansoni*. A light and electron microscopic study. *Arzneimittel-Forschung* 31, 544–554.

Opatowski, Y., Chen, C.-C., Campbell, K.P. and Hirsch, J.A. (2004) Structural analysis of the voltage-dependent Ca^{2+} channel β subunit functional core and its complex with the α1 interaction domain. *Neuron* 42, 387–399.

Pax, R.A., Bennett, J.L., and Fetterer, J.L. (1978) A benzodiazepine derivative and praziquantel: effects on musculature of *Schistosoma mansoni* and *Schistosoma japonicum*. *Naunyn-Schmiedberg's Archives of Pharmacology* 304, 309–315.

Perez-Reyes, E. (2003) Molecular physiology of low-voltage-activated T-type calcium channels. *Physiological Reviews* 83, 117–161.

Petegem, F.V., Clark, K.A., Chatelain, F.C. and Minor, D.L. (2004) Structure of a complex between a voltage-gated calcium channel β subunit and an α-subunit domain. *Nature* 429, 671–675.

Pica-Mattoccia, L. and Cioli, D. (2004) Sex- and stage-related sensitivity of *Schistosoma mansoni* to *in vivo* and *in vitro* praziquantel treatment. *International Journal for Parasitology* 34, 527–533.

Pragnell, M., De Waard, M., Mori, Y., Tanabe, T., Snutch, T.P. and Campbell, K.P. (1994) Calcium channel β-subunit binds to a conserved motif in the I–II cytoplasmic linker of the α_1-subunit. *Nature* 368, 67–70.

Puri, T.S., Gerhardstein, B.L., Zhao, X.-L., Ladner, M.B. and Hosey, M.M. (1997) Differential effects of subunit interactions on protein kinase A- and C-mediated phosphorylation of L-type calcium channels. *Biochemistry* 36, 9605–9615.

Redman, C.A., Robertson, A., Fallon, P.G., Modha, J., Kusel, J.R., Doenhoff, M.J. and Martin, R.J. (1996) Praziquantel: an urgent and exciting challenge. *Parasitology Today* 12, 14–20.

Rossie, S. (1999) Regulation of voltage-sensitive sodium and calcium channels by phosphorylation. *Advances in Second Messenger and Phosphorylation Research* 33, 23–48.

Sabah, A.A., Fletcher, C., Webbe, G. and Doenhoff, M.J. (1986) *Schistosoma mansoni*: chemotherapy of infections of different ages. *Experimental Parasitology* 61, 294–303.

Satin, J., Kyle, J.W., Chen, M., Bell, P., Cribbs, L.L., Fozzard, H.A. and Rogart, R.B. (1992) A mutant of TTX-resistant cardiac sodium channels with TTX-sensitive properties. *Science* 256, 1202–1205.

Staudt, U., Schmahl, G., Blaschke, G. and Mehlhorn, H. (1992) Light and scanning electron microscopy studies on the effects of the enantiomers of praziquantel and its main metabolite on *Schistosoma mansoni in vitro*. *Parasitology Research* 78, 392–397.

Tanaka, M., Ohmae, H., Utsunomiya, H., Nara, T., Irie, Y. and Yasuraoka, K. (1989) A comparison of the antischistosomal effect of levo- and dextro-praziquantel on *Schistosoma japonicum* and *S. mansoni* in mice. *American Journal of Tropical Medicine and Hygiene* 41, 198–203.

Thompson, J.D., Gibson, T.J., Plewniak, F., Jeanmougin, F. and Higgins, D.G. (1997) The ClustalX windows interface: flexible strategies for multiple sequence alignment aided by quality analysis tools. *Nucleic Acids Research* 24, 4876–4882.

Valle, C., Troiani, A.R., Festucci, A., Pica-Mattoccia, L., Liberti, P., Wolstenholme, A., Francklow, K., Doenhoff, M.J. and Cioli, D. (2003) Sequence and level of endogenous expression of calcium channel β subunits in *Schistosoma mansoni* displaying different susceptibilities to praziquantel. *Molecular and Biochemical Parasitology* 130, 111–115.

van der Werf, M.J., de Vlas, S.J., Brooker, S., Looman, C.W., Nagelkerke, N.J., Habbema, J.D. and Engels, D. (2003) Quantification of clinical morbidity associated with schistosome infection in sub-Saharan Africa. *Acta Tropica* 86, 125–139.

Verjovski-Almeida, S., Leite, L.C.C., Dias-Neto, E., Menck, C.F.M. and Wilson, R.A. (2004) Schistosome transcriptome: insights and perspectives for functional genomics. *Trends in Parasitology* 20, 304–308.

Walker, D. and De Waard, M. (1998) Subunit interaction sites in voltage-dependent Ca^{2+} channels: role in channel function. *Trends in Neuroscience* 21, 148–154.

William, S., Sabra, A., Ramzy, F., Mousa, M., Demerdash, Z., Bennett, J.L., Day, T.A., Botros, S. (2001) Stability and reproductive fitness of *Schistosoma mansoni* isolates with decreased sensitivity to praziquantel. *International Journal for Parasitology* 31, 1093–1100.

Wolde Mussie, E., Vande Waa, J., Pax, R.A., Fetterer, R. and Bennett, J.L. (1982) *Schistosoma mansoni*: calcium efflux and effects of calcium-free media on responses of the adult male musculature to praziquantel and other agents inducing contraction. *Experimental Parasitology* 53, 270–278.

Wu, M.H., Wei, C.C., Xu, Z., Yuan, H.C., Lian, W.N., Yang, Q.J., Chen, M., Jiang, Q.W., Wang, C.Z. and Zhang, S.J. (1991) Comparison of the therapeutic efficacy and side effects of a single dose of levo-praziquantel with mixed isomer praziquantel in 278 cases of schistosomiasis japonica. *American Journal of Tropical Medicine and Hygiene* 45, 345–349.

Xiao, S.H. and Catto, B.A. (1989) Comparative *in vitro* and *in vivo* activity of racemic praziquantel and its levorotated isomer on *Schistosoma mansoni*. *Journal of Infectious Diseases* 159, 589–592.

Xiao, S.H., Friedman, P.A., Catto, B.A. and Webster, L.T. (1984) Praziquantel-induced vesicle formation in the tegument of male *Schistosoma mansoni* is calcium dependent. *Journal of Parasitology* 70, 177–179.

Xiao, S.H., Catto, B.A. and Webster, L.T. (1985) Effects of praziquantel on different developmental stages of *Schistosoma mansoni in vitro* and *in vivo*. *Journal of Infectious Diseases* 151, 1130–1137.

Xiao, S.H., Chollet, J., Booth, M., Weiss, N.A. and Tanner, M. (1999) Therapeutic effect of praziquantel enantiomers in mice infected with *Schistosoma mansoni*. *Transactions of the Royal Society of Tropical Medicine and Hygiene* 93, 324–325.

Yue, D.T. (2004) The dawn of high-resolution structure for the queen of the ion channels. *Neuron* 42, 357–361.

15 Cestode Vaccine Development

M.W. Lightowlers
Veterinary Clinical Centre, The University of Melbourne, 250 Princes Highway, Werribee, Victoria 3030, Australia

Introduction

The Cestodes are a group of parasitic flatworms, some of which cause economically or medically important diseases. Those of principal importance belong to the Family Taeniidae, which comprises two genera, *Taenia* and *Echinococcus*. *Taenia* species are the etiological agents of cysticercosis in sheep, cattle, pigs and humans. *Echinococcus* species cause hydatid disease in a variety of animal species, including humans. A number of taeniid cestode species are also natural parasites of laboratory animals and these have been used extensively in early studies of the immunology of *Taenia* and *Echinococcus* parasitism. The taeniid cestode species that are of particular economic importance and those species that have played an important role in immunological investigations are summarized in Table 15.1.

Taeniid cestodes have a prey–predator life cycle involving two mammalian hosts. The definitive host is a carnivore or omnivore, which harbours the adult tapeworm parasite in the small intestine. Mature infective eggs are released with the faeces and, when these are ingested by a suitable species of intermediate host, the oncosphere contained within the egg is liberated through the influence of intestinal secretions, particularly bile. The activated parasite penetrates the mucosa of the small intestine

Table 15.1. Principal medically and economically important taeniid species and those used extensively in laboratory studies.

Species	Obsolete synonyms	Principal intermediate hosts	Metacestode type	Principal definitive host
Taenia solium	*Cysticercus cellulosae*	Pigs (Man)	Cysticercus	Man
T. saginata	*C. bovis*	Cattle	Cysticercus	Man
T. hydatigena	*C. tenuicollis*	Sheep	Cysticercus	Dog
T. ovis	*C. ovis*	Sheep	Cysticercus	Dog
T. multiceps	*Multiceps multiceps*	Sheep (Man)	Coenurus	Dog
T. pisiformis	*C. pisiformis*	Rabbit	Cysticercus	Dog
T. taeniaeformis	*C. fasciolaris*	Rodents	Strobilocercus	Cat
T. crassiceps	–	Rodents	Cysticercus (proliferative)	Dog
Echinococcus granulosus	–	Sheep, goats, cattle, pigs and other herbivores (Man)	Unilocular hydatid cyst	Dog
E. multilocularis	–	Microtine rodents (Man)	Multilocular hydatid cyst	Fox

of the intermediate host and travels by way of the circulating system to its preferred tissue site, in which the parasite encysts. The larval parasite, known as a metacestode, matures to a stage that is infective for the definitive host. The life cycle is completed when the intermediate host tissues containing the mature metacestode are eaten by a suitable species of definitive host. The larval parasite is released and activated by the intestinal secretions of the definitive host and grows into a mature tapeworm. Infection in the definitive hosts rarely causes significant clinical or economic problems. However, infection with cysticercosis or hydatid disease in the intermediate hosts may have severe consequences for health in humans and may cause substantial economic loss when the hosts are livestock species. Typically, the different parasite species are quite restricted in their host range for both definitive and intermediate hosts, however there are some notable exceptions. *Echinococcus granulosus* is the cause of cystic echinococcosis (cystic hydatid disease) and infects a wide range of species as intermediate hosts, including humans. *Taenia solium* is normally transmitted between humans, as the obligate definitive host, and pigs as the intermediate host. However, the eggs of *T. solium* are also infective to humans, if they are ingested, in whom they mature as cysticerci. These frequently occur in the brain and other parts of the nervous system and cause a disease known as neurocysticercosis.

The medical significance of some taeniid cestode species, particularly *E. granulosus*, has stimulated considerable interest worldwide in controlling or eradicating the parasites (Gemmell *et al.*, 2001). Hydatid control campaigns have been waged in many parts of the world. These have relied on the application of anthelmintics to remove tapeworms from domestic dogs, other dog control measures and public education about the disease related particularly to preventing access of dogs to offal infected with hydatid cysts. Where these measures have been able to be applied in a rigorous and sustained manner, hydatid disease has been eradicated, for example, in New Zealand and the island state of Tasmania in Australia (Gemmell *et al.*, 1986). However, few countries have had sufficient resources to commit to these control measures to ensure that effective control or eradication was achieved. The critical problem with control programmes that rely on treatment of the definitive host is the potential for new infections following anthelmintic treatment. Hence, in order to achieve a high level of transmission control it is necessary to re-treat definitive hosts within the pre-patent period of the adult tapeworm

(Gemmell, 1987). In New Zealand, every dog in the country was treated every 6 weeks over a period of years. This strategy was successful in eliminating hydatid disease, however, even this extensive and rigorous campaign failed to eliminate other dog-transmitted cestode species such as *Taenia hydatigena* or *Taenia ovis* (Burridge and Schwabe, 1977; Lawson, 1994). These parasite-control campaigns have highlighted the substantial difficulties associated with control methods directed entirely at the definitive host of cestode parasites. The parasites have a substantial biotic potential such that many years of control efforts can be overturned when only a very small proportion of infections persist in definitive hosts, or due to the immigration of infected hosts into control areas (Gemmell *et al.*, 1987). Indeed, in New Zealand the hydatid control campaign led to a situation where the economic impact of *T. ovis* increased in some areas rather than decreased. In the absence of control, transmission of *T. ovis* was hyperendemic such that most lambs became infected early in life and developed a degree of immunity. In this situation light burdens of infection were common while heavy infestations were relatively rare. However, after several years of 6-weekly dosing of dogs with anthelmintic, transmission rates decreased and the sheep flock comprised many animals that had not been exposed to *T. ovis* and had no immunity to the disease. When an infected dog escaped effective anthelmintic treatment and distributed the parasite's eggs in the environment, many hundreds of sheep over a wide area were found to develop heavy *T. ovis* infections, which became known as 'cysticercosis storms' (Lawson and Gemmell, 1983). The problems that have arisen with the control of transmission of cestode parasites highlight the need for new control measures. One potential option is the use of effective vaccines.

Vaccines to assist with control of taeniid cestode infections would be most easily applied in the parasites' definitive hosts because generally these are fewer in number and more readily accessed than are the intermediate hosts. However, there is little evidence of host-protective immune responses against *Taenia* and *Echinococcus* species in their definitive hosts and attempts to induce protective immune responses have been unreliable and, generally, poorly effective (reviewed by Lightowlers, 1990). On the other hand, immunity to re-infection with these parasites in their intermediate hosts is a prominent feature of the host–parasite relationship (Rickard and Williams, 1982). For this reason, vaccine development efforts have concentrated on the intermediate hosts. Highly effective, defined antigen vaccines have been developed for the prevention of infection with several species of taeniid cestode (summarized in Table 15.2). The greatest success has been achieved using recombinant oncosphere antigens and this chapter focuses on this area of work.

Immunity – Basic Principles

Following an initial infection with taeniid cestode parasites in their intermediate hosts, the infected animals show a substantial degree of immunity to re-infection (Rickard and Williams, 1982). This concomitant immunity, as it is known, has been demonstrated for numerous *Taenia* species and in various host species. From the time of the earliest known experimental investigations, it has been clear that this immunity can be transferred via either passive transfer of serum from infected animals to naïve recipients or via colostral antibodies from an infected dam to her offspring (reviewed by Lightowlers *et al.*, 1992). Hence, antibodies play an important role in concomitant immunity though they are not necessarily the only immune mechanism. Mitchell *et al.* (1980) took advantage of the variable responses of different inbred mouse strains to *Taenia taeniaeformis* infection to undertake a series of elegant experiments that have provided a better understanding of concomitant immunity and differences in 'innate' immunity between different hosts. Mice that were resistant to infection with *T. taeniaeformis* were found to have generated protective antibodies by 9 days post-infection whereas mice that were more susceptible to infection had not. Following an infection with eggs, the early developing parasite is highly susceptible to antibody and complement-mediated host immune responses, however, after about 10 days of development, the parasite transforms from being susceptible

Table 15.2. Recombinant oncosphere antigens of taeniid cestodes, which have been shown to induce host-protective immune responses.

Species	Antigen	Homology group[a]	Protection (%)[b]	Reference
Taenia ovis	To45W	45W	94	Johnson *et al.* (1989)
	To45S	45W	87	Lightowlers *et al.* (1996c)
	To16K	16K	92	Harrison *et al.* (1996)
	To18K	18K	99	Harrison *et al.* (1996)
T. saginata	TSA-9	45W[c]	99	Lightowlers *et al.* (1996b)
	TSA-18	18K[c]	99	Lightowlers *et al.* (1996b)
T. solium	TSOL18	18K	100	Flisser *et al.* (2004)
	TSOL45	45W	97	Flisser *et al.* (2004)
Echinococcus granulosus	EG95	EG95	100	Lightowlers *et al.* (1996a, 1999)
E. multilocularis	EM-95	EG95	83	Gauci *et al.* (2002)

[a]Assignment to a particular homology group, designated by the abbreviation used for the first antigen of the group to be characterized, indicates a high level of amino acid homology between antigens.
[b]Indicates the optimum level of protection achieved in vaccination and challenge trials in the parasite's natural intermediate host species compared to challenge controls.
[c]TSA-9 and TSA-18 were found to act synergistically; results represent those of vaccination trials using the two antigens together.

to immune attack to being insusceptible. Hosts that produce a sufficient amount of protective antibody before this change occurs are able to kill many or all of the developing metacestodes. Conversely, hosts that produce the potentially protective antibodies more slowly, allow many of the developing parasites to survive long enough to undergo the developmental transformation that renders them insusceptible and hence these hosts develop more substantial primary infections. This change in the susceptibility of the developing metacestode to host immune attack seems to coincide with major changes to the parasite's surface structure that occur during their early development (Engelkirk and Williams, 1982, 1983).

Definitive research on the susceptibility of taeniid cestode hosts to primary and secondary infection was undertaken using *T. hydatigena* (Gemmell and Johnstone, 1981). Two different methods to induce infections in sheep were employed, namely oral infection and intramuscular injection of activated oncospheres, thereby enabling the differentiation of parasites arising from primary and secondary infections. They discovered that a high level of immunity was induced following a primary infection that was able to prevent the establishment of parasites from a subsequent secondary challenge infection. However, this immunity did not persist indefinitely and had waned by 18 months after the primary infection, such that at that time it was possible for parasites to establish from a secondary infection despite the continuing presence of viable parasites in the tissues from the primary infection.

Vaccination – Historical Perspectives

In the 1930s extensive studies were undertaken on immunization of laboratory rats and rabbits against challenge infection with *T. taeniaeformis* or *Taenia pisiformis* (Miller, 1931, 1932; Miller and Kerr, 1932; Kan, 1934; Campbell, 1936). These studies established that high levels of protection could be achieved by immunization using non-living extracts derived from different life-cycle stages of the parasite as well as by antigens derived from heterologous cestode species.

During the 1960s and 1970s Gemmell found that exposure of sheep to living oncospheres induced protection against *T. ovis*, *T. hydatigena* or *E. granulosus* (Gemmell, 1962, 1964a,b, 1965, 1966, 1967, 1969, 1970) but that this protection was abrogated if the oncospheres were killed before injection (Gemmell, 1964a, 1969). This led to the belief that protective immune responses were induced only following exposure to living parasites. Rickard and Bell (1971a) investigated this hypothesis by exposing lambs to living oncospheres contained within chambers having a pore size of 0.2 μm that were implanted intraperitoneally. Exposure to the parasites in this way did induce protection from an oral challenge with parasite eggs, implicating the protective antigens as being excretory or secretory products. This was confirmed following the testing of cell-free supernatants collected following *in vitro* culture of activated oncospheres (Rickard and Bell, 1971b). However, it was subsequently revealed that the protective antigens were contained within the unactivated eggs and that activation and culture of the parasites were not required for their production (Rajasekariah *et al.*, 1980). The failure of non-living oncospheres to induce protection in the earlier experiments of Gemmell (1964a, 1969) appears to have been the result of quantitative inadequacies rather than qualitative differences regarding the presence or absence of protective antigens in non-viable parasites. Indeed, the experiments performed initially by Rajasekariah *et al.* (1980), which determined that protection could be achieved with non-living extracts of *T. taeniaeformis* were, to a significant degree, a rediscovery of what had been described some 50 years previously. Following Rickard and Bell's (1971b) discoveries of non-living, host-protective antigens for *T. ovis*, investigations confirmed that oncosphere antigens were highly effective as vaccines against infections with *T. pisiformis* in rabbits (Rajasekariah *et al.*, 1985), *Taenia saginata* in cattle (Rickard and Adolph, 1976), *T. solium* in pigs (Patak and Gaur, 1990) and *E. granulosus* in sheep (Heath *et al.*, 1981; Osborn and Heath, 1982).

Defined Vaccines

The discovery that oncosphere extracts provided a source of potent, protective antigens for both *Taenia* species and *E. granulosus* raised the potential for development of practical vaccines to prevent infections with economic and medically important species. Taeniid cestodes do not proliferate in culture and cannot realistically be obtained from animal infections in sufficient quantities to allow the development of practical vaccines. The application of recombinant DNA procedures in the field of parasitology during the 1980s provided the means to obtain sufficient quantities of vaccine antigens. Indeed, this was a period of great enthusiasm in the field of immunoparasitology, with optimism for the development of effective vaccines against many of the most important parasitic infections. Sadly, this optimism appears to have been misplaced, with very few successes eventuating over the succeeding decades (Mitchell, 1989; Dalton and Mulcahy, 2001). However, the taeniid cestodes have been remarkable exceptions because vaccine development has been spectacularly successful – at least to the point of experimental success (Lightowlers *et al.*, 2003a).

A relatively simple strategy was adopted for the development of vaccines against *Taenia* and *Echinococcus* species (Lightowlers *et al.*, 2003a). Individual host-protective antigens were identified in parasite extracts using immunochemical analyses, fractionation techniques and vaccine trials in the target species. Specific antibodies were prepared against individual host-protective antigens and these were used to identify the antigens expressed by cloned cDNA. Selected clones were sub-cloned into plasmid expression vectors and vaccine trials performed with recombinant proteins in the target host species against an experimental challenge infection.

The taeniid cestodes are a closely related group of species that share many features of their host–parasite immunological relationship. The parasite group includes some members that are relatively easy to work with (e.g. those infecting dogs or cats as definitive hosts and mice, rabbits or sheep as intermediate hosts). The group also includes members that are much more difficult to work with because either the definitive hosts are human (*T. saginata* and *T. solium*), because of the very high cost of undertaking vaccine trials (*T. saginata* in cattle), or because of the difficulties of working with material for challenge infections that can cause serious disease in the scientists involved (*T. solium*, *E. granulosus*, *Echinococcus multilocularis*). Regrettably, the species that are most important from economic or medical perspectives (*T. saginata*, *T. solium*, *Echinococcus* species) fall into the latter category. The strategy that was adopted to produce vaccines against the more important species was to firstly identify defined vaccine antigens in a host–parasite system that could be manipulated relatively easily and inexpensively. The knowledge that was gained with these model parasites was applied to investigate whether there were homologous proteins in the closely related, economically or medically important parasite species, and if so, whether these could be used as vaccines.

The species that was used in the initial work was both an economically important parasite as well as a useful model species. *T. ovis* causes cysticercosis in sheep and is transmitted by dogs acting as the definitive host. This host–parasite system can be maintained relatively inexpensively for laboratory investigations. A successful vaccine was developed for *T. ovis* and the mRNAs encoding the protective antigens of *T. ovis* have been used successfully as laboratory tools to identify highly protective homologous proteins from both *T. saginata* and *T. solium*, as described below. This strategy was not successful for the identification of vaccine antigens for *E. granulosus*, however, a research programme based on the original path taken for *T. ovis* was ultimately successful in developing an effective vaccine for *E. granulosus*.

Taenia ovis

Immunochemical analyses and fractionation studies on *T. ovis* oncosphere proteins identified two groups of host-protective antigens – one having a size of M_r 47–52 kDa and the other of M_r 16–18 kDa (Harrison *et al.*, 1993). Three different recombinant proteins were identified, each of which is capable of inducing very high levels (>90%) of protection against challenge infection in sheep with *T. ovis* eggs (Johnson *et al.*, 1989; Harrison *et al.*, 1996). These proteins are referred to as To45W, To16 and To18. Discovery of the first of these, To45W, has been listed as a Milestone in the History of Parasitology (Cox, 1993) and was the first demonstration of a highly effective defined antigen vaccine against a parasite infection. A practical vaccine based on To45W has been developed and registered for commercial use (Rickard *et al.*, 1995). In addition, the *T. ovis* and *E. granulosus* vaccines have provided useful models for many investigations into the application of recombinant antigen vaccines in the field of parasitology (Table 15.3).

It is interesting to note that initial vaccine trials, which used To45W expressed as a β-galactosidase fusion protein, were unsuccessful (Johnson *et al.*, 1989; Lightowlers *et al.*, 2003a). Following these trials, an alternate *Escherichia coli* expression method was investigated, involving use of a carboxy-terminal fusion with glutathione *S*-transferase (GST). The use of this expression system provided two advantages. First, the vaccine protein could be purified from bacterial lysates using a one-step, non-denaturing affinity purification on glutathione agarose. Second, the oncosphere antigens appear to be relatively toxic for *E. coli* and were expressed at low levels (Johnson *et al.*, 1989; Lightowlers *et al.*, 1996c). Expression together with the abundantly expressed GST protein was found to lead to improved levels of expression. While it is not clear exactly why the GST fusion protein was host-protective when the β-galactosidase fusion protein was not, this 'hiccup' in the development of the *T. ovis* vaccine illustrates an important aspect in the successful identification of this host-protective antigen. No research was undertaken with recombinant *T. ovis* proteins until individual, host-protective antigens had been identified from oncosphere extracts (Harrison *et al.*, 1993). Thus, when initial vaccine trials failed using the recombinant homologue (To45W–β-galactosidase) of a known host-protective (native) antigen, there was sufficient confidence in the potential of the antigen to warrant perseverance, the use of alternative expression system(s) and new vaccine trials. Had the information not been available to indicate that the native protein was host-protective, it is likely that To45W would have been overlooked after the initial vaccine trials, in favour of investigating other cloned antigens. These experiences with *T. ovis* highlight the value of undertaking, wherever possible, detailed investigations into the identification

Table 15.3. Research undertaken using the *Taenia ovis* 45W/18k/16k and *Echinococcus granulosus* EG95 recombinant vaccines, highlighting their application as model anti-parasite vaccines.

Discovery/Research topic	References (*Taenia ovis* 45W/18k/16k)	References (*Echinococcus granulosus* EG95)
Identification of new host-protective antigens	Johnson *et al.* (1989), Harrison *et al.* (1993)	Heath and Lawrence (1996)
Recombinant antigen achieves protection	Johnson *et al.* (1989), Harrison *et al.* (1996)	Lightowlers *et al.* (1996a, 1999)
Duration of immunity and safety	Harrison *et al.* (1999)	Heath *et al.* (2003)
Characterization of the associated gene(s)	Waterkeyn *et al.* (1995, 1997a,b, 1998)	Chow *et al.* (2001)
Stage-specific regulation of expression	Gauci and Lightowlers (1995), Waterkeyn *et al.* (1997a)	Chow *et al.* (2004)
Characterization of the host-protective immune responses	Rothel *et al.* (1996b)	Woollard *et al.* (1998)
Protection against naturally acquired infection	Lawrence *et al.* (1996)	Heath *et al.* (2003)
Immunological studies on vaccine-induced responses	Rothel *et al.* (1996a, 1997c, 1998a, 1998b)	–
Delineation of host-protective fragments and epitopes	Lightowlers *et al.* (1996c), Dadley-Moore *et al.* (1999a)	Woollard *et al.* (1998)
Vaccination with synthetic peptides	Dadley-Moore *et al.* (1999b)	Woollard *et al.* (2000a,b, 2001)
Nucleic acid vaccination	Rothel *et al.* (1997b), Drew *et al.* (1999, 2000a,b, 2001a,b)	Scheerlinck *et al.* (2001)
Vaccination using a recombinant viral construct	Rothel *et al.* (1997a)	–
Identification of host-protective homologues in related parasite species	Gauci *et al.* (1998), Gauci and Lightowlers (2001, 2003), Flisser *et al.* (2004)	Gauci *et al.* (2002)

of protective native antigens before investigations involving recombinant antigens.

Taenia saginata

T. saginata causes cysticercosis in cattle. In the developed world, detection of the parasites in beef carcasses causes the meat to be downgraded or condemned for human consumption. The parasite has a wide distribution but is most prevalent in countries where meat is frequently eaten raw, or incompletely cooked, and where cattle pasture is contaminated with human faeces. Although the obligate definitive host is man, the medical consequences of *T. saginata* taeniasis in humans are usually trivial. Economic losses in the beef industry can, however, be considerable and can have industry-wide impacts such as in many African countries where the parasite is hyperendemic.

Investigations were undertaken to determine if the *T. saginata* genome contained sequences that were closely related to To45W, To16 or To18 cDNAs. Southern blots of *T. saginata* genomic DNA were probed with labelled To45W, To16 or To18 cDNA. These investigations revealed the presence of DNA fragments having close homology to each of the three *T. ovis* cDNAs. Two of these were readily cloned from *T. saginata* oncosphere mRNA and the expressed proteins have been referred to as TSA-9 (the To45W homologue) and TSA-18 (the To18 homologue), respectively. Individually, these proteins were not reliable as vaccines against *T. saginata* infection in cattle. However, in perhaps the most striking example available of vaccine synergy between two recombinant antigens, a combination of TSA-9 and TSA-18 was found to reliably induce very high levels of protection (>95%) in several, independent cattle vaccine trials (Lightowlers *et al.*, 1996b). Not only was this work successful in the development of an effective vaccine against a commercially important parasite, it also validated the strategy whereby the knowledge about host-protective antigens of one taeniid species (*T. ovis*) could be used to rapidly develop an effective vaccine against a related species (*T. saginata*).

Taenia solium

T. solium causes cysticercosis in pigs, and as for *T. saginata*, humans are the obligate definitive hosts. Unlike *T. saginata*, the eggs from the adult *T. solium* tapeworm that are present in the faeces of a tapeworm carrier can infect not only the natural animal intermediate host (pigs) but are infective also for the tapeworm carrier or any other person who might accidentally ingest the eggs. In humans the cysticerci may encyst in the brain, causing neurological disease in many countries in Africa, Asia, Central America and the northern parts of South America (Gemmell *et al.*, 1983). The parasite is largely restricted to the developing world, where free-roaming pigs are present and latrines are either not present or commonly not used.

Analysis of Southern blots of *T. solium* genomic DNA probed with *T. ovis* To45W, To16 or To18 cDNAs identified the presence of related DNA fragments for each of the three probes. The *T. solium* homologues of each have been cloned and the proteins expressed in *E. coli* (designated TSOL45, TSOL18 and TSOL16), respectively. To date, the recombinant TSOL45 and TSOL18 proteins have been tested in four, independent vaccination trials in pigs against a challenge infection with *T. solium* eggs (Flisser *et al.*, 2004; Gonzalez *et al.*, 2005). Two trials have been undertaken in Mexico and one each in Peru and Cameroon. On each occasion TSOL18 has induced 99.5–100% protection in the vaccinated pigs. TSOL45 has also been shown to be highly protective. In the first experiment undertaken in Mexico, TSOL45 was poorly immunogenic and failed to induce a detectable immune response. However, in two subsequent trials, an altered vaccination regime has been used that has improved the level of response to the antigen and in these trials the TSOL45 protein induced 97.1% and 98.6% protection against the challenge infection, with the majority of the vaccinated pigs having no detectable parasites.

The alteration that was made to the *T. solium* vaccination regime related to the use of fusion protein partners in the production of the recombinant oncosphere antigens. All of the protective anti-cestode vaccines summarized in Table 15.2 involve the use of recombinant proteins expressed in *E. coli* as GST fusion proteins. However, a disadvantage of using GST fusion proteins arises because the GST moiety is highly immunogenic (Rothel *et al.*, 1996b; Woollard *et al.*, 2000b; Gonzalez *et al.*, 2005), raising concerns about possible antigenic competition between the GST component and the oncosphere protein.

In the initial *T. solium* vaccination trial described by Flisser *et al.* (2004), pigs were immunized with TSOL45 on two occasions, both times with the GST fusion protein. However, these pigs were neither protected against challenge infection with *T. solium* eggs nor did they produce a detectable serum antibody response to the TSOL45 protein. A second trial was designed to investigate the immune response in pigs given additional immunizations with TSOL45. In this experiment, the pigs did show a detectable, specific antibody response that was boosted by a third injection, however, the antibody titre remained low. A fourth immunization was given using TSOL45 expressed as a fusion protein, not with GST but with maltose-binding protein (MBP). The rationale for use of an MBP fusion protein was to boost the anti-TSOL45 responses without simultaneously boosting the anti-GST responses. This strategy appears to have been successful, with anti-TSOL45 titres rising substantially following the final boost and the vaccinated pigs being protected against a subsequent challenge infection with *T. solium* eggs.

While the use of different fusion proteins in the immunization schedule for TSOL45 was effective in an experimental sense, it is not a practical option for vaccine production. Other modifications to expression constructs have been found to affect the expression levels of recombinant oncosphere antigens, particularly truncations that remove hydrophobic portions of the antigens (Lightowlers *et al.*, 1996b, 2003c). These may allow the production and use of recombinant oncosphere antigens expressed without a fusion partner and, thereby, may overcome any immunological problems inherent with the use of GST or other protein fusions.

Echinococcus granulosus

Infection with *E. granulosus* in intermediate hosts causes cystic hydatid disease (cystic echinococcosis). The infection is transmitted by canids, often domestic dogs, which harbour the small adult tapeworm. Unlike *Taenia* species, *E. granulosus* is able to infect a wide variety of mammalian species as intermediate hosts. Domestic livestock are frequently involved in transmission of the disease, particularly sheep. The metacestode is known as an hydatid cyst and the parasite proliferates internally, often producing large numbers of potentially infective protoscoleces. Hydatid cysts can occur in any body organ but most commonly they are found in the liver and lungs. Infection is acquired in the intermediate host following ingestion of *E. granulosus* eggs from the faeces of an infected definitive host. Accidental ingestion of *E. granulosus* eggs by humans may lead to the development of hydatid cysts in a variety of tissue sites. The clinical consequences of hydatid disease vary according to the number of cysts and their size and location within a patient. The disease often has severe consequences for health. Treatment options are restricted, often of limited effectiveness, and frequently involve extensive surgical procedures (Eckert *et al.*, 2001). The parasite has an almost worldwide distribution, particularly in areas where pastoral industries are prominent. Hydatid disease has been recognized as a major health burden in many countries and substantial effort has been expended in the latter half of the 20th century towards controlling the transmission of the disease (Gemmell *et al.*, 2001).

Following the successful development of the *T. ovis* and *T. saginata* vaccines, investigations were undertaken with *E. granulosus* to determine whether oncosphere antigen homologues of the To45W, To18 or To16 host-protective antigens could be detected. However, Southern blot analyses of *E. granulosus* genomic DNA probed with the *T. ovis* cDNAs failed to identify any genetic homologues. Data published by Heath *et al.* (1981) indicated that oncospheres of *E. granulosus* were a source of potent host-protective antigens, as had been found earlier for *T. taeniaeformis* and *T. ovis*. While it was not possible to short cut the vaccine development process by simply cloning homologues of the host-protective *T. ovis* antigens, it seemed logical that it should be possible to produce a vaccine from first principles, beginning with identification of individual, protective, native antigens in *E. granulosus* oncospheres.

Heath and Lawrence (1996) undertook a series of vaccine trials in sheep to compare oncosphere antigen fractions against a challenge infection with *E. granulosus*. Protective components were identified of M_r values of 23, 25 and 30 kDa. Antisera against these antigens were used to screen a cDNA library prepared using mRNA derived from activated oncospheres of *E. granulosus*. One clone was shown to encode a protein, which had the capacity to induce a high level of protection against *E. granulosus* infection in sheep (Lightowlers *et al.*, 1996a). This protein, referred to as EG95, has now been used successfully in vaccine trials in Australia, New Zealand, Argentina, Chile and China (Lightowlers *et al.*, 1999; Heath *et al.*, 2003). It has also been shown to protect goats and cattle as well as sheep and to be effective in preventing naturally acquired infections in field trials (Heath *et al.*, 2003).

Echinococcus multilocularis

E. multilocularis is the cause of alveolar hydatid disease (alveolar echinococcosis) and is commonly transmitted between foxes as definitive hosts and a number of species of rodents as intermediate hosts. In humans, the metacestode stage of *E. multilocularis* proliferates, invades adjacent tissues and may metastasize. The infection commonly has fatal consequences. *E. multilocularis* is restricted to the northern hemisphere and, in most regions where the disease is endemic, infection in humans is relatively rare (about one case/million population in Europe), although in parts of China the disease occurs more commonly (Eckert *et al.*, 2001). The natural intermediate hosts of the parasite are wild animals and so vaccination of those hosts is not a practical option for the prevention of *E. multilocularis* transmission. However, should an effective vaccine be possible, one option may be the use of vaccination directly in humans. With this in mind, investigations were undertaken to determine whether an oncosphere antigen-based vaccine could be developed against *E. multilocularis*.

Southern blots of *E. multilocularis* genomic DNA were probed with the *E. granulosus* EG95 cDNA, revealing the presence of homologous DNA sequences. These sequences were cloned from mRNA, expressed in *E. coli* and the recombinant protein shown to have the capacity to induce significant protection against challenge infection with *E. multilocularis* eggs in mice (Gauci *et al.*, 2002).

Towards Practical Application of Cestode Vaccines

The remarkable success that has been achieved with identification of protective, recombinant antigens against the *Taenia* and *Echinococcus* species has been in stark contrast to the general lack of success that has been achieved with vaccine development against other parasitic helminth infections (Dalton and Mulcahy, 2001). Hence, very little direction has been available from the experiences of others on how to translate a laboratory success with an anti-parasite vaccine into a practical (commercial) vaccine. First principles would suggest a host of issues that require investigation/optimization, for example:

- dose of antigen;
- number and interval of immunizations;
- route of injection;
- adjuvant;
- duration of protection;
- protection of neonates (active and passive);
- protection against field-derived infection;
- safety;
- commercial-scale production.

In addition to these issues, it would be useful to investigate the potential impact of antigenic variation in the parasite population in relation to resistance to vaccine-induced immune responses. Having now undertaken many of the studies for the *T. ovis*, *T. saginata* and *E. granulosus* vaccines, there may be aspects of what has been learnt along the way that may be useful for others in the future who are faced with a similar vaccine development task.

Clearly there are many different components of a new vaccine's development, which need to be investigated. All of these aspects cannot be investigated simultaneously and the obvious question must be asked: in what priority do we investigate the various components? One of the lessons that can be learnt from the experiences with taeniid cestode vaccines has been that it is critical to consider the issue of vaccine production/productivity very early in the development process. This is because the levels of vaccine productivity that many achieved using methods sufficient to produce vaccine for initial experimental trials may be insufficient to form the basis for a practical vaccine. Modification of vaccine production methods will produce a product that may be significantly different and so vaccine optimization studies (e.g. dose, number of injections, type of adjuvant, etc.) need to be undertaken with the same material as that which will go into the practical vaccine. The *T. ovis*, *T. saginata* and *E. granulosus* vaccines have all required modification of the antigen and/or the antigen production methods in order to produce sufficient quantities of material suitable for commercial scale-up. For example, the *T. ovis* 45W antigen expressed very poorly and in insufficient quantities to form the basis for a commercial vaccine. A series of antigen truncations were investigated in an effort to find a derivative of the To45W protein, which was protective but expressed at a higher level in *E. coli*. One such construct, in which the COOH-terminal 19 amino acids were deleted (To45W-B/X), was found to be both host-protective and expressed at a level several orders of magnitude greater than the original To45W protein (Lightowlers *et al.*, 1996c). Subsequently, it was considered that the affinity purification methodology, which had been used for experimental production of To45W and To45W-B/X, was too expensive for the production of a vaccine for use in livestock animals. Alternative methods were developed for large-scale preparation of vaccine antigen from fermented *E. coli* expressing To45W-B/X, which involved differential solubilization of bacterial inclusion bodies using urea (Dempster *et al.*, 1996). The final product of this process was an antigen preparation differing significantly from the original affinity-purified protein used initially as the host-protective antigen. Clearly the studies which have sought to optimize the levels and duration of immunity induced by the *T. ovis* vaccine would only be relevant if they were carried out with the To45W-B/X antigen prepared from inclusion bodies. With the *T. saginata* and *E. granulosus* vaccines, similar issues need to be addressed in order to be able to produce sufficient quantities of vaccine for practical application.

Protective Immune Responses and Epitope Identification

Early studies with *T. taeniaeformis* in mice (Miller and Gardiner, 1932; Campbell, 1938b; Leid and Williams, 1974; Musoke *et al.*, 1975; Mitchell *et al.*, 1977, 1980), *T. pisiformis* in rabbits (Campbell, 1938a), *T. saginata* in cattle (Lloyd and Soulsby, 1976) and *T. ovis* in sheep (Blundell *et al.*, 1968; Rickard *et al.*, 1977; Heath *et al.*, 1979; Harrison *et al.*, 1993) demonstrated that antibodies in the sera of infected or immunized hosts were capable of transferring passive immunity to naïve hosts. The presence and activities of these protective antibodies were found to be evident *in vitro* through the demonstration that activated oncospheres were killed if cultured in the presence of serum from infected or vaccinated donors (Heath, 1973; Heath and Lawrence, 1981, 1996). This knowledge provided the rationale for the use of antibodies as probes to identify host-protective recombinant proteins. The anti-parasite activity of antisera *in vitro* was found to require a source of complement. Also, inhibition of complement activity *in vivo* by cobra venom factor has been shown to abolish the protective efficacy of passively transferred antibodies against *T. taeniaeformis* in both rats (Musoke and Williams, 1975) and mice (Mitchell *et al.*, 1977), highlighting the dual requirement for both specific antibody and complement in immunity against the oncospheres of both *Echinococcus* and *Taenia*.

For each of the *T. ovis*, *T. saginata* and *E. granulosus* recombinant vaccines, immunity against a challenge infection can be transferred via colostrum from the vaccinated dam to her offspring. Antibodies in the sera of sheep vaccinated with the EG95 recombinant protein have been found to kill *E. granulosus* oncospheres in culture (Heath and Lawrence, 1996; Woollard *et al.*, 2000a,b, 2001). C.T. Kyngdon (Melbourne, 2005, unpublished observation) has confirmed similar lethal effects of ovine, bovine and porcine antisera against their respective recombinant oncosphere antigens (Table 15.2) from *T. ovis*, *T. saginata* and *T. solium*. It can be concluded that a major immune mechanism by which the vaccines against *Taenia* and *Echinococcus* exert their protective effect is through antibody and complement-mediated lysis of the oncosphere or early developing metacestode. While it is possible that other immune mechanisms also exist, the efficacy of specific antibody and complement *in vivo* and *in vitro* would suggest that, while this may not necessarily be the only immune mechanism, it is sufficient to account for the observed protective efficacy of the vaccines.

Precise definition of the host-protection epitope(s) of the *Taenia* and *Echinococcus* vaccine antigens would be valuable if it were to lead to the development of a synthetic vaccine or provide reagents to assist with quality control of vaccine production. Identification of a limited number of host-protective fragments may allow the recombinant antigen to be replaced by synthetic peptides. This would have significant advantages in the production of vaccine and quality control of the product. Even if a synthetic peptide vaccine did not replace the recombinant antigen, knowledge of the critical host-protective epitopes would allow assays to be developed to determine the quality of different vaccine batches. For these reasons, considerable attention has focused on the identification of host-protective epitopes of the To45W and EG95 proteins.

Evidence supporting the important role of complement-fixing antibodies in immunity stimulated by the To45W and EG95 indicated that antibodies would be useful probes for the identification of host-protective epitopes. Synthetic peptides were prepared spanning the full length of the To45W and EG95 proteins and these were reacted in ELISA with sera from sheep, which had been vaccinated and subsequently shown to be immune from parasitic challenge infection (Lightowlers *et al.*, 1996c; Woollard *et al.*, 1998). Prominent immunogenic epitopes were readily identified and those recognized by the majority of sera from individual vaccinated animals were selected, prepared as peptide immunogens and used in vaccination trials. While the peptides themselves were immunogenic and induced specific IgG responses in vaccinated sheep, which displayed identical specificity in Western blots of oncosphere antigens as sera raised to the recombinant proteins, none induced protection against a challenge infection (Dadley-Moore *et al.*, 1999b; Woollard *et al.*, 2000a,b).

The EG95 vaccine protein was also mapped for the location of host-protective epitopes using large, overlapping fragments of the parent protein (Woollard *et al.*, 2001). The EG95 cDNA was sub-cloned as expression constructs representing, approximately, the amino-terminal half of the protein, the carboxy-terminal half and a central portion of the protein overlapping the sequence of the other two fragments. These three fragments, which were expressed using the same methodology used for full-length EG95, were used to raise specific antibodies in sheep that showed an identical pattern in Western blots of oncosphere antigen to that shown by sera raised against the full-length EG95 protein. Similarly, a group of sheep vaccinated with a combination of all three of these EG95 fragments, representing the full protein sequence of EG95, also showed specific recognition of the native oncosphere protein in Western blots. However, none of the groups of sheep vaccinated with fragments, or the pool of fragments, showed any protection against a challenge infection with *E. granulosus* eggs (Woollard *et al.*, 2001).

An analysis of the amino acid sequence of EG95 and its relationship to other proteins (see below) indicated that EG95 may show tertiary structure similar to a fibronectin type III (FnIII) domain. While there is no direct evidence at present either in favour or against this hypothesis, the nature of the folding seen in FnIII domains for which structural information is available (Bork and Doolittle, 1992), would suggest that none of the three EG95 fragments investigated by Woollard and his colleagues would have been able to fold in a manner resembling a portion of the FnIII structure. The current working hypothesis is that the host-protective epitope(s) is/are conformational (Woollard *et al.*, 2001). The lack of effectiveness of antigen fragments or peptides derived from either To45W or EG95, and the existence of sequence similarities between all of the host-protective oncosphere antigens (see below) suggest that the protective responses elicited by all of the recombinant taeniid vaccines may operate via conformational epitopes.

Predicted Protein Structure and Function of Oncosphere Antigens

All of the host-protective oncosphere antigens described to date (for *T. ovis*, *T. saginata*, *T. solium*, *E. granulosus* and *E. multilocularis*) are modular proteins and have common features (Gauci and Lightowlers, 2003; Lightowlers *et al.*, 2003b). All comprise a predicted secretory signal sequence. Much indirect evidence is available to suggest that host-protective oncosphere antigens are secreted. Oncospheres implanted intraperitoneally into sheep in 0.2 μm diffusion chambers elaborate host-protective antigens (Rickard and Bell, 1971a). Supernatants collected from parasites maintained in *in vitro* culture elaborate host-protective antigens (Rickard and Bell, 1971b; Rickard and Adolph, 1977; Osborn *et al.*, 1982). The original recombinant clone encoding To45W was selected using an antiserum raised against *T. ovis* oncosphere secretory products (Johnson *et al.*, 1989). Also, there is evidence that the TSA-18 protein of *T. saginata* (the homologue of To18 and TSOL18 in *T. ovis* and *T. solium*, respectively) is a secreted protein. Benitez *et al.* (1996) published a description of a *T. saginata* protein, which they designated HP6. The nucleotide sequence published at the time suggested that the clone was almost identical to TSA-18, however, the protein that the authors indicated as being expressed by the clone varied from TSA-18 by two frame-shift differences. These differences have subsequently been corrected for HP6 in GenBank, indicating identity between TSA-18 and the clone described as HP6. A cDNA clone representing TSA-18 was selected using a monoclonal antibody raised against *T. saginata* secretory products. The monoclonal antibody revealed localization of the associated antigen in secretory organelles within *T. saginata* oncospheres (Harrison and Parkhouse, 1986; Benitez *et al.*, 1996). Subsequently, Bonay *et al.* (2002) found that the recombinant protein (i.e. TSA-18) enhanced the adhesion of normal rat kidney (NRK) epithelial cells in an *in vitro* assay in a concentration-dependent manner. These data support the suggestion made earlier by Bork and Doolittle (1992) that oncosphere proteins that possess FnIII domains may play functional roles as adhesins.

Evidence to support the functional nature of the secretory signal sequences in oncosphere antigens has been provided by Drew *et al.* (2000a). Cos 7 cells were transiently transfected with a plasmid construct in which the To45W cDNA either included or excluded the predicted secretory signal sequence. The *T. ovis* protein was secreted into the supernatant in a glycosylated form by cells transfected with a construct expressing the full-length To45W. However, cells transfected with a construct from which the signal sequence had been removed, expressed only a somatic, non-glycosylated form of the protein.

The native oncosphere proteins corresponding to both To45W and EG95 are substantially larger than would be predicted by their protein sequences (Johnson *et al.*, 1989; Lightowlers *et al.*, 1996a), and both proteins have a number of N-X-S/T sites for N-linked glycosylation and a high (19% and 20%, respectively) S/T content necessary for O-linked glycosylation. Glycosylation of the To45W native antigen has been confirmed following treatment of the protein with glycosidase PNGase F to remove N-linked carbohydrates (Lightowlers *et al.*, 2003a). It is likely that many or all of the known host-protective oncosphere antigens are glycosylated, or otherwise post-translationally modified, in their native form. However, these modifications are not provided by the prokaryotic expression systems used in production of the protective recombinant proteins. It is clear that post-translational modifications are not required in order for the recombinant antigens to induce protective immune responses and so at least some protective epitopes are associated with the protein component alone. The presence in native oncosphere antigens of additional host-protective epitopes that are associated with their carbohydrate components cannot be excluded; however, epitopes associated with the protein component alone are sufficient to account for the protective efficacy of oncosphere antigens.

All of the oncosphere antigens cloned to date contain either one of two copies of a predicted FnIII domain. These domains are widely distributed in eukaryotic proteins and occur also in some prokaryotic proteins (Bork and Doolittle, 1992). Approximately 2% of animal proteins include FnIII domains. Many, but not all, of these proteins are extracellular and some have roles as adhesins. The structure of this ~100 amino acid domain is highly conserved and consists of two layers with three β strands in one plane and four β strands in another (Potts and Campbell, 1996). Overall, amino acid sequence identity between different FnIII domains is low, even between FnIII repeat domains within fibronectin itself (Plaxco *et al.*, 1997). Nevertheless, certain residues are highly conserved and maintain the tertiary structure of the proteins (Bork and Doolittle, 1992). Other conserved motifs such as an Arg–Gly–Asp (RGD) motif within a loop of some FnIII domains is associated with proteins having cell adhesion properties, as discussed above (Ruoslahti and Pierschbacher, 1987; D'Souza *et al.*, 1991).

Structure of Genes Encoding Oncosphere Antigens

There are many features of the genes encoding oncosphere proteins that are common amongst the genes from different taeniid species. The structure of the genes encoding the various host-protective oncosphere antigens are illustrated in Fig. 15.1. Noteworthy features are the conservation of gene structure between homologues from different *Taenia* species and the conservation in the gene organization corresponding to the modular components of the expressed proteins. The exons of each gene

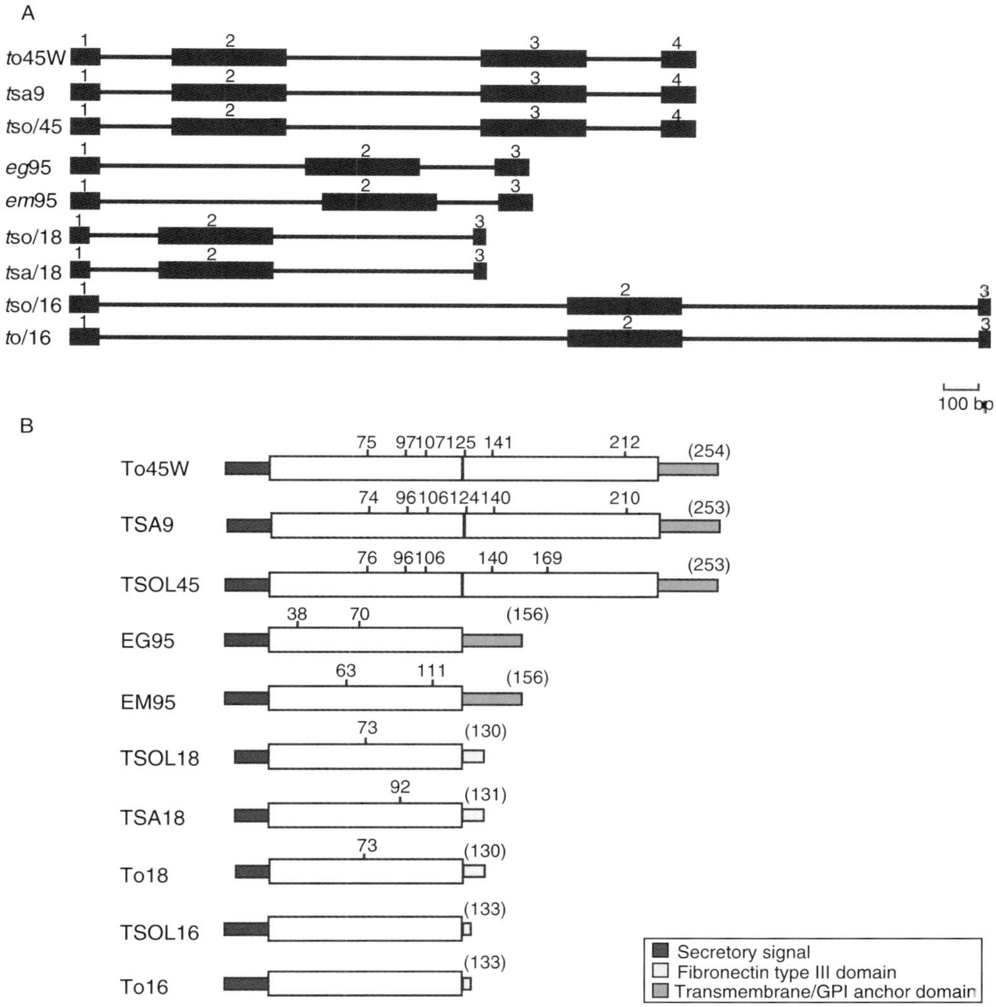

Fig. 15.1. Diagrammatic representation of taeniid cestode genes encoding cloned oncosphere antigens. A. Comparison of gene structures. Exons are represented by black bars labelled 1–4 and introns are shown as black lines. B. Schematic illustration showing the locations of secretory signals, fibronectin type III domains and transmembrane glycosylphosphatidylinositol (GPI) anchor domains of taeniid vaccine antigens. Numbers indicate amino acid position of predicted N-linked glycosylation sites and numbers in brackets represent length of the predicted full-length protein from the initiator methionine. (Reprinted from Lightowlers *et al.*, 2003b with permission from Elsevier.)

correspond to predicted secretory signal sequences, FnIII domain(s) and a small exon encoding the carboxy-terminal portion of the protein. In the case of the 45W homology group, there are two large exons each encoding tandemly expressed FnIII domains (labelled 2 and 3 on the representation of *to*45W in Fig. 15.1A). Some members of the *eg*95 homology group and the 16K and 18K groups also have an intron within the sequence encoding the 3′ untranslated region. Transcripts from the gene(s) encoding the TSOL45 protein in *T. solium* are differentially spliced such that at least three different protein products are potentially expressed by the gene (Gauci and Lightowlers, 2001). It has yet to be confirmed that these differentially spliced transcripts are reflected by the occurrences of the corresponding proteins in the oncosphere. However, Western blots of *T. solium* oncosphere antigens probed with antisera raised to TSOL45 have identified several specifically related proteins (Gonzalez *et al.*, in press), suggesting that the gene does lead to the translation of several different products.

Potential Practical Impact of Vaccination Against Cestode Parasites

While the *T. ovis* vaccine attracted considerable attention as the first, highly successful defined-antigen vaccine against a parasitic disease, and the product was registered for commercial use in New Zealand, regrettably it has never been marketed. Economic and political factors contributed to the decision by the commercial partner in the *T. ovis* vaccine development not to market the product (Rickard *et al.*, 1995).

While *T. ovis* and *T. saginata* have significant economic impact on the meat industry, and *T. solium* and *E. granulosus* have significant medical impacts, often the owner of the associated infected livestock is unaware of the presence of the infection and suffers little or no direct economic loss because of the parasite in their stock. In this situation there is little economic incentive for livestock owners to invest in the use of a vaccine. The final stages of pre-commercial development of the *T. saginata* vaccine are being supported by a sewage treatment authority and it remains to be seen whether the vaccine might see wider usage. The hydatid vaccine has been licensed by a China-based group and a specialized facility has been built for production of the vaccine according to good manufacturing practice (GMP) guidelines. Successful commercial-scale production of the vaccine in China would provide a source of the vaccine for use in China as well as in other parts of the world, if it were deemed to be a valuable adjunct to the other methods being used for hydatid control.

The greatest potential impact for vaccination against the cestode parasites lies with *T. solium*. Unlike *E. granulosus*, *T. solium* transmission is maintained by a relatively restricted number of intermediate hosts. For all practical purposes, *T. solium* is only transmitted by domestic pigs and there is no sylvatic transmission. Also, the number of pigs involved in *T. solium* transmission is often small per production unit, for example, in association with a household or village, and the animals have close contact with their owners or managers. Apart from the breeding stock, most pigs live for less than 1 year. These attributes favour disease control by vaccination and contrast with the situation with hydatid disease in which often large numbers of livestock (several species) may range over extensive pastoral rangelands, and where animals may live for 5, 10 or more years, depending on the host species and circumstances. Another factor, which favours *T. solium* control, is that the obligatory definitive hosts are humans who are more amenable, at least in some ways, to a control campaign than are dogs for hydatid control.

A case has been made that *T. solium* cysticercosis is a potentially eradicable disease. After the development of effective vaccines against *T. ovis* and *T. saginata*, a potential role for vaccination was proposed for control or eradication of *T. solium* (Lightowlers, 1999). Now that an effective vaccine against *T. solium* is a reality (Flisser *et al.*, 2004), the potential for effective control or even eradication of *T. solium* has moved a step closer. However, a modification of the strategy described by Lightowlers (1999) would be suggested now (Table 15.4). As mentioned above, there are numerous problems associated with protecting neonates against disease by vaccination, i.e. protective maternal antibodies interfere with

Table 15.4. Strategy for eradication of *Taenia solium* taeniasis/cysticercosis.

Goal	Intervention
Attack phase	
Remove existing tapeworm infections in humans	Treatment of human population with taeniacide
Prevent the introduction of new sources of infection for humans	Simultaneous oxfendazole treatment and vaccination of all weanling piglets
Consolidation phase	
Prevent new infections in pigs	Simultaneous oxfendazole treatment and vaccination of all weanling piglets
Identification of breakdowns in control	Serological surveillance in humans and pigs; use of sentinal pigs; copro antigen detection in humans

active immunization of the neonate. Also, very young piglets may be immunologically immature and fail to mount a satisfactory (protective) immune response to vaccination. Gonzales *et al.* (1996) showed that a single treatment with oxfendazole killed all *T. solium* cysticerci in treated pigs. Adoption of this treatment in weanling piglets would allow the complex issues surrounding infections in neonatal piglets to be put aside. Oxfendazole treatment, to remove any existing viable parasites, at the same time as vaccination to prevent the development of any new infection, should provide a practical and effective strategy for prevention of *T. solium* transmission through pigs. Mass treatment of the local human population with anthelmintic to remove *Taenia* carriers would attack the parasite in its definitive host. Control would be maintained and reintroduction of the parasite prevented by continuance of the programme of treatment/vaccination of pigs.

Concluding Remarks

The future is bright for anti-cestode vaccines. At present, there seems no 'scientific' reason why the recombinant oncosphere vaccines could not be used effectively in practice. Many of the decisions that will determine whether this comes to fruition will be commercial or political, and these factors are difficult to predict or control. Certainly the tools are now available to have a major impact on the control of hydatid disease and *T. solium* cysticercosis, with the potential to achieve eradication of *T. solium* and provide a major demonstration of the value of vaccination in control of a human parasitic disease.

Remarkably, little is known about the biology of the host-protective oncosphere antigens and their natural functions within the parasites. As the quest for identification of protective antigens comes to a close, with the successful identification of vaccine antigens for *T. ovis*, *T. saginata*, *T. solium* and *E. granulosus*, we now have the opportunity to extend our knowledge of the biology of the host-protective antigens. The application of modern, powerful, investigative techniques such as transient transfection, transgenesis and RNA interference, provide many new opportunities to probe the biology of complex helminth parasites such as the taeniid cestodes.

References

Benitez, L., Garate, T., Harrison, L.J., Kirkham, P., Brookes, S.M. and Parkhouse, R.M. (1996) Cloning and sequencing of the gene encoding the principal 18-kDa secreted antigen of activated oncospheres of *Taenia saginata*. *Molecular and Biochemical Parasitology* 78, 265–268.

Blundell, S.K., Gemmell, M.A. and Macnamara, F.N. (1968) Immunological responses of the mammalian host against tapeworm infections. VI. Demonstration of humoral immunity in sheep induced by the activated embryos of *Taenia hydatigena* and *T. ovis*. *Experimental Parasitology* 23, 79–82.

Bonay, P., Gonzalez, L.M., Benitez, L., Foster, M., Harrison, L.J., Parkhouse, R.M. and Garate, T. (2002) Genomic and functional characterisation of a secreted antigen of *Taenia saginata* oncospheres. *Molecular and Biochemical Parasitology* 121, 269–273.

Bork, P. and Doolittle, R.F. (1992) Proposed acquisition of an animal protein domain by bacteria. *Proceedings of the National Academy of Science USA* 89, 8990–8994.

Burridge, M.J. and Schwabe, C.W. (1977) An epidemiological analysis of factors influencing the increase in *Taenia ovis* prevalence during the New Zealand *Echinococcus granulosus* control program. *Australian Journal for Parasitology* 53, 374–379.

Campbell, D.H. (1936) Active immunization of albino rats with protein fractions from *Taenia taeniaeformis* and its larval form *Cysticercus fasciolaris*. *American Journal of Hygiene* 23, 104–113.

Campbell, D.H. (1938a) The specific absorbability of protective antibodies against *Cysticercus crassicollis* in rats and *C. pisiformis* in rabbits from infected and artificially immunized animals. *Journal of Immunology* 35, 205–216.

Campbell, D.H. (1938b) The specific protective property of serum from rats infected with *Cysticercus crassicollis*. *Journal of Immunology* 35, 195–204.

Chow, C., Gauci, C.G., Cowman, A.F. and Lightowlers, M.W. (2001) A gene family expressing a host-protective antigen of *Echinococcus granulosus*. *Molecular and Biochemical Parasitology* 118, 83–88.

Chow, C., Gauci, C.G., Cowman, A.F. and Lightowlers, M.W. (2004) *Echinococcus granulosus*: oncosphere-specific transcription of genes encoding a host-protective antigen. *Experimental Parasitology* 106, 183–186.

Cox, F.E.G. (1993) Milestones in parasitology. *Parasitology Today* 9, 347–348.

Dadley-Moore, D.L., Lightowlers, M.W., Gauci, C.G., Rothel, J.S. and Jackson, D.C. (1999a) Synthetic peptides as candidate vaccines against *Taenia ovis* infection of sheep. *Letters in Peptide Research* 6, 303–312.

Dadley-Moore, D.L., Lightowlers, M.W., Rothel, J.S. and Jackson, D.C. (1999b) Synthetic peptide antigens induce antibodies to *Taenia ovis* oncospheres. *Vaccine* 17, 1506–1515.

Dalton, J.P. and Mulcahy, G. (2001) Parasite vaccines – a reality? *Veterinary Parasitology* 98, 149–167.

Dempster, R.P., Robinson, C.M. and Harrison, G.B. (1996) Parasite vaccine development: large-scale recovery of immunogenic *Taenia ovis* fusion protein GST-45W(B/X) from *Escherichia coli* inclusion bodies. *Parasitology Research* 82, 291–296.

Drew, D.R., Lightowlers, M. and Strugnell, R.A. (1999) Vaccination with plasmid DNA expressing antigen from genomic or cDNA gene forms induces equivalent humoral immune responses. *Vaccine* 18, 692–702.

Drew, D.R., Lightowlers, M. and Strugnell, R.A. (2000a) Humoral immune responses to DNA vaccines expressing secreted, membrane bound and non-secreted forms of the *Tania ovis* 45W antigen. *Vaccine* 18, 2522–2532.

Drew, D.R., Lightowlers, M.W. and Strugnell, R.A. (2000b) A comparison of DNA vaccines expressing the 45W, 18k and 16k host-protective antigens of *Taenia ovis* in mice and sheep. *Veterinary Immunology and Immunopathology* 76, 171–181.

Drew, D.R., Boyle, J.S., Lew, A.M., Lightowlers, M.W., Chaplin, P.J. and Strugnell, R.A. (2001a) The comparative efficacy of CTLA-4 and L-selectin targeted DNA vaccines in mice and sheep. *Vaccine* 19, 4417–4428.

Drew, D.R., Boyle, J.S., Lew, A.M., Lightowlers, M.W. and Strugnell, R.A. (2001b) The human IgG3 hinge mediates the formation of antigen dimers that enhance humoral immune responses to DNA immunisation. *Vaccine* 19, 4115–4120.

D'Souza, S.E., Ginsberg, M.H. and Plow, E.F. (1991) Arginyl-glycyl-aspartic acid (RGD): a cell adhesion motif. *Trends in Biochemical Science* 16, 246–250.

Eckert, J., Gemmell, M.A., Meslin, F.-X. and Pawlowski, Z.S. (2001) *WHO/OIE Manual on Echinococcosis in Humans and Animals: A Public Health Problem of Global Concern*. WHO/FAO, Paris.

Engelkirk, P.G. and Williams, J.F. (1982) *Taenia taeniaeformis* (Cestoda) in the rat: ultrastructure of the host–parasite interface on days 1 to 7 postinfection. *Journal of Parasitology* 68, 620–633.

Engelkirk, P.G. and Williams, J.F. (1983) *Taenia taeniaeformis* (Cestoda) in the rat: ultrastructure of the host–parasite interface on days 8 to 22 postinfection. *Journal of Parasitology* 69, 828–837.

Flisser, A., Gauci, C.G., Zoli, A., Martinez-Ocana, J., Garza-Rodriguez, A., Dominguez-Alpizar, J.L., Maravilla, P., Rodriguez-Canul, R., Avila, G., Aguilar-Vega, L., Kyngdon, C., Geerts, S. and Lightowlers, M.W. (2004) Induction of protection against porcine cysticercosis by vaccination with recombinant oncosphere antigens. *Infection and Immunity* 72, 5292–5297.

Gauci, C.G. and Lightowlers, M.W. (1995) Developmental regulation of *Taenia ovis* 45W gene expression. *Molecular and Biochemical Parasitology* 73, 263–266.

Gauci, C.G. and Lightowlers, M.W. (2001) Alternative splicing and sequence diversity of transcripts from the oncosphere stage of *Taenia solium* with homology to the 45W antigen of *Taenia ovis*. *Molecular and Biochemical Parasitology* 112, 173–181.

Gauci, C. and Lightowlers, M.W. (2003) Molecular cloning of genes encoding oncosphere proteins reveals conservation of modular protein structure in cestode antigens. *Molecular and Biochemical Parasitology* 127, 193–198.

Gauci, C.G., Flisser, A. and Lightowlers, M.W. (1998) A *Taenia solium* oncosphere protein homologous to host-protective *Taenia ovis* and *Taenia saginata* 18 kDa antigens. *International Journal for Parasitology* 28, 757–760.

Gauci, C., Merli, M., Muller, V., Chow, C., Yagi, K., Mackenstedt, U. and Lightowlers, M.W. (2002) Molecular cloning of a vaccine antigen against infection with the larval stage of *Echinococcus multilocularis*. *Infection and Immunity* 70, 3969–3972.

Gemmell, M.A. (1962) Natural and acquired immunity factors inhibiting penetration of some hexacanth embryos through the intestinal barrier. *Nature* 194, 701–702.

Gemmell, M.A. (1964a) Immunological responses of the mammalian host against tapeworm infections 1. Species specificity of hexacanth embryos in protecting sheep against *Taenia hydatigena*. *Journal of Immunology* 7, 489–499.

Gemmell, M.A. (1964b) Species specificity of the immunogenic complexes of the tapeworm hexacanth embryo. *Nature* 204, 705–707.

Gemmell, M.A. (1965) Immunological responses of the mammalian host against tapeworm infections III. Species specificity of hexacanth embryos in protecting sheep against *Taenia ovis*. *Journal of Immunology* 8, 281–290.

Gemmell, M.A. (1966) Immunological responses of the mammalian host against tapeworm infections. IV. Species specificity of hexacanth embryos in protecting sheep against *Echinococcus granulosus*. *Immunology* 11, 325–335.

Gemmell, M.A. (1967) Species specific and cross-protective functional antigens of the tapeworm embryo. *Nature* 213, 500–501.

Gemmell, M.A. (1969) Immunological responses of the mammalian host against tapeworm infections. X. Immunization of sheep against *Taenia hydatigena* and *T. ovis* with chemically or physically treated embryos. *Experimental Parasitology* 26, 58–66.

Gemmell, M.A. (1970) Hydatidosis and cysticercosis. 3. Induced resistance to the larval phase. *Australian Veterinary Journal* 46, 366–369.

Gemmell, M.A. (1987) A critical approach to the concepts of control and eradication of echinococcosis/hydatidosis and taeniasis/cysticercosis. *International Journal for Parasitology* 17, 465–472.

Gemmell, M.A. and Johnstone, P.D. (1981) Factors regulating tapeworm populations: estimations of the duration of acquired immunity by sheep to *Taenia hydatigena*. *Research in Veterinary Science* 30, 53–56.

Gemmell, M.A., Matyas, Z., Pawlowski, Z. and Soulsby, E.J.L. (1983) *FAO/UNEP/WHO Guidelines for Surveillance, Prevention and Control of Taeniasis/Cysticercosis*. WHO, Geneva, Switzerland.

Gemmell, M.A., Lawson, J.R. and Roberts, M.G. (1986) Control of echinococcosis/hydatidosis: present status of worldwide progress. *Bulletin of the World Health Organization* 64, 333–339.

Gemmell, M.A., Lawson, J.R. and Roberts, M.G. (1987) Population dynamics in echinococcosis and cysticercosis: evaluation of the biological parameters of *Taenia hydatigena* and *T. ovis* and comparison with those of *Echinococcus granulosus*. *Parasitology* 94, 161–180.

Gemmell, M.A., Roberts, M.G., Beard, T.C., Campano Diaz, S., Lawson, J.R. and Nonnemaker, J.M. (2001) In: Eckert, J., Gemmell, M.A., Meslin, F.-X. and Pawlowski, Z.S. (eds) *WHO/OIE Manual on Echinococcosis in Humans and Animals: A Public Health Problem of Global Concern*. WHO/OIE, Paris, pp. 195–237.

Gonzales, A.E., Garcia, H.H., Gilman, R.H., Gavidia, C.M., Tsang, V.C., Bernal, T., Falcon, N., Romero, M. and Lopez-Urbina, M.T. (1996) Effective, single-dose treatment or porcine cysticercosis with oxfendazole. *American Journal of Tropical Medicine and Hygiene* 54, 391–394.

Gonzalez, A.E., Gauci, C.G., Barber, D., Gilman, R.H., Tsang, V.C.W., Garcia, H.H., Verastegui, M. and Lightowlers, M.W. (2005) Vaccination of pigs to control human neurocysticercosis. *American Journal of Tropical Medicine and Hygiene* 72, 837–839.

Harrison, L.J. and Parkhouse, R.M. (1986) Passive protection against *Taenia saginata* infection in cattle by a mouse monoclonal antibody reactive with the surface of the invasive oncosphere. *Parasite Immunology* 8, 319–332.

Harrison, G.B.L., Heath, D.D., Dempster, R.P., Lawrence, S.B., Lightowlers, M.W. and Rickard, M.D. (1993) Identification of host-protective antigens of *Taenia ovis* oncospheres. *International Journal for Parasitology* 23, 41–50.

Harrison, G.B., Heath, D.D., Dempster, R.P., Gauci, C., Newton, S.E., Cameron, W.G., Robinson, C.M., Lawrence, S.B., Lightowlers, M.W. and Rickard, M.D. (1996) Identification and cDNA cloning of two novel low molecular weight host-protective antigens from *Taenia ovis* oncospheres. *International Journal for Parasitology* 26, 195–204.

Harrison, G.B., Shakes, T.R., Robinson, C.M., Lawrence, S.B., Heath, D.D., Dempster, R.P., Lightowlers, M.W. and Rickard, M.D. (1999) Duration of immunity, efficacy and safety in sheep of a recombinant *Taenia ovis* vaccine formulated with saponin or selected adjuvants. *Veterinary Immunology and Immunopathology* 70, 161–172.

Heath, D.D. (1973) Resistance to *Taenia pisiformis* larvae in rabbits. II. Temporal relationships and the development phase affected. *International Journal for Parasitology* 3, 491–498.

Heath, D.D. and Lawrence, S.B. (1981) *Echinococcus granulosus* cysts: early development *in vitro* in the presence of serum from infected sheep. *International Journal for Parasitology* 11, 261–266.

Heath, D.D. and Lawrence, S.B. (1996) Antigenic polypeptides of *Echinococcus granulosus* oncospheres and definition of protective molecules. *Parasite Immunology* 18, 347–357.

Heath, D.D., Yong, W.K., Osborn, P.J., Parmeter, S.N., Lawrence, S.B. and Twaalfhoven, H. (1979) The duration of passive protection against *Taenia ovis* larvae in lambs. *Parasitology* 79, 177–182.

Heath, D.D., Parmeter, S.N., Osborn, P.J. and Lawrence, S.B. (1981) Resistance to *Echinococcus granulosus* infection in lambs. *Journal of Parasitology* 67, 797–799.

Heath, D.D., Jensen, O. and Lightowlers, M.W. (2003) Progress in control of hydatidosis using vaccination – a review of formulation and delivery of the vaccine and recommendations for practical use in control programmes. *Acta Tropica* 85, 133–143.

Johnson, K.S., Harrison, G.B., Lightowlers, M.W., O'Hoy, K.L., Cougle, W.G., Dempster, R.P., Lawrence, S.B., Vinton, J.G., Heath, D.D. and Rickard, M.D. (1989) Vaccination against ovine cysticercosis using a defined recombinant antigen. *Nature* 338, 585–587.

Kan, K. (1934) Immunological studies of *Cysticercus fasciolaris*. *Keio Igaka* 14, 663–687.

Lawrence, S.B., Heath, D.D., Harrison, G.B.L., Robinson, C.M., Dempster, R.P., Gatehouse, T.K., Lightowlers, M.W. and Rickard, M.D. (1996) Pilot field trial of a recombinant *Taenia ovis* vaccine in lambs exposed to natural infection. *New Zealand Veterinary Journal* 44, 155–157.

Lawson, J.R. (1994) Hydatid disease and sheep measles: the history of their control and the economics of a recent change of control policy. *New Zealand Journal of Zoology* 21, 83–89.

Lawson, J.R. and Gemmell, M.A. (1983) Hydatidosis and cysticercosis: the dynamics of transmission. *Advances in Parasitology* 22, 261–308.

Leid, R.W. and Williams, J.F. (1974) The immunological response of the rat to infection with *Taenia taeniaeformis*. I. Immunoglobulin classes involved in passive transfer of resistance. *Immunology* 27, 195–208.

Lightowlers, M.W. (1990) Cestode infections in animals: immunological diagnosis and vaccination. *Revue Scientifique et Technique Office International des Epizooties* 9, 463–487.

Lightowlers, M.W. (1999) Eradication of *Taenia solium* cysticercosis: a role for vaccination of pigs. *International Journal for Parasitology* 29, 811–817.

Lightowlers, M.W., Gemmell, M.A., Harrison, G.B.L., Heath, D.D., Rickard, M.D. and Roberts, M.G. (1992) In: Yong, W.K. (ed.) *Animal Parasite Control Utilizing Biotechnology.* CRC Press, Boca Raton, Florida, pp. 171–198.

Lightowlers, M.W., Lawrence, S.B., Gauci, C.G., Young, J., Ralston, M.J., Maas, D. and Health, D.D. (1996a) Vaccination against hydatidosis using a defined recombinant antigen. *Parasite Immunology* 18, 457–462.

Lightowlers, M.W., Rolfe, R. and Gauci, C.G. (1996b) *Taenia saginata*: vaccination against cysticercosis in cattle with recombinant oncosphere antigens. *Experimental Parasitology* 84, 330–338.

Lightowlers, M.W., Waterkeyn, J.G., Rothel, J.S., Gauci, C.G. and Harrison, G.B. (1996c) Host-protective fragments and antibody binding epitopes of the *Taenia ovis* 45W recombinant antigen. *Parasite Immunology* 18, 507–513.

Lightowlers, M.W., Jensen, O., Fernandez, E., Iriarte, J.A., Woollard, D.J., Gauci, C.G., Jenkins, D.J. and Heath, D.D. (1999) Vaccination trials in Australia and Argentina confirm the effectiveness of the EG95 hydatid vaccine in sheep. *International Journal for Parasitology* 29, 531–534.

Lightowlers, M.W., Colebrook, A.L., Gauci, C.G., Gauci, S.M., Kyngdon, C.T., Monkhouse, J.L., Vallejo Rodriquez, C., Read, A.J., Rolfe, R.A. and Sato, C. (2003a) Vaccination against cestode parasites: anti-helminth vaccines that work and why. *Veterinary Parasitology* 115, 83–123.

Lightowlers, M.W., Gauci, C.G., Chow, C., Drew, D.R., Gauci, S.M., Heath, D.D., Jackson, D.C., Dadley-Moore, D.L. and Read, A.J. (2003b) Molecular and genetic characterisation of the host-protective oncosphere antigens of taeniid cestode parasites. *International Journal for Parasitology* 33, 1207–1217.

Lloyd, S. and Soulsby, E.J.L. (1976) Passive transfer of immunity to neonatal calves against the metacestodes of *Taenia saginata*. *Veterinary Parasitology* 2, 355–362.

Miller, H.M. (1931) The production of artificial immunity in the albino rat to a metazoan parasite. *Journal of Preventive Medicine* 5, 429–452.

Miller, H.M. (1932) Acquired immunity against a metazoan parasite by use of non-specific worm material. *Proceedings of the Society for Experimental Biology* 29, 1125–1126.
Miller, H.M. and Gardiner, M.L. (1932) Passive immunity to infection with a metazoan parasite, *Cysticercus fasciolaris,* in the albino rat. *Journal of Preventive Medicine* 6, 479–496.
Miller, H.M. and Kerr, K.B. (1932) Attempts to immunize rabbits against a larval cestode, *Cysticercus pisiformis. Proceedings of the Society for Experimental Biology and Medicine* 29, 670–671.
Mitchell, G.F. (1989) The Wellcome Trust lecture. Problems specific to parasite vaccines. *Parasitology* 98 Suppl., S19–S28.
Mitchell, G.F., Goding, J.W. and Rickard, M.D. (1977) Studies on immune responses to larval cestodes in mice. Increased susceptibility of certain mouse strains and hypothymic mice to *Taenia taeniaeformis* and analysis of passive transfer of resistance with serum. *Australian Journal of Experimental Biology and Medical Science* 55, 165–186.
Mitchell, G.F., Rajasekariah, G.R. and Rickard, M.D. (1980) A mechanism to account for mouse strain variation in resistance to the larval cestode, *Taenia taeniaeformis. Immunology* 39, 481–489.
Musoke, A.J. and Williams, J.F. (1975) The immunological response of the rat to infection with *Taenia taeniaeformis.* V. Sequence of appearance of protective immunoglobulins and the mechanism of action of 7Sgamma2a antibodies. *Immunology* 29, 855–866.
Musoke, A.J., Williams, J.F., Leid, R.W. and Williams, C.S. (1975) The immunological response of the rat to infection with *Taenia taeniaeformis.* IV. Immunoglobulins involved in passive transfer of resistance from mother to offspring. *Immunology* 29, 845–853.
Osborn, P.J. and Heath, D.D. (1982) Immunisation of lambs against *Echinococcus granulosus* using antigens obtained by incubation of oncospheres *in vitro. Research in Veterinary Science* 33, 132–133.
Osborn, P.J., Heath, D.D. and Roberts, M.G. (1982) Vaccination of sheep against *Taenia ovis.* The response to various dose rates of antigens obtained by incubation of oncospheres *in vitro. Research in Veterinary Science* 32, 351–353.
Patak, K.M.L. and Gaur, S.N.S. (1990) Immunization of pigs with culture antigens of *Taenia solium. Veterinary Parasitology* 34, 353–357.
Plaxco, K.W., Spitzfaden, C., Campbell, I.D. and Dobson, C.M. (1997) A comparison of the folding kinetics and thermodynamics of two homologous fibronectin type III modules. *Journal of Molecular Biology* 270, 763–770.
Potts, J.R. and Campbell, I.D. (1996) Structure and function of fibronectin modules. *Matrix Biology* 15, 313–320.
Rajasekariah, G.R., Rickard, M.D. and Mitchell, G.F. (1980) Immunization of mice against infection with *Taenia taeniaeformis* using various antigens prepared from eggs, oncospheres, developing larvae and strobilocerci. *International Journal for Parasitology* 10, 315–324.
Rajasekariah, G.R., Rickard, M.D. and O'Donnell, I.J. (1985) *Taenia pisiformis*: protective immunization of rabbits with solubilized oncospheral antigens. *Experimental Parasitology* 59, 321–327.
Rickard, M.D. and Adolph, A.J. (1976) Vaccination of calves against *Taenia saginata* infection using a 'parasite-free' vaccine. *Veterinary Parasitology*, 389–392.
Rickard, M.D. and Adolph, A.J. (1977) Vaccination of lambs against infection with *Taenia ovis* using antigens collected during short-term *in vitro* incubation of activated *T. ovis* oncospheres. *Parasitology* 75, 183–188.
Rickard, M.D. and Bell, K.J. (1971a) Immunity produced against *Taenia ovis* and *T. taeniaeformis* infection in lambs and rats following *in vivo* growth of their larvae in filtration membrane diffusion chambers. *Journal of Parasitology* 57, 571–575.
Rickard, M.D. and Bell, K.J. (1971b) Successful vaccination of lambs against infection with *Taenia ovis* using antigens produced during *in vitro* cultivation of the larval stages. *Research in Veterinary Science* 12, 401–402.
Rickard, M.D. and Williams, J.F. (1982) Hydatidosis/cysticercosis: immune mechanisms and immunization against infection. *Advances in Parasitology* 21, 229–296.
Rickard, M.D., Boddington, E.B. and McQuade, N. (1977) Vaccination of lambs against *Taenia ovis* infection using antigens collected during *in vitro* cultivation of larvae: passive protection via colostrum from vaccinated ewes and the duration of immunity from a single vaccination. *Research in Veterinary Science* 23, 368–371.
Rickard, M.D., Harrison, G.B., Heath, D.D. and Lightowlers, M.W. (1995) *Taenia ovis* recombinant vaccine – 'quo vadit'. *Parasitology* 110 Suppl., S5–9.
Rothel, J.S., Corner, L.A., Seow, H.F., Wood, P.R. and Lightowlers, M.W. (1996a) Antigen-specific IgA-secreting cells induced by peripheral vaccination. *Immunology and Cell Biology* 74, 298–300.

Rothel, J.S., Lightowlers, M.W., Seow, H.F., Wood, P.R., Rothel, L.J., Heath, D.D. and Harrison, G.B. (1996b) Immune responses associated with protection in sheep vaccinated with a recombinant antigen from *Taenia ovis*. *Parasite Immunology* 18, 201–208.

Rothel, J.S., Boyle, D.B., Both, G.W., Pye, A.D., Waterkeyn, J.G., Wood, P.R. and Lightowlers, M.W. (1997a) Sequential nucleic acid and recombinant adenovirus vaccination induces host-protective immune responses against *Taenia ovis* infection in sheep. *Parasite Immunology* 19, 221–227.

Rothel, J.S., Waterkeyn, J.G., Strugnell, R.A., Wood, P.R., Seow, H.F., Vadolas, J. and Lightowlers, M.W. (1997b) Nucleic acid vaccination of sheep: use in combination with a conventional adjuvanted vaccine against *Taenia ovis*. *Immunology and Cell Biology* 75, 41–46.

Rothel, J.S., Wood, P.R., Seow, H.F. and Lightowlers, M.W. (1997c) Urea/DTT solubilization of a recombinant *Taenia ovis* antigen, 45W, expressed as a GST fusion protein results in enhanced protective immune response to the 45W moiety. *Vaccine* 15, 469–472.

Rothel, J.S., Corner, L.A., Lightowlers, M.W., Seow, H.F., McWaters, P., Entrican, G. and Wood, P.R. (1998a) Antibody and cytokine responses in efferent lymph following vaccination with different adjuvants. *Veterinary Immunology and Immunopathology* 63, 167–183.

Rothel, J.S., Seow, H.F., Lightowlers, M.W., Parry, B.W., Gauci, C., Hurst, L., Mucha, M.J., Corner, L.A. and Wood, P.R. (1998b) The use of recombinant ovine IL-1beta and TNF-alpha as natural adjuvants and their physiological effects *in vivo*. *Immunology and Cell Biology* 76, 167–172.

Ruoslahti, E. and Pierschbacher, M.D. (1987) New perspectives in cell adhesion: RGD and integrins. *Science* 238, 491–497.

Scheerlinck, J.P., Casey, G., McWaters, P., Kelly, J., Woollard, D., Lightowlers, M.W., Tennent, J.M. and Chaplin, P.J. (2001) The immune response to a DNA vaccine can be modulated by co-delivery of cytokine genes using a DNA prime-protein boost strategy. *Vaccine* 19, 4053–4060.

Waterkeyn, J.G., Lightowlers, M.W., Coppel, R. and Cowman, A.F. (1995) Characterization of the gene family encoding a host-protective antigen of the tapeworm *Taenia ovis*. *Molecular and Biochemical Parasitology* 73, 123–131.

Waterkeyn, J., Gauci, C., Cowman, A. and Lightowlers, M. (1997a) Sequence analysis of a gene family encoding *Taenia ovis* vaccine antigens expressed during embryogenesis of eggs. *Molecular and Biochemical Parasitology* 86, 75–84.

Waterkeyn, J.G., Cowman, A.F. and Lightowlers, M.W. (1997b) *Taenia ovis*: copy number of genes encoding host-protective antigens determined by competitive PCR. *Experimental Parasitology* 86, 84–88.

Waterkeyn, J.G., Gauci, C., Cowman, A.F. and Lightowlers, M.W. (1998) Codon usage in *Taenia* species. *Experimental Parasitology* 88, 76–78.

Woollard, D.J., Gauci, C.G., Heath, D.D. and Lightowlers, M.W. (1998) Epitope specificities and antibody responses to the EG95 hydatid vaccine. *Parasite Immunology* 20, 535–540.

Woollard, D.J., Gauci, C.G. and Lightowlers, M.W. (2000a) Synthetic peptides induce antibody against a host-protective antigen of *Echinococcus granulosus*. *Vaccine* 18, 785–794.

Woollard, D.J., Heath, D.D. and Lightowlers, M.W. (2000b) Assessment of protective immune responses against hydatid disease in sheep by immunization with synthetic peptide antigens. *Parasitology* 121, 145–153.

Woollard, D.J., Gauci, C.G., Heath, D.D. and Lightowlers, M.W. (2001) Protection against hydatid disease induced with the EG95 vaccine is associated with conformational epitopes. *Vaccine* 19, 498–507.

16 The Development of a Schistosome Vaccine

R.A. Wilson and P.S. Coulson
Department of Biology, University of York, York YO10 5YW, UK

Introduction

The World Health Organization (WHO) estimates that schistosomiasis (bilharzia) afflicts some 200 million people in 74 countries, 85% of whom live in sub-Saharan Africa (Chitsulo *et al.*, 2004). *Schistosoma mansoni* and *Schistosoma haematobium* are predominantly human parasites whilst *Schistosoma japonicum* is zoonotic in a wide range of domestic animals as well as humans, and species such as *Schistosoma bovis*, *Schistosoma spindale* and *Schistosoma mattheei* infect domestic livestock. The mammalian host becomes infected when

 Parasitic Flatworms: Molecular Biology, Biochemistry, Immunology and Physiology (eds A.G. Maule and N.J. Marks)

free-swimming larvae penetrate the skin and undertake a complex intravascular migration to reach the site of maturation (the blood vessels of the hepatic portal system for *S. mansoni* and *S. japonicum* and of the bladder wall for *S. haematobium*). The eggs deposited in host tissues by the parasites are the primary agents of disease morbidity because they impair organ function and provoke fibrosis. The disability-linked morbidities, such as diarrhoea, dysuria and anaemia, are widespread in the endemic population. In addition, a recent study has estimated the number of deaths directly attributable to schistosomiasis in sub-Saharan Africa at 280,000 per annum (van der Werf *et al.*, 2003), making it second only to malaria among parasitic diseases as a cause of human suffering.

Approaches to Schistosomiasis Vaccine Development

A notable characteristic of schistosomiasis is the cryptic nature of the infection and associated disease processes. For most people, the immediate manifestations are negligible or indeterminate so that the infected individual is unaware of the accumulating worm burden. The normally slow progression to the pathological state means that 5–10 years may elapse after first exposure before severe pathology becomes apparent, and chemotherapy can do little to reverse late-stage disease. As well as circumventing the risk of resistance due to reliance on a single drug, an effective vaccine would have the advantage of minimizing the chances of progression to the advanced disease state. Given that schistosomes can survive in the hostile environment of the mammalian bloodstream for more than 30 years (Harris *et al.*, 1984), developing a schistosome vaccine was never going to be an easy task. Over the last four decades there have been many attempts but the results have been disappointing. A more detailed rationale for a schistosomiasis vaccine can be found in our earlier publications (Wilson and Coulson, 1998, 1999).

A vaccine that provides protection against infection for an extended period is a powerful weapon for disease control (and ultimately eradication). Most existing vaccines are intended to replicate events after primary exposure that naturally elicits solid immunity. With chronic infections, the vaccinologist starts at a disadvantage in that the existence of a protective immune response is by no means clear or guaranteed. In the case of schistosomiasis this is especially true for children, the group most at risk, where little immunity to reinfection after chemotherapy has been reported (Fulford *et al.*, 1992). As a consequence, there is no obvious paradigm on which to base a vaccine, and researchers have focused their efforts on immune responses in animal models.

Diverse routes have been followed, based on strategies that have succeeded with other pathogens. These include, at their simplest, using crude parasite preparations and individual components purified from them. As an alternative, by analogy with bacteria and viruses, schistosome cercariae have been attenuated with radiation or chemicals rather than repeated passage in culture. Finally, recombinant DNA technology has had a major influence on the selection of candidate antigens. There is no logical sequence, but there is an underlying rationale for each of the various lines of work that have often progressed in parallel. We present here a historical perspective of approaches taken in attempts to attain the elusive goal of a schistosome vaccine. For the last 25 years, there has been considerable optimism that success was tantalisingly close, as more than 40 review articles testify. However, no product is anywhere near to market; indeed only one Phase I/Phase II clinical trial has been performed. After so much effort, it is pertinent to ask why reproducible, sustained, high levels of protection have not been achieved by vaccination of any host.

What Is Meant by Protective Immunity?

In the context of schistosomiasis, where parasites do not multiply in the mammalian host, protective immunity is primarily measured as a reduction in the adult worm burden of vaccinated animals relative to controls. Conventionally, this has been assessed by a terminal procedure involving perfusion of the portal vasculature to flush out the worms. The laboratory mouse has been the mainstay of vaccine testing, with baboons as the favoured non-human primate. The laboratory rat has been used in some instances but this host spontaneously cures around 28 days; perfusions therefore need to be performed earlier, with the danger that parasite migration to the portal system may not be complete. It has been argued that vaccination need not generate sterile immunity because a significant depression in the worm population would reduce both morbidity and transmission. Indeed, as eggs are the cause of morbidity, elimination of worms may not be essential provided that egg production can be drastically reduced by vaccination. Such an anti-fecundity effect would be manifest as a reduction in faecal egg output or tissue egg burden per mature female worm, in vaccinated versus control animals. Similarly, transmission would be reduced if the viability of released eggs (judged by miracidial hatching) was compromised by immune attack.

The above measurements of protection necessitate an interval of at least 5 weeks between challenge and evaluation. Hence, a rapid *in vitro* assay of immune-mediated parasite death that correlated with the protected status of the host would greatly accelerate vaccine development. One such test, involving antibody-dependent cellular cytotoxicity (ADCC), has been widely used to predict the vaccine potential of schistosome antigens (Butterworth *et al.*, 1974, 1982). It has had a major conceptual influence on the selection of vaccine candidates, as the vulnerability of newly transformed schistosomula to killing by various combinations of sera and leucocytes was readily demonstrated. However, even when the *in vitro* mechanisms were being intensively characterized, some researchers advised caution (Smithers, 1987). Indeed, considering the huge interest that ADCC has attracted over the last 30 years, it is remarkable that there is little histopathological evidence for its occurrence *in vivo* to match the detailed ultrastructural analyses of larval killing *in vitro* (McLaren *et al.*, 1978). There has been a gradual acceptance by the majority of researchers that the *in vitro* results are not an accurate predictor of the *in vivo* situation.

Attenuated Parasites Can Induce Protective Immunity

As schistosomes are long-lived and pre-pubertal children show little evidence of acquired immunity, can protection be reliably achieved in a permissive host? Experiments in rodents and primates with cercariae attenuated by radiation provide an emphatic 'yes' in answer to this question. Over a 40-year-period, many laboratories have reproducibly obtained a 60–70% reduction in worm burden after a single vaccination of high responder mouse strains. The attenuated vaccine model has been extensively investigated as a paradigm for a human vaccine (reviewed by Coulson, 1997) and the salient features are summarized below.

An optimum dose of radiation permits parasites to enter mouse skin but their subsequent migration is truncated, restricting them to that site, its draining lymph nodes and the lungs. The prolonged residence of a proportion of attenuated schistosomula in the lymph nodes, compared to normal larvae, provokes a more intense lymphoproliferative response. Schistosome-reactive T lymphocytes, with T helper 1 (Th1) characteristics, enter the circulation and are subsequently recruited to the lungs in response to sequestered attenuated larvae; such cells persist for some time and effectively arm this organ (Smythies *et al.*, 1992). A challenge parasite arriving in the pulmonary vascular bed attracts an inflammatory focus predominantly comprising mononuclear leucocytes. Lung schistosomula are refractory to ADCC *in vitro* and the weight of evidence indicates that the immune effector mechanism acts by blocking onward migration rather than by delivering a lethal blow (Coulson *et al.*, 1998). Production of the Th1 cytokines, interferon gamma and tumour necrosis factor (Street *et al.*, 1999) is integral to the effector response, underlining its cell-mediated basis. This is reinforced by the successful vaccination of μMT mice that cannot make an antibody response (Anderson *et al.*, 1999). The efficacy of cell-mediated immunity does not rule out the potential for a protective humoral response. Indeed, with multiple exposures of mice to attenuated parasites, there is an increasing contribution from antibodies. This is evidenced by the greater success of passive transfer with sera from 3× versus 1× vaccinated mice (Mangold and Dean, 1986; Wilson *et al.*, 1999). Optimum protection of primates requires multiple vaccinations, with five exposures of baboons to attenuated schistosomes resulting in an 84% reduction in worm burden (Kariuki *et al.*, 2004); the correlation between IgG titre at challenge and level of protection suggests a dominant humoral component in the effector mechanism (Soisson *et al.*, 1993; Yole *et al.*, 1996).

Whilst attenuated cercariae can induce protection reliably, their short lifespan (hours) makes them unsuitable for use in the field. Attempts were made to circumvent the problem by cryopreservation of attenuated larvae (Bickle and James, 1978). Although up to 50% protection was achieved in cynomolgus monkeys (Murrell *et al.*, 1979), baboons (Damian *et al.*, 1984), mice (Lewis *et al.*, 1984) and guinea pigs (Xu *et al.*, 1991), the lack of reproducibility in the cryopreservation process (Lewis *et al.*, 1984) meant that this approach eventually petered out. Even with improved cryopreservation, before such a vaccine could become a reality, the ethical and safety issues surrounding the administration to humans of large numbers of attenuated parasites with unknown migration characteristics would need to be addressed.

Unlike the analogous attenuated malaria sporozoite vaccine (Hoffman *et al.*, 2002), the schistosome counterpart has not been tested in humans but our recent demonstration of protection in a small-scale chimpanzee trial (Eberl *et al.*, 2001) attests to its potential as a basis for the development of a recombinant vaccine. However, attempts to replicate the success of the attenuated schistosome vaccine by administration of recombinant antigens have had only limited success. It is notable that, faced with similar problems in replacing the attenuated malaria vaccine with recombinant antigens, it is now proposed to develop a cryopreserved product that meets regulatory, potency and safety requirements, for use in humans (Luke and Hoffman, 2003).

Routes to Antigen Identification

In theory, the administration of crude schistosome extracts could provide a straightforward route to a vaccine (cf. microbial extract vaccines such as pertussis or anthrax). This direct approach has motivated researchers for over 40 years with equivocal results (e.g. Sadun and Lin, 1959). Levels of protection up to 50% (more usually 20–30%; e.g. Hayunga, 1985) have been reported but with a lack of consistency even in the same laboratory (Murrell *et al.*, 1975). A potential drawback of the approach is the extreme heterogeneity of crude preparations coupled with a lack of knowledge about the identity of protective antigens. We now know that the schistosome genome encodes >14,000 genes, with around 7000 expressed in a given life-cycle stage (Verjovski-Almeida *et al.*, 2003). Thus, it is entirely possible that crucial antigens, present in trace amounts, are swamped by abundant and immunogenic but irrelevant proteins. In addition, vaccine strategies that elicit an inappropriate immune response may account for failed experiments.

Faced with these problems, researchers have attempted to define individual parasite proteins as vaccine candidates. This section outlines the various approaches and the major players that emerged.

Mining crude parasite extracts for antigens

A more rational starting point for the use of crude parasite extracts was based on the cell-mediated immune responses associated with the murine attenuated vaccine model. In an attempt to replicate these responses, newly transformed schistosomula, killed by freezing and thawing, were administered intradermally to C57BL/6 strain mice together with live BCG as adjuvant. This non-living vaccine elicited significant, reproducible levels of protection against a cercarial challenge (mean of 50%) that correlated with sensitization for a delayed type hypersensitivity (DTH) rather than a humoral response (James, 1985). A follow-up study revealed that soluble extracts of schistosomula or adult worms were also capable of eliciting protection, provided that the above administration route and adjuvant were used (James *et al.*, 1985). Indeed, it was later shown that the delivery system was crucial to the success of the vaccine (James, 1987; James and Pearce, 1988). Work with inbred strains of mice known to be high or low responders to the attenuated vaccine supplied further evidence that cell-mediated immunity was most likely involved in the protective response in this non-living model (James and DeBlois, 1986). Moreover, a correlation was shown between the level of protection in mice receiving the non-living vaccine and the induction of activated (larvicidal) macrophages, which were considered plausible immune effector cells (James, 1986).

Mice immunized with the non-living vaccine produced little or no antibody against parasite surface membrane antigens and a very restricted response to soluble antigens, recognizing a single polypeptide of M_r 97 kDa (James *et al.*, 1985). When the crude soluble worm antigen extract was separated by gel filtration, the protective capacity was found to be restricted to the fraction containing this molecule (Sher *et al.*, 1986). Following purification by affinity chromatography, Sm-97 (as it was termed) was shown to be minimally glycosylated and localized to the areas just below the tegumental and gut syncytia (Pearce *et al.*, 1986). Purified Sm-97 was capable of eliciting DTH in mice exposed to the non-living vaccine, a finding consistent with its proposed function as a vaccine antigen. A subsequent study identified Sm-97 as paramyosin, an invertebrate muscle protein (Lanar *et al.*, 1986). The purified native antigen administered to mice conferred ~30% protection with BCG (Pearce *et al.*, 1988) and 24–53% without adjuvant (Flanigan *et al.*, 1989).

No further development of Sm-97 seems to have occurred, probably because of difficulties encountered in expressing such a large recombinant protein. However, experiments on *S. japonicum* paramyosin continued throughout the next decade; impressive levels of protection were reported in mice using the native molecule derived from the Philippines strain administered intraperitoneally without adjuvant (62–86%; Ramirez *et al.*, 1996). This was not repeated for the Chinese strain (McManus *et al.*, 1998) although low but significant protection was obtained by subcutaneous administration with Freund's complete adjuvant (FCA). The disparity could not be attributed to strain differences as only a single amino acid exchange occurred between paramyosin from the Philippines and Chinese strains. The full-length *S. japonicum* paramyosin was ultimately expressed as a recombinant protein (Kalinna *et al.*, 1997) but its testing in a variety of hosts (mice, water buffalos, pigs) elicited protection only in the 30–40% range (Chen *et al.*, 2000; McManus *et al.*, 2001); vaccination with a DNA construct achieved similar results (Zhou *et al.*, 2000).

Protective monoclonal antibodies

The development of hybridoma technology in the late 1970s provided an apparent short cut to identify schistosome vaccine candidates. Monoclonal antibodies (mAb), produced by multiple hybridoma lines, could be rapidly screened for efficacy at killing larval parasites *in vitro* and/or passively conferring protection on laboratory rodents. B lymphocytes for fusions were obtained from mice receiving, e.g. an infection (Hazdai *et al.*, 1985), an intravenous injection of eggs (Harn *et al.*, 1984), a vaccination with irradiated cercariae (Bickle *et al.*, 1986) or parasite extracts (Horowitz *et al.*, 1985); alternatively, rats were exposed to cercariae (Grzych *et al.*, 1982) or injected with adult worm antigens (Verwaerde *et al.*, 1987). With hindsight, we can see that the choice of immunogens was in some cases less than ideal. For example, there was no evidence that eggs per se could induce protection (Harrison *et al.*, 1982) even though they were cross-reactive with larval surfaces (Omer Ali *et al.*, 1988). Equally, at the time, evidence was emerging that the immunity displayed by chronically infected mice was a product of pathology (Dean *et al.*, 1981; Wilson *et al.*, 1983). Nonetheless, mAbs were generated that could confer significant protection *in vivo* against cercarial challenge (reviewed by Smithers, 1987). Most notably, the antigen targets of three such mAbs were taken forward to protection experiments.

TPI

A 28 kDa protein was identified as the target of a mAb that conferred partial protection (41–49%) on mice (Harn *et al.*, 1992). Peptide sequencing of the protein revealed it as the *S. mansoni* homologue of mammalian triose phosphate isomerase (TPI), present in all stages of the parasite and transiently on the surface of schistosomula. The T and B cell epitopes of TPI were mapped and multiple antigenic peptide (MAP) constructs synthesized, corresponding to non-conserved regions (Reynolds *et al.*, 1994). There appear to be no actual worm recovery data in the public domain but protection values ranging from 45 to 65% have been reported, following administration of MAP4 (reviewed by Harn *et al.*, 1995). In separate work, *S. japonicum* TPI administered intramuscularly as a DNA construct to mice achieved 14–32% protection (Zhu *et al.*, 2002).

9B-Ag

Another mAb, obtained by fusion of spleen cells from mice with a chronic infection boosted with cercarial glycoprotein showed binding capacity to the surface of newly transformed schistosomula. It was capable of conferring >42% protection upon passive transfer to mice, and recognized an antigen designated 9B-Ag (Hazdai *et al.*, 1985), which appeared to be a multimer consisting of 45 and 30 kDa subunits. The native protein conferred protection on mice, levels depending on the dose and formulation (Hazdai *et al.*, 1985; Mendlovic *et al.*, 1989; Tarrab-Hazdai *et al.*, 1999). It has not been cloned but synthetic (Tarrab-Hazdai *et al.*, 1998) and recombinant (Ben-Yedidia *et al.*, 1999) peptides and a mimotope (Arnon *et al.*, 2000) all conferred approximately 40% protection.

Sm23

This antigen was identified as the target of a mAb produced by fusion of cells from mice primed by intraperitoneal injection of a detergent extract of schistosomula (Harn *et al.*, 1985). The mAb reacted with a molecule on the surface of newly transformed larvae and, surprisingly, 5-day cultured but not *ex vivo* lung worms. The target was a membrane tetraspannin with two highly immunogenic extracellular domains, shown by peptide mapping to contain both B and T cell epitopes in mice (Reynolds *et al.*, 1992). The protective potential of the mAb was not stated but moderate to high levels were reported in a review (Harn *et al.*, 1995), following administration of a MAP. A recent experiment with the recombinant protein in aluminium hydroxide ($Al(OH)_3$) failed to elicit protection (Da'Dara *et al.*, 2003), although DNA vaccine formulations injected alone or with plasmids encoding IL-12 or IL-4 reduced worm burdens by 21–44% (Da'Dara *et al.*, 2001). Other formulations and administration routes failed to better this and results with the *S. japonicum* homologue (Sj23) were just as equivocal (Waine *et al.*, 1999; Zhu *et al.*, 2003).

Anti-idiotypic antibodies

Antibodies used as immunogens elicit antibodies that recognize their idiotype (i.e. antigen-binding site) and hence might mimic the antigenic target of the original immunizing antibody. Such anti-idiotypic antibodies were believed to have vaccine potential, providing an attractive solution to the problem of glycan epitopes not amenable to molecular biology techniques. The first report in schistosome research involved the generation of a rat monoclonal anti-idiotypic antibody against a mAb that recognized a 38 kDa antigen on the surface of schistosomula (Grzych *et al.*, 1985). When used as a vaccine in rats, it stimulated polyclonal antibodies that recognized the original 38 kDa molecule, were cytotoxic for schistosomula in the presence of eosinophils and conferred protection by passive transfer; the rats were also highly protected against cercarial challenge. A second report described an anti-idiotypic antibody raised in rabbits using a mAb that bound a 68 kDa protein from adult worms (Kresina and Olds, 1989). Immunization of mice with this anti-idiotypic antibody generated 16–41% protection and the murine antibodies elicited bound to the surface of schistosomula. There was little further interest until very recently, when a *S. japonicum* anti-idiotypic mAb was described that induced 22–50% protection in mice (Feng *et al.*, 2002) as well as anti-fecundity effects (Feng *et al.*, 2000).

Expression library screening with sera from putatively protected hosts

The advent of recombinant DNA technology in the early 1980s led to a wave of activity and optimism in the quest for a schistosome vaccine. The technology promised to circumvent the problem of scarce antigenic material and, indeed, delivered abundant supplies of defined proteins. It was important to ensure that the gene product detected in the immunological screen, and subsequently transferred to the high-level expression vector, was likely to mediate protective immunity. In this context, key considerations were a source of serum with authenticated protective properties and the parasite stage from which the mRNA was extracted to generate the cDNA library for screening. In the latter case, it is somewhat surprising that most researchers chose to screen libraries from adult worms, which were not considered the targets of protective immunity. The two prominent antigens that emerged from this approach were a muscle protein and a glycolytic enzyme.

IrV5

This molecule was identified using serum, from mice exposed twice to irradiated cercariae, to immunoaffinity purify antigens (Dalton and Strand, 1987), which were then used to vaccinate rabbits (Soisson *et al.*, 1992). One anti-serum (designated IrV5), which bound to the surface of newly transformed schistosomula, was used to clone a 62 kDa fragment of a larger 200 kDa molecule, subsequently shown to be the myosin heavy chain. Administration of the purified recombinant, with the outer membrane protein of meningococcus as adjuvant, elicited up to 75% protection in mice (Soisson *et al.*, 1992), whilst vaccination of baboons with similar formulations elicited a modest level of protection (~25%, range 0–54%; Soisson *et al.*, 1993). Almost a decade later, the *S. japonicum* myosin fragment was cloned and expressed but the recombinant protein did not induce significant protection in mice or rats (Zhang *et al.*, 1998). Furthermore, its formulation in a DNA construct together with a source of CpG motifs did not improve matters (Zhang *et al.*, 2000). Similarly, the myosin fragment administered to pigs failed to protect them, and it has been concluded that the antigen shows little promise as a vaccine candidate (Bickle *et al.*, 2001).

GAPDH

A 37 kDa antigen was identified by Western blotting using sera from humans with low susceptibility to reinfection, suggesting that it was a marker of resistance (Dessein *et al.*, 1988); its identity was subsequently confirmed as glyceraldehyde 3-phosphate dehydrogenase (GAPDH) by screening an adult worm cDNA library with rabbit antisera raised against electroeluted native protein (Goudot-Crozel *et al.*, 1989). (*Note*: The cDNA does not encode a signal sequence or transmembrane domains that could explain the location of the protein on the larval surface; Charrier-Ferrara *et al.*, 1992.) Epitope mapping of the protein sequence led to the identification of a dominant B cell epitope which, coupled to ovalbumin as a carrier in FCA, induced modest protection in mice and rats (Argiro *et al.*, 2000). Incorporation of the cytokine GM-CSF with $Al(OH)_3$ as adjuvant, boosted antibody titers and protection (33–38%; Argiro *et al.*, 1999).

Antigens identified on the basis of their immunogenicity

At least two candidate antigens have been identified primarily as a result of their immunogenic properties when administered to laboratory hosts.

GSTs

The glutathione *S*-transferases (GSTs) emerged independently from work in two laboratories. The presence of a 28 kDa antigen was first demonstrated by the simple expedient of immunizing rats with a soluble extract from adult worms and using the serum generated to immunoprecipitate native protein targets (Balloul *et al.*, 1985). The same antigen was identified on the surface of newly transformed schistosomula and the rat serum was highly cytotoxic for such larvae in the presence of purified eosinophils. It was subsequently cloned and shown to protect rats and hamsters against cercarial challenge (52% and 67%, respectively; Balloul *et al.*, 1987b), and these observations were extended to mice (~40% protection with FCA or $Al(OH)_3$ as adjuvants;

Balloul *et al.*, 1987a). Around this time, the antigen was identified as a 28 kDa GST and shown to be present in various adult worm tissues (Taylor *et al.*, 1988). When baboons were vaccinated twice with Sm28GST in $Al(OH)_3$ plus *Bordetella pertussis*, they had higher worm burdens than controls whereas those receiving three doses of GST in $Al(OH)_3$ alone had a mean protection of 38% (range of 0–80%; Boulanger *et al.*, 1991). In the same study, there was the first mention of an anti-fecundity effect, inferred from the reduced faecal egg output per female worm, paradoxically in the 2× but not 3× vaccinated group. Anti-fecundity effects and a reduced hatching capacity of liver eggs were also reported for mice vaccinated with immunogenic peptides derived from Sm28GST (Xu *et al.*, 1993).

In a separate line of research, the majority of 129 strain mice were shown to be 'resistant' to a primary infection with *S. japonicum* (Mitchell *et al.*, 1984). Moreover, these mice had a dominant antibody response to a 26 kDa antigen from adult worms that was also strongly reactive with sera from rabbits multiply immunized with adult worm extract (Beall and Mitchell, 1986). This allowed the cDNA to be cloned from an expression library and identified as a GST (Smith *et al.*, 1986). The native protein was easily purified on a glutathione affinity column and used for protection experiments, with mostly negative results except for >50% on one occasion in C57BL/6 mice (Mitchell *et al.*, 1988). The natural 'resistance' of 129 strain mice for *S. mansoni* was also demonstrated (Tiu *et al.*, 1986) but unfortunately proved to be the result of a defective portal vasculature rather than acquired immunity (Cox, 1990). This finding undermined the whole concept of immunity induced by GST in 129 strain mice and as a consequence vaccine work with the Philippines strain of *S. japonicum* largely petered out.

There was, however, still momentum for the investigation of GSTs as vaccine candidates. The 26 kDa antigen from the Chinese strain of *S. japonicum* was cloned and shown to elicit a low level of protection in mice (24%) but a stronger anti-fecundity effect (59% reduction in liver eggs/worm pair; Liu *et al.*, 1995). In later experiments, Sj26- and/or Sj28-GSTs were tested in mice (Scott and McManus, 2000), sheep (Xu *et al.*, 1995; Taylor *et al.*, 1998), water buffalo (Shi *et al.*, 2001) and cattle (Shi *et al.*, 2002), some of the trials being performed under field conditions. The results have been variable, with no or low protection in mice versus 30–70% reduction in worm burden but no anti-fecundity effect in sheep. Cattle given Sj28GST as a DNA vaccine were 44% resistant according to worm burden but only 19% on the basis of tissue egg counts.

A vaccine that diminished the fecundity of worms rather than preventing their establishment might represent a more feasible proposition. The patas monkey (*Erythrocebus patas*) has been used as an experimental host for *S. haematobium* because a proportion of worms can reach the blood vessels of the bladder wall. Vaccination with heterologous Sm28GST before challenge with *S. haematobium* achieved a marked and long-lasting reduction in egg excretion in the urine and faeces (55% and 74%, respectively, compared to controls) but worm burdens in the two groups were similar (Boulanger *et al.*, 1995); it is worth noting that only 3–4% of applied parasites matured in control monkeys. A second study using Sh28GST confirmed the earlier observations (Boulanger *et al.*, 1999b) and, together, these data provided the rationale for the clinical trials of Sh28GST in humans (see later). Studies with *S. haematobium* in mice revealed that it was possible to generate far higher levels of anti-GST antibody by vaccination than occurred during a patent infection (Lane *et al.*, 1998). Nevertheless, neither protection nor an anti-fecundity effect could be demonstrated, irrespective of whether the Sh28GST was delivered in microbial vectors, as a DNA construct or as protein in microparticles (A. Lane and R.A. Wilson, unpublished data).

Finally, vaccination of cattle with *S. bovis* GST elicited an anti-fecundity effect, manifest as reduced faecal egg excretion and tissue egg but not adult worm burden (Bushara *et al.*, 1993). Conversely, vaccination of goats (Boulanger *et al.*, 1994) and sheep (Boulanger *et al.*, 1999a) resulted in reduced worm burdens but no anti-fecundity effect.

Sm14

The cross-reactivity of *S. mansoni* and *Fasciola hepatica* antigens was first reported by researchers in Puerto Rico where both parasites were endemic. A major finding was that a 12 kDa *F. hepatica* antigen administered to mice with FCA could induce 52–77% protection against a heterologous challenge with *S. mansoni* cercariae (Hillyer *et al.*, 1988). Independently, a saline extract of adult *S. mansoni* administered to rabbits (not permissive hosts) was shown to induce very high levels of protection (Tendler *et al.*, 1986). A similar result was obtained in mice, with 42% protection achieved even without adjuvant (Tendler *et al.*, 1991). In the same year, fatty acid-binding protein (Sm14) was cloned using a polyclonal antiserum against the saline extract, and immunolocalized to the tubercles, muscle and body of adult worms (Moser *et al.*, 1991). The recombinant Sm14 protein was later shown to provide 37–67% protection in mice and, incidentally, 100% protection against *F. hepatica* challenge, suggesting the possibility of a dual-parasite vaccine (Tendler *et al.*, 1996). The basis for the cross-protection was revealed when the *F. hepatica* antigen was also shown to be a fatty acid-binding protein with 44% sequence identity to Sm14 (Hillyer, 1995).

The *S. japonicum* homologue has been cloned and expressed (91% identity to Sm14; Becker *et al.*, 1994). However, the Sj14 recombinant protein failed to elicit protection in three mouse strains against the Chinese strain of the parasite (Scott *et al.*, 2000). Indeed, it was suggested that its localization to internal tissues and lipid droplets, and absence from the surface or tegument of adult worms (Gobert *et al.*, 1997) makes it an unlikely vaccine candidate (Scott *et al.*, 2000).

An Antigen Cocktail Does Not Improve Vaccine Efficacy

For the most part, each of the antigens described above elicits low to moderate levels of protection. On the assumption that their protective properties might be additive, attempts have been made to improve vaccine efficacy by combining two or more of the antigens in a cocktail. A MAP consisting of epitopes from Sm28GST and TPI was found to stimulate both B and T cell responses in mice but no protection data are available (Ferru *et al.*, 1997). A DNA vaccine encoding epitopes from paramyosin, TPI, Sm23 and Sm28GST (in tandem), the polyprotein encoded by that DNA, or synthetic peptide epitopes on a carrier backbone were administered to three strains of mice but none induced protection (Yang *et al.*, 2000). However, a cocktail DNA vaccine comprising plasmids for Sj28GST, Sj23, Sj14-3-3 and the 62 kDa fragment of *S. japonicum* myosin induced 34% and 45% protection in mice on two out of three occasions; co-administration of the plasmid encoding IL-12 did not improve matters (Zhang *et al.*, 2001a). It appears, therefore, that nothing is gained from delivering two or more candidates simultaneously, and one inference is that the various antigens tested all mediate protection via the same mechanism. Another point of note is that a recent proteomic study revealed many of the candidates (excepting Sm23) to be among the most abundant constituents of the cytosolic and cytoskeletal compartments of the schistosome body, across several life-cycle stages (Curwen *et al.*, 2004). Thus, using a cocktail of the vaccine candidates is little different from using a crude parasite preparation to elicit protection, with approximately the same end result (see section on 'Routes to Antigen Identification').

Antigen Formulation

Another major variable affecting vaccination success is the way in which an antigen is formulated and the conditions under which it is presented. This is well illustrated by the co-administration of recombinant IL-12 with the highly effective attenuated vaccine, which enhances the cell-mediated arm of the immune response and boosts protection levels to >90% (Wynn *et al.*, 1995; Mountford *et al.*, 1996). Replicated in a human vaccine, such an outcome would have a dramatic effect on morbidity and transmission.

In the context of individual candidate antigens, Sm28GST has been the subject of a formidable effort to enhance and sustain immune responses in order to improve efficacy. This has involved the mapping and synthesis of antigenic peptides and their coupling to tetanus toxoid carrier for the vaccination of rats (Auriault *et al.*, 1988), or their assembly into an octamer construct for vaccination of mice, rats and monkeys (Wolowczuk *et al.*, 1991). The encoding cDNA has been incorporated into live vaccine vectors, including *Salmonella typhimurium* (Khan *et al.*, 1994a,b), BCG (Kremer *et al.*, 1996) and *B. pertussis* (Mielcarek *et al.*, 1997, 1998). It has also been formulated as a DNA construct and co-administered with a plasmid encoding IL-18 to boost interferon gamma production (Dupre *et al.*, 2001). The GST protein has been administered intranasally, covalently linked to cholera toxin B subunit in order to promote mucosal immune responses (Sun *et al.*, 1999; Lebens *et al.*, 2003) and as recombinant protein alone entrapped in biodegradable microparticles to provide a long-lasting and stable antibody response (Baras *et al.*, 1999, 2000). Where protection data are given in the above studies, it is fair to say that no spectacular increase in the efficacy of Sm28GST was achieved by any of the protocols, despite abundant evidence for the generation of diverse immune responses. One interpretation is that, as far as the permissive murine host is concerned, there is a ceiling to the level of protection elicited by Sm28GST, irrespective of the intensity and type of immune response generated.

The WHO Vaccine Trials

The optimism within the schistosome research community in the early 1990s that a vaccine was feasible led the WHO to provide funds for independent evaluation of *S. mansoni* candidate antigens, with a view to selecting the most promising for development and clinical trials. Two laboratories were chosen to undertake the testing in mice, according to standard protocols, and researchers worldwide were invited to submit their candidates. In the event, six proteins were tested in trials, Sm28GST, IrV5 and Sm14 (as recombinants), paramyosin (as native protein), TPI and Sm23 (as MAPs); adjuvant formulations were as specified by the antigen donors. Unfortunately, the stated and modest goal of consistent induction of 40% protection or better was not achieved with any of the antigens (Anonymous, 1996). In spite of this disappointing result, further investigations were performed with some of the candidate antigens in what were referred to as human correlate studies. The underlying rationale was presumably that results in mice might not be predictive of protective potential in humans, irrespective of the fact that all of the antigens in the independent trial were selected on the basis of their supposed potential in rodents. The studies in Brazil (Ribeiro de Jesus *et al.*, 2000) and Egypt (Al-Sherbiny *et al.*, 2003) sought to relate putative protected status to cellular and humoral immune responses. In the words of the authors of the Egyptian study 'the immune response profiles produced were unique to each antigen but no clear winner(s) were identified. However, markers for both resistance and susceptibility to reinfection were identified for each molecule, indicating which types of responses to aim for in vaccination'. A similar conclusion was reached in the Brazilian study. On the basis of the independent trials in mice and the above human correlate studies, as things stand at present, there appears to be no obvious case to take any of the WHO candidate antigens forward to clinical trials.

Clinical Trials

A single antigen, Sh28GST, has progressed through Phase I and Phase II trials (Capron *et al.*, 2002). It is important to stress that this was not the antigen used in the independent vaccine trials in mice, or the human correlate studies. Rather, it was selected for development on the basis of vaccination experiments with *S. haematobium* in patas monkeys described earlier (Boulanger *et al.*, 1995, 1999b), where a clear anti-fecundity effect was reported. The product, named Bilhvax, was tested for toxicity in rats, rabbits and dogs before administration in $Al(OH)_3$ to 24 healthy Caucasian volunteers. Following three injections, a strong immune response was elicited in all immunized individuals. Phase Ib trials, performed on healthy Senegalese children, produced a similar result. A follow-up of vaccinated individuals for 4 months after the first immunization showed that Bilhvax induced a high and lasting specific immune response with a Th2-type profile (Capron *et al.*, 2002). Phase II trials on schistosomiasis patients have also been undertaken in Senegal but the results are not in the public domain. To our knowledge, Phase III trials to test vaccine efficacy are not yet underway.

Progress towards human trials raises the thorny question of how to evaluate vaccine success in humans, because protection in experimental animals has almost invariably been measured by worm recovery. One exception was the small-scale attenuated vaccine trial in chimpanzees (Eberl *et al.*, 2001), where estimates of protection were based on faecal egg output and circulating antigen levels but could not be validated against worm burden for ethical reasons. More recently, these indirect estimators of worm burden have been evaluated with respect to actual recoveries in baboons exposed to the attenuated vaccine, revealing their significant overestimation of protection levels (Kariuki *et al.*, 2004). This issue needs to be addressed before we proceed to full-scale vaccine trials for schistosomiasis mansoni but may be less crucial for schistosomiasis haematobia where egg output in the urine is easier to measure.

Why Are High Levels of Protection Difficult to Achieve?

For the most part, the foregoing attempts to vaccinate animals with recombinant antigens have achieved an overall mean of ~30% protection, and there is scope for a meta-analysis of the very large dataset to dissect out the salient features. Setting aside questions about optimum formulation to maximize immune responses, what factors might explain the limited success to date? Almost all of the antigens listed above were described as *transiently* detectable on the surface of newly transformed schistosomula, and some have been shown to elicit antibodies that will mediate ADCC *in vitro*. It is implicit that vaccination with these antigens will stimulate antibody responses against proteins exposed on the surface of newly penetrated larvae in the skin. One of the few pieces of evidence that ADCC can occur *in vivo* is provided by the mAbs that confer a moderate level of protection on mice. Their administration around the time of challenge ensures a very high titre of specific antibody to opsonize incoming parasites. A plausible explanation for the modest protection elicited by vaccination with recombinant antigens is their inability to replicate such titres.

We need to consider the process of larval transformation in relation to the putative dermal effector response. Schistosome larvae rapidly penetrate the epidermis to reach its basement membrane; this is also true for parasites entering hair follicles or sebaceous glands (Wheater and Wilson, 1979; Crabtree and Wilson, 1985). A period of >48 h elapses before most larvae enter the dermis, locate a blood or lymphatic vessel and exit the skin. Transformation of the cercaria into the schistosomulum involves shedding of the tegumental surface and its replacement with a new plasma membrane derived from pre-formed vesicles (Hockley and McLaren, 1973; Skelly and Shoemaker, 1996). This process is completed within <24 h, coincident with both the production of the protective bilayer that overlies the new plasma membrane (Wilson and Barnes, 1977) and the larva becoming refractory to ADCC *in vitro* (Dessein *et al.*, 1981). Intuitively, it seems unlikely that the cytosolic and cytoskeletal proteins, preponderant among the vaccine candidates, would be readily accessible to the immune response in intact parasites. So, how do they mediate ADCC? The most obvious explanation is that they are exposed only during membrane shedding or are associated with the released membrane material, to provide the transient targets for this process. It is in the parasites' interest to minimize the window of vulnerability, and we have suggested that their stay in the epidermis allows them to avoid the immune effector response whilst assembling their disguise (Wilson and Coulson, 1998). The kinetics of the inflammatory response to parasite-released products may simply not be fast enough to catch more than a small proportion of larvae, perhaps only those that are in the most accessible locations.

Other mechanisms have been proposed to explain the limited efficacy of ADCC *in vivo*. Complement components have been implicated in the *in vitro* killing of schistosomula in the presence of effector leucocytes (e.g.

Ramalho-Pinto *et al.*, 1978) but there is evidence that schistosomes, especially the blood-dwelling stages, use several strategies to inhibit membrane attack by both classical and alternative pathways (reviewed by Skelly, 2004). Another possibility involves production of the wrong kind of antibody. Thus, a rat IgG2c mAb could block the cytotoxic activity of an IgG2a mAb directed against the same target (Grzych *et al.*, 1984). Murine IgM mAbs were similarly shown to block the cytotoxic activity of mouse IgG (Yi *et al.*, 1986) and of human infection serum for schistosomula (Dunne *et al.*, 1987). It has also been suggested that in humans, antibodies of the IgG4 isotype can block IgE pathways (Hagan *et al.*, 1991). Yet another alternative involves modulation of immune mechanisms. There is evidence that some components of the cercarial acetabular gland secretions may serve as anti-inflammatory agents that could suppress dermal responses (Sm16; Rao and Ramaswamy, 2000). In addition, it has long been known that the penetrating parasites promote eicosanoid synthesis in the skin (Fusco *et al.*, 1985). It now appears that Sm28GST is responsible for producing prostaglandin D2 (Herve *et al.*, 2003), which impedes the departure of Langerhan's cells from the epidermis and, by extension, antigen presentation in the skin-draining lymph nodes (Angeli *et al.*, 2001). Thus, paradoxically, Sm28GST has been proposed as both a vaccine candidate and an immunomodulator.

With all these caveats, in our personal opinion, most of the current candidates could never be more than a component in a schistosome vaccine because of the intrinsic limitations in the immune responses they provoke. We believe that evolution has equipped schistosomes to deal with the potential threat posed by ADCC in the skin, and we need to broaden our search to other targets, and/or mechanisms of immune attack.

Secreted and Tegument Antigens as Vaccine Targets

Secreted antigens might provide targets for attack, although the immune effector mechanism(s) would necessarily be indirect, e.g. inhibition of crucial enzyme activity or provocation of inflammation that blocked migration. The contents of the cercarial acetabular gland, released during penetration of the epidermis, or proteins from the schistosomular head gland that may facilitate entry into a blood vessel come into this category (Curwen and Wilson, 2003). There is presently a dearth of published information on the proteins contained within these secretions, except for the serine protease of acetabular glands. Referred to as elastase, this protein has been cloned and expressed (McKerrow *et al.*, 1985; Newport *et al.*, 1988) but no protection experiments have been reported.

Analysis of the attenuated vaccine model led several groups to the conclusion that the lung-stage larva, although refractory to ADCC, was a major target of effector responses. Furthermore, in 1× vaccinated mice, $CD4^+$ Th cells were crucial initiators of the focal inflammation that terminated migration. It therefore follows that antigens must either be released from or exposed on the surface of lung schistosomula for processing and presentation by accessory cells to activate the effector T cells. Unfortunately, attempts to identify the components of lung-worm secretions, by raising antisera to screen a lung-stage cDNA library, detected the same abundant cytosolic and cytoskeletal molecules as previous screens of libraries derived from other stages (Harrop *et al.*, 1999), highlighting a limitation of the approach.

A largely unexplored possibility for mediating protection, in view of schistosome longevity, is the material released from the worm gut during the process of blood feeding. Some of the gut proteases involved in haemoglobin degradation have been investigated, and asparaginyl endopeptidase (Sm32, legumain) has been trialled in mice as a DNA construct where, in a single experiment, no significant reduction in worm burden was obtained but an anti-fecundity effect was recorded (Chlichlia *et al.*, 2001). A *S. japonicum* aspartic protease, cathepsin D, has been expressed as an active enzyme in the baculovirus system and tested in mice with Quil A as adjuvant. Challenge worm burdens were significantly reduced (21–38%) but there was no impact on fecundity (Verity *et al.*, 2001). We must assume that the immune responses to these gut proteases would be directed against juvenile and adult worms in the portal vasculature, and successful vaccination against this parasite stage could herald the development of a therapeutic vaccine for delivery to individuals with an existing infection. In this context, the antioxidant enzymes of adult worms have been proposed as vaccine targets, and a mean of 39% protection was obtained after challenge by surgical transfer of adult worms into the portal system of mice vaccinated against superoxide dismutase using a DNA construct (Cook *et al.*, 2004).

The tegument surface is such an obvious target for immune attack that several attempts have been made to induce protection in mice by administration of 'purified' membrane preparations. These were derived from adult worms by a variety of methods, and the degree of enrichment achieved is unclear. However, in at least one case (the freeze-thaw technique; Roberts *et al.*, 1988) we now know that, whilst the final pellet contains membrane proteins, it is dominated by cytosolic and cytoskeletal proteins (Braschi, *et al.*, 2005). Irrespective of this point, a large number of vaccination experiments has been performed, with a variety of adjuvants, giving protection in the range 0–52% but generally less than 30% (Maddison *et al.*, 1978; Roberts *et al.*, 1988; Smithers *et al.*, 1989; Hota-Mitchell *et al.*, 1997). A small number of tegument-associated proteins (including Sm23, see earlier) have been more intensively pursued but they represent only the tip of the iceberg.

Calpain

The native calcium-binding protease, calpain, which translocates to the cytosolic surface of membranes upon calcium binding, was enriched from a membrane preparation and shown to elicit high levels of protection (56–67%) in mice (Hota-Mitchell *et al.*, 1997). This protein was first implicated in protective immunity as the target of a CD4$^+$ T cell clone that could arm peritoneal macrophages to kill schistosomula *in vitro* (Jankovic *et al.*, 1996). The same clone administered intraperitoneally, conferred 65% protection on irradiated mice challenged via this route with cercariae and recombinant IL-2 (this is a very artificial test from which results should be treated with caution). The large subunit of calpain was subsequently expressed in the baculovirus system and used for vaccination experiments in mice, achieving 29–39% protection (Hota-Mitchell *et al.*, 1997). Attempts to improve on this, including the use of calpain as a DNA formulation and plasmids for the co-delivery of immunostimulatory cytokines, increased protection to 42–57% (Siddiqui *et al.*, 2003a,b). Somewhat better results were obtained with gene gun delivery of a DNA construct although a variety of prime-boost strategies involving DNA vector and recombinant vaccinia virus failed to induce protection (Hota-Mitchell *et al.*, 1999). The *S. japonicum* calpain has also been cloned and the recombinant protein shown to induce a moderate reduction in worm burden and a possible anti-fecundity effect (Zhang *et al.*, 2001b).

The 20–22 kDa family

A group of tegumental proteins from *S. mansoni* and *S. japonicum*, with sequence homology to dynein light chain that implies a cytoskeletal function, has also been explored for protective potential. Members of this group have been variously identified as antigens, e.g. the 22.6 kDa protein, which is the allergenic target of IgE responses in human schistosomiasis patients (Santiago *et al.*, 1998). The 20.8 kDa protein from *S. mansoni* administered to mice as a DNA construct elicited approximately 30% protection (Mohamed *et al.*, 1998).

With the exception of Sm23, these proteins appear to be located on the cytoplasmic surface of the tegument membrane, and it is difficult to envisage how they would be accessible to the immune response in intact worms. Do they fall in the category of proteins transiently exposed during cercaria to schistosomulum transformation? If so, would this place the same constraints on their utility as the majority of vaccine candidates we have already considered?

Future Prospects

We start from the premise that the macromolecules most likely to mediate protection are those which the parasite must release or expose on its epithelial surface as part of normal biological processes. Hitherto, these have been difficult to characterize. Fortunately, rapid strides are being made in our understanding of fundamental aspects of schistosome biology that bring the possibility of identifying such molecules within our reach. The developments encompass analyses of the genome, transcriptome and proteome of *S. mansoni*.

The genome

Secreted and surface-exposed molecules were poorly represented in the expressed sequence tag (EST) databases before Autumn 2003 but the completion of a major sequencing project by a Brazilian consortium added a further 125,000 ESTs (Verjovski-Almeida *et al.*, 2003), boosting our ability to find such molecules. Furthermore, the *S. mansoni* genome has been sequenced to 8× coverage, with a draft assembly (v1.0) completed and automatically annotated at the Wellcome Trust Sanger Institute; a genome database, www.SchistoDB.org, was released to the research community in February 2005. Consequently, we should soon have extensive information about the complement of >14,000 genes that comprise the schistosome genome.

Possession of the complete genetic blueprint of the schistosome does not of itself pinpoint new vaccine candidates. Following publication of the *Plasmodium falciparum* genome, a global screen was proposed that involved the formulation of every coding region as a DNA construct for testing in mice (Hoffman, 2000), although it does not seem to have been pursued. The large genome and complex body plan of schistosomes call for a more selective approach. The coding regions of the genome can be searched using a variety of software packages (e.g. SignalP or PSORTII) to identify those sequences encoding putative secretory or membrane-spanning proteins. For a metazoan such as *S. mansoni*, these probably represent 15% of the coding regions, i.e. >2000 genes, still too large a number to contemplate a reverse vaccinology approach as recently promulgated for meningococcus, with a much smaller genome (Rappuoli and Covacci, 2003).

The transcriptome

It is possible that protective antigens are stage-specific, although there is no direct experimental evidence. The advent of microarray technology permits the question of stage specificity to be addressed, and data obtained with the first small-scale schistosome microarray has been published (Hoffmann *et al.*, 2002). At least three more extensive arrays were reported at a recent schistosome genome meeting in August 2004, and use of these to screen gene expression in different life-cycle stages can be expected to identify novel stage-specific genes in the near future. These sequences can also be subjected to motif and domain searching to identify secreted and membrane proteins that might have vaccine potential.

The proteome

The advent of proteomics offers another and more direct route to identify parasite proteins potentially accessible to the immune response (Ashton *et al.*, 2001). Soluble protein preparations can be separated into their individual constituents by 2D electrophoresis, whilst membrane proteins require solubilization in ionic detergents incompatible with isoelectric focusing and so must be processed differently. Ultimately, this involves liquid chromatographic separation of derived peptide mixtures. Irrespective of the separation technique, the resulting peptides are subjected to tandem mass spectrometry to identify the parent protein(s). Best results are obtained by collecting particular secretions (e.g. from cercarial acetabular glands) or using cell fractionation techniques (e.g. tegument membrane isolation) to enrich the target proteins before proteomic analysis. This approach will ultimately provide a complete inventory of the secreted and surface-exposed proteins of larval and adult schistosomes.

The information gleaned from analysis of the genome, transcriptome and proteome can be integrated to prioritize proteins for intensive investigation. Each will need to be expressed as a recombinant or formulated as a DNA construct for rigorous testing of its protective potential. The result should be a new generation of vaccine candidates hopefully emulating or outperforming the high levels of protection achieved by the attenuated vaccine, or inducing self-cure if administered to infected hosts as a therapeutic vaccine.

Concluding Remarks

In writing this review of schistosome vaccine development, with the benefit of hindsight we have attempted to evaluate the published data in a critical but dispassionate manner. We doubt that a vaccine is just around the corner, not least because of our limited understanding of the points in the schistosome life cycle that are susceptible to immune attack. By accident or design, most effort has been directed against the newly transformed schistosomulum but this stage appears to have relatively restricted potential as a vaccine target, and there are other vulnerable stages that have yet to be exploited. The new technologies to analyse the genome, transcriptome and proteome at last provide the way forward to explore a new generation of vaccine candidates firmly rooted in the physiological processes of the parasite. When the inventory of relevant parasite molecules is complete, we shall know the limits of the material from which the elusive schistosome vaccine might be fashioned. By then, we may also have the delivery systems to achieve the appropriate strong, persistent immune responses.

References

Al-Sherbiny, M., Osman, A., Barakat, R., El Morshedy, H., Bergquist, R. and Olds, R. (2003) *In vitro* cellular and humoral responses to *Schistosoma mansoni* vaccine candidate antigens. *Acta Tropica* 88, 117–130.

Anderson, S., Coulson, P.S., Ljubojevic, S., Mountford, A.P. and Wilson, R.A. (1999) The radiation-attenuated schistosome vaccine induces high levels of protective immunity in the absence of B cells. *Immunology* 96, 22–28.

Angeli, V., Faveeuw, C., Roye, O., Fontaine, J., Teissier, E., Capron, A., Wolowczuk, I., Capron, M. and Trottein, F. (2001) Role of the parasite-derived prostaglandin D2 in the inhibition of epidermal Langerhans cell migration during schistosomiasis infection. *The Journal of Experimental Medicine* 193, 1135–1147.

Anonymous (1996) TDR News. The World Health Organization, number 50 June, Geneva, Switzerland.

Argiro, L., Henri, S., Dessein, H., Dessein, A.J. and Bourgois, A. (1999) Induction of a protective immunity against *Schistosoma mansoni* with ovalbumin-coupled Sm37–5 coadsorbed with granulocyte-macrophage colony stimulating factor (GM-CSF) or IL-12 on alum. *Vaccine* 17, 13–18.

Argiro, L.L., Kohlstadt, S.S., Henri, S.S., Dessein, H.H., Matabiau, V.V., Paris, P.P., Bourgois, A.A. and Dessein, A.J. (2000) Identification of a candidate vaccine peptide on the 37 kDa *Schistosoma mansoni* GAPDH. *Vaccine* 18, 2039–2048.

Arnon, R., Tarrab-Hazdai, R. and Steward, M. (2000) A mimotope peptide-based vaccine against *Schistosoma mansoni*: synthesis and characterization. *Immunology* 101, 555–562.

Ashton, P.D., Curwen, R.S. and Wilson, R.A. (2001) Linking proteome and genome: how to identify parasite proteins. *Trends in Parasitology* 17, 198–202.

Auriault, C., Gras-Masse, H., Wolowczuk, I., Pierce, R.J., Balloul, J.M., Neyrinck, J.L., Drobecq, H., Tartar, A. and Capron, A. (1988) Analysis of T and B cell epitopes of the *Schistosoma mansoni* p28 antigen in the rat model by using synthetic peptides. *The Journal of Immunology* 141, 1687–1694.

Balloul, J.M., Pierce, R.J., Grzych, J.M. and Capron, A. (1985) *In vitro* synthesis of a 28 kilodalton antigen present on the surface of the schistosomulum of *Schistosoma mansoni*. *Molecular and Biochemical Parasitology* 17, 105–114.

Balloul, J.M., Grzych, J.M., Pierce, R.J. and Capron, A. (1987a) A purified 28,000 dalton protein from *Schistosoma mansoni* adult worms protects rats and mice against experimental schistosomiasis. *The Journal of Immunology* 138, 3448–3453.

Balloul, J.M., Sondermeyer, P., Dreyer, D., Capron, M., Grzych, J.M., Pierce, R.J., Carvallo, D., Lecocq, J.P. and Capron, A. (1987b) Molecular cloning of a protective antigen of schistosomes. *Nature* 326, 149–153.

Baras, B., Benoit, M.A., Dupre, L., Poulain-Godefroy, O., Schacht, A.M., Capron, A., Gillard, J. and Riveau, G. (1999) Single-dose mucosal immunization with biodegradable microparticles containing a *Schistosoma mansoni* antigen. *Infection and Immunity* 67, 2643–2648.

Baras, B., Benoit, M., Poulain-Godefroy, O., Schacht, A., Capron, A., Gillard, J. and Riveau, G. (2000) Vaccine properties of antigens entrapped in microparticles produced by spray-drying technique and using various polyester polymers. *Vaccine* 18, 1495–1505.

Beall, J.A. and Mitchell, G.F. (1986) Identification of a particular antigen from a parasite cDNA library using antibodies affinity purified from selected portions of Western blots. *The Journal of Immunological Methods* 86, 217–223.

Becker, M.M., Kalinna, B.H., Waine, G.J. and McManus, D.P. (1994) Gene cloning, overproduction and purification of a functionally active cytoplasmic fatty acid-binding protein (Sj-FABPc) from the human blood fluke *Schistosoma japonicum*. *Gene* 148, 321–325.

Ben-Yedidia, T., Tarrab-Hazdai, R., Schechtman, D. and Arnon, R. (1999) Intranasal administration of synthetic recombinant peptide-based vaccine protects mice from infection by *Schistosoma mansoni*. *Infection and Immunity* 67, 4360–4366.

Bickle, Q.D. and James, E.R. (1978) Resistance against *Schistosoma mansoni* induced by immunization of mice with cryopreserved schistosomula. *Transactions of the Royal Society of Tropical Medicine and Hygiene* 72, 677–678.

Bickle, Q.D., Andrews, B.J. and Taylor, M.G. (1986) *Schistosoma mansoni*: characterization of two protective monoclonal antibodies. *Parasite Immunology* 8, 95–107.

Bickle, Q.D., Bogh, H.O., Johansen, M.V. and Zhang, Y. (2001) Comparison of the vaccine efficacy of gamma-irradiated *Schistosoma japonicum* cercariae with the defined antigen Sj62(IrV-5) in pigs. *Veterinary Parasitology* 100, 51–62.

Boulanger, D., Reid, G.D., Sturrock, R.F., Wolowczuk, I., Balloul, J.M., Grezel, D., Pierce, R.J., Otieno, M.F., Guerret, S., Grimaud, J.A., *et al.* (1991) Immunization of mice and baboons with the recombinant Sm28GST affects both worm viability and fecundity after experimental infection with *Schistosoma mansoni*. *Parasite Immunology* 13, 473–490.

Boulanger, D., Trottein, F., Mauny, F., Bremond, P., Couret, D., Pierce, R.J., Kadri, S., Godin, C., Sellin, E., Lecocq, J.P., Sellin, B. and Capron, A (1994) Vaccination of goats against the trematode *Schistosoma bovis* with a recombinant homologous schistosome-derived glutathione S-transferase. *Parasite Immunology* 16, 399–406.

Boulanger, D., Warter, A., Trottein, F., Mauny, F., Bremond, P., Audibert, F., Couret, D., Kadri, S., Godin, C., Sellin, E., Pierce, R.J., Lecocq, J.P., Sellin, B. and Capron, A. (1995) Vaccination of patas monkeys experimentally infected with *Schistosoma haematobium* using a recombinant glutathione S-transferase cloned from *S. mansoni*. *Parasite Immunology* 17, 361–369.

Boulanger, D., Schneider, D., Chippaux, J.P., Sellin, B. and Capron, A. (1999a) Schistosoma bovis: vaccine effects of a recombinant homologous glutathione S-transferase in sheep. *International Journal for Parasitology* 29, 415–418.

Boulanger, D., Warter, A., Sellin, B., Lindner, V., Pierce, R.J., Chippaux, J.P. and Capron, A. (1999b) Vaccine potential of a recombinant S-transferase cloned from *Schistosoma haematobium* in primates experimentally infected with an homologous challenge. *Vaccine* 17, 319–326.

Braschi, S., Curwen, R.S., Ashton, P.D., Verjouski-Almeida, S. and Wilson, R.A. (2005) The tegument surface membranes of the human blood fluke *Schistosoma mansoni*: a proteomic analysis after differential extraction. *Proteomics* (in press).

Bushara, H.O., Bashir, M.E., Malik, K.H., Mukhtar, M.M., Trottein, F., Capron, A. and Taylor, M.G. (1993) Suppression of *Schistosoma bovis* egg production in cattle by vaccination with either glutathione S-transferase or keyhole limpet haemocyanin. *Parasite Immunology* 15, 383–390.

Butterworth, A.E., Sturrock, R.F., Houba, V. and Rees, P.H. (1974) Antibody-dependent cell-mediated damage to schistosomula *in vitro*. *Nature* 252, 503–505.

Butterworth, A.E., Taylor, D.W., Veith, M.C., Vadas, M.A., Dessein, A., Sturrock, R.F. and Wells, E. (1982) Studies on the mechanisms of immunity in human schistosomiasis. *Immunological Reviews* 61, 5–39.

Capron, A., Capron, M. and Riveau, G. (2002) Vaccine development against schistosomiasis from concepts to clinical trials. *British Medical Bulletin* 62, 139–148.

Charrier-Ferrara, S., Caillol, D. and Goudot-Crozel, V. (1992) Complete sequence of the *Schistosoma mansoni* glyceraldehyde-3-phosphate dehydrogenase gene encoding a major surface antigen. *Molecular and Biochemical Parasitology* 56, 339–343.

Chen, H., Nara, T., Zeng, X., Satoh, M., Wu, G., Jiang, W., Yi, F., Kojima, S., Zhang, S. and Hirayama, K. (2000) Vaccination of domestic pig with recombinant paramyosin. Against *Schistosoma japonicum* in China. *Vaccine* 18, 2142–2146.

Chitsulo, L., Loverde, P. and Engels, D. (2004) Schistosomiasis. *Nature Reviews in Microbiology* 2, 12–13.

Chlichlia, K., Bahgat, M., Ruppel, A. and Schirrmacher, V. (2001) DNA vaccination with asparaginyl endopeptidase (Sm32) from the parasite *Schistosoma mansoni*: anti-fecundity effect induced in mice. *Vaccine* 20, 439–447.

Cook, R.M., Carvalho-Queiroz, C., Wilding, G. and LoVerde, P.T. (2004) Nucleic acid vaccination with *Schistosoma mansoni* antioxidant enzyme cytosolic superoxide dismutase and the structural protein filamin confer protection against the adult worm stage. *Infection and Immunity* 72, 6112–6124.

Coulson, P.S. (1997) The radiation-attenuated vaccine against schistosomes in animal models: paradigm for a human vaccine? *Advances in Parasitology* 39, 271–336.

Coulson, P.S., Smythies, L.E., Betts, C., Mabbott, N.A., Sternberg, J.M., Wei, X.G., Liew, F.Y. and Wilson, R.A. (1998) Nitric oxide produced in the lungs of mice immunized with the radiation-attenuated schistosome vaccine is not the major agent causing challenge parasite elimination. *Immunology* 93, 55–63.

Cox, F.E. (1990) Schistosomiasis. The mouse that wasn't immune. *Nature* 345, 18.

Crabtree, J.E. and Wilson, R.A. (1985) *Schistosoma mansoni*: an ultrastructural examination of skin migration in the hamster cheek pouch. *Parasitology* 91, 111–120.

Curwen, R.S. and Wilson, R.A. (2003) Invasion of skin by schistosome cercariae: some neglected facts. *Trends in Parasitology* 19, 63–66.

Curwen, R.S., Ashton, P.D., Johnston, D.A. and Wilson, R.A. (2004) The *Schistosoma mansoni* soluble proteome: a comparison across four life-cycle stages. *Molecular and Biochemical Parasitology* 138, 57–66.

Da'Dara, A.A., Skelly, P.J., Wang, M.M. and Harn, D.A. (2001) Immunization with plasmid DNA encoding the integral membrane protein, Sm23, elicits a protective immune response against schistosome infection in mice. *Vaccine* 20, 359–369.

Da'Dara, A.A., Skelly, P.J., Walker, C.M. and Harn, D.A. (2003) A DNA-prime/protein-boost vaccination regimen enhances Th2 immune responses but not protection following *Schistosoma mansoni* infection. *Parasite Immunology* 25, 429–437.

Dalton, J.P. and Strand, M. (1987) *Schistosoma mansoni* polypeptides immunogenic in mice vaccinated with radiation-attenuated cercariae. *The Journal of Immunology* 139, 2474–2481.

Damian, R.T., Roberts, M.L., Powell, M.R., Clark, J.D., Lewis, F.A. and Stirewalt, M.A. (1984) *Schistosoma mansoni* egg granuloma size reduction in challenged baboons after vaccination with irradiated cryopreserved schistosomula. *Proceedings of the National Academy of Sciences USA* 81, 3552–3556.

Dean, D.A., Bukowski, M.A. and Clark, S.S. (1981) Attempts to transfer the resistance of *Schistosoma mansoni*-infected and irradiated cercaria-immunized mice by means of parabiosis. *American Journal of Tropical Medicine and Hygiene* 30, 113–120.

Dessein, A., Samuelson, J.C., Butterworth, A.E., Hogan, M., Sherry, B.A., Vadas, M.A. and David, J.R. (1981) Immune evasion by *Schistosoma mansoni*: loss of susceptibility to antibody or complement-dependent eosinophil attack by schistosomula cultured in medium free of macromolecules. *Parasitology* 82, 357–374.

Dessein, A.J., Begley, M., Demeure, C., Caillol, D., Fueri, J., dos Reis, M.G., Andrade, Z.A., Prata, A. and Bina, J.C. (1988) Human resistance to *Schistosoma mansoni* is associated with IgG reactivity to a 37-kDa larval surface antigen. *The Journal of Immunology* 140, 2727–2736.

Dunne, D.W., Bickle, Q.D., Butterworth, A.E. and Richardson, B.A. (1987) The blocking of human antibody-dependent, eosinophil-mediated killing of *Schistosoma mansoni* schistosomula by monoclonal antibodies which cross-react with a polysaccharide-containing egg antigen. *Parasitology* 94, 269–280.

Dupre, L., Kremer, L., Wolowczuk, I., Riveau, G., Capron, A. and Locht, C. (2001) Immunostimulatory effect of IL-18-encoding plasmid in DNA vaccination against murine *Schistosoma mansoni* infection. *Vaccine* 19, 1373–1380.

Eberl, M., Langermans, J.A., Frost, P.A., Vervenne, R.A., van Dam, G.J., Deelder, A.M., Thomas, A.W., Coulson, P.S. and Wilson, R.A. (2001) Cellular and humoral immune responses and protection against schistosomes induced by a radiation-attenuated vaccine in chimpanzees. *Infection and Immunity* 69, 5352–5362.

Feng, Z., Li, Y., Qiu, Z., Xue, W., Guan, X. and Wu, G. (2000) Anti-fecundity immunity in mice immunized with anti-idiotypic monoclonal antibodyNP of *Schistosoma japonicum*. *Chinese Medical Journal (Engl.)* 113, 265–268.

Feng, Z., Qiu, Z., Li, Y., Zhu, C., Xue, W. and Guan, X. (2002) Protective immunity induced by the anti-idiotypic monoclonal antibody NP30 of *Schistosoma japonicum*. *Chinese Medical Journal (Engl.)* 115, 576–579.

Ferru, I., Georges, B., Bossus, M., Estaquier, J., Delacre, M., Harn, D.A., Tartar, A., Capron, A., Grassmasse, H. and Auriault, C. (1997) Analysis of the immune response elicited by a multiple antigen peptide (MAP) composed of two distinct protective antigens derived from the parasite *Schistosoma mansoni*. *Parasite Immunology* 19, 1–11.

Flanigan, T.P., King, C.H., Lett, R.R., Nanduri, J. and Mahmoud, A.A. (1989) Induction of resistance to *Schistosoma mansoni* infection in mice by purified parasite paramyosin. *The Journal of Clinical Investigation* 83, 1010–1014.

Fulford, A.J., Butterworth, A.E., Sturrock, R.F. and Ouma, J.H. (1992) On the use of age-intensity data to detect immunity to parasitic infections, with special reference to *Schistosoma mansoni* in Kenya. *Parasitology* 105, 219–227.

Fusco, A.C., Salafsky, B. and Kevin, M.B. (1985) *Schistosoma mansoni*: eicosanoid production by cercariae. *Experimental Parasitology* 59, 44–50.

Gobert, G.N., Stenzel, D.J., Jones, M.K. and McManus, D.P. (1997) Immunolocalization of the fatty acid-binding protein Sj-FABPc within adult *Schistosoma japonicum*. *Parasitology* 115, 33–39.

Goudot-Crozel, V., Caillol, D., Djabali, M. and Dessein, A.J. (1989) The major parasite surface antigen associated with human resistance to schistosomiasis is a 37-kD glyceraldehyde-3-dehydrogenase. *Journal of Experimental Medicine* 170, 2065–2080.

Grzych, J.M., Capron, M., Bazin, H. and Capron, A. (1982) *In vitro* and *in vivo* effector function of rat IgG2a monoclonal anti-*S. mansoni* antibodies. *The Journal of Immunology* 129, 2739–2743.

Grzych, J.M., Capron, M., Dissous, C. and Capron, A. (1984) Blocking activity of rat monoclonal antibodies in experimental schistosomiasis. *The Journal of Immunology* 133, 998–1004.

Grzych, J.M., Capron, M., Lambert, P.H., Dissous, C., Torres, S. and Capron, A. (1985) An anti-idiotype vaccine against experimental schistosomiasis. *Nature* 316, 74–76.

Hagan, P., Blumenthal, U.J., Dunn, D., Simpson, A.J. and Wilkins, H.A. (1991) Human IgE, IgG4 and resistance to reinfection with *Schistosoma haematobium*. *Nature* 349, 243–245.

Harn, D.A., Mitsuyama, M. and David, J.R. (1984) *Schistosoma mansoni*. Anti-egg monoclonal antibodies protect against cercarial challenge *in vivo*. *The Journal of Experimental Medicine* 159, 1371–1387.

Harn, D.A., Mitsuyama, M., Huguenel, E.D. and David, J.R. (1985) *Schistosoma mansoni*: detection by monoclonal antibody of a 22,000-dalton surface membrane antigen, which may be blocked by host molecules on lung stage parasites. *The Journal of Immunology* 135, 2115–2120.

Harn, D.A., Gu, W., Oligino, L.D., Mitsuyama, M., Gebremichael, A. and Richter, D. (1992) A protective monoclonal antibody specifically recognizes and alters the catalytic activity of schistosome triose-phosphate isomerase. *The Journal of Immunology* 148, 562–567.

Harn, D.A., Reynolds, S.R., Chikunguwo, S.M., Furlong, S.T. and Dahl, C.E. (1995) Synthetic peptide vaccines for schistosomiasis. In: Powell, M.F. and Newman, M.J. (eds) *Vaccine Design: The Subunit and Adjuvant Approach*. Plenum Press, New York, pp. 891–905.

Harris, A.R., Russell, R.J. and Charters, A.D. (1984) A review of schistosomiasis in immigrants in Western Australia, demonstrating the unusual longevity of *Schistosoma mansoni*. *Transactions of the Royal Society of Tropical Medicine and Hygiene* 78, 385–388.

Harrison, R.A., Bickle, Q. and Doenhoff, M.J. (1982) Factors affecting the acquisition of resistance against *Schistosoma mansoni* in the mouse. Evidence that the mechanisms, which mediate resistance during early patent infections, may lack immunological specificity. *Parasitology* 84, 93–110.

Harrop, R., Coulson, P.S. and Wilson, R.A. (1999) Characterization, cloning and immunogenicity of antigens released by lung-stage larvae of *Schistosoma mansoni*. *Parasitology* 118, 583–594.

Hayunga, E.G. (1985) Attempted immunization of mice against *S. mansoni* by inoculation with purified glycoprotein antigens from adult worms. *Proceedings of the Helminthological Society of Washington* 52, 184–195.

Hazdai, R.T., Levi-Schaffer, F., Brenner, V., Horowitz, S., Eshhar, Z. and Arnon, R. (1985) Protective monoclonal antibody against *Schistosoma mansoni*: antigen isolation, characterization, and suitability for active immunization. *The Journal of Immunology* 135, 2772–2779.

Herve, M., Angeli, V., Pinzar, E., Wintjens, R., Faveeuw, C., Narumiya, S., Capron, A., Urade, Y., Capron, M., Riveau, G. and Trottein, F. (2003) Pivotal roles of the parasite PGD2 synthase and of the host D prostanoid receptor 1 in schistosome immune evasion. *European Journal of Immunology* 33, 2764–2772.

Hillyer, G.V. (1995) Comparison of purified 12 kDa and recombinant 15 kDa *Fasciola hepatica* antigens related to a *Schistosoma mansoni* fatty acid binding protein. *Memorias do Instituto Oswaldo Cruz* 90, 249–253.

Hillyer, G.V., Garcia Rosa, M.I., Alicea, H. and Hernandez, A. (1988) Successful vaccination against murine *Schistosoma mansoni* infection with a purified 12 kD *Fasciola hepatica* cross-reactive antigen. *American Journal of Tropical Medicine and Hygiene* 38, 103–110.

Hockley, D.J. and McLaren, D.J. (1973) *Schistosoma mansoni*: changes in the outer membrane of the tegument during development from cercaria to adult worm. *International Journal for Parasitology* 3, 13–25.

Hoffman, S.L. (2000) Infectious disease. Research (genomics) is crucial to attacking malaria. *Science* 290, 1509.
Hoffman, S.L., Goh, L.M., Luke, T.C., Schneider, I., Le, T.P., Doolan, D.L., Sacci, J., de la Vega, P., Dowler, M., Paul, C., Gordon, D.M., Stoute, J.A., Church, L.W., Sedegah, M., Heppner, D.G., Ballou, W.R. and Richie, T.L. (2002) Protection of humans against malaria by immunization with radiation-attenuated *Plasmodium falciparum* sporozoites. *Journal of Infectious Diseases* 185, 1155–1164.
Hoffmann, K.F., Johnston, D.A. and Dunne, D.W. (2002) Identification of *Schistosoma mansoni* gender-associated gene transcripts by cDNA microarray profiling. *Genome Biology* 3, research0041.1–0041.12.
Horowitz, S., Brenner, V. and Arnon, R. (1985) *In vivo* protection against *S. mansoni* infection by monoclonal antibodies. *Immunology Letters* 9, 69–73.
Hota-Mitchell, S., Siddiqui, A.A., Dekaban, G.A., Smith, J., Tognon, C. and Podesta, R.B. (1997) Protection against *Schistosoma mansoni* infection with a recombinant baculovirus-expressed subunit of calpain. *Vaccine* 15, 1631–1640.
Hota-Mitchell, S., Clarke, M.W., Podesta, R.B. and Dekaban, G.A. (1999) Recombinant vaccinia viruses and gene gun vectors expressing the large subunit of *Schistosoma mansoni* calpain used in a murine immunization-challenge model. *Vaccine* 17, 1338–1354.
James, S.L. (1985) Induction of protective immunity against *Schistosoma mansoni* by a nonliving vaccine is dependent on the method of antigen presentation. *The Journal of Immunology* 134, 1956–1960.
James, S.L. (1986) Activated macrophages as effector cells of protective immunity to schistosomiasis. *Immunologic Research* 5, 139–148.
James, S.L. (1987) Induction of protective immunity against *Schistosoma mansoni* by a non-living vaccine. V. Effects of varying the immunization and infection schedule and site. *Parasite Immunology* 9, 531–541.
James, S.L. and DeBlois, L.A. (1986) Induction of protective immunity against *Schistosoma mansoni* by a non-living vaccine. II. Response of mouse strains with selective immune defects. *The Journal of Immunology* 136, 3864–3871.
James, S.L. and Pearce, E.J. (1988) The influence of adjuvant on induction of protective immunity by a non-living vaccine against schistosomiasis. *The Journal of Immunology* 140, 2753–2759.
James, S.L., Pearce, E.J. and Sher, A. (1985) Induction of protective immunity against *Schistosoma mansoni* by a nonliving vaccine. I. Partial characterization of antigens recognized by antibodies from mice immunized with soluble schistosome extracts. *The Journal of Immunology* 134, 3432–3438.
Jankovic, D., Aslund, L., Oswald, I.P., Caspar, P., Champion, C., Pearce, E., Coligan, J.E., Strand, M., Sher, A. and James, S.L. (1996) Calpain is the target antigen of a Th1 clone that transfers protective immunity against *Schistosoma mansoni*. *The Journal of Immunology* 157, 806–814.
Kalinna, B.H., Becker, M.M. and McManus, D.P. (1997) Engineering and expression of a full-length cDNA encoding *Schistosoma japonicum* paramyosin. Purification of the recombinant protein and its recognition by infected patient sera. *Acta Tropica* 65, 111–115.
Kariuki, T.M., Farah, I.O., Yole, D.S., Mwenda, J.M., Van Dam, G.J., Deelder, A.M., Wilson, R.A. and Coulson, P.S. (2004) Parameters of the attenuated schistosome vaccine evaluated in the olive baboon. *Infection and Immunity* 72, 5526–5529.
Khan, C.M., Villarreal-Ramos, B., Pierce, R.J., Demarco de Hormaeche, R., McNeill, H., Ali, T., Chatfield, S., Capron, A., Dougan, G. and Hormaeche, C.E. (1994a) Construction, expression, and immunogenicity of multiple tandem copies of the *Schistosoma mansoni* peptide 115-131 of the p28 glutathione S-transferase expressed as C-terminal fusions to tetanus toxin fragment C in a live aro-attenuated vaccine strain of *Salmonella*. *The Journal of Immunology* 153, 5634–5642.
Khan, C.M., Villarreal-Ramos, B., Pierce, R.J., Riveau, G., Demarco de Hormaeche, R., McNeill, H., Ali, T., Fairweather, N., Chatfield, S., Capron, A., Dougan, G. and Hormaeche, C.E. (1994b) Construction, expression, and immunogenicity of the *Schistosoma mansoni* p28 glutathione S-transferase as a genetic fusion to tetanus toxin C in a live aro attenuated vaccine strain of *Salmonella*. *Proceedings of the National Academy of Sciences USA* 91, 11261–11265.
Kremer, L., Riveau, G., Baulard, A., Capron, A. and Locht, C. (1996) Neutralizing antibody responses elicited in mice immunized with recombinant Bacillus Calmette-Guerin producing the *Schistosoma mansoni* glutathione S-transferase. *The Journal of Immunology* 156, 4309–4317.
Kresina, T.F. and Olds, G.R. (1989) Antiidiotypic antibody vaccine in murine schistosomiasis mansoni comprising the internal image of antigen. *The Journal of Clinical Investigation* 83, 912–920.
Lanar, D.E., Pearce, E.J., James, S.L. and Sher, A. (1986) Identification of paramyosin as schistosome antigen recognized by intradermally vaccinated mice. *Science* 234, 593–596.
Lane, A., Boulanger, D., Riveau, G., Capron, A. and Wilson, R.A. (1998) Murine immune responses to *Schistosoma haematobium* and the vaccine candidate rSh28GST. *Parasite Immunology* 20, 359–367.

Lebens, M., Sun, J.B., Sadeghi, H., Backstrom, M., Olsson, I., Mielcarek, N., Li, B.L., Capron, A., Czerkinsky, C. and Holmgren, J. (2003) A mucosally administered recombinant fusion protein vaccine against schistosomiasis protecting against immunopathology and infection. *Vaccine* 21, 514–520.

Lewis, F.A., Stirewalt, M. and Leef, J.L. (1984) *Schistosoma mansoni*: radiation dose and morphologic integrity of schistosomules as factors for an effective cryopreserved live vaccine. *American Journal of Tropical Medicine and Hygiene* 33, 125–131.

Liu, S.X., Song, G.C., Xu, Y.X., Yang, W. and McManus, D.P. (1995) Anti-fecundity immunity induced in pigs vaccinated with recombinant *Schistosoma japonicum* 26 kDa glutathione-S-transferase. *Parasite Immunol* 17, 355–340.

Luke, T.C. and Hoffman, S.L. (2003) Rationale and plans for developing a non-replicating, metabolically active, radiation-attenuated *Plasmodium falciparum* sporozoite vaccine. *Journal of Experimental Biology* 206, 3803–3808.

Maddison, S.E., Slemenda, S.B., Chandler, F.W. and Kagan, I.G. (1978) Studies on putative adult worm-derived vaccines and adjuvants for protection against *Schistosoma mansoni* infection in mice. *Journal of Parasitology* 64, 986–993.

Mangold, B.L. and Dean, D.A. (1986) Passive transfer with serum and IgG antibodies of irradiated cercaria-induced resistance against *Schistosoma mansoni* in mice. *The Journal of Immunology* 136, 2644–2648.

McKerrow, J.H., Pino-Heiss, S., Lindquist, R. and Werb, Z. (1985) Purification and characterization of an elastinolytic proteinase secreted by cercariae of *Schistosoma mansoni*. *The Journal of Biological Chemistry* 260, 3703–3707.

McLaren, D.J., Ramalho-Pinto, F.J. and Smithers, S.R. (1978) Ultrastructural evidence for complement and antibody-dependent damage to schistosomula of *Schistosoma mansoni* by rat eosinophils *in vitro*. *Parasitology* 77, 313–324.

McManus, D.P., Liu, S., Song, G., Xu, Y. and Wong, J.M. (1998) The vaccine efficacy of native paramyosin (Sj-97) against Chinese *Schistosoma japonicum*. *International Journal for Parasitology* 28, 1739–1742.

McManus, D.P., Wong, J.Y., Zhou, J., Cai, C., Zeng, Q., Smyth, D., Li, Y., Kalinna, B.H., Duke, M.J. and Yi, X. (2001) Recombinant paramyosin (rec-Sj-97) tested for immunogenicity and vaccine efficacy against *Schistosoma japonicum* in mice and water buffaloes. *Vaccine* 20, 870–878.

Mendlovic, F., Tarrab-Hazdai, R., Puri, J. and Arnon, R. (1989) Genetic control of immune response to a purified *Schistosoma mansoni* antigen. I. Effect of MHC class II antigens on the cellular, humoral and protective responses. *Parasite Immunology* 11, 667–682.

Mielcarek, N., Cornette, J., Schacht, A.M., Pierce, R.J., Locht, C., Capron, A. and Riveau, G. (1997) Intranasal priming with recombinant *Bordetella pertussis* for the induction of a systemic immune response against a heterologous antigen. *Infection and Immunity* 65, 544–550.

Mielcarek, N., Riveau, G., Remoue, F., Antoine, R., Capron, A. and Locht, C. (1998) Homologous and heterologous protection after single intranasal administration of live attenuated recombinant *Bordetella pertussis*. *Nature Biotechnology* 16, 454–457.

Mitchell, G.F., Cruise, K.M., Garcia, E.G. and Tiu, W.U. (1984) Anti-worm antibody specificities in 129/j mice resistant to infection with *Schistosoma japonicum*. *Journal of Parasitology* 70, 983–985.

Mitchell, G.F., Garcia, E.G., Davern, K.M., Tiu, W.U. and Smith, D.B. (1988) Sensitization against the parasite antigen Sj26 is not sufficient for consistent expression of resistance to *Schistosoma japonicum* in mice. *Transactions of the Royal Society of Tropical Medicine and Hygiene* 82, 885–889.

Mohamed, M.M., Shalaby, K.A., LoVerde, P.T. and Karim, A.M. (1998) Characterization of Sm20.8, a member of a family of schistosome tegumental antigens. *Molecular and Biochemical Parasitology* 96, 15–25.

Moser, D., Tendler, M., Griffiths, G. and Klinkert, M.Q. (1991) A 14-kda *Schistosoma mansoni* polypeptide is homologous to a gene family of fatty acid binding proteins. *The Journal of Biological Chemistry* 266, 8447–8454.

Mountford, A.P., Anderson, S. and Wilson, R.A. (1996) Induction of Th1 cell-mediated protective immunity to *Schistosoma mansoni* by co-administration of larval antigens and IL-12 as an adjuvant. *The Journal of Immunology* 156, 4739–4745.

Murrell, K.D., Dean, D.A. and Stafford, E.E. (1975) Resistance to infection with *Schistosoma mansoni* after immunization with worm extracts or live cercariae: role of cytotoxic antibody in mice and guinea pigs. *The American Journal of Tropical Medicine and Hygiene* 24, 955–962.

Murrell, K.D., Clark, S.S., Dean, D.A. and Vannier, W.E. (1979) *Schistosoma mansoni*: immunization of cynomolgus monkeys by injection of irradiated schistosomula. *Experimental Parasitology* 48, 415–420.

Newport, G.R., McKerrow, J.H., Hedstrom, R., Petitt, M., McGarrigle, L., Barr, P.J. and Agabian, N. (1988) Cloning of the proteinase that facilitates infection by schistosome parasites. *The Journal of Biological Chemistry* 263, 13179–13184.

Omer Ali, P., Smithers, S.R., Bickle, Q., Phillips, S.M., Harn, D. and Simpson, A.J. (1988) Analysis of the anti-*Schistosoma mansoni* surface antibody response during murine infection and its potential contribution to protective immunity. *The Journal of Immunology* 140, 3273–3279.

Pearce, E.J., James, S.L., Dalton, J., Barrall, A., Ramos, C., Strand, M. and Sher, A. (1986) Immunochemical characterization and purification of Sm-97, a *Schistosoma mansoni* antigen monospecifically recognized by antibodies from mice protectively immunized with a nonliving vaccine. *The Journal of Immunology* 137, 3593–3600.

Pearce, E.J., James, S.L., Hieny, S., Lanar, D.E. and Sher, A. (1988) Induction of protective immunity against *Schistosoma mansoni* by vaccination with schistosome paramyosin (Sm97), a nonsurface parasite antigen. *Proceedings of the National Academy of Sciences USA* 85, 5678–5682.

Ramalho-Pinto, F.J. McLaren, D.J. and Smithers, S.R. (1978) Complement-mediated killing of schistosomula of *Schistosoma mansoni* by rat eosinophils *in vitro*. *The Journal of Experimental Medicine* 147, 147–156.

Ramirez, B.L., Kurtis, J.D., Wiest, P.M., Arias, P., Aligui, F., Acosta, L., Peters, P. and Olds, G.R. (1996) Paramyosin: a candidate vaccine antigen against *Schistosoma japonicum*. *Parasite Immunology* 18, 49–52.

Rao, K.V. and Ramaswamy, K. (2000) Cloning and expression of a gene encoding Sm16, an anti-inflammatory protein from *Schistosoma mansoni*. *Molecular and Biochemical Parasitology* 108, 101–108.

Rappuoli, R. and Covacci, A. (2003) Reverse vaccinology and genomics. *Science* 302, 602.

Reynolds, S.R., Shoemaker, C.B. and Harn, D.A. (1992) T and B cell epitope mapping of Sm23, an integral membrane protein of *Schistosoma mansoni*. *The Journal of Immunology* 149, 3995–4001.

Reynolds, S.R., Dahl, C.E. and Harn, D.A. (1994) T and B epitope determination and analysis of multiple antigenic peptides for the *Schistosoma mansoni* experimental vaccine triose-phosphate isomerase. *The Journal of Immunology* 152, 193–200.

Ribeiro de Jesus, A., Araujo, I., Bacellar, O., Magalhaes, A., Pearce, E., Harn, D., Strand, M. and Carvalho, E.M. (2000) Human immune responses to *Schistosoma mansoni* vaccine candidate antigens. *Infection and Immunity* 68, 2797–2803.

Roberts, S.M., Boot, C. and Wilson, R.A. (1988) Antibody responses of rodents to a tegument membrane preparation from adult *Schistosoma mansoni*. *Parasitology* 97, 425–435.

Sadun, E.H. and Lin, S.S. (1959) Studies on the host–parasite relationships to *Schistosoma japonicum*. IV. Resistance acquired by infection, by vaccination and by the injection of immune serum, in monkeys, rabbits and mice. *Journal of Parasitology* 45, 543–548.

Santiago, M.L., Hafalla, J.C., Kurtis, J.D., Aligui, G.L., Wiest, P.M., Olveda, R.M., Olds, G.R., Dunne, D.W. and Ramirez, B.L. (1998) Identification of the *Schistosoma japonicum* 22.6-kDa antigen as a major target of the human IgE response: similarity of IgE-binding epitopes to allergen peptides. *International Archives of Allergy and Immunology* 117, 94–104.

Scott, J.C. and McManus, D.P. (2000) Molecular cloning and enzymatic expression of the 28-kDa glutathione S-transferase of *Schistosoma japonicum*: evidence for sequence variation but lack of consistent vaccine efficacy in the murine host. *Parasitology International* 49, 289–300.

Scott, J.C., Kennedy, M.W. and McManus, D.P. (2000) Molecular and immunological characterisation of a polymorphic cytosolic fatty acid binding protein from the human blood fluke *Schistosoma japonicum*. *Biochimica et Biophysica Acta* 1517, 53–62.

Sher, A., Pearce, E., Hieny, S. and James, S. (1986) Induction of protective immunity against *Schistosoma mansoni* by a nonliving vaccine. IV. Fractionation and antigenic properties of a soluble adult worm immunoprophylactic activity. *The Journal of Immunology* 136, 3878–3883.

Shi, F., Zhang, Y., Ye, P., Lin, J., Cai, Y., Shen, W., Bickle, Q.D. and Taylor, M.G. (2001) Laboratory and field evaluation of *Schistosoma japonicum* DNA vaccines in sheep and water buffalo in China. *Vaccine* 20, 462–467.

Shi, F., Zhang, Y., Lin, J., Zuo, X., Shen, W., Cai, Y., Ye, P., Bickle, Q.D. and Taylor, M.G. (2002) Field testing of *Schistosoma japonicum* DNA vaccines in cattle in China. *Vaccine* 20, 3629–3631.

Siddiqui, A.A., Phillips, T., Charest, H., Podesta, R.B., Quinlin, M.L., Pinkston, J.R., Lloyd, J.D., Paz, M., Villalovos, R.M. and Pompa, J. (2003a) Induction of protective immunity against *Schistosoma mansoni* via DNA priming and boosting with the large subunit of calpain (Sm-p80): adjuvant effects of granulocyte-macrophage colony-stimulating factor and interleukin-4. *Infection and Immunity* 71, 3844–3851.

Siddiqui, A.A., Phillips, T., Charest, H., Podesta, R.B., Quinlin, M.L., Pinkston, J.R., Lloyd, J.D., Pompa, J., Villalovos, R.M. and Paz, M. (2003b) Enhancement of Sm-p80 (large subunit of calpain) induced protective immunity against *Schistosoma mansoni* through co-delivery of interleukin-2 and interleukin-12 in a DNA vaccine formulation. *Vaccine* 21, 2882–2889.

Skelly, P.J. (2004) Intravascular schistosomes and complement. *Trends in Parasitology* 20, 370–374.

Skelly, P.J. and Shoemaker, C.B. (1996) Rapid appearance and asymmetric distribution of glucose transporter SGTP4 at the apical surface of intramammalian-stage *Schistosoma mansoni*. *Proceedings of the National Academy of Sciences USA* 93, 3642–3646.

Smith, D.B., Davern, K.M., Board, P.G., Tiu, W.U., Garcia, E.G. and Mitchell, G.F. (1986) Mr 26,000 antigen of *Schistosoma japonicum* recognized by resistant WEHI 129/j mice is a parasite glutathione S-transferase. *Proceedings of the National Academy of Sciences USA* 83, 8703–8707.

Smithers, S.R. (1987) Vaccination against schistosomes and other systemic helminths. *International Journal for Parasitology* 17, 31–42.

Smithers, S.R., Hackett, F., Ali, P.O. and Simpson, A.J. (1989) Protective immunization of mice against *Schistosoma mansoni* with purified adult worm surface membranes. *Parasite Immunology* 11, 301–318.

Smythies, L.E., Coulson, P.S. and Wilson, R.A. (1992) Monoclonal antibody to Interferon-gamma modifies pulmonary inflammatory responses and abrogates immunity to *Schistosoma mansoni* in mice vaccinated with attenuated cercariae. *The Journal of Immunology* 149, 3654–3658.

Soisson, L.M., Masterson, C.P., Tom, T.D., McNally, M.T., Lowell, G.H. and Strand, M. (1992) Induction of protective immunity in mice using a 62-kDa recombinant fragment of a *Schistosoma mansoni* surface antigen. *The Journal of Immunology* 149, 3612–3620.

Soisson, L.A., Reid, G.D., Farah, I.O., Nyindo, M. and Strand, M. (1993) Protective immunity in baboons vaccinated with a recombinant antigen or radiation-attenuated cercariae of *Schistosoma mansoni* is antibody-dependent. *The Journal of Immunology* 151, 4782–4789.

Street, M., Coulson, P.S., Sadler, C., Warnock, L.J., McLaughlin, D., Bluethmann, H. and Wilson, R.A. (1999) TNF is essential for the cell-mediated protective immunity induced by the radiation-attenuated schistosome vaccine. *The Journal of Immunology* 163, 4489–4494.

Sun, J.B., Mielcarek, N., Lakew, M., Grzych, J.M., Capron, A., Holmgren, J. and Czerkinsky, C. (1999) Intranasal administration of a *Schistosoma mansoni* glutathione S-transferase-cholera toxoid conjugate vaccine evokes antiparasitic and antipathological immunity in mice. *The Journal of Immunology* 163, 1045–1052.

Tarrab-Hazdai, R., Schechtman, D. and Arnon, R. (1998) Synthesis and characterization of a protective peptide-based vaccine against *Schistosoma mansoni*. *Infection and Immunity* 66, 4526–4530.

Tarrab-Hazdai, R., Schechtman, D., Lowell, G., Pirak, E. and Arnon, R. (1999) Proteosome delivery of a protective 9b-antigen against *Schistosoma mansoni*. *International Journal of Immunopharmacology* 21, 205–218.

Taylor, J.B., Vidal, A., Torpier, G., Meyer, D.J., Roitsch, C., Balloul, J.M., Southan, C., Sondermeyer, P., Pemble, S., Lecocq, J.P., Capron, A. and Ketterer, B. (1988) The glutathione transferase activity and tissue distribution of a cloned mr28k protective antigen of *Schistosoma mansoni*. *EMBO Journal* 7, 465–472.

Taylor, M.G., Huggins, M.C., Shi, F., Lin, J., Tian, E., Ye, P., Shen, W., Qian, C.G., Lin, B.F. and Bickle, Q.D. (1998) Production and testing of *Schistosoma japonicum* candidate vaccine antigens in the natural ovine host. *Vaccine* 16, 1290–1298.

Tendler, M., Pinto, R.M., Oliveira Lima, A., Gebara, G. and Katz, N. (1986) *Schistosoma mansoni*: vaccination with adult worm antigens. *International Journal for Parasitology* 16, 347–352.

Tendler, M., Pinto, R.M., Lima Ade, O., Savino, W. and Katz, N. (1991) Vaccination in murine schistosomiasis with adult worm-derived antigens: variables influencing protection in outbred mice. *International Journal for Parasitology* 21, 299–306.

Tendler, M., Brito, C.A., Vilar, M.M., Serra-Freire, N., Diogo, C.M., Almeida, M.S., Delbem, A.C., Da Silva, J.F., Savino, W., Garratt, R.C., Katz, N. and Simpson, A.S. (1996) A *Schistosoma mansoni* fatty acid-binding protein, Sm14, is the potential basis of a dual-purpose anti-helminth vaccine. *Proceedings of the National Academy of Sciences USA* 93, 269–273.

Tiu, W.U., Ehl, A., Walker, J.C., Garcia, E.G., Smith, D.B. and Mitchell, G.F. (1986) Resistance of 129/j mice to *Schistosoma mansoni* infection. *Australian Journal of Experimental Biology and Medical Science* 64, 345–349.

van der Werf, M.J., de Vlas, S.J., Brooker, S., Looman, C.W., Nagelkerke, N.J., Habbema, J.D. and Engels, D. (2003) Quantification of clinical morbidity associated with schistosome infection in sub-Saharan Africa. *Acta Tropica* 86, 125–139.

Verity, C.K., McManus, D.P. and Brindley, P.J. (2001) Vaccine efficacy of recombinant cathepsin D aspartic protease from *Schistosoma japonicum*. *Parasite Immunology* 23, 153–162.

Verjovski-Almeida, S., DeMarco, R., Martins, E.A., Guimaraes, P.E., Ojopi, E.P., Paquola, A.C., Piazza, J.P., Nishiyama, M.Y. Jr, Kitajima, J.P., Adamson, R.E., Ashton, P.D., Bonaldo, M.F., Coulson, P.S., Dillon, G.P., Farias, L.P., Gregorio, S.P., Ho, P.L., Leite, R.A., Malaquias, L.C., Marques, R.C., Miyasato, P.A., Nascimento, A.L., Ohlweiler, F.P., Reis, E.M., Ribeiro, M.A., Sa, R.G., Stukart, G.C., Soares, M.B., Gargioni, C., Kawano, T., Rodrigues, V., Madeira, A.M., Wilson, R.A., Menck, C.F., Setubal, J.C., Leite, L.C. and Dias-Neto, E. (2003) Transcriptome analysis of the acoelomate human parasite *Schistosoma mansoni*. *Nature Genetics* 35, 148–157.

Verwaerde, C., Joseph, M., Capron, M., Pierce, R.J., Damonneville, M., Velge, F., Auriault, C. and Capron, A. (1987) Functional properties of a rat monoclonal IgE antibody specific for *Schistosoma mansoni*. *The Journal of Immunology* 138, 4441–4446.

Waine, G.J., Alarcon, J.B., Qiu, C. and McManus, D.P. (1999) Genetic immunization of mice with DNA encoding the 23 kDa transmembrane surface protein of *Schistosoma japonicum* (Sj23) induces antigen-specific immunoglobulin G antibodies. *Parasite Immunology* 21, 377–381.

Wheater, P.R. and Wilson, R.A. (1979) *Schistosoma mansoni*: a histological study of migration in the laboratory mouse. *Parasitology* 79, 49–62.

Wilson, R.A. and Barnes, P.E. (1977) The formation and turnover of the membranocalyx on the tegument of *Schistosoma mansoni*. *Parasitology* 74, 61–71.

Wilson, R.A. and Coulson, P.S. (1998) Why don't we have a schistosomiasis vaccine? *Parasitology Today* 14, 97–99.

Wilson, R.A. and Coulson, P.S. (1999) Strategies for a schistosome vaccine: can we manipulate the immune response effectively? *Microbes and Infection* 1, 535–543.

Wilson, R.A., Coulson, P.S. and McHugh, S.M. (1983) A significant part of the 'concomitant immunity' of mice to *Schistosoma mansoni* is the consequence of a leaky hepatic portal system, not immune killing. *Parasite Immunology* 5, 595–601.

Wilson, R.A., Coulson, P.S. and Mountford, A.P. (1999) Immune responses to the radiation-attenuated schistosome vaccine: what can we learn from knock-out mice? *Immunological Letters* 65, 117–123.

Wolowczuk, I., Auriault, C., Bossus, M., Boulanger, D., Gras-Masse, H., Mazingue, C., Pierce, R.J., Grezel, D., Reid, G.D. and Tartar, A. (1991) Antigenicity and immunogenicity of a multiple peptidic construction of the *Schistosoma mansoni* Sm-28 GST antigen in rat, mouse, and monkey. 1. Partial protection of Fischer rat after active immunization. *The Journal of Immunology* 146, 1987–1995.

Wynn, T.A., Jankovic, D., Hieny, S., Cheever, A.W., Sher, A. (1995) IL-12 enhances vaccine-induced immunity to Schistosoma mansoni in mice and decreases T helper 2 cytokine expression, IgE production, and tissue eosinophilia. *The Journal of Immunology* 154, 4701–4709.

Xu, S., Shi, F. and Wu, H. (1991) *Schistosoma japonicum*: some parameters affecting the development of protective immunity induced by a cryopreserved, irradiated schistosomula vaccine in guinea-pigs. *Parasitology* 102, 45–47.

Xu, C.B., Verwaerde, C., Gras-Masse, H., Fontaine, J., Bossus, M., Trottein, F., Wolowczuk, I., Tartar, A. and Capron, A. (1993) *Schistosoma mansoni* 28-kDa glutathione S-transferase and immunity against parasite fecundity and egg viability. Role of the amino- and carboxyl-terminal domains. *The Journal of Immunology* 150, 940–949.

Xu, S., Shi, F., Shen, W., Lin, J., Wang, Y., Ye, P., Tian, E., Qian, C., Lin, B., Shi, Y. and Zhang, Z. (1995) Vaccination of sheep against *Schistosoma japonicum* with either glutathione S-transferase, keyhole limpet haemocyanin or the freeze/thaw schistosomula/BCG vaccine. *Veterinary Parasitology* 58, 301–312.

Yang, W., Jackson, D.C., Zeng, Q. and McManus, D.P. (2000) Multi-epitope schistosome vaccine candidates tested for protective immunogenicity in mice. *Vaccine* 19, 103–113.

Yi, X.Y., Omer-Ali, P., Kelly, C., Simpson, A.J. and Smithers, S.R. (1986) IgM antibodies recognizing carbohydrate epitopes shared between schistosomula and miracidia of *Schistosoma mansoni* that block *in vitro* killing. *The Journal of Immunology* 137, 3946–3954.

Yole, D.S., Pemberton, R., Reid, G.D. and Wilson, R.A. (1996) Protective immunity to *Schistosoma mansoni* induced in the olive baboon *Papio anubis* by the irradiated cercaria vaccine. *Parasitology* 112, 37–46.

Zhang, Y., Taylor, M.G. and Bickle, Q.D. (1998) *Schistosoma japonicum* myosin: cloning, expression and vaccination studies with the homologue of the *S. mansoni* myosin fragment IrV-5. *Parasite Immunology* 20, 583–594.

Zhang, Y.Y., Taylor, M.G., Gregoriadis, G., McCrossan, M.V. and Bickle, Q.D. (2000) Immunogenicity of plasmid DNA encoding the 62 kDa fragment of *Schistosoma japonicum* myosin. *Vaccine* 18, 2102–2109.

Zhang, Y., Taylor, M.G., Johansen, M.V. and Bickle, Q.D. (2001a) Vaccination of mice with a cocktail DNA vaccine induces a Th1-type immune response and partial protection against *Schistosoma japonicum* infection. *Vaccine* 20, 724–730.

Zhang, R., Yoshida, A., Kumagai, T., Kawaguchi, H., Maruyama, H., Suzuki, T., Itoh, M., El-Malky, M. and Ohta, N. (2001b) Vaccination with calpain induces a Th1-biased protective immune response against *Schistosoma japonicum*. *Infection and Immunity* 69, 386–391.

Zhou, S., Liu, S., Song, G., Xu, Y. and Sun, W. (2000) Protective immunity induced by the full-length cDNA encoding paramyosin of Chinese *Schistosoma japonicum*. *Vaccine* 18, 3196–3204.

Zhu, Y., Si, J., Ham, D.A., Yu, C., He, W., Hua, W., Yin, X., Liang, Y., Xu, M. and Xu, R. (2002) The protective immunity produced in infected C57BL/6 mice of a DNA vaccine encoding *Schistosoma japonicum* Chinese strain triose-phosphate isomerase. *Southeast Asian Journal of Tropical Medicine and Public Health* 33, 207–213.

Zhu, Y., Ren, J., Harn, D.A., Si, J., Yu, C., Ming, X. and Liang, Y. (2003) Protective immunity induced with 23 kDa membrane protein DNA vaccine of *Schistosoma japonicum* Chinese strain in infected C57BL/6 mice. *Southeast Asian Journal of Tropical Medicine and Public Health* 34, 697–701.

17 Flatworm Parasite Proteomics

R.M. Morphew, J. Barrett and P.M. Brophy
Parasitology Research Group, Institute of Biological Sciences, University of Wales, Aberystwyth, Wales SY23 3DA, UK

Introduction

Genomics is providing a wealth of bioinformatic data on the protein products predicted from gene sequences. To date, the majority of these products have not been assigned a function (Blackstock and Weir, 1999). The number of predicted proteins does not always correlate with the overall complexity of an organism (Van Regenmortel, 2001). Furthermore, genes are switched on and off, and proteins are post-translationally modified during development, and in response to cues from environment such as pathogens, drugs and toxins. Genomics has provided the blueprint for functional genomics studies such as transcriptomics, proteomics and metabolomics.

Proteomics is the 'omics' tool that separates and identifies individual proteins from an initial protein mixture. Proteomics is termed global if it identifies proteins from total preparations, or targeted if, for example, a protein superfamily or protein subset is investigated (Blackstock and Weir, 1999; Pandey and Mann, 2000).

Cell map proteomics reveals the static proteome of a whole organism, tissue, cell or organelle, while expression proteomics investigates changes in a proteome to cellular cues (Godovac-Zimmermann and Brown, 2001). Functional and structural proteomics refers to the investigation of individual proteins such as interactions with ligands.

Identification of proteins using proteomic approaches is usually straightforward if the genome of the organism has been resolved, as proteins can be routinely identified by peptide mass fingerprinting (PMF) against the target blueprint theoretical protein database (see section entitled 'Mass Spectrometry to Unravel the Flatworm Proteome'). Proteomics is more problematic if the target database consists of an expressed sequence tag (EST) collection, individual protein entries or proteins from a related organism. Without a verified genome database, peptide sequencing is often needed to support the identity of the proteins under investigation.

Proteomics provides a powerful new tool in basic and applied parasitology. Reproducible two-dimensional electrophoresis (2DE), mass spectrometry and bioinformatics are becoming commonplace in modern protozoan and helminth parasitology research and teaching laboratories. Proteomics has moved from reviews suggesting the power of the technique to the publication of key experimental papers driving forward our understanding of the host–parasite relationship. Novel proteins have been identified on the surface of malaria-infected erythrocytes (Florens *et al.*, 2004) and a detailed proteomic analysis of the response of *Plasmodium falciparum* to drug stress and cell-cycle changes has been completed (Nirmalan *et al.*, 2004). Proteomics has also uncovered novel proteins in the differentiation of *Leishmania mexicana* (Nugent *et al.*, 2004) and in developmental stages of *Trypanosoma cruzi* (Paba *et al.*, 2004). *Toxoplasma gondii* is also undergoing detailed proteomic analysis (Cohen *et al.*, 2002).

Helminth-based proteomics has also revealed new aspects of the host–parasite relationship, with experimental studies completed and published in parasitic nematodes such as *Haemonchus contortus* (Yatsuda *et al.*, 2003), the cestode *Echinococcus granulosus* (Chemale *et al.*, 2003) and in the parasitic flatworm, *Fasciola hepatica* (Jefferies *et al.*, 2001; Bernal *et al.*, 2004). In addition, the cytosolic proteome of *F. hepatica* has been compared to that of its free-living relative, *Polycelis nigra* to investigate parasite-specific proteins. Both species were originally placed in the same genus of *Fasciola*, such that comparisons could reveal clues to the evolutionary history of the flatworms (Fig. 17.1) (Morphew, 2004, unpublished).

The progress of the *Schistosoma mansoni* genome project, which aims for 95% gene discovery, will provide a platform for routine PMF-based proteomics (Degrave *et al.*, 2001; LoVerde *et al.*, 2004). In addition, there are over 16,000 *S. mansoni* ESTs (LoVerde *et al.*, 2004) and currently over 2000 *F. hepatica* ESTs (D.A. Johnston, London, 2004, personal communication) available for proteomics. These two high-profile gene sequencing initiatives of parasitic flatworms of medical and veterinary importance will also improve our understanding of flatworm evolution. Nevertheless, the vast majority of parasitic flatworms will never have the benefit of being genome-verified or part of an EST programme. Consequently, proteomics investigations may be viewed as being of limited value. However, proteins from non-sequenced flatworms can still be displayed on 2DE arrays, assigned a mass and p*I* fingerprint, and, although not as routine as PMF, protein information can be obtained by direct peptide sequencing.

Parasitic Flatworm Sample Preparation

Parasitic flatworms pose several problematic areas when preparing samples for proteomic analysis. For example, parasites of target species

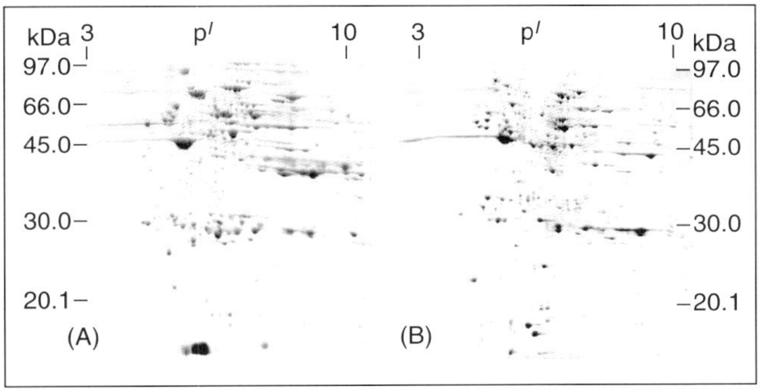

Fig. 17.1. Parasites cannot hide their spots: Two-dimensional electrophoresis gel of the cytosolic fractions of (A) ovine *F. hepatica* and (B) of a free-living close relative *P. nigra* run on 17 cm IPG 3–10 and 12.5% SDS–polyacrylamide gels, Coomassie blue stained. The ability to run comparable 2DE arrays will highlight parasite-specific proteins essential to the parasitic lifestyle (Morphew, 2004, unpublished).

are often in limited availability, flatworms can often have a high concentration of proteases that hinder successful protein array and parasites have to be physically extracted from their normal host environment for analysis.

Protein collection

Proteomics in parasitic flatworms can be completed on intracellular fractions (e.g. microsomal or cytosol) or at the host-interface on excretory–secretory (ES) products. ES analysis can be completed during *in vitro* culture or *in vivo* by, for example, bile or gut content analysis. In all cases, a rapid and careful preparation is vital to prevent altered proteomic profiles due to stress responses (upregulation of heat shock proteins) and action of proteases. Parasitic flatworms are best extracted from fresh host material, washed with a buffered saline solution at approximately the host's body temperature. In *F. hepatica*, for example, this will allow regurgitation of gut contents to remove digested material from, and removal of host material adherent to the outer surfaces of the parasite (Jefferies *et al.*, 2001), both of which can subsequently complicate separation and identification.

Initial proteomic studies in parasitic flatworms have been undertaken using pooled samples. However, proteomic maps can be derived from individual *F. hepatica* worms (Fig. 17.2), although individual variation must be taken into account. The ability to work on individuals allows, for example, the opportunity to more easily observe key cue responses, and assess the overall importance of this individual protein response within populations.

Proteins needed for the survival of flatworms and for host recognition are found as ES products (Sandeman and Howell, 198`). Collection of ES products is usually via *in vitro* culturing systems, which aim to mimic the environmental conditions of the host with respect to pH, oxygen tension, osmotic pressure, temperature and waste removal. By this general *in vitro* approach, ES products from *F. hepatica* were successfully subjected to proteomic analysis by Jefferies *et al.* (2000). Survival for short periods of time outside the host, in a defined system, cannot equate to the complexities of normal physiology or protein expression, and histological and cytological examination should ideally be carried out for culture performance over the time period (usually 2–6 h). However, *in vivo* proteomic analysis of parasitic flatworm ES products is feasible as mass spectrometry can selectively identify host and parasite proteins, respectively. *F. hepatica* cathepsins (cathepsin L-type) have been identified in the bile of *F. hepatica* infected sheep (Fig. 17.3) along with host proteins such as *Ovis aries* carbonic anhydrase II (Morphew, unpublished). In addition, ES products from the cestode *E. granulosus* were removed from hydatid cysts of slaughtered livestock livers

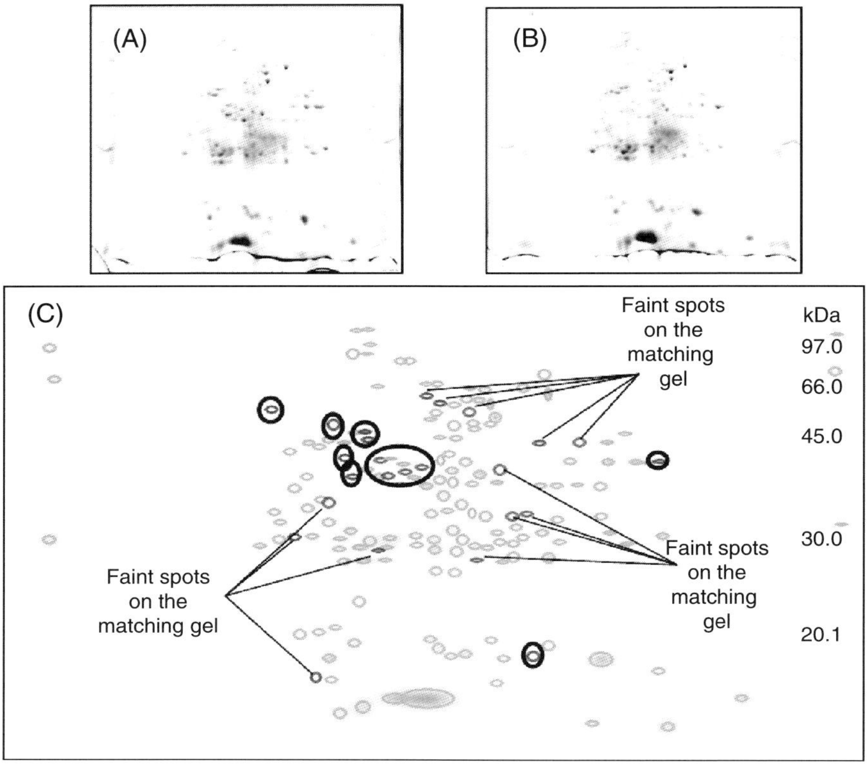

Fig. 17.2. (A) and (B) Two-dimensional electrophoreogram of a cytosol preparation of two single *F. hepatica* (Fluke A and B, respectively) on IPG pH 3–10 and 12.5% SDS–polyacrylamide gel, Coomassie blue stained. (C) Graphic overlay of both single *F. hepatica* electrophoreograms via PDQuest® (Biorad) gel analysis software. Light grey circles represent protein spots present in both trematodes. Discrepancies between both trematodes are indicated, with spots present on both (although not detected by the image software) labelled as faint. Protein spots highlighted via dark black circles represent real differences between the 2DE profiles of the two trematodes highlighting the reproducibility of the technique and indicating natural variation between individual organisms (Morphew, 2004, unpublished data).

and directly analysed without *in vitro* culture by proteomics (Chemale *et al*., 2003).

Solubilization of proteins

As in other organisms, protein extracts derived from flatworms can be obtained by the use of solubilization/denaturation (SD) buffers. These SD buffers are designed to break the interactions among proteins and interfering substances. More importantly they are designed to prevent proteins from re-interacting with one another (Rabilloud, 1996). SD buffers generally consist of chaotrophes, detergents, reductants, ampholytes and protease inhibitors, although Tris buffer may also be added.

Urea, ranging from 5 to 9 M, is the most common choice of chaotropic agent, and disruption of hydrogen and hydrophobic bonding between proteins is the main mode of action. When used in conjunction with thiourea (from 0 to 2 M), the two can aid solubilization of poorly soluble samples. The use of urea or urea–thiourea mixtures in SD buffers has no effect on the charge of the proteins within the sample. This is ideal for use in separating proteins according to their p^{I} in the first

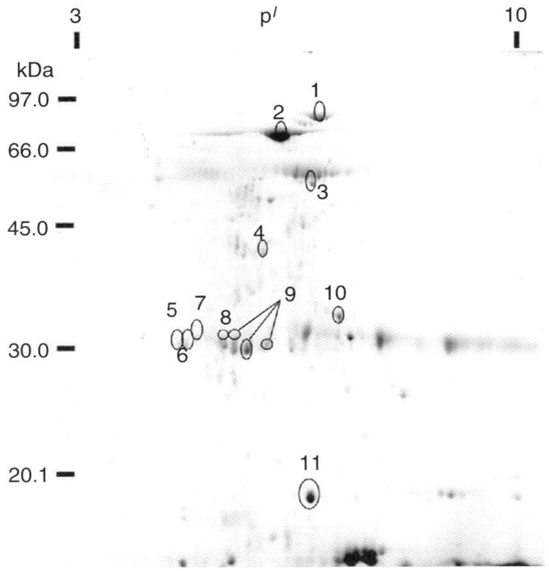

Spot no.	Mowse score	% coverage	M_r	p*I*	Accession no.	Species	Description
1	2.01E+06	24	77753.9	6.75	NP_803450	*B. taurus*	Transferin
2	6.52E+12	49	69188.9	5.8	P14639	*O. aries*	Serum albumin precursor
3	4.63E+10	42	47277.4	6.44	NP_776474	*B. taurus*	Enolase 1
4	1.33E+07	45	33308.1	5.54	NP_776382	*B. taurus*	Regucalcin (senescence marker protein-30)
5	6.10E+05	42	37158.8	5.43	T09259	*F. hepatica*	Cathepsin L-like proteinase
6	1.33E+05	38	37033.7	5.54	AAC47721.1	*F. hepatica*	Secreted Cathepsin L2
7	9.12E+03	23	37149.0	5.55	AAF76330.1	*F. hepatica*	Cathepsin L
8	9.48E+02	26	35234.7	5.79	AAK38169.1	*F. hepatica*	Cathepsin L-like
9	1.05E+04	14–21	35190.5	5.79	AAM111647.1	*F. hepatica*	Cathepsin L
10	3.86E+05	39	29079.9	6.40	P00922	*O. aries*	Carbonic anhydrase II (Carbonate dehydratase II)(CA- I)
11	3.25E+05	65	15563.6	6.16	P09670	*O. aries*	Superoxide dismutase [Cu–Zn]

Fig. 17.3. The infected bile-ome: Two-dimensional electrophoresis gel of *F. hepatica* infected host bile. Run on a 17 cm, pH 3–10, IPG strip in the first dimension and then on 12.5% SDS–PAGE gel in the second dimension with Coomassie blue staining. Host and parasite proteins were identified via their peptide mass fingerprints (Morphew, 2004, unpublished). Valid parasite secreted proteins can only come from *in vivo* proteomics.

dimension. Crystal formation between urea and thiourea can cause blank streaking in the second dimension if handled incorrectly.

The addition of zwitterionic or non-ionic detergents to SD buffers has three important functions. Detergents aim to disrupt membranes and consequently solubilize lipids and delipidate proteins, which are bound to membranes or vesicles. The use of SDS as a detergent will affect the overall net charge and so non-ionic or zwitterionic detergents such as Triton X-100 or CHAPS are used (Shaw and Riederer, 2003). Commonly used CHAPS, which will not affect the overall charge, can be added up to 4% in SD buffers to account for varying sample solubilities. Caprylyl sulphobetaine (SB3-10), belonging to the same class of linear sulphobetaine surfactants as CHAPS, has also provided efficient success for protein separation (Shaw and Riederer, 2003). SB3-10 should not be used in excess of 2%. An important consideration for all detergents is their solubility within urea, the main component of SD buffers, even at high concentrations. Luche *et al.* (2003) have since evaluated several non-ionic and zwitterionic detergents for 2DE analysis.

Reductants provide a means to fully denature proteins within the sample and aid solubilization. To fully break disulphide bonds, reducing conditions with an excess of free thiols are needed. Thiols are the common choice for reductants in SD buffers with dithiothreitol (DTT) generally used. DTT and dithioerythitol (DTE) are effective at producing free thiols and can be used in low concentrations (less than 100 mM) (Rabilloud, 1996). The phosphine tris(2-carboxyethyl) phosphine (TCEP) can be used as a substitute for DTT or DTE but is very acidic. All three migrate to the anode during first dimension separation causing depletion in the basic end of the gel (Shaw and Riederer, 2003).

To aid solubilization, carrier ampholytes are added to the SD buffer. Ampholytes must be added according to the pH range of the first dimension separation (for separation over a pH range of 3 to 10, an ampholyte mix from pH 3 to 10 should be used). Ampholytes are also beneficial in soaking up loose cyanate ions, assisting precipitation of nucleic acids during centrifugation and will inhibit interactions between hydrophobic proteins and immobilines at the basic end of pH gradient strips. Generally concentrations of ampholytes used range from 0.5% to 2%, although excessive use will slow separation (Shaw and Riederer, 2003).

The solubility of proteins from parasitic organisms can often be enhanced by the use of hot sodium dodecyl sulphate (SDS) before solubilization in SD buffers. The ability of SDS to denature proteins also helps in allowing access to hydrophobic proteins, not normally seen after standard preparation procedures. SDS treatment in whole *F. hepatica* preparations has been shown to yield more protein spots visualized on gel analysis than other methods (Jefferies *et al.*, 2000).

Removal of contaminating substances

Many components of whole organisms can interfere with the proteomic process and must be removed to allow good protein separation. Insoluble flatworm proteins have been removed by simple centrifugation (Chemale *et al.*, 2003), reducing the complexity of the sample. Levels of protease activity are generally high within parasitic flatworms and can cause degradation of the collected samples. General buffers for proteomic work are usually sufficient to reduce their activity. However, proteomic buffers may also increase protease activity via protein denaturation. This opens proteins to proteolytic action, so protease inhibitors should be added in all situations. Protease inhibitors are now regularly supplied as a cocktail of many in tablet form. The presence of active proteases is often characterized by a loss of high molecular weight proteins and general smearing of the gel background (Gorg *et al.*, 2000).

Nucleic acids need adequate removal from the sample to prevent binding with proteins and carrier ampholytes in the buffer. If this occurs extensive streaking in the second dimension of electrophoresis is observed (Rabilloud, 1996). Further complications occur with gel pore blocking arising from nucleotide protein binding. To remove these problems, simple ultracentrifugation can be applied to the sample to remove nucleotides based on their densities. In addition, nucleases can be added to digest the nucleotides into smaller oligonucleotides. With this approach, extraneous protein is added and must be taken into account.

Large polysaccharides with low densities can also be removed in conjunction with nucleotide removal by ultracentrifugation as they float. If left, these large molecules have the potential to block the pores of polyacrylamide gels. Other polysaccharide contaminants may need to be removed by the use of SDS or high pH ranges (Rabilloud, 1996). Further to removal of nucleotides and polysaccharides, lipids also need to be eliminated from the sample. This can be done via the use of organic chemicals, most commonly chlorinated, providing strongly denaturing conditions (Rabilloud, 1996), or by the use of centrifugation followed by filtration through glass fibre. The removal of a lipid layer after centrifugation using glass fibre has been effectively shown by Brophy *et al.* (1989), when preparing cytosolic fractions from platyhelminths.

Low concentrations of proteins found in ES products are further complicated by high concentrations of salts. Salt contamination directly interferes with the electrophoresis process. The movement of salt through the first dimension to the extremities of the pH gradient will prolong the time needed to separate proteins and in many cases will draw water with it. This action causes dehydrated regions in the gel pH gradient and in extreme cases can disrupt the gel matrix preventing protein separation. Dialysis of samples can prevent this occurring, although a small loss of proteins may be seen due to the adherence to the membrane surface (Rabilloud, 1996). Centrifugal filters can then be used to concentrate samples before 2DE (Jefferies *et al.*, 2000).

A further suggestion when purifying proteins is precipitation of the soluble fraction, most commonly but not exclusively with trichloroacetic acid (TCA) or ammonium sulphate $(NH_4)_2SO_4$. Samples are then re-suspended in sample buffer ready for 2DE. This has the advantage of not only concentrating and desalting the proteins but will aid sample preparation by removing non-protein materials such as polysaccharides and salts. Pridmore *et al.* (1999) also showed the ability of TCA precipitation to denature endogenous proteases. However, it is important to remember that some proteins may be lost or severely depleted from the sample after TCA precipitation, making quantitative work unreliable (Barrett *et al.*, 2000; Jefferies *et al.*, 2000). Following TCA treatment, proteins are often difficult to re-suspend.

TCA precipitation in the proteomics of *E. granulosus* by Chemale *et al.* (2003) was effective in whole organism preparation, providing superior resolution. In contrast to whole sample preparation using TCA, *E. granulosus* ES products were not improved. However TCA was shown to be very useful in ES products of *F. hepatica* (Jefferies *et al.*, 2001). The addition of phosphotungstic acid (PTA) to TCA precipitation suggested by Yeang *et al.* (1998) improves yields and recovery of proteins. Jiang *et al.* (2004) showed a comparison between four commonly used precipitation methods. Along with TCA precipitation the use of acetone, chloroform/methanol and ammonium sulphate was investigated on human plasma sample preparation. Precipitation with TCA and acetone provided a more efficient sample clean up. However, ammonium sulphate in increasing concentrations provided an excellent method of fractionating a sample, and has the advantage of removing interfering proteins.

Protein Separation before Two-dimensional Electrophoresis

Low-abundance proteins may be visualized more efficiently by increasing the protein load. However, if samples are prepared for a global profile, overcrowding will be seen when the gel is visualized if too much protein is present. Therefore separation before electrophoretic analysis may be required to produce adequate resolution and separation. Overcrowding can be dealt with to some degree via the use of 'zoom' gels. Narrow pH ranges used in the first dimension provide a small section of a global profile with greater resolution of protein spots. Common pH ranges for zoom gels vary, e g. pH 4–7 and pH 5–8. When several are used together, a well-resolved global profile can be achieved.

Along with precipitation with ammonium sulphate to fractionate proteins (Jiang *et al.*, 2004), Jefferies *et al.* (2000) described differential solubilization of samples, using progressively stronger conditions for denaturing of a *F. hepatica* sample and finally with SDS. This allowed enrichment and separation of sample proteins with only a few proteins left in the SDS fraction. This suggests that the remaining protein in the sample provides difficulties for the 2DE process. The ability to fractionate these samples (into soluble proteins, cytosolic proteins, membrane-associated proteins, etc.) drastically reduces the sample complexity. The soluble proteins of *S. mansoni* have already been looked at in this manner allowing an in-depth comparison between life stages (Curwen *et al.*, 2004).

Affinity chromatography (AC) is a fractionation technique widely used in targeted proteomics or protein–protein interaction approaches. This method utilizes an interaction or affinity of a target protein to a substrate (or another protein) immobilized on a support matrix. Figure 17.4 demonstrates that AC can be incorporated into 3D proteomic approaches (Lee and Lee, 2004). Traditionally, AC has been used to purify, for example, carbohydrate-binding proteins from *Diplostomum pseudospathaceum* (Mikes and Man, 2003) and the detoxification superfamily glutathione transferase (GST) from parasitic flatworms (Brophy and Barrett, 1990), specifically shown in

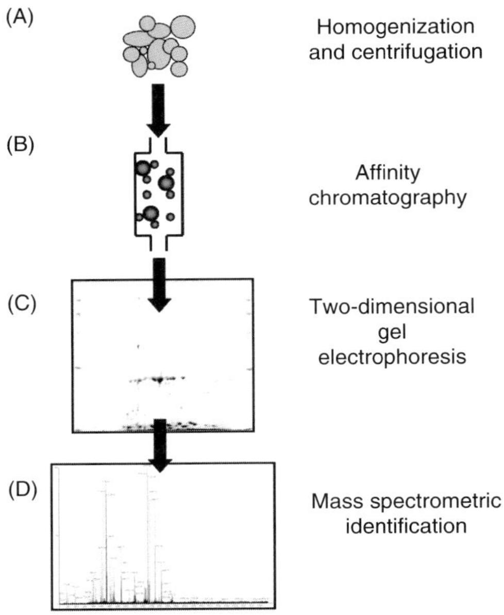

Fig. 17.4. Traditional workflow of two-dimensional electrophoresis and mass spectrometric proteomics with the addition of affinity chromatography. (A) Samples are homogenized and centrifuged to remove insoluble material. (B) Samples are applied to and eluted from an affinity column. (C) The purified fractions are run by standard two-dimensional electrophoresis. (D) Target proteins can then be excised and identified.

F. hepatica (Fig. 17.5). To date, 3D proteomic approaches have characterized the expressed GST-ome in the nematode *Caenorhabditis elegans* (Van Rossum *et al.*, 2001) and the adaptability of GST–protein interaction networks in *C. elegans* under reactive oxygen species stress (Greetham *et al.*, 2004).

Separating Proteins by Two-dimensional Protein Electrophoresis

2DE is a method designed to allow reproducible separation of proteins as an array of spots based on charge and mass (Gorg *et al.*, 2000), as first described by O'Farrell (1975). Gorg *et al.* (2000) provided a comprehensive account of the development of current 2DE analysis. However, 2DE is not a perfect technique, distortions of individual protein spots will occur during casting, polymerization and running of gels, hence the importance of replicates and gel matching analysis.

Isoelectric focusing

The key to the first dimension, isoelectric focusing (IEF), is sample preparation, as IEF

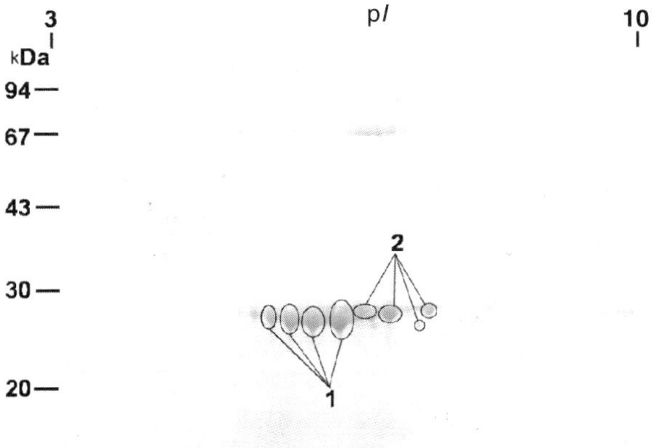

Spot number	Coverage (%)	Mw	p*I*	Accession number	Description
1	31–46	25373.6	5.89	P30112	GST class α
2	46–49	24651.7	6.27	AAA29140	GST class α

Fig. 17.5. The GST-ome: Two-dimensional electrophoresis gel of affinity-purified GSTs from *F. hepatica* ES products run on a 17 cm, pH 3–10, IPG strip in the first dimension and then on 12–14% SDS–PAGE gel in the second dimension with Coomassie blue staining (Jefferies, Morphew and Brophy, 2000, unpublished).

will only be successful if the preparation is salt-free and the sample is in solution. IEF (Fig. 17.6) separates proteins according to their isoelectric point (p*I*). Carrier ampholytes were originally used to create pH gradients, typically pH 3–10 for global 2DE gels. Advances in IEF technology have created pH gradients on plastic backed strips. These immobilized pH gradients (IPGs) allow IEF to be performed without the operator physically making a gel. With this pH gradient firmly fixed (reducing drifting of proteins), the reproducibility is greatly increased from previous IEF technology (Gorg, 1991).

Gorg *et al.* (1999) have adjusted the pH range for IEF using IPGs to increase the width up to pH 12, providing, if required, further separation, but strips based on these ranges are not currently commercially available. Preformed IPG strips allow relatively large quantities of protein to be loaded (up to 4 mg on an 18 cm strip) by using either passive rehydration or active rehydration to assist in high molecular weight protein loading. The ability of loading higher protein levels is especially important in the analysis of low-abundance proteins (Gygi *et al.*, 2000). Unfortunately, membrane proteins do not seem to be represented well after IEF but Santoni *et al.* (2000) have investigated methods to improve this situation, and membrane proteins can be analysed by 1D-based proteomics.

It is important to standardize 2DE for a particular sample and IEF running conditions should remain unchanged from run to run in order to improve reproducibility. An IEF running temperature of 20°C should prevent urea precipitation and carbamylation of proteins, altering protein p*I*. Limiting IEF runs to 40–50 μA per gel strip can often aid successful focusing, and stepping the voltage will allow charge carriers to settle. Due to the robustness of IPG strips, focused strips can be stored at –80°C or used immediately after focusing. Commercial apparatus is available for this first dimension, including Protean IEF (Bio-Rad) and IPGphor (Amersham).

SDS–PAGE: the second dimension

After IEF, IPG strips are equilibrated in an SDS reducing buffer for second dimension separation. Originally used to remove point streaking during silver staining, the method of Gorg *et al.* (1987) is now commonplace although it can be adapted (Vilain *et al.*, 2001). Separation of protein samples in the second dimension is then performed by SDS–polyacrylamide gel electrophoresis (SDS–PAGE). This separates proteins according to their molecular weight (M_r) and many types of commercial apparatus are available.

In the presence of SDS, anionic complexes are formed with proteins with a net negative charge. This allows proteins to be drawn through polyacrylamide gels under the influence of an electric field and to be separated via relative movement through the gel that is based on their respective M_r. For a general protein range of 14 to 100 kDa, 12–12.5% acrylamide, with *N*, *N*-methylenebisacrylamide (Bis) at a 29:1 ratio is used. The percentage acrylamide can be increased to reduce pore size and resolve low M_r proteins more effectively. Precast gradient gels can be purchased or formed in house with a gradient former to provide a superior resolution across the second dimension. For proteins and peptides of interest below 14 kDa, tris-tricine gels are generally recommended.

Staining and extraction

Proteins separated via 2DE need to be visualized for further analysis. Many post-staining

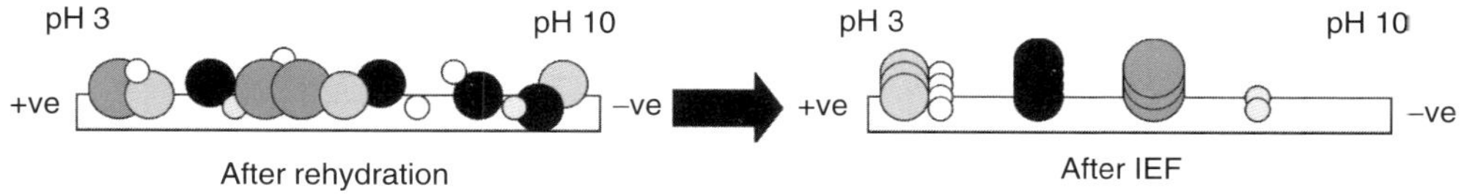

Fig. 17.6. Isoelectric focusing: diagrammatic view of isoelectric focusing (IEF) of a re-hydrated linear immobilized pH 3–10 gradient strip (Biorad, UK).

techniques are available, of which currently the most common are silver and Coomassie brilliant blue (CBB). Silver staining provides sensitivity of detection down to 0.5–1.2 ng of protein. Silver staining methods have been successfully used in parasite proteomics for *F. hepatica* and *Opisthorchis viverrini* (Jefferies *et al.*, 2000; Boonmee *et al.*, 2003). Development of mass spectrometry-compatible silver staining methods now limits protein identification to the ability of modern mass spectrometry equipment (Shevchenko *et al.*, 1996).

CBB provides a simple few-step method for protein visualization and offers detection of proteins down to 10 ng per spot (depending on gel thickness). New commercial colloidal CBB stains do provide a dynamic range of detection, with good compatibility with mass spectrometry for protein identification, but even these new dyes are not linear within the order of magnitude of protein levels found in the cell.

Successful identification of CBB protein spots has been highlighted in both *F. hepatica* and *E. granulosus* (Jefferies *et al.*, 2001; Chemale *et al.*, 2003). Further staining techniques, such as zinc or copper staining, are also available, but currently not generally used. Unfortunately, individual proteins have been found to have different staining characteristics.

Fluorescent post-labelling dyes are also available for protein staining, and for example, SYPRO ruby, provides silver stain range sensitivity (1–2 ng) with compatibility with mass spectrometry, although some report compatibility problems with precast gels. A previously expensive industrial fluorescent proteomics tool, difference-in-gel electrophoresis (DIGE) is now available to academic researchers. First described by Ünlü *et al.* (1997), 2D DIGE overcomes some of the conventional problems associated with comparing two standard 2DE gels, by allowing qualitative comparisons on one gel (Lilley *et al.*, 2002). After isolation, protein samples are minimally labelled with one of three spectrally differing fluorescent dyes; Cyanin-2 (Cy™2), Cy3 or Cy5 (Amersham Biosciences, USA). Labelled protein samples are pooled together and separated within the same 2DE analysis. The separate samples are then visualized individually by scanning at the appropriate wavelengths and images analysed by specialized commercial software (e.g. DeCyder™ Differential Analysis Software, Amersham Biosciences).

Only a small percentage (~3%) of a protein sample is labelled via fluorescent dye, and this addition to each protein alters the molecular mass by approximately 580 Da. In proteins over 30 kDa, this will not significantly alter the position where they resolve by 2DE. However, in smaller proteins this will significantly adjust positioning compared to the unlabelled proteins, but post-electrophoresis staining with a conventional stain allows further analysis. To date, a DIGE analysis of parasite proteins has not been published.

Differential 2DE displays of post-gel labelled silver stained and pre-labelled S-35 proteins also represent a sensitive procedure to visualize and identify synthesized elements of proteomes, respectively.

Image analysis

Replication and reference gel construction are key parts of analysis as 2DE technology is not currently 100% reproducible. This approach is first required to confirm that spots from identical proteins are in the same place. Only then can the relative spot intensities be compared, and any new or missing spots distinguished. Current software that provide analysis of proteins on 2DE gels include PDQuest® (Bio-Rad), Progenesis and Phoretic 2D Evolution (Non-linear), HT Analyser (Genomic Solutions) and Compugen's Z3, and these packages compare images by assigning intensity values to protein spots. Multi-2DE gel analysis is also possible, using landmark proteins, assigning proteins with *x* and *y* coordinates, or by merging gel images. Algorithms within software assess spot intensities on control gels compared to experimental gels, to resolve and determine the expression of individual proteins. If the above software is not available, then standard statistical analysis can be used to assess significance in differential protein expression. Clearly, automated spot matching programs remove subjective error. However, even the most expensive commercial software has difficulty resolving overlapping spots and small position differences due to post-translational modifications (PTMs). Spot shapes can also become distorted during processing, and therefore, human intervention is ultimately required. Image analysis or gel registration remains a relatively neglected area of proteomic development, including the ability to quantify protein expression levels because of limited linearity of stains and scanners.

Mass Spectrometry to Unravel the Flatworm Proteome

Following image analysis, selected protein spots are excised from the gel, either manually by a spotter-pen or automatically by a robot. This is generally followed by enzymatic digestion of the protein(s) in the gel plug. Not all proteases are suitable for PMF, and the most commonly used one is unmodified trypsin which cleaves proteins at lysine or arginine residues (provided the next amino acid is not a proline) (Rosenfeld *et al.*, 1992). The digested peptides are then extracted, and concentrated for subsequent protein identification via mass spectrometry. Cohen and Chait (1997) describe the extraction of whole proteins from gels for protein M_r determination. MS analysis of intact proteins is the principle behind surface-enhanced laser desorption ionization time-of-flight (SELDI ToF), which analyses protein fingerprints following affinity matrix capture.

It is only the recent technological advancements in biological mass spectrometry that has made the analysis of parasite proteomes possible. There are currently two types of mass spectrometric equipment used to identify digested fragments; matrix-assisted laser desorption/ionization time-of-flight mass spectrometry (MALDI ToF) and electrospray ionization mass spectrometry (ESI MS), with the ability to perform tandem mass spectrometry (ESI MS/MS). MALDI ToF, via PMF, makes analysis of genome-verified organisms routine for identification, although a function may not be assigned, e.g. for *Plasmodium*, *C. elegans* and soon the parasitic flatworm *S. mansoni*. ESI MS/MS allows the opportunity to undertake proteomics in parasitic flatworms without a verified genome via generation of peptide sequence information for homology searching. In contrast, Liska and Shevchenko (2003) have developed an approach to analyse MS data for cross-species identification, but to be successful, this approach probably requires at least 80% sequence identity.

Protein identification by mass spectrometry consists of three stages: ionization, separation and detection. However, this will not be discussed in depth here; for an explanation of the technical aspects of modern mass spectrometry please refer to Yates (1998) or Mann *et al.* (2001) amongst others.

Matrix-assisted Laser Desorption/Ionization Time-of-flight Mass Spectrometry (MALDI ToF) for Parasite Analysis

MALDI ToF has revolutionized proteomics with the introduction of PMF. Proteins, not necessarily from gel electrophoresis, are selected and undergo proteolytic digestion. The peptide fragments are then subjected to MALDI ToF analysis to gain accurate mass to charge (*m/z*) data of the peptides within the mixture (Blackstock and Weir, 1999; Liska and Shevchenko, 2003). Individual proteins will fragment on proteolytic digestion in a unique manner, resulting in a peptide 'fingerprint' detectable by mass spectrometry.

Peptide analytes are mixed with a UV (or less common IR) absorbing 'matrix' and allowed to dry, by a variety of methods, to a metal probe. The dried droplet can then be irradiated with a UV/IR source (see Fig. 17.7), to ionize and allow desorption into the vacuum of the mass spectrometer. The generated ions, usually protonated (Zenobi and Knochenmuss, 1998) and singly charged (Gevaert and Vandekerckhove, 2000), are then accelerated into a field-free region to allow separation over a known path length. In the majority of MALDI instruments, a reflector device is incorporated into the ToF tube, effectively doubling the flight path of ions by reversing their trajectories; thereby increasing separation (Fig. 17.7) (Mann and Talbo, 1996). The resulting peptide data is gathered via multichannel plate (MCP) detectors. This process generates peptide *m/z* data with mass accuracies usually

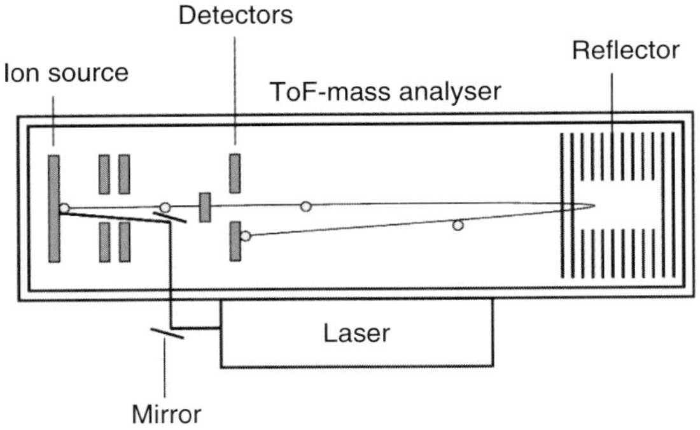

Fig. 17.7. The MALDI: schematic design of a MALDI ToF instrument (Micromass ToF Spec 2E, MMUK) showing ion generation via laser desorption (UV/IR), followed by ion acceleration through the field-free ToF-mass analyser, where the ions are reversed as they enter the reflector to end at the detectors.

around 10–20 ppm, well within 0.2 Da, and with routine sensitivities of 1 pM (Godovac-Zimmermann and Brown, 2001).

MALDI analysis is a highly valuable tool when working on a sequence-verified organism, such as the free-living nematode *C. elegans*. Van Rossum *et al.* (2001) demonstrated the power of PMF by identifying 12 different sequence-related *C. elegans* GSTs. PMFs are also useful in the identification of non-genome verified organisms, provided there are protein sequences in the databases. PMF successfully identified 29 spots of known database proteins from 60 ES products from *F. hepatica* (Jefferies *et al.*, 2001) and enolase was also identified via PMF from *F. hepatica* ES products (Bernal *et al.*, 2004). PMF has also been effective in *E. granulosus* with Chemale *et al.* (2003) identifying 15 database proteins. Four life stages of *S. mansoni* have also been subjected to PMF analysis with 32 proteins identified from database protein entries, all originating from the cytosol (Curwen *et al.*, 2004). Thus, PMF in non-genome verified organisms will possibly identify proteins previously fully characterized, but may suggest new functions for less well-characterized proteins.

Using PMFs for identification allows cross-species identification of highly conserved proteins, suggested by Liska and Shevchenko (2003). Proteins such as regucalcin, transferrin and enolase 1 have been identified from ovine host bile using sequences from bovine sources (Fig. 17.2) in the analysis of the sheep-liver fluke infected bile-ome (Morphew, 2004, unpublished). A single proteomic 2DE array assay may uncover networks for PMF-identified proteins previously only analysed individually in different laboratories.

Electrospray ionization time-of-flight mass spectrometry (ESI ToF MS)

ESI mass spectrometry remains essential to proteomics. The electrospray source introduces analyte into the mass spectrometer by producing a mist containing charged droplets of analyte and solvent. An inert gas, such as nitrogen, is flushed over the ion mist and in the process removes the solvent from the droplets producing desolvated analyte ions. These are often highly charged and is a common characteristic of ESI (Gaskell, 1997; Yates, 1998). The multiple charging of analyte ions effectively reduces the *m/z* value allowing higher resolution and mass accuracy, which therefore aids the analysis of whole proteins as well as standard PMF work. Once produced analyte ions are passed through a series of quadrapole mass analysers and finally to the more common ToF tube for subsequent separation and detection via MCP detectors.

More recently, advances in electrospray technology have helped to look further into proteomics via non-covalent interactions between proteins (Gaskell, 1997). Pramanik *et al.* (1998) have reviewed the ability of ESI to study protein interactions, often a key to triggering many cellular processes, with mass accuracies better than 0.01%. Unfortunately, ESI is less tolerant than MALDI to salt, detergent and buffer contamination and consequently MALDI is used in the majority of cases for PMF analysis due to its ease and speed of use. The real value of ESI MS is highlighted through tandem mass spectrometry (MS/MS) for peptide sequencing.

Analysis of post-translational modifications

Over 400 different types of protein modification have been reported, and every eukaryote protein has usually 8–10 post-translational variants. Protein modifications can be natural or can arise from the 2DE process, e.g. cysteine combines with free acrylamide to give propionamide cysteine.

PTMs, such as methylation or acetylation or addition of complex N- or O-linked glycosylations, are covalently bound modifications to one or several amino acid residues. This can drastically affect the outcome of the altered protein (Mann and Jensen, 2003). Genomic information, at present, cannot predict the processes of PTM and cannot be studied with genomic techniques. With protein modification, being an important area of parasitic flatworm proteomics, the most effective method to research them is with mass spectrometry (Mann *et al.*, 2001).

Typically the protein of interest is first analysed whole and intact using either MALDI or ESI or both if needed to gain an accurate molecular weight. Proteins then undergo proteolytic digestion to provide information related to specific sites on the protein (Roepstorff, 1997). Protein digestion with one or a number of digestion enzymes produces a range of peptide fragments allowing a comparison with a known sequence to identify the site of PTM. For many PTMs enrichment before mass spectrometric analysis can select for specific peptides with the modification of interest. This is often the case with phosphorylation (Mann and Jensen, 2003), which can be enriched by the use of metal AC to enhance phosphopeptides. By eluting enriched samples directly into an instrument capable of sequencing, MS/MS, the site of phosphorylation can be identified. Recently gel-staining techniques have enabled PTMs, such as glycosylation, to be identified whilst still within the gel, allowing proteins of interest to be excised and identified directly.

PTMs are clearly an important area for future flatworm proteomics. Already, proteins are found in multiple 2DE array locations from both *E. granulosus* (heat shock proteins, P 20 and HSP 70) and *F. hepatica* (cathepsins, Cat L type) have been identified (Jefferies *et al.*, 2001; Chemale *et al.*, 2003).

Peptide sequencing via post-source decay and tandem mass spectrometry

MALDI, in general, creates peptide ions via a soft ionization process producing intact peptides and proteins. But when dealing with non-sequenced organisms, peptide sequence data is highly advantageous. In the MALDI process, a number of metastable ions are produced during the ionization process. The action and behaviour of these metastable ions causes the loss of small neutral molecules along with more important fragmentation across the peptide backbone (Gevaert and Vandekerckhove, 2000).

Metastable ion fragmentation is extremely useful when it occurs 'post source'. The reflector device in many MALDI instruments can then be used to scan the selected fragments to produce a post-source decay (PSD) fragmentation spectra. Once correctly interpreted, useful sequence information for non-sequence verified organisms can be gathered. Interpretation can often be complicated due to the presence of many ions, most commonly of *a*, *b* and *y* type. Ions present not only differ in mass by peptide cleavage but also due to loss of neutral molecules and/or the loss of side chains. Coupled with this, poor mass accuracy and limited fragmentation across the peptide backbone make PSD a complex tool for protein sequencing (Mann and Talbo, 1996). However, partial sequence obtained from PSD is often enough to confirm protein identification and so is not to be neglected. For a review of detailed mechanisms of the PSD process and for interpretation procedures, see Spengler (1997).

The fragmentation of peptide ions and subsequent protein sequencing via MS/MS can now be operated in an automated manner. The ability to directly couple ESI sources to HPLC separation allows continuous sequencing of peptides. The experimental design of MS/MS (Fig. 17.8) allows a complete and accurate fragmentation pattern to be produced. Peptides of interest are first selected in a precursor ion scan. The ion of choice, filtered through a quadrapole, is fragmented within a sealed collision cell filled with an inert gas such as nitrogen or argon. This low-energy collision between peptide ions and inert gas causes the peptide bonds to rupture producing fragment ions. The fragments gathered from collision-induced dissociation (CID) are ejected into a standard mass analyser such as a ToF tube with their *m/z* ratios determined. De Hoffmann (1996) has provided comprehensive coverage on the principles of MS/MS.

This technique gives mass accuracies of peptide fragments superior to that of MALDI ToF (Gaskell, 1997). The CID process gives a much more complete fragmentation spectra of *y*, *b* and *a* ions (Mann *et al.*, 2001) and therefore, with the aid of computer software, is easily interpreted and sequenced. Peptide sequence tags from MS/MS experiments allow enhanced database searches with verified organisms or can reveal information on unverified organisms (Barrett *et al.*, 2000; Liska and Shevchenko, 2003). More recently, experimental design of MS/MS instruments, such as ion trapping techniques, is allowing MS/MS experiments to be performed repeatedly on the same fragment (generally three or four times) with the potential to perform MS^n (ability to perform MS/MS to the *n*th time). The ability to investigate fragmentation at this level allows in-depth structural information to be accessed.

Database Mining and Protein Identification

The use of mass spectral data, from either PMF or MS/MS and comparing the obtained value of peptides/fragments with theoretical masses first shown by Henzel *et al.* (1993) is an integral part of data mining. Along with homology modelling to aid X-ray crystallography work in protein structure determination (Smith, 2000), data mining is an essential part of proteomics for identification of flatworm proteomes. Peptide and fragment data, derived from MS work, are entered into easily accessible databases to provide reference information. Three variations of database are available to proteomists to search for identification of proteins; protein, EST and genome databases (Mann *et al.*, 2001).

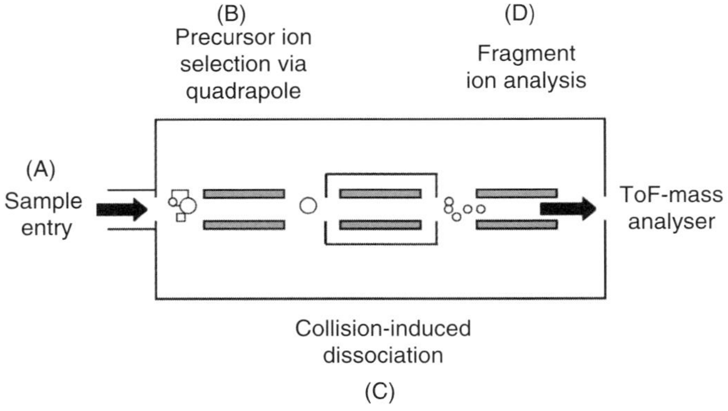

Fig. 17.8. MS/MS: schematic representation of MS/MS by quadrapole mass spectrometers. (A) Sample entry normally via ESI. (B) Precursor ion selection allowing only the ion of interest to continue through the mass spectrometer. (C) Collision-induced dissociation of the ion within a gas cell normally containing argon to produce fragments. (D) The products of fragmentation are focused and analysed via a ToF-mass analyser.

High-speed identification from either PMF or MS/MS data can be achieved from non-redundant protein databases (Mann *et al.*, 2001). These often include valuable annotations giving insight into functions of isolated and identified proteins. Many of these databases, such as NCBInr (National Centre for Biotechnology Information) and Swiss Prot (ExPASy Molecular Biology Server), can be used for protein identification.

EST databases contain sequence data on cDNA sequences, from a number of organisms, made from single random passes. There are millions of ESTs within these databases and although, commonly, they only cover a small section of organisms, genes and proteins, they are still valuable in the identification of proteins (Andersen and Mann, 2000). However, more recent EST projects are believed to cover a greater proportion of an organism's transcriptome.

If PMF and sequence homology searches fail using available protein and EST databases, mass spectrum data can, in principle, be used to search genome databases (GenBank: http://www.ncbi.nlm.nih.gov/). However, only a small percentage of large genomic sequences code for proteins such that bioinformatics still needs to accurately define exon–intron structures (Andersen and Mann, 2000).

Useful databases

EST and protein depository databases are available for assisting proteomics and other post-genomic techniques in parasitic flatworms. Under the umbrella of the World Health Organization Schistosoma Genome Network (http://www.nhm.ac.uk/hosted_sites/schisto/index.html), there are over 180,000 public entries and 140,000 EST input entries for *S. mansoni*, with two major EST projects in progress in Brazil (Verjovski-Almeida *et al.*, 2003). *S. mansoni* genome projects are in progress at The Institute for Genomic Research (TIGR), Rockville, USA (http://www.tigr.org/tdb/e2k1/sma1/) and the Sanger Institute (http://www.sanger.ac.uk/projects/s_mansoni/). *Schistosoma japonicum* has over 45,000 public entries and an equivalent number of input ESTs, with further gene discovery studies (including a genome sequencing project) via collaboration between The Chinese National Human Genome Centre (CHGC) and the Institute of Parasitic Diseases, Chinese Centre for Disease Control and Prevention (http://schistosoma.chgc.sh.cn, Hu *et al.*, 2003).

For *Schistosoma haematobium*, the third major human-infective species, the Sanger Centre, UK, is planning to generate 15,000 ESTs from new cDNA libraries (http://www.sanger.ac.uk/Projects/S_mansoni/). Progress is being made with generating 15,000 *F. hepatica* ESTs at the Sanger Centre, and sequences can currently be downloaded at (ftp://ftp.sanger.ac.uk/pub/databases/Trematode/Fhep/) for the creation of local PMF construction and blast searching facilities. For *E. granulosus*, over 6000 EST entries are available (http://www.ncbi.nlm.nih.gov/). The common PMF searching programs have facilities to search NCBI-listed EST databases. However, this will be problematic and a dedicated PMF searchable clustered EST database should be constructed.

Thus, with the increase in protein databases, many user-friendly internet-based tools have been constructed to search protein, EST and genome databases, of which the technical aspects are often omitted. A variety of search engines and tools are available to identify proteins from PMF data or sequence data, for molecular weight calculation and many other interpretation tools, e.g. Protein Prospector MS-Fit program (http://prospector.ucsf.edu/ UCSF Mass Spectrometry Facility) although a full account will not be given here. Protein modifications can be automatically allowed when searching protein databases to ease identification and some databases identify modifications if they are of specific interest. As with 2DE array, proteolytic cleavage is not ideal, and potential cleavage sites can be 'missed' by the protease. Thus, with most PMF software, one missed cleavage site (or more) can be selected, so that all possible combinations of adjacent peptides are added to the list of theoretical peptides. Furthermore, 100% matching of peptides with a database target is not required for successful identification. To this end, for tryptic digests, 1000–3500 Da sized fragment is a suitable range for searching. Peaks of less than 500 Da are obscured by the matrix and peptides >6000 Da are generally not required in searches. In fact, three- to four-fragment masses can successfully match a database and query protein. Moreover, mass tolerance is allowed in the PMF software for the experimental peptides, and should reflect the accuracy of the mass spectrometer, typically 0.2 Da or 20 ppm or better. Increasing the size of the error window by too much increases the number of false positives.

The results of PMF searches are ordered on a probabilistic basis, including an *E*-value (expectation value), supported by dedicated free-ware programs for statistical analysis (Mascot: http://www.matrix-science.com/). Another confidence indicator for PMF analysis is the percent coverage, i.e. the proportion of the theoretical protein which is covered by the identified peptides. More than 20% coverage of the protein is likely to be significant, but depends on protein size, with small proteins more likely to yield a higher percent coverage. Filtering is also possible in PMF databases, with searches restricted to single species or tissue-specific databases, or within a predefined M_r or p*I* range.

However, it is clear that the ease of collecting proteome data currently exceeds the capacity to analyse it. Therefore, international standards are being sought for proteomic experiments, such as the proteomics standards initiative (PSI, http://psidev.source forge.net/), and the proteomics experiment data repository (PEDRo, http://pedro.man.ac.uk). The aim is to provide proteomic researchers with the opportunity to query and reproduce protocols, analyse raw or metadata from other laboratories, and to link the proteome with the respective transcriptome and the metabolome. The Human Proteome Organization (HUPO) is establishing a defined infrastructure for human data submission, and annotation for the numerous proteomic data platforms. This will also be required to effectively develop parasitic flatworm proteomics.

Proteomics without Two-dimensional Electrophoresis

There are a number of limitations that reduce the efficiency of 2DE. Co-migration of proteins to the same location on a 2DE gel can occur. The advent of zoom gels (narrow pH ranges) can help overcome this problem. The dynamic range offered by 2DE analysis, the ability to label protein over a wide range of concentrations, is up to a maximum of three orders of magnitude. This is directly linked to the capabilities of the staining techniques used and is often not sufficient to account for dynamic ranges common in biological material. Therefore, sample prefractionation or enrichment can be required (Rabilloud, 2002).

More importantly insoluble proteins, such as membrane-bound and nuclear proteins, are under-represented within 2DE analysis. However, membrane proteins can be identified by MS following a 1DE separation. Unfortunately, hydrophobicity is the most common reason for poor representation of abundant proteins and seems set to be with us for quite some time (Santoni *et al.*, 2000; Rabilloud, 2002). Emerging proteomic methods without the use of 2DE are being developed, such as isotope-coded affinity tagging (ICAT) and multi-dimensional protein identification technology (MuDPIT).

Multi-dimensional protein identification technology (MuDPIT)

MuDPIT, alternatively known as 'shotgun' protein identification methods, was developed by Yates (1998) and has the advantage of being based on HPLC (schematically shown in Fig. 17.9). This process involves separation of peptides derived from digested proteins rather than whole proteins and therefore involves proteolytic digestion before analysis. Digestion can be performed with more than one digestion enzyme. The resulting peptides are then subjected to a series of liquid chromatography (LC) separations, most often strong cation exchange (SCX) followed by traditional reverse phase (RP). Peptides, which are eluted from the columns, are then fed directly into a mass spectrometer (ESI) capable of MS/MS (Swart *et al.*, 2003).

This technique offers three main advantages. Firstly, the separation methods of SCX and RP are fairly comparable to that of 2DE. SCX acts in a similar manner to IEF and RP

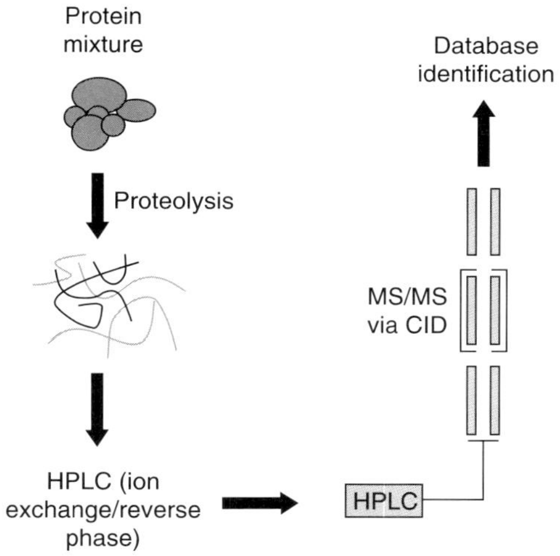

Fig. 17.9. Schematic diagram of MuDPIT proteomics (adapted from Yates, 1998). Protein mixtures of interest are cleaved via proteolytic activity and separated via a series of liquid chromatography separations. Identification of the proteins present relies on sequence tags produced from tandem mass spectrometry. Even low-abundance proteins should be represented at least on one occasion.

shows a trend for increasing retention time correlated with increasing molecular weight, as in the second dimension. Secondly, the two varying solvent types used in both SCX and RP complement one another, when SCX is used as the first dimension. The salt-based solutions for elution from SCX are weak elutants for RP chromatography which requires organic input to elute hydrophobic proteins. Thirdly, compatibility of RP chromatography solvents, water, organic solvents and weak acids, allow this system to be directly coupled to MS/MS systems (Wehr, 2003).

Once peptides have been successfully eluted from multi-dimensional chromatography, they undergo CID within MS/MS instruments. Data-dependent acquisition systems associated with MS/MS allow individual peptides to be fragmented and the resultant spectra to be used to search the databases for possible identification (Wehr, 2003). This method of protein identification, via MuDPIT, is most efficient when working with genome-verified organisms (Yates, 1998). MuDPIT has one distinct advantage in that even low-abundance proteins should be represented at least on one occasion and therefore it has the ability to cover a much larger range of the proteome. However, due to the way MuDPIT is arranged, no information can be gathered regarding relative quantification of proteins, only an absence or presence.

Isotope-coded affinity tagging (ICAT)

To gain quantitative data without the inherent problems associated with gel electrophoresis would seem to be an added advantage. ICAT, (Fig. 17.10), also based on LC, labels reduced cysteine residues from two sample states with a multipart tag. This tag consists of an affinity tag to bind to AC media along with an isotopically coded linker, generally heavy and light forms

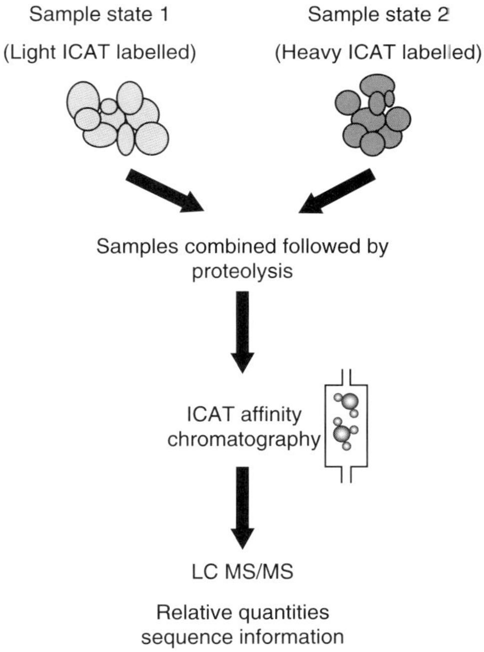

Fig. 17.10. Schematic diagram of the ICAT strategy to proteomics. Samples are labelled, pooled and digested and undergo ICAT affinity chromatography to select cysteine-labelled peptides. Relative quantifications between proteins from both sample states are gained and tandem MS is performed to provide an identification.

of deuterium, one for each sample state (Gygi *et al.*, 1999).

Proteins are first fully denatured and reduced. If not denatured completely, cysteine residues may well be embedded within the protein structure, therefore rendering the ICAT process prone to error due to loss of labelling. Following reduction, protein samples from both sample states are labelled with affinity tags (one isotopic type for each sample state). Once labelled, they undergo proteolytic digestion and AC to select for the labelled cysteine residue peptides. Beneficially, this dramatically decreases the sample complexity. The selected peptides are then fed directly into LC equipment coupled to a MS/MS, where sequence information and relative quantification take place at alternating periods. Quantification, although just relative, is via labelling of identical sequenced peptides from both sample states. One state will be labelled with a heavy form of the tag and the other state will be labelled with the light form of the tag. The two labelled peptides simultaneously elute into the mass spectrometer and will differ in mass from one another by the difference between the two isotopes attached to the heavy and light tags, most commonly between hydrogen and deuterium (Gygi *et al.*, 1999). From this method, a comparison can be made between the two peptides deriving from the two sample states. Commercial ICAT systems are now available with the introduction of cleavable ICAT allowing removal of the affinity tag to improve the performance of MS/MS.

With ICAT providing both sequence information for identification and relative quantification data, it would seem the remaining problems of proteomics should eventually be overcome. Nevertheless, successful proteomics on membrane proteins is still relatively difficult, but not impossible, even with ICAT or MuDPIT, due to their hydrophobicity (Rabilloud, 2003; Wu and Yates, 2003). ICAT requires proteins to contain cysteine residues – one in seven proteins do not (Rabilloud, 2002).

Concluding Remarks

Experimental proteomics studies have been published on several parasitic flatworms including *F. hepatica* and *E. granulosus*, and this will soon be extended to medically important schistosomes. The release of a schistosome theoretical proteome will allow for the routine analysis of global protein expression by PMF in this parasite. Other proteomic studies in parasitic flatworms that are not genome-verified will rely on peptide sequencing, PMF via extensive EST databases that have been clustered, or novel cross-species combined PMF and sequencing strategies. However, to effectively progress flatworm proteomics, standard operating practices (SOPs) and common data deposit procedures are required to be developed by the research community.

Acknowledgements

The authors would like to thank the BBSRC and Pfizer Animal Health (Dr Debra Woods) for a BBSRC CASE studentship supporting R. Morphew and the BBSRC for a project grant (BBC5036381) to support flatworm proteomics.

References

Andersen, J.S. and Mann, M. (2000) Functional genomics by mass spectrometry. *FEBS Letters* 480, 25–31.

Barrett, J., Jefferies, J.R. and Brophy, P.M. (2000) Parasite proteomics. *Parasitology Today* 16, 400–403.

Bernal, D., de la Rubia, J.E., Carrasco-Abad, A.M., Toledo, R., Mas-Coma, S. and Marcilla, A. (2004) Identification of enolase as a plasminogen-binding protein in excretory–secretory products of *Fasciola hepatica*. *FEBS Letters* 563, 203–206.

Blackstock, W.P. and Weir, M.P. (1999) Proteomics: quantitative and physical mapping of cellular proteins. *Trends in Biotechnology* 17, 121–127.

Boonmee, S., Imtawil, K., Wongkham, C. and Wongkham, S. (2003) Comparative proteomic analysis of juvenile and adult liver fluke, *Opisthorchis viverrini*. *Acta Tropica* 88, 233–238.

Brophy, P.M. and Barrett, J. (1990) Glutathione transferase in helminths. *Parasitology* 100, 345–349.

Brophy, P.M., Papadopoulos, A., Touraki, M., Coles, B., Korting, W. and Barrett, J. (1989) Purification of cytosolic glutathione transferases from *Schistocephalus solidus* (plerocercoid): interaction with anthelmintics and products of lipid peroxidation. *Molecular and Biochemical Parasitology* 36, 187–196.

Chemale, G., van Rossum, A.J., Jefferies, J.R., Barrett, J., Brophy, P.M., Ferreira, H.B. and Zaha, A. (2003) Proteomic analysis of the larval stage of the parasite *Echinococcus granulosus*: causative agent of cystic hydatid disease. *Proteomics* 3, 1633–1636.

Cohen, S.L. and Chait, B.T. (1997) Mass spectrometry of whole proteins eluted from sodium dodecyl sulfate-polyacrylamide gel electrophoresis gels. *Analytical Biochemistry* 247, 257–267.

Cohen, A.M., Rumpel, K., Coombs, G.H. and Wastling, J.M. (2002) Characterisation of global protein expression by two-dimensional electrophoresis and mass spectrometry: proteomics of *Toxoplasma gondii*. *International Journal for Parasitology* 32, 39–51.

Curwen, R.S., Ashton, P.D., Johnston, D.A. and Wilson, R.A. (2004) The *Schistosoma mansoni* soluble proteome: a comparison across four life-cycle stages. *Molecular and Biochemical Parasitology* 138, 57–66.

De Hoffmann, E. (1996) Tandem mass spectrometry: a primer. *Journal of Mass Spectrometry* 31, 129–137.

Degrave, W.M., Melville, S., Ivens, A. and Aslett, M. (2001) Parasite genome initiatives. *International Journal for Parasitology* 31, 532–536.

Florens, L., Liu, X., Wang, Y.F., Yang, S.G., Schwartz, O., Peglar, M., Carucci, D.J., Yates, J.R. and Wu, Y.M. (2004) Proteomics approach reveals novel proteins on the surface of malaria-infected erythrocytes. *Molecular and Biochemical Parasitology* 135, 1–11.

Gaskell, S.J. (1997) Electrospray: principles and practice. *Journal of Mass Spectrometry* 32, 677–688.

Gevaert, K. and Vandekerckhove, J. (2000) Protein identification methods in proteomics. *Electrophoresis* 21, 1145–1154.

Godovac-Zimmermann, J. and Brown, L.R. (2001) Perspectives for mass spectrometry and functional proteomics. *Mass Spectrometry Reviews* 20, 1–57.

Gorg, A. (1991) 2-Dimensional electrophoresis. *Nature* 349, 545–546.

Gorg, A., Postel, W., Weser, J., Gunther, S., Strahler, J.R., Hanash, S.M. and Somerlot, L. (1987) Elimination of point streaking on silver stained two-dimensional gels by addition of iodoacetamide to the equilibration buffer. *Electrophoresis* 8, 122–124.

Gorg, A., Obermaier, C., Boguth, G. and Weiss, W. (1999) Recent developments in two-dimensional gel electrophoresis with immobilized pH gradients: wide pH gradients up to pH 12, longer separation distances and simplified procedures. *Electrophoresis* 20, 712–717.

Gorg, A., Obermaier, C., Boguth, G., Harder, A., Scheibe, B., Wildgruber, R. and Weiss, W. (2000) The current state of two-dimensional electrophoresis with immobilized pH gradients. *Electrophoresis* 21, 1037–1053.

Greetham, D., Morgan, C., Campbell, A.M., Van Rossum, A.J., Barrett, J. and Brophy, P.M. (2004) Evidence of glutathione transferase complexing and signaling in the model nematode *Caenorhabditis elegans* using a pull-down proteomic assay. *Proteomics* 4, 1989–1995.

Gygi, S.P., Rist, B., Gerber, S.A., Turecek, F., Gelb, M.H. and Aebersold, R. (1999) Quantitative analysis of complex protein mixtures using isotope-coded affinity tags. *Nature Biotechnology* 17, 994–999.

Gygi, S.P., Corthals, G.L., Zhang, Y., Rochon, Y. and Aebersold, R. (2000) Evaluation of two-dimensional gel electrophoresis-based proteome analysis technology. *Proceedings of the National Academy of Sciences of the United States of America* 97, 9390–9395.

Henzel, W.J., Billeci, T.M., Stults, J.T., Wong, S.C., Grimley, C. and Watanabe, C. (1993) Identifying proteins from 2-dimensional gels by molecular mass searching of peptide-fragments in protein-sequence databases. *Proceedings of the National Academy of Sciences of the United States of America* 90, 5011–5015.

Hu, W., Yan, Q., Shen, D.K., Liu, F., Zhu, Z.D., Song, H.D., Xu, X.R., Wang, Z.J., Rong, Y.P., Zeng, L.C., Wu, J., Zhang, X., Wang, J.J., Xu, X.N., Wang, S.Y., Fu, G., Zhang, X.L., Wang, Z.Q., Brindley, P.J., McManus, D.P., Xue, C.L., Feng, Z., Chen, Z. and Han, Z.G. (2003) Evolutionary and biomedical implications of a *Schistosoma japonicum* complementary DNA resource. *Nature Genetics* 35, 139–147.

Jefferies, J.R., Brophy, P.M. and Barrett, J. (2000) Investigation of *Fasciola hepatica* sample preparation for two-dimensional electrophoresis. *Electrophoresis* 21, 3724–3729.

Jefferies, J.R., Campbell, A.M., Van Rossum, A.J., Barrett, J. and Brophy, P.M. (2001) Proteomic analysis of *Fasciola hepatica* excretory–secretory products. *Proteomics* 1, 1128–1132.

Jiang, L., He, L. and Fountoulakis, M. (2004) Comparison of protein precipitation methods for sample preparation prior to proteomic analysis. *Journal of Chromatography A* 1023, 317–320.

Lee, W.C. and Lee, K.H. (2004) Applications of affinity chromatography in proteomics. *Analytical Biochemistry* 324, 1–10.

Lilley, K.S., Razzaq, A. and Dupree, P. (2002) Two-dimensional gel electrophoresis: recent advances in sample preparation, detection and quantitation. *Current Opinion in Chemical Biology* 6, 46–50.

Liska, A.J. and Shevchenko, A. (2003) Expanding the organismal scope of proteomics: cross-species protein identification by mass spectrometry and its implications. *Proteomics* 3, 19–28.

LoVerde, P.T., Hirai, H., Merrick, J.M., Lee, N.H. and El-Sayed, N. (2004) *Schistosoma mansoni* genome project: an update. *Parasitology International* 53, 183–192.

Luche, S., Santoni, V. and Rabilloud, T. (2003) Evaluation of nonionic and zwitterionic detergents as membrane protein solubilizers in two-dimensional electrophoresis. *Proteomics* 3, 249–253.

Mann, M. and Jensen, O.N. (2003) Proteomic analysis of post-translational modifications. *Nature Biotechnology* 21, 255–261.

Mann, M. and Talbo, G. (1996) Developments in matrix-assisted laser desorption ionization peptide mass spectrometry. *Current Opinion in Biotechnology* 7, 11–19.

Mann, M., Hendrickson, R.C. and Pandey, A. (2001) Analysis of proteins and proteomes by mass spectrometry. *Annual Review of Biochemistry* 70, 437–473.

Mikes, L. and Man, P. (2003) Purification and characterization of a saccharide-binding protein from penetration glands of *Diplostomum pseudospathaceum* – a bifunctional molecule with cysteine protease activity. *Parasitology* 127, 69–77.

Nirmalan, N., Sims, P.F.G. and Hyde, J.E. (2004) Quantitative proteomics of the human malaria parasite *Plasmodium falciparum* and its application to studies of development and inhibition. *Molecular Microbiology* 52, 1187–1199.

Nugent, P.G., Karsani, S.A., Wait, R., Tempero, J. and Smith, D.F. (2004) Proteomic analysis of *Leishmania mexicana* differentiation. *Molecular and Biochemical Parasitology* 136, 51–62.

O'Farrell, P.H. (1975) High-resolution 2-dimensional electrophoresis of proteins. *Journal of Biological Chemistry* 250, 4007–4021.

Paba, J., Ricart, C.A.O., Fontes, W., Santana, J.M., Teixeira, A.R.L., Marchese, J., Williamson, B., Hunt, T., Karger, B.L. and Sousa, M.V. (2004) Proteomic analysis of *Trypanosoma cruzi* developmental stages using isotope-coded affinity tag reagents. *Journal of Proteome Research* 3, 517–524.

Pandey, A. and Mann, M. (2000) Proteomics to study genes and genomes. *Nature* 405, 837–846.

Pramanik, B.N., Bartner, P.L., Mirza, U.A., Liu, Y.H. and Ganguly, A.K. (1998) Electrospray ionization mass spectrometry for the study of non-covalent complexes: an emerging technology. *Journal of Mass Spectrometry* 33, 911–920.

Pridmore, A.M., Devine, D.A., Bonass, W.A. and Silley, P. (1999) Influence of sample preparation technique on two-dimensional gel electrophoresis of proteins from *Porphyromonas gingivalis*. *Letters in Applied Microbiology* 28, 245–249.

Rabilloud, T. (1996) Solubilization of proteins for electrophoretic analyses. *Electrophoresis* 17, 813–829.

Rabilloud, T. (2002) Two-dimensional gel electrophoresis in proteomics: old, old fashioned, but it still climbs up the mountains. *Proteomics* 2, 3–10.

Rabilloud, T. (2003) Membrane proteins ride shotgun – a new mass spectrometry-based approach identifies and characterizes membrane proteins on a large scale. *Nature Biotechnology* 21, 508–510.

Roepstorff, P. (1997) Mass spectrometry in protein studies from genome to function. *Current Opinion in Biotechnology* 8, 6–13.

Rosenfeld, J., Capdevielle, J., Guillemot, J.C. and Ferrara, P. (1992) In-gel digestion of proteins for internal sequence analysis after one- or two-dimensional gel electrophoresis. *Analytical Biochemistry* 203, 173–179.

Sandeman, R.M. and Howell, M.J. (1981) Precipitating antibodies against excretory–secretory antigens of *Fasciola hepatica* in sheep serum. *Veterinary Parasitology* 9, 35–46.

Santoni, V., Molloy, M. and Rabilloud, T. (2000) Membrane proteins and proteomics: un amour impossible? *Electrophoresis* 21, 1054–1070.

Shaw, M.M. and Riederer, B.M. (2003) Sample preparation for two-dimensional gel electrophoresis. *Proteomics* 3, 1408–1417.

Shevchenko, A., Wilm, M., Vorm, O. and Mann, M. (1996) Mass spectrometric sequencing of proteins from silver stained polyacrylamide gels. *Analytical Chemistry* 68, 850–858.

Smith, C.M. (2000) Bioinformatics, genomics, and proteomics. *Scientist* 14, 26–30.

Spengler, B. (1997) Post-source decay analysis in matrix-assisted laser desorption/ionization mass spectrometry of biomolecules. *Journal of Mass Spectrometry* 32, 1019–1036.

Swart, R., Mitulovic, G., Smoluch, M. and Chervet, J.P. (2003) Two dimensional nano LC/MS for the analysis of complex samples in proteomics. *European Journal of Pharmaceutical Sciences* 19, S10.

Ünlü, M., Morgan, M.E. and Minden, J.S. (1997) Difference gel electrophoresis: a single gel method for detecting changes in protein extracts. *Electrophoresis* 18, 2071–2077.
Van Regenmortel, M.H.V. (2001) Proteomics versus genomics. What type of structure–function relationship are we looking for? *Journal of Molecular Recognition* 14, 321–322.
Van Rossum, A.J., Brophy, P.M., Tait, A., Barrett, J. and Jefferies, J.R. (2001) Proteomic identification of glutathione S-transferases from the model nematode *Caenorhabditis elegans*. *Proteomics* 1, 1463–1468.
Verjovski-Almeida, S., DeMarco, R., Martins, E.A.L., Guimaraes, P.E.M., Ojopi, E.P.B., Paquola, A.C M., Piazza, J.P., Nishiyama, M.Y., Kitajima, J.P., Adamson, R.E., Ashton, P.D., Bonaldo, M.F., Coulson, P.S., Dillon, G.P., Farias, L.P., Gregorio, S.P., Ho, P.L., Leite, R.A., Malaquias, L.C.C., Marques, R.C.P., Miyasato, P.A., Nascimento, A., Ohlweiler, F.P., Reis, E.M., Ribeiro, M.A., Sa, R.G., Stukart, C.C., Soares, M.B., Gargioni, C., Kawano, T., Rodrigues, V., Madeira, A., Wilson, R.A., Menck, C.F.M., Setubal, J.C., Leite, L.C.C. and Dias, E. (2003) Transcriptome analysis of the acoelomate human parasite *Schistosoma mansoni*. *Nature Genetics* 35, 148–157.
Vilain, S., Cosette, P., Charlionet, R., Hubert, M., Lange, C., Junter, G.A. and Jouenne, T. (2001) Substituting Coomassie brilliant blue for bromophenol blue in two-dimensional electrophoresis buffers improves the resolution of focusing patterns. *Electrophoresis* 22, 4368–4374.
Wehr, T. (2003) Multidimensional liquid chromatography in proteomic studies. *Lc Gc Europe* 16, 2–8.
Wu, C.C. and Yates, J.R. (2003) The application of mass spectrometry to membrane proteomics. *Nature Biotechnology* 21, 262–267.
Yates, J.R. (1998) Mass spectrometry and the age of the proteome. *Journal of Mass Spectrometry* 33, 1–19.
Yatsuda, A.P., Krijgsveld, J., Cornelissen, A., Heck, A.J.R. and de Vries, E. (2003) Comprehensive analysis of the secreted proteins of the parasite *Haemonchus contortus* reveals extensive sequence variation and differential immune recognition. *Journal of Biological Chemistry* 278, 16941–16951.
Yeang, H.Y., Yusof, F. and Abdullah, L. (1998) Protein purification for the Lowry assay: acid precipitation of proteins in the presence of sodium dodecyl sulfate and other biological detergents. *Analytical Biochemistry* 265, 381–384.
Zenobi, R. and Knochenmuss, R. (1998) Ion formation in MALDI mass spectrometry. *Mass Spectrometry Reviews* 17, 337–366.

18 Proteases in Trematode Biology

J.P. Dalton,[1] C.R. Caffrey,[2] M. Sajid,[2] C. Stack,[1] S. Donnelly,[1] A. Loukas,[3] T. Don,[3] J. McKerrow,[2] D.W. Halton[4] and P.J. Brindley[5]

[1]Institute for the Biotechnology of Infectious Diseases, University of Technology Sydney, Westbourne Street, Gore Hill, Sydney, NSW 2065, Australia; [2]Tropical Disease Research Unit and Sandler Center for Basic Research in Parasitic Diseases, QB3 Building, PO Box 2550, 1700 4th Street, University of California, San Francisco, CA 94143, USA; [3]Helminth Biology Laboratory, Division of Infectious Diseases and Immunology, Queensland Institute of Medical Research, 300 Herston Road, Herston, Brisbane, Queensland 4006, Australia; [4] Parasitology Research Group, School of Biology and Biochemistry, Medical Biology Centre, Queen's University Belfast, 97 Lisburn Road, Belfast BT9 7BL, Northern Ireland, UK; [5]Department of Tropical Medicine, School of Public Health and Tropical Medicine, Tulane University Health Sciences Center, New Orleans, LA 70112, USA

Introduction

Easy access to a readily available and highly nutritious source of food is believed to have been the predominant selection pressure that drove the relationship between the parasite and its host. Endoparasites may have evolved from ectoparasitic monogeneans of fish (Halton and Stranock, 1976; Halton, 1997). Monopisthocotylean members degrade fish skin by secreting enzymes extracorporeally, while further digestion is carried out intracellularly in cells lining a well-differentiated gut. Some polyopisthocotylean monogeneans, however, that resided within the vascularized gill chamber, took advantage of the availability of a more highly nutritious, consistent and renewable diet in the form of blood, and this represented the first major step in the evolution of endoparasitism (Halton and Stranock, 1976; Halton, 1997).

This new nutritional relationship required many alterations to the parasite morphology, particularly at the interfaces of parasite and host where nutrients needed to enter. The external surfaces of endoparasitic flatworms are composed of a multifunctional syncytial tegument permeable to a variety of small organic solutes, and contain sugar and amino acid transporters for the facilitated or active uptake of these molecules and their distribution to various parasite tissues. The internal surfaces, or guts, of flatworm parasites are very well adapted to digesting host tissues, and are often lined with cells extruding long lamellae to increase the surface area for absorption. Within these cells, or gastrodermis, lies highly active machinery for synthesizing and secreting proteolytic enzymes (Dalton *et al.*, 2004).

The relative importance of the tegument and gut in the acquisition of various nutrient molecules likely varies in different flatworms. Regardless of this, the function of the tegument is to take from the host carbohydrates, amino acids and lipids, while that of the gut is primarily the digestion of host tissue macromolecules and the subsequent absorption of small soluble products. Carbohydrates are required for the storage and generation of energy and for structural purposes. Amino acids and fatty acids are used for the synthesis of parasite molecules and egg production but are not metabolized for energy (Tielens, 1997, 1999). Blood provides a rich source of all these nutrients, and its central importance in many physiological functions of the host organisms ensures that its composition, both molecular and cellular, remains consistent and reliable. Hematophagy has its problems, however, including the necessary elimination of residual iron pigment arising from digestion of haemoglobin, the prevention of blood clotting and the inactivation of immunological defence systems – but, as we will see later, parasites have taken all of these into account.

The Trematode Gut

Trematodes are generally suctorial feeders showing relatively little variation in their feeding mechanisms. To some degree, this reflects the uniformity in the physical nature of a diet of fluid and semifluid foods, but it also underscores the efficiency of the oral sucker both as an organ of attachment and as an integral part of the feeding apparatus (Halton, 1997). The majority of our understanding of the structure, physiology and biochemistry of feeding comes from the trematodes of the genera *Schistosoma* and *Fasciola*.

Schistosomes

Schistosomes are in direct contact with blood and therefore do not require a pharynx to draw blood in. Their ingestion of blood relies on the negative pressure created by the contraction of the muscle of the oral sucker and oesophagus. The oesophagus is contiguous with the outer surface and as one moves posteriorly it becomes increasingly convoluted into long parallel folds that act as a sieve to entrap material in the lumen (Senft, 1976). Towards the posterior it becomes more 'glandular' and its large secretory cells synthesize and release into the lumen large numbers of zymogen-like granules. Developmental studies on juvenile schistosomes have shown that these glands first synthesize their secretory material before the final development of the gut proper, and that the first blood meal brings about a marked increase in the rate of granule production in the oesophagus and of granule release into the lumen (Whitfield, 1979). Erythrocyte feeding is initiated by schistosomules during the first week after penetration of the host and intact red blood cells as well as haemoglobin degradation products have been seen in these early developmental stages (Bogitsh and Carter, 1977; Bogitsh and Kirschner, 1986).

Below the oesophagus the bifurcated caecum shows numerous flaps, or lamellae, that sequester the blood material. Using a microelectrode, Senft (1976) estimated the pH of the *Schistosoma mansoni* secreted products to be 5.0–6.0, although within the caecum or between the lamellae it may be lower than this. Therefore, haemoglobin and other proteinaceous material within the caecum may be in the form of a partly or fully denatured precipitate that would be relatively easy to digest with proteases. The gastrodermis is not cellular but syncytial and ultrastructural studies suggest that it may serve primarily as an absorptive tissue, pointing to the possibility of a regional differentiation of digestive–absorptive functions in the alimentary tract of this trematode (Fig. 18.1A).

Fasciola hepatica

The gut of the infective stages of the liver fluke, or newly excysted juvenile (NEJ) is filled with secretory vesicles that are produced and stored in the metacercarial stage. The epithelial cells of this stage are specialized for secretion but not absorption, and studies on the development of the surface lamellae indicate limited absorptive capacity (Bennett, 1975; Fairweather *et al.*, 1999). Following excystment the numbers of stored secretory vesicles dramatically reduce, implying that they are required for excystation and penetration of the intestinal wall. After penetrating the intestine and then the liver capsule (day 3 in mice) a new secretory cell appears and these assume a more adult-like appearance, although true cyclical secretory/absorptive activity is not evident until 2 weeks after infection (Bennett, 1975). True tissue and blood feeding in liver flukes, therefore, does not begin until the parasite is within the liver parenchyma.

Adult liver flukes reside in the bile ducts and feed mainly on blood (and possibly bile duct epithelial cells). To gain access to the blood they must, therefore, puncture the wall of the bile ducts. The alimentary tract of these worms is divided into foregut, comprising mouth, pharynx and oesophagus, and paired and blind-ended intestinal caeca whose lateral diverticula are highly branched in the adult parasite. The coordinated contraction of meridional, equatorial and radial muscles in the suckers in *F. hepatica* generates a powerful suction force that draws a plug of host tissue into the cup of the sucker, effectively bringing the food source to the pharynx (Halton, 1997; Fairweather *et al.*, 1999). Continual localized suction by the sucker and pharynx, aided by the pulsatory action of the musculature in the forebody of the worm, is generally sufficient to breach superficial blood vessels and create a flow of blood that is then forced back into the gut lumen by the peristaltic activity of the pharynx and/or oesophagus (Howell, 1970).

The gastrodermis of liver flukes differs greatly to that of the schistosomes (Fig. 18.1B). The gut is lined by a single layer of epithelial cells of one basic type, which show variations in their fine structure reflecting their functional state (Robinson and Threadgold, 1975; Fairweather *et al.*, 1999). The cells undergo cyclical transformations, involving a secretory phase, in which there is an active Golgi apparatus, numerous mitochondria and many dense zymogen-like secretory bodies, followed by a phase of intense absorption. During the absorptive phase, the surface protrusions or lamellae are longer and more numerous, between which lie exocytosed secretory granules, while inside the cell there are few secretory bodies and the Golgi complexes are largely inactive. However, numerous cytoplasmic bodies indicative of endocytosis and autophagy are present (Robinson and Threadgold, 1975; see Fig. 18.1B). The events in any one cell are not synchronized with those of adjacent cells so that secretion and absorption take place more or less simultaneously throughout the gastrodermis. A third group of cells, with a more developed sub-epithelial musculature, is associated with the movement of ingested material in the gut lumen and mixing with the contents of the secretory vesicles (Fairweather *et al.*, 1999).

Blood as a Source of Amino Acids

Trematode flukes require ingested host proteins not just as a source of nutrition for somatic

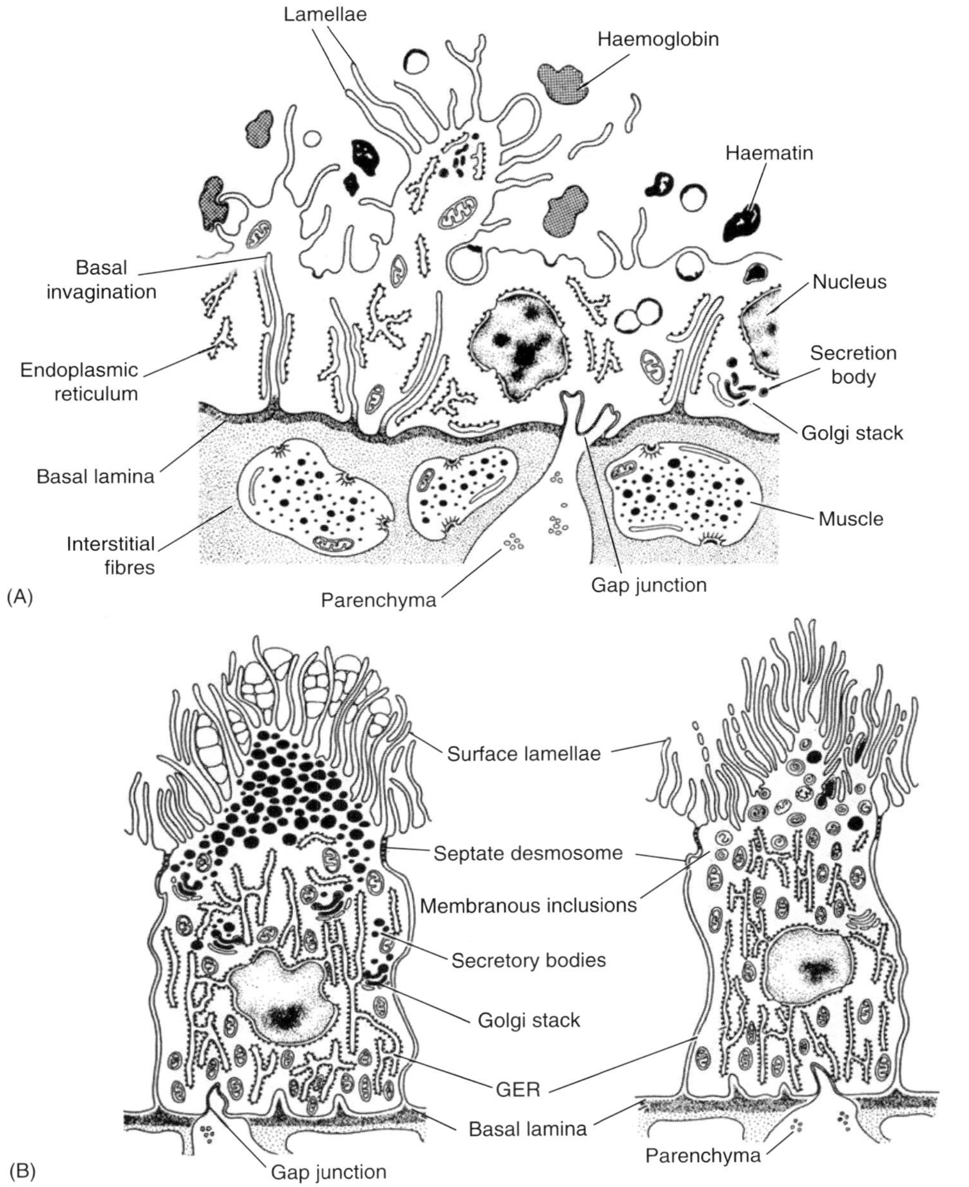

Fig. 18.1. Schematics demonstrating the syncytial nature of the gastrodermis of *Schistosoma mansoni* (A) which contrasts to the cellular nature of the gastrodermis of *Fasciola hepatica* (B). The *F. hepatica* cells go through cycles of secretion and absorption.

growth but also for egg production. In 1959, Timms and Bueding showed that schistosomes could be maintained *in vitro* in the presence of a defined medium that only included haemoglobin as the protein source (Timms and Bueding, 1959). Haemoglobin, which is present in blood at 16 g/100 ml, therefore, appears to be the primary source of amino acids, although serum proteins (6–8 g/100 ml) and cell debris must also be important. Amino acid transporters have been identified in the tegument of schistosomes indicating that at least a proportion of the amino acid requirement of trematodes may enter through the surface (Dalton *et al.*, 2004).

Egg production in trematodes puts a heavy demand on the parasite for nutrients. Female *S. mansoni* worms ingest ten times more red blood corpuscles (330,000/h) than males in order to produce the estimated 100–200 eggs/day/female (Loker, 1983). *Schistosoma japonicum* females are even more prodigious producing 500–2000 eggs/day (Loker, 1983; Fan and Kang, 2003). The hermaphroditic adult liver fluke, on the other hand, consumes 0.03 ml of blood (approximately 350×10^6 red blood cells)/h and produces from 20,000 to 50,000 eggs/day (Whitfield, 1979; Andrews, 1999; Dalton *et al.*, 2004).

Free haemoglobin is a dangerous substance as both free haemoglobin and haem released following protease degradation can act as a 'Fenton reagent' with oxygen to generate reactive oxygen species that can damage membrane proteins and lipids (McGonigle *et al.*, 1998). It is generally believed that the parasite neutralizes free haem by complexation to form haematin and hence avoid oxidative damage (Oliveira and Oliveira, 2002). The guts of trematodes become filled with a dark pigment following digestion of blood (particularly obvious in female schistosomes and adult liver flukes). However, as digestion in trematodes is predominantly extracellular, any residual waste, including haematin pigment, appears in the lumen of the blind-ended gut and is eliminated regularly by the simple process of regurgitation (*F. hepatica* empties its gut every 3 h) (Hanna, 1975; Dalton *et al.*, 2004).

Prevention of Blood Clotting

Preventing the clotting of ingested blood in the gut is an obvious concern that blood-feeding parasites need to address. Recently, a schistosome anticoagulant protein (SAP, Sm 10.3), which does not resemble any previously described invertebrate anticoagulant protein, has been described (Bentley *et al.*, 2002). The parent protein is processed to several smaller proteins of 24–22.4 kDa, which contain tandem repeats of 27 amino acids. Immunolocalization studies showed that SAP is produced by the sub-tegumental cell bodies of the oesophageal 'gland' and is packaged into secretory granules before being transported through the luminal syncytium to the lumen where it mixes with the blood meal. Recombinant SAP significantly prolongs the activated partial thromboplastin time (APTT) *in vitro,* but the precise mechanism of anticoagulation still remains unclear. Inhibition of Factor X, a central component in both the intrinsic and extrinsic coagulation cascades, is a possibility (Bentley *et al.*, 2002).

Considerably more is known about anticoagulants from the blood-feeding hookworm, *Ancylostoma caninum*. Several families of small, anti-clotting peptides have been described in *A. caninum*. AcAP, an 8.7 kDa peptide secreted by adult worms that does not exhibit homology to other anticoagulants, but is a potent inhibitor of Factor Xa, was described by Cappello *et al.* (1995). A family of related peptides, one of which uniquely inhibited a complex of blood coagulation factor VIIA and tissue factor, was also reported (Stanssens *et al.*, 1996).

In addition to inhibitors of coagulation factors, adult *A. caninum* also secrete a protein that inhibits adherence of platelets to fibrinogen and collagen (Chadderdon and Cappello, 1999). Cloning of the platelet inhibitor (Del Valle *et al.*, 2003) revealed that this is a member of the pathogenesis-related protein (PRP) superfamily of hookworms, a family of secreted proteins with diverse biological activities (Zhan *et al.*, 2003). While similar proteins have yet to be characterized in schistosomes, multiple PRPs are present in the *S. mansoni* transcriptome (Verjovski-Almeida *et al.*, 2003). These anticoagulant proteins are not only potential targets for anti-helminth vaccines and specific inhibitors but are of interest in the development of new anti-thrombotic therapies for human and veterinary medicine (Verjovski-Almeida *et al.*, 2003; Hotez *et al.*, 2004).

Lysis of Blood Cells

Despite our knowledge of proteolytic digestion of haemoglobin by haematophagous parasites, the step immediately preceding this event, haemolysis, has received far less attention. Schistosomes, liver flukes and other blood-feeding helminths, including hookworms and trichostrongyle nematodes of livestock, ingest intact erythrocytes. These cells are then lysed by haemolytic factors released by cells that line the gastrointestinal tracts of these parasites.

The contribution of the oesophagus to the overall digestive process in schistosomes is unclear. As mentioned in the section on 'Schistosomes' this tissue is lined by gland cells that secrete granules that were suggested to contain cytolytic and/or digestive enzymes and it was considered that both lysis and digestion of blood cells began in the oesophagus (Ernst, 1975; Bogitsh and Carter, 1977). However, the speed with which a blood meal bolus passes through the oesophagous (~1 s) would limit its capacity as a digestive organ per se. In addition, digestive enzymes were not detected in the oesophagus, indicating that digestion takes place solely within the gut caecum (Bogitsh and Dresden, 1983). Accordingly, it is likely that the secretory granules observed in the cells lining the oesophagus secrete lytic and anticoagulant substances rather than peptidases.

The nature of the red blood cell lytic factors is only beginning to be elucidated although haemolytic activity was described in oesophageal and intestinal homogenates of adult *S. mansoni* almost 20 years ago (Kasschau *et al.*, 1986). This lytic activity was thought to be pore-forming in nature because it caused red blood cell ghost formation and responded in a similar way to the addition of sugars and salts as other known small-membrane pore-formers like saponin (Kasschau and Prill, 1988). More recently, mRNAs encoding small, pore-forming proteins with sequence similarity to the amoebapores of *Entamoeba histolytica* (Young *et al.*, 1982) were reported for the flukes *Clonorchis sinensis* (Lee *et al.*, 2002) and *F. hepatica* (Reed *et al.*, 2000; Espino and Hillyer, 2003). Both recombinant proteins lysed red cells and the former, termed clonorin, is expressed in the gut of adult worms. Amoebapores are a subfamily of saposin-like pore-forming proteins (Zhai and Saier, 2000) and amoebapore-like expressed sequence tags (ESTs) have been identified in schistosomes, hookworms and *Paragonimus westermani*. Full-length gene cloning, functional expression and haemolytic activity are currently being explored in our laboratories (T. Don and A. Loukas, Brisbane, 2004, personal communication; Fig. 18.2).

The function of haemolytic factors in trematodes may not, however, be exclusive to red cell lysis. A heat-stable haemolytic activity was also found in *S. mansoni* egg extracts, suggestive of a surfactant type haemolysin that might play a role in the movement of eggs through host tissues (Asahi *et al.*, 1984). Also, adherence of red blood cell ghosts to the tegument of schistosomula has been observed *in vitro*, indicating the presence of a surface-bound or excreted haemolysin (Caulfield and Cianci, 1985).

Peptidases Involved in the Digestion of Blood and Tissue Proteins

Given the demands of worm nutrition and fecundity, it is not surprising that considerable anabolic effort is invested in producing alimentary peptidases (proteolytic enzymes). Indeed, ultrastructural analysis of the gastrodermal cells of trematodes reveals evidence of intense protein synthesis, trafficking, storage and delivery, i.e. high numbers of ribosomes, Golgi apparatus and secretory vesicles, that reflects the production of digestive enzymes. It has been estimated that adult *F. hepatica* secrete 0.5–1.0 μg cathepsin L enzymes/parasite/h, representing 5% of the total protein of the adult worm (Dalton *et al.*, 2004). Further, quantitative data from the ongoing *S. japonicum* EST project demonstrates that the genes for the alimentary peptidases, cathepsins B and D (described later), rank third and fourth, respectively, as the most highly expressed genes in female worms (http://schistosoma.chgc.sh.com.cn/summary.general/index.htm), representing approximately 2% of all the cDNAs thus far identified. Finally, almost 10% of the ESTs identified in adult *F. hepatica* and *C. sinensis* cDNA libraries encode cathepsin L proteases indicating their pivotal role in parasite biology (Lee *et al.*, 2003; J.P. Dalton, unpublished data). The proteases of trematodes have been extensively reviewed elsewhere (Brindley *et al.*, 1997; Tort *et al.*, 1999; Sajid and McKerrow, 2002; Caffrey *et al.*, 2004) and, hence, only a brief overview is provided below.

Digestion in schistosomes

The schistosome gastrodermis produces an array of both endopeptidases (that cleave within a substrate protein) and exopeptidases (cleavage occurring at either end of a substrate), and their concerted action allows for the complete

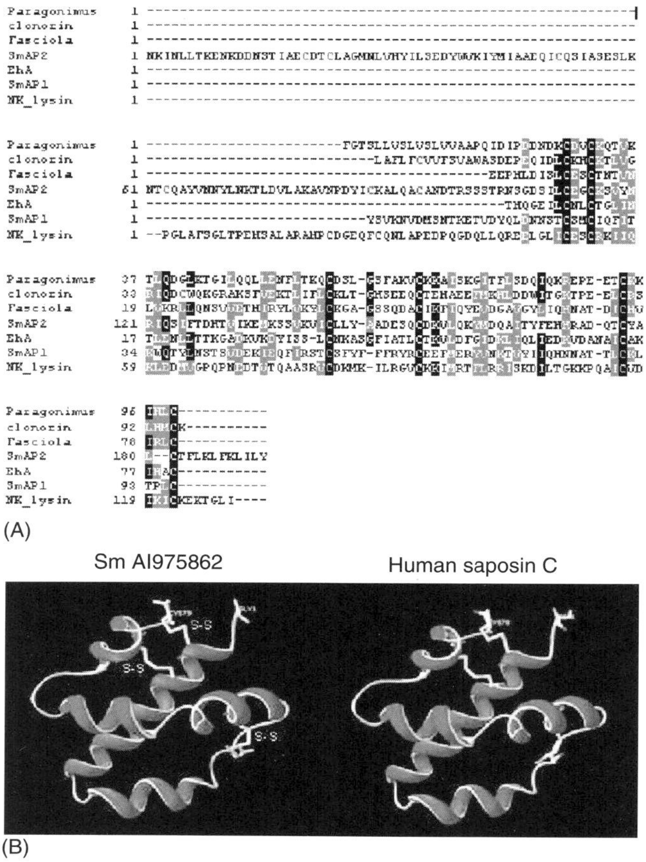

Fig. 18.2. Sequence alignment comparing single and double domain (SmAP2) saposin-like proteins from various trematodes, *Entamoeba histolytica* and human NK lysin (A). SmAP1 and SmAP2 are from *Schistosoma mansoni* (T. Don and A. Loukas, Brisbane, 2004, personal communication). Note the six conserved cysteine residues in each domain and the additional six N-terminal cysteines (bold type) in the N-terminal domain of SmAP2. Molecular model of another *S. mansoni* saposin-like protein (Protein Data Bank (PDB) Accession Number AI975862) based on the known NMR structure of human saposin C (PDB Accession Number 1M12) (B). S-S denotes predicted Cys residues involved in disulphide bond formation based on the known disulphides from saposin C. Note the similar folds adopted between the two sequences despite less than 50% identity at the primary sequence level. Modelling was performed using the Swiss Model server and structures were viewed with Swiss pdb Viewer.

hydrolysis of ingested proteins to amino acids (Tort *et al.*, 1999). The predominant peptidase class comprises the cysteine peptidases, which mediate peptide bond hydrolysis through a reactive sulphydryl group (Barrett and Rawlings, 2001). Cysteine endoproteases orthologous to the clan CA mammalian cathepsins, including cathepsins B, L and F, are found in gastrodermal cells and are secreted into the gut lumen (a comprehensive description of the classes of peptidases can be found in the MEROPS database published by Rawlings *et al.*, 2004). Each of these has overlapping but distinct preferences for peptide bonds and, hence, together can affect the efficient breakdown of proteins. As cysteine proteases generally operate at slightly acid pH (5.0–6.5), they are suitable for functioning within the low pH of the gut lumen (for *S. mansoni*) where initial cleavage steps take place (Dresden *et al.*, 1981; Dalton *et al.*, 1996; Brindley *et al.*, 1997; Tort *et al.*, 1999; Sajid and McKerrow, 2002).

Aspartic proteases orthologous to mammalian cathepsin D, which mediate peptide bond hydrolysis via two carboxylate groups (Barrett and Rawlings, 2001), likely act alongside cysteine proteases to achieve further digestion of proteins (Brindley *et al.*, 2001). A member of the clan CD cysteine peptidases, represented by an asparaginyl endopeptidase or 'legumain', is also an important protease secreted by the gastrodermis and may function directly in protein degradation or indirectly by processing and activating cysteine and aspartic proteases from the inactive zymogens (see section on 'Regulation of the Digestive Process'; Tort *et al.*, 1999; Caffrey *et al.*, 2004).

Final degradation of substrates to oligopeptides and free amino acids may involve gastrodermal exopeptidases such as a cathepsin C (Caffrey *et al.*, 2004), which removes dipeptides from the N-terminus of proteins, and a leucine aminopeptidase (LAP; McCarthy *et al.*, 2004), which is capable of releasing free amino acids from peptides and dipeptides. However, it is notable that cathepsin B also exhibits carboxydipeptidase activity and, therefore, may well play a dual role (Tort *et al.*, 1999; Caffrey *et al.*, 2004).

Digestion in liver flukes

Digestion of host tissues in the gut of *F. hepatica* differs to that of schistosomes as it appears to rely mainly on cathepsin L cysteine proteases (Dalton and Heffernan, 1989; Dalton *et al.*, 2003). Adult parasites release copious amounts of these proteases into the gut lumen from the surrounding gastrodermal epithelial cells during their secretory phase (Fig. 18.1; Yamasaki *et al.*, 1992; Smith *et al.*, 1993; Grams *et al.*, 2001; Dalton *et al.*, 2003). Cathepsin L proteases efficiently cleave interstitial matrix proteins such as haemoglobin, fibronectin, laminin and native collagen (Berasain *et al.*, 1997) and are suited to the proteolysis of host proteins as they have broad specificities, are active over a wide pH range (pH 3.0–8.0) and exhibit high stability at 37°C (Dalton *et al.*, 2003). Cathepsin B cysteine proteases may be more important in the early migratory stages of the parasite, particularly when they are migrating through the intestinal wall and entering the liver (Law *et al.*, 2003), while cathepsin Ls take over the function of tissue penetration and feeding in the stages that migrate from the liver into the bile duct stages (Dalton *et al.*, 2003).

Liver flukes also possess cathepsin C and LAP exopeptidases that are orthologous to the schistosome enzymes. These exopeptidases most likely complete the digestive process to yield free dipeptides and amino acids, respectively, from peptides generated by endoproteolytic cysteine protease activity on host proteins. Both cathepsin C and LAP have been immunolocalized to gastrodermal cells (Carmona *et al.*, 1994; Acosta *et al.*, 1998; J.P. Dalton, unpublished data).

Regulation of the Digestive Process

Studies by Halton and Stranock (1976) indicate that the presence of the blood meal in the gut and the resulting distension of the lumen is the trigger that induces an immediate outflow of digestive secretions from gastrodermal cells. These enzymes are secreted into an environment that is slightly acidic (Senft, 1976), which is important in several ways for regulating the digestive process. First, cysteine proteases and aspartic proteases are produced as inactive precursors or zymogens and require a slightly acidic pH to undergo autoprocessing and activation. Recent studies by Collins *et al.* (2004) have shown that the inactive cathepsin L zymogens are stored within the secretory vesicles of the gastrodermal cells and that processing to the fully active mature protease likely occurs in the acidic environment of the gut lumen. Second, the activity of the various proteases involved in digestion is optimal at acidic pHs. However, while the cathepsin L and B cysteine endoproteases function in a slightly acid environment (pH 4.0–6.0), aspartic proteases appear to be most active at even lower pH values (pH 2.5–4.6). Third, proteins such as haemoglobin are more easily digested at lower pH probably due to the relaxing of their tertiary structure (Becker *et al.*, 1995; J.P. Dalton, unpublished data). The low pH of the gut may be maintained by proton pumps in the lining of the gut wall and/or by the release of free protons following cleavage of peptide bonds by the proteases (Dalton *et al.*, 2004).

Although, as mentioned above, proteases undergo autoprocessing at low pH, it is possible that this process is initiated or facilitated by asparaginyl endopeptidases that cleave cathepsin zymogens at the junction between their propeptides and mature enzymes (Dalton *et al.*, 1995; Dalton and Brindley, 1996). Recently, Sajid *et al.* (2003) demonstrated that *S. mansoni* cathepsin B1 precursor could be *trans*-processed and activated by recombinant *S. mansoni* asparaginyl endopeptidase and suggested that this might be one level at which schistosomes regulate protease activity in the gut. Therefore, at the time of secretion of cathepsins into the lumen, some molecules may be already activated by *trans*-processing, which would accelerate the autoprocessing and activation of additional cathepsin zymogen due to the low environmental pH.

It is not clear whether catabolism of proteins by trematode gastrodermal peptidases is biochemically ordered, whereby some enzymes are responsible for the initial cleavage of substrates, thus facilitating subsequent digestion by other enzymes (Caffrey *et al.*, 2004; Dalton *et al.*, 2004). However, some order may be introduced by the compartmentalization of enzymes and their substrates. Caffrey *et al.* (2004) proposed that proteins partially degraded by cysteine protease activity may be taken up by the gastrodermal syncytium via pino- or endocytosis for further hydrolysis in a presumptive (phago)lysosomal compartment. The lower pH (4.0–5.0) of these vesicles would allow activity of cathepsin L, B and C cysteine proteases and cathepsin D aspartic proteases to produce small peptides and dipeptides (Tort *et al.*, 1999; Caffrey *et al.*, 2004). These may then be transported into the gastrodermal syncytium (schistosomes) or epithelial cells (*Fasciola*) for final breakdown to amino acids by LAP, which is functionally active only above pH 6.0 (McCarthy *et al.*, 2004). Finally, free amino acids would be distributed to the various parasite tissues via amino acid transporters (permeases) at the basal layer of the syncytium or tegumental cells (Dalton *et al.*, 2004).

Non-feeding Functions Attributed to Peptidases

A recurring theme regarding the study of parasite peptidases, including those from trematodes, has been an economy of effort whereby the same peptidase is found in different tissues or life stages of the parasite, presumably to perform different functions. For example, the cysteine proteases of adult *S. mansoni*, cathepsin B1 and asparaginyl endopeptidase, are found not only in the gut but also in the flame cells of cercariae where they possibly contribute to osmoregulation and/or excretion (Skelly and Shoemaker, 2001). Similarly, *S. japonicum* cathepsins D and L are present in eggs and miracidia as well as the gastrodermis of adults (Day *et al.*, 1995; Verity *et al.*, 1999). Finally, the cathepsin B (cat-B1) of adult *Fasciola gigantica* is expressed in the gut and most other tissues of the parasite and is, therefore, believed to function in multiple cellular processes (Meemon *et al.*, 2004).

Invasion and migration through host tissue

In addition to operating in internal tissues, many trematode peptidases are released extracorporeally where they work to the advantage of the parasite by degrading host tissues. This is especially relevant for juvenile parasites, which must forcibly penetrate through a number of tissues before arriving at the predilection site. Key to this adaptability is that many of the parasite cysteine proteases are catalytically active and stable at neutral pH (in contrast to their mammalian homologues which tend to function at low pH) and are capable of cleaving a wide range of native macromolecular substrates. For *F. gigantica*, two cathepsin Bs (FG cat-B2 and B3) are produced by metacercariae, NEJs and young juveniles, but not by adults, thus implying a selective function in tissue migration (Meemon *et al.*, 2004). Both cathepsins are released from the juvenile gut, which is capable of secretion, but not absorption (Bennett, 1975; Fairweather *et al.*, 1999). Similarly, *F. hepatica* expresses some cathepsin L proteases only in the NEJ stage and, these therefore, have been suggested to facilitate both excystment and migration of the juvenile parasites to the liver (Irving *et al.*, 2003). *F. hepatica* cysteine proteases (including cathepsins L1 and L2) degrade the relevant extracellular matrix proteins including fibrillar types I and III collagen, basement membrane type IV collagen, fibronectin and laminin *in vitro* (Berasain *et al.*, 1997) and it is likely that cysteine proteases released by the migratory stages of other trematodes such as *Paragonimus* and *Clonorchis* also degrade these substrates.

Aside from gut peptidases, it is possible that tegumental enzymes either actively released or sloughed off also contribute to the hydrolysis of tissue proteins, thereby facilitating migration. In addition, surface peptidases may contribute to the turnover and renewal of tegumental proteins or glycocalyx (Lammas and Duffus, 1983), the digestion of endocytosed proteins or assist in the evasion of host immune responses (see below). Consistent with these functions is the presence of cathepsins B and F in the tegument of schistosomes (Skelly and Shoemaker, 2001; Caffrey *et al.*, 2002; Meemon *et al.*, 2004).

The cercarial life stage of schistosomes is uniquely adapted for invasion of and migration through host tissues (Fig. 18.3; McKerrow and Salter, 2002; Curwen and Wilson, 2003). For *S. mansoni*, secretory vesicles, some of which contain peptidases, are produced in a number of specialized compartments including the acetabular gland cells and head gland cells (Curwen and Wilson, 2003). Little is known about the peptidases present in the head gland cells, which secrete their contents later than the acetabular cells and may assist in invading the blood vasculature (McKerrow and Salter, 2002; Curwen and Wilson, 2003). Within the acetabular gland cells, however, is a well-characterized 28 kDa serine peptidase termed cercarial protease or elastase due to its sequence similarity with mammalian pancreatic elastase (Marikovsky *et al.*, 1990). This enzyme is capable of degrading the relevant connective tissue molecules, including elastin, collagen and glycoproteins (McKerrow *et al.*, 1983). Some debate exists as to exactly when and for how long after skin invasion the acetabular gland cells release their contents (including cercarial elastase; McKerrow and Salter, 2002; Curwen and Wilson, 2003). However, recent studies with human skin have demonstrated the presence of cercariae in the dermis 10 min after invasion (Whitfield *et al.*, 2003), well within the time frame of release of acetabular gland cell contents (Fig. 18.3). Two of five genes encoding *S. mansoni* elastase isoforms account for greater than 90% of the activity measured (Salter *et al.*, 2002). During penetration, the elastase may also serve to remove the cercarial glycocalyx furthering the transformation of cercariae to schistosomulae (Fishelson *et al.*, 1992). Functionally similar enzymes have been identified in the invasive stages of other trematodes including *Schistosomatium douthitii* (Amiri *et al.*, 1988) and *Trichobilharzia ocellate* (Bahgat and Ruppel, 2002).

Excystment of juvenile parasites

A number of reports have suggested a relationship between cysteine protease activity and excystment of metacercariae. Cysteine proteases may act alone or in concert with extrinsic host factors such as trypsin in the host gut (Intapan and Maleewong, 2001; Li *et al.*, 2004). *In vitro*, *Paragonimus ohirai* metacercarial cysteine protease activity is induced by addition of the bile salt, sodium cholate and parasite

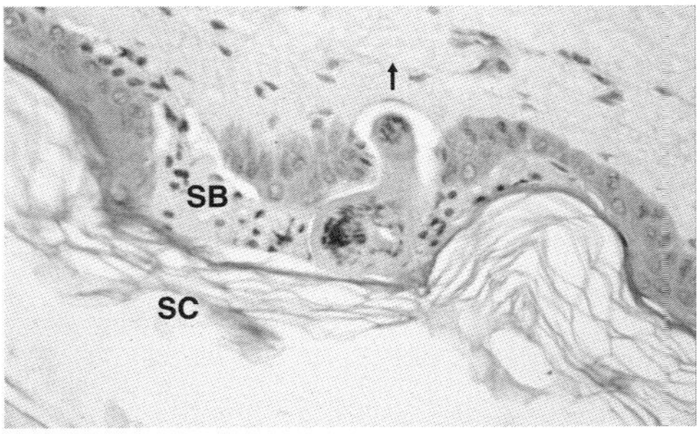

Fig. 18.3. A cercaria of *Schistosoma mansoni* penetrating the skin epidermis. The parasite has traversed the outer skin layer, stratum corneum (SC), and is halfway through the underlying dermis, stratum basale (SB). Movement through these layers is facilitated by proteases, most particularly an elastase, which is secreted from the cercarial acetabular glands. The cercaria is moving, as indicated by the arrow, into the inner skin layers where it will meet and enter a venule.

excystment is inhibited by inhibitors of cysteine proteases, but not serine peptidases (Ikeda, 2003). A possible source of these proteases is the excretory bladder where two *P. westermani* metacercarial cysteine proteases of 27 and 28 kDa have been localized (Yang *et al.*, 2002). The syncytial cells of the excretory bladder contain a large number of Golgi complexes, consistent with a function as a secretory organ (Orido, 1990). The specific activities of the 27 and 28 kDa peptidases are highest in metacercariae and NEJ, but decline thereafter, suggesting their direct involvement in excystation and subsequent migration (Chung *et al.*, 1997). This downregulation of expression is similar to that reported for the *F. gigantica* metacercarial cathepsins B2 and B3 described in the previous section (Meemon *et al.*, 2004).

Hatching of eggs

Trematode eggs are known to contain and release peptidases. For *S. mansoni,* it has been proposed that peptidases facilitate the migration of eggs through tissues and eventually into the external environment (Asch and Dresden, 1979; Day *et al.*, 1995). The same reports identified cysteine protease activity in *S. mansoni* egg extracts. Subsequently, three cysteine protease activities were partially purified from eggs, two of which were cathepsin B-like on the basis of substrate preference (Sung and Dresden, 1986). A monoclonal antibody localized one of these enzymes to the 'penetration' glands of unhatched miracidia (Dresden *et al.*, 1983), therefore, suggesting that the enzyme(s) may function in hatching and/or penetration of the intermediate snail host. The latter function was also proposed for cysteine proteases isolated from miracidial extracts and culture media containing transforming miracidia (Yoshino *et al.*, 1993). Consistent with these propositions is a report that transforming miracidia release peptidases that degrade the glycoprotein portion of the extracellular matrix (Pino-Heiss *et al.*, 1986). Finally, two peptidases of unknown classification were identified in the excretory–secretory products of *S. mansoni* eggs (Ashton *et al.*, 2001), as was a fibrinolytic activity, which was suggested to counteract the thrombogenic potential of the egg surface (Doenhoff *et al.*, 2003).

Protease Phylogeny

Proteases are believed to have arisen by a process of gene duplication followed by divergence from primitive enzymes with broad specificity (Neurath, 1984). As proteases evolved, their temporal and spatial expression changed, facilitating the development of more specialist functions. In helminth parasites, gene duplication and divergence occurred frequently in proteases even within particular species (Tort *et al.*, 1999). The net results are that parasites express several or many paralogous enzymes that perform different functions in different tissues at various stages of their life cycle, as highlighted in the previous section. Phylogenetic studies are important for understanding the evolution of proteases, relating structure to function (revealed by divergent primary structures and substrate-specificity studies) and identifying critical differences between host and parasite enzymes that may be exploited in anti-parasite drug design efforts.

Cathepsin B

Trematode cathepsin B-like peptidases exhibit less divergence (>65% similarity to one another) than the cathepsin L cysteine proteases (Fig. 18.4). A phylogenetic comparison of trematode cathepsin Bs identifies two main clades, termed here as cathepsin B1 and cathepsin B2. This observation is interesting as these two enzymes exhibit distinct differences in their localization in schistosome tissue; the former is associated with the gut, while the latter is expressed in both gut and tegument. Accordingly, this phylogenetic divergence may provide an example of the evolution of distinct functions for helminth cathepsin Bs, although the role of the cathepsin B2 in the tegument needs to be fully elucidated. It can also be seen from Fig. 18.3 that cathepsin C peptidases diverged early from the cathepsin B lineage and through increasing changes in structure became exclusively exopeptidic (see also Tort *et al.*, 1999).

The trematode cathepsin B cysteine proteases possess a 20-amino-acid insert termed the 'occluding loop', a feature that is used to classify them. The intact loop is found in all

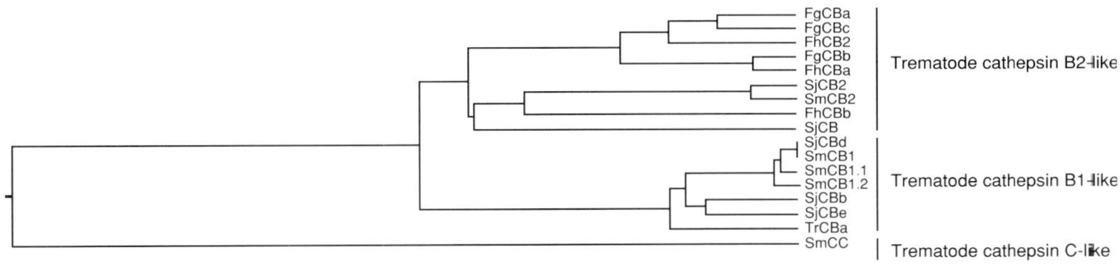

Fig. 18.4. Phylogenetic comparison of the trematode cathepsin B-like family generated with ClustalW and using the full-length peptidase sequences. The suffices a, b, c, etc. were added to differentiate the sequences retrieved. All sequences represent individual polypeptides; any duplicate or partial sequences were omitted. Keys: Fg, *Fasciola gigantica*; Fh, *Fasciola hepatica*; Sj, *Schistosoma japonicum*; Sm, *Schistosoma mansoni*; Tr, *Trichobilharzia regenti*; CB, cathepsin B; CC, cathepsin C. The peptidase acronyms correspond to the following Protein Data Bank Accession Numbers (shown in parentheses): FgCBa (AAO73004), FgCBb (AAO73003), FgCBc (AAO37002), FhCBa (AAD11445), FhCBb (CDAA80449), FhCB2 (CAD32937), SmCC (2123443A), SjCB (AAP05883), SjCBb (P43157), SjCB2 (AAO59414), SjCBd (P25792), SjCBe (S31909), SmCB1 (P25792), SmCB1.1 (CAD44624), SmCB1.2 (CAD44625), SmCB2 (CAC85211) and TrCBa (J. Dvorák, Prague, 2004, personal communication).

vertebrate, as well as many parasite cathepsin Bs, and contains two histidine residues at positions 110 and 111 (human cathepsin B numbering; Fig. 18.5), which confer exopeptidase activity, specifically dipeptidyl carboxypeptidase activity. While the schistosome cathepsins B1 and B2 possess these two histidine residues, and would therefore be expected to exhibit dipeptidyl carboxypeptidase activity, the *Fasciola* cathepsin B has lost His[111] (Tort *et al.*, 1999). However, the loss of His[111] may not necessarily result in the loss of exopeptidase activity as Krupa *et al.* (2002) showed that human cathepsin B with this residue altered, nevertheless, retained such activity.

Cathepsins L and F

Phylogenetic analysis, based on the primary sequences of the catalytic domains of cysteine proteases, has demonstrated a number of different clades within the trematode non-cathepsin B-like papain superfamily (Fig. 18.6; Tort *et al.*, 1999; Park *et al.*, 2002; Kang *et al.*, 2004). The parasite enzymes fall into two main

```
                       **
SjCB2    CQPYEFPPCEHHTLG-PLPVCDGDV-ETPPC
SmCB2    CQPYEFPPCEHHVIG-PLPSCDGDV-ETPSC
SjCBd    CEPYPFPKCEHHTKG-KYPPCGSKIYNTPRC
SmCB1    CEPYPFPKCEHHTKG-KYPPCGSKIYNTPRC
SmCB1a   CEPYPFPKCEHHTKG-KYPPCGSKIYKTPRC
SmCB1b   CEPYPFPKCEHHTKG-KYPPCGSKIYKTPRC
SjCBb    CQPYPFPKCEHHTKG-KYPACGTKIYKTPQC
SjCBf    CQPYPFPKCEHHTKG-KYPACGTKIYKTPQC
SjCBi    CQPYPFPKCEHHTKG-KYPACGTKIYKTPQC
TrCB1a   CQPYPFPKCEHHTTG-KYPECGEKIYKTPKC
FgCBa    CLPYPFPKCSHGVVTPGLPPCPRDIYPTPKC
FgCBc    CLPYPFPKCSHLEETPGLAPCPRELYATPKC
FhCB2    CLPYPFPQCRHPGSRSQLNPCPRYTYPTPSC
FgCBb    CQPWMFTKCDHVGDSRKYSRCPHYTYPTPPC
FhCBa    CQPWMFTKCDHVGDSRKYSRCPHYTYPTPPC
```

Fig. 18.5. Sequence alignment of the occluding loop of trematode cathepsin Bs. The predicted loop region is indicated by a black bar. The histidine residues conferring exopeptidase activity are marked with asterisks.

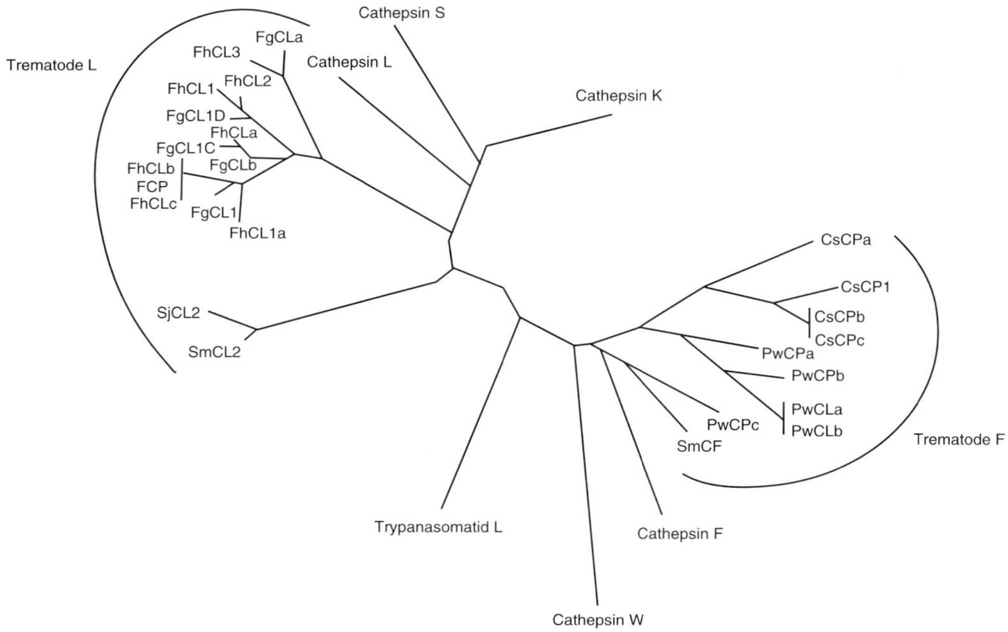

Fig. 18.6. Rooted phylogenetic comparison of the cathepsin L superfamily. Keys: Cs, *Clonorchis sinensis*; Pw, *Paragonimus westermani*; Fg, *Fasciola gigantica*; Fh, *Fasciola hepatica*; Sj, *Schistosoma japonicum*; Sm, *Schistosoma mansoni*; CL, cathepsin L; CF, cathepsin F; CP, cysteine peptidase. The terms cathepsin, trypanosomatid and filarial refer to mammalian, *Trypanosoma cruzi* and *Onchorcerca volvulus* peptidase orthologues, respectively. The peptidase acronyms correspond to the following Protein Data Bank Accession Numbers (shown in parentheses): Cathepsin F (NM_003793), cathepsin K (NM_000396), cathepsin L (M20496), cathepsin S (M90696), cathepsin W (NM_001335), CsCP1 (AF093243), CsCPa (U85984), CsCPb (U85983), CsCPc (AB020036), PwCLa (AF362769), PwCPa (D21124), PwCPb (AF71801), PwCLb (U70537), PwCPc (U69120), SmCF (U07345), trypanasomatid L (M84342), FgCLa (AF419329), FgCLb (AF239264), FgCL1 D and C (AF112566), FhCLa (AF271385), FhCLb (AB009306), FhCLc (L33772), FhCL1 (Z22765), FhCL1a (U62288), FhCL2 (U62280), FhCL3 (AJ270093), FCP (S70380), SjCL2 (U38476) and SmCL2 (Z32529).

clades: (i) cathespin L and (ii) cathepsin F. *Fasciola* and *Schistosoma* gut proteases belong to the cathepsin L lineage, which also includes the digestive enzymes of insects and crustaceans, and hence, these enzymes may have evolved to perform this function (Tort *et al.*, 1999). Ultimately, this lineage gave rise to the mammalian lysosomal cathepsins L, S and K, with which the trematode orthologues show 40–55% amino acid similarity (Tort *et al.*, 1999).

It is interesting that within the cathepsin L clade of *Fasciola*, significant allelic diversification has taken place. Thus far, 6 and 17 different cathepsin L sequences have been identified for *F. gigantica* and *F. hepatica*, respectively, but all belong to a monophyletic group (Fig. 18.6; Irving *et al.*, 2003). The possession of a large gene family may allow the simultaneous expression of several genes and enable the production of greater quantities of proteolytic enzyme. As a result, the parasite might be more efficient at penetrating and feeding on host tissues (Dalton *et al.*, 2004). The approximate time of divergence of the *Fasciola* sequences (~19 million years ago (Mya)) using molecular clock determinations coincides with a period of radiation of host pecoran species (ancestors of cow, sheep, etc.). Accordingly, it was suggested that the diversification of cathepsin Ls was an adaptive response to the availability of new host species (Irving *et al.*, 2003).

The cathepsin F clade of cysteine proteases has been identified in *Clonorchis*, *Paragonimus* and *Schistosoma*; but, thus far, not in *Fasciola* (Fig. 18.6). Also, members of this clade are more homologous to one another (60–76%) than those of the cathepsin L clade (36–37%) (Park *et al.*, 2002; Kang *et al.*, 2004), perhaps indicating a greater conservation of function. This lineage ultimately gave rise to the mammalian cathepsin F clade, which is ubiquitously expressed in cells and most likely performs household functions. Significant, however, is that most trematode cathepsin Fs (except *P. westermani* (Protein Data Bank (PDB) Accession Number U69120; Park *et al.*, 2001)) lack the cystatin-like domain situated upstream of the propeptide domains of mammalian, *Brugia malayi* and *Caenorhabditis elegans* cathepsin Fs. This might suggest an alternate function(s) and/or cellular localization for the trematode enzymes. Although *S. mansoni* cathepsin F has been localized to the gastrodermis of the adult it may have a function (or operate in a separate organelle) distinct from that of cathepsin L (Brady *et al.*, 2000a,b; Bogitsh *et al.*, 2001).

Aspartic proteases

Phylogenetic analysis has demonstrated that the schistosome aspartic protease is more related in primary amino acid sequence to mammalian cathepsin D (clan AA proteases, peptidase family A1) than to enzymes of the other major categories of vertebrate aspartic proteases (52% identical and 69% similar at the amino acid level). The parasite and mammalian cathepsin Ds exhibit remarkably similar secondary and three-dimensional structures, and share some cross-reactive epitopes (Brinkworth *et al.*, 2001; Morales *et al.*, 2004). However, they differ in their fine substrate specificity and by the presence of a long COOH-terminal extension in the schistosome orthologue (Brindley *et al.*, 2001). Homology modelling studies identified significant differences in the active site grooves of the schistosome and human cathepsin Ds, suggesting that they could be exploited for novel anti-parasitic agents (Brinkworth *et al.*, 2001).

Gene Structure, Orthology and Molecular Evolution

While primary structure analysis can provide information on phylogeny, the orthology between parasite and mammalian proteases can be substantiated by gene structure analysis. For example, the *S. mansoni* cathepsin D aspartic protease gene locus is ~13 kb in length with seven exons interrupted by six introns. The exons range in length from 49 to 294 bp and the introns from 30 to 5025 bp (Morales *et al.*, 2004). The genomic organization of *S. mansoni* cathepsin D is remarkably similar in sequence, structure and complexity to human cathepsin D (although the latter possesses a nine exon structure; Holm *et al.*, 1984), including a conservation of all six exon–intron boundaries of the schistosome gene (Fig. 18.7). By contrast, the schistosome cathepsin D gene organization is less similar to aspartic protease genes of the nematodes *C. elegans* (Tcherepanova *et al.*, 2000) and *Haemonchus contortus* (Longbottom *et al.*, 1997), indicating that it belongs to a separate aspartic protease lineage that eventually gave rise to human cathepsin D. Retrotransposons, which were detected in the introns of the cathepsin D gene, and are found in high copy number in the schistosome genome (Brindley *et al.*, 2003), may have influenced the evolutionary divergence of the organization of the schistosome aspartic protease gene away from the vertebrate nine-exon paradigm, through their effects on unequal crossing over (homologous recombination; Morales *et al.*, 2004).

The gene structure of a *F. gigantica* cathepsin L (GenBank Accession Number AB010923) has been described by Yamasaki *et al.* (2002). The gene locus spans ~2 kb, with four exons interrupted by three introns. The intron breakpoints do not correlate with the junctions of the prepeptide, propeptide and mature enzyme domain, indicating that gene structure does not correspond to the functional units of the protease. A second cathepsin L gene (GenBank Accession Number AB010924) exhibited a similar structure with the exception of the size of the third intron. Nevertheless, the organization of both *F. gigantica* cathepsin L genes is more compact and interrupted by fewer introns than their mammalian counter-

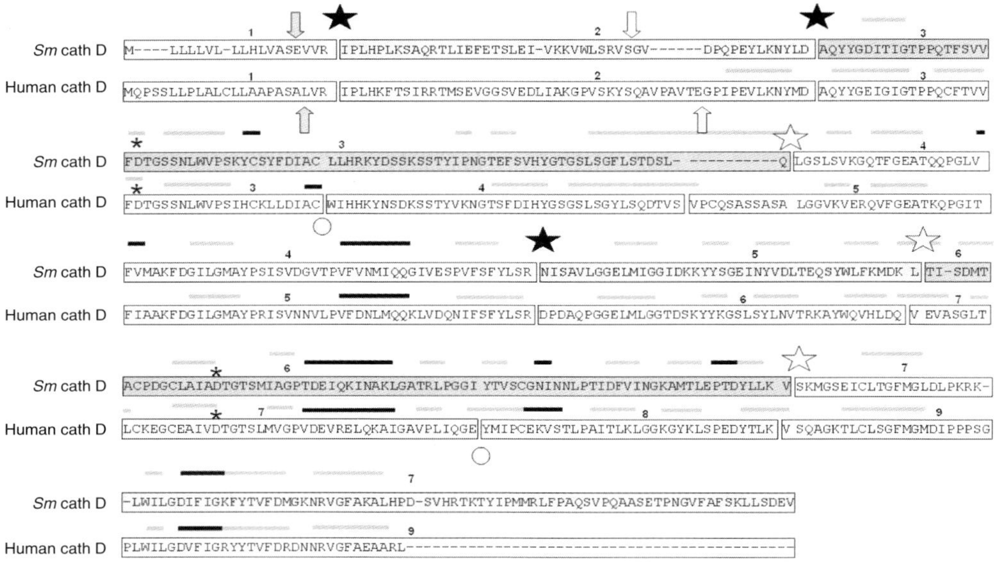

Fig. 18.7. Comparison of the exon–intron structures of *Schistosoma mansoni* cathepsin D (*Sm* cath D) and human cathepsin D (Human cath D), using the system of Jean *et al.* (2001). The two enzymes were aligned for maximal homology and the sequence corresponding to an exon for each protein was separated and boxed. Exon numbering is shown above the boxes. The position of the active site dyad of DTG motifs is indicated with asterisks (*). Black stars indicate where intron positions are conserved between the two enzymes, while white stars indicate where introns occur in similar, but not exactly the same, positions. Grey circles indicate positions of introns in the human cathepsin D gene that are absent from the schistosome cathepsin D gene. The secondary structure of the schistosome enzyme, as predicted by the Swiss Model First Approach mode, is indicated above the primary amino acid sequence, where grey bars indicate beta sheets and black bars indicate alpha helices. In like fashion, the secondary structure of human cathepsin D is indicated above the amino acid sequence. The position of cleavage of signal peptides is indicated with grey arrows, and that of cleavage of the propeptides from mature enzymes is indicated with white arrows. (After Morales *et al.*, 2004, with permission.)

part (Ishidoh *et al.*, 1989; Yamasaki *et al.*, 2002). However, the conserved intron insertion boundaries support the orthology of the parasite with crustacean and mammalian cathepsin Ls. By contrast, the structures of the *F. gigantica* genes are less similar to a cathepsin L-like gene reported from the tapeworm *Spirometra erinacei* (Yamasaki *et al.*, 2002).

Although similar gene comparisons are not yet available for the other platyhelminth proteases, these investigations should soon be feasible given that the human genome sequence is available and that the entire genome sequences of both *S. japonicum* and *S. mansoni* will be available in the near future (Hu *et al.*, 2003; El-Sayed *et al.*, 2004). Indeed, the gene structure of ostensible host orthologues is already available for some of the enzymes, e.g. mammalian cathepsins B and L (Ishidoh *et al.*, 1989; Ferrara *et al.*, 1990; Chauhan *et al.*, 1993).

Concluding Remarks

Proteases have played a pivotal role in the development of parasitism. By the time flatworms had emerged, members of all the major families of proteases had evolved such that parasites exhibit a complete profile of exo- and endoproteases. The best characterized of these proteases, including cathepsins B, L, C, D and LAP, are vital in the complex process of nutrient uptake from the host by sequentially degrading blood tissue proteins such as haemoglobin to free amino acids. However, other proteases are involved in separate essential functions, including tegumental turnover, parasite excystment, egg hatching and host penetration.

Our interest in trematode peptidases stems not just from a desire to understand their molecular and cellular functions, but to use this knowledge to develop practical applications in diagnostics, chemotherapy and vaccination. To date, cysteine proteases have proven sensitive and selective markers for the diagnosis of trematode infection in humans (Ruppel *et al.*, 1990; O'Neill *et al.*, 1999; Nagano *et al.*, 2004) and farm animals (Cornelissen *et al.*, 2001; Rokni *et al.*, 2002). Cathepsin B1 is a primary target for small-molecule cysteine protease inhibitors that decrease *S. mansoni* worm burden and fecundity in infected mice (Wasilewski *et al.*, 1996). In addition, significant success has been achieved in protecting animals against helminth infections due to *F. hepatica*, *S. mansoni*, *H. contortus* and *Trichostrongylus* sp., using digestive proteases as recombinant protein or DNA vaccines (Kofta *et al.*, 2000; Chlichlia *et al.*, 2001; Dalton and Mulcahy, 2001).

All of the above applications will benefit further from the widespread and increasing use of recombinant protein expression technologies, such as those utilizing the yeasts *Saccharomyces cerevisiae* and *Pichia pastoris* (Dowd *et al.*, 1997; Caffrey *et al.*, 2004; Collins *et al.*, 2004). The availability of pure and standardized quantities of recombinant protein opens the way for large-scale vaccine trials and reliable antigen-based diagnoses of trematode disease. The standardization of protein production is also important for anti-helminth drug design involving combinatorial chemical libraries and medium- to high-throughput screening systems. Such an approach has already been successfully employed in the discovery of potent inhibitors of peptidases of parasitic protozoa (Shenai *et al.*, 2003).

References

Acosta, D., Goni, F. and Carmona, C. (1998) Characterization and partial purification of a leucine aminopeptidase from *Fasciola hepatica*. *Journal of Parasitology* 84, 1–7.

Amiri, P., Sakanari, J., Basch, P., Newport, G. and McKerrow J.H. (1988) The *Schistosomatium douthitti* cercarial elastase is biochemically and structurally distinct from that of *Schistosoma mansoni*. *Molecular and Biochemical Parasitology* 28, 113–120.

Andrews, S.J. (1999) The Life cycle of *Fasciola hepatica*. In: Dalton, J.P. (ed.) *Fasciolosis*. CAB International, Wallingford, UK, pp. 1–29.

Asahi, H., Moribayashi, A., Sendo, F. and Kobayakawa, T. (1984) Hemolytic factors in *Schistosoma japonicum* eggs. *Infection and Immunity* 46, 514–518.

Asch, H.L. and Dresden, M.H. (1979) Acidic thiol proteinase activity of *Schistosoma mansoni* egg extracts. *Journal of Parasitology* 65, 543–549.

Ashton, P.D., Harrop, R., Shah, B. and Wilson, R.A. (2001) The schistosome egg: development and secretions. *Parasitology* 122, 329–338.

Bahgat, M. and Ruppel, A. (2002) Biochemical comparison of the serine protease (elastase) activities in cercarial secretions from *Trichobilharzia ocellata* and *Schistosoma mansoni*. *Parasitology Research* 88, 495–500.

Barrett, A.J. and Rawlings, N.D. (2001) Evolutionary lines of cysteine peptidases. *Biological Chemistry* 382, 727–733.

Becker, M.M., Harrop, S.A., Dalton, J.P., Kalinna, B.H., McManus, D.P. and Brindley, P.J. (1995) Cloning and characterization of the *Schistosoma japonicum* aspartic proteinase involved in haemoglobin degradation. *Journal of Biological Chemistry* 270, 24496–24501.

Bennett, C.E. (1975) *Fasciola hepatica*: development of the caecal epithelium during migration in the mouse. *Experimental Parasitology* 37, 426–441.

Bentley, G., Gutsmann, V., Pasi, J. and Agnew, A. (2002) Characterisation of a schistosome anti-coagulant protein, SAP. Molecular and Cellular Biology of Helminth Parasites Conference, September 2002, Hydra, Greece.

Berasain, P., Goni, F., McGonigle, S., Dowd, A., Dalton, J.P., Frangione, B. and Carmona, C. (1997) Proteinases secreted by *Fasciola hepatica* degrade extracellular matrix and basement membrane components. *Journal of Parasitology* 83, 1–5.

Bogitsh, B.J. and Carter, O.S. (1977) *Schistosoma mansoni*: ultrastructural studies on the esophageal secretory granules. *Journal of Parasitology* 63, 681–686.

Bogitsh, B.J. and Dresden, M.D. (1983) *Schistosoma japonicum*: biochemistry and cytochemistry of dipeptidyl aminopeptidase-II-like activity in adults. *Experimental Parasitology* 60, 163–170.

Bogitsh, B.J. and Kirschner, K.F. (1986) *Schistosoma japonicum*: ultrastructural localization of a haemoglobinase using mercury labelled pepstatin. *Experimental Parasitology* 62, 211–215.

Bogitsh, B.J., Dalton, J.P., Brady, C.P. and Brindley, P.J. (2001) Gut-associated immunolocalisation of the *Schistosoma mansoni* cysteine proteases, SmCL1 and SmCL2. *Journal of Parasitology* 87, 237–241.

Brady, C.P., Brindley, P.J., Dowd, A.J. and Dalton, J.P. (2000a) *Schistosoma mansoni*: differential expression of cathepsin L1 and cathepsin L2 suggests discrete biological functions for each enzyme. *Experimental Parasitology* 94, 75–83.

Brady, C.P., Brinkworth, R.I., Dalton, J.P., Dowd, A.J., Verity, C.K. and Brindley, P.J. (2000b) Molecular modelling and substrate specificity of discrete cruzipain-like and cathepsinL-like cysteine proteinases of human blood fluke *Schistosoma mansoni*. *Archives in Biochemistry and Biophysics* 380, 46–55.

Brindley, P.J., Kalinna, B.M., Dalton, J.P., Day, S.R., Wong, J.Y.M., Smythe, M.L. and McManus, D.P. (1997) Proteolytic degradation of host haemoglobin by schistosomes. *Molecular and Biochemical Parasitology* 89, 1–9.

Brindley, P.J., Kalinna, B.H., Wong, J.Y.M., Bogitsh, B.J., King, L.T., Smyth, D.J., Verity, C.K., Abbenante, G., Brinkworth, R.I., Fairlie, D.P., Smythe, M.L., Milburn, P.J., Bielefiedt-Ohmann, H., Zheng, Y. and McManus, D.P. (2001) Proteolysis of human haemoglobin by schistosome cathepsin D. *Molecular and Biochemical Parasitology* 112, 103–112.

Brindley, P.J., Laha, T., McManus, D.P. and Loukas, A. (2003) Mobile genetic elements colonizing the genomes of metazoan parasites. *Trends in Parasitology* 19, 79–87.

Brinkworth, R.I., Prociv, P., Loukas, A. and Brindley, P.J. (2001) Haemoglobin-degrading, aspartic protease of blood-feeding parasites: substrate specificity revealed by homology models. *The Journal of Biological Chemistry* 276, 38844–38851.

Caffrey, C.R., Salter, J.P., Lucas, K.D., Khiem, D., Hsieh, I., Lim, K.C., Ruppel, A., McKerrow, J.H. and Sajid, M. (2002) SmCB2, a novel tegumental cathepsin B from adult *Schistosoma mansoni*. *Molecular and Biochemical Parasitology* 121, 49–61.

Caffrey, C.R., McKerrow, J.H., Salter, J.P. and Sajid, M. (2004) Blood 'n' Guts: an update on schistosome digestive peptidases. *Trends in Parasitology* 20, 241–248.

Cappello, M., Vlasuk, G.P., Bergum, P.W., Huang, S. and Hotez, P.J. (1995) *Ancylostoma caninum* anticoagulant peptide: a hookworm-derived inhibitor of human coagulation factor Xa. *Proceedings of the National Academy of Sciences USA* 92, 6152–6156.

Carmona, C., McGonigle, S., Dowd, A.J., Smith, A.M., Coughlan, S., MacGowran, E. and Dalton, J.P. (1994) A dipeptidylpeptidase secreted by *Fasciola hepatica*. *Parasitology* 109, 113–118.

Caulfield, J.P. and Cianci, C.M. (1985) Human erythrocytes adhering to schistosomula of *Schistosoma mansoni* lyse and fail to transfer membrane components to the parasite. *Journal of Cell Biology* 101, 158–166.

Chadderdon, R.C. and Cappello, M. (1999) The hookworm platelet inhibitor: functional blockade of integrins GPIIb/IIIa (alphaIIbbeta3) and GPIa/IIa (alpha2beta1) inhibits platelet aggregation and adhesion *in vitro*. *Journal of Infectious Diseases* 179, 1235–1241.

Chauhan, S.S., Popescu, N.C., Ray, D., Fleischmann, R., Gottesman, M.M. and Troen, B.R. (1993) Cloning, genomic organization, and chromosomal localization of human cathepsin L. *Journal of Biological Chemistry* 268, 1039–1045.

Chlichlia, K., Bahgat, M., Ruppel, A. and Schirrmacher, V. (2001) DNA vaccination with asparaginyl endopeptidase (Sm32) from the parasite *Schistosoma mansoni*: anti-fecundity effect induced in mice. *Vaccine* 20, 439–447.

Chung, Y.B., Yang, H.J., Kang, S.Y., Kong, Y. and Cho, S.Y. (1997) Activities of different cysteine proteases of *Paragonimus westermani* in cleaving human IgG. *Korean Journal of Parasitology* 35, 139–142.

Collins, P.R., Stack, C.M., O'Neill, S.M., Doyle, S., Ryan, T., Brennan, G.P., Mousley, A., Stewart, M., Maule, A.G., Dalton, J.P. and Donnelly, S. (2004) Cathepsin L1, the major protease involved in liver fluke (*Fasciola hepatica*) virulence: propeptide cleavage sites and autoactivation of the zymogen secreted from gastrodermal cells. *Journal of Biological Chemistry* 279, 17038–17046.

Cornelissen, J.B., Gaasenbeek, C.P., Borgsteede, F.H., Holland, W.G., Harmsen, M.M. and Boersma, W.J. (2001) Early immunodiagnosis of fasciolosis in ruminants using recombinant *Fasciola hepatica* cathepsin L-like protease. *International Journal of Parasitology* 31, 728–737.

Curwen, R.S. and Wilson, R.A. (2003) Invasion of skin by schistosome cercariae: some neglected facts. *Trends in Parasitology* 19, 63–66.
Dalton, J.P. and Brindley, P.J. (1996) Schistosome asparaginyl endopeptidase Sm32 in haemoglobin digestion. *Parasitology Today* 12, 125.
Dalton, J.P. and Heffernan, M. (1989) Thiol proteases released *in vitro* by *Fasciola hepatica. Molecular and Biochemical Parasitology* 35, 161–166.
Dalton, J.P. and Mulcahy, G. (2001) Parasite vaccines – a reality? *Veterinary Parasitology* 2079, 1–19.
Dalton, J.P., Hola-Jamriska, L. and Brindley, P.J. (1995) Asparaginyl endopeptidase activity in adult *Schistosoma mansoni. Parasitology* 111, 575–580.
Dalton, J.P., Clough, K.A., Jones, M.K. and Brindley, P.J. (1996) Characterization of the cathepsin-like cysteine proteinases of *Schistosoma mansoni. Infection and Immunity* 64, 1328–1334.
Dalton, J.P., O'Neill, S., Stack, C., Collins, P., Walshe, A., Sekiya, M., Doyle, S., Mulcahy, G., Hoyle, D., Khaznadji, E., Moire, N., Brennan, G., Mousley, A., Kreshchenko, N., Maule, A.G. and Donnelly, S.M. (2003) *Fasciola hepatica* cathepsin L-like proteases: biology, function, and potential in the development of first generation liver fluke vaccines. *International Journal of Parasitology* 33, 1173–1181.
Dalton, J.P., Skelly, P. and Halton, D.W. (2004) Role of the tegument and gut in nutrient uptake by parasitic platyhelminths. *Canadian Journal of Zoology* 82, 211–232.
Day, S.R., Dalton, J.P., Clough, K.A., Leonardo, L., Tiu, W.U. and Brindley, P.J. (1995) Cloning and characterization of cathepsin L proteinases of *Schistosoma japonicum. Biochemical and Biophysical Research Communications* 217, 1–9.
Del Valle, A., Jones, B.F., Harrison, L.M., Chadderdon, R.C. and Cappello, M. (2003) Isolation and molecular cloning of a secreted hookworm platelet inhibitor from adult *Ancylostoma caninum. Molecular and Biochemical Parasitology* 129, 167–177.
Doenhoff, M.J., Stanley, R.G., Pryce, D., Curtis, R.C., Parry, H., Griffiths, K. and Jackson, C.L. (2003) Identification of a fibrinolytic enzyme in *Schistosoma mansoni* eggs and modulated blood fibrinogen metabolism in *S. mansoni*-infected mice. *Parasitology* 126, 231–234.
Dowd, A.J., Tort, J., Roche, L. and Dalton, J.P. (1997) Isolation of a cDNA encoding *Fasciola hepatica* cathepsin L2 and functional expression in *Saccharomyces cerevisiae. Molecular and Biochemical Parasitology* 88, 163–174.
Dresden, M.H., Rutledge, M.L. and Chappell, C.L. (1981) Properties of the acid thiol proteinase from *Schistosoma mansoni* adults. *Molecular Biochemical Parasitology* 4, 61–66.
Dresden, M.H., Sung, C.K. and Deelder, A.M. (1983) A monoclonal antibody from infected mice to a *Schistosoma mansoni* egg proteinase. *Journal of Immunology*, 130, 1–3.
El-Sayed, N.M., Bartholomeu, D., Ivens, A., Johnston, D.A. and LoVerde, P.T. (2004) Advances in schistosome genomics. *Trends in Parasitology* 20, 154–157.
Ernst, S.C. (1975) Biochemical and cytochemical studies of digestive–absorptive functions of esophagus, cecum, and tegument in *Schistosoma mansoni*: acid phosphatase and tracer studies. *Journal of Parasitology* 61, 633–647.
Espino, A.M. and Hillyer, G.V. (2003) Molecular cloning of a member of the *Fasciola hepatica* saposin-like protein family. *Journal of Parasitology* 89, 545–552.
Fairweather, I., Threadgold, L.T. and Hanna, R.E.B. (1999) Development of *Fasciola hepatica* in the mammalian host. In: Dalton, J.P. (ed.) *Fasciolosis*. CAB International, Wallingford, UK, pp. 47–111.
Fan, P.C. and Kang, Y.C. (2003) Egg production capacity of one-pair worms of *Schistosoma japonicum* in albino mice. *Southeast Asian Journal of Tropical Medicine and Public Health* 34, 708–712.
Ferrara, M., Wojcik, F., Rhaissi, H., Mordier, S., Roux, M.P. and Bechet, D. (1990) Gene structure of mouse cathepsin B. *FEBS Letters* 273, 195–199.
Fishelson, Z., Amiri, P., Friend, D.S., Marikovsky, M., Petitt, M., Newport, G. and McKerrow, J.H. (1992) *Schistosoma mansoni*: cell-specific expression and secretion of a serine protease during development of cercariae. *Experimental Parasitology* 75, 87–98.
Grams, R., Vichasri-Grams, S., Sobhorn, P., Upatham, E.S. and Viyanant, V. (2001) Molecular cloning and characterisation of cathepsin L encoding genes from *Fasciola gigantica. Parasitology International* 50, 105–114.
Halton, D.W. (1997) Nutritional adaptations to parasitism within the platyhelminths. *International Journal for Parasitology* 27, 693–704.
Halton, D.W. and Stranock, S.D. (1976) The fine structure and histochemistry of the caecal epithelium of *Calicotyle kröyeri* (Monogenea: Monopoisthocotylea). *International Journal for Parasitology* 6, 253–263.
Hanna, R.E.B. (1975) *Fasciola hepatica*: a light and electron microscope autoradiographic study of protein synthesis and secretion by gut cells in tissue slices. *Experimental Parasitology* 39, 18–28.

Holm, I., Ollo, R., Panthier, J.J. and Rougeon, F. (1984) Evolution of aspartyl proteases by gene duplication: the mouse renin gene is organized by two homologous clusters of four exons. *EMBO Journal* 3, 557–562.

Hotez, P.J., Brooker, S., Bethony, J.M., Bottazzi, M.E., Loukas, A. and Xiao, S. (2004) Hookworm infection. *New England Journal of Medicine* 351, 799–807.

Howell, M.J. (1970) Ingestion by two species of digenetic trematode. *Journal of Parasitology* 56, 391–392.

Hu, W., Yan, Q., Shen, D.K., Liu, F., Zhu, Z.D., Song, H.D., Wang, Z.J., Rong, Y.P., Zeng, L.C., Wu, J., Zhang, X., Wang, J.J., Xu, X.N., Wang, S.Y., Fu, G., Xu, X.R., Zhang, X.L., Wang, Z.Q., Brindley, P.J., McManus, D.P., Xue, C.L., Zheng, F., Chen, Z. and Han, Z.G. (2003) Evolutionary and biomedical implications of a *Schistosoma japonicum* complementary DNA resource. *Nature Genetics* 35, 139–147.

Ikeda, T. (2003) Involvement of cysteine proteinases in excystment of *Paragonimus ohirai* metacercariae induced by sodium cholate and A23187. *Journal of Helminthology* 77, 21–26.

Intapan, P. and Maleewong, W. (2001) *In vitro* excystation of *Paragonimus heterotremus* metacercariae. *Journal of Parasitology* 87, 1184–1186.

Irving, J.A., Spithill, T.W., Pike, R.N., Whisstock, J.C. and Smooker, P.M. (2003) The evolution of enzyme specificity in *Fasciola* spp. *Journal of Molecular Evolution* 57, 1–15.

Ishidoh, K., Kominami, E., Suzuki, K. and Katsunuma, N. (1989) Gene structure and 5′-upstream sequence of rat cathepsin L. *FEBS Letters* 259, 71–74.

Jean, L., Long, M., Young, J., Pery, P. and Tomley, F. (2001) Aspartyl proteinase genes from apicomplexan parasites: evidence for evolution of the gene structure. *Trends in Parasitology* 17, 491–498.

Kang, T.H., Yun, D.H., Lee, E.H., Chung, Y.B., Bae, Y.A., Chung, J.Y., Kang, I., Kim, J., Cho, S.Y. and Kong, Y. (2004) A cathepsin F of adult *Clonorchis sinensis* and its phylogenetic conservation in trematodes. *Parasitology* 128, 195–207.

Kasschau, M.R. and Prill, M.S. (1988) Comparative effects of sugars on the haemolytic activity of *Schistosoma mansoni* and other haemolytic agents. *Comparative Biochemistry and Physiology* 90, 453–458.

Kasschau, M.R., Robinson, D.C. and Dresden, M.H. (1986) *Schistosoma mansoni*: activation of hemolytic activity in homogenates from live adult worms. *Experimental Parasitology* 62, 442–449.

Kofta, W., Mieszczanek, J., Plucienniczak, G. and Wedrychowicz, H. (2000) Successful DNA immunisation of rats against fasciolosis. *Vaccine* 18, 2985–2990.

Krupa, J.C., Hasnain, S., Nagler, D.K., Menard, R. and Mort, J.S. (2002) S2′ substrate specificity and the role of His110 and His111 in the exopeptidase activity of human cathepsin B. *Biochemical Journal* 361, 613–619.

Lammas, D.A. and Duffus, W.P. (1983) The shedding of the outer glycocalyx of juvenile *Fasciola hepatica*. *Veterinary Parasitology* 12, 165–178.

Law, R.H., Smooker, P.M., Irving, J.A., Piedrafita, D., Ponting, R., Kennedy, N.J., Whisstock, J.C., Pike, R.N. and Spithill, T.W. (2003) Cloning and expression of the major secreted cathepsin B-like protein from juvenile *Fasciola hepatica* and analysis of immunogenicity following liver fluke infection. *Infection and Immunity* 71, 6921–6932.

Lee, J.Y., Cho, P.Y., Kim, T.Y., Kang, S.Y., Song, K.Y. and Hong, S.J. (2002) Hemolytic activity and developmental expression of pore forming peptide, clonorin. *Biochemical and Biophysical Research Communications* 296, 1238–1244.

Lee, J.S., Lee, J., Park, S.J. and Yung, T.S. (2003) Analysis of the genes expressed in *Clonorchis sinensis* adults using the expressed sequence tag approach. *Parasitology Research* 91, 283–289.

Li, S., Chung, Y.B., Chung, B.S., Choi, M.H., Yu, J.R. and Hong, S.T. (2004) The involvement of the cysteine proteases of *Clonorchis sinensis* metacercariae in excystment. *Parasitology Research* 93, 36–40.

Loker, E.S. (1983) A comparative study of the life-histories of mammalian schistosomes. *Parasitology* 87, 343–369.

Longbottom, D., Redmond, D.L., Russell, M., Liddell, S., Smith, W.D. and Knox, D.P. (1997) Molecular cloning and characterisation of a putative aspartate proteinase associated with a gut membrane protein complex from adult *Haemonchus contortus*. *Molecular Biochemical Parasitology* 88, 63–72.

Marikovsky, M., Arnon, R. and Fishelson, Z. (1990) *Schistosoma mansoni*: localization of the 28 kDa secreted protease in cercaria. *Parasite Immunology* 12, 389–401.

McCarthy, E., Stack, C., Donnelly, S.M., Doyle, S., Mann, V.H., Brindley, P.J., Stewart, M., Day, T.A., Maule, A.G. and Dalton, J.P. (2004) Leucine aminopeptidase of the human blood flukes, *Schistosoma mansoni* and *Schistosoma japonicum*. *International Journal of Parasitology* 34, 703–714.

McGonigle, S., Dalton, J.P. and James, E.R. (1998) Peroxidoxins. *Parasitology Today* 14, 139–145.

McKerrow, J.H. and Salter, J. (2002) Invasion of skin by *Schistosoma* cercariae. *Trends in Parasitology* 18, 193–195.

McKerrow, J.H., Keene, W.E., Jeong, K.H. and Werb, Z. (1983) Degradation of extracellular matrix by larvae of *Schistosoma mansoni*. I. Degradation by cercariae as a model for initial parasite invasion of host. *Laboratory Investigations* 49, 195–200.
Meemon, K., Grams, R., Vichasri-Grams, S., Hofmann, A., Korge, G., Viyanant, V., Upatham, E.S., Habe, S. and Sobhon, P. (2004) Molecular cloning and analysis of stage- and tissue-specific expression of cathepsin B encoding genes from *Fasciola gigantica. Molecular and Biochemical Parasitology* 136, 1–10.
Morales, M.E., Kalinna, B.H., Heyers, O., Schulmeister, A., Mann, V.H., Copeland, C.S. Loukas, A. and Brindley, P.J. (2004) Genomic organization of the *Schistosoma mansoni* aspartic protease gene, a platyhelminth orthologue of mammalian lysosomal cathepsin D. *Gene* 338, 99–109.
Nagano, I., Pei, F., Wu, Z., Wu, J., Cui, H., Boonmars, T. and Takahashi, Y. (2004) Molecular expression of a cysteine proteinase of *Clonorchis sinensis* and its application to an enzyme-linked immunosorbent assay for immunodiagnosis of clonorchiasis. *Clinical, Diagnostic and Laboratory Immunology* 11, 411–416.
Neurath, H. (1984) Evolution of proteolytic enzymes. *Science* 224, 350–357.
Oliveira, P.L. and Oliveira, M.F. (2002) Vampires, Pastuer and reactive oxygen species: is the switch from aerobic to anaerobic metabolism a preventive anti-oxidant defence in blood-feeding parasites. *FEBS Letters* 525, 3–6.
O'Neill, S.M., Parkinson, M., Dowd, A.J., Strauss, W., Angles, R. and Dalton, J.P. (1999) Immunodiagnosis of human fasciolosis using recombinant *Fasciola hepatica* cathepsin L1 cysteine proteinase. *American Journal of Tropical Medicine and Hygiene* 60, 749–751.
Orido, Y. (1990) Development of the excretory bladder of the lung fluke *Paragonimus ohirai* (Trematoda: Troglotrematidae). *Journal of Parasitology* 76, 205–211.
Park, H., Hong, K.M., Sakanari, J.A., Choi, J.H., Park, S.K., Kim, K.Y., Hwang, H.A., Paik, M.K., Yun, K.J., Shin, C.H., Lee, J.B., Ryu, J.S. and Min, D.Y. (2001) *Paragonimus westermani*: cloning of a cathepsin F-like cysteine proteinase from the adult worm. *Experimental Parasitology* 98, 223–227.
Park, H., Kim, S.I., Hong, K.M., Kim, M.J., Shinm, C.H., Ryu, J.S., Min, D.Y., Lee, J.B. and Hwang, U.W. (2002) Characterization and classification of five cysteine proteinases expressed by *Paragonimus westermani* adult worm. *Experimental Parasitology* 102, 143–149.
Pino-Heiss, S., Petitt, M., Beckstead, J.H. and McKerrow, J.H. (1986) Preparation of mouse monoclonal antibodies and evidence for a host-immune response to the preacetabular gland proteinase of *Schistosoma mansoni* cercariae. *American Journal of Tropical Medicine and Hygiene* 35, 536–543.
Rawlings, N.D., Tolle, D.P. and Barrett, A.J. (2004) MEROPS: the peptidase database. *Nucleic Acids Research* 32, 160–164.
Reed, M.B., Strugnell, R.A., Panaccio, M. and Spithill, T.W. (2000) A novel member of the NK-lysin protein family is developmentally regulated and secreted by *Fasciola hepatica. Molecular Biochemical Parasitology* 105, 297–303.
Robinson, G. and Threadgold, L.T. (1975) Electron microscope studies of *Fasciola hepatica*. XII. The fine structure of the gastrodermis. *Experimental Parasitology* 37, 20–36.
Rokni, M.B., Massoud, J., O'Neill, S.M., Parkinson, M. and Dalton, J.P (2002) Diagnosis of human fasciolosis in the Gilan province of Northern Iran: application of cathepsin L-ELISA. *Diagnosis of Microbiological and Infectious Diseases* 44, 175–179.
Ruppel, A., Idris, M.A., Sulaiman, S.M. and Hilali, A.M. (1990) *Schistosoma mansoni* diagnostic antigens (Sm 31/12): a sero-epidemiological study in the Sudan. *Tropical and Medical Parasitology* 41, 127–130.
Sajid, M. and McKerrow, J.H. (2002) Cysteine proteases of parasitic organisms. *Molecular and Biochemical Parasitology* 120, 1–21.
Sajid, M., McKerrow, J.H., Hansell, E., Mathieu, M.A., Lucas, K.D., Hsieh, I., Greenbaum, D., Bogyo, M., Salter, J.P., Lim, K.C., Franklin, C., Kim, J.H. and Caffrey, C.R. (2003) Functional expression and characterization of *Schistosoma mansoni* cathepsin B and its trans-activation by an endogenous asparaginyl endopeptidase. *Molecular Biochemical Parasitology* 131, 65–75.
Salter, J.P., Choe, Y., Albrecht, H., Franklin, C., Lim, K.C., Craik, C.S. and McKerrow, J.H. (2002) Cercarial elastase is encoded by a functionally conserved gene family across multiple species of schistosomes. *Journal of Biological Chemistry* 277, 24618–24624.
Senft, A.W. (1976) Observations on the physiology of the gut of *Schistosoma mansoni*. In: Van dev Bossche, H. (ed.) *The Biochemistry of Parasites and Host–Parasite Relationships*. North-Holland Publishing Company, Oxford, UK.
Shenai, B.R., Lee, B.J., Alvarez-Hernandez, A., Chong, P.Y., Emal, C.D., Neitz, R.J., Roush, W.R. and Rosenthal, P.J. (2003) Structure–activity relationships for inhibition of cysteine protease activity and development

of Plasmodium falciparum by peptidyl vinyl sulfones. *Antimicrobial Agents and Chemotherapy* 47, 154–160.

Skelly, P.J. and Shoemaker, C.B. (2001) *Schistosoma mansoni* proteases Sm31 (cathepsin B) and Sm32 (legumain) are expressed in the cecum and protonephridia of cercariae. *Journal of Parasitology* 87, 1218–1221.

Smith, A.M., Dalton, J.P., Clough, K.A., Kilbane, C.L., Harrop, S.A., Hole, N. and Brindley, P.J. (1993) Adult *Schistosoma mansoni* express cathepsin L proteinase. *Molecular and Biochemical Parasitology* 67, 11–19.

Stanssens, P., Bergum, P.W., Gansemans, Y., Jespers, L., Laroche, Y., Huang, S., Maki, S., Messens, J., Lauwereys, M., Cappello, M., Hotez, P.J., Lasters, I. and Vlasuk, G.P. (1996) Anticoagulant repertoire of the hookworm *Ancylostoma caninum*. *Proceedings of the National Academy of Sciences USA* 93, 2149–2154.

Sung, C.K. and Dresden, M.H. (1986) Cysteinyl proteinases of *Schistosoma mansoni* eggs: purification and partial characterization. *Journal of Parasitology* 72, 891–900.

Tcherepanova, I., Bhattacharyya, L., Rubin, C. and Freedman, J. (2000) Aspartic proteases from the nematode *Caenorhabditis elegans*: structural organization and developmental and cell-specific expression of *asp-1*. *Journal of Biological Chemistry* 275, 26359–26369.

Tielens, A.G.M. (1997) Biochemistry of trematodes. In: Fried, B and Graczyk, T.K. (eds) *Advances in Trematode Biology*. CRC Press, Boca Raton, Florida, pp. 309–344.

Tielens, A.G.M. (1999) Metabolism. In: Dalton, J.P. (ed.) *Fasciolosis*. CAB International, Wallingford, UK, pp. 277–306.

Timms, A.R. and Bueding, E. (1959) Studies of a proteolytic enzyme from *Schistosoma mansoni*. *British Journal of Pharmacology* 14, 68–73.

Tort, J., Brindley, P.J., Knox, D., Wolfe, K.H. and Dalton, J.P. (1999) Helminth proteinases and their associated genes. *Advances in Parasitology* 43, 161–266.

Verity, C.K., McManus, D.P. and Brindley, P.J. (1999) Developmental expression of cathepsin D aspartic protease in *Schistosoma japonicum*. *International Journal of Parasitology* 29, 1819–1824.

Verjovski-Almeida, S., DeMarco, R., Martins, E.A., Guimaraes, P.E., Ojopi, E.P., Paquola, A.C., Piazza, J.P., Nishiyama, M.Y. Jr, Kitajima, J.P., Adamson, R.E., Ashton, P.D., Bonaldo, M.F., Coulson, P.S., Dillon, G.P., Farias, L.P., Gregorio, S.P., Ho, P.L., Leite, R.A., Malaquias, L.C., Marques, R.C., Miyasato, P.A., Nascimento, A.L., Ohlweiler, F.P., Reis, E.M., Ribeiro, M.A., Sa, R.G., Stukart, G.C., Soares, M.B., Gargioni, C., Kawano, T., Rodrigues, V., Madeira, A.M., Wilson, R.A., Menck, C.F., Setubal, J.C., Leite, L.C. and Dias-Neto, E. (2003) Transcriptome analysis of the acoelomate human parasite *Schistosoma mansoni*. *Nature Genetics* 35, 148–157.

Wasilewski, M.M., Lim, K.C., Phillips, J. and McKerrow, J.H. (1996) Cysteine protease inhibitors block schistosome haemoglobin degradation *in vitro* and decrease worm burden and egg production *in vivo*. *Molecular and Biochemical Parasitology* 81, 179–189.

Whitfield, P.J. (1979) *The Biology of Parasitism: An Introduction to the Study of Associating Organisms*. Edward Arnold, London.

Whitfield, P.J., Bartlett, A., Brown, M.B. and Marriot, C. (2003) Invasion by schistosome cercariae: studies with human skin explants. *Trends in Parasitology* 19, 339–340.

Yamasaki, H., Kominami, E. and Aoki, T. (1992) Immunocytochemical localization of a cysteine protease in adult worms of the liver fluke *Fasciola* sp. *Parasitology Research* 78, 574–580.

Yamasaki, H., Mineki, R., Murayama, K., Ito, A. and Aoki, T. (2002) Characterisation and expression of the *Fasciola gigantica* cathepsin L gene. *International Journal for Parasitology* 32, 1031–1042.

Yang, H.J., Chung, Y.B., Kang, S.Y. and Cho, S.Y. (2002) Excretory bladder: the source of cysteine protease in *Paragonimus westermani* metaceraciae. *Korean Journal of Parasitology* 40, 89–92.

Yoshino, T.P., Lodes, M.J., Rege, A.A. and Chappell, C.L. (1993) Proteinase activity in miracidia, transformation excretory–secretory products, and primary sporocysts of *Schistosoma mansoni*. *Journal of Parasitology* 79, 23–31.

Young, J.D., Young, T.M., Lu, L.P., Unkeless, J.C. and Cohn, Z.A. (1982) Characterization of a membrane pore-forming protein from *Entamoeba histolytica*. *Journal of Experimental Medicine* 156, 1677–1690.

Zhai, Y. and Saier, M.H. Jr (2000) The amoebapore superfamily. *Biochimica et Biophysica Acta* 1469, 87–99.

Zhan, B., Liu, Y., Badamchian, M., Williamson, A., Feng, J., Loukas, A., Hawdon, J.M. and Hotez, P.J. (2003) Molecular characterisation of the *Ancylostoma*-secreted protein family from the adult stage of *Ancylostoma caninum*. *International Journal for Parasitology* 33, 897–907.

19 Signalling Molecules and Nerve–Muscle Function

A.G. Maule[1], N.J. Marks[1] and T.A. Day[2]

[1]*Parasitology Research Group, Queen's University Belfast, 97, Lisburn Road, Belfast BT9 7BL, Northern Ireland, UK;* [2]*Department of Biomedical Sciences, Veterinary Medicine Building, Iowa State University, Ames, IA 50011, USA*

Introduction

The anthelmintics at the forefront of parasitic helminth control were identified through empirical screening processes whereby an active compound was chosen for its detrimental effect on selected parasites. The majority of these appear to interfere with or disrupt helminth neuromuscular function. Therefore, the random processes involved in empirical screening selected neuromuscular control as an *a priori* target for anthelmintics. Although this situation is most pronounced in the nematodes, where the majority of leading drugs target receptors or channels associated with neuromuscular function, this phenomenon is also apparent within the flatworms where, for example, praziquantel disrupts motor control. The key components of motor control within helminths are sensory neurons, their associated interneurons, motorneurons and muscle, all of which interplay to generate coordinated motor activity that is critical to parasite survival. The need for highly coordinated motor activity is probably best displayed by many free-living stages of parasites that must migrate within the environment to access the host. However, adult parasites need motor activity to migrate to their preferred location within the host, to maintain that position within the host or perform migrations such as the diurnal movements of some tapeworms, to control the entry and exit of food materials into their gut (excluding tapeworms), to perform the coordinated muscle movements involved in egg formation and release, to facilitate reproduction and, in the case of dioecious species such as schistosomes, to anchor the female worm in the gynaecophoric

 Parasitic Flatworms: Molecular Biology, Biochemistry, Immunology and Physiology (eds A.G. Maule and N.J. Marks)

canal. The central role played by motor coordination in many aspects of parasitic helminth biology provides a strong incentive to continue to broaden our knowledge of this system and, in the process, uncover drug targets of potential therapeutic value.

Current understanding of nerve–muscle integration and function in flatworms is very limited, especially in comparison to their rounded cousins, the nematodes. A major impediment to advancing this knowledge is the absence of a free-living model that can be easily obtained, maintained and manipulated in the laboratory – there is no *Caenorhabditis elegans* equivalent in the phylum Platyhelminthes and, at this time, the strength of evidence to support the use of free-living species such as *Schmidtea mediterranea*, *Girardia* (*Dugesia*) *tigrina* or *Procerodes littoralis* is limited. Most molecular genetic progress has been made with *S. mediterranea* (Sanchez Alvarado *et al.*, 2002), although its utility as a model for parasitic flatworms has not been established. Nevertheless, these free-living species do offer opportunities to investigate nerve–muscle systems that are believed to be extremely similar between free-living and parasitic worms. A case in point is the high levels of similarity seen in neuropeptide signalling molecules across parasites from the nematode clades where parasitism has arisen, independently, several times.

So, most commonly, research on parasitic helminth nerve and muscle is aimed at the discovery and characterization of novel drug targets. However, there is some added value to platyhelminth neurobiology research in that flatworms are relatively simple animals that are close to the base of metazoan evolution. In this way, flatworms may provide a potential source for key discoveries in the evolution of metazoan nervous systems. Unfortunately, flatworms are not very amenable to neurophysiology and current understanding of inter- and intracellular signalling molecules involved in the control and modulation of nerve and muscle function is scant. Nevertheless, some progress continues to be made in several areas of flatworm neurobiology research and these are likely to develop quickly with the relatively new resource of expressed sequence tags (ESTs) for selected flatworm parasites and the schistosome genome projects. This chapter aims to highlight the most salient features of flatworm nerve–muscle function and, where possible, to draw upon data from free-living flatworms and evaluate their potential in providing useful information on parasitic flatworms.

Nervous System Structure

The organization of the nervous system in flatworms incorporates: a peripheral nerve net-like arrangement of sub-body wall nerves that are believed to provide a conduit for sensory inputs and for the control of body wall muscle activity; a central ganglionic brain that connects with a series of longitudinal nerve cords that are interlinked by transverse commissures to form the typical orthogonal arrangement; and, nerve complexes associated with muscular structures of the reproductive and intestinal tracts and attachment organs. Gross depictions of the neuroanatomy of flatworm parasites are available in Halton and Maule (2004). Much of our understanding of the organization of the nervous system in parasitic helminths has emanated from immunocytochemical studies using fluorescence and confocal scanning laser microscopy to map the distributions of selected neuronal signalling molecules (see Halton, 2004). For an overview of the organization of flatworm, nerve and muscle structures readers are directed to a number of recent reviews (see Halton and Gustafsson, 1996; Halton *et al.*, 1998; Gustafsson *et al.*, 2002; Halton and Maule, 2004).

Briefly, the flatworm brain is most commonly a ring-like or semicircular structure formed by the connection of two concentrations of nerve axons and nerve cell bodies by a well-developed commissure or commissures. These bilobed accumulations of nerves form the cerebral ganglia that occupy the lateral margins of the brain and comprise a core of nerve axons that are enveloped by a loose gathering of the associated cell bodies; at least some flatworm brains appear to be further surrounded by multilayer sheaths of mesenchyme that represent primitive glia-like cells (Sukhdeo and Sukhdeo, 1994).

The cerebral ganglia are the point of origin for the longitudinal nerve cords that form the vertical struts for the ladder-like central nervous system. Digeneans and monogeneans display longitudinal cord asymmetry in that the ventral nerve cords are the most highly developed, an adaptation believed to provide for ventral feeding, reproductive and attachment organs. In contrast, the gutless cestodes usually display both bilateral and dorsoventral nerve cord symmetry. Where attachment organs exist, these are invariably highly muscular and well innervated as their fine control is critical to parasite survival. Furthermore, the pharynx, intestine and reproductive organs are richly endowed with muscle and associated innervation that provides for deft motor coordination and which is largely derived from central nerve components. Apart from the modulation of body wall muscle by peripheral nerves, sensory organs provide input to this system so that it can trigger the motor responses needed for appropriate behavioural changes. Diminution of the numbers of sensory nerve endings in parasitic flatworms provides a significant contrast to their free-living relatives. Nevertheless, most parasitic flatworms have sensory organs that relay information that is critical to their host location and intrahost migratory activities.

Muscle Structure and Organization

Just below the basement membrane of the outer epithelial (free-living flatworms) or tegumental (parasitic flatworms) layer, flatworm body wall is invested with well-developed somatic muscle layers that provide for motor functions associated with locomotion. Most commonly, flatworm somatic muscle comprises distinct bands of circular, longitudinal and diagonally arranged fibres that form an encasing lattice of muscle, which facilitates complex movements associated with locomotion (peristaltic movements, twisting, turning, crawling and swimming). Many trematodes and monogeneans possess an outer layer of tightly packed circular muscle fibres that overlie the longitudinal muscle fibre layer as well as an inner layer of more loosely arranged diagonal muscle fibres; the diagonal muscle fibre layer is reduced or absent in cestodes. Indeed, cestodes possess additional longitudinal somatic muscle layers that are not part of their body wall but which lie deeper within their body tissue. In some monogenean fish gill flukes, large, well-developed somatic muscle bundles run longitudinally between the pharynx and posterior attachment apparatus (haptor) and these have been proposed to play a role in reflex-like contractile responses that help maintain attachment during sudden water turbulence (Maule *et al.*, 1989; Halton *et al.*, 1998). Additional dorsoventral muscle bundles have been observed in trematodes and monogeneans and are believed to help maintain the flattened body shape – although a role in locomotion has not been ruled out. It has been proposed that the multiple layers of flatworm somatic muscle allow for functional segregation through physiological or pharmacological differences (Pax *et al.*, 1996).

Researchers have been dogged by the inability to delineate between myoactivity and neuroactivity, mainly because the isolation of individual tissues by dissection is extremely difficult. This obstruction precipitated the development of methods to generate dispersed muscle fibres, and these preparations provided the first visual information on individual muscle fibres and/or muscle fibre bundles; three morphologically distinct fibre types were recorded from *Schistosoma mansoni*, frayed, spindle-shaped and crescent-shaped fibres (Day *et al.*, 1993). Although the location of all three fibre types within the worm is not clear, it has been suggested that the frayed fibres originate from the longitudinal muscle layer.

At the ultrastructural level, flatworm muscle resembles smooth muscle with individual, non-striated myofibrils being delimited by the sarcolemma and interconnected by gap junctions. Also, flatworm muscles lack a T-tubule system that is characteristic of striated muscle in other animal groups. The contractile portion of flatworm myofibrils contains thick myosin and thin actin filaments that connect with the sarcolemma via attachment plaques or desmosomes. Actomyosin cross-bridges have been reported and where overlap has been observed, ratios that vary from 9:1 to 12:1 have been observed. Although flatworm muscle is mostly non-striated, pseudo-striated (e.g. in the tail of schistosome cercariae; Dorsey *et al.*, 2002; Mair *et al.*, 2003) and obliquely striated (e.g. tentacular bulb of the trypanorhynch, *Grillotia erinaceus*; Ward *et al.*, 1986) muscles have been reported. It is presumed that the role played by these structures has demanded the development of a more efficient muscle type that is best served in the more striated format.

It is believed that fine cytoplasmic arms connect the nuclei-containing portion of the myocytons with the contractile components such that many parenchymal cells are in fact myocytons (Lumsden and Hildreth, 1983). Also, some cytoplasmic arms are believed to form multiple neuromuscular junctions that provide for neuronal control/modulation of muscle activity. Limited information is available on the functional relationship of trematode nerve and muscle although the situation in cestodes has been described as polyneuronal and polyterminal in that each neuron can form synapse-like contacts with multiple myocytons and each myocyton can synapse with several neurons (Webb, 1987). Morphologically atypical synapses and paracrine release sites have been described in the relationship between muscle and nerve in flatworms; details on the role they play are not available, although they are believed to facilitate nerve–muscle communication.

Neuronal Signalling Molecules

Signalling molecules are commonly divided into the so-called classical transmitters and the neuropeptides. The former include small molecules that are usually recycled at the synapse and which, amongst other things, serve to induce profound and fast alterations in nerve or muscle activity. In contrast, neuropeptides are synthesized in the nerve cell body and are not recycled, usually act through second messenger pathways that serve to amplify the signal and generally modulate the activities of nerve and muscle in a more subtle and long-lasting manner. These distinctions may be helpful as a general rule of thumb, but such rigid delineation of the actions of the classical transmitters and the neuropeptides is not always appropriate as some classical transmitters have subtle and slow effects and some neuropeptides have fast and profound actions on their target cells. In fact, it has been proposed that neuropeptides originally served as fast transmitters in early-diverging animals and only later assumed the more modulatory role with which they are commonly associated in mammals (Greenberg and Price, 1992).

Classical transmitters, their synthesis, degradation and receptors

Acetylcholine

Numerous histochemical staining methods have shown the widespread presence of cholinesterase/acetylcholinesterase (ChE/AChE) in the nervous systems of flatworms and have been used to support the presence of acetylcholine (ACh); ACh is degraded by the action of AChE. Furthermore, antisera to ACh have revealed its occurrence in the nervous systems of both a cestode (*Hymenolepis diminuta*; Samii and Webb, 1990) and a trematode (*S. mansoni*; Hillman, 1983). More recently, molecular cloning of AChE from *Schistosoma haematobium* facilitated the generation of an antiserum to the recombinant enzyme and immunocytochemistry confirmed its location to neuronal structures in this worm, but also revealed its presence in the tegument (Jones *et al.*, 2002); homologues of the *S. haematobium* AChE have been cloned from *S. mansoni* and *Schistosoma bovis* (Bentley *et al.*, 2003). A consistent theme of these studies has been the widespread nature of cholinergic innervation in both central and peripheral nerve elements and the association of cholinergic innervation with all muscular organs, pointing to a key role in muscle action.

ACh, cholinomimetics and acetylcholinesterase inhibitors have been shown to inhibit the movement of flatworms and to cause flaccid paralysis (for reviews, see Pax *et al.*, 1996; Halton *et al.*, 1997; Halton and Maule, 2004; Ribeiro *et al.*, 2005). This is in contrast to the mainly myoexcitatory effects of ACh at the vertebrate neuromuscular junction and in many other invertebrates including nematodes. Although ACh has predominantly inhibitory actions on flatworm muscle it has been shown to have inconsistent effects on the muscle of the monogenean, *Diclidophora merlangi* (Maule *et al.*, 1989), to induce the contraction of muscle fibres dispersed from planarians (Blair and Anderson, 1994; Moneypenny *et al.*, 2001) and, to cause increased muscle activity and shortening in intact planarians (Buttarelli *et al.*, 2000). These data could indicate inconsistency across the phylum Platyhelminthes or a distinction in the actions of cholinergic drugs in the different flatworm classes. Nevertheless, the most consistent picture of ACh action in parasitic flatworms is as a muscle relaxant. This is supported by work on individual muscle fibres from *S. mansoni* that indicate the occurrence of muscle-based, inhibitory ACh receptors (Day *et al.*, 1996).

There are two recognized types of ACh receptors, directly ligand-gated ion channels that are designated nicotinic ACh receptors (nAChR), and G-protein-coupled receptors (GPCR) that are designated muscarinic ACh receptors (mAChR). It is unknown if such a delineation, based on the actions of nicotine and muscarine at the ACh receptors of higher vertebrates, will hold for flatworms. Both muscarinic and nicotinic agonists have been shown to have inhibitory actions on flatworms (Pax *et al.*, 1981, 1984; Mellin *et al.*, 1983). However, selective muscarinic and nicotinic agonists (apart from the nicotinic agonist, α-bungarotoxin) failed to inhibit contraction of dispersed muscle fibres from *S. mansoni* (Day *et al.*, 1996). These results led to the conclusion that the schistosome muscle-based ACh receptor was pharmacologically distinct from the mammalian homologue.

Molecular support for the presence of neuronal-type nAChRs in flatworms was provided by Cebria *et al.* (2002) and Bentley *et al.* (2004) who cloned nAChR subunits from the planarian *Dugesia japonica* and the trematode *S. haematobium*, respectively. The two novel nAChR subunit-expressing genes from *S. haematobium* appear to encode putative alpha (ShAR1alpha) and non-alpha (ShAR1beta) subunits. Although these putative AChR subunits possess the key structural features common to all nAChR subunits, they have an unusually large cytoplasmic domain between the M3 and M4 transmembrane regions. Interestingly, ShAR1alpha was expressed on the surface membranes whereas ShAR1beta was localized to the musculature and some cells within the connective parenchyma. Unfortunately, the functional channel has not been expressed as ShAR1alpha did not form functional channels on its own or in combination with ShAR1beta or the chick beta2 subunit in *Xenopus laevis* oocytes. The planarian AChR subunits were localized to exclusively neuronal tissues using *in situ* hybridization (Cebria *et al.*, 2002). Clearly, functional expression is needed to ascertain the physiology of these channels in trematodes and planarians and to establish if they are, indeed, functionally distinct. Examination of the available schistosome ESTs indicates the occurrence of a putative muscarinic ACh receptor.

5-Hydroxytryptamine (5-HT; serotonin)

Serotonin is the most widely studied classical transmitter within flatworms. Specific antisera have identified serotonin-immunoreactivity within the central and peripheral nervous system of all flatworm parasites that have been investigated (see Pax *et al.*, 1996; Halton and Maule, 2004; Ribeiro *et al.*, 2005). Reports on the synthesis of serotonin in flatworms are inconsistent, with some indicating *de novo* synthesis and others suggesting that worms are unable to synthesize serotonin from tryptophan, although synthesis from 5-hydroxytryptophan was possible (see Ribeiro *et al.*, 2005). A gene encoding tryptophan hydroxylase, which catalyses the conversion of tryptophan to 5-hydroxytryptophan, has been characterized from *S. mansoni* (Hamdan and Ribeiro, 1999) and is expressed in all life stages (Boyle *et al.*, 2003). Monoamine oxidase (MAO), which converts biogenic amines to their corresponding aldehydes, is widely involved in the degradation of serotonin. In this respect, the degradation of serotonin has been recorded in *H. diminuta* and *S. mansoni* (Nimmo-Smith and Raison, 1968; Ribeiro and Webb, 1984) and known MAO inhibitors stimulate flatworm motor activity (Holmes and Fairweather, 1984). It is also apparent that many flatworm parasites possess mechanisms of serotonin uptake, a means of exploiting the often serotonin-rich environment they inhabit (see Boyle *et al.*, 2003). Indeed, an EST that encodes a serotonin transporter protein has been identified in schistosomes (Hu *et al.*, 2003; Verjovski-Almeida *et al.*, 2003). Serotonin reuptake is the main mechanism of signal termination at the neuronal synapse and there is evidence for such a serotonin transport system in nerve and muscle in the cestode *H. diminuta*

(Osloobi and Webb, 1999). Although serotonin receptors (5HTLpla1 and 5HTLpla4) have only been cloned from the planarian, *G. tigrina* (Saitoh *et al.*, 1997), there is a plethora of physiological, biochemical and receptor binding data that support their presence in parasitic flatworms (see Halton *et al.*, 1997; Halton and Maule 2004; Ribeiro *et al.*, 2005).

Physiologically, there is a long established link between serotonin and the stimulation of cyclic adenosine monophosphate (cAMP) levels in flatworms, consistent with the hypothesis that serotonin activates a Gs-linked guanosine triphosphate binding GPCR (McNall and Mansour, 1984; Mansour and Mansour, 1986, 1989; Ribeiro and Webb, 1987). Heterotrimeric G-proteins relay signals from GPCRs to cytosolic signalling molecules; the Gs class of G-protein α-subunit most commonly stimulates adenylate cyclase, which generates cytosolic cAMP. As well as stimulating the production of cAMP, serotonin has been shown to increase glucose absorption, glycogenolysis and glycolysis (see Halton *et al.*, 1997). As early as the 1950s, serotonin was shown to have excitatory effects on the movement of *Fasciola hepatica* (Mansour, 1957) and, since that time, a raft of studies have reported the excitatory actions of this biogenic amine on intact worms and/or muscle strip preparations from cestodes, monogeneans and trematodes (see Halton *et al.*, 1997). The pharmacology of the serotonin response in *S. mansoni* did not match that known for vertebrate serotonin receptors (Willcockson and Hillman, 1984). Although the exposure of *S. mansoni* dispersed muscle fibres to serotonin did not stimulate muscle activity, its presence was a prerequisite for depolarization-induced contraction (Day *et al.*, 1994a). In contrast to the apparent absence of a direct contractile effect of serotonin on schistosome muscle fibres, it did induce contraction of muscle fibres from two planarian species (see Blair and Anderson, 1996; Moneypenny *et al.*, 2001). It is not known if this is a planarian/ trematode distinction or simply due to differences in the dispersed fibre assays employed. It is also unclear if the serotonin-induced elevation of energy production and cAMP levels seen in flatworm homogenates is responsible for the increased motor activity or whether they represent distinct signalling pathways (see Pax *et al.*, 1996). Available evidence indicates that the pharmacology of the receptors for both responses is similar.

Catecholamines

The data supporting the catecholamines (dopamine and noradrenaline) as signalling molecules involved in motor function in flatworms is much less compelling than that for serotonin and ACh. Nevertheless, there is biochemical support for the presence of noradrenaline and dopamine in cestodes and trematodes, although localization data is largely limited to planarians where there is significant variation in their neuronal distribution patterns between species (see Halton and Maule, 2004). Evidence for the appropriate synthetic pathways in flatworms comes from the ability of *H. diminuta* to synthesize dopamine from tyrosine and the characterization of a tyrosine hydroxylase (the rate-limiting enzyme in the biosynthesis of catecholamines) cDNA from *S. mansoni* (Hamdan and Ribeiro, 1998). There is no consistent pattern to the effects of catecholamines on the motor activity of parasitic flatworms. For example, dopamine inhibits contractility in *S. mansoni* (Pax *et al.*, 1984), whereas it stimulates muscle contractions in *F. hepatica* (Holmes and Fairweather, 1984) and the monogenean, *D. merlangi* (Maule *et al.*, 1989). Coupled with this inhibition of schistosome contractility, dopamine enhanced the excitatory effects of serotonin and muscle responses to electrical stimulation (Pax *et al.*, 1981; Willcockson and Hillman, 1984). Both noradrenaline and adrenaline caused lengthening in *S. mansoni* (Pax *et al.*, 1984), mild inhibitory effects on *F. hepatica* (Holmes and Fairweather, 1984) and excitatory effects on *D. merlangi* (Maule *et al.*, 1989). The actions of catecholamines on flatworm motor function appear to be variable and species dependent.

Histamine

Although there is biochemical evidence for the occurrence of histamine in trematodes and cestodes, there is very limited data indicating a role in the modulation of nerve or muscle.

Histamine has been localized to the nervous system of the frog-lung trematode, *Haplometra cylindracea* (Eriksson *et al.*, 1996) and in larval stages of the fish tapeworm, *Diphyllobothrium dendriticum* (Wikgren *et al.*, 1990). Histamine has been reported to inhibit the locomotory movements of *S. mansoni* cercariae (Ercoli *et al.*, 1985) and to modulate the motor activity of *H. diminuta* (Sukhdeo *et al.*, 1984). Following heterologous expression of a schistosome GPCR (SmGPCR), histamine was identified as a selective agonist suggesting that this may be a schistosome histamine receptor (Hamdan *et al.*, 2002). One peculiarity of SmGPCR is its tolerance (maintenance of histamine responsiveness) of an asparagine[111] to aspartic acid[111] mutation within the third transmembrane domain; most biogenic amine GPCRs require the asparagine[111] to facilitate ligand activation. A role for the parasite histaminergic system in the host–parasite interaction has been proposed (Ribeiro *et al.*, 2005), but a role in neuromuscular function has not been elucidated.

γ-Aminobutyric acid (GABA)

There is evidence for the biosynthesis of GABA in cestodes and trematodes (Eriksson *et al.*, 1995; Mendonça-Silva *et al.*, 2004). Furthermore, GABA has been localized to the nervous system and has been detected biochemically in a trematode (*F. hepatica*) and a cestode (*Moniezia expansa*) (Eriksson *et al.*, 1995). One of the best studied mammalian brain receptors is the GABA A-type and three putative GABA A-like receptors have been reported in the schistosome EST database (see Ribeiro *et al.*, 2005), although functional confirmation and distribution await investigation. Unfortunately, nerve–muscle modulatory effects of GABA in parasitic flatworms have not been reported and the only known physiological effects in the phylum relate to its picrotoxin-sensitive depression of electrical activity in the nerve cord of the turbellarian, *Notoplana acticola* (Keenan *et al.*, 1979). Picrotoxin (a GABA A-subtype selective antagonist) reduced locomotory activities in *S. mansoni*, consistent with its inhibition of a GABA A-like ion channel (Mendonça-Silva *et al.*, 2004).

Glutamate

In the early 1980s, measurements of *in situ* fluorescent spectra suggested that an amino acid, possibly glutamate, occurred in the nervous system of the fish tapeworm, *Gyrocotyle fumbriata* (Keenan and Koopowitz, 1982). Subsequent immunolocalization studies have identified the amino acid glutamate in the nervous systems of two cestodes, *H. diminuta* (Webb and Eklove, 1989) and *Mesocestoides corti*, and in three trematodes, *S. mansoni*, *Trichobilharzia ocellata* cercariae (Solis-Soto and Brink, 1994) and *F. hepatica* (Brownlee and Fairweather, 1996). Glutamate has excitatory effects in cestodes and was reported to stimulate nerve cord activity in *G. fumbriata* (Keenan and Koopowitz, 1982) and induce muscle contraction in *H. diminuta* (Webb, 1988; Thompson and Mettrick, 1989). A range of other glutamate-related observations have been made in *H. diminuta*: glutamate-triggered inositol triphosphate production (Webb, 1997), implicating the involvement of a glutamate-activated GPCR; at different concentrations, glutamate was reported to elevate or depress cAMP levels in *H. diminuta* homogenates (Eklove and Webb, 1981); the depolarization of *H. diminuta* tissue slices by elevated extracellular potassium or by electrical stimulation triggered the release of glutamate, and, there is evidence that this worm has a high- and low-affinity glutamate transport system. Glutamate-induced contraction of *S. mansoni* dispersed muscle fibres is believed to be related to the actions of a high-affinity glutamate transporter in the muscle, rather than a glutamate receptor (Miller *et al.*, 1996; Pax *et al.*, 1996). The glutamate transporter is believed to be sodium-dependent and electrogenic, meaning that its activities would induce muscle depolarization and contraction, consistent with the physiological observations. More recently, *S. mansoni* membrane fractions have been found to possess a glutamate receptor that has pharmacological characteristics of the kainate receptor subfamily (Mendonça-Silva *et al.*, 2002). Kainate has a profound effect on the behaviour of adult *S. mansoni*, causing them to adopt a corkscrew-like posture consistent with somatic muscle contraction (Mendonça-Silva *et al.*, 2002). ESTs for two ionotropic glutamate receptor subunits were identified in a planarian and, using *in situ* hybridization, were shown to be expressed in different regions of the brain (Cebria *et al.*, 2002). Examination of schistosome ESTs identified multiple candidate ligand-gated glutamate receptors as well as putative glutamate GPCRs (Ribeiro *et al.*, 2005). The interplay, if any, between the proposed muscle-based glutamate transporter and the uncharacterized putative glutamate receptors in flatworm neuromuscular function awaits investigation.

Octopamine

There is very limited evidence for octopamine in flatworms and there is no evidence to support a role in motor function. It has been detected in *H. diminuta*, which was shown to be able to synthesize octopamine from tyrosine (Ribeiro and Webb, 1983).

Nitric oxide

The diffusible gas, nitric oxide (NO), is now a well-established neuronal signalling molecule in vertebrates and invertebrates (Colasanti and Venturini, 1998). NADPH (the reduced form of nicotinamide adenine dinucleotide phosphate)-diaphorase activity is widely believed to be representative of nitric oxide synthase (NOS) activity and, along with specific antibodies, has been used to indicate the presence of NOS in cestode and trematode nervous systems (Gustafsson *et al.*, 1996; Lindholm *et al.*, 1998; Terenina *et al.*, 1999; Gustafsson *et al.*, 2001; Kohn *et al.*, 2001a; Tandon *et al.*, 2001; Terenina and Gustafsson, 2003). NO has been shown to interact with the heme group of soluble guanylate cyclase to stimulate the production of cyclic guanosine monophosphate (cGMP). Analysis of cGMP immunostaining in nerves of *H. diminuta* indicated an increase in cAMP levels following exposure to the NO donor, sodium nitroprusside (Gustafsson *et al.*, 2003). More recently, a radiometric assay was used to reveal an increase and then decrease in cGMP levels in *H. diminuta* (Onufriev *et al.*, 2005) upon exposure to sodium nitroprusside. Clearly, there is evidence to support an NO signalling system in flatworm nervous systems, but the influence that NO has on motor function, if any, is unknown.

Neuropeptides

Unlike the situation in nematodes and arthropods, where hundreds of neuropeptides have been elucidated using biochemical, physiological and molecular techniques, flatworm neuropeptide discovery is still in its infancy. This is despite the characterization of the first flatworm neuropeptide almost 15 years ago (Maule *et al.*, 1991). Progress has been hindered by the inability to obtain large quantities of flatworm neuronal tissues and the absence, until recently, of a significant body of genomic and/or EST data for flatworms. Nevertheless, the available evidence does provide a snapshot of what appear to be the most abundant and widespread flatworm neuropeptides.

Neuropeptide F

Neuropeptide F (NPF) was first identified from the cestode, *M. expansa* as a 39-amino acid, C-terminally amidated neuropeptide that had some of the structural characteristics of mammalian neuropeptide Y (NPY), a highly abundant brain peptide in all higher animals (Maule *et al.*, 1991). NPF antisera then revealed that this peptide was abundantly expressed throughout the nervous systems of all classes and stages of flatworm (see Halton *et al.*, 1994; Halton and Gustafsson, 1996; Day and Maule, 1999). A year later, the second known NPF was structurally characterized from the land planarian, *Artioposthia triangulata* (= *Arthurdendyus triangulatus*) (Curry *et al.*, 1992); it had the same structural features as *M. expansa* NPF (mxNPF), although it was 36-amino-acids long, as are members of the NPY family of peptides in higher animals. The presence of an Arg-X-Arg-aromatic-amide C-terminus and signature tyrosyl residues 10 and 17 amino acids from the C-terminus, which are some of the defining features of NPYs, indicated that the flatworm peptides were *bonafide* NPY orthologues. However, vertebrate NPYs also had a poly-prolyl N-terminus, a feature that was absent from the flatworm peptides and which therefore brought into question their relationship to NPY. Further support for an evolutionary link between NPY and NPF came from the characterization of the mxNPF-encoding gene (Mair *et al.*, 2000). It had an unusual intron–exon organization that matched that seen in vertebrate NPYs. Nuclear magnetic resonance examination of the solution structure of mxNPF revealed the random structure of the N-(Pro^1–Asn^{16}) and C-(Gly^{35}–Phe^{39}) termini either side of an alpha helix, and that its structure was similar to that of porcine NPY (Miskolzie and Kotovych, 2002). Recently, the first trematode NPF was identified on an *S. japonicum* EST and the full-length cDNAs were cloned from *S. japonicum* (*sjnpf*) and *S. mansoni* (*smnpf*; Humphries *et al.*, 2004). Both of these genes were found to encode 36-amino-acid peptides that had identical C-terminal pentapeptides (Gly-Arg-Pro-Arg-Phe-amide) to the two known flatworm NPFs as well as the position-matched tyrosyl residues (see Table 19.1).

A role for NPF in motor function was first intimated when the C-terminal nonapeptide of mxNPF was found to increase muscle contractility in muscle strips from juvenile *F. hepatica* (Marks *et al.*, 1996). Subsequently, whole mxNPF was shown to stimulate motility in *M. corti* larvae (Hrckova *et al.*, 2004), although only at comparatively high concentrations (10^{-4} M). A series of signalling pathway and calcium movement inhibitors suggested that the excitatory effects of NPF required the activity of a GPCR, extra- and intracellular calcium, protein kinase A (PKA) and adenylate cyclase. The work did not establish which of these components were directly linked to NPF-induced excitatory effects but suggested that the actions of NPF in flatworms were quite distinct from those seen in vertebrates where inhibitory effects linked to adenylate cyclase inhibition predominate. In contrast, smNPF has been shown to inhibit the forskolin-induced stimulation of cAMP levels in homogenates of *S. mansoni* (Humphries *et al.*, 2004). These inhibitory effects of smNPF in *S. mansoni* were still evident at concentrations as low as 10^{-11} M; mxNPF had similar effects but only at concentrations of 10^{-7} M and above. It is possible that the comparatively high concentrations of mxNPF needed to trigger increased motor activity in *M. corti* and *F. hepatica* are due to the non-specific interaction of NPF with other endogenous receptors, although poor tissue penetration and the use of non-native NPF ligands could have contributed to the need for higher concentrations. The fact that smNPF

Table 19.1. Amino acid sequences of neuropeptides identified from platyhelminths.

Peptide	Amino acid sequence
Neuropeptide F (NPF)/Neuropeptide Y (NPY)	
Arthurdendyus triangulatus NPF	KVVHLRPRSSFSSEDEYQI**Y**LRNVSK**Y**IQLYG**R**P**RFa**
Moniezia expansa NPF	PDQDSIVNPSDLVLDNKAALRD**Y**LRQINE**Y**FAIIG**R**P**RFa**
Schistosoma japonicum NPF	AQALAKLMTLFYTSDAFNK**Y**MENLDA**Y**YMLRG**R**P**RFa**
Schistosoma mansoni NPF	AQALAKLMSLFYTSDAFNK**Y**MENLDA**Y**YMLRG**R**P**RFa**
Aplysia californica NPF	DNSEMLAPPPRPEEFTSAQQLRQ**Y**LAALNE**Y**YSIMG**R**P**RFa**
Lymnaea stagnalis NPF	MLTPPQRPEEFKNPNELRK**Y**LKALNE**Y**YAIVG**R**P**RFa**
Drosophila melanogaster NPF	SNSRPPRKNDVNTMADAYK**F**LQDLDT**Y**YGDRA**R**V**RFa**
Lamprey NPY	FPNKPDSPGEDAPAEDLAR**Y**LSAVRH**Y**INLIT**R**Q**RYa**
Homo sapiens NPY	YPSKPDNPGEDAPAEDMAR**Y**YSALRH**Y**INLIT**R**Q**RYa**
FMRFamide-related peptides (FaRPs)	
Arthurdendyus triangulates	RYI**Rfa**
Bdelloura candida	GYI**Rfa**
	YI**Rfa**
Moniezia expansa	GNFF**Rfa**
Girardia tigrina	GYI**Rfa**
Procerodes littoralis	GYI**Rfa**
Macrocallista nimbosa	FM**Rfa**

Selected sequelogs are shown for comparison. Single letter annotation for amino acids used throughout. Amino acids that are characteristic of family members are highlighted in bold and underlined.

appears to depress cAMP levels in *S. mansoni* consistent with NPY actions in vertebrates points to the remarkable conservation of this signalling pathway from flatworms to humans. There is unpublished evidence that forskolin-stimulated cAMP levels are also inhibited in planarian homogenates by mxNPF suggesting that this pathway may be conserved across flatworms (H.H. Omar and T.A. Day, Iowa State University, unpublished observations). It is not clear that these biochemical actions observed in homogenates are taking place in the muscle, but the muscle is indeed highly innervated with NPF-containing neurons.

FMRFamide-related peptides

FMRFamide-related peptides (FaRPs) are the best known and most widely studied family of invertebrate neuropeptides and have a characteristic C-terminal tetrapeptide comprising an aromatic amino acid, hydrophobic amino acid, arginyl-phenylalanylamide. In some other invertebrate phyla, where there has been an immense proliferation of FaRP peptides (e.g. nematodes possess >60 distinct FaRPs), the signature requirements have been relaxed such that they need only possess an aromatic-variable dipeptide or a variable-hydrophobic dipeptide prior to the Arg-Phe-amide. In complete contrast to the diversity of FaRPs seen in nematodes, flatworms appear to have only a very small FaRP complement. FaRPs have been structurally characterized from four planarians (*A. triangulata, Bdelloura candida, G. tigrina, P. littoralis*) and a cestode (*M. expansa*), and all conform to the most rigorous signature requirements. The planarian peptides include the tetra- and pentapeptides, YIRFamide (*B. candida*; Johnston *et al.*, 1996), GYIRFamide (*B. candida, G. tigrina* and *P. littoralis*; Johnston *et al.*, 1995, 1996; R.N. Johnston, Queen's University Belfast, unpublished observations) and RYIRFamide (*A. triangulata*; Maule *et al.*, 1994), whereas the only known parasitic flatworm FaRP is the cestode hexapeptide, GNFFRFamide (*M. expansa*; Maule *et al.*, 1993) (see Table 19.1). Although FaRP-immunoreactive species have been identified in extracts of trematodes, none have been structurally characterized.

FaRP immunostaining patterns in flatworms have been widely reported and are consistent between the classes in that all major central and peripheral nerve elements appear to contain these peptides (see Halton *et al.*, 1994; Halton and Gustafsson, 1996; Shaw *et al.*, 1996; Day and Maule, 1999; Halton and Maule, 2004). Another key feature of FaRP distribution is its occurrence in nerves that innervate muscle systems, from attachment organs to feeding apparatus and reproductive ducts (see Halton and Maule, 2004). The localization data alone implicate FaRPs in pivotal brain functions including motor modulation.

FaRPs have been shown to enhance the contractility of whole worm (*M. corti* larvae; Hrckova *et al.*, 2002) and muscle strip/trimmed preparations (*F. hepatica, D. merlangi, S. mansoni*; Marks *et al.*, 1996; Graham *et al.*, 1997; Moneypenny *et al.*, 1997; T.A. Day, Iowa State University, unpublished observations) from trematodes and cestodes and to induce contraction in dispersed muscle fibres from planarians and a trematode (*S. mansoni, B. candida, P. littoralis*; Day *et al.*, 1994b; Johnston *et al.*, 1996; Moneypenny *et al.*, 2001). It should be noted that whole flatworm preparations are most commonly unresponsive to neuropeptides such that the effects on *M. corti* larvae are unusual; FaRPs have no effects on intact *S. mansoni* or *F. hepatica* (Pax *et al.*, 1996; N.J. Marks, Queen's University Belfast, unpublished observations). The consistent theme for FaRP effects in flatworms is muscle contraction and, unlike the situation for some of the classical transmitters and NPF, the effects appear to be consistent across the flatworm classes. The potent contraction of dispersed muscle fibres that is caused by FaRPs indicates the occurrence of a muscle-based receptor. The presence or absence of FaRP receptors on flatworm nerves has not been examined.

Structure–activity data on the effects of flatworm FaRPs on *S. mansoni* muscle fibres reveal that the replacement of any residue in the YIRFamide signature with the amino acid alanine abolished peptide activity (Day *et al.*, 1997). Although no trematode FaRP has been structurally characterized, evidence from physiology studies reveals that the YIRFamide-containing peptides induce contraction of dispersed *S. mansoni* muscle fibres with high potency (GYIRFamide had an EC_{50} of 10^{-9} M) whereas GNFFRFamide was much less effective (EC_{50} of 10^{-6} M) (Day *et al.*, 1994b). Similarly, GNFFRFamide was about 1000-fold less effective than the YIRFamides at stimulating muscle contractility in trimmed *F. hepatica* juveniles (Marks *et al.*, 1996). These data would point to the endogenous trematode FaRPs being structurally more similar to the planarian peptides than the cestode variant. Furthermore, the effects of all these flatworm FaRPs on *P. littoralis* muscle fibres were abolished by GYIR(d)Famide (an GYIRFamide analogue with the D-isomer of Phe in the C-terminal position), suggesting that they all interact at a single muscle receptor to elicit muscle contraction. Although supporting data is limited, cestode preparations appear to be most responsive to GNFFRFamide amongst the flatworm FaRPs. FaRP-induced excitation of *S. mansoni* muscle fibres was shown to be dependent on calcium in the extracellular medium, although these effects were not abolished by known calcium channel blockers (Day *et al.*, 1994b).

Attempts have been made to unravel the mechanism of FaRP action on flatworm muscle. To this end, data is available for the effects of selected inhibitors of known signalling pathways on the FaRP-induced excitation of muscle strips prepared from *F. hepatica* (Graham *et al.*, 2000). The effects of GYIRFamide on *F. hepatica* muscle strips were reduced but not abolished by the G-protein inhibitor, guanosine 5′-2-*O*-(thio)diphosphate (GDPβS), implicating a GPCR in the FaRP-induced excitatory effects. This was not surprising since GPCRs are the most common receptor targets of neuropeptides, even though some molluscan and nematode FaRPs have been shown to activate directly ligand-gated ion channels (Linguеglia *et al.*, 1995; Maule *et al.*, 1995; Purcell *et al.*, 2002a,b). A common consequence of GPCR activation is modulation of adenylate cyclase activity that results in the regulation of cytosolic cAMP. Known adenylate cyclase and PKA inhibitors failed to give meaningful results due to direct effects on motor activity. However, inhibition of phospholipase C (PLC; with neomycin sulphate) and protein kinase C (PKC; with chelerytherine chloride) blunted the GYIRFamide response, suggesting a pathway involving G-protein activation of PLC- and PKC-dependent signalling pathways. Since it is unknown if the site of action of flatworm FaRPs is confined to muscle, it is difficult to make definitive conclusions on the signalling processes involved where muscle and nerve occur together. However, similar studies carried out on dispersed muscle fibres of *S. mansoni* and *P. littoralis* have confirmed that FaRP-induced contraction is blocked by inhibitors of PKC (T.A. Day, M.I.J. Totten and A.G. Maule, unpublished observations).

Neuropeptide processing enzymes

Further support for the occurrence and activity of neuropeptides in flatworms stems from the presence of enzymes necessary for the production of the mature peptides. For example, the first step in the generation of a mature neuropeptide is often the excision of a propeptide from a longer pre-propeptide. This is achieved by the activity of prohormone convertases, and a number of putative platyhelminth convertase-encoding cDNAs are represented within the EST databases.

Critical to the bioactivity of both the NPF and FaRPs mentioned above are the enzymes that produce the amidated carboxy terminus from a glycine-extended propeptide. Specifically, the Arg-Phe-Gly carboxy terminus is converted to Arg-Phe-amide by a sequential reaction that normally requires two enzymes, peptidylglycine alpha-hydroxylating monooxygenase (PHM) and peptidyl-alphahydroxyglycine-alpha-amidating lyase (PAL). *S. mansoni* FaRPs and NPFs which have unamidated C-termini have been shown to lack bioactivity in schistosomes (Day *et al.*, 1997; Humphries *et al.*, 2004). *S. mansoni* has been shown to possess a PHM that has the requisite amidating activity and is abundantly expressed throughout the nervous system (Mair *et al.*, 2004). In fact, the distribution of endogenous PHM is more widespread than both FaRPs and NPFs, suggesting that there may be more yet-undiscovered amidated peptides within the nervous system.

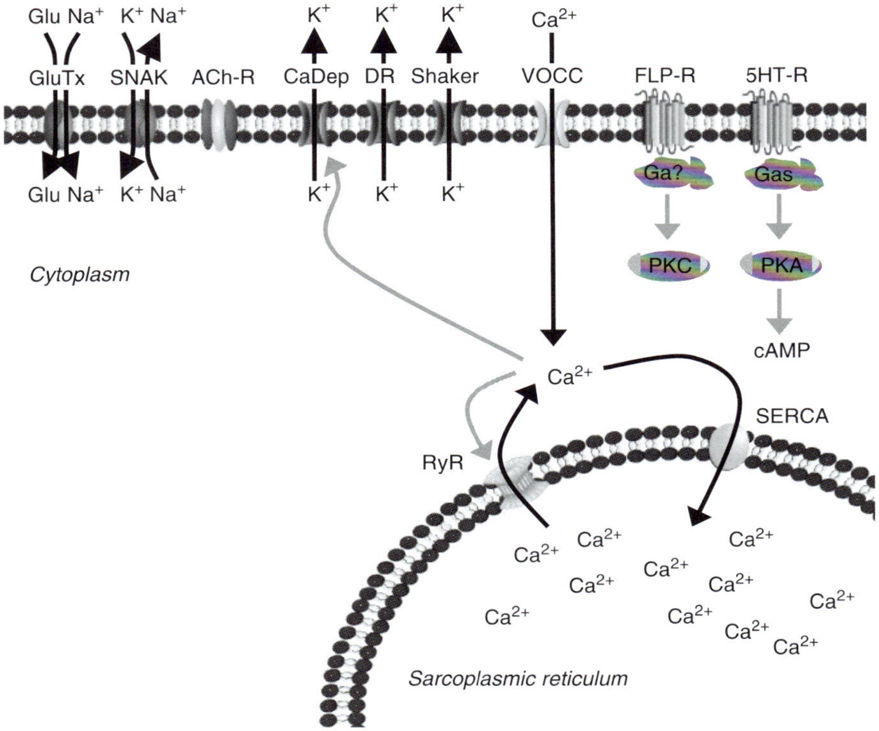

Fig. 19.1. A composite schematic of molecules demonstrated to be functional in flatworm muscle physiology. More than 50 years ago, Chance and Mansour (1953) demonstrated the myoexcitatory effects of exogenous serotonin on *F. hepatica*. Since then, much evidence has accumulated in support of serotonin's central role in muscle control, including demonstration that serotonin is coupled to the stimulation of protein kinase A (PKA) and the production of cAMP (Mansour, 1984; Ribiero and Webb 1987). Serotonin has direct effects on flatworm muscle fibres (Day *et al.*, 1994a) that are also linked to cAMP. A stimulatory trimeric G-protein (Gαs) has been cloned from *S. mansoni* and has been associated with the serotonin receptor (Mansour and Mansour, 1986, 1989). Specific FaRPs have been shown to elicit contraction of individual muscle fibres from a number of flatworm species (Day *et al.*, 1994b, 1997; Johnston *et al.*, 1996; Moneypenny *et al.*, 2001), demonstrating the presence of FaRP receptors (FMRF amide-like peptide receptors; FLP-R) on the muscle. Excitatory FaRP receptors in flatworms have been linked to protein kinase C (PKC) activation (Graham *et al.*, 2000; Day and Maule, unpublished observations). The presence of voltage-operated calcium channels (VOCC) in flatworms have been demonstrated by the dihydropyridine blockade of depolarization-induced contractions in whole worms and isolated muscle (Fetterer *et al.*, 1980; Day *et al.*, 1994b; Graham *et al.*, 1999), by the recording of voltage-gated calcium currents in *G. tigrina* (Cobbett and Day 2003), and by the molecular cloning of various channel subunits (Kohn *et al.*, 2001b,c). At least three different potassium channels are present in the muscle: calcium-dependent potassium channels (CaDep); delayed rectifier channels (DR); and fast shaker-type channels (Shaker) (Blair *et al.*, 1991; Day *et al.*, 1993, 1995; Kim *et al.*, 1995a,b; Kumar *et al.*, 2004). Flatworm muscle stores calcium in the sarcoplasmic reticulum and ryanodine receptors (RyR) mediate a Ca^{2+}-triggered release of intracellular Ca^{2+} (Graham *et al.*, 1999; Day *et al.*, 2000); the other endogenous regulators of these channels in flatworms are not known. Thapsigargin-sensitive sarcoendoplasmic reticular calcium ATPases (SERCA) are involved in pumping cytosolic calcium into the sarcoplasmic reticulum, and a number of these pumps have been identified at a molecular level (de Mendonca *et al.*, 1995; Talla *et al.*, 1998). Cholinergic compounds have effects on the musculature of whole animals and isolated muscle fibers (Blair and Anderson, 1994; Day *et al.*, 1996; Buttarelli *et al.*, 2000; Moneypenny *et al.*, 2001), indicating the presence of acetylcholine receptors (ACh-R) on the muscle. However, the vague pharmacology of the cholinergic effects makes it difficult to conclude if they are of the same nature as the acetylcholine receptors recently cloned from flatworms (Cebria *et al.*, 2002; Bentley *et al.*, 2004). An electrogenic glutamate transporter (GluTx; Miller *et al.*, 1996) and sodium–potassium ATPase (SNAK; Skelly *et al.*, 2001) are also present in some flatworm muscle.

Concluding Remarks

Current understanding of the role played by classical transmitters and neuropeptides in the regulation of motor function in flatworms is, at best, fragmentary. Nevertheless, there is compelling evidence that serotonin, ACh, glutamate, GABA, FaRPs and, to a lesser extent, catecholamines, NO and NPF, all contribute to nerve–muscle function in flatworms. It should be noted that definitive proof of a neurotransmitter/neuromodulator function has not been established for any of these signalling molecules. However, flatworm EST data is now providing resources for the identification of the appropriate synthetic, processing and degradative enzymes, transporters and receptors that will help confirm the involvement of these signalling molecules in motor function. The most significant hurdles to the characterization of these signalling systems and their interplay in the modulation of motor behaviours in parasitic flatworms focus on our inability to silence the appropriate genes in the adult stages, adequately monitor any phenotypic changes, and perform the appropriate membrane level physiology on nerve and muscle cells independently. Progress in the application of RNA interference (RNAi) procedures to parasitic flatworms has been slow and, although more significant progress has been made using free-living flatworms, their utility as models has not been established. It is clear that there are neuronal signalling systems that have been conserved between planarians and parasitic flatworms, but others show some differences between these groupings and more detailed comparative studies are warranted if we are to exploit the advances made in planarian research. Regardless of these facts, many of the neuronal and muscle proteins that are being uncovered have potential as drug targets. Of particular interest are those signalling molecules associated with muscle (Fig. 19.1) as these are believed to be most amenable/accessible to drug intervention. The discovery of flatworm ACh, serotonin, histamine and glutamate receptors, coupled with recent breakthroughs in FaRP receptor identification in insects and nematodes, provide an impetus to flatworm receptor research, which is likely to represent a significant component of the next phase of discovery in this area. These receptors provide the most obvious candidate drug targets that warrant functional characterization and exploitation in drug screening programmes.

Acknowledgement

The work in the laboratories of A.G.M. and T.A.D. is supported in part by NIH grant AI049162. N.J.M. and A.G.M. also acknowledge the support of the Department of Agriculture and Rural Development for Northern Ireland.

References

Bentley, G.N., Jones, A.K. and Agnew, A. (2003) Mapping and sequencing of acetylcholinesterase genes from the platyhelminth blood fluke *Schistosoma*. *Gene* 314, 103–112.

Bentley, G.N., Jones, A.K., Parra, W.G.O. and Agnew, A. (2004) ShAR1alpha and ShAR1beta: novel putative nicotinic acetylcholine receptor subunits from the platyhelminth blood fluke *Schistosoma*. *Gene* 329, 27–38.

Blair, K.L. and Anderson, P.A.V. (1994) Physiological and pharmacological properties of muscle cells isolated from the flatworm *Bdelloura candida* (Tricladida). *Parasitology* 109, 325–335.

Blair, K.L. and Anderson, P.A.V. (1996) Physiology and pharmacology of turbellarian neuromuscular systems. *Parasitology* 113, S73–S82.

Blair, K.L., Day, T.A., Lewis, M.C., Bennett, J.L. and Pax, R.A. (1991) Studies on muscle cells isolated from *Schistosoma mansoni*: a Ca^{2+}-dependent K^{+} channel. *Parasitology* 102, 251–258.

Boyle, J.P., Hillyer, J.F. and Yoshino, T.P. (2003) Pharmacological and autoradiographical characterization of serotonin transporter-like activity in sporocysts of the human blood fluke, *Schistosoma mansoni Journal of Comparative Physiology A* 189, 631–641.

Brownlee, D.J.A. and Fairweather, I. (1996) Immunocytochemical localization of glutamate-like immunoreactivity within the nervous system of the cestode, *Mesocestoides corti* and the trematode, *Fasciola hepatica*. *Parasitology Research* 82, 423–427.

Buttarelli, F.R., Pontieri, F.E., Margotta, V. and Palladini, G. (2000) Acetylcholine/dopamine interaction in planaria. *Comparative Biochemistry and Physiology C* 125, 225–231.

Cebria, F., Kudome, T., Nakazawa, M., Mineta, K., Ikeo, K., Gojobori, T. and Agata, K. (2002) The expression of neural-specific genes reveals the structural and molecular complexity of the planarian central nervous system. *Mechanisms of Development* 116, 199–204.

Chance, M.R. and Mansour, T.E. (1953) A contribution to the pharmacology of movement in the liver fluke. *British Journal of Pharmacology* 8, 134–138.

Cobbett, P. and Day, T.A. (2003) Functional voltage-gated Ca^{2+} channels in muscle fibers of the platyhelminth *Dugesia tigrina*. *Comparative Biochemistry and Physiology A* 134, 593–605.

Colasanti, M. and Venturini, G. (1998) Nitric oxide in invertebrates. *Molecular Neurobiology* 17, 157–174.

Curry, W.J., Shaw, C., Johnston, C.F., Thim, L. and Buchanan, K.D. (1992) Neuropeptide F: primary structure from the tubellarian, *Artioposthia triangulata*. *Comparative Biochemistry and Physiology C* 101, 269–274.

Day, T.A. and Maule, A.G. (1999) Parasitic peptides! The structure and function of neuropeptides in parasitic worms. *Peptides* 20, 999–1019.

Day, T.A., Orr, N., Bennett, J.L. and Pax, R.A. (1993) Voltage-gated currents in muscle cells of *Schistosoma mansoni*. *Parasitology* 106, 471–477.

Day, T.A., Bennett, J.L. and Pax, R.A. (1994a) Serotonin and its requirement for maintenance of contractility in muscle fibres isolated from *Schistosoma mansoni*. *Parasitology* 108, 425–432.

Day, T.A., Maule, A.G., Shaw, C., Halton, D.W., Moore, S., Bennett, J.L. and Pax, R.A. (1994b) Platyhelminth FMRFamide-related peptides (FaRPs) contract *Schistosoma mansoni* (Trematoda: Digenea) muscle fibres *in vitro*. *Parasitology* 109, 455–459.

Day, T.A., Kim, E., Bennett, J.L. and Pax, R.A. (1995) Analysis of the kinetics and voltage-dependency of transient and delayed K^+ currents in muscle fibres isolated from the flatworm *Schistosoma mansoni*. *Comparative Biochemistry and Physiology* 111, 79–87.

Day, T.A., Chen, G.Z., Miller, C., Tian, M., Bennett, J.L. and Pax, R.A. (1996) Cholinergic inhibition of muscle fibres isolated from *Schistosoma mansoni* (Trematoda: Digenea). *Parasitology* 113, 55–61.

Day, T.A., Maule, A.G., Shaw, C. and Pax, R.A. (1997) Structure–activity relationships of FMRFamide-related peptides contracting *Schistosoma mansoni* muscle. *Peptides* 18, 917–921.

Day, T.A., Haithcock, J., Kimber, M. and Maule, A.G. (2000) Functional ryanodine receptor channels in flatworm muscle fibres. *Parasitology* 120, 417–422.

de Mendonca, R.L., Beck, E., Rumjanek, F.D. and Goffeau, A. (1995) Cloning and characterization of a putative calcium-transporting ATPase gene from *Schistosoma mansoni*. *Molecular and Biochemical Parasitology* 72, 129–139.

Dorsey, C.H., Cousin, C.E., Lewis, F.A. and Stirewalt, M.A. (2002) Ultrastructure of the *Schistosoma mansoni* cercaria. *Micron* 33, 279–323.

Eklove, H. and Webb, R.A. (1981) The effect of L-glutamate and related agents on adenylate cyclase in the cestode *Hymenolepis diminuta*. *Canadian Journal of Physiology and Pharmacology* 69, 28–36.

Ercoli, N., Payares, G. and Nunez, D. (1985) *Schistosoma mansoni*: neurotransmitters and the mobility of cercaria and schistosomules. *Experimental Parasitology* 59, 204–216.

Eriksson, K.S., Maule, A.G., Halton, D.W., Panula, P.A.J. and Shaw, C. (1995) GABA in the nervous system of parasitic flatworms. *Parasitology* 110, 339–346.

Eriksson, K.S., Johnston, R.N., Shaw, C., Halton, D.W. and Panula, P.A.J. (1996) Widespread distribution of histamine in the nervous system of a trematode flatworm. *Journal of Comparative Neurology* 373, 220–227.

Fetterer, R.H., Pax, R.A. and Bennett, J.L. (1980) Praziquantel, potassium and 2,4-dinitrophenol: analysis of their action on the musculature of *Schistosoma mansoni*. *European Journal of Pharmacology* 64, 31–38.

Graham, M.K., Fairweather, I. and McGeown, J.G. (1997) The effects of FaRPs on the motility of isolated muscle strips from the liver fluke, *Fasciola hepatica*. *Parasitology* 114, 455–465.

Graham, M.K., McGeown, J.G. and Fairweather, I. (1999) Ionic mechanisms underlying spontaneous muscle contractions in the liver fluke, *Fasciola hepatica*. *American Journal of Physiology* 277, R374–R383.

Graham, M.K., Fairweather, I. and McGeown, J.G. (2000) Second messengers mediating mechanical responses to the FARP GYIRFamide in the fluke *Fasciola hepatica*. *American Journal of Physiology* 279, R2089–R2094.

Greenberg, M.J. and Price, D.A. (1992) Relationships among the FMRFamide-like peptides. *Progress in Brain Research* 92, 25–37.

Gustafsson, M.K.S., Lindholm, A.M., Terenina, N. and Reuter, M. (1996) NO nerves in a tapeworm! NADPDH diaphorase histochemistry in adult *Hymenolepis diminuta*. *Parasitology* 113, 559–565.

Gustafsson, M.K.S., Terenina, N.B., Kreshchenko, N.D., Reuter, M., Maule, A.G. and Halton, D.W. (2001) Comparative study of the spatial relationship between nicotinamide adenine dinucleotide phosphate-diaphorase activity, serotonin-immunoreactivity, and GYIRFamide-immunoreactivity and the musculature of the adult liver fluke, *Fasciola hepatica*. *Journal of Comparative Neurology* 429, 71–79.

Gustafsson, M.K., Halton, D.W., Kreshchenko, N.D., Movsessian, S.O., Raikova, O.I., Reuter, M. and Terenina, N.B. (2002) Neuropeptides in flatworms. *Peptides* 23, 2053–2061.

Gustafsson, M.K., Terenina, N.B., Reuter, M. and Movsessian, S.O. (2003) NO nerves and their targets in a tapeworm: an immunocytochemical study of cGMP in *Hymenolepis diminuta*. *Parasitology Research* 90, 148–152.

Halton, D.W. (2004) Microscopy and the helminth parasite. *Micron* 35, 361–390.

Halton, D.W. and Gustafsson, M.K.S. (1996) Functional morphology of the platyhelminth nervous system. *Parasitology* 113, S47–S72.

Halton, D.W. and Maule, A.G. (2004) Flatworm nerve–muscle: structural and functional analysis. *Canadian Journal of Zoology* 82, 316–333.

Halton, D.W., Shaw, C., Maule, A.G. and Smart, D.S. (1994) Regulatory peptides in helminth parasites. *Advances in Parasitology* 34, 163–227.

Halton, D.W., Maule, A.G. and Shaw, C. (1997) Trematode neurobiology. In: Fried, B. and Graczyk, T.K. (eds) *Advances in Trematode Biology*. CRC Press, Boca Raton, New York, pp. 345–382.

Halton, D.W., Maule, A.G., Mair, G.R. and Shaw, C. (1998) Monogenean neuromusculature: some structural and functional correlates. *International Journal for Parasitology* 28, 1609–1623.

Hamdan, F. and Ribeiro, P. (1998) Cloning, functional expression and characterization of a novel form of tyrosine hydroxylase from the human parasite, *Schistosoma mansoni*. *Journal of Neurochemistry* 71, 1369–1380.

Hamdan, F. and Ribeiro, P. (1999) Characterization of a stable form of tryptophan hydroxylase from the human parasite, *Schistosoma mansoni*. *Journal of Biological Chemistry* 274, 21746–21754.

Hamdan, F., Abramovitz, M., Mousa, A., Xie, J., Durocher, Y. and Ribeiro, P. (2002) A novel *Schistosoma mansoni* G protein-coupled receptor is responsive to histamine. *Molecular Biochemical Parasitology* 119, 75–86.

Hillman, G.R. (1983) The neuropharmacology of schistosomes. *Pharmacology and Therapeutics* 22, 103–115.

Holmes, S.D. and Fairweather, I. (1984) *Fasciola hepatica*: the effects of neuropharmacological agents upon *in vitro* motility. *Experimental Parasitology* 58, 194–208.

Hrckova, G., Velebny, S., Halton, D.W. and Maule, A.G. (2002) *Mesocestoides corti* (syn. *M. vogae*): modulation of larval motility by neuropeptides, serotonin and acetylcholine. *Parasitology* 124, 409–421.

Hrckova, G., Velebny, S., Halton, D.W., Day, T.A. and Maule, A.G. (2004) Pharmacological characterisation of neuropeptide F (NPF)-induced effects on the motility of *Mesocestoides corti* (syn. *M. vogae*) larvae. *International Journal for Parasitology* 34, 83–93.

Hu, W., Yan, Q., Shen, D.K., Liu, F., Zhu, Z.D., Song, H.D., Xu, X.R., Wang, Z.J., Rong, Y.P., Zeng, L.C., Wu, J., Zhang, X., Wang, J.J., Xu, X.N., Wang, S.Y., Fu, G., Zhang, X.L., Wang, Z.Q., Brindley, P.J., Mcmanus, D.P., Xue, C.L., Feng, Z., Chen, Z. and Han, Z.G. (2003) Evolutionary and biomedical implications of a *Schistosoma japonicum* complementary DNA resource. *Nature Genetics* 35, 139–147.

Humphries, J.E., Kimber, M.J., Barton, E., Hsu, W., Marks, N.J., Greer, B., Harriott, P., Maule, A.G. and Day, T.A. (2004) Structure and bioactivity of neuropeptide F from the human parasites *Schistosoma mansoni* and *Schistosoma japonicum*. *The Journal of Biological Chemistry* 279, 39880–39885.

Johnston, R.N., Shaw, C., Halton, D.W., Verhaert, P. and Baguna, J. (1995) GYIRFamide: a novel FMRFamide-related peptide (FaRP) from the triclad turbellarian, *Dugesia tigrina*. *Biochemical and Biophysical Research Communications* 209, 689–697.

Johnston, R.N., Shaw, C., Halton, D.W., Verhaert, P., Blair, K.L., Brennan, G.P., Price, D.A. and Anderson, P.A.V. (1996) Isolation, localisation, and bioactivity of the FMRFamide-related neuropeptides GYIRFamide and YIRFamide from the marine turbellarian *Bdelloura candida*. *Journal of Neurochemistry* 67, 814–821.

Jones, A.K., Bentley, G.N., Parra, W.G.O. and Agnew, A. (2002) Molecular characterization of an acetylcholinesterase implicated in the regulation of glucose scavenging by the parasite *Schistosoma*. *The FASEB Journal* 16, 51–69.

Keenan, L. and Koopowitz, H. (1982) Physiology and *in situ* identification of putative aminergic transmitters in the nervous system of *Gyrocotyle fimbriata*, a parasitic flatworm. *Journal of Neurobiology* 13, 9–21.

Keenan, L., Koopowitz, H. and Bernardo, K. (1979) Primitive nervous systems: action of aminergic drugs and blocking agents on activity in the ventral nerve cord of the flatworm *Notoplana acticola*. *Journal of Neurobiology* 10, 397–407.

Kim, E., Day, T.A., Bennett, J.L. and Pax, R.A. (1995a) Cloning and functional expression of a Shaker-related voltage-gated potassium channel gene from *Schistosoma mansoni* (Trematoda: Digenea). *Parasitology* 110, 171–180.

Kim, E., Day, T.A., Marks, N.J., Johnston, R.N., Halton, D.W., Shaw, C., Chen, G.Z., Bennett, J.L. and Pax, R.A. (1995b) Immunohistochemical localization of a Shaker-related voltage-gated potassium channel protein in *Schistosoma mansoni* (Trematoda: Digenea). *Experimental Parasitology* 81, 421–429.

Kohn, A.B., Moroz, L.L., Lea, J.M. and Greenberg, R.M. (2001a) Distribution of nitric oxide synthase immunoreactivity in the nervous system and peripheral tissues of *Schistosoma mansoni*. *Parasitology* 122, 87–92.

Kohn, A.B., Anderson, P.A., Roberts-Misterly, J.M. and Greenberg, R.M. (2001b) Schistosome calcium channel beta subunits. Unusual modulatory effects and potential role in the action of the antischistosomal drug praziquantel. *Journal of Biological Chemistry* 276, 36873–36876.

Kohn, A.B., Lea, J., Roberts-Misterly, J.M., Anderson, P.A. and Greenberg, R.M. (2001c) Structure of three high voltage-activated calcium channel alpha1 subunits from *Schistosoma mansoni*. *Parasitology* 123, 489–497.

Kumar, D., White, C., Fairweather, I. and McGeown, J.G. (2004) Electrophysiological and pharmacological characterization of K^+-currents in muscle fibres isolated from the ventral sucker of *Fasciola hepatica*. *Parasitology* 129, 779–793.

Lindholm, A.M., Reuter, M. and Gustafsson, M.K. (1998) The NADPH-diaphorase staining reaction in relation to the aminergic and peptidergic nervous system and the musculature of adult *Diphyllobothrium dentriticum*. *Parasitology* 117, 283–292.

Lingueglia, E., Champigny, G., Lazdunski, M. and Barbry, P. (1995) Cloning of the amiloride-sensitive FMRFamide peptide-gated sodium channel. *Nature* 378, 730–733.

Lumsden, R.D. and Hildreth, M.B. (1983) The fine structure of adult tapeworms. In: Arme, C. and Pappas, W.C. (eds) *Biology of the Eucestoda*. Academic Press, New York, pp. 177–234.

Mair, G.R., Halton, D.W., Shaw, C. and Maule, A.G. (2000) The neuropeptide F (NPF) encoding gene from the cestode, *Moniezia expansa*. *Parasitology* 120, 71–77.

Mair, G.R., Maule, A.G., Fried, B., Day, T.A. and Halton, D.W. (2003) Organization of the musculature of schistosome cercariae. *Journal of Parasitology* 89, 623–625.

Mair, G.R., Niciu, M.J., Stewart, M.T., Brennan, G., Omar, H., Halton, D.W., Mains, R., Eipper, B.A., Maule, A.G. and Day, T.A. (2004) A functionally atypical amidating enzyme from the human parasite *Schistosoma mansoni*. *The FASEB Journal* 18, 114–121.

Mansour, T.E. (1957) The effect of lysergic acid diethylamide, 5-hydroxytryptamine and related compounds on the liver fluke *Fasciola hepatica*. *British Journal of Pharmacology* 12, 406–409.

Mansour, T.E. (1984) Serotonin receptors in parasitic worms. *Advances in Parasitology* 23, 1–36.

Mansour, J.M. and Mansour T.E. (1986) GTP-binding proteins associated with serotonin activated adenylate cyclase in *Fasciola hepatica*. *Molecular and Biochemical Parasitology* 21, 139–149.

Mansour, J.M. and Mansour T.E. (1989) Identification of GTP-binding proteins in *Fasciola hepatica* and *Schistosoma mansoni* by immunoblotting. *Molecular and Biochemical Parasitology* 36, 11–18.

Marks, N.J., Johnston, S.S., Maule, A.G., Halton, D.W., Shaw, C., Geary, T.G., Moore, S. and Thompson, D.P. (1996) Physiological effects of platyhelminth RFamide peptides on muscle-strip preparations of *Fasciola hepatica* (Trematoda: Digenea). *Parasitology* 113, 393–401.

Maule, A.G., Halton, D.W., Allen, J.M. and Fairweather, I. (1989) Studies on motility *in vitro* of an ectoparasitic monogenean *Diclidophora merlangi Parasitology* 98, 85–93.

Maule, A.G., Shaw, C., Halton, D.W., Thim, L., Johnston, C.F., Fairweather, I. and Buchanan, K.D. (1991) Neuropeptide F: a novel parasitic flatworm regulatory peptide from *Moniezia expansa* (Cestoda: Cyclophyllidea). *Parasitology* 102, 309–316.

Maule, A.G., Shaw, C., Halton, D.W. and Thim, L. (1993) GNFFRFamide: a novel FMRFamide-immunoreactive peptide isolated from the sheep tapeworm, *Moniezia expansa*. *Biochemical and Biophysical Research Communications* 193, 1054–1060.

Maule, A.G., Shaw, C., Halton, D.W., Curry, W.J. and Thim, L. (1994) RYIRFamide: a turbellarian FMRFamide-related peptide (FaRP). *Regulatory Peptides* 50, 37–43.

Maule, A.G., Geary, T.G., Bowman, J.W., Marks, N.J., Blair, K.L., Halton, D.W., Shaw, C. and Thompson, D.P. (1995) Inhibitory effects of nematode FMRFamide-related peptides (FaRPs) on muscle strips from *Ascaris suum*. *Invertebrate Neuroscience* 1, 255–265.

McNall, S.J. and Mansour, T.E. (1984) Novel serotonin receptors in *Fasciola*: characterization by studies on adenylate cyclase activation and [^{3}H]LSD binding. *Biochemical Pharmacology* 33, 2789–2797.

Mellin, T., Busch, R., Wang, C. and Kath, G. (1983) Neuropharmacology of the parasitic trematode *Schistosoma mansoni*. *American Journal of Tropical Medicine and Hygiene* 32, 83–93.

Mendonça-Silva, D.L., Pessoa, R.F. and Noel, F. (2002) Evidence for the presence of glutamatergic receptors in adult *Schistosoma mansoni*. *Biochemical Pharmacology* 64, 1337–1344.

Mendonça-Silva, D.L., Gardino, P.F., Kubrusly, R.C.C., Mello, F.G. and Noel, F. (2004) Characterization of GABAergic neurotransmission in adult *Schistosoma mansoni*. *Parasitology* 129, 137–146.

Miller, C.L., Day, T.A., Bennett, J.L. and Pax, R.A. (1996) *Schistosoma mansoni*: L glutamate-induced contractions in isolated muscle fibers; evidence for a glutamate transporter. *Experimental Parasitology* 84, 410–419.

Miskolzie, M. and Kotovych, G. (2002) The NMR-derived conformation of neuropeptide F from *Moniezia expansa*. *Journal of Biomolecular Structure and Dynamics* 19, 991–998.

Moneypenny, C.G., Maule, A.G., Shaw, C., Day, T.A., Pax, R.A. and Halton, D.W. (1997) Physiological effects of platyhelminth FMRFamide-related peptides (FaRPs) on the motility of the monogenean *Diclidophora merlangi*. *Parasitology* 115, 281–288.

Moneypenny, C.G., Kreshchenko, N., Moffett, C.L., Halton, D.W., Day, T.A. and Maule, A.G. (2001) Physiological effects of FMRFamide-related peptides and classical transmitters on dispersed muscle fibres of the turbellarian, *Procerodes littoralis*. *Parasitology* 122, 447–455.

Nimmo-Smith, R.H. and Raison, C.G. (1968) Monoamine oxidase activity of *Schistosoma mansoni*. *Comparative Biochemistry and Physiology C* 62, 403–416.

Onufriev, M.V., Gulyaeva, N.V., Terenina, N.B., Tolstenkov, O.O. and Gustafsson, M.K. (2005) The effect of a nitric oxide donor on the synthesis of cGMP in *Hymenolepis diminuta*: a radiometric study. *Parasitology Research* 95, 22–24.

Osloobi, N. and Webb, R.A. (1999) Localization of a sodium-dependent high-affinity serotonin transporter and recruitment of exogenous serotonin by the cestode *Hymenolepis diminuta*: an autoradiographic and immunohistochemical study. *Canadian Journal of Zoology* 77, 1265–1277.

Pax, R.A., Siefker, C., Hickox, T. and Bennett, J.L. (1981) *Schistosoma mansoni*: neurotransmitters, longitudinal musculature and effects of electrical stimulation. *Experimental Parasitology* 52, 346–355.

Pax, R.A., Siefker, C. and Bennett, J.L. (1984) *Schistosoma mansoni*: differences in acetylcholine, dopamine and serotonin control of circular and longitudinal parasite muscles. *Experimental Parasitology* 58, 312–324.

Pax, R.A., Day, T.A., Miller, C.L. and Bennett, J.L. (1996) Neuromuscular physiology and pharmacology of parasitic flatworms. *Parasitology* 113, S83–S96.

Purcell, J., Robertson, A.P., Thompson, D.P. and Martin, R.J. (2002a) The time-course of the response to the FMRFamide-related peptide PF4 in *Ascaris suum* muscle cells indicates direct gating of a chloride ion-channel. *Parasitology* 124, 649–656.

Purcell, J., Robertson, A.P., Thompson, D.P. and Martin, R.J. (2002b) PF4, a FMRFamide-related peptide, gates low-conductance Cl^- channels in *Ascaris suum*. *European Journal of Pharmacology* 456, 11–17.

Ribeiro, P. and Webb, R.A. (1983) The occurrence and synthesis of octopamine and other biogenic amines in the cestode *Hymenolepis diminuta*. *Molecular and Biochemical Parasitology* 7, 53–62.

Ribeiro, P. and Webb, R.A. (1984) The occurrence, synthesis and metabolism of 5-hydroxytryptamine and 5-hydroxytryptophan in the cestode *Hymenolepis diminuta*. *Comparative Biochemistry and Physiology C* 79, 159–164.

Ribeiro, P. and Webb, R.A. (1987) Characterization of a serotonin transporter and an adenylate cyclase-linked serotonin receptor in the cestode *Hymenolepis diminuta*. *Life Sciences* 40, 755–768.

Ribeiro, P., El-Shehabi, N. and Patockas, N. (2005) Classical transmitters and their receptors in flatworms. Parasitology (in press).

Saitoh, O., Yuruzume, E., Watanabe, K. and Nakata, H. (1997) Molecular identification of a G protein-coupled receptor family which is expressed in planarians. *Gene* 195, 55–61.

Samii, S.I. and Webb, R.A. (1990) Acetylcholine-like immunoreactivity in the cestode *Hymenolepis diminuta*. *Brain Research* 513, 161–165.

Sanchez Alvarado, A., Newmark, P.A., Robb, S.M. and Juste, R. (2002) The *Schmidtea mediterranea* database as a molecular resource for studying platyhelminthes, stem cells and regeneration. *Development* 129, 5659–5665.

Shaw, C., Maule, A.G. and Halton, D.W. (1996) Platyhelminth FMRFamide-related peptides (FaRPs). *International Journal for Parasitology* 26, 335–346.

Skelly, P.J., Dougan, P.M., Maule, A.G., Day, T.A. and Shoemaker, C.B. (2001) Cloning and characterization of a muscle isoform of a Na-, K-ATPase alpha subunit (SNaK1) from *Schistosoma mansoni*. *Parasitology* 123, 277–284.

Solis-Soto, J.M. and Brink, M.D. (1994) Immunocytochemical study on biologically active neurosubstances in daughter sporocysts and cercaria of *Trichobilharzia ocellata* and *Schistosoma mansoni*. *Parasitology* 108, 301–311.

Sukhdeo, S.C. and Sukhdeo, M.V. (1994) Mesenchyme cells in *Fasciola hepatica* (Platyhelminthes): primitive glia? *Tissue and Cell* 26, 123–131.

Sukhdeo, M.V.K., Hsu, S.C., Thompson, C.S. and Mettrick, D.F. (1984) *Hymenolepis diminuta*: behavioural effects of 5-hydroxytryptamine, acetylcholine, histamine and somatostatin. *Journal of Parasitology* 70, 682–688.

Talla, E., de Mendonca, R.L., Degand, I., Goffeau, A. and Ghislain, M. (1998) *Schistosoma mansoni* Ca^{2+}-ATPase SMA2 restores viability to yeast Ca^{2+}-ATPase-deficient strains and functions in calcineurin-mediated Ca^{2+} tolerance. *Journal of Biological Chemistry* 273, 27831–27840.

Tandon, V., Kar, P.K. and Saha, N. (2001) NO nerves in trematodes, too! NADPH-diaphorase activity in adult *Fasciolopsis buski*. *Parasitology International* 50, 157–163.

Terenina, N.B. and Gustafsson, M.K. (2003) Nitric oxide and its target cells in cercaria of *Diplostomum chromatophorum*: a histochemical and immunocytochemical study. *Parasitology Research* 89, 199–206.

Terenina, N.B., Reuter, M. and Gustafsson, M.K. (1999) An experimental, NADPH-diaphorase histochemical and immunocytochemical study of *Mesocestoides vogae* tetrathyridia. *International Journal for Parasitology* 29, 787–793.

Thompson, C.S. and Mettrick, D.F. (1989) The effects of 5-hydroxytryptamine and glutamate on muscle contraction in *Hymenolepis diminuta* (Cestoda). *Canadian Journal of Zoology* 67, 1257–1262.

Verjovski-Almeida, S., DeMarco, R., Martins, E.A, Guimaraes, P.E, Ojopi, E.P., Paquola, A.C, Piazza, J.P., Nishiyama, M.Y. Jr, Kitajima, J.P., Adamson, R.E., Ashton, P.D., Bonaldo, M.F., Coulson, P.S., Dillon, G.P., Farias, L.P., Gregorio, S.P., Ho, P.L., Leite, R.A., Malaquias, L.C., Marques, R.C., Miyasato, P.A., Nascimento, A.L., Ohlweiler, F.P., Reis, E.M., Ribeiro, M.A., Sa, R.G., Stukart, G.C., Soares, M.B., Gargionic, C., Kawano, T., Rodrigues, V., Madeira, A.M., Wilson, R.A., Menck, C.F., Setubal, J.C., Leite, L.C. and Dias-Neto, E. (2003) Transcriptome analysis of the acoelomate human parasite *Schistosoma mansoni*. *Nature Genetics* 35, 148–157.

Ward, S.M., McKerr, G. and Allen, J.M. (1986) Structure and ultrastructure of muscle systems within *Grillotia erinaceus* metacestodes (Cestoda: Trypanorhyncha). *Parasitology* 93, 587–597.

Webb, R.A. (1987) Innervation of muscle in the cestode *Hymenolepis microstoma*. *Canadian Journal of Zoology* 65, 928–935.

Webb, R.A. (1988) Release of exogenously supplied L-[3H]-glutamate and endogenous glutamate from tissue slices of the cestode *Hymenolepis diminuta*. *Canadian Journal of Physiology and Pharmacology* 66, 889–894.

Webb, R.A. (1997) Glutamatergic inositol 1,4,5-triphosphate production in tissue slices of the cestode *Hymenolepis diminuta*. *Comparative Biochemistry and Physiology C* 118, 19–26.

Webb, R.A. and Eklove, H. (1989) Demonstration of intense glutamate-like immunoreactivity in the longitudinal nerve cords of the cestode, *Hymenolepis diminuta*. *Parasitology Research* 75, 545–548.

Wikgren, M., Reuter, M., Gustafsson, M.K.S. and Lindroos, P. (1990) Immunocytochemical localization of histamine in flatworms. *Cell Tissue Research* 260, 479–484.

Willcockson, W.S. and Hillman, G.R. (1984) Drug effects on the 5-HT response of *Schistosoma mansoni*. *Comparative Biochemistry and Physiology C* 77, 199–203.

20 Unusual Aspects of Metabolism in Flatworm Parasites

A.G.M. Tielens and J.J. van Hellemond
Faculty of Veterinary Medicine, Department of Biochemistry and Cell Biology, Utrecht University, PO Box 801765, The Netherlands

Introduction

The phylum Platyhelminthes includes various dorsoventrally flattened, bilaterally symmetrical animals, which are often described as flatworms. They are present in all aquatic environments, and parasitic as well as non-parasitic species of flatworms exist. Metabolically speaking, all flatworms are very much alike, but from a more biological point of view some differences can be seen and will, therefore, be introduced here. Within the phylum Platyhelminthes, the class Turbellaria contains mainly free-living flatworms, but most platyhelminths are parasites of vertebrates, and all vertebrate species are prone to infection by one or more species of flatworms. The great majority of parasitic flatworms belong to the classes Aspidogastrea,

Monogenea, Digenea and Cestoda. The first three classes are often together referred to as 'trematodes'. The aspidogastrids are typically endoparasites of many molluscs and turtles, and are not very host-specific. The monogenea form a diverse group of platyhelminths parasitic primarily on the external surfaces and gills of marine and freshwater fish. In contrast to aspidogastrids and monogeneans, digeneans (flukes) have indirect and complex life cycles. They all have stages in their life cycle that are parasitic in molluscs, and often have in addition other stages in their life cycle that are parasitic in other hosts. Digeneans are widespread endoparasites, infecting all classes of vertebrates and they may inhabit a large variety of organs of their host. Cestodes, on the other hand, all live in the intestines of vertebrate hosts. They are also called tapeworms because as adults they are extremely flattened, segmented and often very long.

This chapter will be restricted to digeneans and cestodes because the metabolism of parasites from the other classes is hardly studied, if at all. The liver fluke *Fasciola hepatica* and the blood-dwelling *Schistosoma mansoni* are the best-studied digeneans, and most metabolic studies on tapeworms have been performed on *Hymenolepis diminuta*. These parasites will, therefore, play the leading part in this chapter. On the other hand it should be noted that the metabolism of most parasitic worms is very similar, even in the unusual features. All parasitic worms are completely adapted to a parasitic way of life, and these adaptations are very comparable in the various types of platyhelminths, and the same adaptations are even found in parasitic nematodes (roundworms). It is these adaptations, some of which are remarkable, that will be discussed further here. These unusual aspects of metabolism are the result of adaptations, either to the environmental changes that occur during their life cycle, the rather unusual niche many parasites inhabit (for instance the dark world in the intestine of the host where food is plenty but oxygen scarce), or simply their parasitic lifestyles, i.e. 'theft'. Substrates for energy metabolism as well as building blocks for the synthesis of macromolecules are simply taken from the host, and this has resulted, ultimately, in the loss of many obsolete biosynthetic capacities.

Parasitic versus Free-living Stages

Commonly, during their complex life cycle, parasitic helminths alternate between free-living and parasitic stages, and the availability of substrates during these diverse stages varies considerably. The free-living stages, which are simply vehicles to go from one host to the next, do not multiply, and they possess limited or no biosynthetic capacities. The free-living stages of parasitic flatworms, like miracidia, cercariae and metacercariae, are self-supporting. They usually live in water, do not obtain substrates other than oxygen from the environment and are completely dependent on the energy reserves they acquired in their previous host. These stages will die if their endogenous reserves are exhausted before they have entered the next (intermediate) host.

On the other hand, once inside the host, substrates are plentiful, and the only concern of these parasitic stages is to produce offspring and avoid being destroyed by the immune system of the host. Parasitic flatworms often live in the 'land of milk and honey'. They do not have to hunt for food like predators, nor do they have to search and gather their food like scavengers and grazers. To obtain food, they only have to use their 'mouth' (oral sucker) or their tegument (see section on 'Parasitic Stages'). During their stay in the host, which will last for the rest of their life, they are pampered by the host, but against its will. To multiply and produce eggs, these parasitic stages in fact use energy as well as building blocks provided by the host.

Nutrition

Free-living stages

Parasitic flatworms have distinct stages in their life cycle with a large variation in the availability of substrates. When they reside inside their intermediate or definitive host, the parasite can and will obtain substrates from this host, as the term 'parasite' itself signifies. The free-living stages, (meta)cercariae, eggs and miracidia, on the other hand, live on their endogenous reserves, which they obtained during their previous parasitic stage. Glycogen is present in these stages and is used to span the time before a new host is entered and new substrates become available. As these reserves are limited, the free-living stages (miracidia, cercariae) are either short-lived and search frantically for a new host, or they enter into a dormant stage (metacercariae and eggs), waiting to be reactivated by external triggers.

In eggs from most flatworm parasites, like those of, for example, *F. hepatica*, the vitelline cells inside the egg function as food reserves and these eggs, which embryonate outside the mammalian host, require no extra nutrients for the developing embryo. Eggs from schistosomes, on the other hand, develop to maturity inside human tissues and only mature eggs are excreted. In this environment, substrates are abundant and the uptake of a wide variety of substrates has been reported. The eggshells are porous and the eggs take up glucose and amino acids, and excrete proteins (Kawanaka *et al.*, 1986). These eggs cannot mature without the uptake of nutrients. Studies on *Schistosoma japonicum* eggs showed that maturation could be achieved in a complex medium. Eggs of schistosomes, however, appear to be the exception; in the eggs of most parasitic flatworms a large number of vitelline cells provide all the nutritional requirements for survival (prior to a parasitic lifestyle).

Parasitic stages

The parasitic stages inside their respective hosts need substrates for the biosynthesis of structural elements as well as for energy metabolism, and all of these nutrients are taken from the host. Feeding on blood is the main feeding strategy of digeneans. The liver fluke *F. hepatica* lives in the bile ducts of its host and uses the oral sucker both for attachment and for feeding on the blood of its host. Schistosomes, on the other hand, literally live in blood, inside the vasculature of their hosts (birds and mammals), being bathed continuously in the medium that provides all their nutritional requirements. In contrast to these digenean worms, cestodes reside in the alimentary tracts of vertebrates, a niche with a wealth of predigested low molecular weight compounds. See Halton (1997) for an extensive review on nutritional adaptations to parasitism within the plathyhelminths.

The modes of uptake of nutrients by parasitic flatworms are not principally different from those used by free-living organisms. Within the phylum Platyhelminthes, there are three main variations on the mode of feeding.

1. Flatworms with an alimentary tract ingest food into this canal, where it is first digested and then absorbed by the intestinal epithelium. Ectoparasitic flatworms like monogenea are restricted to this type of food uptake.
2. Endoparasitic digeneans, on the other hand, possess two surfaces that can potentially absorb nutrients: the intestinal epithelium and the external surface (tegument). The relative contribution of both surfaces in nutrition is still unknown, but both intestine and tegument are considered to be important in nutrition in these endoparasitic helminths. The incomplete digestive tract of digeneans must first be filled with nutrients and after digestion the undigested remains must be regurgitated through the same oral opening.
3. For cestodes, uptake via the external surface, the syncitial tegument, is the sole means of obtaining nutrients, as they do not have an alimentary tract. This absence of an alimentary tract is a striking feature of all cestodes. In this respect, it should be realized that all cestodes are parasites.

The intestines of trematodes are lined with an absorptive epithelium, which takes up nutrients after they have been degraded in the cul-de-sac intestine by digestive enzymes. *Schistosoma* sp. and *Fasciola* sp. utilize proteases for the degradation of host globin in their intestines (Carmona *et al.*, 1994; Yamasaki *et al.*, 2002; Chapter 18, this volume). Cysteine proteases (cathepsin L and cathepsin B) and aspartyl proteases play a role in the digestion of host haemoglobin. These proteases have been characterized and their respective roles analysed (Klinkert *et al.*, 1994; Dalton *et al.*, 1995; Skelly and Shoemaker, 2001). Schistosomes, and in particular the female, ingest red blood cells, most likely as a source of polypeptides and amino acids to facilitate egg laying. Proteolytic enzymes are not only important for trematodes to digest proteins in order to obtain nutrients, but they also play a crucial role in the host–parasite interaction, in penetration of the host skin, in egg secretion and in immune evasion. For instance, in the invasion of schistosomula through host tissue, serine proteases are involved, which also protect the parasites from complement attack by cleaving complement fragments (Chapter 18, this volume).

The tegument of trematodes plays an important role in nutrient absorption and the numerous pits, channels and invaginations effectively increase the surface area. This outer membrane contains glucose-transporter proteins and enzymes, like peptidases, involved in amino acid absorption. Absorption of simple substrates that are abundant in the blood of the host, like glucose and amino acids, occurs primarily via the tegument instead of the intestinal epithelium. Glucose transporters, homologues of the human glucose transporter protein GLUT 1, are evenly distributed throughout the tegumental surfaces (ventral and dorsal) of male and female *S. mansoni*, and this finding implies that both sexes have the ability to take up glucose across the entire tegument (Skelly and Shoemaker, 1996). These studies showed that one of these glucose transporters is expressed only in the mammalian stages of the life cycle and is specifically localized within the apical membranes of the tegument, which is a further indication that the uptake of glucose occurs almost exclusively via the tegument. Cholesterol and other lipids are absorbed via the tegument as well (Moffat and Kusel, 1992; Rogers *et al.*, 1993). The outer membrane of schistosomes also contains receptors, which bind human low-density lipoproteins (LDL) (Rumjanek *et al.*, 1988). Uptake of lipids via bound LDL has, however, never been demonstrated and it is suggested that binding of these lipoprotein complexes to the receptors on the surface of the parasite is only used as a disguise, to shield the parasite antigens from the host antibodies (Xu and Caulfield, 1992).

Energy Metabolism

Substrates of energy metabolism

In contrast to the host, in parasitic flatworms a clear distinction exists between the substrates for energy metabolism and those for biosynthetic purposes. The mammalian host can adapt its metabolism to the availability of substrates. Depending on the supply, carbohydrates, lipids or proteins can be the main source of energy. Parasitic flatworms, on the other hand, are completely dependent on carbohydrates for their energy metabolism. Because of the very limited aerobic capacity of the adult parasites (see sections on 'Free-living versus Parasitic Stages' and 'Transitions in Energy Metabolism During the Life Cycle'), their metabolism has to be mainly fermentative. Carbohydrates are suitable substrates for fermentation, as both oxidation and reduction of this substrate can occur (see section on 'Anaerobic Glycolysis versus Malate Dismutation'). Lipids are too reduced for this purpose, and therefore, cannot be used for the production of ATP by parasitic helminths in general. Most free-living stages of parasitic flatworms probably have a classical aerobic metabolism and could, therefore, in principle be able to utilize lipids for the production of ATP. Very little research has been done on lipid metabolism in these stages of parasitic helminths, and up to now no evidence has been demonstrated for the degradation of lipids by *F. hepatica*, *S. mansoni*, or any other parasitic flatworm.

Generally, free-living stages of parasitic worms are supposed to depend on their endogenous glycogen reserves that are used to span the gap in food supply until the next host is entered. Endogenous glycogen reserves are also used by intestinal parasites, this time to span the time between host meals when glucose concentrations in the intestine are low. In the rat tapeworm *H. diminuta*, it was shown that glycogen degradation occurs when the host is fasting and that synthesis of glycogen occurs after host refeeding (reviewed in Andreassen *et al.*, 1999). Most parasitic flatworms contain glycogen reserves, and even the blood-dwelling adult schistosomes, which are in an environment with a steady supply of glucose, contain large and metabolically active glycogen stores. Studies on individual *S. mansoni* worm pairs, living inside the veins of the final host, showed that these glycogen stores are continuously synthesized by the parasite. Glycogen is consumed probably intermittently, for short, intensive, bursts of activity, for instance during the muscle activity necessary for crawling. Replenishment of the glycogen reserves is, however, not induced by a marked decrease in the glycogen levels, but occurs in *S. mansoni in vivo* slowly and continuously (Tielens *et al*., 1990).

The oxidation of amino acids is considered to contribute little to the overall energy metabolism in parasitic flatworms, but amino acids are necessary as precursors of protein synthesis for normal growth and reproduction.

Free-living versus parasitic stages

It is assumed that all free-living stages of parasitic helminths have an aerobic energy metabolism and degrade their endogenous reserves to CO_2 via the classical pathways of glycolysis and the Krebs cycle (Fig. 20.1). The stored carbohydrate reserves of free-living stages (glycogen) are degraded to pyruvate via the classic glycolytic pathway within the cytosol. The end product of the cytosolic steps, pyruvate, is transported into the mitochondria where it is converted into acetyl-CoA. Acetyl-CoA is then further degraded via the Krebs cycle and most ATP is produced via the mitochondrial respiratory chain and oxidative phosphorylation (Fig. 20.1). Oxygen acts as the final electron acceptor to oxidize the generated reduced coenzyme (NADH). This degradation of carbohydrates is not different from the well-known aerobic mammalian-type of metabolism. Although not studied extensively in parasitic flatworms, this aerobic type of metabolism has been shown to occur in the free-living stages of *S. mansoni*, cercariae and miracidia, and in the miracidia of the liver fluke, *F. hepatica* (Tielens *et al*., 1984; Boyunaga *et al*., 2001).

Adult parasitic helminths have an energy metabolism strikingly different from that in the free-living stages. Although the end products of carbohydrate metabolism vary greatly between different species of adult parasitic flatworms, none of them degrades carbohydrates completely to CO_2, as the free-living stages do. In general, parasitic helminths do not use oxygen as final electron acceptor, but have a fermentative metabolism instead. When oxygen cannot function as terminal electron acceptor, the degradation of substrates must be in redox balance, i.e. the number of NADH-producing reactions must equal the number of NADH-consuming ones (without the use of oxygen). In adult parasitic platyhelminths, two pathways occur to maintain this balance between NADH formation and usage: (i) anaerobic glycolysis and (ii) malate dismutation.

Anaerobic glycolysis versus malate dismutation

Some adult flatworms ferment their carbohydrate substrates to lactate, a classical adaptation to metabolism without oxygen, although these parasites may perform this process for a reason other than lack of oxygen (see section on 'Transitions in Energy Metabolism during the Life Cycle'). This so-called anaerobic glycolysis, which yields two molecules of ATP per molecule of glucose degraded, occurs via the well-known Embden–Meyerhof pathway (Fig. 20.1). *Schistosoma* spp., *Clonorchis sinensis, Echinococcus granulosis* and *Taenia* spp. are examples of parasitic flatworms that excrete large amounts of lactate. All parasitic helminths use anaerobic glycolysis to a certain extent, but most use mainly a different pathway to ferment carbohydrates: malate dismutation.

In this malate dismutation pathway, carbohydrates are degraded to phosphoenolpyruvate (PEP) via the classical glycolytic pathway. This PEP is then carboxylated by PEP carboxykinase (PEPCK) to oxaloacetate, which is subsequently reduced to malate. This malate is transported into the mitochondria and is degraded in a split pathway. A portion of the malate is oxidized to acetate and another portion is reduced to succinate, which can then be further metabolized to propionate (Fig. 20.1).

In the oxidative branch of malate dismutation, malic enzyme oxidizes malate to pyruvate, which is then further oxidized to acetyl-CoA by pyruvate dehydrogenase, an enzyme complex specially adapted to anaerobic functioning in *Ascaris suum* and possibly in other parasitic helminths like the trematode *F. hepatica* and the cestode *Dipylidium caninum* (Diaz and Komuniecki, 1994; Klingbeil *et al*., 1996). Parasitic helminths like *F. hepatica* use an acetate:succinate CoA-transferase (ASCT) for

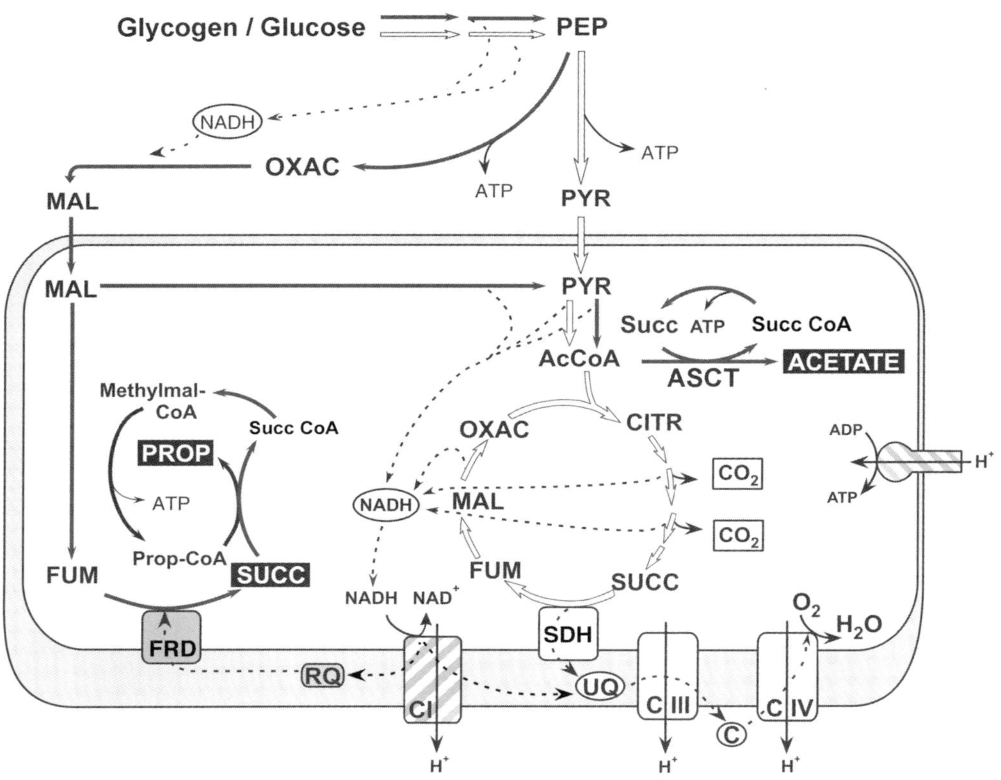

Fig. 20.1. Generalized scheme of the main pathways of aerobic and anaerobic carbohydrate degradation in parasitic flatworms. The aerobic pathway is indicated by open arrows, whereas the anaerobic pathway (malate dismutation) is indicated by solid arrows. Abbreviations: AcCoA, acetyl-CoA; ASCT, acetate:succinate CoA-transferase; C, cytochrome *c*; CI–CIV, complexes I–IV of the respiratory chain; CITR, citrate; FRD, fumarate reductase; FUM, fumarate; MAL, malate; Methylmal-CoA, methylmalonyl-CoA; OXAC, oxaloacetate; PEP, phosphoenolpyruvate; PROP, propionate; Prop-CoA, propionyl-CoA; PYR, pyruvate; RQ, rhodoquinone; SDH, succinate dehydrogenase; SUCC, succinate; Succ CoA, succinyl CoA; UQ, ubiquinone.

the production of acetate from acetyl-CoA (Van Vugt *et al.*, 1979; Saz *et al.*, 1996; Van Hellemond *et al.*, 1998). ATP is formed when subsequently the succinyl-CoA, which is produced during acetate formation by ASCT, is recycled to succinate by succinyl-CoA synthetase (SCS), an enzyme usually participating in the Krebs cycle (Fig. 20.1).

The production of acetate from part of the incoming malate, which results in the formation of NADH, is balanced by the reduction of another portion of the malate to succinate. This reduction of malate to succinate occurs via two reactions that reverse part of the Krebs cycle (Fig. 20.1). In *F. hepatica*, succinate is not a main end product as it is further metabolized to propionate via a reversal of the reactions that occur in mammals during the formation of succinyl-CoA from propionate. This decarboxylation of succinate to propionate is accompanied by substrate-level phosphorylation (Köhler *et al.*, 1978; Pietrzak and Saz, 1981). As can be seen in Fig. 20.1, redox balance inside the mitochondria is maintained when twice as much propionate as acetate is produced. The reduction of fumarate to succinate is coupled to an electron-transport-linked phosphorylation of ADP at site I of the respiratory chain (Figs 20.1 and 20.2). Apart from the electron-transport associated ATP formation in

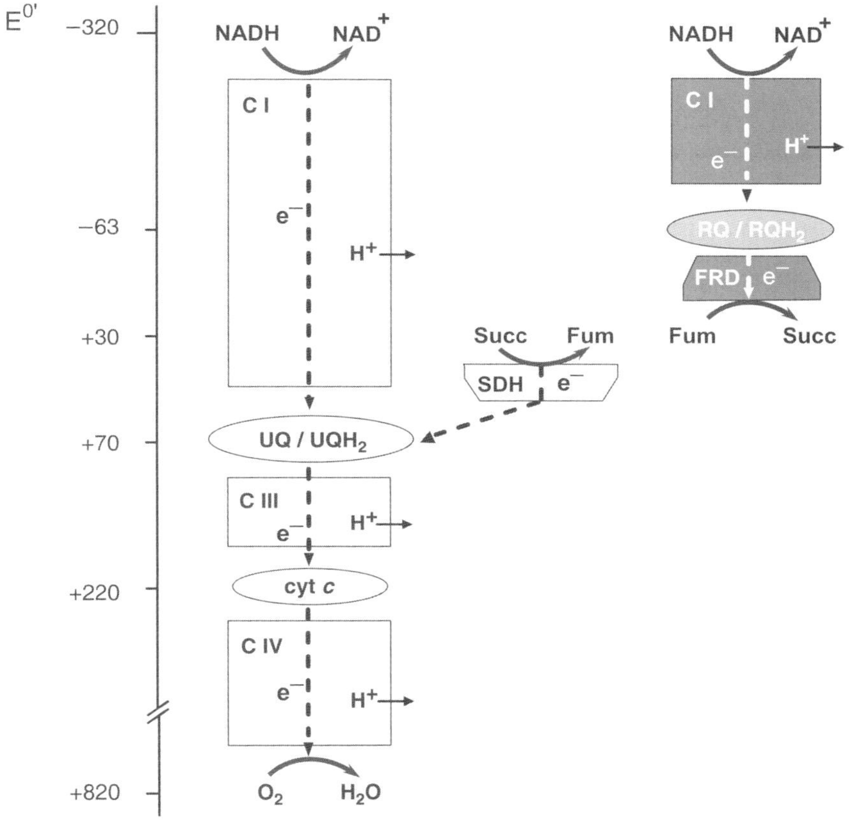

Fig. 20.2. Schematic representation of the mitochondrial respiratory chains of parasitic flatworms. Boxes indicate electron-transport chain complexes, whereas ovals represent the electron transporters UQ, RQ and cytochrome *c*. The open boxes represent complexes involved in the classical aerobic respiratory chain, whereas grey boxes represent complexes involved in malate dismutation. The vertical bar represents a scale for the standard redox potentials in mV. Translocation of protons by the complexes is indicated by H^+ →. Abbreviations: CI, CIII and CIV, complexes I, III and IV of the respiratory chain; cyt *c*, cytochrome *c*; FRD, fumarate reductase; Fum, fumarate; SDH, succinate dehydrogenase; Succ, succinate; RQ, rhodoquinone; UQ, ubiquinone.

the reduction of fumarate, this so-called malate dismutation is accompanied by substrate-level phosphorylations (Fig. 20.1). In total, the anaerobic degradation of glucose to propionate and acetate yields about 5 mol of ATP/mol of glucose degraded, which compares favourably with the 2 mol of ATP produced during lactate production.

In the trematode *F. hepatica* as well as in the cestodes *H. diminuta* and *Hymenolepis microstoma*, malic enzyme is in fact not NAD but NADP-dependent, producing NADPH during the formation of pyruvate (Tielens *et al.*, 1987; Fioravanti *et al.*, 1992). However, NADPH cannot be used by the enzyme fumarate reductase (FRD) in the reductive branch of malate dismutation. Cestodes have been shown to possess an active NADPH:NAD transhydrogenase associated with the inner-mitochondrial membrane. This enzyme couples the malic enzyme reaction to fumarate reduction in these organisms. Presumably, the same enzyme is used by other parasitic flatworms such as *F. hepatica* with an NADP-dependent malic enzyme that produce fumarate.

Unusual aspects of glycolysis

Several glycolytic enzymes in parasitic helminths have been purified and studied in great detail. These studies were stimulated by the early observation that the chemotherapeutic action of antimonials on schistosomes was associated with an inhibition of its phosphofructokinase (PFK), a glycolytic enzyme with a high flux control. Helminth PFKs appear to be much more sensitive to these anti-schistosomal drugs than the PFK of the host. Attack of the ATP-producing systems of parasites is considered to be one of the more rational and promising ways of combating parasitic diseases, as energy (ATP) is one of the few things parasites cannot directly obtain from the host. Special attention has been, of course, directed towards enzymes that are absent in the host or where differences are observed between host and parasite enzymes.

The initial enzyme in the glycolytic sequence, hexokinase, controls the entry of glucose into the pathway and is critical in the regulation of carbohydrate utilization. In mammals four isoenzymes are present. Isoenzymes I–III, monomers with a molecular mass of about 100,000 Da, have a high affinity for glucose and are strongly inhibited by their end product, glucose-6-phosphate. Type IV, also called glucokinase, is a monomer of about 50,000, has a low affinity for glucose and is weakly inhibited by glucose-6-phosphate. Only a few hexokinases from parasitic helminths have been studied in detail. In *S. mansoni*, hexokinase appears to be an enzyme with a high flux control of glycolysis, and cercariae and adults contain only a single isoform. The schistosomal enzyme is structurally related to other members of the hexokinase family and the amino acid sequence shows a significant homology to the mammalian hexokinases (Tielens *et al.*, 1994). The evolutionary relation of this schistosomal enzyme with other hexokinases is interesting, as the schistosomal hexokinase with its relatively high affinity for glucose and sensitivity for inhibition by glucose-6-phosphate, kinetically resembles the 100 kDa mammalian type hexokinases, while its molecular mass is only 50 kDa. Hexokinase appears to play a distinctive role in the rapid transition in energy metabolism that occurs in schistosomes when free-living cercariae enter the mammalian host (see section on 'Transitions in Energy Metabolism during the Life Cycle').

PFK is usually the main regulatory site of glycolysis, and *F. hepatica* PFK has been studied extensively. In general, the activity of PFK is regulated by several effectors. The most potent activators of mammalian as well as *Fasciola* PFK are fructose-2,6-bisphosphate and AMP, whereas ATP is inhibitory (Van Schaftingen *et al.*, 1981; Kamemoto *et al.*, 1987). In contrast to mammalian PFKs, the *Fasciola* enzyme is relatively insensitive to inhibition by citrate (Kamemoto *et al.*, 1987). However, the most remarkable difference with mammalian PFKs is that the kinetic properties of *Fasciola* PFK are strongly influenced by phosphorylation. Phosphorylation by a cAMP-dependent protein kinase results in PFK activation and it was proposed that this phosphorylation plays, at least in part, a functional role in the liver fluke regulation of PFK *in vivo* (Kamemoto *et al.*, 1989). Fasciola PFK is phosphorylated with cAMP-dependent protein kinase to 0.2 mol P/mol subunit and this results in a threefold increase in maximal enzyme activity compared to the unphosphorylated enzyme (Mahrenholz *et al.*, 1991). This phosphorylation occurs at a threonine residue in a phosphorylation site, which shows no homology to the phosphorylation site of *Ascaris* PFK an enzyme also known to be stimulated by phosphorylation (Kulkarni *et al.*, 1987; Mahrenholz *et al.*, 1991). Further, structural data to determine the differences between mammalian PFKs and those of parasitic helminths are not yet available.

Fructose-1,6-bisphosphatase (FBPase) is also active in adult *F. hepatica* and, together with the activity of PFK, can trigger extensive substrate cycling. ^{13}C-NMR experiments showed that substrate cycling is prominent in the glycolytic pathway of adult flukes, at least *in vitro* (Matthews *et al.*, 1986). Although this cycling consumes ATP, it is generally assumed that the occurrence of futile cycling enables the cell to regulate its net glycolytic flux more efficiently.

Pyruvate kinase (PK) from *F. hepatica* closely resembles L-type PK from mammalian liver (Behm and Bryant, 1980). It shows cooperative kinetics with PEP, but in the presence of fructose 1,6-bisphosphate (FBP) this changes to Michaelis–Menten kinetics. The enzyme is inhibited by ATP and this inhibition can be relieved by FBP. Regulation of PK from *Fasciola* by more recently discovered effectors of mammalian-type PKs, like glucose 1,6-bisphosphate and phosphorylation/dephosphorylation, has not yet been reported.

PEPCK functions in *F. hepatica* and *H. diminuta*, like in other parasitic helminths, as a CO_2-fixing enzyme. This is remarkably different from the decarboxylating anabolic role of PEPCK in mammalian gluconeogenesis. Parasitic helminths are not unique, however, in this respect, because PEPCK functions as a CO_2-fixing enzyme in the glycolytic degradation of glucose in many other invertebrates, such as some marine organisms. PEPCK activity appears to be controlled primarily by the concentration of enzyme, substrates and products, in contrast to PK activity, which is under tight allosteric control (Behm and Bryant, 1982).

PEPCK is a key enzyme in the cytosolic production of malate, an important process in the anaerobic energy metabolism of *F. hepatica* adults, whereas PK is a key enzyme in the glycolytic pathway to pyruvate, which is operative in the aerobic functioning juvenile liver fluke (see section on 'Transitions in Energy Metabolism during the Life Cycle'). Therefore, during the change from an aerobic to an anaerobic energy metabolism that occurs during the development of *F. hepatica* in the final host, a change in the relative importance of PK and PEPCK may be expected. Such a change was indeed observed, as the development of the fluke in the host liver was accompanied by an almost complete disappearance of PK activity (Tielens *et al.*, 1987). A comparable transition in PEPCK/PK ratios was observed during the development of the cestode *H. diminuta* (Fioravanti *et al.*, 1998).

One should bear in mind, however, that the fate of PEP at the PK/PEPCK branch point cannot simply be estimated by measuring the ratio of PK and PEPCK activities *in vitro*, as the use of PEP in each reaction is determined by K_m values and substrate concentrations, by the extent of phosphorylation of PK, by the presence of (in)activators and by the rate of the subsequent reactions.

Unusual aspects of mitochondrial metabolism

Malate dismutation is a mitochondrial adaptation to anaerobic functioning. As a consequence, and contrary to classical mitochondria, these mitochondria do not use oxygen as the terminal acceptor of the electrons in the electron-transport chain. Parasitic flatworms utilizing anaerobic malate dismutation produce a wide range of metabolic end products of carbohydrate metabolism, but their mitochondrial pathways are surprisingly similar given their diverse phylogenetic origins. Typically, these organisms operate a portion of the Krebs cycle (from oxaloacetate to succinate) and, in some cases β-oxidation, in a direction opposite to that found in aerobic organisms.

All parasitic flatworms capable of anaerobic metabolism favour malate as the primary mitochondrial substrate and the oxidative decarboxylations of first malate and then pyruvate generate intramitochondrial reducing power in the form of NADH (Fig. 20.1). In contrast, the pathways used to reoxidize intramitochondrial NADH are quite diverse and depend on the stage or species of parasite under examination, but in all cases, redox balance is maintained and electron-transport associated ATP is generated by the NADH-reduction of fumarate to succinate. In the cestode, *H. diminuta*, succinate and acetate are the major end products of anaerobic malate dismutation and are excreted in the predicted 2:1 ratio. In the trematode *F. hepatica*, succinate is then further decarboxylated to propionate with an additional substrate level phosphorylation coupled to the decarboxylation of methylmalonyl CoA. *F. hepatica* forms primarily propionate and acetate as end products, again in a ratio of 2:1 to maintain redox balance.

Compared with typical aerobic mitochondria, the three main distinctions of these anaerobic mitochondria are (i) the enzyme catalyzing the conversion of fumarate to succinate; (ii) the quinone that connects this electron transfer to the enzyme complex in the electron-transport chain; and (iii) the presence of ASCT, which converts acetyl-CoA into acetate. These characteristic features of anaerobically functioning mitochondria are further discussed in the three following sections.

Succinate dehydrogenase versus fumarate reductase

In juvenile liver fluke and miracidia, a respiratory chain up to cytochrome *c* oxidase is active and all evidence obtained so far indicates that in *F. hepatica* at least this electron-transport chain is not different from the classical one present in mammalian mitochondria (Figs 20.1 and 20.2). In the aerobically functioning stages, electrons are transferred from NADH and succinate to ubiquinone via complex I and II of the respiratory chain, respectively. Subsequently, these electrons are transferred from the formed ubiquinol to oxygen via the complexes III and IV of the respiratory chain.

In the mitochondria of anaerobically functioning parasitic flatworms such as adult *F. hepatica* and *H. diminuta*, however, this electron-transport chain is altered, as oxygen is not used as the terminal electron acceptor. During malate dismutation, endogenously produced fumarate functions as the terminal electron acceptor. In this case, electrons are transferred from NADH to fumarate via complex I and FRD (Fig. 20.2). This implies that during the development of *F. hepatica* in its final host, a transition occurs from succinate oxidation via succinate dehydrogenase (SDH) in the Krebs cycle of the juvenile liver fluke to the reverse reaction: reduction of fumarate to succinate in the adult. Bacteria contain two homologous but distinct enzyme complexes, one to oxidize succinate (SDH) and one to reduce fumarate (FRD), although each enzyme will catalyse both reactions *in vitro*. *In vivo*, distinct enzyme complexes are needed for these opposite reactions as the electron flow through the complex is reversed, which implies differences in the affinity for electrons (standard electron potential) of the electron-binding domains of these enzyme complexes (Ackrell *et al.*, 1992; Van Hellemond and Tielens, 1994). Distinct enzyme complexes have been described in the parasitic nematodes *Haemonchus contortus* (Roos and Tielens, 1994) and *A. suum* (Saruta *et al.*, 1995). These complexes were shown to be differentially expressed during the life cycle of the parasites and are suggested to function either as a SDH or as a FRD. Firm evidence for differential expression of two different enzyme complexes in parasitic flatworms is still lacking, but it has been shown that the kinetic properties of the complex from adult *F. hepatica* resemble those of other organisms known to reduce fumarate (Van Hellemond *et al.*, 1995). Mitochondria of adult *F. hepatica* had a low SDH/FRD ratio, which corresponds to their *in vivo* function. On the other hand, significant differences in activity ratios were not detected between different stages of *F. hepatica* (adult and metacercariae), although *in vivo*, the free-living stage oxidizes succinate, whereas the adult reduces fumarate. Apparently, the observed change in the activity ratio that was observed between free-living and parasitic stages of *A. suum* does not occur in *F. hepatica* (Van Hellemond *et al.*, 1995; Saruta *et al.*, 1995). The low SDH/FRD activity ratio, which was observed in *F. hepatica* and in all other fumarate-reducing eukaryotes (parasitic helminths as well as lower marine organisms), is caused by a markedly increased fumarate reduction activity compared with mammalian-type mitochondria, which correlates with the significant capacity for fumarate reduction of the mitochondria from fumarate-reducing eukaryotes. SDHs or FRDs from cestodes have not yet been studied at the protein or DNA level.

Ubiquinone versus rhodoquinone

In addition to distinct enzyme complexes for succinate oxidation and fumarate reduction, distinct quinones are involved in these processes in parasitic flatworms. The electron-transport chains of many bacteria employ menaquinone when fumarate is the final electron acceptor (Collins and Jones, 1981). In parasitic helminths, however, the presence of rhodoquinone was demonstrated and because rhodoquinone is present mainly in anaerobic, fumarate-reducing stages, it was suggested that rhodoquinol functions as an electron donor in fumarate reduction, similar to menaquinol in bacterial fumarate reduction (Allen, 1973). It was shown that rhodoquinone is essential for efficient electron transport in the anaerobic electron-transport chain of all eukaryotic organisms known to reduce fumarate during anoxia (Van Hellemond *et al.*, 1995). Rhodoquinone, with its relatively low redox potential (E'^0 = −63 mV), is comparable to menaquinone in this respect and transfers electrons via FRD to fumarate (E'^0 = 30 mV), whereas ubiquinone (E'^0 = 110 mV) preferentially donates electrons to complex III of the electron-transport chain (Fig. 20.2).

The essential role of rhodoquinone in anaerobic energy metabolism has been demonstrated during the entire life cycle of *F. hepatica*; the amount of rhodoquinone present reflected the importance of fumarate reduction in various stages (Van Hellemond *et al.*, 1995, 1996). Throughout the life cycle of *F. hepatica*, a strong correlation is found between the quinone composition and the type of metabolism: the amount of ubiquinone is correlated with the use of the aerobic respiratory chain, and the amount of rhodoquinone with the use of fumarate reduction. Both quinones are synthesized *de novo* by *F. hepatica* and the rate of synthesis of both quinones appears to be related to varying needs during the life cycle. Furthermore, it was demonstrated that in adult *F. hepatica*, rhodoquinone is not produced by modification of ubiquinone obtained from the host, but that rhodoquinone and ubiquinone are synthesized *de novo* via the mevalonate pathway (Fig. 20.3). Further analyses of rhodoquinone synthesis prompted the suggestion that rhodoquinone is synthesized by *F. hepatica* via a pathway nearly identical to that of ubiquinone biosynthesis; possibly only the last reaction differs (Van Hellemond *et al.*, 1996).

As expected, rhodoquinone is also present in *H. diminuta*, a cestode known to produce succinate as end product. Surprisingly, rhodoquinone is also present in the homolactic fermenter, *S. mansoni*, where it is especially present in sporocysts, the snail stage of this parasite, which produces succinate under anaerobic conditions (see section on 'Transitions in Energy Metabolism during the Life Cycle').

Acetate:succinate CoA-transferase (ASCT)

In the oxidative branch of malate dismutation, malic enzyme oxidizes malate to pyruvate, which is then further oxidized to acetyl-CoA. Subsequently, the CoA-moiety of acetyl-CoA is transferred to succinate by an ASCT, after which the resulting acetate is excreted as end product (Saz *et al.*, 1996; Van Hellemond *et al.*, 1998). The produced succinyl-CoA is subsequently recycled to succinate by SCS,

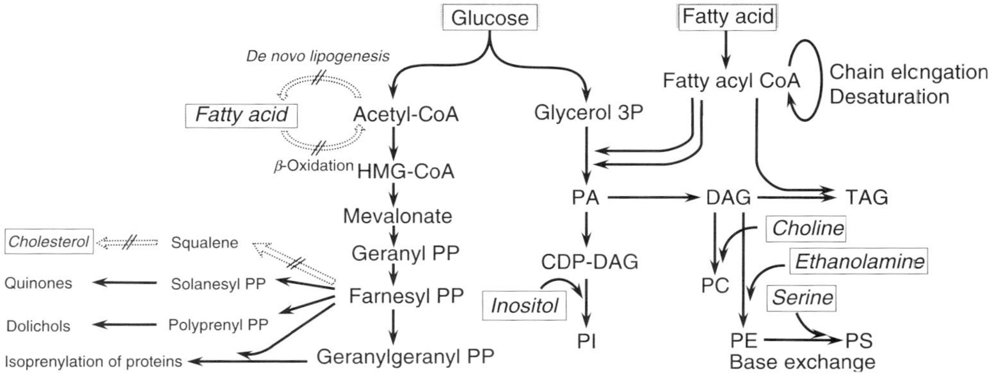

Fig. 20.3. Schematic representation of the main pathways in the lipid metabolism of parasitic flatworms. Boxed substrates are supplied by the host. Pathways present in mammalian systems but absent in parasitic flatworms are shown by open arrows. Abbreviations: DAG, diacylglycerol; CDP-DAG, cytidine diphosphodiacylglycerol; Farnesyl PP, farnesyl pyrophosphate; Geranyl PP, geranylpyrophosphate; Geranylgeranyl PP, geranylgeranylpyrophosphate; HMG-CoA, hydroxymethylglutaryl-CoA; TAG, triacylglycerol; PA, phosphatidic acid; PC, phosphatidylcholine; PE, phosphatidylethanolamine; PI, phosphatidylinositol; PS, phosphatidylserine.

a process that produces ATP from ADP (or GTP from GDP). ASCT is not only present in anaerobic mitochondria but also in the strictly aerobically functioning mitochondria of the trypanosomatids *Trypanosoma brucei* and *Leishmania* (Van Hellemond *et al.*, 1998) as well as in the hydrogenosomes of trichomonads (Steinbuchel and Müller, 1986). The localization of ASCT in these mitochondria identified the first metabolic pathway common to mitochondria and hydrogenosomes: an ASCT/SCS cycle, which is further evidence for the common evolutionary origin of mitochondria and hydrogenosomes (Van Hellemond *et al.*, 1998). Interestingly, organisms lacking compartmentation of metabolism, such as *Giardia* and *Entamoeba*, also lack this cycle and instead produce acetate from acetyl-CoA by a single step reaction catalysed by acetyl-CoA synthetase. This suggests that compartmentalization and the presence of the acetyl/succinyl-CoA cycle are linked.

ASCT is not present in the mitochondria of all organisms, and has so far only been detected in anaerobically functioning mitochondria of parasitic worms and the aerobically functioning mitochondria of trypanosomatids. These latter mitochondria are interesting, as these protozoa are considered to be the earliest-branching eukaryotes that contain mitochondria. Hence, these mitochondria could contain enzymes present in the ancestral mitochondrial organelle. The presence of ASCT in hydrogenosomes on the one hand, and on the other hand in aerobically as well as anaerobically functioning mitochondria, is very intriguing. Future phylogenetic analyses of ASCT genes of various mitochondria and hydrogenosomes could yield further information on the evolutionary relationships between the various ATP-producing organelles.

Evolutionary origin of anaerobic mitochondria

Mitochondria evolved by an endosymbiotic event between an anaerobically functioning archaebacterial host and an aerobic α-proteobacterium. However, true anaerobically functioning mitochondria, such as those found in parasitic flatworms and some marine organisms, most likely did not originate directly from the pluripotent ancestral mitochondrion, but arose later in evolution from the aerobic type of mitochondria after these had already adapted to an aerobic way of life by losing their anaerobic capacities.

It could be envisaged that the true anaerobically functioning mitochondria, such as those present in most adult parasitic flatworms, originated from the pluripotent ancestral organism by adaptation to an anaerobic environment, like the hydrogenosomes of trichomonads. In our opinion, however, these anaerobically functioning mitochondria evolved later in evolution from the more conventional aerobic type of mitochondria after these had adapted to an aerobic way of life by losing their anaerobic capacities (Tielens *et al.*, 2002). It should be noted that these adaptations or unusual aspects of metabolism are not unique to parasitic flatworms, or even to parasites in general. Anaerobic mitochondria, although a hallmark of parasites, are also found in other organisms that live (temporarily) in the absence of oxygen.

Several observations indicate that these anaerobically functioning mitochondria evolved from classical mitochondria, and did not originate (by adaptation to an anaerobic environment) directly from the facultative anaerobic pluripotent ancestral cell that was the result of the endosymbiosis of an α-proteobacterial symbiont and the anaerobic host. First, all sequence data available up to now on the fumarate-reducing enzymes (FRDs) of these organisms demonstrate that these enzymes are closely related to the SDHs of classical aerobic mitochondria (Tielens *et al.*, 2002). These classical mitochondria are supposed to have evolved by losing the anaerobic capacities of the pluripotent ancestral cell, which probably already possessed a comparable enzyme for anaerobic fumarate reduction. Therefore, it is more likely that this adaptation towards anaerobic metabolism did not occur immediately after the symbiotic event, but after the earlier prokaryotic FRD for anaerobic functioning was lost. Second, the same argument holds true for the quinone used for this anaerobic fumarate reduction. All anaerobically functioning, fumarate reducing eukaryotes investigated so far use rhodoquinone for the transport of electrons from complex I to the FRD, while the more ancient prokaryotic systems use menaquinone. Again, the anaerobically functioning mitochondria use a molecule, which is structurally more related (a benzoquinone) to the molecule of the classical mitochondria, ubiquinone (also a benzoquinone), than to the molecule used by the anaerobic prokaryotes, menaquinone (a naphthoquinone). As the synthesis of rhodoquinone probably differs only in the last step from the synthesis of ubiquinone, this indicates that also the electron transporter was adapted to anaerobic functioning after the synthesis of the original transporter (menaquinone) was lost by adaptation to an aerobic environment. Moreover, the fact that both classical and anaerobically functioning mitochondria are known to occur within the same organism, for instance, parasitic flatworms, points towards the later evolution of classical aerobic mitochondria to an anaerobic environment rather than an earlier adaptation to an anaerobic environment. This is consistent with the belief that parasitic flatworms evolved from free-living worms that displayed the aerobic metabolism seen in free-living worms today.

For these reasons we suggest that anaerobically functioning mitochondria evolved from classical mitochondria by adaptation to an anaerobic environment. This adaptation could have occurred by small structural modifications of the SDH enzyme such as changes in the redox potentials of the iron–sulphur clusters, making it better suited to fumarate reduction, and resulting in an enzyme that is structurally more related to the prototype SDH, but which is functionally more related to the older FRDs of prokaryotes. In addition, an adaptation could have occurred in the synthesis of the quinone involved in fumarate reduction. The biosynthetic pathway of rhodoquinone is still unknown, as is the origin of the enzyme(s) involved in the last steps in this biosynthesis. It is possible that it was acquired by lateral gene transfer from, for instance, a *Rhodospirillum*-like prokaryotic organism, which is known to synthesize rhodoquinone.

Anaerobic metabolism is usually considered to be an early-evolved and rather primitive way of life. However the above scenario reveals that these anaerobically functioning mitochondria are not primitive, but represent an adaptation of the traditional aerobic mitochondria to anaerobic environments. In this way, these anaerobic mitochondria appear to have undergone further evolution rather than be representative of a more primitive form.

Transitions in energy metabolism during the life cycle

In the different stages of its life cycle, *F. hepatica* has to adapt to different environments. Among other chemical and physical parameters, the availability of oxygen and substrates varies widely during the life cycle. Therefore, transitions have to occur, for instance, from the fully aerobic functioning of the juvenile liver fluke to the almost completely anaerobic functioning of the adult parasite in the bile duct. Most likely a comparable switch from an aerobic to an anaerobic energy metabolism will also occur during the development of miracidia into sporocysts and rediae, but this switch has not yet been studied at the metabolic level. It should be realized that these transitions in *F. hepatica* are definitive, as the same organism will not encounter its previous environment again. The change to a fermentative metabolism is permanent and is not forced by a temporarily hypoxic environment or sudden, intense activity. Therefore, these metabolic switches are not comparable to the transient switches that occur in, for example, skeletal muscle or marine organisms, like the mussel *Mytilus edulis*.

After emergence from the metacercarial cyst, the juvenile liver fluke, which is almost exclusively dependent on Krebs cycle activity, develops gradually into a fermenting adult living in the bile ducts of the definitive host (Tielens, 1994). During this development, three different pathways of glucose breakdown successively produce the majority of ATP. Krebs cycle activity, which provides the vast majority of ATP in the juvenile fluke, gradually decreases during the development of *F. hepatica* in the liver parenchyma. Concomitantly, acetate becomes the major end product of the late parenchymal and early bile-duct stages of the worm. This formation of acetate requires oxygen for the reoxidation of the formed NADH, and is the most important source of energy for the developing fluke in the liver parenchyma. Finally, in the bile ducts, the anaerobic production of propionate and acetate provides essentially all the ATP required by the adult worm (Tielens *et al.*, 1984). Apparently, the aerobic capacity of *F. hepatica* decreases during its development in the host liver. The observed decrease in Krebs cycle activity per mg total protein cannot be explained by a slow decrease in the number of remnant mitochondria from an earlier, more aerobic stage. It was shown that, when calculated per fluke, the Krebs cycle activity actually increases immensely during development; it was found to be directly proportional to the surface area of the fluke (Tielens *et al.*, 1984). This correlation holds true for the entire development of newly excysted juvenile to mature fluke in the bile ducts. These observations support the view that Krebs cycle activity is limited by the diffusion of oxygen and occurs only in the outer layer of the parasite. Parasitic flatworms possess neither respiratory organs nor a circulatory system such that the availability of oxygen inside the tissues is dependent on its slow diffusion, and parasite growth limits its aerobic capacity. In the adult stage, the aerobically functioning outer layer represents less than 1% of the total volume of the worm, such that its metabolism appears to be completely anaerobic.

The metabolism of the developing liver fluke is thus altered by a lack of oxygen. This happens in two steps. First, as outlined above, in the aerobic liver parenchyma of the host the limited diffusion of oxygen in the growing fluke will result in a lack of oxygen in the innermost tissues of the parasite. Second, after arrival in the bile ducts, the entire fluke will have to function anaerobically because of the very low oxygen content of the bile. The prolonged stay in this anaerobic environment ultimately resulted in metabolic adaptations both in the cytosol and in the mitochondria (Lloyd, 1986; Tielens *et al.*, 1987). A striking example of such an adaptation is the change in quinone composition that occurs during the development of *F. hepatica* (Van Hellemond *et al.*, 1995, 1996). Mitochondria of adult *Fasciola gigantica* showed weak or no cytochrome *c* oxidase activity, indicating that this parasite is also adapted to the anaerobic environment in the bile duct (Fujino *et al.*, 1995).

Similarly, a switch from aerobic metabolism in free-living stages to anaerobic metabolism in parasitic stages occurs in schistosomes. The free-living cercariae and miracidia of *S. mansoni* possess an aerobic energy metabolism in which their endogenous glycogen reserves are degraded mainly to CO_2. Adult schistosomes, on the other hand, live in the blood stream of their host and, despite their small size and life in an aerobic environment, they have a fermentative metabolism and degrade glucose mainly to lactate. When cercariae penetrate the skin of the final host and transform into schistosomula, they switch rapidly from Krebs cycle activity to lactate production via glycolysis. This metabolic switch was shown to be initiated by the sudden presence of external glucose when the free-living stages penetrate the new host, and is not linked to a decreased availability of oxygen, like in *F. hepatica*. The mere presence of external glucose results in an increased glycolytic flux, probably caused by the rapid uptake of glucose that occurs upon expression at the surface of a specific schistosomal glucose transporter protein, SGTP4 (Skelly and Shoemaker, 1996). This increased glycolysis is maintained as a result of the specific kinetic properties of schistosomal hexokinase, the first enzyme in glucose catabolism. The observed rapid switch to lactate production occurs only in cercarial heads, the region of the larvae that develops into the mature parasite. The tail of the cercaria, which propels the organism through water, is fully dependent on the degradation of endogenous glycogen reserves as it contains little or no SGTP4 and hexokinase, and it degenerates following its separation from the penetrating schistosomulum. In contrast to *Ascaris*, *F. hepatica* and several other helminth species, no significant changes in the energy metabolism occur during the further development of schistosomes. Lactate remains the main end product, although Krebs cycle activity and oxidative phosphorylation also contribute significantly to ATP production, even in adults.

In the transition from free-living *S. mansoni* miracidia to the parasitic stage in the intermediate snail host (sporocysts), a change from aerobic to anaerobic metabolism occurs. Whereas the miracidia degrade their endogenous glycogen reserves mainly aerobically via the Krebs cycle to CO_2, the sporocysts are facultative anaerobes, fully equipped to adjust their energy metabolism to the variable conditions inside the snail and alternate between aerobic and anaerobic functioning. Although sporocysts produce some lactate under aerobic conditions in the intermediate host, they are largely dependent on Krebs cycle activity for ATP production, and they switch to lactate and succinate production under anaerobic conditions (Van Hellemond *et al.*, 1997). The succinate is produced via the pathway already known from other parasites (discussed in section on 'Anaerobic Glycolysis versus Malate Dismutation') that uses fumarate reduction with rhodoquinone as the electron carrier in the electron-transport chain (Fig. 20.1).

The cestode *H. diminuta* is predominantly anaerobic throughout its life cycle, but nevertheless also shows metabolic variation during its development inside the final host. During this development from cysticercoid to adult, a transition occurs from lactate formation towards electron-transport-dependent succinate formation. Differences in PEP-utilizing / succinate-forming enzymatic activities were noted in comparisons of immature, mature and pregravid–gravid segments of *H. diminuta* (Fioravanti *et al.*, 1998). In fact, research on *H. diminuta* has demonstrated that several variations in metabolism exist within this tapeworm (reviewed in Andreassen *et al.*, 1999). Metabolic differences between various strains have also been observed. Furthermore, differences in metabolism within an individual tapeworm have been noted. For example, a metabolic gradient was found along the length of adult *H. diminuta*, with the first 2 cm, including the scolex, producing relatively more lactate and acetate, and less succinate, than the remainder of the worm. To complicate matters even further, the individual host was found to influence metabolism. Metabolic variation not only existed between worms from the same strain obtained from different rats (Bennet *et al.*, 1993), but also changing the beetle intermediate host influenced the metabolism of adults in their rat final host.

Biosynthetic Capacities

In accordance with their opportunistic way of life, parasitic flatworms have limited biosynthetic capacities; as described in 'Introduction', they obtain many simple substrates from their hosts. More complex molecules that the parasite cannot obtain directly from the host are synthesized from these simpler building blocks. Obviously, the parasite has to synthesize complex structures like proteins and DNA. In general, the biosynthetic pathways of parasitic helminths bear a close resemblance to those of their mammalian hosts. However, the enzymes of these pathways often possess different properties, and in cases where parasites produce unique end products, there may be distinct pathway components that involve unique enzymes that are absent from the host.

Carbohydrates, like glucose, are not synthesized *de novo* by parasitic flatworms, and gluconeogenesis has never been demonstrated except from intermediates at the level of triose phosphates (Bryant and Behm, 1989). Simple carbohydrates are obtained from the host and are then used by the parasite to synthesize complex ones, for instance, production of the glycocalyx glycoproteins or the synthesis of glycogen, the storage carbohydrate used by all parasitic flatworms.

As in other organisms, parasitic flatworms require the 20 amino acids for protein synthesis and the formation of other biomolecules; the majority of amino acids are essential for the parasite and are host-derived. However, parasites also have the capacity to synthesize some amino acids from simple carbon precursors, usually via pathways common to most other organisms. The carbon skeletons of several amino acids are derived from glycolytic or Krebs cycle intermediates, while transamination reactions provide the amino group, which is usually donated by glutamate. Glutamate is formed by α-ketoglutarate dehydrogenase from the Krebs cycle intermediate α-ketoglutarate and ammonia. A remarkable feature of the amino acid metabolism in some parasitic flatworms is the production and excretion of large quantities of proline. The extensive proline production by *F. hepatica*, schistosomes and other trematodes correlates with an extremely active proline biosynthetic pathway in these parasitic flatworms. In contrast to higher organisms, in which glutamate serves as the precursor for proline synthesis, the liver fluke uses host-derived arginine as substrate for the formation of this amino acid. Arginine is first cleaved to urea and ornithine, and this ornithine is then converted via glutamate-γ-semialdehyde to proline. Although this pathway is present in most animals, in proline-producing parasitic flatworms the enzymes involved are several times more active than in mammalian tissues. Combined with the fact that proline oxidase activity is absent or very low in these parasites, this serves to explain the high levels of proline produced by these helminths. Excessive proline production has been implicated in the pathogenesis of trematode infections, including bile duct hyperplasia in fasciolosis and fibrosis in schistosomiasis (Wolf-Spengler and Isseroff, 1983).

Lipids

Lipids like phospholipids, triacylglycerols and cholesterol play various important roles in living organisms. Phospholipids and cholesterol are the main components of biological membranes, which not only form the boundaries of the various compartments inside the cells and organisms, but also act as the interface between the organism and the environment, and in case of parasitic flatworms, the site of interaction between parasite and host.

Very little is known about lipid metabolism in parasites, although in several aspects, it is quite different from that of the mammalian host. Lipids, like phospholipids, triacylglycerols and cholesterol, are not synthesized *de novo* by parasitic flatworms, but are obtained directly from the host (cholesterol) or are synthesized from building blocks obtained from the host (fatty acids and in case of phospholipids, also the head groups). A schematic and partly hypothetical overview of the main pathways of lipid metabolism in parasitic flatworms is shown in Fig. 20.3. Fatty acids are not degraded and thus not used for ATP production (see section on 'Substrates of Energy Metabolism'). *De novo* fatty acid synthesis does also not occur, and therefore, fatty acids must be obtained from the host. Nevertheless, a comparison of host and parasite fatty acids revealed the presence of some fatty acids in adult flukes that are virtually absent from the host (Oldenborg *et al.*, 1975). These parasite-specific fatty acids are not synthesized *de novo*, but are produced by the modification of fatty acids obtained from the host. *F. hepatica* is unable to desaturate fatty acids, but can use acetate for the elongation of host-derived fatty acids (Oldenborg *et al.*, 1975). This chain elongation has specificity for certain fatty acids, of which the elongation of 18:1 to 20:1 is quantitatively the most important. In general, two different pathways exist for the elongation of fatty acids, one being a reversal of β-oxidation and in the other acetyl-CoA is carboxylated to malonyl-CoA and subsequently added to pre-existing fatty acids. It has not yet been fully established which pathway occurs in *F. hepatica*, but acetyl-CoA carboxylase is present in this parasite whereas there is no evidence for the occurrence of a reversal of β-oxidation (Brouwers *et al.*, 1997). The reason for the persistence during evolution of chain elongation of large amounts of fatty acids, while all other synthesizing and catabolic pathways of fatty acids are lost, is unclear.

Fatty acids, with or without prior chain elongation, are incorporated into phospholipids and triacylglycerols (Fig. 20.3). Phosphatidylcholine (PC) and phosphatidylethanolamine (PE) are the major phospholipid classes present in *F. hepatica* and *S. mansoni*, and both the head groups and the fatty acids are obtained from the host and used as precursors. Pathways for synthesis of the other phospholipids have not been thoroughly investigated in parasitic flatworms. Generally, the phospholipid classes present in the host are also found in parasitic helminths, although sometimes in different ratios. Also, *F. hepatica* contains a large amount of glycerophosphocholine, a degradation product of PC (Mansour *et al.*, 1982; Tielens *et al.*, 1982). While this indicates a high phospholipid turnover in *F. hepatica*, its role in lipid metabolism has not yet been further investigated.

Although sterols like cholesterol are not synthesized *de novo* by parasitic flatworms, they do possess an active mevalonate pathway (Fig. 20.3) (reviewed in Coppens and Courtoy, 1996). This pathway has been studied in *S. mansoni*, and all available evidence indicates that it is similar to the lipid metabolism seen in *F. hepatica*. The mevalonate pathway was shown to be used by *S. mansoni* for the synthesis of dolichols for protein glycosylation, of quinones as electron transporters in the respiratory chain and of farnesyl and geranylgeranyl pyrophosphates as substrates for the isoprenylation of proteins (Chen and Bennett, 1993; Foster *et al.*, 1993). A key enzyme in the mevalonate pathway is 3-hydroxymethylglutaryl-CoA reductase (HMG-CoA reductase) and it was shown that the schistosomal enzyme differs from the mammalian type, both structurally and in its regulatory properties (Rajkovic *et al.*, 1989; Chen *et al.*, 1991). Farnesyl pyrophosphate plays a key role in the mevalonate pathway as it is the last common substrate for the synthesis of all end products (Fig. 20.3). As mentioned already, the branch leading from farnesyl pyrophosphate via squalene to cholesterol is not operative in parasitic flatworms, whereas the other branches are active, at least in *S. mansoni* and probably also in *F. hepatica* and *H. diminuta*.

One of the branches from farnesyl pyrophosphate leads to the formation of quinones, essential lipids in electron-transfer chains where they carry electrons from one protein complex to the next. In aerobic energy metabolism where the quinone has to transfer electrons to complex III of this chain, a ubiquinone is required. However, *F. hepatica* and *H. diminuta* have an anaerobic energy metabolism and this requires a quinone with a lower standard redox potential (see Fig. 20.2). Rhodoquinone was shown to play an essential role in anaerobic energy metabolism during the entire life cycle of *F. hepatica* (Van Hellemond *et al.*, 1996). It was also demonstrated that in *F. hepatica* and *S. mansoni*, this rhodoquinone was not produced by modification of the ubiquinone obtained from the host but was synthesized *de novo* by the parasite. Most likely, ubiquinone and rhodoquinone are derived from the same precursor and only the last step in rhodoquinone synthesis differs from that in ubiquinone synthesis. As rhodoquinone is essential for the anaerobic energy metabolism of parasitic helminths and is absent from their hosts, rhodoquinone synthesis (in particular the enzyme catalysing the last step) is a key target for anthelmintic drug design.

It is not only fatty acids that are modified (see section on 'Lipids') in parasitic flatworms, but also cholesterol obtained from the host is not only used directly as a component of membranes in *F. hepatica* and *S. mansoni*, but it is also used as a substrate for the synthesis of other compounds such as ecdysteroids (Foster *et al.*, 1992).

Apparently, parasitic flatworms have discarded some pathways of *de novo* lipid synthesis, but have selectively retained several biosynthetic pathways that modify host lipids. Although lipids like fatty acids and cholesterol are obtained from the host, less abundant lipids that are more difficult to acquire because of their low concentration in the host (e.g. specific unsaturated fatty acids, eicosanoids, ecdysteroids and quinones) are synthesized *de novo* by the parasite, usually by the modification of more abundant substrates. In this way, lipid metabolism of parasitic flatworms is adapted to an opportunistic way of life, just like their energy metabolism.

Purines, pyrimidines and polyamines

Purine and pyrimidine nucleotides are essential components of many biochemical molecules, from DNA and RNA to ATP and NAD. In recent years, the pyrimidine and especially the purine metabolism of parasitic helminths have been investigated extensively, mainly because they are different from the pathways in the mammalian host such that they have potential as targets for chemotherapeutic attack. For a review of purine and pyrimidine pathways in parasitic helminths and protozoa, see Berens *et al.* (1995). Although parasitic helminths do not synthesize purines *de novo*, but obtain them from the host, they do possess elaborate purine salvage pathways for a more economical management of this resource. Pyrimidines, on the other hand, are synthesized *de novo* by all parasitic flatworms studied so far and, as with mammalian cells, their ability to salvage pyrimidines is limited.

Polyamines like spermidine and spermine, which bind tightly to nucleic acids and are abundant in rapidly proliferating cells, are present in parasitic helminths in amounts comparable to those in vertebrate cells. As the enzymes necessary for their synthesis are lacking in adult parasitic flatworms, it is assumed that these compounds are obtained from the host (Bacchi and Yarlett, 1995). In *F. hepatica* a polyamine N-acetyltransferase has been characterized, that is thought to play a major role in the polyamine metabolism of this parasite by inactivating excess amines (Aisien and Walter, 1993).

Concluding Remarks

In the different stages of their life cycle, parasitic flatworms have to adapt their behaviour and metabolism to the different environments they encounter (Table 20.1). The most important variations in the environment are the availability of food and oxygen. Free-living stages do not gather food and are therefore completely dependent on the endogenous energy stores they acquired from the previous host. Oxygen, on the other hand, is mostly present in the environment of free-living stages. This enables the free-living stages to live very economically; they use Krebs cycle activity and oxidative phosphorylation to obtain as much energy as possible from their endogenous glycogen stores. This thrifty metabolism very well serves their only mission, becoming a parasitic stage in the next host.

Parasitic stages, on the other hand, generally do not use oxygen as the final electron acceptor but use fermentative processes to obtain most of their ATP. For these stages, an uneconomical energy metabolism is not detrimental, as the host provides the nutrients. Most adult flatworms inside the final host produce end products of a fermentative carbohydrate breakdown, such as succinate, acetate, propionate and lactate. These end products are formed via malate dismutation, a fermentative pathway, which is present in all types of parasitic worms (flatworms as well as many nematodes), but which is also present in animals like freshwater snails, mussels, oysters and other marine organisms. Malate dismutation is linked to a specially

Table 20.1. Generalized scheme of the main differences in behaviour and metabolism between free-living and parasitic stages of parasitic flatworms.

	Free-living stages	Parasitic stages[a]
Feeding	–	+
Reproduction	–	+
Biosynthetic capacities	–	+/–
Motility	++[b]	+/–
Energy metabolism	Aerobic	Anaerobic
Substrate	Glycogen	Glucose
Catabolic pathways used	Glycolysis, Krebs cycle	Glycolysis, malate dismutation
Main end products	CO_2	Succinate, propionate, acetate, lactate
Electron-transfer chain		
Protein components	Complexes I, II, III and IV	Complex I and fumarate reductase
Electron carriers	Ubiquinone, cytochrome *c*	Rhodoquinone
Final electron acceptor	Oxygen	Fumarate
ATP production	Mainly oxidative phosphorylation	Substrate-level and oxidative phosphorylation

[a]Mainly studied in the definitive host only.
[b]Often substantial motility, e.g. cercariae and miracidia of digeneans.

adapted electron-transfer chain that generates ATP via oxidative phosphorylation without the use of oxygen. The energy metabolism of parasitic flatworms is clearly and substantially different from that of the host and, therefore, is a suitable target for anti-parasitic drugs.

Apparently, parasitic flatworms have adapted to exploit anaerobic niches and the surplus of substrates that are commonly present in their environment. This enables them not only to use low-efficiency catabolic pathways, but also to lose several anabolic pathways such as those involved in the biosynthesis of fatty acids and sterols. Obviously, parasitic flatworms are exquisitely adapted to an opportunistic way of life.

References

Ackrell, B.A.C., Johnson, M.K., Gunsalus, R.P. and Cecchini, G. (1992) Structure and function of succinate dehydrogenase and fumarate reductase. In: Muller, F. (ed.) *Chemistry and Biochemistry of Flavoenzymes*, Vol. III. CRC Press, Boca Raton, Florida, pp. 229–297.

Aisien, S.O. and Walter, R.D. (1993) Biogenic-amine acetylation: an additional function of the N-acetyltransferase from *Fasciola hepatica*. *Biochemical Journal* 291, 733–737.

Allen, P.C. (1973) Helminths: comparison of their rhodoquinone. *Experimental Parasitogy* 34, 211–219.

Andreassen, J., Bennet-Jenkins, E.M. and Bryant, C. (1999) Immunology and biochemistry of *Hymenolepis diminuta*. *Advances in Parasitology* 42, 223–275.

Bacchi, C.J. and Yarlett, N. (1995) Polyamine metabolism. In: Marr, J.J. and Muller, M. (eds) *Biochemistry and Molecular Biology of Parasites*. Academic Press, New York, pp. 119–131.

Behm, C.A. and Bryant, C. (1980) Regulatory properties of a partially purified preparation of pyruvate kinase from *Fasciola hepatica*. *International Journal for Parasitology* 10, 107–114.

Behm, C.A. and Bryant, C. (1982) Phosphoenolpyruvate carboxykinase from *Fasciola hepatica*. *International Journal for Parasitology* 12, 271–278.

Bennet, E.M., Heath, P.A. and Bryant, C. (1993) The effects of changes in the definitive host environment on the metabolism of *Hymenolepis diminuta* during growth and maturation. *International Journal for Parasitology* 23, 57–68.

Berens, R.L., Krug, E.C. and Marr, J.J. (1995) Purine and pyrimidine metabolism. In: Marr, J. and Muller, M. (eds) *Biochemistry and Molecular Biology of Parasites*. Academic Press, New York, pp. 89–117.

Boyunaga, H., Schmitz, M.G., Brouwers, J.F., Van Hellemond, J.J. and Tielens, A.G.M. (2001) *Fasciola hepatica* miracidia are dependent on respiration and endogenous glycogen degradation for their energy generation. *Parasitology* 122, 169–173.

Brouwers, J.F., Smeenk, I.M., Van Golde, L.M.G. and Tielens, A.G.M. (1997) The incorporation, modification and turnover of fatty acids in adult *Schistosoma mansoni*. *Molecular and Biochemical Parasitology* 88, 175–185.

Bryant, C. and Behm, C. (1989) *Biochemical Adaptations in Parasites*. Chapman & Hall, London.

Carmona, C., McGonigle, S., Dowd, A.J., Smith, A.M., Coughlan, S., McGowran, E. and Dalton, J.P. (1994) A dipeptidylpeptidase secreted by *Fasciola hepatica*. *Parasitology* 109, 113–118.

Chen, G.Z. and Bennett, J.L. (1993) Characterization of mevalonate-labeled lipids isolated from parasite proteins in *Schistosoma mansoni*. *Molecular and Biochemical Parasitology* 59, 287–292.

Chen, G.Z., Foster, L. and Bennett, J.L. (1991) Purification and characterization of 3-hydroxymethylglutaryl-coenzyme A reductase of *Schistosoma mansoni*: regulation of parasite enzyme activity differs from mammalian host. *Experimental Parasitology* 73, 82–92.

Collins, M.D. and Jones, D. (1981) Distribution of isoprenoid quinone structural types in bacteria and their taxonomic implication. *Microbiological Reviews* 45, 316–354.

Coppens, I. and Courtoy, P.J. (1996) The mevalonate pathway in parasitic protozoa and helminths. *Experimental Parasitology* 82, 76–85.

Dalton, J.P., Smith, A.M., Clough, K.A. and Brindley, P.J. (1995) Digestion of haemoglobin by schistosomes: 35 years on. *Parasitology Today* 11, 299–303.

Diaz, F. and Komuniecki, R.W. (1994) Pyruvate dehydrogenase complexes from the equine nematode, *Parascaris equorum*, and the canine cestode, *Dipylidium caninum*, helminths exhibiting anaerobic mitochondrial metabolism. *Molecular and Biochemical Parasitology* 67, 289–299.

Fioravanti, C.F., McKelvey, J.R. and Reisig, J.M. (1992) Energy-linked mitochondrial pyridine nucleotide transhydrogenase of adult *Hymenolepis diminuta*. *Journal of Parasitology* 78, 774–778.

Fioravanti, C.F., Walker, D.J. and Sandhu, P.S. (1998) Metabolic transition in the development of *Hymenolepis diminuta* (Cestoda). *Parasitological Research* 84, 777–782.

Foster, J.M., Mercer, J.G. and Rees, H.H. (1992) Analysis of ecdysteroids in the trematodes, *Schistosoma mansoni* and *Fasciola hepatica*. *Tropical Medicine and Parasitology* 43, 239–244.

Foster, J.M., Pennock, J.F., Marshall, I. and Rees, H.H. (1993) Biosynthesis of isoprenoid compounds in *Schistosoma mansoni*. *Molecular and Biochemical Parasitology* 61, 275–284.

Fujino, T., Fried, B. and Takamiya, S. (1995) Cytochemical localization of cytochrome *c* oxidase activity in mitochondria in the tegument and tegumental and parenchymal cells of the trematodes *Echinostoma trivolvis*, *Zygocotyle lunata*, *Schistosoma mansoni*, *Fasciola gigantica* and *Paragonimus ohirai*. *Journal of Helminthology* 69, 195–201.

Halton, D.W. (1997) Nutritional adaptations to parasitism within the platyhelminthes. *International Journal for Parasitology* 27, 693–704.

Kamemoto, E.S., Iltzsch, M.H., Lan, L. and Mansour, T.E. (1987) Phosphofructokinase from *Fasciola hepatica*: activation by phosphorylation and other regulatory properties distinct from the mammalian enzyme. *Archives in Biochemistry and Biophysics* 258, 101–111.

Kamemoto, E.S., Lan, L. and Mansour, T.E. (1989) *In vivo* regulation of phosphorylation level and activity of phosphofructokinase by serotonin in *Fasciola hepatica*. *Archives in Biochemistry and Biophysics* 271, 553–559.

Kawanaka, M., Hayashi, S. and Carter, C.E. (1986) Uptake and excretion of amino acids and utilization of glucose by *Schistosoma japonicum* eggs. *Japanese Journal of Medical Sciences and Biology* 39, 199–206.

Klingbeil, M.M., Walker, D.J., Arnette, R., Sidawy, E., Hayton, K., Komuniecki, P.R. and Komuniecki, R. (1996) Identification of a novel dihydrolipoyl dehydrogenase-binding protein in the pyruvate dehydrogenase complex of the anaerobic parasitic nematode, *Ascaris suum*. *Journal of Biological Chemistry* 271, 5451–5457.

Klinkert, M.Q., Cioli, D., Shaw, E., Turk, V., Bode, W. and Butler, R. (1994) Sequence and structure similarities of cathepsin B from the parasite *Schistosoma mansoni* and human liver. *FEBS Letters* 351, 397–400.

Köhler, P., Bryant, C. and Behm, C.A. (1978) ATP synthesis in a succinate decarboxylase system from *Fasciola hepatica* mitochondria. *International Journal for Parasitology* 8, 399–404.

Kulkarni, G., Rao, G.S., Srinivasan, N.G., Hofer, H.W., Yuan, P.M. and Harris, B.G. (1987) *Ascaris suum* phosphofructokinase. Phosphorylation by protein kinase and sequence of the phosphopeptide. *Journal of Biological Chemistry* 262, 32–34.

Lloyd, G.M. (1986) Energy metabolism and its regulation in the adult liver fluke *Fasciola hepatica*. *Parasitology* 93, 217–248.

Mahrenholz, A.M., Hefta, S.A. and Mansour, T.E. (1991) Phosphofructokinase from *Fasciola hepatica*: sequence of the cAMP-dependent protein kinase phosphorylation site. *Archives in Biochemistry and Biophysics* 288, 463–467.

Mansour, T.E., Morris, P.G., Feeney, J. and Roberts, G.C. (1982) A ^{31}P-NMR study of the intact liver fluke *Fasciola hepatica*. *Biochimica Biophysica Acta* 721, 336–340.

Matthews, P.M., Foxall, D., Shen, L. and Mansour, T.E. (1986) Nuclear magnetic resonance studies of carbohydrate metabolism and substrate cycling in *Fasciola hepatica*. *Molecular Pharmacology* 29, 65–73.

Moffat, D. and Kusel, J.R. (1992) Fluorescent lipid uptake and transport in adult *Schistosoma mansoni*. *Parasitology* 105, 81–89.

Oldenborg, V., Van Vugt, F. and Van Golde, L.M.G. (1975) Composition and metabolism of phospholipids of *Fasciola hepatica*, the common liver fluke. *Biochimica Biophysica Acta* 398, 101–110.

Pietrzak, S.M. and Saz, H.J. (1981) Succinate decarboxylation to propionate and the associated phosphorylation in *Fasciola hepatica* and *Spirometra mansonoides*. *Molecular and Biochemical Parasitology* 3, 61–70.

Rajkovic, A., Simonsen, J.N., Davis, R.E. and Rottman, F.M. (1989) Molecular cloning and sequence analysis of 3-hydroxy-3-methylglutaryl-coenzyme A reductase from the human parasite *Schistosoma mansoni*. *Proceedings of the National Academy of Sciences USA* 86, 8217–8221.

Rogers, R.A., Jack, R.M. and Furlong, S.T. (1993) Lipid and membrane protein transfer from human neutrophils to schistosomes is mediated by ligand binding. *Journal of Cell Science* 106 (Pt 2), 485–491.

Roos, M.H. and Tielens, A.G.M. (1994) Differential expression of two succinate dehydrogenase subunit-B genes and a transition in energy metabolism during the development of the parasitic nematode *Haemonchus contortus*. *Molecular and Biochemical Parasitology* 66, 273–281.

Rumjanek, F.D., Campos, E.G. and Afonso, L.C. (1988) Evidence for the occurrence of LDL receptors in extracts of schistosomula of *Schistosoma mansoni*. *Molecular and Biochemical Parasitology* 28, 145–152.

Saruta, F., Kuramochi, T., Nakamura, K., Takamiya, S., Yu, Y., Aoki, T., Sekimizu, K., Kojima, S. and Kita, K. (1995) Stage-specific isoforms of complex II (succinate-ubiquinone oxidoreductase) in mitochondria from the parasitic nematode, *Ascaris suum*. *Journal of Biological Chemistry* 270, 928–932.

Saz, H.J., deBruyn, B. and de Mata, Z. (1996) Acyl-CoA transferase activities in homogenates of *Fasciola hepatica* adults. *Journal of Parasitology* 82, 694–696.

Skelly, P.J. and Shoemaker, C.B. (1996) Rapid appearance and asymmetric distribution of glucose transporter SGTP4 at the apical surface of intramammalian-stage *Schistosoma mansoni*. *Proceedings of the National Academy of Sciences USA* 93, 3642–3646.

Skelly, P.J. and Shoemaker, C.B. (2001) *Schistosoma mansoni* proteases Sm31 (cathepsin B) and Sm32 (legumain) are expressed in the cecum and protonephridia of cercariae. *Journal of Parasitology* 87, 1218–1221.

Steinbuchel, A. and Müller, M. (1986) Anaerobic pyruvate metabolism of *Tritrichomonas foetus* and *Trichomonas vaginalis* hydrogenosomes. *Molecular and Biochemical Parasitology* 20, 57–65.

Tielens, A.G.M. (1994) Energy generation in parasitic helminths. *Parasitology Today* 10, 346–352.

Tielens, A.G.M., Nicolay, K. and van den Bergh, S.G. (1982) ^{31}P-NMR studies of pH homeostasis in intact adult *Fasciola hepatica*. *Molecular and Biochemical Parasitology* 6, 175–180.

Tielens, A.G.M., van den Heuvel, J.M. and van den Bergh, S.G. (1984) The energy metabolism of *Fasciola hepatica* during its development in the final host. *Molecular and Biochemical Parasitology* 13, 301–307.

Tielens, A.G.M., van den Heuvel, J.M. and van den Bergh, S.G. (1987) Differences in intermediary energy metabolism between juvenile and adult *Fasciola hepatica*. *Molecular and Biochemical Parasitology* 24, 273–281.

Tielens, A.G.M., van den Heuvel, J.M. and van den Bergh, S.G. (1990) Continuous synthesis of glycogen by individual worm pairs of *Schistosoma mansoni* inside the veins of the final host. *Molecular and Biochemical Parasitology* 39, 195–201.

Tielens, A.G.M., van den Heuvel, J.M., van Mazijk, H.J., Wilson, J.E. and Shoemaker, C.B. (1994) The 50-kDa glucose 6-phosphate-sensitive hexokinase of *Schistosoma mansoni*. *Journal of Biological Chemistry* 269, 24736–24741.

Tielens, A.G.M., Rotte, C., Van Hellemond, J.J. and Martin, W. (2002) Mitochondria as we don't know them. *Trends in Biochemical Sciences* 27, 564–572.

Van Hellemond, J.J. and Tielens, A.G.M. (1994) Expression and functional properties of fumarate reductase. *Biochemical Journal* 304, 321–331.

Van Hellemond, J.J., Klockiewicz, M., Gaasenbeek, C.P., Roos, M.H. and Tielens, A.G.M. (1995) Rhodoquinone and complex II of the electron transport chain in anaerobically functioning eukaryotes. *Journal of Biological Chemistry* 270, 31065–31070.

Van Hellemond, J.J., Luijten, M., Flesch, F.M., Gaasenbeek, C.P. and Tielens, A.G.M. (1996) Rhodoquinone is synthesized *de novo* by *Fasciola hepatica*. *Molecular and Biochemical Parasitology* 82, 217–226.

Van Hellemond, J.J., Van Remoortere, A. and Tielens, A.G.M. (1997) *Schistosoma mansoni* sporocysts contain rhodoquinone and produce succinate by fumarate reduction. *Parasitology* 115, 177–182.

Van Hellemond, J.J., Opperdoes, F.R. and Tielens, A.G.M. (1998) Trypanosomatidae produce acetate via a mitochondrial acetate: succinate CoA-transferase. *Proceedings of the National Academy of Sciences USA* 95, 3036–3041.

Van Schaftingen, E., Jett, M.F., Hue, L. and Hers, H.G. (1981) Control of liver 6-phosphofructokinase by fructose 2,6-bisphosphate and other effectors. *Proceedings of the National Academy of Sciences USA* 78, 3483–3486.

Van Vugt, F., van der Meer, P. and van den Bergh, S.G. (1979) The formation of propionate and acetate as terminal processes in the energy metabolism of the adult liver fluke *Fasciola hepatica*. *International Journal of Biochemistry* 10, 11–18.

Wolf-Spengler, M.L. and Isseroff, H. (1983) Fascioliasis: bile duct collagen induced by proline from the worm. *Journal of Parasitology* 69, 290–294.

Xu, X. and Caulfield, J.P. (1992) Characterization of human low density lipoprotein binding proteins on the surface of schistosomula of *Schistosoma mansoni*. *European Journal of Cell Biology* 57, 229–235.

Yamasaki, H., Mineki, R., Murayama, K., Ito, A. and Aoki, T. (2002) Characterisation and expression of the *Fasciola gigantica* cathepsin L gene. *International Journal for Parasitology* 32, 1031–1042.

21 Glycoconjugate Structures

M. Wuhrer[1] and R. Geyer[2]
[1]*Biomolecular Mass Spectrometry Unit, Department of Parasitology, Leiden University Medical Center, PO Box 9600, 2300 RC Leiden, The Netherlands;* [2]*Institute of Biochemistry, University of Giessen, Friedrichstrasse 24, 35392 Giessen, Germany*

Introduction

Glycosphingolipids and glycoproteins of parasitic helminths have been shown to express multiple biological activities ranging from molecular mimicry in order to avoid immune recognition to modulation of the host's immune response. In particular, the oligosaccharide moieties of these glycoconjugates are known to play important roles in host–parasite interactions. Despite this fact, however, knowledge on the precise carbohydrate structures of these molecules is still fragmentary in many cases. The aim of this chapter is to summarize current structural data on glycoproteins and glycolipids expressed by parasitic metazoa focusing exclusively on trematodes and cestodes.

Trematodes (Digenea)

Schistosoma

In terms of its glycobiology, *Schistosoma mansoni* is the most intensely studied parasitic helminth, which reflects the importance of schistosome glycans in host–parasite relations. In schistosomiasis, carbohydrate determinants are a major target of the immune response (Eberl *et al.*, 2001). Schistosomes express a large variety of carbohydrates, ranging from mammalian-type structures to unique, often highly fucosylated schistosome-specific glycan motifs. Schistosome glycoconjugates are often surface-exposed or secreted (Deelder *et al.*, 1976; Bergwerff *et al.*, 1994; van Dam *et al.*, 1994) and provide a basis for direct diagnostic methods (see, for example, de Jonge *et al.*, 1991). Furthermore, some of the carbohydrate determinants expressed by schistosomes may induce protective immunity (Omer-Ali *et al.*, 1988; van der Kleij *et al.*, 1999; Eberl *et al.*, 2001).

The glycobiology of schistosomes has been reviewed several times, with a special focus on glycoconjugate structures (Cummings and Nyame, 1996, 1999; Khoo, 2001) or antigenicity (van Dam and Deelder, 1996; Hokke and Deelder, 2001; Nyame *et al.*, 2004). In this chapter, we will give a comprehensive overview of carbohydrate structures of schistosomes attached to proteins and lipids with a focus on the stage-specific expression of glycans and the sharing of carbohydrate determinants between parasite and mammalian hosts.

Glycosphingolipids of schistosomes

While elongated mammalian glycosphingolipids are based on lactosylceramide (Gal(β1-4)Glc(β1-1)ceramide) (Kolter *et al.*, 2002), *S. mansoni* expresses a unique GalNAc(β1-4)Glcβ1-disaccharide moiety attached to ceramide which has been termed the 'schisto' core (Makaaru *et al.*, 1992; Wuhrer *et al.*, 2000b). Besides glucosylceramide and galactosylceramide, this schisto-specific glycosphingolipid GalNAc(β1-4)Glc(β1-1)ceramide seems to be expressed in all life-cycle stages, although considerable variation in the ceramide structures is exhibited (Wuhrer *et al.*, 2000b; Figs. 21.1 and 21.2).

For more complex *S. mansoni* glycolipids, epitope typing with monoclonal antibodies and infection sera revealed a marked antigenicity and stage-specific expression of glycolipid species (Weiss *et al.*, 1986; Wuhrer *et al.*, 1999). Elongated cercarial glycolipids, for example, were shown to be based on the GalNAc(β1-4)Glc(β1-1)ceramide 'schisto' core and contained mainly terminal Lewis X trisaccharide units (Wuhrer *et al.*, 2000c; Fig. 21.1). Some of these glycolipids carried an additional fucose (α1-3)-linked to galactose, thus forming a pseudo-Lewis Y motif (Fig. 21.1) named in analogy to the Lewis Y blood group determinant.

Gal(β1-1)Cer [1]

Glc(β1-1)Cer [1]

GalNAc(β1-4)Glc(β1-1)Cer [1,2]

GlcNAc(β1-3)GalNAc(β1-4)Glc(β1-1)Cer [2]

Gal(β1-4)GlcNAc(β1-3)GalNAc(β1-4)Glc(β1-1)Cer [2]

Lewis X: Gal(β1-4)GlcNAc(β1-3)GalNAc(β1-4)Glc(β1-1)Cer [2]
Fuc(α1-3)

Lewis X: Gal(β1-4)GlcNAc(β1-3)GlcNAc(β1-3)GalNAc(β1-4)Glc(β1-1)Cer [2]
Fuc(α1-3)

Pseudo-Lewis Y: Gal(β1-4)GlcNAc(β1-3)GalNAc(β1-4)Glc(β1-1)Cer [2]
Fuc(α1-3) Fuc(a1-3)

Fig. 21.1. Glycosphingolipids of *S. mansoni* cercariae. Data on the glycosphingolipid structures are compiled as described in Wuhrer *et al.* (2000b) [1] and (2000c) [2]. Lewis X and pseudo-Lewis Y epitopes are boxed. Cer, ceramide.

Fig. 21.2. Glycosphingolipids of *S. mansoni* eggs and adults. The elucidation of the glycosphingolipid structures is reported in Makaaru *et al.* (1992) [1], Wuhrer *et al.* (2000b) [2] and (2002) [3], and Khoo *et al.* (1997a) [4]. Cer, ceramide.

Both Lewis Y and pseudo-Lewis Y are based on the Lewis X trisaccharide unit and differ only in the (α1-2)- or (α1-3)-linkage of a second fucose to galactose, respectively.

While the majority of *S. mansoni* cercarial glycolipids carry Lewis X or pseudo-Lewis Y units, another group of cercarial glycolipids is highly antigenic and shares antigenic motifs with the large glycolipids of eggs (Weiss *et al.*, 1986; Wuhrer *et al.*, 1999). In contrast to a structural model of Levery *et al.* (1992), which postulated a backbone structure of alternating HexNAc and fucose residues, further analysis by Khoo *et al.* (1997a) and Wuhrer *et al.* (2002) revealed for *S. mansoni* egg glycolipids a backbone of *N*-acetylhexosamine residues decorated with oligofucosyl side chains (Fig. 21.2). *S. mansoni* egg glycolipids exhibit significant biological activities as they were shown to induce cytokine production in human monocytes, thus leading to elevated interleukin-10, interleukin-6 and tumour necrosis factor alpha levels (van der Kleij *et al.*, 2002). Furthermore, both human and animal *S. mansoni* infection sera strongly react with *S. mansoni* egg glycosphingolipids (Weiss *et al.*, 1986; Wuhrer *et al.*, 1999). The major antigenic determinant of these glycosphingolipids is recognized by the monoclonal antibody (mAb) M2D3H (Wuhrer *et al.*, 1999, 2000a), and has been identified as Fuc(α1-3)GalNAcβ1- (Kantelhardt *et al.*, 2002).

Schistosome infection sera as well as the mAb M2D3H cross-react with glycans of keyhole limpet haemocyanin (KLH), the oxygen-transporting protein of the haemolymph of the marine snail *Megathura crenulata* (Hamilton *et al.*, 1999; Wuhrer *et al.*, 1999, 2000a; Kantelhardt *et al.*, 2002), a phenomenon which allows the use of KLH in antibody-based diagnosis of schistosomiasis (Mansour *et al.*, 1989; Alves-Brito *et al.*, 1992; Yuesheng *et al.*, 1994) as well as vaccination trials (Bashir *et al.*, 1994; Taylor *et al.*, 1998). The putative target epitope on KLH, which is bound by schisto-reactive antibodies, has been assigned to *N*-glycans of KLH, which exhibit a Fuc(α1-3)GalNAc(β1-4)[Fuc(α1-3)]GlcNAc-β1-antenna (Geyer *et al.*, 2004).

Besides KLH, other molluscan glycoproteins were shown to be recognized by schistosome-directed antibodies due to a shared cross-reactive determinant defined by the mAb IPLSm1 (Dissous *et al.*, 1986). The mAbs IPLSm1 and M2D3H likewise seem to recognize this cross-reactive determinant. This cross-reactivity might be of biological relevance, since the haemolymph of the *S. mansoni* intermediate host, *Biomphalaria glabrata*, also exhibits cross-reactivity with parasite glycoproteins and glycolipids (Dissous *et al.*, 1986; Wuhrer *et al.*, 2000a). This sharing of carbohydrate determinant(s) between parasite and intermediate host could be interpreted as an attempt to evade defence mechanisms of the snail by camouflage and fits into the concept of 'molecular mimicry' (Damian, 1989).

Protein glycosylation of schistosomes

CERCARIAL STAGE. The first report about protein glycosylation in cercariae described the structures of the major *O*-glycans of its highly immunogenic glycocalyx (Khoo *et al.*, 1995; Fig. 21.3). These glycans share the motif of oligofucosyl chains attached to a GalNAc(β1-4)GlcNAc(β1-(LacdiNAc)) unit with schistosome egg glycosphingolipids (Khoo *et al.*, 1997a; Fig. 21.2). While these cercarial *O*-glycans are based on type 1 and 2 cores, a second group of cercarial *O*-glycans is based on a novel core structure comprising an *O*-linked GalNAc with β-linked Gal residues in both the 3rd and 6th positions (Huang *et al.*, 2001). This *O*-glycan core may carry Lewis X trisaccharide units (Fig. 21.3). *N*-glycans of *S. mansoni* cercariae are core-(α1-6)-fucosylated throughout and show often additional (β1-2)-core xylosylation (Khoo *et al.*, 2001). The *N*-glycans possess mainly hybrid-type or diantennary structures ending in Lewis X trisaccharide units (Fig. 21.3). Taken together, Lewis X has been found as a major epitope on cercarial glycolipids (Fig. 21.1) as well as *N*-glycans and has also been described for certain *O*-glycans of this life-cycle stage (Fig. 21.3). Hence, this carbohydrate unit seems to be a general motif of cercarial glycosylation.

ADULT STAGE. *N*-glycans of *S. mansoni* adult worms have been structurally characterized by various groups applying a wide range of techniques such as metabolic labelling, lectin- or immuno-affinity chromatography, mass spectrometry and NMR. A large proportion of the *N*-glycans has been shown to be non-fucosylated, oligomannosidic or core-(α1-6)-fucosylated, paucimannosidic structures (Nyame *et al.*, 1988; Khoo *et al.*, 2001). Xylosylated species were not detected. Moreover, a great variety of complex-type structures were described (Khoo *et al.*, 2001). Complex-type species may

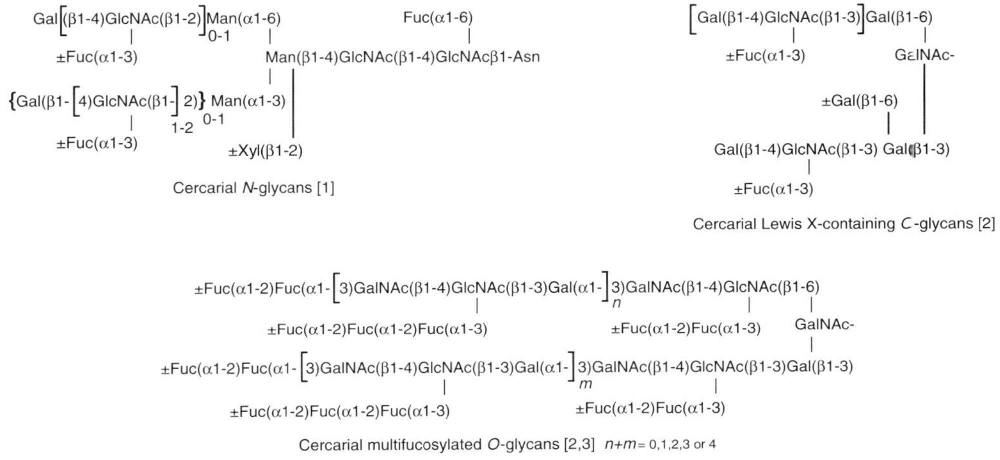

Fig. 21.3. Protein glycosylation of *S. mansoni* cercariae. Cercarial *N*-glycan structures were reported by Khoo *et al.* (2001) [1]. *O*-glycan structures were elucidated by Huang *et al.* (2001) [2] and Khoo *et al.* (1995) [3].

contain two LacdiNAc antennae, which may be fucosylated (Srivatsan *et al.*, 1992a; Fig. 21.4). Alternatively, *N*-glycans may contain two to four *N*-acetyllactosamine antennae with additional fucosylated *N*-acetyllactosamine repeats, resulting in oligomeric Lewis X structures (Srivatsan *et al.*, 1992b; Fig. 21.4). Furthermore, studies on potential *O*-glycosylation in adult worms revealed the presence of *O*-linked Tn-antigen (GalNAcα1-Ser/Thr), T-antigen (Gal(β1-3)GalNAcα1-Ser/Thr), and *O*-linked GlcNAc (Nyame *et al.*, 1987; Cummings and Nyame, 1996).

In contrast to the studies mentioned above, in which the glycosylation of complex protein extracts from adult worms has been analysed, there are two publications focusing on the carbohydrate structures of two purified, highly antigenic glycoproteins from schistosome adults: the circulating cathodic antigen (CCA) and the circulating anodic antigen (CAA). CCA exhibits large *O*-glycans with multimeric Lewis X entities (van Dam *et al.*, 1994; Fig. 21.4), whereas CAA carries highly negatively charged *O*-glycans constructed of a polymeric (β1-6)-linked GalNAc backbone substituted with (β1-3)-linked glucuronic acid residues (Bergwerff *et al.*, 1994; Fig. 21.4). These two examples indicate that a more detailed look at the carbohydrate structures of specific sets of *S. mansoni* antigens might reveal additional, novel glycosylation features, which are easily overlooked when analysing complex mixtures.

EGG STAGE. Egg extracts are known to contain many glycoproteins, which differentially express carbohydrate determinants (Weiss and Strand, 1985). While the glycosylation of soluble egg antigen (SEA) is thought to have a role in pivotal immunological processes in schistosomiasis infection (Faveeuw *et al.*, 2002, 2003) and the induction of hepatic granulomas (Weiss *et al.*, 1987; van de Vijver *et al.*, 2004), data regarding the precise structures of the glycoprotein-glycans expressed in this life-cycle stage are still incomplete. It is known, however, that both *N*-glycans and *O*-glycans contain fucosylated and non-fucosylated variants

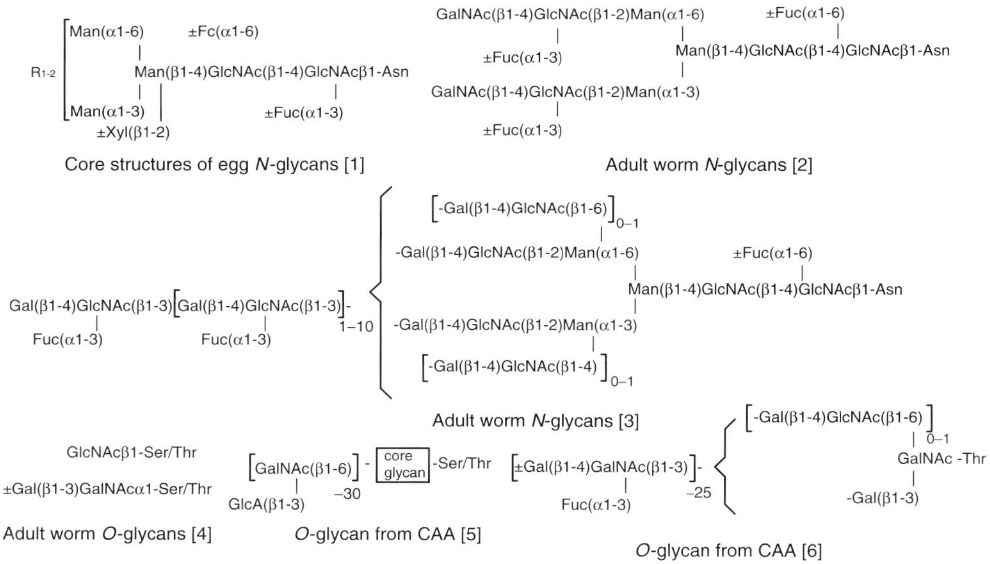

Fig. 21.4. Protein glycosylation of *S. mansoni* adults and eggs. *N*-glycans of egg glycoproteins were characterized by Khoo *et al.* (1997b) [1]. The presented adult worm *N*-glycans were published by Srivatsan *et al.* (1992a) [2] and (1992b) [3]. *O*-linked GlcNAc as well as T- and Tn-antigen were reported by Nyame *et al.* (1987) [4]. The repetitive *O*-glycan structures of the circulating antigens CAA and CCA were described by Bergwerff *et al.* (1994) [5] and van Dam *et al.* (1994) [6], respectively. R, terminal sequences (see text).

of HexNAc, *N*-acetyllactosamine (Gal(β1-4)GlcNAcβ1-) and LacdiNAc (GalNAc(β1-4)GlcNAcβ1-) units at their non-reducing ends (Khoo *et al.*, 1997b; abbreviated as 'R' in Fig. 21.4). Respective *N*-glycan core structures were found to contain optionally an (α1-6)-linked fucose residue. A portion of the core-fucosylated structures was found to contain xylose(β1-2)-linked to the β-mannose, and/or an additional (α1-3)-fucose unit at their innermost GlcNAc to yield highly unusual difucosylated core structures (Khoo *et al.*, 1997b; Fig. 21.4). In the case of *O*-glycans, the data indicated a heterogeneous pattern of core structures, which have not yet been studied in detail (Khoo *et al.*, 1997b).

The Lewis X epitope and the host–parasite relationship

Several observations indicate a major role for schistosome Lewis X-containing glycoconjugates in the interaction with the mammalian host:

(i) The temporal and spatial expression pattern of Lewis X in schistosomes favours this interaction. While anti-Lewis X monoclonal antibodies (mAbs) allow the detection of this carbohydrate epitope only around gland openings of the cercaria (van Remoortere *et al.*, 2000), this pattern changes dramatically when the parasite enters the mammalian host. Upon transformation, the highly antigenic glycocalyx is stripped off, the tail is lost, and Lewis X-positive patches can be observed after antibody staining all over the newly formed surface of the schistosomulum (Ko *et al.*, 1990; Köster and Strand, 1994; van Remoortere *et al.*, 2000). In later life-cycle stages, Lewis X remains constantly exposed at the site of parasite–host interactions. Adult schistosomes, for example, weakly express Lewis X epitopes in their own tegument and gut. In addition, they secrete Lewis X-containing gut antigens a major one of which is CCA. Likewise, the shell of *S. mansoni* eggs is recognized by anti-Lewis X mAbs (van Remoortere *et al.*, 2000), thus verifying again the exposure of this epitope.

(ii) Lewis X, which is also termed CD15 or stage-specific embryonic antigen I (SSEA I), is shared between the parasite and the mammalian host. Lewis X is expressed in a wide range of mouse and human tissues (Fox *et al.*, 1983). On a structural basis, Lewis X has been detected, often in parallel to sialyl-Lewis X, on glycolipids (Fukuda *et al.*, 1985; Symington *et al.*, 1985) and glycoproteins (Fukuda *et al.*, 1984; Spooncer *et al.*, 1984) of human granulocytes.

(iii) The Lewis X epitope is the target of an immune response in schistosomiasis, leading to autoimmunity (Nyame *et al.*, 1996, 1997, 1998). Anti-Lewis X autoantibodies have been shown to mediate complement-dependent cytolysis of Lewis X-presenting host cells *in vitro* (Nyame *et al.*, 1996, 1997; van Dam *et al.*, 1996), an effect which might cause the mild neutropenia observed in schistosomiasis (Borojevic *et al.*, 1983).

(iv) Parasitic Lewis X also seems to be targeted by the innate immune system through the C-type lectin dendritic cell-specific ICAM-3-grabbing non-integrin (DC-SIGN), which binds to SEA of *S. mansoni* as well as to Lewis X-containing neoglycoconjugates (van Die *et al.*, 2003) and might be involved in the early cellular immune reaction to schistosomula.

(v) Finally, Lewis X-containing glycoconjugates have an immunomodulatory effect in murine and human schistosomiasis as demonstrated by the induction of cellular secretion of interleukin 10 *in vitro* (Velupillai and Harn, 1994; Velupillai *et al.*, 2000).

Fasciola

So far, no structural studies on protein glycosylation in *Fasciola* have been reported yet, whereas neutral and acidic glycosphingolipid structures have been recently characterized. The results revealed that both *Fasciola hepatica* and *Fasciola gigantica* contain galactosylceramide, glucosylceramide, lactosylceramide and globotriaosylceramide, thus exhibiting glycolipids of the globo-series (Wuhrer *et al.*, 2001; Fig. 21.5). Schisto-series glycosphingolipids could not be detected. Besides globotriaosylceramide, adult *F. hepatica* flukes have been further shown to contain isoglobotriaosylceramide as well as Forssman-type glycosphingolipids with terminal GalNAc(α1-3)GalNAc β1-motifs (Wuhrer *et al.*, 2004; Fig. 21.5). Histochemical studies indicated Forssman

Neutral glycolipids

Gal(β1-1)Cer [1]

Glc(β1-1)Cer [1]

Gal(β1-4)Glc(β1-1)Cer [1]

Gal(α1-4)Gal(β1-4)Glc(β1-1)Cer [1,3]

Gal(α1-3)Gal(β1-4)Glc(β1-1)Cer [3]

Gal(β1-6)Gal(β1-4)Glc(β1-1)Cer [3]

Gal(β1-6)Gal(α1-3/4)Glc(β1-4)Glc(β1-1)Cer [3]

Gal(β1-6)Gal(β1-6)Gal(α1-3/4)Gal(β1-4)Glc(β1-1)Cer [3]

GalNAc(α1-3)GalNAc(β1-3/4)Gal(α1-3/4)Gal(β1-4)Glc(β1-1)Cer [3]

Acidic glycolipid

GlcNAc(α1-HPO_3-6)Gal(β1-1)Cer [2]

Fig. 21.5. Glycosphingolipids of *F. hepatica* adults. The elucidation of the glycosphingolipid structures is reported in Wuhrer *et al.* (2001) [1], (2003) [2] and (2004) [3]. Gal(β1-6)-epitopes mediating cross-reactivity with glycosphingolipids from cestodes are boxed. Cer, ceramide.

glycolipids to occur mainly in the gut of the parasite (Wuhrer *et al.*, 2004). Analyses of lipid moieties of the Forssman glycolipids purified from the parasite indicated that the Forssman glycosphingolipids found in *F. hepatica* seem to be taken up from sheep (Wuhrer *et al.*, 2004). In contrast, the smaller glycosphingolipids (glucosylceramide, galactosylceramide, lactosylceramide and globotriaosylceramide) exhibit, in part, ceramide structures with 4-hydroxysphinganine derivatives (phytosphingosines, t18:0 and t20:0) and 2-hydroxy fatty acids (Wuhrer *et al.*, 2001), which resemble the ceramide structures of schistosome glycosphingolipids (Khoo *et al.*, 1997a; Wuhrer *et al.*, 2000b). Hence, these glycolipids may be assumed to be of parasitic origin. Intriguingly, *F. hepatica* further exhibits glycosphingolipids with unusual Gal(β1-6)Galβ1-motifs (Wuhrer *et al.*, 2004; boxed structures in Fig. 21.5). This finding is remarkable as the same structural units are present on neogala-series glycosphingolipids of Cyclophyllida (see below). Corresponding neogala-reactive antibodies were detected in *F. hepatica* infection sera based on their binding to *Echinococcus granulosus* and *Taenia crassiceps* neogala glycosphingolipids. These antibodies might contribute to the known serological cross-reactivity between *F. hepatica* and cestode infection sera.

In addition to the neutral glycosphingolipid species described above, *F. hepatica* was found to express a novel type of acidic, antigenic glycolipid with a structure of GlcNAc(α1-HPO_3-6)Gal(1-1)ceramide (Wuhrer *et al.*, 2003; Fig. 21.5). In terms of ceramide composition, this acidic glycolipid seems to be highly related to the *F. hepatica* ceramide monohexosides comprising mainly phytosphingosines and 2-hydroxy fatty acids. Due to its recognition by *F. hepatica* infection sera, the GlcNAc(α1-phosphate) moiety might be of potential relevance in the serodiagnosis of fasciolosis. Furthermore, an inositol-containing species was characterized as 1-*O*-hexadecyl-*sn*-glycerol-3-phosphoinositol (Wuhrer *et al.*, 2003), which might be a member of a group of phosphatidylinositol lipids in *F. hepatica*. Since this molecule does not represent a glycolipid, it is not listed in Fig. 21.5.

Cestodes

Glycosphingolipids of cestode species (tapeworms) analysed so far can be structurally divided into two different groups and are, therefore, separately discussed. As far as available, structural data on glycoprotein-glycans are presented independently.

Pseudophyllida

Structural analyses of the neutral glycosphingolipids from plerocercoids of the parasites *Spirometra erinacei* (Kawakami *et al.*, 1993, 1995, 1996), *Spirometra mansonoides* (Singh *et al.*, 1987), *Diplogonoporus balaenopterae* (Kawakami *et al.*, 1997) as well as adult worms and plerocercoids of *Diphyllobothrium hottai* (Iriko *et al.*, 2002) revealed the presence of a novel and so far unique biosynthetic series (spirometo-series) which is based on galactosylceramide and characterized by the monosaccharide sequence Gal(β1-4)Glc(β1-3)Gal-ceramide (Fig. 21.6). Glycolipids of this type have been also termed 'spirometosides'. As a characteristic feature, these compounds are, in part, further substituted by (α1-3)-linked fucosyl and/or (β1-6)-linked galactosyl residues. Although the presence of Gal(β1-6)Gal units (marked by boxes in Fig. 21.6) provides some structural similarity to the neogala-series glycosphingolipids found in Cyclophyllida (see below), the occurrence of penultimate glucose residues clearly differentiates these compounds from all other glycosphingolipids analysed so far. Hence, these glycosphingolipids might be of taxonomic significance. Since spirometosides have been predominantly detected at the external surface of the worm, it is further suspected that these glycolipids might play a functionally important role in the host–parasite interaction (Yanagisawa *et al.*, 1999).

Glc(β1-1)Cer [1,6]

Gal(β1-1)Cer [1,4,6]

Gal(1-3)Gal(1-1)Cer [6]

Glc(β1-3)Gal(β1-1)Cer [6]

```
                        [Gal(β1-6)]
                                 |
          Glc(β1-3)Gal(β1-1)Cer   [6]

Gal(β1-4)Glc(β1-3)Gal(β1-1)Cer   [2,5,6]
         |
Fuc(α1-3)

                        [Gal(β1-6)]
                                 |
Gal(β1-4)Glc(β1-3)Gal(β1-1)Cer   [3,5]
         |
Fuc(α1-3)
```

Fig. 21.6. Glycosphingolipids of Pseudophyllida. Glycolipids of *Spirometra erinacei* were analysed by Kawakami *et al.* (1995) [1], (1993) [2] and (1996) [3]. Glycolipids of *S. mansonoides* were analysed by Singh *et al.* (1987) [4]. Glycolipids of *D. balaenopterae* and *D. hottai* were characterized by Kawakami *et al.* (1997) [5] and Iriko *et al.* (2002) [6], respectively. Gal(β1-6)-units are marked by boxes. Cer, ceramide.

In contrast to the striking divergences in the carbohydrate parts, respective ceramide moieties resemble those of other cestode glycosphingolipids in so far as sphinganine (d18:0) and 4-hydroxy-sphingoid bases (t18:0, t20:0) as well as, in part hydroxylated, fatty acids with 16 to 28 carbon atoms, in particular C26:0 and C28:0, represent predominant ceramide constituents, which might be an indication of similar biosynthetic routes.

Cyclophyllida

Glycosphingolipids

Neutral glycosphingolipids derived from metacestodes of *Echinococcus multilocularis*

(Persat *et al.*, 1990, 1992) or *T. crassiceps* (Dennis *et al.*, 1992), *Taenia solium* cysts (Lopez-Marin *et al.*, 2002) as well as adult cestodes of *Metroliasthes coturnix* (Nishimura *et al.*, 1991) express almost exclusively neogala-series carbohydrate moieties (Fig. 21.7) despite originating from different parasites, different life-cycle stages and different vertebrate hosts. The only exceptions are the ceramide disaccharide from *M. coturnix* which corresponds to the gala-series (Gal(α1-4)Gal(β1-1)Cer) and a ceramide tetrasaccharide derivative of *T. crassiceps* terminating in a Gal(α1-4)Gal(β1-6)-unit. A fraction of ceramide disaccharide and ceramide trisaccharide components of *E. multilocularis* is further fucosylated, thus forming a Fuc(α1-3)[±Gal(β1-6)] Gal(β1-6)-moiety. Intriguingly, such neogala-series glycosphingolipids inhibit the proliferation of human peripheral blood mononuclear cells (PBMCs), which suggests that this type of glycolipid may play an immunologically relevant role in alveolar hydatid disease (Persat *et al.*, 1996).

The ceramide moieties of these glycolipids display again considerable heterogeneity in both sphingoid-base and fatty acid composition. The long-chain bases are typified by the predominance of dihydrosphingosine (sphinganine; d18:0) and phytosphingosine (4-hydroxysphinganine; t18:0) in glycosphingolipids from *T. crassiceps* (Dennis *et al.*, 1992) and *E. multilocularis* (Persat *et al.*, 1990, 1992), whilst *M. coturnix* ceramide moieties are composed of C18–C20 sphinganine (d18:0, d20:0) and 4-hydroxysphinganine (Nishimura *et al.*, 1991). As a characteristic feature, glycolipids from all sources carry amide-bound long-chain fatty acids like hexacosanoic (C26:0) and octacosanoic acids (C28:0), along with short-chain, partially hydroxylated fatty acids with 16 or 18 carbon atoms in the case of *E. multilocularis* glycolipids, or long-chain 2-hydroxy fatty acids (e.g. C26h:0) in

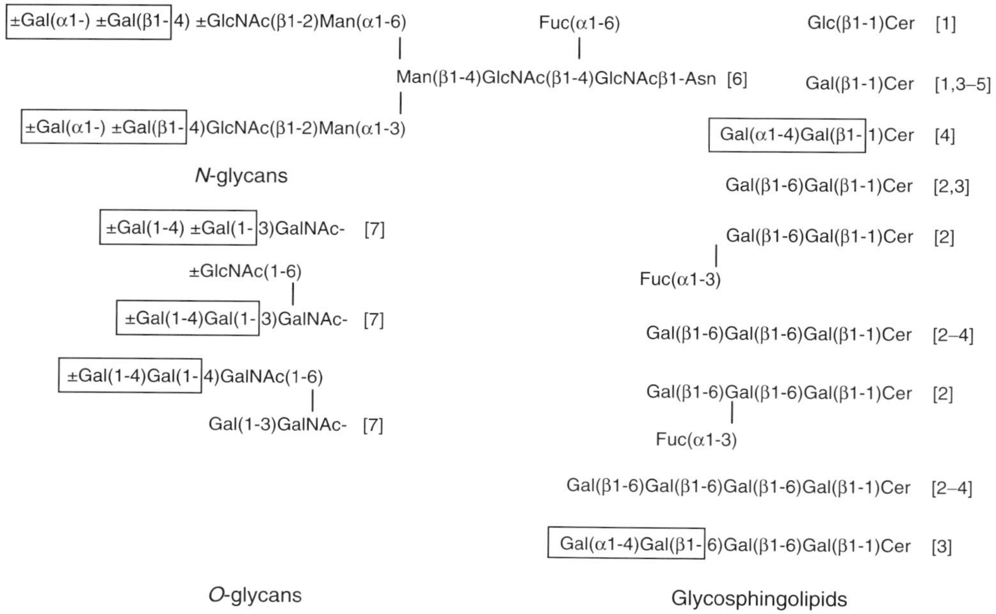

Fig. 21.7. Glycosphingolipids and protein glycosylation of Cyclophyllida. Glycolipids of *E. multilocularis* were analysed by Persat *et al.* (1990) [1] and (1992) [2]. Glycolipids of *T. crassiceps* and *M. coturnix* were characterized by Dennis *et al.* (1992) [3] and Nishimura *et al.* (1991) [4], respectively. The major glycolipid of *T. solium* was analysed by Lopez-Marin *et al.* (2002) [5]. Cer, ceramide. *N*-glycans of *E. granulosus* and *O*-glycans of *E. multilocularis* were described by Khoo *et al.* (1997c) [6] and Hülsmeier *et al.* (2002) [7], respectively. Possible Galα1-4)Gal1- structural elements are boxed.

M. coturnix ceramide. Intriguingly, fucosylated glycolipids obtained from *E. multilocularis* were found to possess only short-chain fatty acids with 16 and 18 carbon atoms (mainly C16:0 and C18h:0), which might indicate that the (α1-3)-fucosyltransferase involved in the biosynthesis of such compounds requires an acceptor substrate with shorter fatty acids, whereas (β1-6)-galactosyltransferase from this source obviously accepts all kinds of galactosylceramide precursors (Persat *et al.*, 1992).

In addition to neutral compounds, cestode-derived glycosphingolipids have been reported to comprise also small amounts of gangliosides, i.e. acidic, sialic acid-containing glycolipid molecules (Dennis *et al.*, 1993; Persat *et al.*, 1994). Structural studies performed on the acidic glycosphingolipids from *E. multilocularis* led to the identification of G_{M1}, G_{M3}, G_{D1a} and, to a lesser degree, G_{M2}, all belonging to the same a-ganglio-series (Persat *et al.*, 1994). Structural analyses of the respective ceramide moieties, however, revealed that *E. multilocularis* gangliosides contain completely different ceramide moieties characterized by the presence of sphingosine (d18:1) and non-hydroxylated fatty acids with 16 to 24 carbon atoms. Since the ceramide composition and the structures of the respective carbohydrate chains resemble that of mammalian gangliosides, it has been concluded that these minor components of *E. multilocularis* metacestode-derived glycosphingolipids are probably taken up from the host (Persat *et al.*, 1994, 1995). Therefore, these structures are not listed in Fig. 21.7.

Glycoproteins

Carbohydrates have been shown to contribute significantly to the antigenicity of cestode glycoproteins (Obregón-Henao *et al.*, 2001; Elayoubi *et al.*, 2003). Furthermore, there is evidence that the laminated layer of *E. multilocularis* metacestodes is largely composed of carbohydrates (Ingold *et al.*, 2000) which may be involved in immunosuppressive events that occur at the host–parasite interface (Walker *et al.*, 2004). At the structural level, however, little is known about the carbohydrate moieties of cestode glycoproteins in general. One exception is the highly species-specific *E. multilocularis* antigen referred to as Em2(G11) which is a major component of the laminated layer. Carbohydrate structure analyses (Hülsmeier *et al.*, 2002) revealed that Em2(G11) is a mucin-type glycoprotein carrying *O*-linked glycans with core 1 (Gal(β1-3)GalNAc) or core 2 (GlcNAc(β1-6)[Gal(β1-3)]GalNAc) structures (Fig. 21.7). As a characteristic feature, these glycans are capped, in part, by Gal(1-4)Gal-units, thus forming novel *O*-linked oligosaccharide structures. Although the anomeric configurations of the respective monosaccharide constituents have not been determined, this Gal(1-4)Gal-epitope appears to be homologous to the Gal(α1-4)Galβ-determinant found, at least in part, in cestode glycosphingolipids (Fig. 21.7). The identification of *O*-linked Tn antigen (GalNAc-*O*-Ser/Thr) in this cestode glycoprotein is in agreement with lectin-based studies which demonstrated the presence of this epitope in both larval and adult worm extracts of *E. granulosus* (Alvarez Errico *et al.*, 2001).

Structural characterization of *N*-linked glycans, enzymatically released from crude extracts of *E. granulosus* hydatid cyst membranes and protoscoleces, demonstrated that carbohydrates from the cyst membranes represent predominantly non-charged oligosaccharides with partially truncated *N*-acetyllactosamine (Gal(β1-4)GlcNAc) antennae and core-fucosylation (Khoo *et al.*, 1997c). A significant proportion of these antennae were found to be capped by α-galactose, thus forming again Galα-Galβ-terminal structures (Fig. 21.7). Although the exact linkage position of α-Gal has not been established, this finding supports the hypothesis that Gal(α1-4)Galβ-determinants represent a carbohydrate epitope which may be shared by both cestode glycosphingolipids as well as *N*- and *O*-linked glycoprotein-glycans. In addition to the complex-type carbohydrates mentioned above, *N*-glycans recovered from protoscoleces extracts comprised small proportions of high mannose and truncated core structures, as well. In principle, analogous results were obtained for *N*-linked glycans from *T. solium* metacestodes (Haslam *et al.*, 2003). In this case, however, sugar chains with terminal Galα–Galβ structures have not been reported.

Concluding Remarks

Flatworms express a vast variety of carbohydrate structures. While fucosylation makes a major contribution to glycan diversity and antigenicity in schistosomes (e.g. core-α1-3/6-fucosylation, oligofucosyl chains, Lewis X, pseudo-Lewis Y), the trematode *F. hepatica* as well as cestodes of the genera *Echinococcus* and *Taenia* exhibit a variety of antigenic galactosylated glycoconjugates (e.g. neogala- and globo-series type of structures). Besides the production of highly antigenic glycoconjugates, which often function as major targets of the host's immune response, presentation of host-like glycans occurs. This phenomenon is well described for schistosomes (expression of Lewis X epitopes), but seems to be more general and has been also reported for *F. hepatica* (presence of Forssman antigen and globotriaosylceramide). Surface-expressed carbohydrate determinants, which are shared between parasite and host, can be interpreted as a camouflage coat according to the concept of 'molecular mimicry'. In addition, they may have an immunomodulatory role.

While future work will certainly come up with many more structural features of flatworm glycoconjugates, the rapid progress of the flatworm genome projects should lead to the identification and allow the characterization of many enzymes of the glycosylation machinery. Together with immunological studies and the expected establishment of methods for the genetic manipulation of flatworms, future research should be able to generate a much more detailed picture of flatworm glycobiology.

References

Alvarez Errico, D., Medeiros, A., Miguez, M., Casaravilla, C., Malgor, R., Carmona, C., Nieto, A. and Osinaga, E. (2001) *O*-glycosylation in *Echinococcus granulosus*: identification and characterization of the carcinoma-associated Tn antigen. *Experimental Parasitology* 98, 100–109.

Alves-Brito, C.F., Simpson, A.J.G., Bahia-Oliveira, L.M.G., Rabello, A.L.T., Rocha, R.S., Lambertucci, J.R., Gazzinelli, G., Katz, N. and Correa-Oliveira, R. (1992) Analysis of anti-keyhole limpet haemocyanin antibody in Brazilians supports its use for the diagnosis of acute schistosomiasis mansoni. *Transactions of the Royal Society of Tropical Medicine and Hygiene* 86, 53–56.

Bashir, M., Bushara, H., Cook, L., Fuhui, S., He, D., Huggins, M., Jiaojiao, L., Malik, K., Moloney, A., Mukhtar, M., Ping, Y., Shoutai, X., Taylor, M. and Yaochuan, S. (1994) Evaluation of defined antigen vaccines against *Schistosoma bovis* and *S. japonicum* in bovines. *Tropical and Geographical Medicine* 46, 255–258.

Bergwerff, A.A., van Dam, G.J., Rotmans, J.P., Deelder, A.M., Kamerling, J.P. and Vliegenthart, J.F.G. (1994) The immunologically reactive part of immunopurified circulating anodic antigen from *Schistosoma mansoni* is a threonine-linked polysaccharide consisting of -6)-(β-D-Glc*p*A-(1–3))-β-D-Gal*p*NAc-(1-repeating units. *Journal of Biological Chemistry* 269, 31510–31517.

Borojevic, R., Santos-da-Silva, C. and Carvalho, E.A. (1983) Chronic schistosomiasis mansoni: splenic myelopoiesis and inhibition of neutrophil granulocytopoiesis mediated by the sera of patients. *Journal of Infectious Diseases* 148, 422–426.

Cummings, R.D. and Nyame, A.K. (1996) Glycobiology of schistosomiasis. *FASEB Journal* 10, 838–848.

Cummings, R.D. and Nyame, A.K. (1999) Schistosome glycoconjugates. *Biochimica et Biophysica Acta* 1455, 363–374.

Damian, R.T. (1989) Molecular mimicry. *Current Topics in Microbiological Immunology* 145, 101–115.

de Jonge, N., Rabello, A.L., Krijger, F.W., Kremsner, P.G., Rocha, R.S., Katz, N. and Deelder, A.M. (1991) Levels of the schistosome circulating anodic and cathodic antigens in serum of schistosomiasis patients from Brazil. *Transactions of the Royal Society of Tropical Medicine and Hygiene* 85, 756–759.

Deelder, A.M., Klappe, H.T., van den Aardweg, G.J. and van Meerbeke, E.H. (1976) *Schistosoma mansoni*: demonstration of two circulating antigens in infected hamsters. *Experimental Parasitology* 40, 189–197.

Dennis, R.D., Baumeister, S., Geyer, R., Peter-Katalinic, J., Hartmann, R., Egge, H., Geyer, E. and Wiegandt, H. (1992) Glycosphingolipids in cestodes – chemical structures of ceramide monosaccharide, disaccharide, trisaccharide and tetrasaccharide from metacestodes of the fox tapeworm, *Taenia crassiceps* (Cestoda: Cyclophyllidea). *European Journal of Biochemistry* 207, 1053–1062.

Dennis, R.D., Baumeister, S., Irmer, G., Gasser, R.B. and Geyer, E. (1993) Chromatographic and antigenic properties of *Echinococcus granulosus* hydatid cyst-derived glycolipids. *Parasite Immunology* 15, 669–681.

Dissous, C., Grzych, J.M. and Capron, A. (1986) *Schistosoma mansoni* shares a protective carbohydrate epitope with freshwater and marine snails. *Nature* 323, 443–445.

Eberl, M., Langermans, J.A., Vervenne, R.A., Nyame, A.K., Cummings, R.D., Thomas, A.W., Coulson, P.S. and Wilson, R.A. (2001) Antibodies to glycans dominate the host response to schistosome larvae and eggs: is their role protective or subversive? *Journal of Infectious Diseases* 183, 1238–1247.

Elayoubi, F.A., Fraser, A., Jenkins, D.J. and Craig, P.S. (2003) Partial characterisation of carbohydrate-rich *Echinococcus granulosus* coproantigens. *International Journal of Parasitology* 33, 1553–1559.

Faveeuw, C., Angeli, V., Fontaine, J., Maliszewski, C., Capron, A., Van Kaer, L., Moser, M., Capron, M. and Trottein, F. (2002) Antigen presentation by CD1d contributes to the amplification of Th2 responses to *Schistosoma mansoni* glycoconjugates in mice. *Journal of Immunology* 169, 906–912.

Faveeuw, C., Mallevaey, T., Paschinger, K., Wilson, I.B., Fontaine, J., Mollicone, R., Oriol, R., Altmann, F., Lerouge, P., Capron, M. and Trottein, F. (2003) Schistosome *N*-glycans containing core alpha 3-fucose and core beta 2-xylose epitopes are strong inducers of Th2 responses in mice. *European Journal of Immunology* 33, 1271–1281.

Fox, N., Damjanov, I., Knowles, B.B. and Solter, D. (1983) Immunohistochemical localization of the mouse stage-specific embryonic antigen 1 in human tissues and tumors. *Cancer Research* 43, 669–678.

Fukuda, M., Spooncer, E., Oates, J.E., Dell, A. and Klock, J.C. (1984) Structure of sialylated fucosyl lactosaminoglycan isolated from human granulocytes. *Journal of Biological Chemistry* 259, 10925–10935.

Fukuda, M.N., Dell, A., Oates, J.E., Wu, P., Klock, J.C. and Fukuda, M. (1985) Structures of glycosphingolipids isolated from human granulocytes. The presence of a series of linear poly-*N*-acetyllactosaminylceramide and its significance in glycolipids of whole blood cells. *Journal of Biological Chemistry* 260, 1067–1082.

Geyer, H., Wuhrer, M., Kurokawa, T. and Geyer, R. (2004) Characterization of keyhole limpet hemocyanin (KLH) glycans sharing a carbohydrate epitope with *Schistosoma mansoni* glycoconjugates. *Micron* 35, 105–106.

Hamilton, J.V., Chiodini, P.L., Fallon, P.G. and Doenhoff, M.J. (1999) Periodate-sensitive immunological cross-reactivity between keyhole limpet haemocyanin (KLH) and serodiagnostic *Schistosoma mansoni* egg antigens. *Parasitology* 118, 83–90.

Haslam, S.M., Restrepo, B.I., Obregón-Henao, A., Teale, J.M., Morris, H.R. and Dell, A. (2003) Structural characterization of the *N*-linked glycans from *Taenia solium* metacestodes. *Molecular and Biochemical Parasitology* 126, 103–107.

Hokke, C.H. and Deelder, A.M. (2001) Schistosome glycoconjugates in host–parasite interplay. *Glycoconjugate Journal* 18, 573–587.

Huang, H.H., Tsai, P.L. and Khoo, K.H. (2001) Selective expression of different fucosylated epitopes on two distinct sets of *Schistosoma mansoni* cercarial *O*-glycans: identification of a novel core type and Lewis X structure. *Glycobiology* 11, 395–406.

Hülsmeier, A.J., Gehrig, P.M., Geyer, R., Sack, R., Gottstein, B., Deplazes, P. and Kohler, P. (2002) A major *Echinococcus multilocularis* antigen is a mucin-type glycoprotein. *Journal of Biological Chemistry* 277, 5742–5748.

Ingold, K., Gottstein, B. and Hemphill, A. (2000) High molecular mass glycans are major structural elements associated with the laminated layer of *in vitro* cultivated *Echinococcus multilocularis* metacestodes. *International Journal of Parasitology* 30, 207–214.

Iriko, H., Nakamura, K., Kojima, H., Iida-Tanaka, N., Kasama, T., Kawakami, Y., Ishizuka, I., Uchida, A., Murata, Y. and Tamai, Y. (2002) Chemical structures and immunolocalization of glycosphingolipids isolated from *Diphyllobothrium hottai* adult worms and plerocercoids. *European Journal of Biochemistry* 269, 3549–3559.

Kantelhardt, S.R., Wuhrer, M., Dennis, R.D., Doenhoff, M.J., Bickle, Q. and Geyer, R. (2002) Fuc(alpha1-3) GalNAc-: major antigenic motif of *Schistosoma mansoni* glycolipids implicated in infection sera and keyhole limpet hemocyanin cross-reactivity. *Biochemical Journal* 366, 217–223.

Kawakami, Y., Nakamura, K., Kojima, H., Suzuki, M., Inagaki, F., Suzuki, A., Sonoki, S., Uchida, A., Murata, Y. and Tamai, Y. (1993) Glycosphingolipids of plerocercoids of parasite, *Spirometra erinacei*. *Glycoconjugate Journal* 10, 339.

Kawakami, Y., Kojima, H., Nakamura, K., Suzuki, M., Uchida, A., Murata, Y. and Tamai, Y. (1995) Monohexosylceramides of larval and adult forms of the tapeworm, *Spirometra erinacei*. *Lipids* 30, 333–337.

Kawakami, Y., Nakamura, K., Kojima, H., Suzuki, M., Inagaki, F., Suzuki, A., Ikuta, J., Uchida, A., Murata, Y. and Tamai, Y. (1996) A novel fucosyltetrahexosylceramide in plerocercoids of the parasite *Spirometra erinacei*. *European Journal of Biochemistry* 239, 905–911.

Kawakami, Y., Yanagisawa, M., Kojima, H., Nakamura, H., Araki, J., Uchida, A., Murata, Y. and Tamai, Y. (1997) Glycosphingolipids of parasite, *Diplogonoporus balaenopterae*. *Glycoconjugate Journal* 14, S83.

Khoo, K.H. (2001) Structural variations in schistosomal glycans. *Trends in Glycoscience and Glycotechnology* 31, 493–506.

Khoo, K.H., Sarda, S., Xu, X., Caulfield, J.P., McNeil, M.R., Homans, S.W., Morris, H.R. and Dell, A. (1995) A unique multifucosylated -3GalNAcβ1-4GlcNAcβ1-3Galα1- motif constitutes the repeating unit of the complex *O*-glycans derived from the cercarial glycocalyx of *Schistosoma mansoni*. *Journal of Biological Chemistry* 270, 17114–17123.

Khoo, K.H., Chatterjee, D., Caulfield, J.P., Morris, H.R. and Dell, A. (1997a) Structural characterization of glycosphingolipids from the eggs of *Schistosoma mansoni* and *Schistosoma japonicum*. *Glycobiology* 7, 653–661.

Khoo, K.H., Chatterjee, D., Caulfield, J.P., Morris, H.R. and Dell, A. (1997b) Structural mapping of the glycans from the egg glycoproteins of *Schistosoma mansoni* and *Schistosoma japonicum*: identification of novel core structures and terminal sequences. *Glycobiology* 7, 663–677.

Khoo, K.H., Nieto, A., Morris, H.R. and Dell, A. (1997c) Structural characterization of the *N*-glycans from *Echinococcus granulosus* cyst membrane and protoscoleces. *Molecular and Biochemical Parasitology* 86, 237–248.

Khoo, K.H., Huang, H.H. and Lee, K.M. (2001) Characteristic structural features of schistosome cercarial *N*-glycans: expression of Lewis X and core xylosylation. *Glycobiology* 11, 149–163.

Ko, A.I., Dräger, U.C. and Harn, D.A. (1990) A *Schistosoma mansoni* epitope recognized by a protective monoclonal antibody is identical to the stage-specific embryonic antigen 1. *Proceedings of the National Academy of Sciences of the United States of America* 87, 4159–4163.

Kolter, T., Proia, R.L. and Sandhoff, K. (2002) Combinatorial ganglioside biosynthesis. *Journal of Biological Chemistry* 277, 25859–25862.

Köster, B. and Strand, M. (1994) *Schistosoma mansoni*: immunolocalization of two different fucose-containing carbohydrate epitopes. *Parasitology* 108, 433–446.

Levery, S.B., Weiss, J.B., Salyan, M.E.K., Roberts, C.E., Hakomori, S., Magnani, J.L. and Strand, M. (1992) Characterization of a series of novel fucose-containing glycosphingolipid immunogens from eggs of *Schistosoma mansoni*. *Journal of Biological Chemistry* 267, 5542–5551.

Lopez-Marin, L.M., Montrozier, H., Lemassu, A., Garcia, E., Segura, E. and Daffe, M. (2002) Structure and antigenicity of the major glycolipid from *Taenia solium cysticerci*. *Molecular and Biochemical Parasitology* 119, 33–42.

Makaaru, C.K., Damian, R.T., Smith, D.F. and Cummings, R.D. (1992) The human blood fluke *Schistosoma mansoni* synthesizes a novel type of glycosphingolipid. *Journal of Biological Chemistry* 267, 2251–2257.

Mansour, M.A., Ali, P.O., Farid, Z., Simpson, A.J.G. and Woody, J.W. (1989) Serological differentiation of acute and chronic schistosomiasis mansoni by antibody responses to keyhole limpet hemocyanin. *American Journal of Tropical Medicine and Hygiene* 41, 338–344.

Nishimura, K., Suzuki, A. and Kino, H. (1991) Sphingolipids of a cestode *Metroliasthes coturnix*. *Biochimica et Biophysica Acta* 1086, 141–150.

Nyame, K., Cummings, R.D. and Damian, R.T. (1987) *Schistosoma mansoni* synthesizes glycoproteins containing terminal *O*-linked *N*-acetylglucosamine residues. *Journal of Biological Chemistry* 262, 7990–7995.

Nyame, K., Cummings, R.D. and Damian, R.T. (1988) Characterization of the *N*- and *O*-linked oligosaccharides in glycoproteins synthesized by *Schistosoma mansoni* schistosomula. *Journal of Parasitology* 74, 562–572.

Nyame, A.K., Pilcher, J.B., Tsang, V.C.W. and Cummings, R.D. (1996) *Schistosoma mansoni* infection in humans and primates induces cytolytic antibodies to surface Le^x determinants on myeloid cells. *Experimental Parasitology* 82, 191–200.

Nyame, A.K., Pilcher, J.B., Tsang, V.C.W. and Cummings, R.D. (1997) Rodents infected with *Schistosoma mansoni* produce cytolytic IgG and IgM antibodies to the Lewis X antigen. *Glycobiology* 7, 207–215.

Nyame, A.K., DeBose-Boyd, R., Long, T.D., Tsang, V.C. and Cummings, R.D. (1998) Expression of Le^x antigen in *Schistosoma japonicum* and *S. haematobium* and immune responses to Le^x in infected animals: lack of Le^x expression in other trematodes and nematodes. *Glycobiology* 8, 615–624.

Nyame, A.K., Kawar, Z.S. and Cummings, R.D. (2004) Antigenic glycans in parasitic infections: implications for vaccines and diagnostics. *Archives of Biochemistry and Biophysics* 426, 182–200.

Obregón-Henao, A., Gil, D.L., Gomez, D.I., Sanzon, F., Teale, J.M. and Restrepo, B.I. (2001) The role of *N*-linked carbohydrates in the antigenicity of *Taenia solium* metacestode glycoproteins of 12, 16 and 18 kD. *Molecular and Biochemical Parasitology* 114, 209–215.

Omer-Ali, P., Smithers, S.R., Bickle, Q., Phillips, S.M., Harn, D. and Simpson, A.J.G. (1988) Analysis of the anti-*Schistosoma mansoni* surface antibody response during murine infection and its potential contribution to protective immunity. *Journal of Immunology* 140, 3273–3279.

Persat, F., Bouhours, J.F., Mojon, M. and Petavy, A.F. (1990) Analysis of the monohexosylceramide fraction of *Echinococcus multilocularis* metacestodes. *Molecular and Biochemical Parasitology* 41, 1–6.

Persat, F., Bouhours, J.-F., Mojon, M. and Petavy, A.-F. (1992) Glycosphingolipids with Galβ1-6Gal sequences in metacestodes of the parasite *Echinococcus multilocularis*. *Journal of Biological Chemistry* 267, 8764–8769.

Persat, F., Bouhours, J.F., Petavy, A.F. and Mojon, M. (1994) Gangliosides of *Echinococcus multilocularis* metacestodes. *Biochimica et Biophysica Acta* 1225, 297–303.

Persat, F., Bouhours, J.F., Petavy, A.F. and Mojon, M. (1995) Free ceramides of *Echinococcus multilocularis* metacestodes. *Biochimica et Biophysica Acta* 1255, 280–284.

Persat, F., Vincent, C., Schmitt, D. and Mojon, M. (1996) Inhibition of human peripheral blood mononuclear cell proliferative response by glycosphingolipids from metacestodes of *Echinococcus multilocularis*. *Infection and Immunity* 64, 3682–3687.

Singh, B.N., Beach, D.H., Walenga, R.W., Mueller, J.F. and Holz, G.G. (1987) Neutral glycosphingolipids of *Spirometra* life-cycle stages. In: MacInnis, A.J. (ed.) *Molecular Paradigms of Eradicating Helminthic Parasites*. A.R. Liss, New York, pp. 493–506.

Spooncer, E., Fukuda, M., Klock, J.C., Oates, J.E. and Dell, A. (1984) Isolation and characterization of polyfucosylated lactosaminoglycan from human granulocytes. *Journal of Biological Chemistry* 259, 4792–4801.

Srivatsan, J., Smith, D.F. and Cummings, R.D. (1992a) *Schistosoma mansoni* synthesizes novel biantennary Asn-linked oligosaccharides containing terminal β-linked *N*-acetylgalactosamine. *Glycobiology* 2, 445–452.

Srivatsan, J., Smith, D.F. and Cummings, R.D. (1992b) The human blood fluke *Schistosoma mansoni* synthesizes glycoproteins containing the Lewis X antigen. *Journal of Biological Chemistry* 267, 20196–20203.

Symington, F.W., Hedges, D.L. and Hakomori, S. (1985) Glycolipid antigens of human polymorphonuclear neutrophils and the inducible HL-60 myeloid leukemia line. *Journal of Immunology* 134, 2498–2506.

Taylor, M.G., Huggins, M.C., Shi, F., Lin, J., Tian, E., Ye, P., Shen, W., Qian, C.G., Lin, B.F. and Bickle, Q.D. (1998) Production and testing of *Schistosoma japonicum* candidate vaccine antigens in the natural ovine host. *Vaccine* 16, 1290–1298.

van Dam, G.J. and Deelder, A.M. (1996) Glycoproteins of parasites – schistosome glycoconjugates and their role in host–parasite pathological interactions. In: Montreuil, J., Vliegenthart, J.F.G. and Schachter, H. (eds) *Glycoproteins and Disease*, Vol. I. Elsevier, Amsterdam, pp. 123–142.

van Dam, G.J., Bergwerff, A.A., Thomas-Oates, J.E., Rotmans, J.P., Kamerling, J.P., Vliegenthart, J.F. and Deelder, A.M. (1994) The immunologically reactive *O*-linked polysaccharide chains derived from circulating cathodic antigen isolated from the human blood fluke *Schistosoma mansoni* have Lewis *x* as repeating unit. *European Journal of Biochemistry* 225, 467–482.

van Dam, G.J., Claas, F.H., Yazdanbakhsh, M., Kruize, Y.C., van Keulen, A.C., Ferreira, S.T., Rotmans, J.P. and Deelder, A.M. (1996) *Schistosoma mansoni* excretory circulating cathodic antigen shares Lewis-*x* epitopes with a human granulocyte surface antigen and evokes host antibodies mediating complement-dependent lysis of granulocytes. *Blood* 88, 4246–4251.

van der Kleij, D., Tielens, A.G. and Yazdanbakhsh, M. (1999) Recognition of schistosome glycolipids by immunoglobulin E: possible role in immunity. *Infection and Immunity* 67, 5946–5950.

van der Kleij, D., van Remoortere, A., Schuitemaker, J.H., Kapsenberg, M.L., Deelder, A.M., Tielens, A.G., Hokke, C.H. and Yazdanbakhsh, M. (2002) Triggering of innate immune responses by schistosome egg glycolipids and their carbohydrate epitope GalNAcβ1-4(Fucα1-2Fucα1-3)GlcNAc. *Journal of Infectious Diseases* 185, 531–539.

van de Vijver, K.K., Hokke, C.H., van Remoortere, A., Jacobs, W., Deelder, A.M. and Van Marck, E.A. (2004) Glycans of *Schistosoma mansoni* and keyhole limpet haemocyanin induce hepatic granulomas *in vivo*. *International Journal for Parasitology* 34, 951–961.

van Die, I., van Vliet, S.J., Nyame, A.K., Cummings, R.D., Bank, C.M., Appelmelk, B., Geijtenbeek, T.B. and van Kooyk, Y. (2003) The dendritic cell-specific C-type lectin DC-SIGN is a receptor for *Schistosoma mansoni* egg antigens and recognizes the glycan antigen Lewis *x*. *Glycobiology* 13, 471–478.

van Remoortere, A., Hokke, C.H., van Dam, G.J., van Die, I., Deelder, A.M. and van den Eijnden, D.H. (2000) Various stages of *Schistosoma* express Lewis^X, LacdiNAc, GalNAcβ1-4(Fucα1-3)GlcNAc and

GalNAcβ1-4(Fucα1-2Fucα1-3)GlcNAc carbohydrate epitopes: detection with monoclonal antibodies that are characterized by enzymatically synthesized neoglycoproteins. *Glycobiology* 10, 601–609.

Velupillai, P. and Harn, D.A. (1994) Oligosaccharide-specific induction of interleukin 10 production by B220+ cells from schistosome-infected mice: a mechanism for regulation of CD4+ T-cell subsets. *Proceedings of the National Academy of Sciences of the United States of America* 91, 18–22.

Velupillai, P., dos Reis, E.A., dos Reis, M.G. and Harn, D.A. (2000) Lewis(x)-containing oligosaccharide attenuates schistosome egg antigen-induced immune depression in human schistosomiasis. *Human Immunology* 61, 225–232.

Walker, M., Baz, A., Dematteis, S., Stettler, M., Gottstein, B., Schaller, J. and Hemphill, A. (2004) Isolation and characterization of a secretory component of *Echinococcus multilocularis* metacestodes potentially involved in modulating the host–parasite interface. *Infection and Immunity* 72, 527–536.

Weiss, J.B. and Strand, M. (1985) Characterization of developmentally regulated epitopes of *Schistosoma mansoni* egg glycoprotein antigens. *Journal of Immunology* 135, 1421–1429.

Weiss, J.B., Magnani, J.L. and Strand, M. (1986) Identification of *Schistosoma mansoni* glycolipids that share immunogenic carbohydrate epitopes with glycoproteins. *Journal of Immunology* 136, 4275–4282.

Weiss, J.B., Aronstein, W.S. and Strand, M. (1987) *Schistosoma mansoni*: stimulation of artificial granuloma formation *in vivo* by carbohydrate determinants. *Experimental Parasitology* 64, 228–236.

Wuhrer, M., Dennis, R.D., Doenhoff, M.J., Bickle, Q., Lochnit, G. and Geyer, R. (1999) Immunochemical characterisation of *Schistosoma mansoni* glycolipid antigens. *Molecular and Biochemical Parasitology* 103, 155–169.

Wuhrer, M., Dennis, R.D., Doenhoff, M.J. and Geyer, R. (2000a) A fucose-containing epitope is shared by keyhole limpet haemocyanin and *Schistosoma mansoni* glycosphingolipids. *Molecular and Biochemical Parasitology* 110, 237–246.

Wuhrer, M., Dennis, R.D., Doenhoff, M.J. and Geyer, R. (2000b) Stage-associated expression of ceramide structures in glycosphingolipids from the human trematode parasite *Schistosoma mansoni*. *Biochimica et Biophysica Acta* 1524, 155–161.

Wuhrer, M., Dennis, R.D., Doenhoff, M.J., Lochnit, G. and Geyer, R. (2000c) *Schistosoma mansoni* cercarial glycolipids are dominated by Lewis X and pseudo-Lewis Y structures. *Glycobiology* 10, 89–101.

Wuhrer, M., Berkefeld, C., Dennis, R.D., Idris, M.A. and Geyer, R. (2001) The liver flukes *Fasciola gigantica* and *Fasciola hepatica* express the leucocyte cluster of differentiation marker CD77 (globotriaosylceramide) in their tegument. *Biological Chemistry* 382, 195–207.

Wuhrer, M., Kantelhardt, S.R., Dennis, R.D., Doenhoff, M.J., Lochnit, G. and Geyer, R. (2002) Characterization of glycosphingolipids from *Schistosoma mansoni* eggs carrying Fuc(α1-3)GalNAc-, GalNAc(β1-4)[Fuc(α1-3)]GlcNAc- and Gal(β1-4)[Fuc(α1-3)]GlcNAc- (Lewis X) terminal structures. *European Journal of Biochemistry* 269, 481–493.

Wuhrer, M., Grimm, C., Zahringer, U., Dennis, R.D., Berkefeld, C.M., Idris, M.A. and Geyer, R. (2003) A novel GlcNAcalpha1-HPO3-6Gal(1-1)ceramide antigen and alkylated inositol-phosphoglycerolipids expressed by the liver fluke *Fasciola hepatica*. *Glycobiology* 13, 129–137.

Wuhrer, M., Grimm, C., Dennis, R.D., Idris, M.A. and Geyer, R. (2004) The parasitic trematode *Fasciola hepatica* exhibits mammalian-type glycolipids as well as Gal(beta1-6)Gal-terminating glycolipids that account for cestode serological cross-reactivity. *Glycobiology* 14, 115–126.

Yanagisawa, M., Kojima, H., Kawakami, Y., Iriko, H., Nakamura, T., Nakamura, K., Uchida, A., Murata, Y. and Tamai, Y. (1999) A monoclonal antibody against a glycolipid SEGLx from *Spirometra erinaceieuropaei* plerocercoid. *Molecular and Biochemical Parasitology* 102, 225–235.

Yuesheng, L., Rabello, A.L.T., Simpson, A.J.G. and Katz, N. (1994) The serological differentiation of acute and chronic *Schistosoma japonicum* infection by ELISA using keyhole limpet haemocyanin as antigen. *Transactions of the Royal Society of Tropical Medicine and Hygiene* 88, 249–251.

22 Gene Silencing in Flatworms using RNA Interference

P.J. Skelly
Department of Biomedical Sciences, Division of Infectious Diseases, Tufts University School of Veterinary Medicine, 200 Westboro Road, Grafton, MA 01536, USA

Introduction

RNA interference (RNAi) is defined as the mechanism by which gene-specific double-stranded RNA (dsRNA) triggers degradation of homologous mRNA transcripts. One paper refers to this as 'killing the messenger'; the result is sequence-specific gene silencing (Tabara *et al.*, 1998; Hunter, 1999; Grishok and Mello, 2002; Malhotra *et al.*, 2002; Dykxhoorn *et al.*, 2003). The phenomenon was first discovered in the free-living nematode, *Caenorhabditis elegans* when it was observed that dsRNA was at least 10-fold more potent at suppressing the expression of a target gene than either sense or antisense RNAs alone (Fire *et al.*, 1998). This silencing by dsRNA in *C. elegans* has since been found to have a number of remarkable properties. First, RNAi can be induced either by injecting dsRNA into the worm or by introducing dsRNA through feeding either the dsRNA itself or bacteria engineered to express it (Tabara *et al.*, 1998; Timmons *et al.*, 2001). Second, RNAi is potent since only a small number of molecules of dsRNA per cell can trigger gene silencing throughout the treated animal (Grishok *et al.*, 2001; Zamore, 2001). Third, the phenomenon can be relatively long-lasting and can often be detected in the F1, or first generation, progeny of treated animals, and beyond (Grishok and Mello, 2002).

Molecular Mechanisms of Silencing

Our present understanding of the mechanisms underlying dsRNA-induced gene silencing is derived largely from genetic studies in *C. elegans* and plants and from biochemical studies of *Drosophila* extracts (Klink and Wolniak, 2000; Maine, 2001; Worby *et al.*, 2001; Hutvagner and Zamore, 2002a,b; Schmid *et al.*, 2002). A current model for RNAi and related silencing phenomena is shown diagrammatically in Fig. 22.1. Silencing first involves the recognition of dsRNAs by RNAse III ribonucleases that convert them into 20–26 nucleotide-long RNAs (Step 1, Fig. 22.1) (Carthew, 2001). These small interfering RNAs (siRNAs) are double-stranded and contain 5′-phosphorylated termini and two-nucleotide 3′-overhangs that are a signature of the silencing pathways of plants and animals (Fjose *et al.*, 2001; Lipardi *et al.*, 2001). One class of RNAse III ribonucleases, called the Dicer enzymes, process dsRNA into siRNAs (Grishok *et al.*, 2001). Dicer enzymes are multidomain, evolutionarily conserved proteins comprising helicase, RNAse III and dsRNA-binding regions (Hannon, 2002). Dicer enzymes also possess a PAZ domain, which is named after the proteins Piwi, Argonaut and Zwille; the PAZ domain is about 150 amino acids long and contains an RNA-binding fold (Lingel *et al.*, 2003). siRNAs, generated through dsRNA cleavage by Dicer, are proposed to next join with an effector nuclease, known as the RNA-induced silencing complex (RISC) (Step 2, Fig. 22.1) (Grishok and Mello, 2002). In an ATP-dependent step, siRNAs are unwound and RISC is activated (Step 3, Fig. 22.1) (Nykanen *et al.*, 2001). The siRNAs are presumed to identify homologous mRNA substrates through Watson–Crick base-pairing (Step 4, Fig. 22.1). The RISC recognizes and destroys such target mRNAs in an endonucleolytic manner (Step 5, Fig. 22.1) (Grishok *et al.*, 2001; Grishok and Mello, 2002). Cleavage requires extensive complementarity between the target mRNA and an antisense siRNA and occurs near the middle of the region bounded by the siRNA strand (Schwarz *et al.*, 2003).

The RISC assembly pathway and the full composition of the silencing complex are under active investigation and several proteins have been identified as being RISC-associated (Pham *et al.*, 2004). In *C. elegans*, RDE proteins bind to long dsRNAs and are thought to present these to Dicer for processing (Tabara *et al.*, 1999, 2002). Argonaute protein family members appear to be essential components of the RISC (Tabara *et al.*, 1999; Hammond *et al.*, 2001; Shi *et al.*, 2004). These proteins have conserved structural features; they are basic proteins of about 100 kDa that possess two characteristic domains: PAZ and Piwi. The ~150 amino acid PAZ domain is also located in Dicer and, as noted above, contains an RNA-binding fold (Lingel *et al.*, 2003). The Piwi domain is approximately 330 amino acids long and is found at the carboxyl-terminus of Argonaute-family proteins where it functions to interact with the RNAse III domain of Dicer (Tahbaz *et al.*, 2004). In some organisms (plants, fungi, *C. elegans*), an RNA-dependent RNA polymerase is also essential for RNAi but in other organisms (*Drosophila*, mammals) this is not the case (Baulcombe, 1999; Smardon *et al.*, 2000; Volpe *et al.*, 2002). In a similar manner, the RNAi response can spread throughout some organisms (plants, *C. elegans*) but not others (*Drosophila*, mammals) and this may be related to the presence or absence of dsRNA uptake or transport systems in some creatures (Feinberg and Hunter, 2003). In *C. elegans*, the putative transmembrane protein SID-1 has been implicated as a channel for dsRNA that facilitates the spread of RNAi (Winston *et al.*, 2002).

Biological Functions of Gene Silencing

A variety of functions have been proposed for the endogenous role of the RNAi machinery within cells. These include a role in the destruction of viral RNAs. The fact that some viruses encode inhibitors of RNAi supports the contention that RNAi represents a viral defence system in some organisms (Silhavy *et al.*, 2002). RNAi may also suppress the expression of potentially harmful segments of the genome such as transposons which might otherwise destabilize the genome by acting as insertional mutagens (Hannon, 2002). RNAi has also been suggested to participate

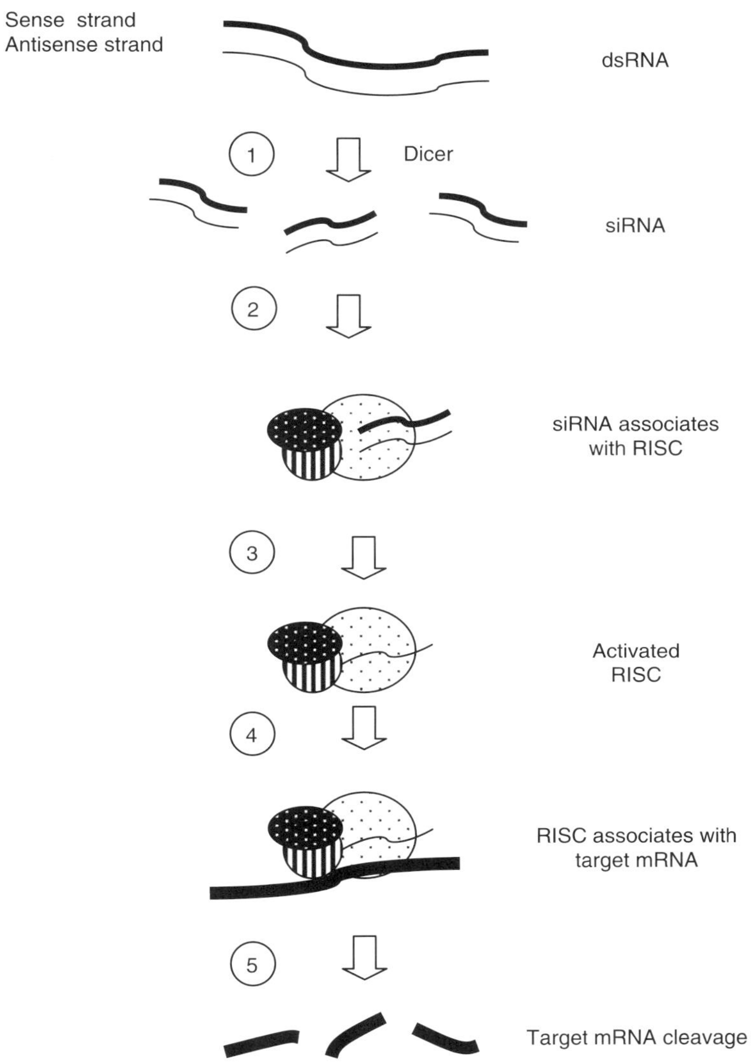

Fig. 22.1. A model of gene silencing. Long, double-stranded RNAs (dsRNAs) are processed into 20–26 nucleotide small interfering RNAs (siRNAs) by Dicer (Step 1). The siRNAs associate with an RNA-induced silencing complex (RISC, Step 2), unwinding and activating in the process (Step 3). The antisense strand of the siRNA guides the RISC to complementary mRNA molecules (Step 4), which are cleaved and destroyed (Step 5). Sense and antisense RNA strands are indicated by thick and thin lines, respectively.

in the regulation of mRNA stability or in nuclear microdomain organization. It has been further suggested that components of the RNAi machinery may function in several pathways of gene regulation (Djikeng *et al.*, 2001; Allshire, 2002; Hannon, 2002; Hu *et al.*, 2002; Zilberman *et al.*, 2003; Shi *et al.*, 2004). *C. elegans* worms lacking Dicer are sterile and show alterations in developmental timing, highlighting the important role this enzyme plays in nematode biology (Knight and Bass, 2001).

Over the past several years, many groups have described endogenous RNA substrates that enter silencing pathways (Lagos-Quintana *et al.*, 2002; Reinhart *et al.*, 2002; Ambros *et al.*, 2003; Lai *et al.*, 2003). These substrates are called microRNAs (miRNAs) and are short non-coding RNAs that form hairpin structures. This makes the stem of the miRNA hairpins largely double-stranded but the sequences contain mismatches. These miRNAs enter silencing pathways, leading to gene silencing through either mRNA destruction or the suppression of protein synthesis (Hutvagner and Zamore, 2002a,b). miRNAs are transcribed from the genomes of a variety of organisms. In the cytoplasm, pre-miRNAs are processed by Dicer into mature, ~22 nucleotide miRNAs. Such molecules were first described as controlling the timing of *C. elegans* development (Lee *et al.*, 1993).

Introducing long dsRNA (>30 nucleotides) into mammalian cells initiates a potent antiviral response, exemplified by the non-specific inhibition of protein synthesis and RNA degradation. The mammalian antiviral response can be bypassed, however, by the introduction or expression of smaller (generally 21–23 nucleotide long) dsRNAs (Elbashir *et al.*, 2001). These siRNAs can suppress mammalian gene expression in a highly specific manner without eliciting a vigorous, non-specific antiviral response.

RNAi as a Molecular and Therapeutic Tool

In addition to the considerable interest in understanding the biology of endogenous gene-silencing phenomena, there is awareness that RNAi is an immensely valuable molecular tool for analysing gene function. One of the most effective ways to determine the biological function of a protein is to examine the phenotype of organisms that have reduced levels of the protein. Classically, this has been achieved by mutating the gene encoding the protein using either molecular genetics or chemical mutagens. Alternative methods for suppressing specific genes, such as RNAi, have more recently proved very powerful. In recent years, RNAi has permeated all fields of eukaryotic biology as a central technology of molecular analysis and this is particularly true in those organisms that are not amenable to classical genetic approaches. For key model organisms too, RNAi has proven invaluable. For instance, in *C. elegans* testing the functions of individual genes by RNAi has extended to analysis of all of the worms predicted approximately 19,000 genes (Carpenter and Sabatini, 2004). A similar genome-wide functional screen in *Drosophila melanogaster* has also been undertaken and similar strategies are being pursued in other organisms (Carpenter and Sabatini, 2004).

The potency and specificity of RNAi has resulted in a focus on its therapeutic use for the prevention and treatment of a variety of conditions, including infectious diseases (McCaffrey *et al.*, 2003; Park *et al.*, 2003; Song *et al.*, 2003; Zhang *et al.*, 2003). For instance, intravenous siRNA administration was used to successfully suppress gene expression in brain cancer *in vivo* in adult rats (Zhang *et al.*, 2003). Local gene knockdown in specific brain tissues of adult mice was earlier reported (Hommel *et al.*, 2003). Treatment of mice with siRNA duplexes targeting the Tnfrsf6 gene (encoding the Fas receptor) was able to protect mice from liver failure and fibrosis in two models of autoimmune hepatitis (Song *et al.*, 2003). Work with RNAi in the development of treatments for infectious disease has, so far, focused on viral infection. For example, administration of siRNA targeting the HIV-1 *env* gene resulted in this gene being specifically silenced in peripheral blood mononuclear cells (Park *et al.*, 2003). In experiments involving hepatitis B virus (HBV) infection, RNA administration *in vivo* blocked replication of HBV in hepatocytes of infected mice (McCaffrey *et al.*, 2003). In this case, immunohistochemical detection of HBV core antigen revealed more than 99% reduction in stained hepatocytes upon RNAi treatment, demonstrating the validity of the approach in the treatment of viral diseases.

RNAi Protocols

It is important to note that RNAi suppresses gene expression but does not abolish gene function. The degree of suppression observed depends on the amount of dsRNA that enters the RNAi pathway and on the half life and abundance of the corresponding protein. In addition, tissue types can differ in their degree of RNAi penetrance such that suppression can vary between different tissues (Timmons *et al.*, 2001). A number of experimental strategies exist for inducing RNAi and several are illustrated in Fig. 22.2. Target organisms can be exposed to long dsRNA or short, siRNA. To generate long dsRNA, a fragment of the target's coding DNA is first obtained, routinely by PCR. The PCR may be carried out using primers encoding part of the target as well as terminal RNA polymerase promoter sites. In this way, the resultant DNA product is bounded by RNA polymerase promoter sequences (depicted by 'flags' in Fig. 22.2, upper left); the PCR product can then be purified and used as a template *in vitro* to generate single-stranded RNAs, using appropriate RNA polymerases (Fig. 22.2, A). Complementary, single-stranded RNAs are annealed *in vitro* to generate the final, dsRNA product (Fig. 22.2, B). This dsRNA may be cleaved by Dicer to generate siRNAs (Fig. 22.2, E) that can participate in gene silencing as illustrated in Fig. 22.1.

Synthesis, purification and annealing of siRNA by industrial chemical processes, for direct use in gene suppression experiments, is becoming increasingly popular. The characteristics of those sequences that constitute the most inhibitory siRNAs are unresolved; it is not currently possible to predict the degree of gene suppression a particular siRNA will produce. It is recommended that the siRNA target site be located >100 nucleotides from the initiation codon and >50 nucleotides from the termination codon (Sui *et al.*, 2002). For optimal siRNA secondary structure, the GC ratio should ideally be between 45% and 55%. It is

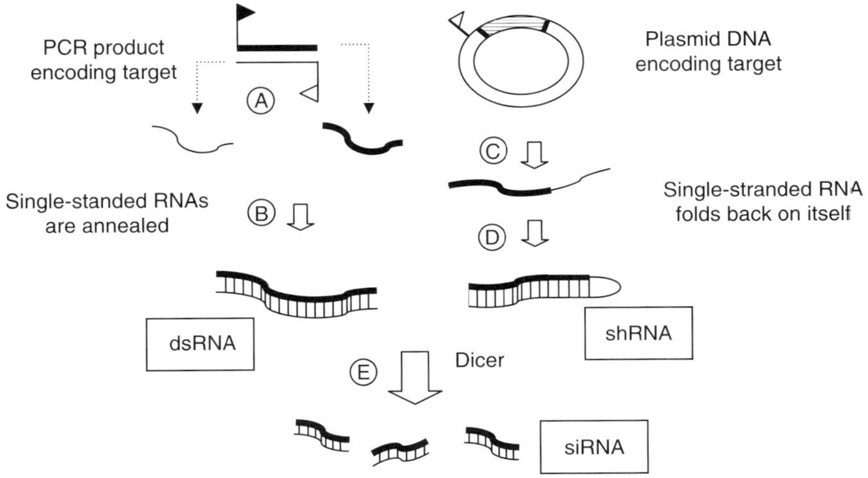

Fig. 22.2. Protocols for gene silencing. To induce gene silencing, target organisms are exposed to double-stranded RNA (dsRNA). Sense and antisense RNA sequences are indicated by thick and thin lines, respectively. To generate long dsRNA, a fragment of the target's coding DNA is obtained, routinely by PCR using primers that contain RNA polymerase promoter sites (black and white 'flags'). The PCR product is used as a template to generate single-stranded RNAs *in vitro*, using appropriate RNA polymerases (A). The complementary, sense and antisense RNAs are annealed *in vitro* to generate a dsRNA product (B). This dsRNA may be cleaved by Dicer to generate siRNAs (E). Small interfering RNAs (siRNAs) can be synthesized chemically for direct use in gene silencing studies. siRNAs can also be produced with DNA plasmids or expression cassettes (upper right) that are used to generate RNA molecules containing inverted repeats (C). By folding back on themselves, these RNA molecules effectively become double-stranded and are then termed short hairpin RNAs (shRNAs) (D). As is the case for longer dsRNAs, these shRNAs may also be processed by Dicer into siRNAs (E) to participate in gene silencing, as illustrated in Fig. 22.1.

important to ensure that the sequence is specific to the target gene by performing a database search. This minimizes the chance that non-target genes will be suppressed by the introduced siRNA.

In some cases, long dsRNAs have been cleaved using recombinant ribonuclease III to produce multiple siRNAs (Yang *et al.*, 2002). siRNAs can also be produced using DNA plasmids or expression cassettes (Fig. 22.2, upper right). These constructs are used to generate inverted repeats that are separated by a spacer of three to nine nucleotides (Fig. 22.2, C). By folding back on themselves, the RNA molecules effectively become double-stranded and are termed short hairpin RNAs (shRNAs) (Fig. 22.2, D). These shRNAs are processed by Dicer into siRNAs (Fig. 22.2, E) and these can participate in gene silencing as illustrated in Fig. 22.1.

RNAi in Flatworms

RNAi has been described in organisms of diverse phylogeny – plants, fungi, insects, nematodes, African trypanosomes, *Dictyostelium discoideum*, spiders and vertebrates (Wianny and Zernicka-Goetz, 2000; Schoppmeier and Damen, 2001; Caplen *et al.*, 2002; Hutvagner and Zamore, 2002a,b; Martens *et al.*, 2002; McRobert and McConkey, 2002; Morris *et al.*, 2002; Urwin *et al.*, 2002; Ullu *et al.*, 2004). However, the phenomenon is not universal; several organisms have been described that do not display RNAi. These include the unicellular parasites *Leishmania major*, *Trypanosoma cruzi* and the yeast *Saccharomyces cerevisiae* (Robinson and Beverley, 2003; DaRocha *et al.*, 2004; Ullu *et al.*, 2004). In the case of flatworms however, the machinery for targeted gene suppression through RNAi is known to exist for both free-living and parasite forms (Sanchez Alvarado and Newmark, 1999; Boyle *et al.*, 2003; Sanchez Alvarado, 2003; Skelly *et al.*, 2003).

Planaria

In flatworms, RNAi was first described in the free-living planarian *Schmidtea mediterranea* (Sanchez Alvarado and Newmark, 1999). Planarians have long been used as models to study the development and the regeneration (Salo and Baguna, 2002; Sanchez Alvarado, 2003). Small fragments of planarians are capable of regenerating complete individuals and recent effort has focused on investigating the molecular mechanisms underlying such extraordinary regenerative capabilities. To investigate the role of body wall myosin in regeneration, individual *S. mediterranea* were cut transversely into several pieces and myosin dsRNA was injected into the exposed parenchyma and/or the gastrovascular system (Sanchez Alvarado and Newmark, 1999). The fragments were then allowed to heal and regenerate at room temperature. After 3 days, immunofluorescence analysis revealed that animals injected with the myosin dsRNA exhibited little, if any, regeneration of the body wall musculature within their blastemas relative to controls (injected with water or single-stranded RNA) (Sanchez Alvarado and Newmark, 1999). Similarly, 1 day after injecting head fragments with opsin dsRNA, opsin mRNA was undetectable in the retinular cells of the planarian photoreceptor, as determined by *in situ* hybridization. In contrast, the expression level of opsin was unchanged in the photoreceptors of control heads (injected with water or an irrelevant (α-tubulin) dsRNA) (Sanchez Alvarado and Newmark, 1999). These experiments suggest that dsRNA in planarians acts in the manner described for other organisms – by reducing the normal accumulation of endogenous RNA transcripts. In addition, since RNAi was observed away from the site of dsRNA injection, this implies the presence of a dsRNA delivery system in planarians.

Rather than introducing dsRNA by microinjection, other workers simply soaked the flatworms in dsRNA (Orii *et al.*, 2003). Fragments of *Dugesia japonica* were soaked in water containing dsRNA encoding the DjIFb protein that is expressed in the body margin. Immunofluorescence analysis showed that this treatment suppressed DjIFb expression in regenerated worms (Orii *et al.*, 2003). This soaking method requires more RNA than microinjection and the RNAi effect is dependent on the concentration of dsRNA used. However, the effect observed at high concentration could also be achieved by repeated treatment at a lower concentration of dsRNA. RNAi was never seen in worms fragmented after dsRNA treatment (Orii *et al.*, 2003). This suggests that the intact, mucous-covered, body surface is impermeable to dsRNA.

Ingestion of bacteria expressing dsRNA by planarians can also trigger the RNAi response, leading to specific gene inhibition, albeit with lower penetrance (Newmark *et al.*, 2003). The method was used to target genes in a variety of cell types including secretory cells, cells of the gastrovascular system and neurons (Newmark *et al.*, 2003). This is in contrast to *C. elegans* whose mature neurons are refractory to RNAi (Timmons *et al.*, 2001). From the experiments reported so far, the inhibitory effect observed in planarians following ingestion or injection of dsRNA has been put at about 3 weeks (Pineda *et al.*, 2000; Newmark *et al.*, 2003).

Given that flatworms are sensitive to RNAi following dsRNA treatment by soaking, feeding or microinjection, the technology is being used to investigate aspects of planarian biology. For instance, expression of the *sine oculis* gene (*Gtso*) of *Girardia tigrina* has been specifically suppressed by dsRNA treatment (Pineda *et al.*, 2000). The gene is expressed exclusively in photoreceptor cells. Loss of function of *Gtso* by RNAi during planarian regeneration completely inhibits eye regeneration. *Gtso* is also essential for maintaining the differentiated state of photoreceptor cells (Pineda *et al.*, 2000). Other work involves the *nou-darake* (*ndk*) gene which encodes a fibroblast growth factor receptor-related protein of *D. japonica*. RNAi of *ndk* expression results in ectopic brain formation throughout the planarian (Cebria *et al.*, 2002). This work suggests that *ndk* modulates fibroblast growth factor signalling in planarian stem cells (neoblasts) to restrict brain tissue to the head region (Cebria *et al.*, 2002).

Schistosomes

RNAi has also been used to suppress gene expression in another type of flatworm, specifically the trematode *Schistosoma mansoni* (Boyle *et al.*, 2003; Skelly *et al.*, 2003). Schistosomes are parasitic platyhelminths that cause a chronic, debilitating disease called schistosomiasis that afflicts over 200 million people globally. The disease is characterized by the presence of adult worms, or blood flukes, within the portal and mesenteric veins or within the veins of the b adder (depending on the precise schistosome species). These worms, living as male/female pairs, can survive for many years during which time the female produces hundreds of eggs per day. The primary pathological consequences of schistosome infection are the host's immunological response to these eggs within host tissues.

The schistosome life cycle is extraordinary for its complexity and dramatic transformations. Eggs produced by adult *S. mansoni* pass through the intestinal wall into the environment. Once in fresh water, the egg hatches to release a multicellular, free-swimming, ciliated organism called a miracidium. Using chemotaxis, the miracidium finds and infects a freshwater snail intermediate host and undergoes a complicated maturation. The miracidium first transforms into a mother sporocyst (in a process called miracidial transformation) and later spawns numerous daughter sporocysts that migrate to different sites within the snail. These sporocysts asexually produce schistosome larvae called cercariae that are released daily into the environment. Water-borne cercariae use their propulsive tails to find and infect vertebrate hosts. With the aid of secreted enzymes, the cercariae penetrate the skin and transform into juvenile forms called schistosomula, in a process called cercarial transformation. Transformation involves complex biochemical and morphological changes including the shedding of the cercarial tail, the release of the cercarial tegument and the biogenesis of a new, intra-vertebrate tegumental covering. Over the next 6 weeks or so, the schistosomula migrate via the bloodstream to the lungs, then to the liver and eventually to the mesenteric veins where the male/female pairs begin to produce eggs.

Two different developmental stages of *S. mansoni* have been tested for RNAi – the intra-vertebrate larval form called the schistosomulum and the intra-molluscan larval stage called the sporocyst (Boyle *et al.*, 2003; Skelly *et al.*, 2003). In schistosomula, the cathepsin

B gene was targeted for suppression by RNAi. Cathepsins are proteolytic enzymes that are involved in haemoglobin digestion; the haemoglobin is obtained from red blood cells that are eaten by the blood-dwelling schistosomes. A family of cathepsins has been described in *S. mansoni* including the major gut-associated papain-like cysteine proteinase cathepsin B (also known as Sm31) (Gotz and Klinkert, 1993; Tort *et al.*, 1999). In these RNAi experiments, cercariae recovered from infected snails were vortexed to separate tails from bodies and the bodies were recovered by centrifugation. Cathepsin B dsRNA was then added to the parasites that were incubated in RPMI culture medium at 37°C. Under these conditions the parasites undergo transformation; it was hypothesized that the dsRNA may gain easier access into the schistosomes during this period when their external membranes are in a state of flux (Skelly *et al.*, 2003). It was shown that after 6-day incubation, the expression of the cathepsin B gene was suppressed in parasites exposed to cathepsin B dsRNA but not in parasites exposed to control dsRNAs. Immunofluorescence analysis using anti-cathepsin B antibodies showed that the enzyme was undetectable in most cathepsin B dsRNA-treated parasites (63%), while the remainder of these parasites (37%) did stain for the protein. In contrast, in the majority of parasites treated with control dsRNAs, cathepsin B was evident (74–85%) (Skelly *et al.*, 2003). Enzyme activity measurements in these parasite populations also revealed the suppressive effect of cathepsin B dsRNA treatment; activity levels detected in soluble extracts of treated parasites were five- to sixfold lower than those of controls (Skelly *et al.*, 2003). Finally, using reverse transcriptase-PCR, gene suppression was evidenced by the inability to detect cathepsin B cDNA using RNA obtained from treated parasites. Despite the substantial suppression of expression of the cathepsin B protease, treated parasites did not exhibit an overtly altered phenotype and all groups exhibited equivalent viability at the end of the 6-day culture period. This means that exposure to cathepsin B dsRNA is not lethal for cultured parasites at the developmental stage examined. This is not surprising given the proposed primary role of cathepsin B in degrading haemoglobin in the host bloodstream and the fact that treated parasites were young schistosomula maintained *in vitro* in nutrient-rich medium.

To examine RNAi at the sporocyst stage of the *S. mansoni* life cycle, two genes were targeted for suppression; the glycolytic enzyme glyceraldehyde-3-phosphate dehydrogenase (GAPDH) and a glucose transporter protein (SGTP1) (Charrier-Ferrara *et al.*, 1992; Skelly *et al.*, 1994). For these experiments, miracidia were cultured with GAPDH or SGTP1 dsRNA in a balanced salt solution. This mimics the salt concentration of the intermediate snail host of *S. mansoni* and promotes miracidial transformation and sporocyst development. After 6 days in culture, significant reductions (60–90%) in target gene transcript levels were detected by quantitative PCR analysis while non-target transcript levels were unaltered (Boyle *et al.*, 2003). As little as 10 nM SGTP1 dsRNA was sufficient to reduce SGTP1 transcript levels. Additionally, the glucose uptake capacity of sporocysts treated with SGTP1 dsRNA was significantly diminished relative to controls (Boyle *et al.*, 2003). In some experiments, sporocysts were washed following 6-day exposure to dsRNA and were then further cultured for 14 or 28 additional days in the absence of dsRNA. The level of SGTP1 transcripts was significantly reduced at both time points, while non-target transcript levels were unaltered. This demonstrates that the effects of SGTP1 dsRNA treatment persisted for up to 4 weeks (Boyle *et al.*, 2003). The time frame for persistence of an RNAi effect in schistosomes is comparable to that reported above for planarians (Pineda *et al.*, 2000; Newmark *et al.*, 2003).

RNAi in larval schistosomes was only observed if dsRNA was administered at, or shortly after, the initiation of miracidial transformation. As the time between transformation initiation and the addition of dsRNA increased, the effectiveness of the treatment decreased; by 24 h, dsRNA treatment was completely ineffective at reducing target transcript levels (Boyle *et al.*, 2003). As noted earlier, both miracidial transformation and cercarial transformation involve complex outer tegumental membrane reorganizations; it may be that the parasites are most susceptible to RNAi when dsRNA is administered during these processes. The dsRNA may most efficiently gain entry into the larvae at these times. To test this, miracidia were exposed to rhodamine-labelled SGTP1 dsRNA either at the start of transformation or 20 h later. After 4 h exposure, larvae were fixed and the degree of labelled dsRNA entry into both groups was assessed. It was reported that qualitatively both the fluorescence intensity and the proportion of larvae that were fluorescent did not differ between the groups and label was clearly detected within the bodies of some larvae in both groups (Boyle *et al.*, 2003). This means that it is unlikely that dsRNA entry is responsible for differences in the effects of dsRNA in the two life stages examined. However, it remains possible that dsRNA cannot enter in sufficient amounts to exert a demonstrable effect at lengthy time points beyond miracidial transformation (Boyle *et al.*, 2003). As noted, dsRNAs have been introduced into flatworms by soaking the organisms in dsRNA-containing solutions, by microinjection or by feeding some species with bacteria expressing dsRNA. In other systems, transfection reagents (mostly liposomal or amine-based) have been used to promote nucleic acid uptake. Attempts to optimize RNAi in *S. mansoni* with one such reagent (Lipofectamine) were not successful (Skelly *et al.*, 2003).

Concluding Remarks

These studies examining the impact of exposing schistosomula and sporocysts to dsRNA have demonstrated the susceptibility of these *S. mansoni* life stages to dsRNA-mediated interference. This ability to manipulate gene expression through RNAi represents a powerful new tool for investigating gene function in these debilitating human parasites and adds to the growing arsenal of gene manipulation techniques being developed for use with parasitic flatworms (Davis *et al.*, 1999; Wippersteg *et al.*, 2002; and see Chapter 7, this volume). With suitable molecular genetic tools such as RNAi, it will now be possible to more fully exploit flatworm genome databases that are currently under development (Hu *et al.*, 2003; Verjovski-Almeida *et al.*, 2003). In this manner, it will be possible to identify worm genes and gain a fuller understanding of platyhelminth cellular and molecular biology. All of which should help to highlight the vulnerabilities of parasitic forms to immunological, chemotherapeutic or other interventions.

Acknowledgements

This work was supported by NIH Grant R01 AI056272. Thanks to Chuck Shoemaker for critically reading the manuscript.

References

Allshire, R. (2002) RNAi and heterochromatin – a hushed-up affair. *Science* 297, 1818–1819.

Ambros, V., Lee, R.C., Lavanway, A., Williams, P.T. and Jewell, D. (2003) MicroRNAs and other tiny endogenous RNAs in *C. elegans*. *Current Biology* 13, 807–818.

Baulcombe, D.C. (1999) Gene silencing: RNA makes no protein. *Currrent Biology* 9, R599–R601.

Boyle, J.P., Wu, X.J., Shoemaker, C.B. and Yoshino, T.P. (2003) Using RNA interference to manipulate endogenous gene expression in *Schistosoma mansoni* sporocysts. *Molecular and Biochemical Parasitology* 128, 205–215.

Caplen, N., Zheng, Z., Falgout, B. and Morgan, R. (2002) Inhibition of viral gene expression and replication in mosquito cells by dsRNA-triggered RNA interference. *Molecular Therapy* 6, 243.

Carpenter, A.E. and Sabatini, D.M. (2004) Systematic genome-wide screens of gene function. *Nature Reviews Genetics* 5, 11–22.

Carthew, R.W. (2001) Gene silencing by double-stranded RNA. *Current Opinion in Cell Biology* 13, 244–248.

Cebria, F., Kobayashi, C., Umesono, Y., Nakazawa, M., Mineta, K., Ikeo, K., Gojobori, T., Itoh, M., Taira, M., Sanchez Alvarado, A. and Agata, K. (2002) FGFR-related gene *nou-darake* restricts brain tissues to the head region of planarians. *Nature* 419, 620–624.

Charrier-Ferrara, S., Caillol, D. and Goudot-Crozel, V. (1992) Complete sequence of the *Schistosoma mansoni* glyceraldehyde-3-phosphate dehydrogenase gene encoding a major surface antigen. *Molecular and Biochemical Parasitology* 56, 339–343.
DaRocha, W.D., Otsu, K., Teixeira, S.M. and Donelson, J.E. (2004) Tests of cytoplasmic RNA interference (RNAi) and construction of a tetracycline-inducible T7 promoter system in *Trypanosoma cruzi*. *Molecular and Biochemical Parasitology* 133, 175–186.
Davis, R.E., Parra, A., LoVerde, P.T., Ribeiro, E., Glorioso, G. and Hodgson, S. (1999) Transient expression of DNA and RNA in parasitic helminths by using particle bombardment. *Proceedings of the National Academy of Sciences USA* 96, 8687–8692.
Djikeng, A., Shi, H., Tschudi, C. and Ullu, E. (2001) RNA interference in *Trypanosoma brucei*: cloning of small interfering RNAs provides evidence for retroposon-derived 24–26-nucleotide RNAs. *RNA* 7, 1522–1530.
Dykxhoorn, D.M., Novina, C.D. and Sharp, P.A. (2003) Killing the messenger: short RNAs that silence gene expression. *Nature Reviews Molecular Cell Biology* 4, 457–467.
Elbashir, S.M., Harborth, J., Lendeckel, W., Yalcin, A., Weber, K. and Tuschl, T. (2001) Duplexes of 21-nucleotide RNAs mediate RNA interference in cultured mammalian cells. *Nature* 411, 494–498.
Feinberg, E.H. and Hunter, C.P. (2003) Transport of dsRNA into cells by the transmembrane protein SID-1. *Science* 301, 1545–1547.
Fire, A., Xu, S., Montgomery, M.K., Kostas, S.A., Driver, S.E. and Mello, C.C. (1998) Potent and specific genetic interference by double-stranded RNA in *Caenorhabditis elegans*. *Nature* 391, 806–811.
Fjose, A., Ellingsen, S., Wargelius, A. and Seo, H.C. (2001) RNA interference: mechanisms and applications. *Biotechnology Annual Review* 7, 31–57.
Gotz, B. and Klinkert, M.Q. (1993) Expression and partial characterization of a cathepsin B-like enzyme (Sm31) and a proposed 'haemoglobinase' (Sm32) from *Schistosoma mansoni*. *Biochemical Journal* 290, 801–806.
Grishok, A. and Mello, C.C. (2002) RNAi (Nematodes: *Caenorhabditis elegans*). *Advances in Genetics* 46, 339–360.
Grishok, A., Pasquinelli, A.E., Conte, D., Li, N., Parrish, S., Ha, I., Baillie, D.L., Fire, A., Ruvkun, G. and Mello, C.C. (2001) Genes and mechanisms related to RNA interference regulate expression of the small temporal RNAs that control *C. elegans* developmental timing. *Cell* 106, 23–34.
Hammond, S.M., Boettcher, S., Caudy, A.A., Kobayashi, R. and Hannon, G.J. (2001) Argonaute2, a link between genetic and biochemical analyses of RNAi. *Science* 293, 1146–1150.
Hannon, G.J. (2002) RNA interference. *Nature* 418, 244–251.
Hommel, J.D., Sears, R.M., Georgescu, D., Simmons, D.L. and DiLeone, R.J. (2003) Local gene knockdown in the brain using viral-mediated RNA interference. *Nature Medicine* 9, 1539–1544.
Hu, W., Myers, C., Kilzer, J., Pfaff, S. and Bushman, F. (2002) Inhibition of retroviral pathogenesis by RNA interference. *Current Biology* 12, 1301.
Hu, W., Yan, Q., Shen, D.K., Liu, F., Zhu, Z.D., Song, H.D., Xu, X.R., Wang, Z.J., Rong, Y.P., Zeng, L.C., Wu, J., Zhang, X., Wang, J.J., Xu, X.N., Wang, S.Y., Fu, G., Zhang, X.L., Wang, Z.Q., Brindley, P.J., McManus, D.P., Xue, C.L., Feng, Z., Chen, Z. and Han, Z.G. (2003) Evolutionary and biomedical implications of a *Schistosoma japonicum* complementary DNA resource. *Nature Genetics* 35, 139–147.
Hunter, C.P. (1999) Genetics: a touch of elegance with RNAi. *Current Biology* 9, R440–R442.
Hutvagner, G. and Zamore, P.D. (2002a) A microRNA in a multiple-turnover RNAi enzyme complex. *Science* 297, 2056–2060.
Hutvagner, G. and Zamore, P.D. (2002b) RNAi: nature abhors a double-strand. *Current Opinion in Genetics and Development* 12, 225–232.
Klink, V.P. and Wolniak, S.M. (2000) The efficacy of RNAi in the study of the plant cytoskeleton. *Journal of Plant Growth Regulation* 19, 371–384.
Knight, S.W. and Bass, B.L. (2001) A role for the RNase III enzyme DCR-1 in RNA interference and germ line development in *Caenorhabditis elegans*. *Science* 293, 2269–2271.
Lagos-Quintana, M., Rauhut, R., Yalcin, A., Meyer, J., Lendeckel, W. and Tuschl, T. (2002) Identification of tissue-specific microRNAs from mouse. *Current Biology* 12, 735–739.
Lai, E.C., Tomancak, P., Williams, R.W. and Rubin, G.M. (2003) Computational identification of *Drosophila* microRNA genes. *Genome Biology* 4, R42.
Lee, R.C., Feinbaum, R.L. and Ambros, V. (1993) The *C. elegans* heterochronic gene lin-4 encodes small RNAs with antisense complementarity to lin-14. *Cell* 75, 843–854.
Lingel, A., Simon, B., Izaurralde, E. and Sattler, M. (2003) Structure and nucleic acid binding of the *Drosophila* Argonaute 2 PAZ domain. *Nature* 426, 465–469.

Lipardi, C., Wei, Q. and Paterson, B.M. (2001) RNAi as random degradative PCR: siRNA primers convert mRNA into dsRNAs that are degraded to generate new siRNAs. *Cell* 107, 297–307.

Maine, E.M. (2001) RNAi as a tool for understanding germline development in *Caenorhabditis elegans*: uses and cautions. *Developmental Biology* 239, 177–189.

Malhotra, P., Dasaradhi, P.V., Kumar, A., Mohmmed, A., Agrawal, N., Bhatnagar, R.K. and Chauhan, V.S. (2002) Double-stranded RNA-mediated gene silencing of cysteine proteases (falcipain-1 and -2) of *Plasmodium falciparum*. *Molecular Microbiology* 45, 1245–1254.

Martens, H., Novotny, J., Oberstrass, J., Steck, T.L., Postlethwait, P. and Nellen, W. (2002) RNAi in *Dictyostelium*: the role of RNA-directed RNA polymerases and double-stranded RNase. *Molecular Biology of the Cell* 13, 445–453.

McCaffrey, A.P., Nakai, H., Pandey, K., Huang, Z., Salazar, F.H., Xu, H., Wieland, S.F., Marion, P.L. and Kay, M.A. (2003) Inhibition of hepatitis B virus in mice by RNA interference. *Nature Biotechnology* 21, 639–644.

McRobert, L. and McConkey, G.A. (2002) RNA interference (RNAi) inhibits growth of *Plasmodium falciparum*. *Molecular and Biochemical Parasitology* 119, 273–278.

Morris, J.C., Wang, Z., Drew, M.E. and Englund, P.T. (2002) Glycolysis modulates trypanosome glycoprotein expression as revealed by an RNAi library. *EMBO Journal* 21, 4429–4438.

Newmark, P.A., Reddien, P.W., Cebria, F. and Sanchez Alvarado, A. (2003) Ingestion of bacterially expressed double-stranded RNA inhibits gene expression in planarians. *Proceedings of the National Academy of Sciences USA* 100(Suppl. 1), 11861–11865.

Nykanen, A., Haley, B. and Zamore, P.D. (2001) ATP requirements and small interfering RNA structure in the RNA interference pathway. *Cell* 107, 309–321.

Orii, H., Mochii, M. and Watanabe, K. (2003) A simple "soaking method" for RNA interference in the planarian *Dugesia japonica*. *Developmental Genes and Evolution* 213, 138–141.

Park, W.S., Hayafune, M., Miyano-Kurosaki, N. and Takaku, H. (2003) Specific HIV-1 *env* gene silencing by small interfering RNAs in human peripheral blood mononuclear cells. *Gene Therapy* 10, 2046–2050.

Pham, J.W., Pellino, J.L., Lee, Y.S., Carthew, R.W. and Sontheimer, E.J. (2004) A Dicer-2-dependent 80s complex cleaves targeted mRNAs during RNAi in *Drosophila*. *Cell* 117, 83–94.

Pineda, D., Gonzalez, J., Callaerts, P., Ikeo, K., Gehring, W.J. and Salo, E. (2000) Searching for the prototypic eye genetic network: sine oculis is essential for eye regeneration in planarians. *Proceedings of the National Academy of Sciences USA* 97, 4525–4529.

Reinhart, B.J., Weinstein, E.G., Rhoades, M.W., Bartel, B. and Bartel, D.P. (2002) MicroRNAs in plants. *Genes and Development* 16, 1616–1626.

Robinson, K.A. and Beverley, S.M. (2003) Improvements in transfection efficiency and tests of RNA interference (RNAi) approaches in the protozoan parasite *Leishmania*. *Molecular and Biochemical Parasitology* 128, 217–228.

Salo, E. and Baguna, J. (2002) Regeneration in planarians and other worms: new findings, new tools, and new perspectives. *Journal of Experimental Zoology* 292, 528–539.

Sanchez Alvarado, A. (2003) The freshwater planarian *Schmidtea mediterranea*: embryogenesis, stem cells and regeneration. *Current Opinion in Genetics and Development* 13, 438–444.

Sanchez Alvarado, A. and Newmark, P.A. (1999) Double-stranded RNA specifically disrupts gene expression during planarian regeneration. *Proceedings of the National Academy of Sciences USA* 96, 5049–5054.

Schmid, A., Schindelholz, B. and Zinn, K. (2002) Combinatorial RNAi: a method for evaluating the functions of gene families in *Drosophila*. *Trends in Neuroscience* 25, 71–74.

Schoppmeier, M. and Damen, W.G. (2001) Double-stranded RNA interference in the spider *Cupiennius salei*: the role of Distal-less is evolutionarily conserved in arthropod appendage formation. *Developmental Genes and Evolution* 211, 76–82.

Schwarz, D.S., Hutvagner, G., Du, T., Xu, Z., Aronin, N. and Zamore, P.D. (2003) Asymmetry in the assembly of the RNAi enzyme complex. *Cell* 115, 199–208.

Shi, H., Djikeng, A., Tschudi, C. and Ullu, E. (2004) Argonaute protein in the early divergent eukaryote *Trypanosoma brucei*: control of small interfering RNA accumulation and retroposon transcript abundance. *Molecular and Cellular Biology* 24, 420–427.

Silhavy, D., Molnar, A., Lucioli, A., Szittya, G., Hornyik, C., Tavazza, M. and Burgyan, J. (2002) A viral protein suppresses RNA silencing and binds silencing-generated, 21- to 25-nucleotide double-stranded RNAs. *EMBO Journal* 21, 3070–3080.

Skelly, P.J., Kim, J.W., Cunningham, J. and Shoemaker, C.B. (1994) Cloning, characterization, and functional expression of cDNAs encoding glucose transporter proteins from the human parasite *Schistosoma mansoni*. *Journal of Biological Chemistry* 269, 4247–4253.

Skelly, P.J., Da'dara, A. and Harn, D.A. (2003) Suppression of cathepsin B expression in *Schistosoma mansoni* by RNA interference. *International Journal for Parasitology* 33, 363–369.

Smardon, A., Spoerke, J.M., Stacey, S.C., Klein, M.E., Mackin, N. and Maine, E.M. (2000) EGO-1 is related to RNA-directed RNA polymerase and functions in germ-line development and RNA interference in *C. elegans*. *Current Biology* 10, 169–178.

Song, E., Lee, S.K., Wang, J., Ince, N., Ouyang, N., Min, J., Chen, J., Shankar, P. and Lieberman, J. (2003) RNA interference targeting Fas protects mice from fulminant hepatitis. *Nature Medicine* 9, 347–351.

Sui, G., Soohoo, C., Affar el, B., Gay, F., Shi, Y. and Forrester, W.C. (2002) A DNA vector-based RNAi technology to suppress gene expression in mammalian cells. *Proceedings of the National Academy of Sciences USA* 99, 5515–5520.

Tabara, H., Grishok, A. and Mello, C.C. (1998) RNAi in *C. elegans*: soaking in the genome sequence. *Science* 282, 430–431.

Tabara, H., Sarkissian, M., Kelly, W.G., Fleenor, J., Grishok, A., Timmons, L., Fire, A. and Mello, C.C. (1999) The rde-1 gene, RNA interference, and transposon silencing in *C. elegans*. *Cell* 99, 123–132.

Tabara, H., Yigit, E., Siomi, H. and Mello, C.C. (2002) The dsRNA binding protein RDE-4 interacts with RDE-1, DCR-1, and a DExH-box helicase to direct RNAi in *C. elegans*. *Cell* 109, 861–871.

Tahbaz, N., Kolb, F.A., Zhang, H., Jaronczyk, K., Filipowicz, W. and Hobman, T.C. (2004) Characterization of the interactions between mammalian PAZ PIWI domain proteins and Dicer. *EMBO Reports* 5, 189–194.

Timmons, L., Court, D.L. and Fire, A. (2001) Ingestion of bacterially expressed dsRNAs can produce specific and potent genetic interference in *Caenorhabditis elegans*. *Gene* 263, 103–112.

Tort, J., Brindley, P.J., Knox, D., Wolfe, K.H. and Dalton, J.P. (1999) Proteinases and associated genes of parasitic helminths. *Advances in Parasitology* 43, 161–266.

Ullu, E., Tschudi, C. and Chakraborty, T. (2004) RNA interference in protozoan parasites. *Cellular Microbiology* 6, 509–519.

Urwin, P.E., Lilley, C.J. and Atkinson, H.J. (2002) Ingestion of double-stranded RNA by preparasitic juvenile cyst nematodes leads to RNA interference. *Molecular Plant–Microbe Interactions* 15, 747–752.

Verjovski-Almeida, S., DeMarco, R., Martins, E.A., Guimaraes, P.E., Ojopi, E.P., Paquola, A.C., Piazza, J.P., Nishiyama, M.Y., Kitajima, J.P., Adamson, R.E., Ashton, P.D., Bonaldo, M.F., Coulson, P.S., Dillon, G.P., Farias, L.P., Gregorio, S.P., Ho, P.L., Leite, R.A., Malaquias, L.C., Marques, R.C., Miyasato, P.A., Nascimento, A.L., Ohlweiler, F.P., Reis, E.M., Ribeiro, M.A., Sa, R.G., Stukart, G.C., Soares, M.B., Gargioni, C., Kawano, T., Rodrigues, V., Madeira, A.M., Wilson, R.A., Menck, C.F., Setubal, J.C., Leite, L.C. and Dias-Neto, E. (2003) Transcriptome analysis of the acoelomate human parasite *Schistosoma mansoni*. *Nature Genetics* 35, 148–157.

Volpe, T.A., Kidner, C., Hall, I.M., Teng, G., Grewal, S.I. and Martienssen, R.A. (2002) Regulation of heterochromatic silencing and histone H3 lysine-9 methylation by RNAi. *Science* 297, 1833–1837.

Wianny, F. and Zernicka-Goetz, M. (2000) Specific interference with gene function by double-stranded RNA in early mouse development. *Nature Cell Biology* 2, 70–75.

Winston, W.M., Molodowitch, C. and Hunter, C.P. (2002) Systemic RNAi in *C. elegans* requires the putative transmembrane protein SID-1. *Science* 295, 2456–2459.

Wippersteg, V., Kapp, K., Kunz, W., Jackstadt, W.P., Zahner, H. and Grevelding, C.G. (2002) HSP70-controlled GFP expression in transiently transformed schistosomes. *Molecular and Biochemical Parasitology* 120, 141–150.

Worby, C.A., Simonson-Leff, N. and Dixon, J.E. (2001) RNA interference of gene expression (RNAi) in cultured *Drosophila* cells. *Science STKE* 2001, pl1.

Yang, D., Buchholz, F., Huang, Z., Goga, A., Chen, C.Y., Brodsky, F.M. and Bishop, J.M. (2002) Short RNA duplexes produced by hydrolysis with *Escherichia coli* RNase III mediate effective RNA interference in mammalian cells. *Proceedings of the National Academy of Sciences USA* 99, 9942–9947.

Zamore, P.D. (2001) RNA interference: listening to the sound of silence. *Nature Structural Biology* 8, 746–750.

Zhang, Y., Boado, R.J. and Pardridge, W.M. (2003) *In vivo* knockdown of gene expression in brain cancer with intravenous RNAi in adult rats. *The Journal of Gene Medicine* 5, 1039–1045.

Zilberman, D., Cao, X. and Jacobsen, S.E. (2003) ARGONAUTE4 control of locus-specific siRNA accumulation and DNA and histone methylation. *Science* 299, 716–719.

Index

Page numbers in *italics* refer to figures and tables